# Shock Wave and High Pressure Phenomena

More information about this series at http://www.springer.com/series/1774

James R. Asay • Lalit C. Chhabildas
R. Jeffery Lawrence • Mary Ann Sweeney

# Impactful Times

## Memories of 60 Years of Shock Wave Research at Sandia National Laboratories

Springer

James R. Asay, retired
Sandia National Laboratories,
Albuquerque, NM, USA

Lalit C. Chhabildas, retired
Sandia National Laboratories,
Albuquerque, NM, USA

R. Jeffery Lawrence, retired
Sandia National Laboratories,
Albuquerque, NM, USA

Mary Ann Sweeney
Sandia National Laboratories,
Albuquerque, NM, USA

ISSN 2197-9529 ISSN 2197-9537 (electronic)
Shock Wave and High Pressure Phenomena
ISBN 978-3-319-33345-8 ISBN 978-3-319-33347-2 (eBook)
DOI 10.1007/978-3-319-33347-2

Library of Congress Control Number: 2016944359



Printed on acid-free paper

This Springer imprint is published by Springer Nature
The registered company is Springer International Publishing AG
The registered company address is: Gewerbestrasse 11, 6330 Cham, Switzerland

# Preface

The effort to document this history of shock wave research, entitled *Impactful Times: Memories of 60 Years of Shock Wave Research at Sandia National Laboratories*, began in the early fall of 2011. James (Jim) Asay and Lalit Chhabildas had been queried by many new staff who joined Sandia in the late 1990s and early 2000s about providing a history of shock wave technology development. To make this a comprehensive document, we decided to start from the mid 1950s, when shock wave research originated at Sandia, and carry the development of the technology and Sandia's accomplishments through to the present. *Impactful Times* includes the review of experiments in hypervelocity impact and magnetic loading to study the high-pressure response of materials, advanced models and *ab initio* theories to describe that response, and state-of-the-art computations applied to a wide range of real-world problems.

The two of us had a good knowledge of the shock wave research conducted at Sandia from the 1970s through to the present because we had joined the laboratory in the early to mid 1970s. We believed, however, that it was important to involve people who had participated in the research prior to that time as well. This was done in two ways. One was to ask those directly involved in the shock wave research back to the 1950s and 1960s to provide their personal recollections. We also talked with many of the early participants, including Lynn Barker, Al Chabai, Dennis Hayes, Roy (Red) Hollenbach, Orval Jones, Charles Karnes, Don Lundergan, Darrell Munson, and Ray Reed, to get their perspective on those early times. The other way was to invite Bob Graham to coauthor the publication, since he had initiated many of the early experimental techniques in the mid 1950s and has been a leader of shock wave research at Sandia and throughout the scientific community. Bob originally agreed to do this; however, other commitments prevented his long-term involvement. Later, R. Jeffery (Jeff) Lawrence and Mary Ann Sweeney joined the effort to produce this history. Jeff's involvement with shock wave research at Sandia dates back to 1963 when he was engaged in nuclear weapon vulnerability testing as an officer at Kirtland Air Force Base. His participation therefore provided an important part of the early history. Jeff was also closely connected to many efforts to develop the original models describing dynamic material behavior and the early computer codes for analyzing shock wave problems. A little later, Mary Ann Sweeney joined the team. As a member of the technical staff of the Pulsed Power Sciences Center at Sandia since the mid 1970s, who used Sam Thompson's computer codes in the 1970s and 1980s, and who more recently is also serving as the editor-in-chief of the annual Department of Energy (DOE) National Nuclear Security Administration (NNSA) *Stockpile Stewardship and Management Plan*, she provided knowledge of the DOE and NNSA research programs, along with technical and editing skills that not only added depth and breadth to the book but also resulted in a concise and balanced presentation.

We organized our book into two parts. Part I, "Building Shock Wave Capabilities," discusses the development of new experimental platforms at Sandia to produce precise loading conditions and novel diagnostics to probe the behavior of shocked materials, starting in the 1950s and continuing to the present. Topics also include the complementary development of theoretical and modeling activities, experimental shock wave drivers, and diagnostics development. Throughout the technical discussions, we have attempted to identify all the key players as well as the major technologies that were developed. A brief discussion of each advance is presented, along with its assessment and attributes.

To provide a more personal account of these developments, a large number of recollections are provided in Part II, "Memories of Shock Wave Research." These individual recollections present a window into the personal perspective and experience of researchers who participated in the shock wave program at Sandia. We made a strong effort to include as many people as possible. Over 80 were contacted, with around half providing their personal experiences. Each contributor was asked to summarize his or her role in shock wave research and to highlight interesting events and anecdotes that happened along the way. We purposely gave little guidance on style and format, which is why some recollections are written as a stream of consciousness while others are more technical and consist of short annotated summaries of major papers. The result is a rich and interesting mix that highlights individual personalities, personal struggles, and technical successes. The dates provided for each person represent their hiring and retirement date at Sandia; we have also made an attempt to identify the actual dates an individual participated in shock wave research. Quotations from the individual recollections are liberally sprinkled throughout the text to bring out the perspective of the research and to provide a pointer to further reading of the recollections. In some cases (e.g., Bob Graham, Walter Herrmann, Orval Jones, George Samara, and Sam Thompson), the recollections of others were used to capture the seminal contributions that individuals made to shock wave research. In summary, the technical discussions and recollections offer a unique insight into the shock wave program that covers six decades at Sandia.

A bibliography of almost 1000 references is provided for those interested in pursuing in more detail specific areas of shock wave research and technology at Sandia. The bibliography was developed by asking the contributors to the recollections to identify 20 or so of their key publications. These were used to prepare the discussion of experimental, modeling, and computational developments that occurred over each decade. In cases where people who made major contributions during that time did not provide their recollections or could not be contacted, a literature search was done to identify their significant contributions.

We also felt that attaching faces to those who participated in the shock wave program would add interest. Each of Chapters 2 through 7, which are organized by decade, contains a section at the end called "People and Places." The photographs in this section illustrate key facilities developed during the specific decade and the individuals who participated in shock wave research. It was not always possible to find a photograph for a person corresponding to the decade in which he or she did shock wave research; however, we were able to find photographs for most people at some point in their career, typically when they had achieved a specific milepost such as a promotion or had received an award or honor. While the photographs showing specific accomplishments are not complete, they are representative of the group's achievements. All photographs have been annotated, which helps put them in perspective. The photographs and the associated captions often tell a story in themselves.

The shock wave group as a whole has had a major impact in both management and technical arenas at Sandia National Laboratories and throughout the broader scientific community. Two individuals became executive vice presidents at Sandia, and one was appointed president of the Nevada Test Site (now known as the Nevada National Security Site), an organization involved with nuclear weapon testing activities. Many others were promoted to middle management positions within Sandia. Scientific accomplishments are equally wide-ranging. Three individuals within the Sandia shock wave program were elected to the National Academy of Engineering. A great many were promoted to top levels of scientific or engineering achievement at Sandia, either as a laboratory fellow, a distinguished member of technical staff, or a senior scientist; one retired from Sandia and became a Senior Technologist, a major scientific senior executive leadership and advisory position at the Air Force Research Laboratory, a large Department of Defense organization. A large number of individuals were also appointed fellows of various scientific societies and many received top awards for technical achievement in their respective scientific or engineering organizations. Another became a high-level program administrator at NNSA. These individual achievements are too extensive to report here, but the photographs in the People and Places sections provide a small glimpse into some of them. In summary, Sandia's shock wave research program has had a significant impact not only in managing technical activities but also in scientific accomplishments as part of the national and international community.

As will become apparent in our book, two major shock wave efforts have lasted for most of Sandia's history. One focused primarily on scientific aspects of shock wave phenomena, while the other emphasized engineering applications. Three of us (Asay, Chhabildas, and Lawrence) are from the engineering side, so there is a built-in bias toward this aspect of research. However, we tried to present a balanced picture of shock wave research, especially in identifying key shock wave technologies.

Also, some contributors to the recollections were involved with the science aspects of shock wave research, while others were involved with the engineering aspects of shock wave research; this gives a balanced perspective in many cases. Although we tried to be as objective as possible, we wish to apologize for any oversights that may have occurred by not recognizing specific individuals and their research.

Albuquerque, NM, USA

James R. Asay
Lalit C. Chhabildas
R. Jeffery Lawrence
Mary Ann Sweeney

# Acknowledgments

Many individuals contributed to the successful completion of this book. First and foremost, we are grateful to our families and particularly to our spouses, Patricia (Pat) Asay, Annette Chhabildas, Jane Lawrence, and Edward Ricco, who patiently supported not only our careers but strongly encouraged the preparation of this book over several years. Secondly, the book would not have been possible without the contributions from researchers involved in shock wave studies who dedicated their time to prepare detailed recollections of their Sandia careers. They provided a personal touch, which became the focus for our descriptions of the technical developments in shock compression science at Sandia over the past 60 years.

Several individuals provided personal accounts that were crucial to our investigations into the early days of shock wave research at Sandia. Interviews with C. Donald (Don) Lundergan, combined with his recollections, clarified how shock wave research was organized in the 1950s; his integration of experiment, theory, and computation ultimately set the stage for the early pioneering challenges that researchers faced in bringing shock wave technology to a high-precision science. This vision for an integrated approach to shock wave research persists to the present day. Orval E. Jones complemented that early history by providing a high-level view of how shock compression science became one of the major research thrusts at Sandia. Early perspectives were also described by several researchers who helped shape the shock wave field through their innovative efforts and resulted in major advances throughout the shock wave community, while aiding Sandia's program as well. In particular, we recognize Robert (Bob) A. Graham for his groundbreaking contributions, which include the first time-resolved stress gauge that enabled many scientific and applied research applications, the initiation of research in shock chemistry at Sandia, and the founding of the Topical Group on Shock Compression of Condensed Matter as part of the American Physical Society. This topical group and its biennial conference have become the premier organization and forum for presenting and discussing shock wave research. Without these early visionary developments, our book would not have been possible. Bob was also instrumental to the development of this book by recommending initial ideas and suggestions on how a history of shock wave technologies should be organized. In addition, Lynn

M. Barker and Roy (Red) E. Hollenbach offered a unique perspective of the early history of shock wave research at Sandia by exemplifying how their innovative efforts in gun impact technologies and groundbreaking instrumentation, such as the VISAR, enabled Sandia National Laboratories to obtain and maintain leadership in the field. Albert (Al) J. Chabai, Dennis B. Hayes, and Ray P. Reed afforded additional personal perspectives of how the shock wave programs in the 1950s led to an expansion of Sandia's role to include nuclear testing at the Nevada Test Site (now known as the Nevada National Security Site).

Many others helped to investigate the early development of shock wave research at Sandia. The Sandia corporate archivist, Myra L. O'Canna, produced invaluable information by researching *Sandia Lab News* articles and archival photos of contributors who worked in the shock wave program at Sandia over the past 60 years. Her dedicated efforts clarified several issues regarding many of the individual research efforts. Marguerite E. Hess, Diana S. Gonzales, Alice Parsons, and Robert Martinez also assisted in resolving several historical questions relating to Sandia reports and copyright issues. Two corporate historians, Alan Carr at Los Alamos National Laboratory and Rebecca Ullrich at Sandia, provided information about the early years at both national security laboratories. Rebecca's efforts were also essential in locating additional archival photos.

Several investigators helped to make the manuscript more comprehensive, including (in alphabetical order) (1) Melvin (Mel) R. Baer for a descriptive rendition on the evolution of multiphase modeling of energetic materials at Sandia and a perceptive discussion of Sandia's role to help resolve the cause of the turret explosion on board the USS Battleship Iowa; (2) Mark B. Boslough for a fascinating personal account of the events leading up to the impact of the Shoemaker-Levy-9 comet on Jupiter and Sandia's participation in interpreting Hubble Space Telescope images of the event; (3) Michael (Mike) D. Furnish for a concise perspective of subcritical experiments at the Nevada National Security Site and also for his careful and diligent efforts in resolving many technical, programmatic, and publication issues; (4) Eugene (Gene) S. Hertel for an insightful and comprehensive discussion of computer code developments at Sandia National Laboratories; (5) James (Jim) E. Kennedy for laying out the events leading to an energetic materials research program at Sandia; (6) J. Michael (Mike) McGlaun for an explanation of computational developments and a firsthand narrative of Samuel (Sam) L. Thompson's contributions to computer code development at Sandia, which underlie many of the contributions to solving national and international shock wave research problems; and (7) Bruno Morosin for a detailed description of the research program initiated on solid-state shock chemistry in the 1980s.

Several researchers from other laboratories were also extremely helpful in preparing the book, including (in alphabetical order) (1) Eric L. Christiansen of NASA, Johnson Space Center, Houston, TX, for assistance in providing information on the space debris program; (2) John R. Cogar, Corvid Technologies, Mooresville, NC, and Brian L. Kiser, Naval Surface Warfare Center, Dahlgren Division, VA, for advice and for providing a graphic illustrating the technical issues related to high-velocity engagements of weapon systems; (3) Jerry W. Forbes, senior scientist at

Energetics Technology Center, St. Charles, MD, for advice on publishing shock wave physics books; (4) Christopher (Kit) H. Neel and David E. Lambert, Munitions Directorate, Eglin Air Force Base, Eglin, FL, who researched several technical issues regarding shock wave publications; and (5) William (Bill) J. Nellis, Harvard University, Cambridge, MA, for providing another perspective of Bob Graham's leadership and seminal contributions to the field of shock compression science.

The document could not have been completed without the support of many people, including (in alphabetical order) (1) Laveryn L. Apodaca for assistance in preparing parts of the manuscript, (2) Steven R. Asay for providing graphics support and preparing several of the figures used in the book, (3) Michael Beckett and Madelynne J. Farber for legal assistance in resolving copyright issues and providing advice on our negotiation of the contractual agreement with the Springer publishing company, (4) Darren L. Buie and Luis Paz for carefully reviewing all the individual recollections and chapters for classification and other sensitive issues, and (5) Amy L. Lucero for her assistance in organizing the bibliography for use in the manuscript.

Finally, we would like to acknowledge John M. Taylor for a thorough reading of the manuscript and for making concise recommendations to improve the accuracy of presentation and to clarify historical events. We also express our deep appreciation to the management and staff of Sandia National Laboratories, particularly in the Pulsed Power Sciences Center, for providing support and exhibiting strong interest in the completion of this book.

# Contents

**Part I Building Shock Wave Capabilities**

**1 Introduction** . . . 3
1.1 Sandia's Roots . . . 3
1.2 Science and Engineering . . . 6
1.3 Building Capability in Shock Wave Research . . . 9
1.3.1 Advances in Experimental, Diagnostic, and Modeling Capabilities . . . 10
1.3.2 Advances and Applications of Computational Capabilities . . . 13
1.4 Book Organization . . . 16

**2 The 1950s: Origins** . . . 19
2.1 Background . . . 19
2.2 Component and System Requirements . . . 27
2.3 Nuclear Testing and Shock Wave Research . . . 29
2.4 Explosive Methods for Shock Wave Research . . . 31
2.5 Development of Precision Impact Launchers . . . 34
2.6 A Family of Innovative Impact Launchers at Sandia . . . 41
2.7 People and Places of the 1950s . . . 44

**3 The 1960s: Explosive Growth** . . . 49
3.1 Background . . . 49
3.2 Time-Resolved Stress Measurements . . . 51
3.3 Time-Resolved Particle Velocity Measurements . . . 56
3.4 Elastic–Plastic Materials . . . 63
3.5 Shock-Induced Spallation . . . 65
3.6 Viscoelastic Materials . . . 67
3.7 Porous Materials (Foams) . . . 69
3.8 Computational Capabilities . . . 72

3.9 Answering DOD's Needs .................................... 75
3.10 People and Places of the 1960s ........................... 79

**4 The 1970s: New Opportunities** .............................. 85
4.1 Background ............................................... 85
4.2 Velocity Interferometer System for Any Reflector ......... 86
4.3 Shock-Induced Phase Transitions .......................... 90
4.4 Two-Dimensional Computer Codes ........................... 93
4.5 Composites and Mixtures .................................. 95
4.6 Damage-Based Spallation Models ........................... 97
4.7 Geological Materials: Oil Shale .......................... 99
4.8 Piezoelectric and Ferroelectric Materials ................ 100
4.9 Third-Order Elastic Constants ............................ 101
4.10 Pressure–Shear Loading .................................. 102
4.11 Shock Rise Time and the Fourth Power Law ................ 105
4.12 Acceleration Waves ...................................... 107
4.13 Shock Thermodynamics Applied Research Facility .......... 108
4.14 Mass Ejecta from Shocked Surfaces ....................... 112
4.15 Energetic Materials ..................................... 113
4.16 People and Places of the 1970s .......................... 118

**5 The 1980s: Heady Times** ..................................... 127
5.1 Background ............................................... 127
5.2 High-Pressure Material Strength .......................... 128
5.3 Three-Stage Railgun for the Strategic Defense Initiative . 131
5.4 Metallization of Hydrogen on the Two-Stage Gun ........... 134
5.5 Graded-Density Impactor for "Soft," or Ramp, Impact ...... 135
5.6 The Sandia HyperVelocity Launcher ........................ 138
5.7 Generalized Fragmentation Theories ....................... 142
5.8 Laser Windows for Mbar Profile Studies ................... 146
5.9 Shock-Induced Solid State Chemistry ...................... 149
5.10 Piezoelectric Polymers for the Bauer Shock Gauge ........ 151
5.11 High-Fidelity Ferroelectric Models ...................... 153
5.12 CTH: A Robust 3-D Hydrodynamic Code ..................... 155
5.13 Turret Explosion Onboard the USS Iowa ................... 158
5.14 People and Places of the 1980s .......................... 161

**6 The 1990s: Black Monday** .................................... 171
6.1 Background ............................................... 171
6.2 Kinetic Energy Kill for Theater Missile Defense .......... 176
6.3 Space Debris Impact on the International Space Station ... 180
6.4 The DOD/DOE Memorandum of Understanding .................. 182
6.5 Mesoscale Modeling ....................................... 183
6.6 Line VISAR for Mesoscale Studies ......................... 187
6.7 MAVEN: Model Accreditation via Experimental
Sciences for Nuclear Weapons ................................. 190

6.8 Underground Testing on Subcritical Experiments .............. 192
6.9 The Role of the Accelerated Strategic Computing Initiative in Shock Wave Research .............. 196
6.10 The Shoemaker–Levy Comet Impact on Jupiter at 60 km/s .............. 198
6.11 ALEGRA: The Next-Generation Hydrodynamic Code .............. 204
6.12 People and Places of the 1990s .............. 208

**7 The 2000s: A New Millennium** .............. 215
7.1 Background .............. 215
7.2 Development of Shock Wave Capabilities on Z .............. 220
7.3 Ramp Loading to Multi-Megabars .............. 223
7.4 Magnetically-Driven Hypervelocity Flyer Plates .............. 228
7.5 *Ab Initio* Equation of State Theories .............. 236
7.6 Containment of Toxic Materials .............. 239
7.7 Compact Pulser: Veloce .............. 243
7.8 Magnetically Applied Pressure Shear (MAPS) .............. 245
7.9 STAR in the New Millennium .............. 247
7.9.1 Effects of Shear Stress on Granular Material Compaction .............. 247
7.9.2 Reverse Taylor Impact Studies .............. 249
7.9.3 Shock-Induced Vaporization: Kinetic Effects .............. 252
7.10 People and Places of the 2000s .............. 254

**8 Looking to the Future** .............. 263
8.1 Looking Back .............. 263
8.2 Looking Forward .............. 269

**Part II Memories of Shock Wave Research**

**9 Memories of Shock Wave Research at Sandia** .............. 275

**James A. Ang** .............. 276

**James R. Asay** .............. 280

**Melvin R. Baer** .............. 306

**Lynn M. Barker** .............. 320

**Mark B. Boslough** .............. 329

**Barry M. Butcher** .............. 341

**Albert J. Chabai** .............. 350

**Lalit C. Chhabildas** .............. 353

**Michael P. Desjarlais** .............. 377

**George E. Duvall (A Tribute by the Editors)** .............. 379

**Michael D. Furnish** . . . . . . . . . . 382

**Dennis E. Grady** . . . . . . . . . . 393

**Robert A. Graham (Recollections of William J. Nellis and Tribute by the Editors)** . . . . . . . . . . 396

**Thomas A. Haill** . . . . . . . . . . 401

**Clint A. Hall** . . . . . . . . . . 408

**Dennis B. Hayes** . . . . . . . . . . 420

**Walter Herrmann (*in Memoriam* by Orval E. Jones)** . . . . . . . . . . 424

**Eugene S. Hertel, Jr.** . . . . . . . . . . 427

**Roy (Red) E. Hollenbach** . . . . . . . . . . 434

**James N. Johnson** . . . . . . . . . . 435

**Orval E. Jones (A Tribute by the Editors)** . . . . . . . . . . 438

**Charles H. Karnes** . . . . . . . . . . 441

**James E. Kennedy** . . . . . . . . . . 442

**Marlin E. Kipp** . . . . . . . . . . 449

**Marcus D. Knudson** . . . . . . . . . . 462

**Carl H. Konrad** . . . . . . . . . . 470

**R. Jeffery Lawrence** . . . . . . . . . . 480

**Raymond W. Lemke** . . . . . . . . . . 488

**C. Donald Lundergan** . . . . . . . . . . 495

**Peter C. Lysne** . . . . . . . . . . 502

**J. Michael McGlaun** . . . . . . . . . . 503

**Stephen T. Montgomery** . . . . . . . . . . 513

**Bruno Morosin** . . . . . . . . . . 521

**Darrell E. Munson** . . . . . . . . . . 527

**Ray P. Reed** . . . . . . . . . . 536

**William D. Reinhart** . . . . . . . . . . 547

**George A. Samara (*in Memoriam* by Alton D. Romig, Jr. and Robert A. Graham)** . . . . . . . . . . 551

**Karl W. Schuler** . . . . . . . . . . 556

**Herbert J. Sutherland** . . . . . . . . . . 562

**Samuel L. Thompson (A Tribute by J. Michael McGlaun: An Enduring Nationwide Impact)** . . . . . 564

**Wayne M. Trott** . . . . . 567

**Timothy G. Trucano** . . . . . 572

**Tracy J. Vogler** . . . . . 583

**Jack L. Wise** . . . . . 589

**List of Acronyms and Abbreviations** . . . . . 595

**Bibliography** . . . . . 601

**Index of Names of Individuals** . . . . . 647

**Index of Terms** . . . . . 657

# Part I
# Building Shock Wave Capabilities

# Chapter 1
# Introduction

## 1.1 Sandia's Roots

The origin of Sandia National Laboratories began with World War II and the Manhattan Project.[1] Prior to the United States entering the war, the U.S. Army leased land then known as Oxnard Field on the desert outskirts of Albuquerque, New Mexico, to refuel and service Army and Navy aircraft in transit. In January 1941, the construction began on the Albuquerque Army Air Base, leading to the establishment near the end of the year of the "Bombardier School—Army Advanced Flying School." Shortly afterward the base was renamed Kirtland Field, after the Army military pilot Colonel Roy S. Kirtland and, in mid 1942, the Army acquired the installation. During the war, Kirtland Field expanded and served as a major Army Air Corps training installation.

General Leslie Groves and Dr. Robert Oppenheimer codirected the Manhattan Project.[2] Two major components of the Manhattan Project were Project Trinity and Project Alberta. Project Trinity prepared for the detonation, on July 16, 1945, of the first nuclear bomb (nicknamed "the Gadget") at the Trinity Site in New Mexico. Project Alberta involved the assembly, testing, arming, and delivery of the first airborne nuclear weapon (nicknamed "Little Boy"), a uranium gun-type nuclear

---

[1] For additional details about Sandia's roots and early years, see the first externally published volume of the history of Sandia, by Necah Stewart Furman, *Sandia National Laboratories: The Postwar Decade* (University of New Mexico Press, Albuquerque, 1990). That 858-page treatise includes extensive notes and a substantial bibliography. See also Leland Johnson, "Sandia National Laboratories: A history of exceptional service in the national interest," ed. by C. Mora, J. Taylor, R. Ullrich, Sandia National Laboratories Report SAND97-1029, Albuquerque, NM, 1997.

[2] *Los Alamos National Laboratory: A Proud Past, an Exciting Future* appeared in 1995 as a special issue of *Dateline Los Alamos*, a monthly publication of the Public Affairs Office of Los Alamos National Laboratory. That 50-year Trinity anniversary issue provides an interesting retrospective with many recollections and photographs.



J.R. Asay et al., *Impactful Times*, Shock Wave and High Pressure Phenomena,
DOI 10.1007/978-3-319-33347-2_1

device, to Hiroshima on August 6, 1945. "Fat Man," a plutonium implosion-type device, was delivered to Nagasaki 3 days later.

During the Manhattan Project, engineering activities at Project Y or Site Y,[3] the high-mesa hideaway in the Jemez Mountains northwest of Santa Fe, were carried out by "Z Division." Z Division was named for the division head, Jerrold Zacharias, a professor from the Radiation Laboratory at the Massachusetts Institute of Technology. The division was conceived as an ordnance design, testing, and assembly arm of the nuclear bomb program. In July 1945, the site of this forerunner of Sandia National Laboratories was established in Albuquerque to handle weapon development, testing, and bomb assembly.

Near the end of the war, Groves and Oppenheimer faced the challenge of turning the nuclear bomb effort into an operation to produce and maintain the Nation's stockpile and develop peaceful uses of nuclear energy. Space at Site Y was then at a premium. Moreover, the members of Z Division needed to work closely with the military, and the remoteness of the site made transport of weapon components difficult. In addition, to encourage personnel to stay after the end of the war, Groves and Oppenheimer decided to focus what became the Los Alamos Scientific Laboratory on weapon design and to relocate the weapon production and assembly elsewhere. In the several months before July 16, Oppenheimer began looking for a location to continue that weapon engineering work—especially its non-nuclear aspects. Kirtland Field had served the transportation needs for Projects Trinity and Alberta, hence, the decision to relocate Z Division there permanently. The military base was therefore transferred from the Army Air Corps to the U.S. Army Service Forces Chief of Engineering District and assigned to the Manhattan Engineer District of the War Department.

By the close of the war, Z Division had begun to consolidate weapon assembly at the Albuquerque site. By 1946, the site was known as "Sandia Base," after the nearby Sandia Mountains.[4] Figure 1.1 is an early photograph that includes the Albuquerque site. From 1947 to 1971, Sandia Base was the principal nuclear weapon installation of the U.S. Department of Defense (DOD). Over that period, nuclear weapon research that included development, design, testing, and training initiated by the Manhattan Project was conducted there and its subsidiary, Manzano Base. Fabrication, assembly, and storage of weapons were also performed at Sandia Base. In 1971, Sandia Base was merged into Kirtland Air Force Base.

By April 1, 1948, Z Division had grown to about 500 people and was renamed Sandia Laboratory. By mid 1948, it grew to around 1000 employees. On May 13, 1949,

[3] During the war, the term "Los Alamos" was not used to ensure secrecy. The earliest official reference to Los Alamos Scientific Laboratory, according to Alan Carr, the Los Alamos National Laboratory historian, was in mid October 1945. The name was changed to Los Alamos National Laboratory on January 1, 1981.

[4] These mountains were named for the Spanish word for watermelon because, as viewed from the west, they turn that color at sunset in the winter.

**Fig. 1.1** Kirtland Army Air Base is in the foreground with Sandia Base in the background (1945). The Sandia mountains are visible in the far left background (Reprinted with permission of Sandia National Laboratories)

President Harry S. Truman sent a terse letter to Leroy Wilson, the president of the American Telephone and Telegraph Company (AT&T).

> I am informed that the Atomic Energy Commission intends to ask that the Bell Telephone Laboratories accept under contract the direction of the Sandia Laboratory at Albuquerque, New Mexico. This operation, which is a vital segment of the atomic weapons program, is of extreme importance and urgency in the national defense, and should have the best possible technical direction.... In my opinion, you have here an opportunity to render an exceptional service in the national interest....

On November 1, 1949, a no-fee contract was established with AT&T to manage Sandia Laboratory through Sandia Corporation, an AT&T subsidiary. In 1979, the U.S. Congress designated Sandia as a national laboratory. The AT&T contract to operate Sandia remained in effect until October 1993. At present, Sandia National Laboratories[5] (SNL) is managed and operated by Sandia Corporation, a wholly owned subsidiary of Lockheed Martin Corporation.

## 1.2 Science and Engineering[6]

Sandia's urgent task after the war was to build a stockpile of nuclear weapons. The early nuclear weapons were carried by subsonic aircraft. At that time, the structural and environmental requirements did not differ substantially from conventional weapons. Hence, the existing engineering procedures and materials for system design were sufficient, but nuclear weapons had stringent reliability and safety requirements. The engineering groups at Sandia actively developed innovative safety concepts and conducted extensive tests for operational reliability during this period. Parachutes to permit delayed delivery, weak-link and strong-link safety systems, and reliable barometric and contact fuzing were among the new subsystems that had to operate in all environments.

The advent of supersonic and exoatmospheric missile delivery systems imposed more demands. Addressing these issues required understanding the effects of adverse environments on components and subsystems. A major issue was the effect of radiation-produced shocks during a nuclear burst. The low-energy x-ray radiation often caused material ablation, and the ensuing shock waves could cause major damage. The resulting stress waves propagated through materials and often produced component or subsystem failure, which could preclude proper weapon reentry and operation. Resolving this issue required knowledge of the stress levels produced by the radiation pulse and the tensile failure strength (referred to as "spallation strength") of a broad range of materials. Knowledge of mechanical response, such as elastic–plastic behavior, did not exist at the stress and loading rates (usually referred to as "strain rates") experienced in these environments. Extrapolation of material response from low strain rates and stresses using the highest loading rates available at the time (typically provided by Hopkinson bars[7]) was not reliable. The extensive shock wave research at Los Alamos during and after the Manhattan Project provided information on the high-pressure equation of state (EOS) properties but essentially no

[5] Sandia operates laboratories and testing facilities in Albuquerque, New Mexico, in Livermore, California, at the Tonopah Test Range in Nevada and at the Kauai Test Facility in Hawaii. Sandia also has offices in Carlsbad, New Mexico (the Waste Isolation Pilot Plant), in Mercury, Nevada (to support the Nevada National Security Site), in Amarillo, Texas (the Weapons Evaluation Test Laboratory), and a program office in Washington, D.C.

[6] Much of the material in this section was summarized from the recollections of C.D. Lundergan, B.M. Butcher, A.J. Chabai, and R.P. Reed in Part II.

[7] Hopkinson bars provided data at loading stresses of a few kbars and strain rates up to about 1000/s.

information on mechanical properties in the tens of kbars[8] range and at loading times of less than 1 μs.

The response of materials subjected to impacts associated with delivery velocities and defensive countermeasures was also important. For reliable operation, the subsystem or component had to retain its integrity for several microseconds after impact with the ground or other structures or after activation of contact and other weapon fuzes, to prevent deformation and destruction. Hence, this issue concerned how large amplitude stress waves propagated through many materials.

Weapon systems were also becoming more complex as component miniaturization evolved. Furthermore, operation of a weapon was sequential, with each component operating fully before the next one was started. Hence, the size and timing were critical to weapon delivery. Coupled with miniaturization, another critical factor was the energy to power components such as gyros, triggers, and radar. All required power, either electrically or through mechanical actuation. Weapons would usually be stored for long periods and then have to function without failure. Small, on-board explosive devices were attractive since they could be stored and then detonated to produce significant power on demand.

The propagation of stress waves generated by the detonation of explosives can cause damage to adjacent components and subsystems. How the stress waves were generated and spread through the weapon system had to be determined. A basic understanding of solid material response at a few kbars and at high loading rates for engineering models did not exist.

The myriad issues facing weapon designers and the need to understand material properties prompted shock wave research on the following topics:

1. Dynamic response of mixtures, composites, and polymers in explosive power supplies to predict stress limits and operational times
2. Electromechanical effects and operational limits from shock loading piezoelectric and ferroelectric (FE) materials
3. Energetic materials for detonators and explosive power supplies
4. Pulsed radiation effects on weapon structures and components
5. Ground shock effects from nuclear explosions in geological media and the effects on above- and below-ground structures
6. Compaction behavior of porous materials for shock cushioning of components and subsystems
7. Time-resolved gauges to quantify material response to pulsed radiation

By the mid 1950s, Sandia management decided to establish a research program modeled after AT&T Bell Laboratories to support these engineering applications. An immediate outcome was the formation of the Physical Sciences Department and the Physical Research Department in the Research Directorate managed by Stuart C. Hight. By 1957, each department had 30 to 40 researchers in several scientific

---

[8] 1 kbar = 1000 atm or 14,500 lb per square inch (psi). The pressure units of kbar and Mbar (1 Mbar = 1000 kbars) are used throughout the text, although some figures show pressure in gigapascals or GPa (1 GPa = 10 kbars = 0.01 Mbar).

disciplines. The Physical Sciences Department under Richard Claassen was engaged in fundamental research, and the Physical Research Department under George Hansche was engaged in applied research. Other departments in the Research Directorate included the technical areas of weapon effects, mathematics, and aerodynamics.[9]

Two visionary individuals profoundly influenced the early and subsequent development of shock wave research at Sandia. One was Frank W. Neilson, and the other was C. Donald Lundergan. Neilson initiated a research effort in the mid 1950s to understand the response of FE ceramics to shock loading. That effort resulted in the ground-breaking development of time-resolved stress gauges.[10] As a division supervisor, Neilson recruited Orval Jones from the California Institute of Technology. Jones then played a major role in the development and application of shock wave research at Sandia.

On the engineering side, Charles Bild, Director of Materials and Process Development, also encouraged and strongly supported shock wave research for weapon component and system development. At the request of Leon Smith, manager of the Electrical Systems Department in that directorate, Lundergan was asked in 1957 to perform projectile impact experiments with an air-driven gun on contact fuzes that were being developed. This work laid the groundwork for a new program to measure the dynamic response of materials under precisely controlled impact conditions (see Chap. 2). Lundergan recognized the critical importance of the emerging shock wave technology to Sandia's mission and proposed a comprehensive plan for a new department focused on all aspects of the technology (theoretical modeling, computational capability, and experimental research). In 1959, he recruited Lynn Barker who, along with Roy (Red) Hollenbach, had a profound influence on the development and application of experimental shock wave techniques.[11] Later, in the mid 1960s, Lundergan recruited Walt Herrmann from the Massachusetts Institute of Technology (MIT) to initiate a program in material modeling and the development of hydrodynamic computer codes. Herrmann was successful in this challenging assignment and in closely coupling these activities to a strong experimental program, as discussed in the following chapters.

A third technical area at Sandia in which shock wave technology has played an important role is geological materials. Since its formation, Sandia has participated with the other two national security laboratories, Los Alamos National Laboratory (LANL) and Lawrence Livermore National Laboratory (LLNL),[12] in field testing

---

[9] These details are from a private discussion with Orval Jones in 2012. A tribute to Orval is included in Part II.

[10] This information is from Bob Graham's commemoration of George Samara's work at Sandia (George Samara Memorial Symposium, Sandia National Laboratories, on May 18, 2007. Bob's *in Memoriam* is included in Part II.

[11] Lynn Barker took a leave of absence to study for a PhD in physics at Cornell University in 1961 but returned to Sandia in 1962 after his wife became very ill.

[12] The laboratories have been known by those names since 1981. These acronyms are used in the rest of the text.

nuclear weapons. Aboveground testing was common until October 10, 1963, when the United States, Britain, and the Soviet Union signed a treaty on banning nuclear tests in the atmosphere, oceans, and space. Sandia continued to participate in underground tests at the Nevada Test Site[13] (NTS) led by its two sister laboratories.

Several Sandians made significant contributions to the national defense posture through nuclear testing in the early years, including Bob Bass, Carter Broyles, Ronald Carlson, Albert Chabai, Hunter DeVault, Doris Hankins, Bill Perrett, Carl T. Smith, and Luke Vortman. An important contribution was a new gauge technique to estimate the yield of nuclear weapons. In the early 1960s, Bass and Chabai adapted the SLIFER[14] technique, originally developed at LLNL, to obtain continuous shock wave attenuation data in geological materials (e.g., volcanic tuff, granite, desert alluvium, and salt). Bass and Chabai discovered a universal power–law relationship for shock position versus arrival time as a function of yield that was independent of the geological materials. Bob Brownlee of Los Alamos used the Sandia SLIFER technique and this relationship on many nuclear events at NTS, since it could provide yield estimates within an hour after the detonation of a device.

With the end of the Cold War and the Comprehensive Test Ban Treaty (CTBT), which has been signed but not yet ratified and so is not yet in force, all underground nuclear testing was discontinued. Since then the National Nuclear Security Administration (NNSA) has relied mainly on the Stockpile Stewardship Program (SSP) to sustain and assess the safety, security, and effectiveness of the stockpile through advanced computing and development of complex models based on laboratory experiments, without the use of underground nuclear tests. An extraordinary set of science, technology, and engineering facilities supports the SSP. At Sandia, the Z pulsed power facility and the Shock Thermodynamics and Applied Research (STAR) facility, in particular, support shock physics studies for materials of interest. These contributions are discussed in more detail in Chaps. 6 and 7.

## 1.3 Building Capability in Shock Wave Research

The decision to initiate a shock wave research program at Sandia was instrumental in the rapid development of capabilities to address a wide range of weapon science and basic science problems. In the mid 1950s, a small ongoing effort involving shock waves was directed toward understanding how the electrical output of FE crystals responded to dynamic impact, such as shock loading by small explosive charges. The motivation for that research was the need to develop various components, including explosively actuated power supplies, to detonate the nuclear weapon.

[13] The Nevada Test Site was renamed the Nevada National Security Site (NNSS) in August 2010.

[14] SLIFER is an acronym for Shorted Location Indicator by Frequency of Electrical Resonance.

### 1.3.1 *Advances in Experimental, Diagnostic, and Modeling Capabilities*

The pioneering team of Frank Neilson, Bill Benedick, and Bob Graham began by investigating the electrical output of FE ceramics subjected to shock loading. Barium titanate was one of the first materials studied to understand how loading conditions, such as peak stress and loading times, generated the electrical output. Understanding the coupled mechanical and electrical response required experiments with carefully controlled loading conditions. Precision diagnostics were also needed to measure the input stress and electrical time histories; controlled shock wave techniques and instrumentation were just beginning to become available in the mid 1950s.

The simplest configuration to produce well-controlled shock compression is planar loading of a thin disk. In this configuration, a planar shock wave propagates through the disk and is detected when it arrives at the back surface, thereby determining the shock velocity. In general, measurements are made of the shock velocity and either the input shock pressure or particle velocity produced by the shock wave.[15] These two quantities can then be applied in the equations for planar shock motion to determine the pressure and density, or the specific volume (inverse density), produced in a sample. By performing experiments at different initial shock pressures, a locus of pressure–density points is obtained, which is referred to as the Hugoniot curve for the material. This curve is used to develop material models to describe the dynamic response for various applications.

Facilities for laboratory shock wave experiments were extremely limited in the 1950s. The first experiments at Sandia were conducted with the high explosive (HE) loading technique developed as part of the Manhattan Project. Explosive shock wave generators, known as plane-wave lenses, were used to generate the high shock pressures. Similar techniques, but scaled to lower pressures, were used by Neilson's team. Accurate diagnostics, mainly time-of-arrival shorting pins that had been developed to detect the arrival of the shock wave at the back surface of a sample, were also used. In addition, the shock or mechanical response and the electrical output of FE materials were determined to quantify the coupled electromechanical behavior.

The HE experiments were conducted outdoors at a remote site in Tech Area III, about five miles south of Tech Area I at Sandia. In the mid 1960s, this effort evolved into a major thrust within the Physical Research Department that focused on the physics and chemistry of shock compression for a broad range of materials. The development and application of this work are discussed in more detail in Chaps. 3, 4, and 5.

---

[15] A longitudinal stress state is produced in a solid under planar loading. In a fluid, this is the pressure. In a solid, the difference between the two states is two-thirds of the yield stress of the solid, which is generally small, so stress and pressure are similar in most cases. Stress and pressure are used interchangeably throughout this book unless it is necessary to distinguish between the two in some discussions.

In parallel with shock wave research on FE materials, a similar program began in another organization to understand the response of contact fuzes and other weapon components and systems. This experimental effort under Don Lundergan began a few years after the explosives work on FEs and at first focused on the mechanical behavior and engineering aspects of component response. Along with Lundergan, Lynn Barker and Red Hollenbach were the principal participants in that parallel effort, which began around 1957.

Instead of explosive charges, in 1958 Lundergan took a different approach that had a lasting influence on shock wave research, not only at Sandia but in the entire shock wave community. He used a 100-mm-bore compressed air gun to accelerate a projectile with controlled velocities up to about 0.3 km/s that, upon impact, produced planar shock waves in a flat-disk target. The air gun was a "hand-me-down" that had been used to test the effects of impacts on weapon components. Precise perpendicular impact with minimal "tilt" was achieved by accurate boring and polishing of the gun bore, designing projectiles with flat, normal impact plates, and precisely aligning flat target samples to the normal impact plates. In addition, the use of compressed air as an accelerant rather than explosives avoided several operational restrictions, such as the location of the experiment and safety regulations associated with explosives. Later this gun was upgraded to use helium gas, which increased the impact velocity, and was moved to more comfortable quarters in Tech Area I as an all-weather indoor facility.

Lundergan's initial air-gun launcher marked the start of a family of precision guns for material property studies and also prompted their development at other institutions, as described in Chap. 2. Figure 1.2 summarizes the development of gun capabilities at Sandia. In 1961, Bob Graham developed a 40-mm-bore powder gun for increased velocity (Graham, Ingram, and Ingram 1961). Somewhat later, Graham developed a helium gun to launch projectiles to about 1 km/s (Graham 1961a). This was a major achievement in shock wave experiments at the time. In 1969, Darrell Munson and Ray Reed implemented an 89-mm-bore powder gun that doubled the launch velocities to over 2 km/s, providing impact pressures of several hundred kbars. A few years later, Bob May achieved a major advance in launcher capability with the construction of a two-stage light-gas gun (Munson and May 1975), which tripled the velocity to over 7 km/s and gave impact pressures of 3 to 4 Mbars. Around 1990, Lalit Chhabildas invented a three-stage launcher, the HyperVelocity Launcher (HVL), which more than doubled the velocity to 16 km/s, as described in Chap. 5 (Chhabildas et al. 1992, 1995).

Along with the rapid advances in facilities, diagnostics for precision measurements advanced quickly. The initial instrumentation for measuring shock wave properties consisted primarily of electrical shorting pins of different heights on the back surface of a shocked sample. Shorting of the pins determined the displacement versus time history produced by a shock wave and thus the shock velocity (Smith and Barker 1962). This simple technique was not optimal for determining the structure of the shock wave. To provide more detail about the shock structure, Barker and Hollenbach devoted intense efforts to measuring the free-surface displacement continuously. Both a slant-wire resistor (Barker 1961, 1962; Barker and Hollenbach 1964) and displacement

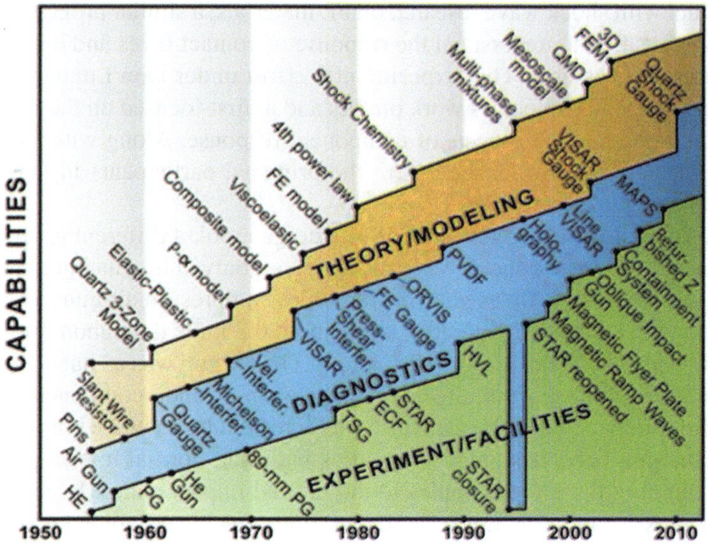


**Fig. 1.2** Development of shock wave capabilities at Sandia. The large drop in experimental capabilities in 1994 resulted from the management decision to dismantle the STAR facility and eliminate all associated experimental activities. Acronyms used in the figure are: ECF (Explosives Component Facility), FE (ferroelectric), FEM (ferroelectric model), HE (high explosive), HVL (HyperVelocity Launcher), MAPS (magnetically applied pressure shear), ORVIS (Optically Recorded Velocity Interferometer), P-α model (model for response of porous materials), PG (powder gun), PVDF (polyvinylidene difluoride), QMD (quantum molecular dynamics), STAR (Shock Thermodynamics Applied Research), TSG (Two-Stage light gas Gun), VISAR (Velocity Interferometer System for Any Reflector)

interferometer techniques (Barker and Hollenbach 1965) were developed, as illustrated in Fig. 1.2. These developments are discussed in Chap. 3.

Two ground-breaking innovations in the 1960s discussed in Chap. 3 were the X-cut quartz gauge to measure stress directly (Neilson and Benedick 1960) and an interferometer technique to measure the particle velocity produced by a propagating shock wave (Barker 1968). Neilson, Benedick, and Graham's quartz gauge advance (Neilson and Benedick 1960; Neilson, Benedick, Brooks et al. 1962; Graham 1961b, 1975) provided a continuous measure of shock pressure (which is also referred to here as stress) versus time (i.e., a time-resolved shock profile). A game-changing advance in instrumentation was the 1972 invention of the VISAR (Velocity Interferometer System for Any Reflector), which allowed detailed shock wave studies of any material (Barker and Hollenbach 1972). These two innovations impacted shock wave research worldwide.

These applications include:

- dynamic elastic yielding,
- dynamic compressive and tensile strength,
- determination of phase changes produced by shock compression and the associated transformation kinetics,
- initiation and growth of energetic reactions,

- optical properties under shock compression, and
- effects of nonhomogeneous material properties on planar shock compression.

Other major advances followed the VISAR, including

- ORVIS (Optically Recorded Velocity Interferometer System),
- the piezoelectric polymer PVDF (polyvinylidene difluoride),
- the Line VISAR,
- dynamic holography for measuring nonuniform material motions, and
- the use of X-cut quartz as a shock velocity vs. time profile gauge, rather than as a stress vs. time (or particle velocity vs. time) profile gauge, at pressures exceeding 3 Mbars to measure shock velocity directly (Knudson and Desjarlais 2009).

All these developments are discussed further in subsequent chapters

Dynamic material models to predict the response for materials of interest to the Sandia weapons program (e.g., metals, viscoelastic materials, geological materials, mixtures and composites, ferroelectric and piezoelectric materials, porous materials, and explosives) were not available in the mid 1950s and early 1960s. The rapid advance in experimental capabilities led to new models to understand the fundamental deformation mechanisms and to predict how a material or component would respond to shock loading. The modeling deficiency was explicitly addressed in the 1960s, beginning with Neilson's three-zone model of the electrical response of shocked piezoelectrics (Neilson and Benedick 1960). That model had a major impact on the development of materials for shock wave gauges. Other models developed in this period were elastic–plastic models for metals, composite models for mixtures and laminated materials, models for energetic materials, and models for dynamic failure and fragmentation. Moreover, in the 2000s *ab initio* models, such as quantum molecular dynamics (QMD), began to predict very high pressure EOS response. Figure 1.2 above summarizes the progress in developing a theoretical understanding of shocked materials. That progress required integration of three activities: experimental advances, diagnostic development, and advances in theoretical modeling, as discussed in subsequent chapters.

### *1.3.2 Advances and Applications of Computational Capabilities*

The development of realistic models for shock wave response allowed detailed analysis and prediction of the performance of weapon components and systems during dynamic loading. However, this step required advanced computer codes to simulate configurations in three dimensions and computers with sufficient speed and memory to perform simulations of complex configurations. Fortunately, both tools were developed in the same time frame as the shock wave capabilities in Fig. 1.2, thereby allowing increasingly larger and more realistic simulations. Figure 1.3 summarizes the major advances in codes and computer capabilities from the 1950s to the present.

In the 1950s and to some extent the 1960s, the computer capabilities and simulations for modeling shock wave propagation had not advanced to the point where com-

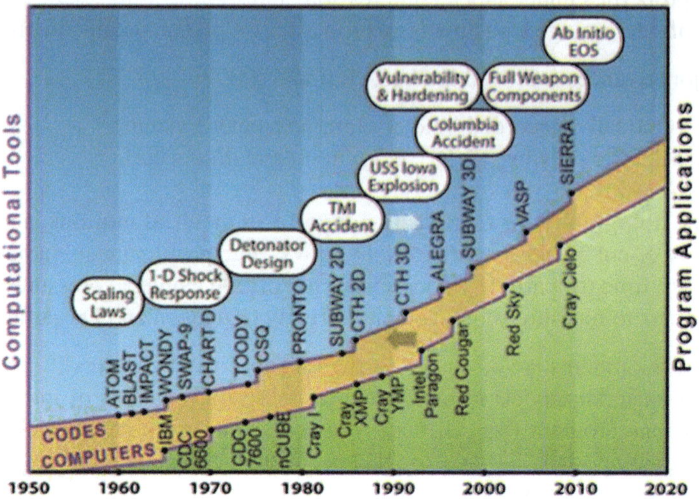


**Fig. 1.3** Hydrodynamic computer codes (blue background) and main frame computers (green background) used for shock wave simulations. Acronyms used in the figure are: 1-D (one-dimensional), ALEGRA (Arbitrary Lagrangian Eulerian General Research Applications), CDC (Control Data Corporation), EOS (equation of state), IBM (International Business Machines), TMI (Three Mile Island). Balloons with white backgrounds identify a few major applications of Sandia analytical and computational capabilities used during the corresponding time periods

plex problems—such as crater formation from an explosive blast, underground shock propagation and interaction with a structure, or coupled radiation-shock response—could be simulated in realistic geometries. In the early years, many of the problems had to be addressed with a variety of scaling laws. With the rapid advances in simulation codes and more powerful computers, the situation began to change in the 1960s.

During this period, mainframe International Business Machines (IBM) computers were available, although they were limited by speed and memory that restricted their use to simple problems such as the design and analysis of planar shock wave experiments or one-dimensional (1-D) approximations of weapon components. In addition, the material models and the codes (usually referred to as "hydrodynamic codes" but sometimes as hydrocodes) were not available. In the mid 1960s, the only codes to simulate shock wave problems were the 1-D code WONDY, a finite-difference Lagrangian code, and SWAP-9, a characteristics-type code that tracked shock waves.

From the early 1950s to the early 1970s, the capability of mainframe computers grew a thousandfold, consistent with Moore's law,[16] which predicts a doubling of computing capability every 18 months. The Control Data Corporation (CDC) machines introduced in the early 1970s provided a major advance in computing power to develop the next-generation hydrodynamic codes, namely, the two-dimensional (2-D) codes TOODY and CSQ, as identified in Fig. 1.3. Sandia purchased a CDC 6600, considered to be the first supercomputer, around 1970. By

[16] What became known as Moore's law was published in the April 19, 1965, issue of the monthly magazine *Electronics*, by Intel co-founder Gordon Moore.

1976 two CDC 6600-class computers were available. Later, a CDC 7600-class computer was acquired, which was about ten times faster than the CDC-6600 and had 512 kbytes of memory. With these computers, the hydrocodes TOODY and CSQ solved increasingly complex problems, such as the hypervelocity impact of a nylon ball onto a steel plate or analysis of concrete rupture during the Three Mile Island (TMI) accident in 1979. The purchase of a Cray 1, with 8 Mbytes of memory, in the early 1980s significantly increased the capability to perform large shock wave problems.[17] The next large computer at Sandia was the Cray XMP, followed by a Cray YMP in the late 1980s. These machines had the capability to store the very large databases required for large numerical simulations.

The Cray YMP was followed by the massively parallel processing (MPP) computers in use today. The first MPP computer at Sandia was the Intel Paragon, which figured prominently in solving several large shock wave problems. The first production MPP computer for solving shock wave applications was the Accelerated Strategic Computing Initiative (ASCI) Red machine. Sandia researchers now routinely use the Cray Cielo supercomputer at LANL for the largest shock wave calculations. Cielo is ten million times faster than the Cray 1.[18]

The enormous increase in computing power over the last several decades has had a profound impact on the complexity, size, and type of shock wave problems that can be analyzed. The top part of Fig. 1.3 highlights a few of the many applications possible because of the increased capabilities. In the 1950s and early 1960s, scaling solutions were often used to predict the effects of complex shock wave propagation in geological media or the damage to structures by large explosions. In the mid-to-late 1960s, WONDY and SWAP-9 began to be used for 1-D shock response simulations, including the design and analysis of experiments and some weapon applications. Porous material crush up was of particular interest to the weapons program and could be predicted with WONDY, as discussed in Chap. 3.

In the 1970s, CSQ and TOODY simulated several important weapons problems as well as problems of national import, including the TMI nuclear reactor accident (discussed briefly in Chap. 4); in the late 1980s, as discussed in Chap. 5, the CTH code was used to analyze the turret explosion aboard the USS Iowa. The combined development in the early 1990s of three-dimensional (3-D) hydrodynamic codes, such as CTH, and the advent of fast MPP computing enabled a greater range and increased complexity of applications that could be addressed. Notable examples include analysis of the Shoemaker-Levy comet impact on Jupiter and missile intercept lethality, both of which are discussed in Chap. 6. The Jupiter impact simulation provided estimates of crater size and impact plumes, allowing observers on the Earth to interpret the astronomical observations obtained with the Hubble telescope. In the 2000s, these capabilities determined that the cause of the Columbia shuttle disaster was foam debris from the fuel tank impacting the orbiter wing.

---

[17] An interesting aside is that the 2011 Smart Phone is equivalent in computer power to the early 1980s Cray 1.

[18] The fastest computer now available, LLNL's Sequoia, is about a thousand times faster than that predicted by Moore's law (Bob Schmitt, Sandia National Laboratories, 2013).

Code development efforts continued into the late 1990s, resulting in increased capabilities to describe complex material phenomena, including the 3-D SUBWAY code to simulate the electromechanical response of weapon components (as discussed in Chap. 6). The 3-D magnetohydrodynamic code ALEGRA[19] further expanded the types of phenomena that could be studied (see Chap. 6). The *ab initio* calculations of material properties that are now possible are discussed in Chap. 7. Several other model development efforts and examples of simulations are discussed in the chapters. The remarkable upshot is the enormous increase in the complexity of problems that are solvable through the concerted efforts of shock wave research, hydrodynamics code development, and increased computing capability.

To accomplish these major advances, the number of Sandia staff involved in shock wave research grew rapidly in the 1960s. By the end of the 1950s, about a dozen staff were associated with that research. By the end of the 1960s, about four dozen researchers were involved. Hiring of shock wave researchers continued into the 1970s and beyond, but at a slower rate. In the mid-to-late 1970s, a number of researchers left the Physics of Solids Research and Solid Dynamics Research Departments, primarily because of other growing programs at Sandia, such as the Energy Initiative in the U.S. Department of Energy (DOE). A resurgence of hiring in shock wave research by the Pulsed Power Sciences Center occurred in the early 2000s with the discovery of magnetic loading, which considerably expanded the material response regions accessible through shock compression. This new area of shock physics is discussed in Chap. 7.

## 1.4 Book Organization

The book is divided into two parts, Part I, Building Shock Wave Capabilities, and Part II, Memories of Shock Wave Research. The rest of Part I contains seven chapters. Chapters 2 to 7, organized by decade, highlight the major developments and activities during each specific decade. Chapter 2, "Origins," describes the programmatic needs at Sandia that led to the emerging shock wave effort in the 1950s. Chapter 3, "Explosive Growth," denotes the rapid increase of shock wave capabilities in the 1960s. Chapter 4, "New Opportunities," emphasizes the wealth of opportunities that emerged in the 1970s and the computational tools that were developed concurrently and applied to important problems. Chapter 5, "Heady Times," documents the breadth of applications that used these capabilities in the 1980s. Chapter 6, "Black Monday," describes the near elimination and resurgence of Sandia's experimental shock wave capabilities and the impact of computational advances to solve complex shock wave problems. Chapter 7, "A New Millennium," discusses the techniques and initiatives that have reenergized shock wave research at Sandia. The final chapter, "Looking to the Future," provides some possible directions for

[19] ALEGRA is the acronym for the Arbitrary Lagrangian Eulerian General Research Applications code.

future shock wave research. Part II presents the individual recollections of over 40 shock wave researchers at Sandia; selected quotes from those recollections are included in the chapters to highlight the personal struggles and successes of individual researchers. Finally, an extensive set of nearly 1000 technical publications is listed in the Bibliography for those interested in pursuing in more detail the shock wave research conducted at Sandia National Laboratories over the last 60 years.

# Chapter 2
# The 1950s: Origins

## 2.1 Background

Shock compression science, in its modern form, evolved from the early work conducted at Los Alamos resulting from the Manhattan Project. That work, of course, was an important aspect of the development of nuclear weapons, first the fission bomb toward the end of World War II and then the fusion bomb, both of which were mainstays of the Cold War era.

Work on nuclear weapons and in many ancillary areas continued after the war as high-priority military research and development (R&D) programs. However, to keep overall control of nuclear energy research in civilian hands, in particular for developing nuclear weapons, the Atomic Energy Commission (AEC) was established in 1946. The AEC was subsequently transformed into the Energy Research and Development Administration (ERDA) in 1974 and then into the Department of Energy (DOE) in 1977, where it remains today. Most of the postwar work on shock compression science was sponsored by these organizations and conducted at the national security labs, by many civilian contractors of DOE and the Department of Defense (DOD), as well as at universities, including (but not limited to) Brown University, California Institute of Technology (Caltech), Georgia Tech, Harvard University, Massachusetts Institute of Technology (MIT), University of California at San Diego (UCSD), and Washington State University (WSU). These follow-on efforts involved somewhat different paths. The largest effort concentrated on improving the understanding and capabilities of newer and safer nuclear weapons, which were more tailored to specific military objectives. In addition, scientists within DOD and the national security labs recognized the need to study and understand the effects of these new weapons on military and civilian facilities and systems.

An urgent task of AEC and DOD after World War II was to build a stockpile of nuclear weapons. For Sandia, a principal requirement was that components and structures of nuclear weapon systems meet operational requirements and that the weapons reliably operate when deployed. However, the weapons also needed to be



J.R. Asay et al., *Impactful Times*, Shock Wave and High Pressure Phenomena,
DOI 10.1007/978-3-319-33347-2_2

unquestionably safe from unintentional nuclear detonation under all other conditions. The early nuclear weapons were bombs carried by subsonic aircraft, simply because their structural and environmental requirements did not substantially differ from conventional weapons. Accordingly, the use of existing engineering procedures and materials in the design of these early systems was sufficient, although they required faultless application to meet the more stringent reliability and safety requirements. Sandia engineering groups were starting active development of innovative safety concepts and extensive testing of weapons for operational reliability during this period. New concepts for the components and subsystems included parachutes to permit delayed delivery, weak-link and strong-link safety systems, and reliable barometric and contact fuzing.

In addition to understanding the response of these newly designed components and systems to shock or dynamic loading, Sandia's engineers needed to measure and predict effects of shock waves produced during a nuclear explosion, both on the Nation's military systems and on intended targets. These requirements led to an extensive program of nuclear testing in the atmosphere, on the ground, underwater, and much later to fully contained underground tests. Early on, these studies mostly involved large-scale phenomena such as cratering and the destruction of buildings and other structures. The effects of these tests were often observed with late-time diagnostics that led to scaling analyses. At the same time, scientists were requesting more sophisticated data related to the actual testing. In particular, better determinations of the actual nuclear yield, as well as the dynamic loads on the targets, were required to improve the large-scale effects analyses. Analysis of the specific response of the weapon systems and their components generally required detailed, time-resolved data. This led to an increasing awareness of the need for complementary R&D in the area of shock generation and propagation. These laboratory investigations, however, were at pressure and size scales significantly smaller than the early LANL work, which was more directly connected with "bomb" physics, i.e., the operational response of actual nuclear packages and weapon components during implosion or gun assembly of the nuclear material.

The national security laboratories, and in particular Sandia, already had efforts underway in several related fields. These were typically in areas needed for Sandia's core mission at the time to "weaponize," or add, all the required non-nuclear "bells and whistles" to the new weapons in the development queue. More specifically, the engineering groups at Sandia were tasked to develop and analyze the subsystems and components that would be subjected to impulsive pressures and stresses. In many operational scenarios, these dynamic loads could reach many tens of kilobars in times of less than 1 μs (one millionth of a second). As these operational requirements were identified and formulated, Sandia realized that it did not have the required experimental capabilities and that they were not available anywhere else for that matter. Because of this deficiency, there was a compelling need to develop and implement the appropriate capabilities. As a result, rapid growth of shock wave capabilities began in the late 1950s and early 1960s to allow studies of dynamic behavior on materials of interest to the weapons program. This led to the design of diagnostics and loading techniques for the precision measurement of time-dependent material response at relevant stress and time regimes. Development of these capabilities was essential to Sandia's mission. In addition, the fundamental experimental and modeling approaches implemented during that time laid the groundwork for research programs that expanded to an investigation at stresses of tens of Mbars, which are still of interest today for weapon and other applications.

In the late 1950s and early-to-mid 1960s, at least four principal R&D efforts were in place related to dynamic material response. A primary objective was to evaluate the behavior of weapon components subjected to shock wave loads—that is, the dynamic response of contact fuzes to impact or the performance of explosive devices being developed to generate electrical current for various weapon components. Support of the Nation's nuclear weapon testing activities was another early driver of the shock wave research. Both programmatic drivers motivated basic research focused on determining the properties and detailed behavior of many classes of materials under dynamic loading. This information was urgently needed to develop components and subsystems that could be reliably used in weapon applications. The research approach emphasized the development of theories and models to predict how shock physics and weapon effects phenomena would contribute to understanding the operation of weapon components, their vulnerability to hostile threats, and how to improve the survivability and effectiveness of these systems.

Sandia's early work in support of nuclear testing was connected to aboveground weapon tests and to underground tests (UGTs). Two areas in which Sandia, along with Los Alamos, played important roles were (1) the short-time determination of nuclear yields for early UGTs and (2) the development of scaling laws and prediction methods to describe the phenomena associated with crater formation for surface and near-surface bursts. Al Chabai and others were developing scaling laws to describe these phenomena. As Dennis Hayes remembers, "Al Chabai was a skilled user and proponent of dimensional analysis as a tool to organize measurements that did not lend themselves to modeling. For instance, his work on scaling of explosive cratering data gained international recognition and is still in use today."

Sandia made yield measurements of underground nuclear explosions primarily by using SLIFER (Shorted Location Indicator by Frequency of Electrical Resonance) cables. Prior to LLNL's development of this real-time technology, the yields from UGTs took weeks, or more often months, to generate using radiochemical methods. After he joined Sandia in the late 1970s, Hayes made an important discovery that was to have a far-reaching impact on the nuclear test program. In his words,

> Being a good student of Al [Chabai]'s work led me to try some similar analysis of my own circa 1964. This Sandia group, along with their LANL colleagues, had amassed hundreds of time-of-arrival measurements made very close-in to several dozen underground nuclear explosions. But the data remained largely unanalyzed owing to a lack of sophisticated modeling capability. It turns out that each underground test was assigned a yield based on radioactive isotope ratios gathered in post-shot gases, i.e., a radiochemical analysis. Scaling both distance from and time-of-arrival of the shock by the cube-root of the radiochemical yield produced a remarkable, low scatter universal curve time-of-arrival graph that I proudly tacked above my desk. I was just a technician and too naïve to understand the importance of publication.

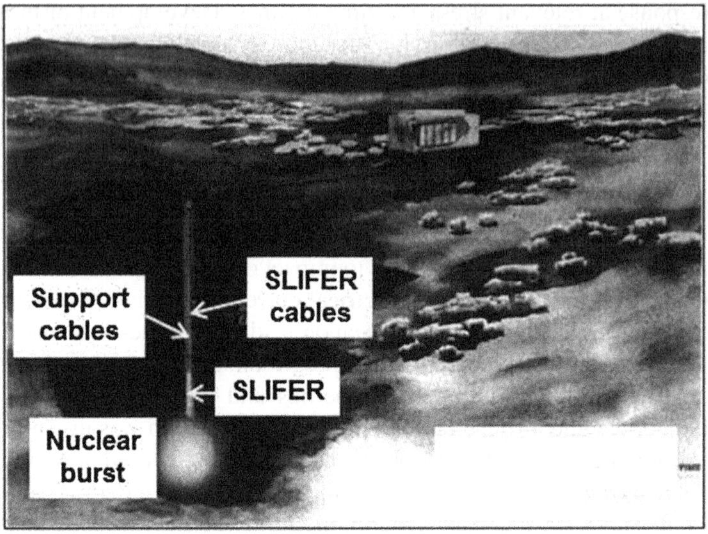


**Fig. 2.1** Artist's concept of an emplaced SLIFER gauge. (Reprinted with permission from Bass et al. 1976, Sandia National Laboratories)

SLIFER cable data could provide good yield estimates, typically with uncertainties less than 20 %, in hours or less. The technology was extremely valuable to weapon developers and other experimenters, who could rapidly feed the data back into testing and design projects. The SLIFER concept, shown schematically in Fig. 2.1, operates by measuring the resonant frequency of a progressively shorted cable emplaced radially from the nuclear detonation point. The hardware is simple, consisting of a length of air or solid dielectric coaxial cable as a component in an oscillating circuit. The wavelengths used, typically less than one quarter as long as the resonant frequency, are inductive. As a strong shock wave expands outward along the cable from the detonation point, the cable is crushed, thereby decreasing the inductance and frequency of the circuit. The frequency is then measured directly via the length of the cable as it is continually shorted by the ground shock. Thus, measurement of the frequency, along with the time of arrival, determines the shock front position versus time. Around that time, Hayes wrote a code that converted frequency vs. time to distance along the cable vs. time for the SLIFER experiments. The experimenters could then triangulate the results to get the actual distance vs. time data and infer yield. LANL investigators, including Bob Brownlee, found the data particularly useful. As Al Chabai notes in his recollections:

> This power law relationship was noted by Bob Brownlee of Los Alamos National Laboratory (LANL), who engaged Sandia in having SLIFER cable measurements made on many LANL weapon development events at the Nevada Test Site[1] (NTS). Adjacent to the device emplacement hole, one to three vertical holes (at about one million dollars per hole) were

[1] Editors' note: Now called the Nevada National Security Site.

> drilled, into which we placed the SLIFER cables. Accurate placement surveys were made of cable locations relative to the detonation center.
> This technique allowed us to estimate nuclear yield within an hour after detonation, which was extremely valuable to the nuclear device engineers, allowing them to obtain data immediately on the performance of their weapon designs. On most of these tests, a post-shot hole was drilled down to the camouflet[2] produced by the explosion and a radiochemistry sample was obtained to determine the yield and performance of the device. But these results were not available for months after the test, and this is why our SLIFER measurements were of such value to LANL. This universal relationship that we discovered was later refined by Don Eilers of LANL using more data and has been adopted as the standard for possible use in on-site yield measurements of nuclear detonation tests conducted by foreign nations.

More details of the SLIFER technique are in Al Chabai's recollections and in Bass et al. 1976. The scaling relation initially developed by Hayes was refined and broadly applied in UGTs; it played an important role in estimating nuclear yield and became one of the primary tools for monitoring compliance to the 1974 Threshold Test Ban Treaty.

Sandia's work on cratering phenomenology continued into the 1980s and involved analytic and scaling law techniques and, subsequently, single- and multi-dimensional hydrocode efforts. In the late 1950s and early 1960s, there was significant interest in predicting the size of the craters formed by conventional and nuclear explosions. Since the dynamic properties of geological materials, hydrodynamic computer codes, and large mainframe computers were not available to solve these problems numerically, Al Chabai conducted extensive investigations of conventional and nuclear cratering using dimensional analyses to distinguish which scaling relations were fundamental to crater formation (Chabai 1959, 1965; Chabai and Hankins 1960). His work established the appropriate scaling rules for these two classes of explosives and laid the groundwork for experiments to validate the relations that were developed. The models and calculations were validated through detailed comparisons with the experimental data. The latter involved full-scale nuclear and non-nuclear field tests, as well as high-quality laboratory-scale investigations. Two of the craters from the early 1960s are shown in Fig. 2.2.

The Scooter event was fired at the Nevada Test Site in October 1960, with 500 tons of conventional high explosive detonated below the surface. It was the largest event of its type ever exploded in the United States. The crater (Fig. 2.2a) was approximately 300 ft in diameter and 80 ft deep at its deepest point. Figure 2.2b shows the Sedan crater, which was formed when a 104-kt nuclear explosive buried under 635 ft of desert alluvium was fired at the Nevada Test Site on July 6, 1962, displacing 12 million tons of earth. The Sedan crater (Fig. 2.2b) is 320 ft deep and 1280 ft in diameter. Both events were part of Operation Plowshare, a U.S. program to develop techniques to use nuclear explosives for peaceful purposes and part of what are called Peaceful Nuclear Explosions. Over the ensuing years, this work led to major improvements in Sandia's hydrocode capabilities. Al Chabai, Bob Bass, and Paul Yarrington made important contributions to these efforts. Many of these

[2] Editors' note: An underground cavity formed by a nuclear explosion.

**Fig. 2.2** Craters at the Nevada Test Site. (**a**) Crater from the Scooter event. (**b**) Sedan crater. (Photos reprinted with permission of Sandia National Laboratories)

early results and the data began to appear in the open literature in the late 1960s and 1970s.

The first major research thrust in shock compression science at Sandia was based on techniques developed at Los Alamos during the Manhattan Project. Starting in the 1950s, Sandia's Physical Research and Physical Sciences departments began to study the properties and dynamic behavior of ferroelectric materials at pressures of a few tens of kbars and high temperatures. The motivation was to understand the electrical properties under shock loading to develop "one-shot" power supplies for weapon applications. By the late 1950s, these groups comprised an important expansion of the "engineering" emphasis at Sandia. Led by George Hansche and Dick Claassen, these two departments eventually included such leading scientists as Frank Neilson, Bob Graham, Fred Vook, George Samara, and Orval Jones, as well as the future Sandia president Al Narath. Projects important to the future directions of shock wave research included the study of ferromagnetic metals, ferroelectric ceramics, and piezoelectric quartz crystals. Although these groups were oriented toward basic research, the investigations of the shock response of single (X-cut) crystal quartz evolved into the first shock-wave-sensing gauge that could measure the submicrosecond, high-pressure, shock data needed by the nuclear weapon effects community in the early 1960s (Neilson and Benedick 1960; Graham 1961b, 1962; Graham et al. 1961, 1965a, b; Neilson et al. 1962). The extreme electrical and mechanical environments being investigated in studies of shocked components and UGT experiments required robust quartz stress gauges to resolve the fine detail of shock wave structures. This development was a game changer for a wide variety of material studies, including UGT weapon effects testing capabilities for defense programs. It was one of the many innovations and advances developed by this research group.

George Samara was one of the managers who provided a long-term vision in this organization; he continued to manage the Physics of Solids Research Department from 1971 until his death in 2006. Also, beginning in the 1950s, another group of Sandians in the fledgling stages was directed more toward applied research. The principal effort of that new group was part of a section in the Systems Engineering Division initially supervised by two of the earliest proponents of shock wave engineering at the Laboratory, Lynn Barker and then Don Lundergan. This group played a leading role in developing innovative experimental tools and facilities and identifying the analysis capabilities to investigate questions about the dynamic response of weapon components and the materials contained within them.

These shock wave technologies were increasingly called upon to help define the effects of various weapon outputs on full-scale systems, as well as their components. These included the vulnerability and survivability of U.S. systems, as well as potential countermeasures to enhance their capabilities in hostile environments. As time progressed, the data from these efforts assisted in the design and implementation of many new components and related systems. A vastly improved understanding of system and component development, and their vulnerabilities and survivability in adverse weapon environments, was also an important outcome. A better definition of the capabilities and lethality of U.S. weapons against opposing systems and targets was a parallel undertaking. These major accomplishments resulted from the synergistic combination of the experimental, theoretical, and modeling capabilities that were continuously evolving and improving at Sandia throughout these early years.

An example of the foresight of researchers working in this arena was illustrated in a 1957 report (Lundergan 1957) that appears to be one of the earliest references to Sandia's role in using shock wave technology to generate reliable, high quality fuze systems for nuclear weapons. In this report, Lundergan noted that existing contact fuze designs were all based on empirical evaluations of limited observations from full-scale testing. As such, they contained no real guidance for future designs, which would surely require reliability and survivability under vastly different operational environments. His opinion was that the only viable approach for obtaining the needed data was to invest in laboratory-scale testing capabilities that could provide highly accurate time-resolved pressure measurements in a repeatable manner and with configurations amenable to modeling and future design calculations. He concluded that detailed material equations of state and constitutive relations were essential and that gun launcher technologies would be the best tools for that purpose. Based on earlier LANL work, he identified the requirements for obtaining shock Hugoniot (curves describing the characteristic dynamic pressure–density response of materials under shock loading) for many new and untested materials as well as for obtaining the finer details of material behavior, such as elastic–plastic response. These details influence the complex wave reflections at material interfaces and wave fronts and many other features that are important factors for optimizing fuze designs. These phenomena would become areas for major research at Sandia and other organizations for decades. With this broad wisdom and insight, Don Lundergan can probably be credited with implementing applied shock wave R&D technologies at Sandia, with the research still being conducted in the manner he proposed.

Developing a basic understanding of experimental and modeling technologies was needed to address myriad design problems. In fact, these technologies represented the core of shock wave R&D that would occupy Sandia for the next five decades and more. The research areas that began in the 1950s included:

- dynamic response of mixtures, composites, and polymers used in explosive power supplies to predict operational stress limits and times,
- dynamic failure properties of materials used in components and subsystems to predict stress wave effects and operational capabilities of weapon components and subsystems,
- electromechanical effects produced by shock loading of piezoelectric and ferroelectric materials used for power generation required to predict their operational limits,
- energetic materials to predict the operation of detonators and explosively actuated ferroelectric power supplies,
- compaction behavior of porous materials used for shock cushioning of components and subsystems,
- time-resolved stress gauges to provide a fundamental understanding of material response to pulsed radiation loads,
- effects of pulsed radiation on weapon structures and components to predict operational conditions for weapon components, and
- the ground shock produced by nuclear explosions in geological media to predict the damage incurred by above- and below-ground structures.

The following sections will discuss these developments:

- Component and System Requirements (Sect. 2.2)
- Nuclear Testing and Shock Wave Research (Sect. 2.3)
- Explosive Methods for Shock Wave Research (Sect. 2.4)
- Development of Precision Impact Launchers (Sect. 2.5)
- A Family of Innovative Impact Launchers at Sandia (Sect. 2.6)

## 2.2 Component and System Requirements

The advent of supersonic and exoatmospheric missile delivery systems and their associated environments imposed additional demands on components and subsystems. To address these issues, it was necessary to understand the effects of adverse environments. A major problem was the effect of radiation-produced shock loading that occurred during a high-altitude exoatmospheric nuclear burst, which is where most missile and reentry vehicle engagement scenarios take place. The high-fluence, low-energy x-ray radiation at the largest output of these detonations often caused rapid surface ablation of materials, producing high-intensity shock waves that could cause internal damage. These stress waves propagated through materials in a system, often causing failure of materials and components, which in turn could preclude proper weapon reentry and operation. Resolving this issue required knowledge of the stresses produced by the radiation pulse, the duration of the pulse, and the tensile failure strength (referred to as spallation or spall strength) of materials. The overall impulsive load on the aeroshell could also be large enough to generate substantial buckling of the outer reentry vehicle structure. Knowledge of mechanical response, such as elastic–plastic behavior, did not exist at the appropriate stress and loading rates (usually referred to as strain rates) needed for the applications. Extrapolation of material response available at low strain rates and stresses using the highest loading rates available at the time (typically Hopkinson bars located at a few universities) was not reliable because of the extremely nonlinear response of the materials. Hopkinson bars provided material data at loading stresses of a few kbars and strain rates up to about $10^3$/s. The extensive shock wave research at Los Alamos during and after the Manhattan Project provided information on hydrodynamic high-pressure EOS properties, but essentially nothing on the mechanical and constitutive properties in the lower tens of kbar range. Missing was the range of strain rates associated with Sandia's design needs, which typically exceeded $10^5$/s (loading times of less than 1 μs) and loading stresses of several tens of kbars and above.

In addition to shock waves produced by nuclear weapon radiation, other problems involved the response of materials subjected to impacts associated with the delivery velocities and the effects of defensive countermeasures. A critical requirement for reliable operation was that the structure and, in particular, the arming, fuzing, and firing components had to retain integrity for a period of time (typically several milliseconds) after impact with the ground or other structures or after activation of contact and other weapon fuzes, in order to prevent deformation and destruction of these subsystems before detonation. The solution to this problem required knowledge of how large-amplitude stress waves, including the detailed wave structure, propagate through the many materials of the system.

Another motivation was that weapon systems were becoming more complex as component miniaturization evolved: the first weapons had been assembled with little emphasis on size requirements. Furthermore, operation of a weapon was sequential, with each component fully operating before the next started. These requirements made it clear that size and timing were critical to nuclear weapon delivery. Coupled with miniaturization, another factor was the energy needed to power the weapon components; these include the gyros, triggers, and radar, all of which are needed to make the system work. All required power, either electrically or through mechanical actuation. Similarly, weapons would generally be stored for long periods of time and then have to function without failure. This made onboard energy produced by small explosive devices, such as explosively actuated ferroelectric power supplies, attractive since they could be stored for long times and then detonated to produce significant electrical power on demand. Other small explosive charges were used to activate a variety of components in the weapon system.

However, the energy released by explosives in the early stages of nuclear detonation often causes damage to adjacent components as the large-amplitude stress waves propagate into surrounding components and subsystems. It was therefore necessary to determine exactly how these stress waves were generated and spread through the system. Stress wave propagation in liquids was well understood at the time, as well as that for small-amplitude elastic stress (sound) waves in solids. However, a basic understanding of solid material response at high stresses and high loading rates did not exist for engineering applications and models.

A pressing urgency was to understand this wide range of conditions and resolve the related uncertainties in device performance via fundamental engineering research in shock loading. Existing experimental facilities for these studies were severely limited; hence, experimental efforts were being formulated in the mid 1950s to provide the data. Early shock wave experiments to examine the ability to produce large electrical currents on demand focused on ferroelectric materials using small explosive charges to produce controlled planar loading of the samples. The output current could then be directly correlated with the loading stress to provide data on electromechanical response that was useful in developing predictive models. Frank Neilson, Bob Graham, and Bill Benedick played crucial roles in this endeavor. Within a short time, they realized that more precise control of the sample loading conditions was needed and could be achieved with precisely controlled plate impact techniques. This new approach established one of the main thrusts for shock wave experiments that persisted over the next 60 years.

The major shock wave techniques initiated in the 1950s and refined in the following years to meet the engineering requirements of components and systems are summarized below:

- use of explosives for the first studies of sample materials subjected to controlled one-dimensional or planar impact loading,
- characterization of piezoelectric and ferroelectric materials, specifically barium titanate, with explosively driven shock wave methods,

- development of the first precision plate impact launcher to study shock wave response under planar loading, which replaced the explosive technique for most studies, and
- initiation of experimental efforts to develop time-resolved instrumentation for detailed studies of mechanical response and dynamic tensile failure.

Two research groups were involved in these efforts. One focused on the physical processes occurring during shock deformation and was centered in what was then the Physical Research Department, initially under the management of G. E. Hansche. Principal individuals in the late 1950s and the 1960s included Dave Anderson, Bill Benedick, Bob Graham, W. J. Halpin, J. R. (Reid) Holland, George Ingram, Jerry Kennedy, Jim Kennedy, Pete Lysne, Frank Neilson, Mark Percival, Ray Reed, Dick Rohde, T. J. Tucker, and Floyd Tuler. Many of these researchers went on to make major accomplishments in shock wave research.

The shock wave group in the Physical Research Department began studying explosive loading to subject weapon components to planar stress loading. The central goal of these early studies was to understand electromechanical effects in ferroelectric materials, such as barium titanate, subjected to shock loading by explosive charges. Neilson provided visionary leadership in directing much of the experimental work and recognizing fundamental principles for producing shock-induced electromechanical response. He proposed use of the displacement current generated during rapid changes of electric polarization produced by shock loading as a power source for weapon subsystems. This development has had a profound impact on virtually all Sandia components requiring large electrical current on demand. These ferroelectric studies led to several important discoveries crucial to weapon and scientific applications and placed Sandia in a leadership role for this class of materials—in short, national and international recognition for Sandia shock wave research was jump started by this effort.

The other research group, in the Electrical Systems Department at this time, focused more on engineering applications, particularly the mechanical aspects of material response appropriate to weapon components and subsystems. Principal researchers included Don Lundergan, Lynn Barker, and Red Hollenbach. That group significantly expanded from the mid 1950s to the mid 1960s. Some additional individuals enlisted in this effort by the end of the 1960s included Larry Bertholf, Barry Butcher, Peter Chen, Doug Drumheller, Tommy Guess, Walter Herrmann, Darrell Hicks, James Johnson, Charles Karnes, Sam Key, Jeff Lawrence, Larry Lee, Bob May, Darrell Munson, Jace Nunziato, Karl Schuler, Billy Thorne, Bob Walsh, and Emily Young. As with the Physical Research Department, these individuals also made major contributions to the research.

## 2.3 Nuclear Testing and Shock Wave Research

Leading up to and after the Limited Test Ban Treaty in 1963, UGTs became the only technique that could be used to validate the many material and component nuclear response models being developed at the three laboratories. Although not a mainstream effort of the Sandia shock wave research program, computational and experimental support of the engineering and design groups directly involved in UGTs was a major thrust during the 1950s and 1960s. Underground experiments were considerably more expensive than aboveground nuclear weapons tests. However, the UGTs provided better controlled and reproducible conditions and, with their large-scale vacuum systems, represented the operational exoatmospheric environments far better. The UGT program was thus an essential backbone of nuclear weapon effects studies of vulnerability, survivability, and lethality for the strategic defense community, both at Sandia and other organizations.

New materials were being developed at a rapid rate for heat shields, radar windows, hardening and countermeasures, and many other applications. All had to be tested in operational nuclear detonation environments. However, before definitive analyses of such experiments could be performed, the equations of state and shock response properties of these new materials had to be determined. The experimental data and modeling capabilities were being developed in parallel and used to address these issues by Sandia's two shock research departments. Furthermore, the nuclear tests and their increasingly sophisticated diagnostics were beginning to yield the only data that could be used to validate the newly developed models in extreme environments. At the same time, laboratory-based characterization of materials and model developments were occurring. The result of these concurrent parallel paths was model validation, which gave confidence in the use of analysis techniques to design newer, safer, more capable operational systems. This large-scale effort, centered on UGTs, was an early driver for the shock engineering capabilities that Sandia continues to develop today.

This was the situation through the 1960s and 1970s until (1) the Threshold Test Ban Treaty (TTBT) was signed in 1974 and entered into force in 1990 and (2) the development of large-scale pulsed power accelerators capable of reproducing many extreme conditions in aboveground tests (AGTs) in the laboratory. The TTBT, which had a 150-kt limit, slowed the rate of the full-scale UGT program until its end in 1992. The new AGT facilities, which had been in the development queue for several years, led to a series of pulsed power accelerators at Sandia, represented by HERMES, the Particle Beam Fusion Accelerator (PBFA), Saturn, and more recently the Z machine (which will be discussed in Chap. 7). The end of the UGT effort was also the beginning of the subcritical experiment (SCE) effort in 1996. SCEs typically involve high explosives and nuclear materials, but the latter never reach a critical mass; hence, no nuclear yield is generated. However, SCEs provide extensive shock data on plutonium and other materials needed by weapon designers. More details of this research are in Chap. 6. These facilities and programs have assumed many of the roles that UGTs historically supplied in the post-World War II decades.

Nuclear field testing identified several materials issues to be addressed through shock wave research at Sandia:

- dynamic material properties, including phase transitions of various geological materials and structural materials of interest to ground shock studies,
- very high pressure and temperature response of materials to characterize the impulse produced by radiation deposition in regimes where vaporization could occur,

- measurement of stress pulses produced by radiation deposition in components and subsystems for determination of radiation coupling coefficients, and
- failure properties of materials under impulsive loading, particularly spallation strength.

## 2.4 Explosive Methods for Shock Wave Research

The shock wave group in the Physical Research Department began using explosive loading in the mid 1950s to subject weapon components to planar stresses. The motivation was to understand the electromechanical effects produced in ferroelectric materials, such as barium titanate, subjected to shock loading by explosive charges. Early experiments were performed at a remote Sandia test area on Kirtland Air Force Base close to the present-day Area Y located near Area III, where a major shock wave facility is still located. Nielson and colleagues conducted many experiments on these materials (Kulterman et al. 1958; Neilson and Benedick 1960), leading to electromechanical models and, in particular, to the renowned three-zone model of electric output from shocked piezoelectrics. This model accounted for elastic yielding in single crystal quartz, with the resulting formation of elastic and plastic waves that changed the polarization state of the material and the generated electric current. A notable feature of the model prediction and the output of the quartz was that the output current was proportional to the stress or pressure produced at the specimen-gauge interface. Neilson also proposed the use of the displacement current generated during rapid changes of electric polarization when certain ferroelectrics are shock loaded. These effects allowed the use of these materials as power sources in weapon subsystems, which were widely used in Sandia components that required large amounts of electrical current during pulsed operation.

A principal requirement of shock wave experiments is to produce a well-defined thermodynamic state for obtaining material property information, such as the Hugoniot response of the material. Figure 2.3 illustrates the use of planar shock loading to obtain data on high-pressure dynamic properties. The figure shows the planar loading of a constant pressure for fluids (or longitudinal stress for solids) on the left side of a large flat specimen. In the simplest case of a fluid, the instantaneous application of pressure produces a planar shock wave that propagates through the sample with velocity $U_S$. For constant pressure loading, the shock wave will travel as a steady front, transitioning the material from an initial state of density (or specific volume, which is the inverse of density), specific internal energy, particle velocity $u_p$ (assumed to be zero if the material is at rest), and initial pressure (also usually assumed to be zero) to a final uniform state of these variables. To determine the final pressure, density, energy, shock, and particle velocities, two kinematic properties of the shock wave must be measured and used with the equations for conservation of momentum, mass, and energy (Davison and Graham 1979; Boslough

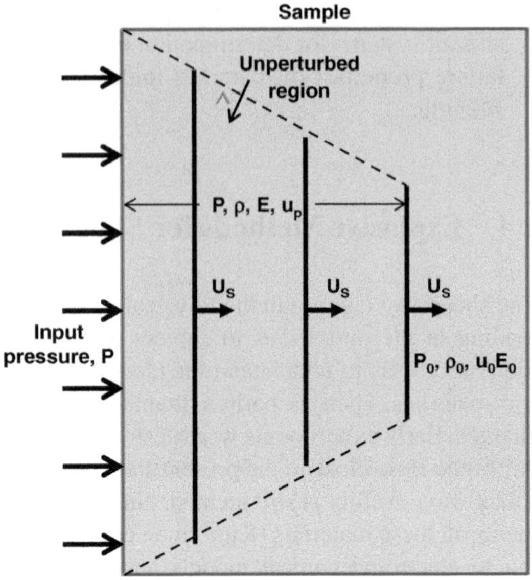


**Fig. 2.3** Concept of shock loading, showing the propagation of a planar shock front with velocity $U_s$. That shock compresses the material from an initial pressure $P_0$ (usually zero) and an initial particle velocity $u_0$ (usually zero), density $\rho_0$, and internal energy $E_0$. Behind the shock, the final states are pressure P, density ρ, particle velocity $u_p$, and internal energy E. Only the unperturbed region at the sample center, unaffected by release waves from the sample sides produces good data

and Asay 1993; Drumheller 1998; Davison 2008; Forbes 2012). The resulting pressure and density states define a single point on the Hugoniot curve, which is the locus of states produced by shock loading to different final pressures. Typically, the shock velocity is measured, along with either the pressure or the particle velocity induced by the shock wave. The shock velocity is determined by the transit time across the specimen if the peak pressure is constant and the shock wave structure does not change with propagation distance. The shock pressure can be measured with an appropriate gauge, such as the quartz gauge discussed earlier, or other types of stress gauges, or inferred from particle velocity gauges. The particle velocity induced by the shock can be determined by measuring the free surface velocity (which is nearly twice the particle velocity) using various techniques that will be discussed in the following sections. In addition, the lateral dimensions of the sample must be large enough that waves emanating from the edges[3] do not perturb the measurement of the two variables, i.e., the motion remains planar, or uniaxial, in the direction of motion over the measurement region.

In the mid 1950s, techniques for producing planar shock loading were limited. The main method was the explosive loading technique developed at Los Alamos for determining the hydrodynamic response of materials at extremely high pressures of a few hundred kbars to several Mbars. The basic technique for using explosives to produce planar loading is shown schematically in Fig. 2.4. A combination of explosives with different reaction rates are configured so that the final loading of a flat specimen is planar, i.e., the loading pressure is applied nearly instantaneously

[3]The waves emanating from the edges are represented in Fig. 2.3 as dashed lines that define the remaining region of uniaxial strain as the shock front propagates through the material.

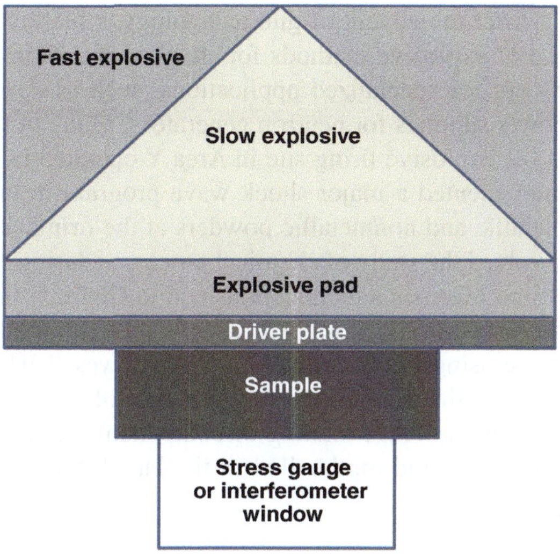


**Fig. 2.4** Typical configuration for using an explosive plane wave generator for shock wave experiments

over a flat surface of the sample. The configuration is referred to as a "plane wave generator." This approach was adapted for lower pressure work at Sandia by Neilson and his colleagues using the appropriate explosives and driver plates. Several electrical arrival pins on the driver plate are typically used to determine the starting time for the shock in the specimen. Either a gauge or arrival pins determine the arrival time of the shock at the back surface and, hence, the shock velocity. If a time-resolved stress gauge is used on the back surface, the shock stress can be determined as a function of time, which is useful for inferring material properties in the shocked state. If electrical shorting pins offset by different distances from the back surface are used to detect the shocked surface (i.e., the gauge is removed), the displacement of the surface can be determined versus time and differentiated to obtain the free surface velocity. From these two measurements, the shock velocity and either the stress or the particle velocity are used with the equations of motion to determine the thermodynamic state.

Although the use of small explosive charges to study planar shock loading of ferroelectric and other materials for one-dimensional (uniaxial strain) shock wave studies was a major development early in the Sandia shock wave program, the method had several disadvantages: the inherent safety hazards of using explosives, the limitations for producing controlled loading and unloading, the nonconstant peak drive pressure, and the inability to produce controlled unloading from the shocked state. To mitigate the safety risks, the experiments were conducted in a remote desert location of Tech Area III, which was exposed to rattlesnakes, scorpions, and other unpleasant environmental conditions (e.g., the New Mexico dust storms in March). Although satisfying safety concerns, this was not an ideal environment for shock wave research. However, the technique was used for a variety of experimental studies and yielded excellent results (Jones et al. 1962; Jones and Holland 1964, 1968; Jones 1967; Jones and Mote 1969).

After the advent of gun technology, which is discussed in the next section, the use of explosive methods for shock wave studies substantially declined at Sandia except for specialized applications, such as developing improved detonators and power supplies for neutron generators. Many of these experiments were conducted at the explosive firing site in Area Y operated by Bill Benedick. Bob Graham also implemented a major shock wave program in the mid 1980s to compress porous metallic and nonmetallic powders at the firing site; Graham's investigation, which involved the study of chemical processes during shock compression, is discussed in Bruno Morosin's recollections and in Chap. 5. In addition to the use of explosives for precision shock wave measurements, Bob Benham developed an explosive technique using light-initiated high explosives (LIHE) to expose full weapon components to shock impulses using a sheet of explosives referred to as a simulation of nuclear explosive loading of components (Benham 1976). The LIHE facility was shut down and mothballed in the late 1980s and reactivated later for specialized applications.

## 2.5 Development of Precision Impact Launchers

A short while after he arrived at Sandia in 1955, Don Lundergan initiated a shock wave program separate from the explosive work conducted in the Physical Research Department. In 1957, at the request of Leon Smith, manager of the Electrical Systems Department, Lundergan used a cast-off ballistic gun for projectile impact experiments on contact fuzes. At that time, Al Beck, an engineer in one of the weapon components organizations, had been using a 4-inch-bore military gun in Area III to study the response of weapon components subjected to projectile impact. A 1956 photo of the control room for the Beck gun, which was located near the Drop Tower in Area III, is shown in Sect. 2.7, People and Places of the 1950s. Lundergan initially used the gun for experiments on contact fuzes and did rudimentary calculations of the impact pressures produced in the components (Lundergan 1957). The impact configuration for each shot was not carefully controlled in the experiments, so the electrical response of the fuze was not predictable. Lundergan quickly realized that well-controlled impacts were needed to produce precise states to correlate impact stress to fuze output. Slightly earlier, Bill Benedick had performed analyses to evaluate the feasibility of using air guns for impact studies (Benedick 1956), so the time was ripe for taking ballistic gun technology used in military applications to the next level of refinement and precision. By around 1958, the gun originally used for contact fuze studies was converted to a precision launcher for shock wave material studies.

In 1955, after Lynn Barker finished his military service in the Korean War (where he received the U.S. Navy Distinguished Flying Cross) and also finished a master's degree in physics from the University of Arizona, he joined the engineering staff at Sandia. His initial responsibilities were to develop safe nuclear weapon delivery and escape methods for aircraft dropping nuclear weapons. While engaged in this critical problem, Barker made his first notable accomplishment by inventing the first Sandia analog computer, which he used to analyze heat loads produced on aircraft structures during the delivery and detonation of nuclear weapons.[4] By early 1957, both Barker and Lundergan were in the Electrical Systems Department under Leon Smith. Later that year, Barker was promoted to Section Supervisor in that department. Lundergan then reported to him, until Barker stepped down from management in 1960[5] and Lundergan was promoted to his position. In early 1962, Lundergan was further promoted to supervisor of the Materials Processing Division with Barker and Red Hollenbach as direct reports, making the three of them the core group to develop shock wave techniques for engineering applications.

To achieve the precision impact conditions required for planar shock wave studies, the Beck gun was accurately re-bored to more precise dimensions, refitted with a breech that could use compressed air to accelerate projectiles with flat impactors to different impact velocities, and engineered with a target alignment system that allowed precision plate impact with low angular misalignment, or tilt. A target chamber that could be evacuated was used to provide access to the new recording instrumentation being developed to measure shock wave properties accurately (Lundergan 1960). With his innovative mechanical and electrical talents, Hollenbach played a leading role in developing this gun into a precision impact facility. The facility became operational around 1958 for material property studies and was the beginning of a family of precision guns for impact experiments. Charles Bild, Director of Materials and Process Development, aided the emerging shock wave program on the engineering side. He encouraged and strongly supported efforts for weapon component and system development. With management support, the capability grew rapidly. As Hollenbach states in his recollections,

> We started with nothing—a cast off shock tube, no instrumentation, two people—and I'm not sure we were on the project full time. We progressed from firing the gun out across the mesa to a sawdust-filled bunker that challenged us to find the test piece after each shot. Early test results were obtained from postmortem visual (eyeball and microscope) examination of the 'target.' Somewhere about this time our group seemed to really grow—more people, better facilities, lots of instrumentation and much more sophisticated techniques (pins, slanted resistor) for obtaining and reducing the data from each shot.

Figure 2.5 shows a photograph of the original Lundergan precision impact gun. The gun facility was remotely located in the open high desert (in Tech Area III) amid the tumbleweeds, common dust storms, heat, and cold, as well as an occasional rattlesnake. An interesting detail is that, to our knowledge, this was the first precision impact facility in the United States (and probably anywhere) for material property studies. The muzzle of the gun was precision lapped to be perpendicular with the bore axis, thereby ensuring near normal impact of a projectile with an impact surface perpendicular to the projectile axis. Its unprecedented capability for

[4] L. M. Barker, private communication to J. R. Asay, 2011.

[5] Lynn Barker left Sandia in 1961 to pursue a PhD degree in physics at Columbia University in New York. He returned in 1962 because of his wife's ill health and did not complete the research and a dissertation for the degree.

**Fig. 2.5** The first gun launcher constructed for precision shock wave research, 1958. (Photo taken on June 27, 1958, reprinted with permission of Sandia National Laboratories)

obtaining previously unavailable data paved the way for additional gun facilities at Sandia and other institutions throughout the world.[6]

In addition to providing safer operating conditions for producing shock wave experiments than possible with explosive techniques, gun launchers produce better impact conditions for studying material response. The shock wave structures produced in fluids by explosive loading and plate impact are contrasted in Fig. 2.6. The shock structure produced by plane explosive loading consists of a sharp increase in pressure to a peak value at the shock front, which propagates through the fluid sample. For most explosive configurations, the peak driving pressure on the fluid is not constant, so the rarefaction wave produced by the explosive driver travels faster than the shock front and attenuates its amplitude, thereby producing a triangular wave and resulting in a phenomenon referred to as a Taylor wave attenuation. This varying shock pressure throughout the sample makes it difficult to study material

[6]In a private conversation with Jim Asay (November, 2012), Lundergan indicated that Dick Fowles, who was at Stanford Research Institute (SRI) at the time the gun facility was being developed, requested construction details for building a similar facility at SRI. After Fowles moved to Washington State University in the mid 1960s, he and Prof. George Duvall constructed a 4-inch-bore helium gun at the university. This facility was used by a large number of graduate students for fundamental material property studies in support of graduate degrees, including one of the authors, Jim Asay, who used it, along with Sandia quartz gauges, for his PhD research on the dynamic yielding of single-crystal lithium fluoride.

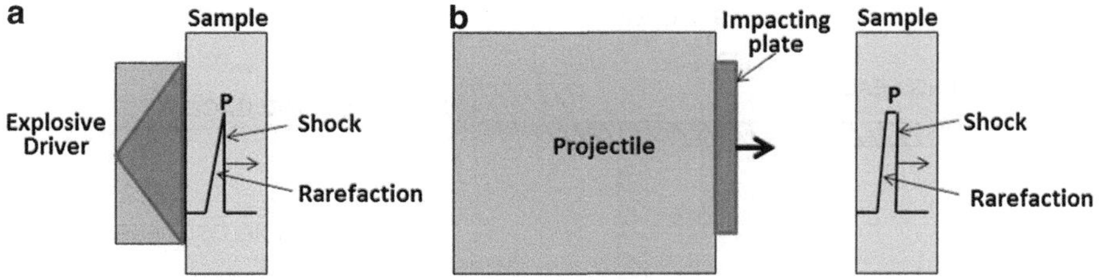


**Fig. 2.6** Explosive and plate impact techniques for shock loading a fluid sample. (**a**) Explosive loading produces an abrupt increase in pressure at the left boundary of a sample, resulting in a shock wave that travels to the right with amplitude P. Explosive loading also produces an immediate unloading of the shock pressure, resulting in a rarefaction wave that follows the shock and attenuates the peak shock pressure as the wave propagates across the specimen (**b**) Plate impact produces a shock wave with a constant duration until the rarefaction wave created by reflection of the shock from the back side of the impactor finally catches up with the right-going shock and attenuates it. This configuration allows study of materials at constant shock loading and also their study for controlled unloading

properties at a given pressure and therefore to generate an accurate Hugoniot curve for the material. The problem of attenuation was eliminated in later advances that used an explosive lens to launch a flyer plate to impact a stationary plate—very similar to the gun-launched techniques. With impact loading, the input pressure to the sample remains constant until the shock wave traveling to the left in the impactor plate reflects from its left boundary and returns to the sample interface. For many impactor configurations, this also produces a rarefaction wave, which will ultimately catch the shock front and attenuate it, as with explosive loading. However, the experiment can be designed so the driving pressure, and therefore the shock amplitude, is constant during its propagation so that material properties can be accurately measured at a specific shock pressure. In addition, a time-resolved measurement of the structure of the shock wave and the rarefaction wave provides important information about material properties, particularly the unloading response of the material, as will be illustrated in the following chapters.

The use of flat plate impact was transformational for shock wave research, but techniques had to be developed for precisely characterizing all aspects of the shock event, which required new instrumentation to measure specific variables of the shock wave and impact conditions. In particular, the stress and particle velocity states produced by shock loading were the principal variables to be recorded. Barry Butcher captured this in his recollections:

> Guns had previously been used for impact experiments, but these were rudimentary in terms of our air gun. The control room was full of state-of-the-art oscilloscopes and counters. Results of the shots were recorded with Polaroid® film by taking pictures of the single sweeps of the scopes triggered by impact. This film was then digitized with a rather costly digitizer, with the results, as I remember, printed out on IBM punch cards.

Although development of specialized impact techniques necessary for precision measurements seems trivial in light of today's capabilities, innovative diagnostics had to be invented for several purposes. These purposes included accurate characterization of the time of impact of the center of the sample to within nanoseconds

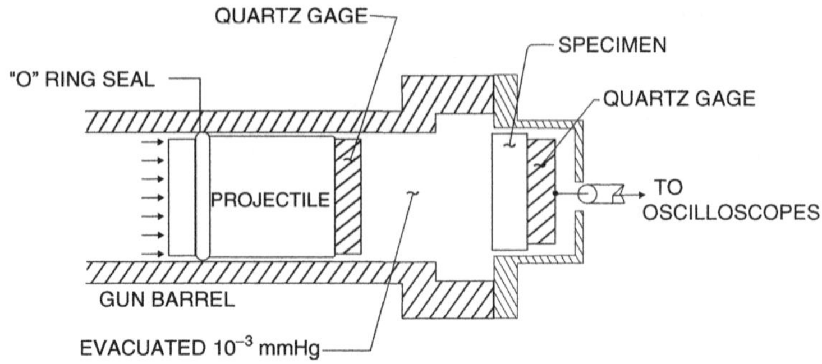


**Fig. 2.7** Typical plate impact experiment for gas-driven gun experiments. (Reprinted with permission from Graham et al. 1967a, Copyright 1967, AIP Publishing LLC)

(ns), control of misalignment (or tilt) between the impactor and target to less than 1/20$^{th}$ of a degree, and measurement of the projectile velocity to better than 1 % uncertainty. Moreover, timing instrumentation with ns resolution was needed to measure the projectile velocity and propagation velocity accurately for the shock and rarefaction waves produced after unloading from the impact stress. These measurements, particularly for the unloading waves, required development of time-resolved instrumentation to determine the changes in the wave structure. In the 1950s, these techniques did not exist anywhere.

Sandia researchers began to develop the appropriate methods for precision gun and time-resolved wave structure measurements. Intensive efforts to characterize impact conditions and measure shock properties were initiated in the 1950s and became routine by the late 1960s. These technical advances were published in a series of papers (Lundergan 1957, 1960, 1961; Graham 1958, 1961a, b; Graham et al. 1961; Neilson and Benedick 1960; Barker 1961, 1962; Smith and Barker 1962). In addition to measurement diagnostics, Graham developed a highly precise impact facility in the early 1960s for producing routine impacts with misalignments of about 1/50th of a degree, which resulted in extremely good time resolution when using quartz gauges to determine shock wave structure (Graham 1967). A schematic of one of his experimental configurations is shown in Fig. 2.7. More details of time-resolved shock wave structure measurements and how they provided a new understanding of dynamic material response are discussed in Chap. 3.

In the early 1960s, the gun from the desert of Tech Area III was moved to Tech Area I and modernized into a facility capable of performing a wide variety of shock wave experiments. A photograph of the original Lundergan gun is shown in Fig. 2.8 after its relocation to more comfortable quarters, and the gun was reengineered for more flexible use and a broader range of material studies. An unknown investigator is shown in the photograph inspecting a projectile to be used in the gun. Also shown is a schematic view of the structural components. The gun was used for general-purpose material studies until a second 100-mm gas gun, using helium as the pressurized gas, was built later in the 1960s and housed in the same building. The original Lundergan gun was then configured for exclusive use with beryllium,

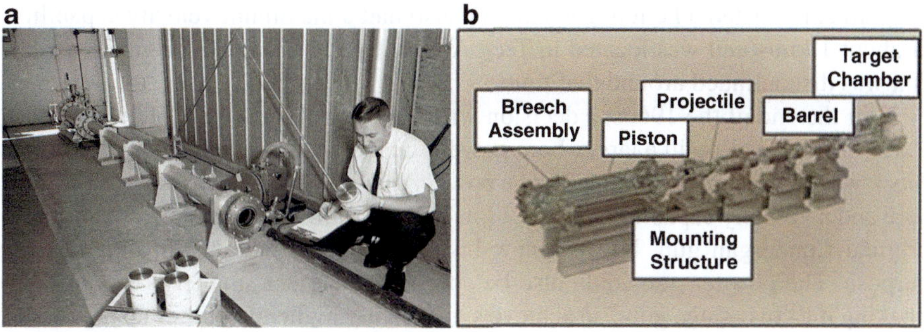


**Fig. 2.8** (**a**) The original 100-mm-diameter air gun relocated in Tech Area I, 1960, and then (**b**) a second gas gun, as shown schematically, was engineered for improved performance and experimental access. ((**a**) was provided by Sandia National Laboratories, (**b**) is from the private collection of B. M. Butcher)

a highly toxic material. Later in the 1980s, both guns were moved back to the desert, but this time to a large building in Area Y, which is located next to Area III where the first gun was built. This facility later became known as the STAR facility, which will be discussed in Chap. 4. Lundergan's original precision gun launcher was also modified for oblique impact studies; these studies will be discussed further in Chap. 7.

Darrell Munson joined Sandia at a time when significant changes were occurring in gun technology as he notes in his recollections,

> …I was offered employment and started in 1961 with the Metallurgy group under Charles Bild, the Director, Bill O'Neill, the Department Manager, Cecil Russell, the Division Supervisor,[7] and Keith Mead, the Section Leader. For two years this involved routine metallurgical problems, one of which was to obtain pure metal specimens for a group headed by Donald Lundergan, the Division Supervisor, by operating a compressed gas gun facility to study impact effects on bomb components, especially contact fuzes. This group was probably the first group to recognize the requirements for shock wave material studies to support the weapons programs…. I was given a section supervisory position in Donald Lundergan's division, under Charles Bild, to oversee the 4-inch compressed gas gun operations. While the gun facility was in Area III, the offices and laboratory activities were in Building 806 in Area I…. 1966 was an interesting year; a massive reorganization took place with the elimination of all section supervisors and my return to staff member status, at my own request. In 1967, another major rearrangement took place to accommodate the increased demand for shock wave material studies. Bild had an old Building (855) in Area I modified to accommodate both the 4-inch gun from Area III and a new 4-inch gun built for use in engineering studies. Butcher was in charge of the former and Larry Lee the latter….

After Lundergan's demonstration of gas guns for precision impact experiments in 1958, the technology quickly evolved at Sandia. In the early 1960s, Bob Graham converted a 40-mm ballistic gun into a precision powder gun capable of launching projectiles to about 1 km/s for shock wave studies (Graham 1961a). It was initially used for developing the quartz stress gauge that was to have widespread use. This gun was located in a remote area, as was the original Lundergan gun. Slightly later,

[7] Editors' note: Division Managers were called Division Supervisors at that time in Sandia's history.

Graham constructed a helium gas gun that also had a maximum velocity capability of about 1 km/s and was located in Tech Area I. Use of Graham's original powder gun was discontinued around that time.

Near the end of the 1960s, a decision was made by Walter (Walt) Herrmann and Darrell Munson to build a larger-bore powder gun and a two-stage light gas gun for shock wave experiments in Area Y. The powder gun was designed to have an 89-mm bore and a velocity capability to over 2 km/s. The location was in the vicinity of the original Lundergan gun, but in a new building constructed specifically for that purpose. The powder gun could also be coupled to another smaller-bore gun tube (making it a "two-stage gun," to achieve significantly higher velocities up to 7 km/s. Darrell Munson recalls the events leading to this major advance:

> In 1968 we became aware of some new innovative technology based on two-stage light gas guns… that were being developed and were much superior to explosive methods for high speed impacts. Some interest developed at Sandia in the new technology, which I promoted, and as a result of a meeting with the Vice President of Research (name not recalled), Al Narath, and others, I made a pitch to acquire a two-stage light gas gun facility. Over the next year I guided the procurement of a 3.5-inch gun from General Motors (GM) in Goleta, California, based on a gun constructed and operated by Sid Green and Arfon Jones. I was also responsible for both the conceptual and the final design (called 'Title III' at the time) of Building 9956 to house it. The project appeared as a line item in the Congressional Budget at something well over $100,000. The gun arrived in November 1969, was installed by GM staff, and fired its first test shot successfully before Christmas.
>
> The gun building, 9956, was about 100 ft long and 20 ft wide and high enough to permit an overhead crane. Butting up to this building was a bomb proof instrument room, which contained the oscilloscopes, pulse generators, and firing circuits. The 60-ft-long gun was aligned along the center line of the building and included an added 6-ft-diameter impact chamber. Sandia staff included Bob May, Ray Reed, Paul Matson, Bob Biesecker, and Charles Kinsey, and the support of others who still manned the 9950 facility as well as some from other divisions. Somewhat later, Carl Konrad also joined the group. The gun could be used either as a two-stage light gas gun with a final stage diameter of 1.125 inch, or as a simple powder gun with a 3.5-inch diameter. Joining these gun tubes was a tapered diameter reduction coupler section.

The 89-mm powder gun was capable of launching projectiles to over 2 km/s; that improvement quadrupled the peak shock pressures possible with the helium gun (up to about 1 Mbar in high density materials). Furthermore, the two-stage configuration nearly quadrupled the velocity possible with the single-stage powder gun, allowing peak shock pressures of 3 Mbars in high density materials. This capability was important to Sandia applications, such as vulnerability studies and other applications to be discussed later, because it allowed the first studies of shock-induced vaporization that changes the physical processes as well as the amount of momentum transfer produced by impact or radiation deposition.

Later in the 1980s, the powder gun and the two-stage light gas gun combination had grown into a large research facility (see Lynn Barker's recollections and Chap. 4), internationally known as the Shock Thermodynamics Applied Research (STAR) facility. The facility is still operational and presently houses five separate research guns, including the original Lundergan gun that underwent another substantial modification with the milling of a slot in the bore to allow slanted plate, i.e., oblique, impact that produces both longitudinal and shear states of stress upon impact (see Chap. 7).

Don Lundergan's vision of a strong interdisciplinary program in shock wave modeling and experimental research was taking root on the experimental side in the 1950s and would blossom with the development of modeling and computational tools in the following decades.

## 2.6 A Family of Innovative Impact Launchers at Sandia

The use of smooth bore guns for plane shock wave studies was a paradigm shift from the use of explosives. After the successful demonstration of Lundergan's precision air gun, the technology quickly evolved at Sandia and elsewhere. Figure 2.9 presents a schematic of the three major types of gun impact systems developed at Sandia for material studies. Lundergan's gun, which is referred to as a gas gun in the figure, used compressed air in the gun breech to launch a projectile containing a flat

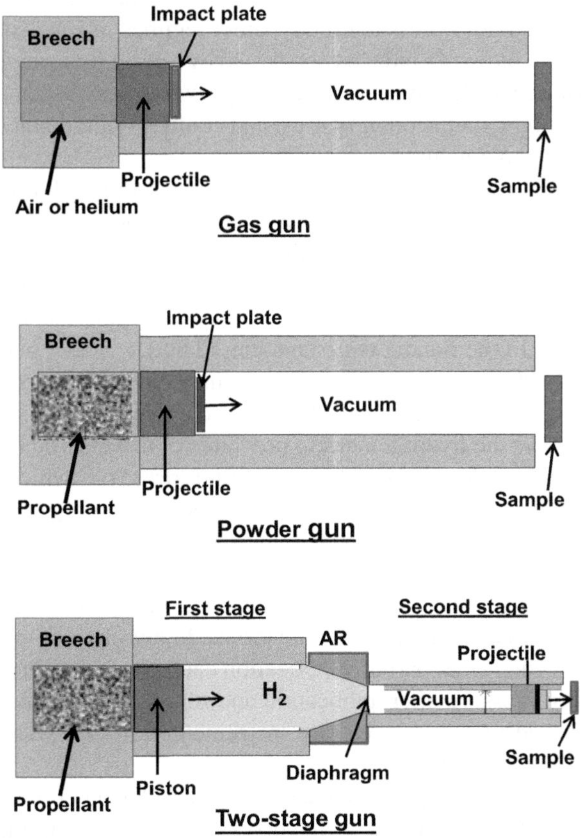


**Fig. 2.9** Gun launcher configurations developed to achieve increasing impact pressures. In the two-stage gun, the piston compresses the gas in a conical section called the acceleration reservoir, denoted as AR in this figure

impactor up to about 0.3 km/s. The impact pressures produced at this velocity were typically in the range of a few tens of kbars for most materials of interest. The capability enabled studies of elastic–plastic response at low pressure, but there was a need to increase the projectile velocity for both practical and scientific applications. By the early 1960s, helium gas compressed to several thousand psi in the breech of the gun replaced pressurized air, allowing projectile velocities to over 1 km/s, because of the higher sound speed of helium, and resulted in impact pressures exceeding 100 kbars for a wide variety of materials. This met the basic need of several weapon design efforts in progress, but even higher impact velocities were required to study phenomena of interest to the weapons program, in particular, material failure and fragmentation and the full compaction of porous materials being used to mitigate shocks in weapon systems.

Higher impact velocities were achieved by replacing pressurized gas with gunpowder, which resulted in a new plate impact launcher that became known as the powder gun. An 89-mm-bore powder gun was first operational in late 1969 for material studies and achieved a peak impact velocity of 2.3 km/s, resulting in impact pressures of a few hundred kbars in aluminum and up to 1 Mbar in high-density materials like tantalum. This gun was extensively used for material studies beginning in the early 1970s and is still in use today. A shortcoming of the gun was that it could not access thermodynamic regimes of interest for weapon problems, such as the blow-off-induced shock waves caused by x-ray deposition and vaporization. These shock waves were typically several tens of kbars in magnitude and of short duration, on the order of 1 μs, and could do significant damage to weapon materials and components.

The thermodynamic regime of shock-induced vaporization could be partially accessed with a new gun technology developed at General Motors in the late 1950s, referred to as a two-stage light gas gun, or often referred to only as the two-stage gun. Sandia acquired this technology in the late 1970s and perfected it for use in material studies, including shock-induced vaporization in low-melting-point materials. In the Sandia two-stage gun (TSG), a heavy piston, typically about 20 lb of polyethylene, is accelerated in the first stage, which consists of a large-bore powder gun, and compresses hydrogen gas contained in this stage. The piston further compresses the hydrogen gas to pressures of around 100,000 psi in a conical section of the gun referred to as an acceleration reservoir, which connects the first stage to a smaller-bore second-stage launch tube. The highly compressed gas ruptures a burst diaphragm at the entrance to the second stage and accelerates a lightweight projectile, typically a few tens of grams, to velocities up to 8 km/s. In most applications, the upper practical velocity is around 7 km/s, which results in impact pressures of a few Mbars in high-density materials. In low-melting-point materials, like cadmium, zinc, and lead, partial vaporization can occur when the impacted material returns to zero pressure after application and release of the shock wave. These states can be probed with laser velocity instrumentation that was developed at Sandia in the following decades.

There was a programmatic need to develop an even higher launch velocity capability, often referred to as hypervelocity, for full vaporization studies in support of x-ray vulnerability analyses and to access a broader range of weapon physics problems. As the advent of precision gun launchers represented a paradigm shift from using explosive for shock wave studies, another breakthrough was required to achieve higher impact velocities. The first attempt was through development of a three-stage railgun consisting of a two-stage gun that injected projectiles into a third-stage section, referred to as a railgun, at around 6 km/s. The railgun section used electromagnetic acceleration rather than gas pressure to achieve an additional velocity boost. This technique is discussed in Chap. 5 and was partially successful, but it did not meet the intended goal of 15 km/s.

An alternate attempt to achieve hypervelocity was successful and required another paradigm shift in propulsion techniques. This approach used the application of a very intense pressure (typically 1 Mbar or higher as opposed to the few thousand psi used in low-velocity guns) ramped up in pressure over about a microsecond to avoid shock formation. This impulse is used to accelerate a thin flyer plate to hypervelocity. The key feature was the smooth increase in applied driving pressure that accelerated the flyer plate, but prevented shock heating and subsequent melting or vaporization of the plate. This approach resulted in an effective extension to the TSG, resulting in a three-stage launcher that was called the HyperVelocity Launcher (HVL), which attained an impact velocity of about 12 km/s for EOS studies; it is described in more detail in Chap. 5. An improved version of this concept, referred to as the Enhanced HyperVelocity Launcher (EHVL), was able to launch material fragments to about 20 km/s for hypervelocity lethality studies.

A similar approach of using magnetic pressure produced by a fast pulsed power machine to achieve higher loading pressure over similar loading time scales resulted in even higher flyer plate velocities, successfully demonstrating 46 km/s. This technique, referred to as the magnetic flyer (MF), is discussed in Chap. 7. The higher pressures accessible with magnetically-driven flyer plates produce impact pressures as high as 40 Mbars. Furthermore, by using combined shock and ramp loading, the full spectrum of pressure–density–temperature states on the EOS surface is accessible, as discussed in Chap. 7.

The phenomenal increase in plate impact technology achieved over the past 60 years in the Sandia shock wave program is illustrated in Fig. 2.10. All of these capabilities still exist at Sandia and are routinely used for a wide variety of weapons and other scientific problems. Each will be discussed in more detail in the following chapters.

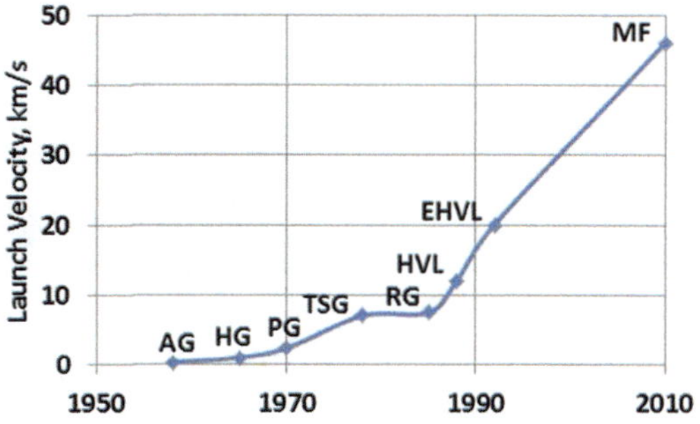


**Fig. 2.10** Maximum launch velocities of flyer plate techniques developed at Sandia. AG air gun, HG helium gun, PG powder gun, TSG two-stage gun, RG railgun, HVL HyperVelocity Launcher, EHVL Enhanced HyperVelocity launcher, MF magnetic flyer

## 2.7 People and Places of the 1950s

**Red Hollenbach, promoted to Section Supervisor, 1952 (Reprinted with permission of Sandia National Laboratories)**

**Luke Vortman, promoted to Division Supervisor, 1951 (Reprinted with permission of Sandia National Laboratories)**

**Dennis Hayes attending UNM, working night shift at Sandia, 1959 (Reprinted with permission of Sandia National Laboratories)**

**Lynn Barker, promoted to Section Supervisor in the Electrical Systems Division, 1957 (Reprinted with permission of Sandia National Laboratories)**

**Bill Perrett performing field tests, circa 1950, private collection, Bob Perrett (Bill Perrett's son)**

**Don Lundergan, promoted to Section Supervisor in the Systems Engineering Division, 1960 (Reprinted with permission of Sandia National Laboratories)**

**One of Sandia's first permanent buildings (Building 800), 1949 (Reprinted with permission of Sandia National Laboratories)**

**Operation Scooter, with 500 tons of high explosive that was exploded in the Nevada desert as part of Operation Plowshare, 1960 (Reprinted with permission of Sandia National Laboratories)**

**Artillery gun used for contact fuze studies by Al Beck, 1956 (Photo requested by Allan Beck 9/14/56, reprinted with permission of Sandia National Laboratories)**

**Al Schwarz (*right*) explains weapon specifications to Henry Winchell and Evelyn Evans, 1956 (Reprinted with permission of Sandia National Laboratories)**

# Chapter 3
# The 1960s: Explosive Growth

## 3.1 Background

The 1960s witnessed phenomenal growth of shock wave capabilities at Sandia. Along with rapidly evolving techniques for producing precisely controlled loading at ever-increasing shock pressures, pivotal improvements in precision diagnostics were occurring at a record rate. The goal was to probe the detailed structure of shock waves to understand specific aspects of dynamic material response, such as the ubiquitous two-wave structure observed in materials that exhibit both elastic and plastic response under shock loading. This information was needed for the new material models being developed. It was important, as well, to apply this new technology to the pressing requirements of the Nation's defense community. In particular, it was necessary to understand the stress response of materials in nuclear environments, such as the stress wave response of materials subjected to pulsed radiation sources, so that appropriate experimental techniques and material models could be developed to simulate effects on weapon components and systems.

Significant resources were invested in facilities and instrumentation, and the number of staff dramatically increased during the 1960s. Staff members with experimental expertise and with strong theoretical, modeling, and numerical analysis background were also recruited. As Lundergan notes in his reminiscences,

> .... the time of transition required the addition of staff, schooled in the appropriate disciplines. Digital computing was coming of age, and problems requiring the results of shock wave research were mounting. During the early 1960s Larry Bertholf, Charles Karnes, Walter Herrmann, Peter Chen, Barry Butcher, Karl Schuler, Darrell Munson, Floyd Tuler, Sam Key, and other well qualified individuals were added to the organization. Their contributions added to the understanding and quantification of the time dependency for spallation and the propagation of stress waves in composites, laminates, and foams. These material characteristics and associated wave propagation codes provided the input and calculational techniques to solve problems in the weapons program, nuclear reactor safety, armor design, and many other areas.



J.R. Asay et al., *Impactful Times*, Shock Wave and High Pressure Phenomena,
DOI 10.1007/978-3-319-33347-2_3

The collective expertise and the new techniques quickly advanced the understanding of basic material phenomena and led to state-of-the-art material models for the digital computer codes that were also rapidly being developed to simulate weapon components and subsystems. The 1960s decade represented a major leap forward in achieving Lundergan's vision, formed in the previous decade, of an interdisciplinary shock wave capability at Sandia.[1] With the new lab-wide emphasis on research, abundant financial resources were available that led to the corporate commitment to hire researchers with advanced degrees in the appropriate fields. By the end of the 1950s, about a dozen staff were associated with shock wave research. By the end of the 1960s, about four dozen researchers were involved. Hiring continued into the 1970s and beyond, but at a slower rate. Red Hollenbach captured this explosive growth in his recollections: "[W]e went from rags to riches." Frank Neilson, the visionary genius of the research organization, was also quoted regarding Sandia research as saying, "[H]ere money is like oil—you squirt it on to make things go faster."[2] By the 1970s, these tightly focused efforts led to Sandia's recognition as one of the Nation's leading organizations for shock wave research.

In 1967 Herrmann became manager of the newly established Mechanical Behavior Department, with four divisions engaged in material response. After a major Sandia reorganization the following year, the department became the Solid Dynamics Research Department, with Barry Butcher, Charles Karnes, Darrell Munson, and Olden Burchett as the division supervisors. Lundergan had stepped down from management to become a staff member in Munson's division. Herrmann's management strategy was to integrate the experimental, modeling, and computational divisions vertically within a single department, in keeping with Lundergan's original vision. This organizational structure allowed rapid solution of complex technical problems through integrated research in each area. Herrmann's management philosophy differed from that of most other organizations, which kept these functions separate, thereby requiring intergroup negotiations to set priorities that tended to slow progress in solving problems. For example, in many program reviews among Sandia, Los Alamos, and Livermore, Sandia's integrated approach often produced results *first* when the three laboratories were addressing a common technical problem. This interdisciplinary approach persisted for over 20 years until the Solid Dynamics Research Department was dissolved because of the Sandia-wide reorganization in the early 1990s, which is discussed in more detail in Chap. 6. After that structural change, the individual divisions (called departments in the new structure) reported to a director, rather than to a second-level department manager as in the previous arrangement. The new arrangement, which was meant to increase operational and fiscal efficiencies, was not optimal for the shock wave effort and disrupted the differentiating advantage of the integrated, interdisciplinary approach of Lundergan's original vision.

The staffing level and organizational structure of the Physics of Solids Research Department were also undergoing rapid changes at this time. Orval Jones had joined Sandia in 1961 as a staff member working with Frank Neilson and Bill Benedick on ferroelectric materials. During his early career, he made several superb contributions

[1] Private conversation between Don Lundergan and Jim Asay, December 2012.

[2] Private communication from Orval Jones to Jeff Lawrence, August 2013.

to the development of shock compression science at Sandia, which included teaching a course on shock wave physics to the growing staff in this specialty and participating in strategic planning for the growing research sciences. Those efforts resulted in the formation of several research departments at Sandia and, in particular, a department dedicated to shock wave research. Strategic planning by Orval Jones and others to define key research thrusts at Sandia in the 1960s created a solid foundation for the strong and continuing management commitment to shock wave research, which attracted top scientists into this discipline for decades to come. Furthermore, Orval made ground-breaking technical contributions to shock wave research. These included early studies of elastic yielding in metals with the quartz gauge (Jones et al. 1962), the first time-resolved measurements of elastic yielding in metals using quartz gauges (Jones and Mote 1969), and developing a dislocation-based theory of elastic precursor decay with Jim Johnson and Tom Michaels of Washington State University (Johnson et al. 1970; Jones 1971), in addition to several other technical accomplishments (Jones and Graham 1971; Jones 1972, 1973). His visionary accomplishments set the tone and direction for many research programs at Sandia and elsewhere for many years to come.

Jones quickly moved up in the organization and became a division supervisor by 1965. George Samara, who was to make an indelible imprint on the research and reputation of that department, joined Sandia in 1962 but was obligated to an immediate 3-year tour in the Army. He returned to the same department in 1965 and initiated a research program in static high pressure. Samara became a division supervisor in 1967, and in 1973 he was promoted to manager of the Physics of Solids Research Department under Orval Jones, who was then the Director of Solid State Sciences Research. Samara continued as the Department Manager until the Sandia-wide reorganization in the early 1990s. Walt Herrmann's department was also placed under Jones' directorate in the new organizational structure.

The major shock wave research efforts of the 1960s are discussed in the following sections:

- Time-Resolved Stress Measurements (Sect. 3.2)
- Time-Resolved Particle Velocity Measurements (Sect. 3.3)
- Elastic–Plastic Materials (Sect. 3.4)
- Shock-Induced Spallation (Sect. 3.5)
- Viscoelastic Materials (Sect. 3.6)
- Porous Materials (Foams) (Sect. 3.7)
- Computational Capabilities (Sect. 3.8)
- Answering DOD's Needs (Sect. 3.9)

## 3.2 Time-Resolved Stress Measurements

As the decade began, diagnostic development proceeded at a rapid rate in the Applied Research Division under George Anderson and in the Systems Engineering Division under Don Lundergan. Anderson's division concentrated on developing a basic understanding of ferroelectric and piezoelectric materials, initially using HE plane-wave drivers but converting to gun launchers by the mid 1960s. Lundergan's division focused on refining precision impact techniques for the air gun launcher fabricated in 1958 and on developing new instrumentation to diagnose shock wave structure. Detailed shock wave structure measurements were needed because studies of weapon components and systems and underground nuclear tests indicated that such details mattered for successful operation of the components or systems. The staff in both divisions aggressively pursued development of time-resolved diagnostics to measure shock waves produced by plate impact or explosive loading. The technical directions of each group were initially different, but the considerable overlap in research efforts eventually led to competition for resources and responsibilities.

The overriding challenge was the urgent need for diagnostics that could precisely probe the detailed structure of shock waves with nanosecond resolution. In the fuzing of a weapon, for example, both elastic and plastic waves are produced upon impact of the contact fuze, and how each of these waves propagates through the weapon system and component must be predicted. This information can be obtained by analyzing the fine structure of shock waves under well controlled impact conditions. The detailed structure of a shock wave provides crucial information about physical mechanisms, such as the critical longitudinal stress that produces elastic yielding (referred to as the Hugoniot elastic limit, or HEL) and that controls the loading process. A typical metal remains elastic up to the HEL, followed by plastic (or fluid-like) behavior to the peak loading stress. This loading response produces two waves with different velocities at low shock pressures: the "elastic wave" travels at the elastic velocity, which is nearly the ambient ultrasonic longitudinal velocity, and the "plastic wave" travels at a velocity, $U_s$, dependent on bulk properties of the material and is strongly stress dependent (Davison 2008; Forbes 2012). Figure 3.1 illustrates the two-wave structure, with the longitudinal stress denoted as $\sigma_x$ (for wave propagation in the x direction) and the associated particle velocity as $u_p$. How the detailed measurements of wave structure result in, or alternatively are caused by, the stress–strain response is shown in Fig. 3.1b. Stress or particle velocity measurements of each wave and the two-wave velocities determine the stress–strain curve through the conservation equations for planar shock propagation (Davison and Graham 1979; Boslough and Asay 1993; Drumheller 1998; Davison 2008; Forbes 2012). The Hugoniot, the locus of points obtained by single-shock loading, is the primary information for analyzing the

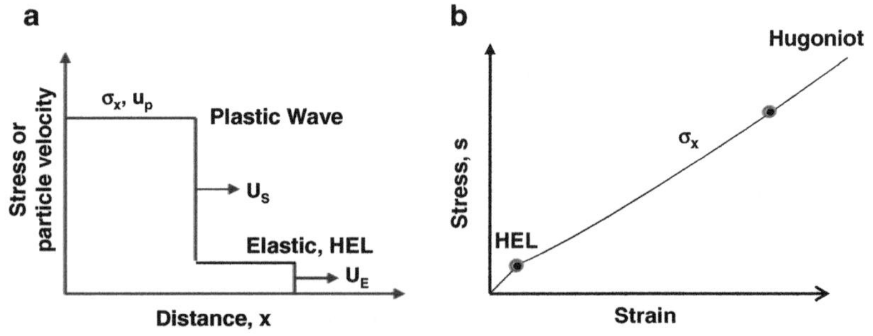


**Fig. 3.1** (**a**) Two-wave structure produced in an elastic–plastic material at low impact stresses. (**b**) Corresponding Hugoniot curve generated from several shock wave experiments

material response to shock loading. For many shock wave applications, especially those needing rapid solutions, the Hugoniot provides the basic material characterization that permits at least first-cut answers to the questions at hand.

To measure the fine structure of shock waves, Sandia researchers strategically invested extensive resources to develop diagnostics for measuring either the time-resolved stress or the time-resolved particle velocity profile produced by shock loading. As described in Chap. 2, using explosive loading, Neilson and his colleagues rapidly advanced the understanding of piezoelectric and ferroelectric response under shock compression (Kulterman et al. 1958; Neilson and Benedick 1960). This focus had an immediate payoff in the renowned "Three-Zone Model" for predicting the current produced from piezoelectric crystals by shock compression (Neilson and Benedick 1960). The model recognized the formation of elastic and plastic waves for quartz shocked above the HEL and the polarization and electric field states produced behind each wave. It provided the theoretical foundation for developing X-cut crystalline quartz into a high-quality stress wave gauge. By placing a quartz gauge on the back surface of a flat sample and recording the output current versus time as a shock wave arrives at the interface, a time-resolved stress profile related to the response of the material is recorded. An important attribute of the gauge is that the current profile produced at the back surface depends only on the stress profile at the interface between the sample and the gauge.

Figure 3.2 shows the three principal quartz gauge configurations that were developed for stress wave measurements. The most widely used one is the shunted guard ring

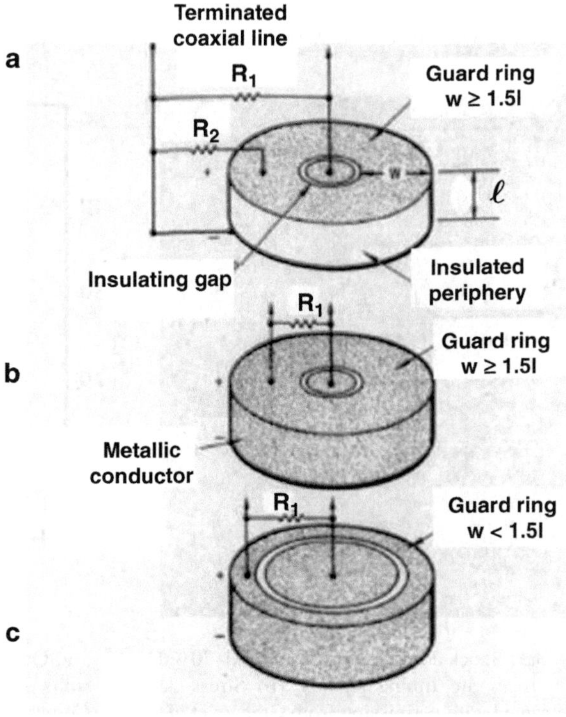


**Fig. 3.2** Quartz gauge configurations for shock wave experiments: (**a**) shunted guard ring, (**b**) shorted guard ring, and (**c**) wraparound guard ring (Reprinted with permission from Graham 1975, Copyright 1975, AIP Publishing LLC)

configuration, which has a universal response determined only by the material properties of quartz at fixed strain. The shorted guard ring and wraparound guard ring configurations have also been employed for specific applications. Graham discusses the specific requirements and calibrations for the three configurations (Graham 1975).

By the early 1960s, significant progress had been made in fully understanding ferroelectric and piezoelectric materials. Jones, Neilson, and Benedick had further developed a linear piezoelectric model to quantify the time-resolved structure of shock waves (Jones et al. 1962). However, the model had limited applicability since it assumed linear piezoelectric, dielectric, and elastic behavior, whereas these responses were known to be highly nonlinear. Higher-order electrostrictive effects and unusual dielectric breakdown behavior in quartz were not well understood and were not incorporated into the model at the time. Graham and his colleagues conducted comprehensive studies to determine these effects. These characterizations enabled a fundamental understanding of crystalline quartz to shock loading and led to the development of X-cut quartz for shock wave experiments to about 30 kbars where dynamic yielding occurs (Graham 1961b, 1962, 1972a, b, 1975, 1976, 1977, 1979a, b; Graham et al. 1961, 1965b; Neilson et al. 1962; Graham and Jacobson 1973; Graham and Chen 1975; Graham and Yang 1975; Graham and Reed 1978).

The quartz gauge had a transformational effect on shock wave research. Its main advantage at that time was that a principal shock wave variable, namely, the stress state, could be directly measured for the first time, rather than derived from other measurements such as the particle velocity obtained by differentiating displacement data. Furthermore, the time resolution of quartz gauge measurements was inherently very good, typically in the nanosecond range. At Sandia and other institutions, the gauge was used in a wide variety of experiments. Figure 3.3 shows an example

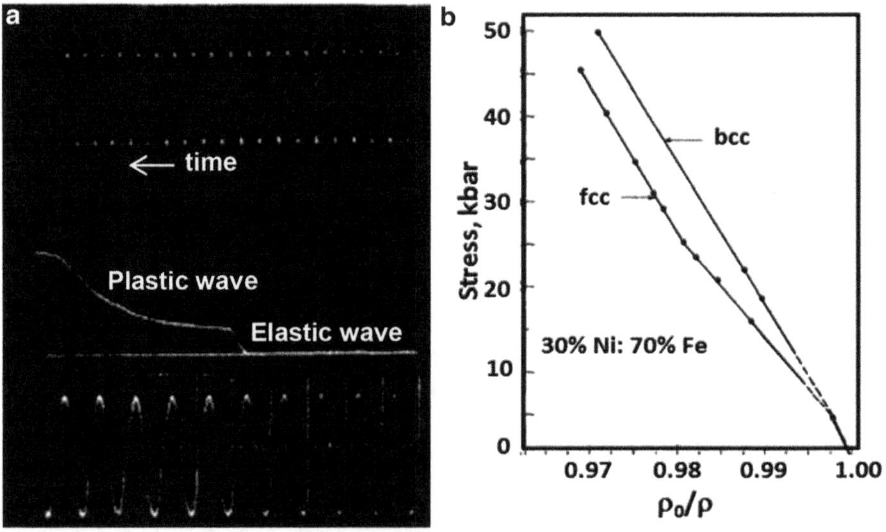


**Fig. 3.3** Shock compression of 30 % Ni–70 % Fe alloy. (**a**) Quartz gauge record. The *top* and *bottom traces* are timing signals. (**b**) Stress–density results show the bcc-fcc phase transition (Reprinted with permission from Graham et al. 1967a, Copyright 1967, AIP Publishing LLC)

**Fig. 3.4** Frank Neilson, Bill Benedick, and Bob Graham (*left* to *right*) share a patent for the quartz gauge (1967). Graham holds a quartz gauge disk (Reprinted with permission of Sandia National Laboratories)

of its use for the study of the bcc-fcc (body-centered cubic to face-centered cubic) phase transformation in the 30 % Ni–70 % Fe alloy (Graham et al. 1967a). The oscilloscope trace in part (a) clearly shows the two-wave elastic–plastic structure, and the stress–density plot in part (b) shows the elastic–plastic compression behavior and the effects of the phase transition. The initial rapid increase in stress with increasing density represents the elastic response. The 1979 report by Davison and Graham provides a comprehensive summary of the extensive applications of quartz gauge technology (Davison and Graham 1979).

Graham's pioneering research moved piezoelectric science into the large-strain, high-electric-field regime and, combined with the research of Neilson and Benedick, resulted in a versatile, highly precise stress wave gauge for use in a wide variety of applications. This groundbreaking work also garnered a patent, as shown in Fig. 3.4.

The quartz gauge was a transformational development for many areas of shock wave research and has been used in a large number of scientific and weapon applications, including nuclear explosive testing (Hayes and Kennedy 1969), energetic reactions (Kennedy 1970; Hayes and Mitchell 1978), and elastic yielding (Jones et al. 1962). Because of its robust behavior in severe environments, it was the principal gauge for measuring the stresses produced in weapon materials by pulsed x-ray radiation in the highly specialized application of UGTs (Graham and Ingram 1968; Graham and Jacobson 1973). Al Chabai, who participated in many of these events, notes in his recollections that

> .... The quartz gauge provided, with high accuracy, the wave shape and peak shock wave amplitude produced in the sample by blowoff from x-ray exposure. Among the many measurements made (e.g., impulse, temperature, shock pressure, displacement, etc.), it is fair to say that the quartz gauge provided the bulk of the most important quantitative data, allowing the U.S. to develop warheads for the Navy and the Air Force that were hardened to x rays, thereby providing a credible deterrent.

Research on ferroelectric and piezoelectric materials was continued into the 1970s and early 1980s by Graham and several other researchers who further improved the basic understanding through an unprecedented number of investigations. This significantly increased the range of practical applications (Lysne 1972b, 1973, 1975, 1976, 1977, 1978a, b; Lysne and Bartel 1975; Graham 1961b, 1962, 1972a, b, 1975, 1976, 1977; Graham and Yang 1975; Graham and Reed 1978; Graham et al. 1992; Stanton and Graham 1977; Chen et al. 1976a, b, 1978; Chen and Montgomery 1977, 1978, 1980).

Later in the 1980s, a polymer, polyvinylidene difluoride (PVDF), was observed to be piezoelectric. PVDF was exploited for use as a stress wave gauge by sandwiching a thin layer between thick sample layers with electrical leads extending to the outside of the sample (Graham 1979b, 1980; Bauer et al. 1992; Lee, Williams, Graham, and Bauer 1986; Lee, Graham, Bauer, and Reed 1988; Lee, Hyndman, Reed, and Bauer 1990; Lee, Johnson, Bauer et al. 1992; Reed and Greenwoll 1989; Reed et al. 1990). This configuration allowed determination of the *rate of change of stress* as a shock wave passes through the gauge, rather than the *magnitude of stress*, as for the quartz gauge. The PVDF gauge was another major advance in shock wave instrumentation and is discussed in more detail in Chap. 5, Sect. 5.10.

## 3.3 Time-Resolved Particle Velocity Measurements

Don Lundergan's[3] Engineering Systems Division took a different approach to developing precision shock wave instrumentation. His group focused on techniques to measure the particle velocity accurately at the back surface of a shocked material (usually referred to as the free surface velocity if a gauge is not used). As with the quartz gauge, the goal was to obtain a time-resolved measurement of the particle velocity, which would provide detailed information about material response. Time-resolved particle velocity and stress profiles are equivalent through the conservation equations of motion, and both can be used to derive the same material property information. In the earliest approach to determine particle velocity, a large array of electrical shorting pins (30 or more closely spaced pins) were offset at different heights from the surface and used to track the surface displacement during shock arrival. Displacements could be measured with a few micron resolution and time to a few nanoseconds. Differentiation of the displacement data produced a free surface velocity vs. time or velocity-time profile (Lundergan 1960, 1961). Lundergan used the pin technique to detect elastic and plastic waves and determine the elastic–plastic response of aluminum to 37 kbars (Lundergan 1960, 1961), which was later confirmed with more accurate particle velocity data.

The next advance was continuous measurement of free surface displacement. The first step in that direction was the slant-wire resistor gauge (Barker 1961, 1962; Barker and Hollenbach 1964). Figure 3.5 shows a schematic of the configuration. The technique employed a resistive wire mounted at a slight angle on the back free

[3] Lundergan was a section supervisor in 1960 and became Division Supervisor of the Materials Properties Test Division in 1962.

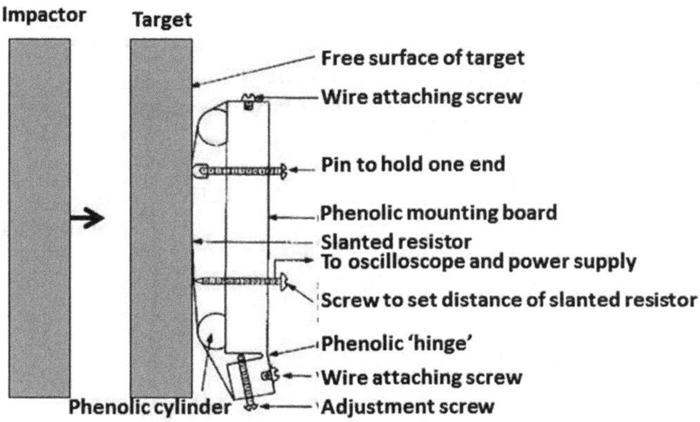


**Fig. 3.5** Slant-wire resistor for continuous measurement of free surface displacement (Reprinted with permission from Barker and Hollenbach 1964, Copyright 1964, AIP Publishing LLC)

surface of a conducting sample or on a metal plating of a nonconducting sample. The wire was continuously shorted as the shock wave arrived, thereby changing the wire resistance, which was recorded as a voltage change. The technique continuously determined the time-resolved free surface displacement with about 3 μm resolution and the particle velocity through differentiation (Barker 1961, 1962; Smith and Barker 1962; Barker and Hollenbach 1964). It yielded accurate Hugoniot curve measurements in aluminum (Lundergan 1961; Lundergan and Herrmann 1963; Barker et al. 1964) and, later, in aluminum, copper, and lead to peak pressures of 116 kbars (Munson and Barker 1966).

The next significant advance was the displacement interferometer, which increased the precision by allowing surface displacement measurements with sub-wavelength resolution. Although this interferometer had the same limitations as the slant-wire resistor in that it measured surface displacement rather than particle velocity directly, it was the start of a new approach using laser interferometry that would pay huge dividends in shock wave research. Barker and Hollenbach began their investigation of using interferometry for shock wave studies in late 1963 when Charlie Karnes had just arrived at Sandia. As Karnes recalls,

> …. I reported to Sandia in June 1963, and I believe it was in September 1963 that Lynn had the idea of using a Michelson interferometer to measure the free surface displacement history of an impacted target. He and Red Hollenbach set up an experiment in our lab on the second floor of Building 805. It was a simple, but elegant experiment.
>
> They used an aluminum tube about three to four feet long and four inches inside diameter having a wall thickness of 0.5 inch. A lapped aluminum impact plate was bonded to a polyurethane foam cylinder and placed in one end of the tube, the breech end. Another lapped aluminum plate, the target, was put on the other end of the cylinder. The mating surface of the target end of the tube was precision ground and lapped square. The plastic cylinder with the impact plate was held at one end of the aluminum tube, the barrel, with a thin wire, which was attached to something fixed in the lab. The target plate with the interferometer components was at the other end of the barrel. The barrel was soft evacuated, creating an atmospheric-pressure gas gun, a 'vacuum gun' if you will, using atmospheric pressure as the propellant. The vacuum held the target plate firmly against the end of the barrel. A photomultiplier was set up to monitor the fringes, with the output being displayed on an oscilloscope. A low-powered He-Ne laser was used to illuminate the interferometer.

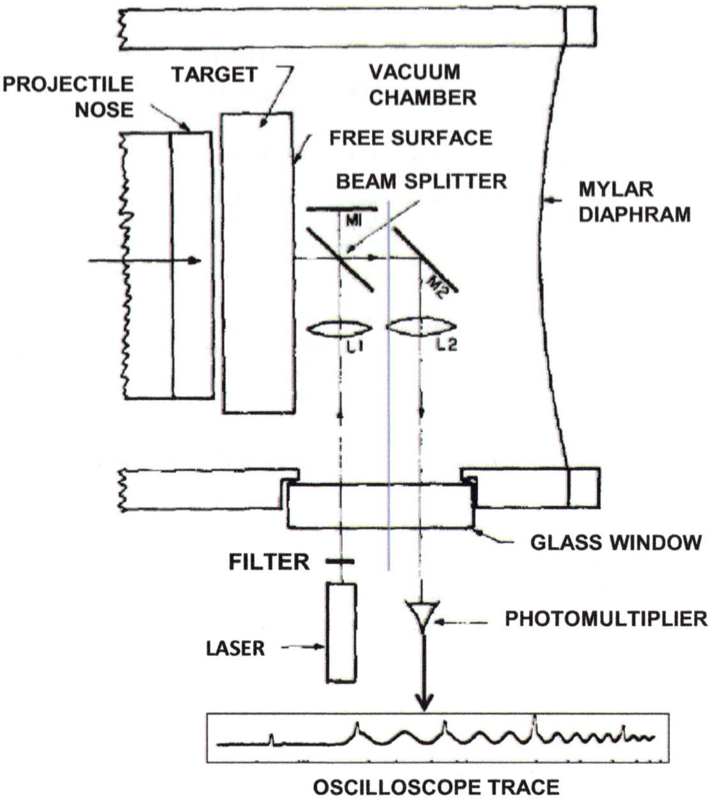


**Fig. 3.6** Schematic of a Michelson interferometer setup on a shock wave experiment. The oscilloscope record shows a frequency trace (with timing marks) that is differentiated to produce a particle velocity profile of the shock waves arriving at the back surface (Reprinted with permission from Barker and Hollenbach 1965, Copyright 1965, AIP Publishing LLC)

> Another thin wire was attached to the target plate with the interferometer setup. Lynn held onto that 'target wire' to keep the components from hitting the floor. To fire the 'gun,' Red cut the thin wire holding the projectile in place, and the projectile flew down the tube and impacted the target plate. That was the first shock wave interferometer experiment. I was in the lab watching it.

Refinement quickly followed for application to shock wave studies, and in November 1965, the first paper on Michelson interferometry for shock wave experiments was published (Barker and Hollenbach 1965). After differentiation of the continuous displacement data, the elastic–plastic wave structure was clearly observable. Figure 3.6 shows a schematic of the interferometer configuration. The fringe data shown at the bottom of the figure produced a time-resolved record of the displacement with submicron resolution.

The displacement interferometer was a major advance over the slant-wire resistor because of increased spatial resolution, but was limited to relatively low surface velocities because of the limited frequency response of recording oscilloscopes.

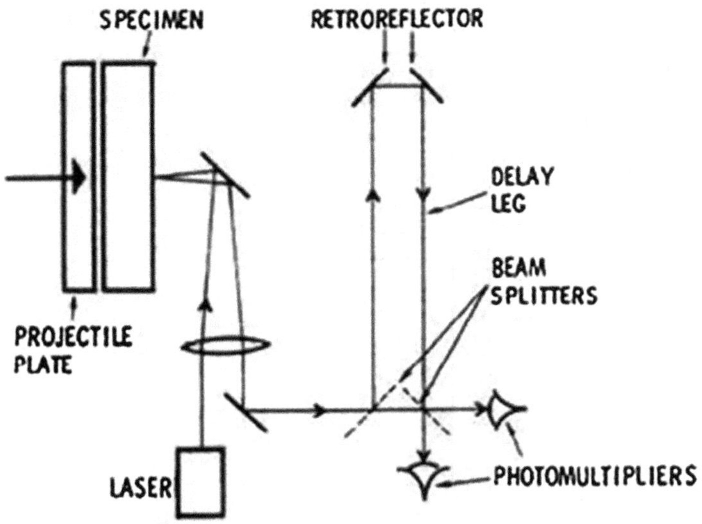


**Fig. 3.7** Velocity interferometer for free surface velocity measurements (Reprinted with permission from Barker 2000a, Copyright 2000, AIP Publishing LLC)

However, the technique was quickly superseded by the Sandia velocity interferometer, another Barker brainchild and a pivotal advance in shock wave instrumentation, which used optical differentiation to provide a direct measure of particle velocity with better accuracy and with time resolution of a few nanoseconds (Barker 1968, 2000a). The velocity interferometer considerably extended the stress range possible with the quartz gauge, since it could access shock stresses exceeding 100 kbars. Its main limitation was that the reflecting surface had to remain specular (i.e., mirror like) during the measurement. For this reason, it was not generally used for nonhomogeneous materials, such as porous materials or materials undergoing phase transitions.

Figure 3.7 shows a schematic of the interferometer, which was referred to as the Sandia velocity interferometer. This technique uses an optical delay to generate an optical signal a few ns before the direct signal from the shocked sample. When combined, a beat frequency is obtained, which is directly proportional to the particle velocity of the recorded surface (Barker and Hollenbach 1972). As with the quartz gauge, the velocity interferometer was a groundbreaking advance that allowed direct determination of a shock wave variable, namely, the particle velocity, with high accuracy. This was another example of the seminal advances Barker and Hollenbach made during the period 1964–1974 in their laboratory, referred to as the "Serendipity Lab." Jim Johnson has an interesting remembrance.

> The partnership of Lynn Barker and Red Hollenbach in the 'Serendipity Lab' set a standard for experimental shock wave research that has never been equaled in my opinion and experience. I have never forgotten the fact that the speed of light is very nearly one foot per nanosecond after Lynn explained to me that cable lengths were important in the new nanosecond regime in which shock wave experiments were being conducted at Sandia. …

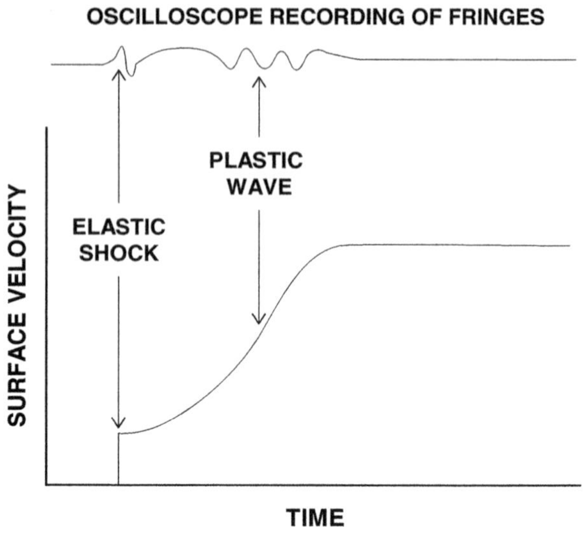


**Fig. 3.8** Particle velocity profile obtained on an Al sample with a fused silica window mounted on its back surface, showing the interferometer fringe record on *top* and the velocity profile on the *bottom*. Time scale is arbitrary (Reprinted with permission from Johnson and Barker 1969, Copyright 1969, AIP Publishing LLC)

Figure 3.8 illustrates how a direct measurement of the fringe shift in the interferometer relates to particle velocity. The top part of the graph shows an actual fringe record obtained on a 6061-T6 aluminum alloy using a fused silica window on the back surface of the sample arrangement shown in Fig. 3.7 (Johnson and Barker 1969).

The interferometer found use in a wide range of applications, including measurement of the fine structure of shock waves in aluminum (Barker 1968), which allowed the first steady wave analyses of propagating shock waves (Johnson and Barker 1969), plate laminates (Lundergan 1970; Lundergan and Drumheller 1971a, b; Barker 1971b), viscoelastic materials (Schuler 1970; Schuler and Nunziato 1974, 1976; Nunziato et al. 1972), and elastic precursor decay (Asay et al. 1975).

One of the earliest uses of the interferometer for evaluating physical mechanisms of shock compression was made by Johnson and Barker on the aluminum alloy, 6061-T6 (Johnson and Barker 1969). Shock loading of sufficiently thick samples produces a steady shock wave in most materials with positive curvature in their stress–strain response that propagates without change in shape. This effect is due to the balance between dissipative mechanisms that tend to spread the wave and the concave nonlinear response that tends to steepen the wave. Johnson and Barker were the first to perform a careful study of steady plastic wave propagation in aluminum. The calculations were done for impact stresses ranging from about 9 to 80 kbars and are illustrated in Fig. 3.9. The initial elastic shock is observed in each case, which is represented by a constant velocity plateau, followed by the plastic compression wave. The experimental results show that the risetime of the plastic wave decreases rapidly as the shock amplitude increases. Experiments performed on different sample thicknesses demonstrate that the plastic shock risetime remains unchanged for each peak stress and is thus steady. Also shown in the figure are calculations of the steady waves using a dislocation theory of plastic dissipation and

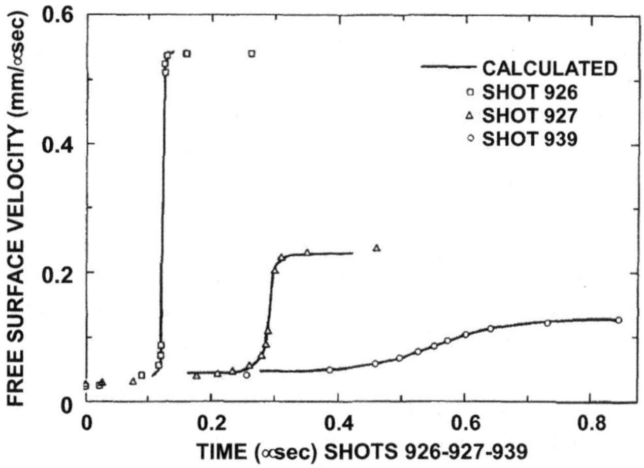


**Fig. 3.9** Comparison of experimental profiles with a dislocation theory of dynamic yielding (Reprinted with permission from Johnson and Barker 1969, Copyright 1969, AIP Publishing LLC)

a steady wave analysis method (Johnson and Barker 1969). The dislocation model included a description of the initial dislocation density of the material, a factor for multiplication of dislocations during shock compression, and a function to describe the stress dependence of dislocation velocity. Excellent agreement is obtained between experiment and theory, which represented a major advance in modeling capability in the 1960s and helped to establish the advantages of making time-resolved wave profile measurements for investigating physical mechanisms of dynamic material response.

A notable result of these experiments is the strong dependence of strain rate or equivalently risetime of the plastic wave on peak loading stress, which became important for establishing a general relationship between shock strain rate and shock stress. Later, Lalit Chhabildas plotted these results and found a relationship of maximum strain rate with peak stress (Chhabildas and Asay 1979), which was further determined to be a fourth power relationship (Grady 1981b). As will be discussed in Chap. 4, this response was formalized in the late 1970s through additional experiments by Chhabildas and colleagues and with theoretical analyses performed by Dennis Grady in the early 1980s, who found that the fourth power functional behavior applied generally to a broad class of solid materials. That observation has significant implications for the fundamental physical mechanisms that control shock dissipation (Grady 2015).

In addition to studies of compressive material response, free-surface interferometry allowed studies of the tensile failure stress (often referred to as "spall" strength), which is discussed in more detail in Sect. 3.5 and often occurs when a shock wave, followed by an unloading wave, reflects from a free surface. As first noted by Lundergan and colleagues, the ability to measure spall strength accurately was an important advance, since this information was critical to weapon applications involving material or component failure (Lundergan and Smith 1962; Lundergan 1963).

Another major advance in interferometry was the use of optical windows placed on the back surface of a sample. As with the quartz gauge, a calibrated and closely impedance-matched window provided a better estimate of the actual shock wave structure in the sample and permitted studies of the unloading behavior, as shown in Fig. 3.8. In 1970 Barker and Hollenbach reported the first optical calibration of Z-cut sapphire, fused silica, and polymethyl methacrylate (PMMA), windows that could be used for interferometry experiments (Barker and Hollenbach 1970). With the exception of the PMMA, which was viscoelastic in its usable stress range, the two other window materials remained elastic during use.

The development of laser windows considerably extended interferometer capabilities and provided additional insight into physical mechanisms of shock compression. A notable observation first noted by Barker in the 1960s showed that the initial elastic loading is sharp, but the elastic unloading is spread out (Barker 1968). This effect, which is observed in essentially all metals, has been speculated to occur for different reasons, including heterogeneous effects (Asay and Lipkin 1978) or dislocation pileups and pinned loops that produce reverse plastic flow immediately upon release from the shocked state (Johnson 1993). Recent experiments on single-crystal aluminum suggest the latter mechanism may dominate in most cases (Winey et al. 2012).

Several other window materials have been explored since Barker and Hollenbach's early work. Single-crystal <100> LiF remains transparent and has been optically calibrated for shock loading to 1.2 Mbars by Wise and Chhabildas (1986). Furnish and coworkers extended the use of LiF to shock stresses of about 1.8 Mbars (Furnish et al. 1999), where it quickly loses transparency after initial shock loading. However, LiF remains transparent to at least 8 Mbars during ramp loading (Fratanduono et al. 2011), indicating a strong dependence of transparency on temperature. LiF is useful in a wide variety of shock wave applications with conventional gun launchers and both laser and magnetic compression experiments. It has become the "workhorse" for shock wave studies at high pressure and is the most extensively used laser window for conventional velocity interferometry experiments.

As mentioned earlier, X-cut quartz yields at about 30 kbars under shock compression and is not used as a stress gauge beyond that. However, for shock stresses exceeding 5 Mbars, X-cut quartz has again become a standard shock wave gauge. In this case, the quartz becomes conductive immediately behind the shock front, so that the propagating shock front in the material is a reflective surface that moves at the shock velocity. Using interferometry, shock velocity can then be directly measured accurately, so the gauge is actually a time-resolved *shock velocity* or stress gauge (since the EOS of the window is usually known) rather than a time-resolved *particle velocity* gauge. This unique application is an outcome of a game-changing development in interferometry, namely, the VISAR (Velocity Interferometer System for Any Reflector), which Barker and Hollenbach developed in the 1970s. The VISAR and several of its applications are discussed in Chap. 4.

## 3.4 Elastic–Plastic Materials

The operation of many Sandia components and subsystems is moderated by the mechanical response of metals and other materials, which depends on elastic and plastic deformation at high loading rates and high stresses. This behavior is usually referred to as elastic–plastic response. The elastic and plastic waves produced by shock loading above a material's elastic limit travel with different velocities, as indicated in Fig. 3.1, and thus arrive at the recording position at different times, as indicated in Fig. 3.8. The elastic velocity is constant (for non-decaying elastic waves), but the shock velocity of the plastic wave is stress dependent, causing a decreasing temporal separation between the waves as the stress increases. Hence, for applications the material response must be measured accurately over the complete stress region of interest. Understanding this behavior in the wide range of materials used in weapon systems was one of the principal motivations for Don Lundergan's original objective of establishing a comprehensive shock wave research program. Lundergan recalls the programmatic motivation for starting a research program on elastic–plastic materials.

> .... data existed on propagation of stress waves and material behavior at low strain rate boundaries such as those investigated using Hopkinson bars and comparable university laboratory equipment. The work at the national laboratories, in investigating the hydrodynamic behavior of materials, provided an upper boundary. Missing was the range of strain rates associated with the design needs of Sandia subsystems. Accordingly, the investigation of the elastoplastic behavior of materials was initiated at our laboratory.

Knowledge of the elastic yield stress under shock compression was a basic requirement for models in computer simulations and was urgently pursued. Walt Herrmann's principal research as a professor at MIT prior to joining Sandia in 1964 involved the development of dynamic material models (e.g., Herrmann et al. 1962). Lundergan's precision impact facility also began generating material property data in the early 1960s. Lundergan therefore collaborated with Herrmann to integrate Sandia's emerging experimental results with the evolving material models. After Lundergan published his first experimental paper on the shock loading response of aluminum (Lundergan 1961), a broader paper on aluminum was published with Herrmann a couple years later (Lundergan and Herrmann 1963). When Lundergan recruited Herrmann to his division in 1964, the experimental and modeling efforts significantly accelerated, and several more modeling and experimental papers on elastic–plastic response were published (Barker et al. 1964, 1966; Herrmann et al. 1970; Herrmann 1973, 1974).

The first experimental paper (Lundergan 1961), reporting on the elastic–plastic response of aluminum to about 37 kbars, used an array of contact pins to determine the free surface displacement and thus the particle velocity. Although limited in time and spatial resolution, the compression response of aluminum was accurately determined and clearly showed the elastic–plastic behavior over this stress range. This work was followed by more accurate measurements of shocked aluminum using the slant-wire resistor (Lundergan and Herrmann 1963) and even more accurate

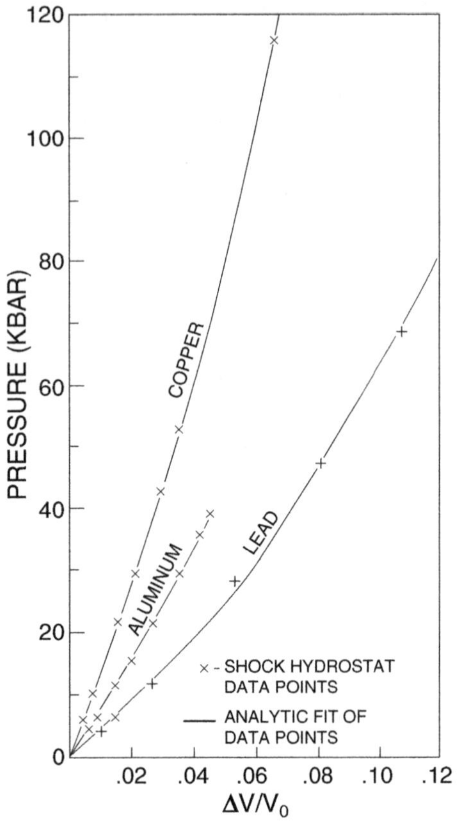


**Fig. 3.10** Pressure vs. change in volume data for copper, aluminum, and lead determined from shock wave experiments (Reprinted with permission from Munson and Barker 1966, Copyright 1966, AIP Publishing LLC)

measurements using the Michelson displacement interferometer technique (Barker and Hollenbach 1965; Barker et al. 1966; Butcher and Karnes 1966; Munson and Barker 1966). To expand the range of materials and critically test techniques to obtain basic EOS data, Munson and Barker published shock compression results for copper, aluminum, and lead that demonstrated consistency between the slant-wire techniques and the Michelson interferometer and also showed good agreement with hydrostatic pressure data when corrections for temperature effects were taken into account. Figure 3.10 shows these results. The figure provides pressure versus engineering strain, $\Delta V/V_0$, for shock loading. The convex upward nature of each curve ensures that arbitrary plastic waves produced by shock compression will evolve into steady shock waves with a risetime balanced by dissipative processes and the convexity of the compression curve.

The rapidly evolving capabilities of the quartz and interferometer diagnostics in the early 1960s provided a much needed database for initial elastic yield properties for the models being used to describe high rate yielding phenomena. In particular, data were emerging that suggested elastic yielding was rate dependent and the elastic limit changed with propagation distance. Jones, Neilson, and Benedick were among the first to investigate elastic yielding systematically in several metals with the newly invented quartz gauge (Jones et al. 1962).

Dislocation theories of dynamic yielding were also being developed at this time and just beginning to be used to predict shock-induced yielding. Barry Butcher and Darrell Munson explored the use of dislocation theories similar to those of Johnson and Barker (1969) to explain dynamic yielding in polycrystalline metals (Butcher and Munson 1967). Orval Jones and Jim Mote extended the experimental research on dynamic yielding to single-crystal copper (Jones and Mote 1969), laying the groundwork for new theoretical descriptions of how dislocation motion on different slip plane systems affects dynamic yielding. The theory for orientation dependence of yielding under shock loading was developed independently by Jim Johnson at Sandia and Tom Michaels, a graduate student at Washington State University, resulting in the classic joint paper (Johnson, Jones, and Michaels 1970). The theory was then applied to new experimental results on single-crystal copper (Jones and Mote 1969) and single-crystal tungsten (Michaels 1972). The theory was a major advance for describing dynamic yielding and is still being used. Other research on the dynamic yielding of LiF, conducted about the same time using both quartz gauges and velocity interferometry, concerned how minute amounts of impurities can have an overriding influence on dynamic elastic yielding (Asay et al. 1972, 1975).

Throughout the 1960s, the Sandia velocity interferometer technique and the quartz gauge instrumentation were used to study dynamic yielding in a variety of other materials (Jones and Holland 1964, 1968; Rohde 1969; Rohde and Jones 1968; Rohde and Graham 1969, 1973; Rohde et al. 1972). Based on the extensive work of both groups, Graham and Jones published a summary of the accumulated data on dynamic elastic limits (Graham and Jones 1968).

Under Orval Jones, the Physics of Solids Research Department also hired several staff with expertise in materials science to put shock wave studies of dynamic yielding on a firm materials science basis. Al Stevens, Dick Rohde, and Larry Pope were enlisted for their expertise in this area and performed several investigations of a number of material issues that Johnson then incorporated into the material models. These issues included the effects of multi-dimensional flow (Johnson 1968b), twinning on dynamic yielding (Johnson and Rohde 1971), and crystal anisotropy for arbitrary directions of wave propagation (Johnson 1971, 1972a, b, 1974b; Johnson and Pope 1975). Another notable paper (Johnson and Barker 1969) illustrated the role of dislocation mechanisms on the evolution of steady shock waves, i.e., shock waves that retain their temporal structure as the wave propagates through a material.

## 3.5 Shock-Induced Spallation

Complementary to dynamic yielding during shock compression, material failure often occurs in experiments through the tensile stresses produced by interaction of reflected stress waves from surfaces followed by unloading waves (also called a rarefaction wave). The tensile stress at which this occurs is referred to as the spallation stress or spall strength. Don Lundergan recalled the weapons motivation for developing an understanding of material failure.

> The advent of supersonic and exoatmospheric missile delivery systems and their associated environments imposed new demands on the components and subsystems for which Sandia was responsible. To meet these demands required the understanding of the effects of these severe environments. One such demand was understanding the behavior of materials subjected to impacts at velocities associated with the new delivery systems and to radiation associated with space and countermeasures. One problem related to high velocity impacts was that the structure had to retain its integrity for a period after impact and fuze actuation to prevent deformation of critical structures or destruction of critical components. The problem associated with high energy radiation was the ablation of materials and the resulting high intensity shock waves that produced spallation that could preclude [successful] reentry.

Because this effect was so critical to the operational integrity of components and subsystems in weapons, especially for vulnerability analyses, a focused research effort was initiated in the Solid Dynamics Research Department to determine spall strength thresholds and develop material models to predict when spallation would occur. After joining Sandia in 1962, Barry Butcher was asked to initiate an experimental research program on the spallation behavior of metals. He recalls that

> …. A phenomenon known as spall was of concern. Spall occurs when a compressive shock wave is reflected from a free surface. As the wave travels back into the material and interacts with a rarefaction wave, it grows in tension until it is sufficiently strong to fracture the material. The consequence is that the material between the surface and the fracture flies off at a high velocity. This is the same phenomenon as when a marble hits two inline marbles at rest. The impacting marble stops and the marble farthest away flies off at the impact velocity. The possibility of spall in systems activated by explosive energy is rather unsettling to designers because fast moving debris could fly through space and impact special parts before they have a chance to function.

The new gun impact methods developed by Lundergan in the late 1950s and the evolving interferometry developed by Barker and Hollenbach were ideal tools for determining the spall strength by configuring the impacting plate and sample dimensions and then measuring the free surface velocity history. Figure 3.11 illustrates a typical spall signal for iron. The first increase in free surface velocity at 2.0 μs represents the elastic wave, with an HEL of 2.2 GPa. The sudden decrease, or "pullback," in the free surface velocity after the peak velocity is achieved is designated as $\Delta u_{fs}$ in the figure. It results from the internal splitting of the sample into two pieces caused by internal wave interactions as Butcher explains above. The subsequent ringing in free surface velocity is caused by reverberations in the spalled sample, i.e., the split piece. By measuring the free surface velocity signature, the spall strength is easily determined to be 3.8 GPa.

In the 1960s and 1970s, several metal spallation studies were conducted (Lundergan and Smith 1962; Lundergan 1963; Butcher 1967, 1968; Guess and Lee 1968; Tuler and Butcher 1968; Davison and Johnson 1970; Davison 1974; Davison and Stevens 1972, 1973; Davison et al. 1972, 1977; Davison and Kipp 1978; Stevens and Tuler 1971; Stevens and Pope 1973; Stevens et al. 1973). These systematic experiments at different peak loading stresses and varying tensile loading rates

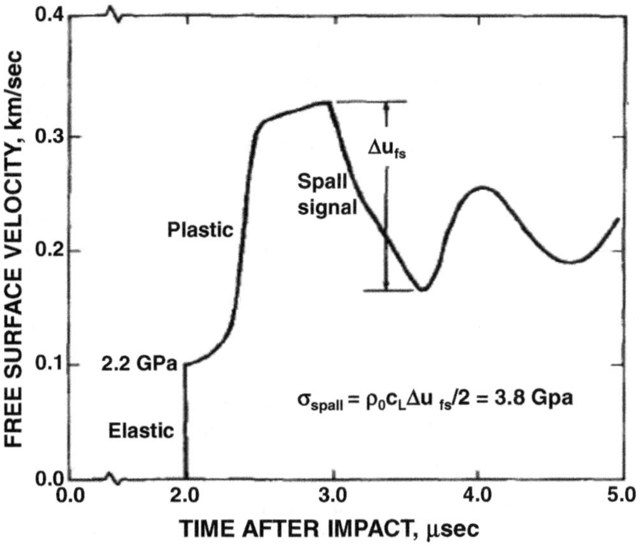


**Fig. 3.11** Free surface velocity of impact for iron showing the pullback in velocity $\Delta u_{fs}$, which denotes spallation. The spall strength $\sigma_{sp}$ is determined with the formula shown, where $\rho_0$ is density and $C_L$ is elastic wave speed (Reprinted with permission from Bertholf et al. 1975, Copyright 1975, AIP Publishing LLC)

(controlled by sample thicknesses) provided a unique database for developing theoretical descriptions of spallation. Herrmann's 1962 paper on the dynamic response of aluminum was one of the first to discuss the wave interactions and material properties important to spallation (Herrmann et al. 1962). Tuler and Butcher (1968) were the first to incorporate time-dependent material failure into a rate model describing metal spall. The early work, initiated at Sandia by Lee Davison and coworkers in the Physics of Solids Research Department, motivated more physically based nucleation and growth models for spallation. To put the theoretical description on a firm physical basis, a host of materials science studies were conducted to investigate the metallurgical effects and damage morphology produced during spallation. This dedicated research resulted in a significant advance in the ability to engineer weapon systems for use in a large variety of environments.

## 3.6 Viscoelastic Materials

Another class of materials of interest to weapon designers included polymeric materials. They have a significantly different loading response from metals and exhibit a strong time dependence in their elastic response, which is usually referred to as viscoelasticity. Probing the detailed shock wave structure under well-controlled loading conditions is necessary to understand the dynamic response of these materials. Upon joining Sandia in 1967, Karl Schuler initiated a systematic study of polymethyl methacrylate (PMMA), also known by the commercial name Plexiglas®. PMMA exhibits viscoelastic behavior during shock loading with a unique wave structure. At the time of these studies, the theoretical underpinnings of viscoelastic phenomena had not been established for the high-rate, high-pressure regime. As Schuler states in his recollections,

> I was hired by Don Lundergan, who wanted me to work on wave propagation in polymeric materials. There were two reasons for this: the metal behavior at low shock pressures (<50 kbars) was being adequately covered but many polymers involved in nuclear weapons had only been studied quasi-statically. Under shock loading their characterization was lacking, particularly for time-dependent effects. Secondly, Lynn Barker was developing the laser interferometer, and Plexiglas® looked like an ideal low-impedance window material.

Schuler was one of the first users of the Sandia velocity interferometer. His progress in applying this diagnostic to PMMA greatly benefited from close interaction with Lynn Barker. As Schuler remembers,

> I guess because of both the physical proximity and our joint interest in Plexiglas®, I worked closely with Lynn Barker. He was definitely a great mentor to me. I'll always remember that in his cubicle there was a blackboard and along the top of the blackboard was written 'I'm perfectly willing to be wrong.' Very seldom was Barker wrong.

Schuler's experimental program quickly produced major results concerning the viscoelastic response of PMMA that were published in the early 1970s (Schuler 1970, 1971).

When the highly talented theoretician, Jace Nunziato, joined Sandia under Walt Herrmann around 1970, he and Schuler teamed together. Through this collaboration, the PMMA data were rapidly integrated into a rational mechanics framework and published in an extensive series of papers (Nunziato et al. 1974a, b, 1975; Nunziato and Schuler 1973a, b; Walsh and Schuler 1973; Schuler and Nunziato 1974, 1976; Barker et al. 1974b). Schuler comments in his recollections:

> My weak point was doing the paperwork at the end. I was not a good or fast writer. I always had a cartoon on my office wall depicting an outhouse with the caption 'The job's not done till the paperwork's finished.' Fortunately, management recognized this and teamed me up with Jace Nunziato, who was a superb writer. Between the two of us we pumped out quite a few papers. I'd do the experiments, reduce the data, make up some figures and tables, and discuss them with Jace. Together we would test published theories with the data and scope out a paper….

The viscoelastic model of PMMA developed through their collaboration was incorporated into the WONDY code, which was also well developed at that time. That code provided the ideal test bed to compare the new theories with experimental measurements and performed admirably in that role. Figure 3.12 shows a PMMA shock loading experiment and a 1-D simulation using Nunziato's viscoelastic model. The initial sudden increase in particle velocity at zero time represents the arrival of the fully elastic response of the material. The gradual increase in particle velocity after that time results from viscoelastic behavior. The tight link of experiment and theory provided a quick comparison of the unique data with the superb theoretical research of Nunziato and Ed Walsh, a University of Florida professor.

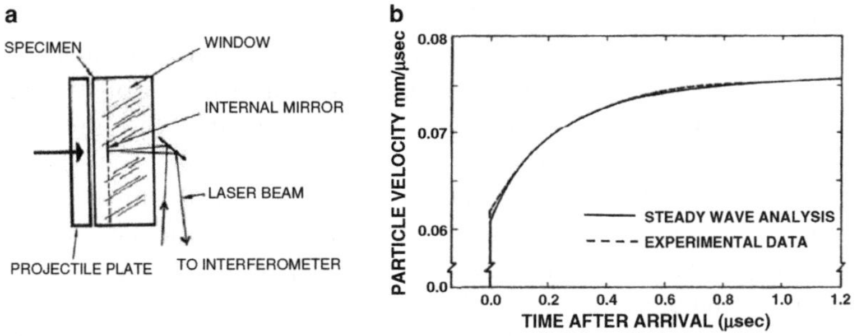


**Fig. 3.12** Viscoelastic wave profiles in polymethyl methacrylate. (**a**) Experimental technique. (**b**) Comparison of experimental and simulated profiles with arbitrary time scale. Note the contrast to elastic–plastic behavior exhibited by metals in Fig. 3.8 (Reprinted from Nunziato et al. 1974b, Figs. 8 and 26, Copyright 1974, with permission of Springer Publishing)

The significant progress in developing the theoretical framework in the 1970s led to publication of a widely distributed monograph in the prestigious Handbuch der Physik book, *Mechanics of Solids* (Nunziato et al. 1974b and second edition, published in 1984).

## 3.7 Porous Materials (Foams)

Porous solids are materials with less than theoretical density, usually referred to as "distended" materials or "foams." Sandia research on shock compaction of porous materials is another example of how integrated experimental and theoretical efforts produced significant results in a short time. Knowledge of the dynamic response of porous materials was required for a number of weapon applications, resulting in considerable theoretical and experimental resources being devoted by the Solid Dynamics Research Department to developing a basic understanding of these materials. As Butcher notes in his recollections,

> Energy absorption by porous materials of metal and plastic (that is, foam) was another area I was asked to investigate. Energy is dissipated as a fragment moves through the foam and the foam is severely deformed. Actually, in an explosive environment, an awful lot of foam is needed to slow down the rapidly moving fragments, but the presence of the foam is certainly better than no foam at all since it helps keep things in place. Foam is also useful when very high energy from x rays is deposited within it. With foam, the vaporized material has room to expand rather than being pushed like a piston. Some of my work concerned this phenomenon. With underground testing in progress during this time, I evaluated the effectiveness of this material response on several underground nuclear tests to see if we could duplicate the results mathematically. Those tests were indeed big physics, and I am still overwhelmed by them….

The early attempts to treat the compaction of porous materials assumed that under shock loading, the initial porous or distended material is compressed without strength effects to a fully dense solid, resulting in higher temperatures caused by the work done in crushing the pores. This model is referred to as the "snow plow" model and does very well describing the pressure, density, and temperature of compacted porous materials at pressures far exceeding the strength of the material. Forbes presents this thermodynamic model that describes the high-pressure compaction of porous materials (Forbes 2012).

However, Sandia applications of porous materials usually involve the low-pressure partial compaction regime where strength effects cannot be ignored. Walt Herrmann recognized that the dynamic response of a porous material would remain elastic up to a longitudinal stress at which the applied stress is sufficient to begin collapsing internal voids; for stresses above that, the material would gradually compact to solid density. For a stress exceeding that for complete compaction, the response is governed by the EOS of a fully dense solid, but at a higher temperature, as described by the snow plow model. The irreversible work done in collapsing the pores produces internal heating and results in significantly higher temperatures than compressing a fully dense solid to the same stress.

Beginning in the late 1960s, Herrmann initiated a comprehensive research program to describe shock wave propagation in porous materials. As leader of this effort, he developed a model to describe the compaction process that is still used by researchers worldwide because of its simplicity, even though more sophisticated models have since been developed. He devised a parameter $\alpha$, which is the distended or actual volume divided by the ambient volume of the solid (or, equivalently, the solid density divided by the initial distended density), to describe the compaction process, thereby establishing the name "P-$\alpha$ model" (Herrmann 1968, 1969a, b, 1971, 1972).[4] The compaction to solid density usually occurs over a stress range of a few to tens of kbars, depending on material strength, since internal stresses cause all the pores to collapse. The theoretical framework of porous material compaction was published in the classic paper "Constitutive equation for the dynamic compaction of ductile porous materials" (Herrmann 1969a) and in the precursor Sandia Report (Herrmann 1968). A typical compaction curve for porous iron is shown in Fig. 3.13. The sudden increase in pressure for each of the curves represents elastic response. The following curves merging with the single curve called "solid" represent the compaction behavior of the material. The P-$\alpha$ model was developed to describe this response specifically. The same parameter is used in the model for all of the different initial porosities shown.

Extensive experiments were conducted in the late 1960s to develop and verify the model (Lee 1967, 1968, 1972; Lysne and Halpin 1968; Butcher and Karnes 1969; Butcher 1971, 1973; Butcher et al. 1973, 1974). Later theoretical papers dealt with various refinements (Herrmann 1971, 1972). The P-$\alpha$ model has been incorporated into several codes and consistently provides a good estimate of material response during shock-induced compaction (Herrmann 1969b, 1971).

To mitigate shock damage to materials in real systems, porosity was seized on as an ideal countermeasure or hardening approach. For materials where intrinsic strength was not a direct or prime requisite, porosity could be introduced to attenuate shock amplitudes that propagate and cause damage to other material layers.

[4] Rigorously, longitudinal stress is in the model description, but historically this was referred to as pressure, hence the term P-$\alpha$ model.

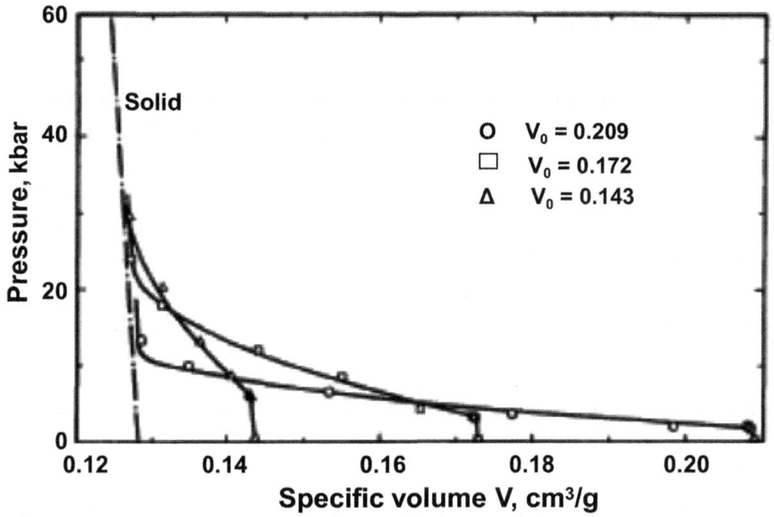


**Fig. 3.13** Compaction curve for porous iron of different initial densities with comparison of the P-α curves to experimental data. Specific volume is inverse density (From Herrmann 1972, reprinted with permission of Sandia National Laboratories)

Because part of the work is done in crushing the porous material voids, the shock amplitudes are rapidly reduced by the porosity. In fact, it is the first percent or so of the void content that yields by far the largest shock attenuation exhibited by this class of materials (Herrmann and Lawrence 1978).

Several other models, including rate-dependent pore compaction, were developed to describe porous materials (Davison 1971; Johnson 1968a; Butcher 1971; Butcher et al. 1974; Holt et al. 1974), but the P-α model is the simplest and most widely used. It was pervasive throughout the scientific and weapons communities at that time and is still in use today. More recently, Grady and colleagues developed a generalized version of the P-α model, called the "P-λ model," which covers a wider class of materials, such as mixtures and materials undergoing phase transformations (Grady et al. 2000; Grady and Winfree 2001; Fenton et al. 2012). The combined theoretical and experimental research on porous materials established Sandia's leadership role for modeling this class of materials. The investigations conducted on porous materials are one of the crowning achievements of shock wave research at Sandia.

In an extension to the initial studies of porous material compaction for longitudinal stress alone, recent experiments on the oblique-impact gun (i.e., the original Lundergan gun modified to produce combined longitudinal and shear stress states) showed that the final material state after compaction cannot be described by a single parameter such as α. Vogler found that granular particles fractured during the compaction process and the particle size were significantly affected by imposing a shear stress, in addition to the longitudinal stress (Vogler et al. 2011). The added shearing force strongly influenced the process, resulting in smaller particle sizes for the compacted material.

## 3.8 Computational Capabilities

Up to the 1960s, practical computational technology for analyzing shock wave propagation consisted of paper, pencil, and mechanical calculating machines. For example, in the Manhattan Project, a large number of people were needed to design and analyze the response of nuclear weapons using hand calculations and the limited calculating machines available at the time. As Don Lundergan notes in his recollections,

> The initial calculations, around 1957, to determine the propagation of stress waves in a weapon structure were done by using known elastic wave velocities to hand plot the progression of stress waves and estimate the times of critical events. The calculations were both tedious and inaccurate because of the lack of computational techniques and material properties. The calculations, however, did indicate that a problem existed.... A year or so later, we had progressed to using limited data for a few materials and Marchant mechanical machines to do the calculations.

As noted in his recollections, Lundergan quickly recognized this deficiency and proposed a comprehensive research effort to develop integrated capabilities that would allow the solution of complex weapons problems. Ray Reed remembers:

> Don had long recognized the emerging critical importance of shock wave technology to the Sandia mission. He had formally developed and proposed a comprehensive and very detailed plan for an entirely new department focused on all aspects—theoretical, computational, and experimental—of shock wave technology critically related to the Sandia mission. I joined the newly formed department Don had conceived and promoted rather than my prior R&D department. The new organization, as conceived by Lundergan, had emerged as a department of four divisions soon headed by Walter Herrmann, who was newly hired from the Massachusetts Institute of Technology. The department expanded quickly with its divisions led by supervisors Larry Bertholf, Olden Burchett, Barry Butcher, and Darrell Munson. That new department, the Solid Dynamics Department, as designed by Lundergan, was created especially for applied computational and experimental shock wave technology in a weapons directorate to complement the theoretical research efforts of the Physical Research Department.

The 1960s witnessed the beginning of computer codes at Sandia that could employ the material models being developed during this time to predict the behavior of stress wave propagation in a wide range of materials and components subjected to a variety of dynamic loading conditions.

One of the earliest and most widely available hydrocodes, WONDY (Herrmann et al. 1967), was developed in the 1960s and 1970s, initially by Herrmann at MIT, and then much more extensively at Sandia (e.g., Lawrence 1970; Kipp and Lawrence 1982). A one-dimensional Lagrangian[5] wave propagation code was designed in modular fashion to allow easy changes to the material response models and driving conditions. Its major advantages are that it is stand-alone, including both input and output, and can run single cases on current-generation PCs and laptops in just seconds. Because of its flexibility, WONDY is suited for large parameter studies involving

[5]The Lagrangian frame of reference is fixed in material coordinates.

variations in equations of state and constitutive models, as well as for a large variety of initial and boundary conditions.

The code was initially used to solve wave propagation in elastic–plastic materials, but its modular construction allowed application to a wide variety of different material models. WONDY underwent several modifications as it was refined for specific technical applications and for running in the evolving Sandia computing environment (see Fig. 1.3) in the 1960s and 1970s (Herrmann et al. 1967, 1970; Herrmann 1969c; Lawrence 1970; Kipp and Lawrence 1982; Herrmann and Bertholf 1983). Jeff Lawrence describes the major features of the code.

> Because of its flexibility and modular design, WONDY was used during these years as a 'test bed' for many new models, new numerical techniques, and other computer-oriented tasks (e.g., Input/Output). Examples of new equations of state and constitutive relations included material porosity, cumulative-damage fracture and spall, and various rate-dependent material responses. Rate dependence has its own characteristic time constant, different from that associated with the numerical stability of standard shock-propagation analyses. The new technique of sub-cycling, necessary for stability of dual time-constant solutions, was therefore developed. The rate-dependent models were also an effective approach to studying complex, layered composites. Other innovations treated the electrical response of piezoelectric and ferroelectric materials as well as mesh rezoning for improved computational efficiency.

A good example of WONDY's capabilities is the analysis of shock attenuation in rate-dependent materials (Herrmann and Lawrence 1978). During the development and use of experimental and modeling capabilities, several classes of materials exhibited significant rate-dependent behavior. Since the effects of rate dependence on pulse attenuation were not known, but are important for weapon applications, WONDY was used to evaluate these effects. A notional example of the attenuation effects in a rate-dependent viscoelastic solid is shown in Fig. 3.14. Although it is notional, this example is a representative of real material response. Once the specific parameters and rate effects are quantified for a given material, say a composite, the curves can be used to determine the amplitude of the shock stress at a given location. The significant result is that the effects of shock attenuation can be tailored for engineered materials to the anticipated loading by adjusting the relaxation time. For composite materials, the tailoring can be done by using various configurations of layered materials, and the composite can be modeled as viscoelastic (Barker 1971b; Lawrence 1973), as discussed in Chap. 4 (see Fig. 4.7). This technique allows fabrication of composite configurations to ensure protection of weapon components by limiting their peak stress exposure. The modeling capability has been used in various applications, including the effects of pulsed radiation on weapon components.

The next step in code development included 2-D hydrodynamic codes, which were needed to simulate complex weapon components and systems. The development of 2-D Eulerian[6]-Lagrangian hydrodynamic codes was also initiated in the late 1960s. Lawrence captures the evolution of Eulerian hydrodynamic codes in the following recollection.

[6] The Eulerian frame of reference is fixed in the laboratory.

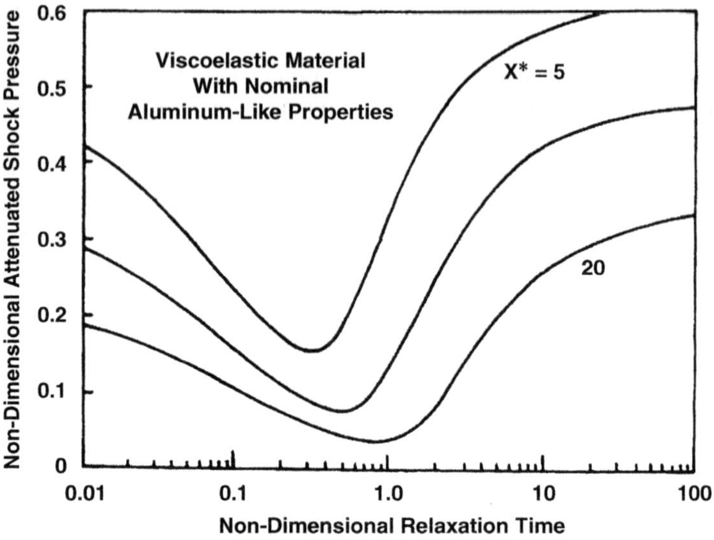


**Fig. 3.14** Viscoelasticity effect on shock attenuation of Al-like material for triangular shock with peak pressure ~1/3 the material bulk modulus. Relaxation time is non-dimensionalized by initial pulse width; attenuated shock pressure is non-dimensionalized by peak load. Curves are propagation distances times initial spatial shock pulse width. Figure represents a 250-kbar peak pressure and 0.1-$\mu$s pulse width (typical of a high-explosive load for ~0.6-mm spatial pulse width) for 1 ns to 10 $\mu$s relaxation times. Near figure center, relaxation time yields shock attenuation 0.1–0.2 times applied load, depending on propagation distance (~0.3–1 cm here). Short relaxation times (*left*) produce attenuation equivalent to a rate-independent elastic-plastic material; long relaxation times (*right*) yield responses nearly equivalent to purely elastic behavior, with considerably less attenuation (Reprinted and modified with permission from Herrmann and Lawrence 1978, Copyright 1978, ASME)

In the 1970s and 1980s, Sam Thompson and his colleagues began migrating from the group that supported the Nuclear Regulatory Commission to the applied shock physics group. This was still a period of rapid shock physics code development at Sandia. Thompson brought his independently developed set of hydrocodes, which started with the 1-D versions, CHART and CHARTD. These evolved into the 2-D version, CSQ (CHART-SQuared), and later into the 3-D code, CTH (CSQ to the Three-Halves power). In contrast to WONDY and TOODY, these codes were Eulerian, as opposed to Lagrangian, and included radiation-transport capabilities and their own equation-of-state formulations. In addition to Thompson, the developers included Mike McGlaun, Hank Lauson, James Peery, Steve Rottler, and others.

However, this family of codes, although incorporating internal zone-to-zone radiation transport, did not include explicit treatment for the time-dependent deposition of external radiation sources, such as that which would be required for the pulsed x-ray sources from nuclear weapons. These sources were generally included in the analyses through external routines developed for that task.

Regarding Thompson's codes, Mike McGlaun also recalls,

> Sam developed an integrated suite of software that analyzed one-dimensional (1-D) and two-dimensional (2-D) strong shock problems. One of Sam's shock codes was the 1-D finite-difference Lagrangian radiation-shock physics code CHARTD.... Sam also developed a powerful suite of pre- and post-processing graphical software to analyze the equations of state, time histories of Lagrangian and Eulerian points, and 1-D and 2-D calculations. Sam's software was powerful, widely used, and had a reputation for being accurate, reliable, and well supported.

In the mid 1960s, Lynn Barker began working on another approach for performing 1-D shock wave simulations. This method decomposed the shock wave into small stress jumps that propagate through the material in accordance with the 1-D equations of motion, similar to the approach used in characteristics codes (Asay et al. 1975). The latest version of Barker's SWAP (Stress Wave Applications Program) code was called SWAP-9 (Barker and Young 1974). As Lawrence further states,

> At the same time, Lynn Barker, with the assistance of Emily Young, was developing the SWAP code. SWAP, a method-of-characteristics code, ran much faster than any of the finite-difference codes and yielded clear mappings of propagation and attenuation of the various shock and pressure waves throughout the computational space. However, it was more difficult to interpret for complex problems where wave interactions often became very complicated. Multi-dimensional characteristics codes were flirted with, but received essentially no long-term support.

SWAP was not used to any extent in weapon applications, but was extremely useful to experimentalists and modelers for designing and analyzing shock wave experiments.

## 3.9 Answering DOD's Needs

Much of the increase in numerical modeling activity, especially those caused by pulsed x-ray shock generation, was prompted by the tremendous increase in the rate of underground nuclear weapon effects testing. There was a great need for analytical capabilities to take advantage of the stream of data being generated by these tests. This activity started in the late 1950s, but mainly in the early-to-mid 1960s, there was an increasing and broader interest in the nuclear vulnerability and survivability of missile systems, reentry vehicles, and even satellites. The principal threat to our weapon systems was exposure to exoatmospheric nuclear detonations and specifically to the unrestricted output of ~3/4 of their total output energy, which was in the form of low-energy but high fluencies of pulsed thermal (i.e., black body) x-rays. This was probably the chief rationale for the increased weapon effects UGT activity. At that time, the Sandia group headed by Al Chabai (with Dennis Hayes as a member of that group) was beginning to discuss many of these issues with the researchers at the Air Force Weapons Laboratory (AFWL). It was clear from these interactions that the main tools for investigating these issues were shock wave oriented, directly related either to shock generation and propagation or to the structural response resulting from impulsive loads produced by pulsed radiation loads. In fact, it was Chabai and his team who pointed toward the shock physics work underway at Sandia, which was at the time headed by Don Lundergan and included Lynn Barker, Barry Butcher, and their colleagues.

Concurrently, in the early 1960s, there was a push from the DOD to study and understand these issues as they impacted the detailed operational capabilities of missiles and other systems. The main group driving this effort was the Defense Atomic Support Agency (DASA), which subsequently evolved into the Defense Nuclear Agency (DNA), and is now the Defense Threat Reduction Agency (DTRA). This involved studying the key areas of vulnerability and hardening of the Nation's own systems, as well as similar studies from the somewhat different perspective of the lethality of these weapons against opposing systems. After the cessation of atmospheric and high-altitude nuclear tests in the early 1960s, UGTs became the main source for these data. UGT activities were generally of two types: weapon development tests, mostly run by Los Alamos and Lawrence Livermore National Laboratories, and weapon effects tests, mostly conducted under the auspices of DOD/DASA. There was some overlap, and often some effects experiments would be conducted on development shots and vice versa.

AFWL was the focus of much of the weapon effects test preparation, design, and execution, as well as post-test analysis. With the proximity of Sandia Labs on Kirtland AFB (called Sandia Base until 1971), along with Sandia's involvement in most phases of the UGT program, it was natural to establish a close working relationship. In this small part of the "big picture," the main players at Sandia were Carter Broyles and his group, who supported many UGT activities at the Nevada Test Site.

On the other side of the base, at AFWL, Air Force Capt. Don Lamberson headed a group of young Air Force officers who did most of the pre-test design and post-test analysis of one class of those tests, that is, the one-dimensional, pulsed x-ray-driven, shock wave investigations. Those experiments were typically conducted on one-dimensional coupons of of basic, special, and composite materials of great interest to the Air Force. Typically the tests consisted of active, time-resolved, shock wave profile measurements. They employed, for the first time, the Sandia-developed quartz gauges.

The objectives of these and related experiments were to obtain actual dynamic pressure histories for basic, special, and composites materials that would support establishing vulnerability and lethality levels for missiles and weapon systems. Validation of the theoretical calculations of dynamic material and system responses was thus an important driver. An additional goal was to evaluate early concepts of countermeasures for hardening military systems against nuclear weapon effects. Another important aim was to evaluate new active and passive experimental techniques, especially under the extreme conditions of the UGT environment. Because of the uncertainties associated with the UGT environments, it was learned early on that, if space were available, tests of the samples should be replicated whenever possible.

The rapid growth of computer codes for shock wave problems, especially in the 1960s, enabled detailed analyses of these pulsed x-ray effects for complex weapon engagements. These codes were necessary because of the extreme nonlinearity of the shock-propagation phenomena: the simple or closed-form analysis techniques could not provide accurate solutions.

Dennis Hayes wrote one of the first 1-D computer codes, VANDAL, to evaluate the absorption and deposition of x-rays in potential targets. Herrmann and his colleagues modified the one-dimensional finite-difference code, WONDY, to treat these short-pulse x-ray deposition problems. Other codes implemented during this period included the 1-D radiation hydrodynamic code CHARTD, which modeled internal radiation transport (Thompson 1969, 1970, 1973; Thompson and Lauson 1972). The emergence of 2-D computer capabilities for shock wave problems also appeared in the 1960s with the first 2-D Lagrangian hydrodynamic code TOODY (Thorne and Herrmann 1967) and the 2-D Eulerian follow on to CHARTD, called CSQ, which like its 1-D predecessor included radiation transport using the diffusion approximation (Thompson 1979).

Many experiments were conducted at the Nevada Test Site to quantify the stress waves produced by nuclear weapon explosions and compare the results with numerical predictions. The quartz gauge played a principal role in quantifying radiation-produced stress pulses for x-ray vulnerability studies. As Al Chabai recalls,

> Different nuclear sources were utilized to provide x rays with various spectral distributions, and experiments were conducted within the pipes to measure the effects of x-ray deposition on materials. Often the level of x-ray deposition was high enough that the surface of exposed material samples was heated above the vaporization level, producing 'blowoff' from the sample and generation of a consequent shock wave in the remaining solid material. The quartz gauge developed by Frank Neilson at Sandia was a very significant instrument for measuring this shock wave. This measurement technique was utilized extensively on all x-ray vulnerability tests at NTS, not only by Sandia, but by many other experimenters. The quartz gauge provided, with high accuracy, the wave shape and peak shock wave amplitude produced in the sample by blowoff from x-ray exposure. Among the many measurements made (e.g., impulse, temperature, shock pressure, displacement, etc.), it is fair to say that the quartz gauge provided the bulk of the most important quantitative data, allowing the U.S. to develop warheads for the Navy and the Air Force that were hardened to x rays, thereby providing a credible deterrent.

Emphasizing the game-changing capability of the quartz gauge, Ray Reed, who participated in many of the experiments, also recalls:

> … I learned that the patented Sandia Quartz Gauge design for laboratory use could not be used in nuclear field tests.… Building on those arbitrary designs, I standardized a family of 'Sandia Field Test Quartz Gauges,' different from the ideal laboratory 'Sandia Quartz Gauge.' … In the ideal 'Sandia Quartz Gauge,' only slightly tilted and sustained planar impact by a thick impactor results in a very narrow step to an ideal sustained flat plateau as the stress wave progresses through the quartz disk. That demonstrates the needed linear response. Rather, in a 'Sandia Field Test Quartz Gauge,' a sustained, nearly planar

impact also results in an initial narrow step but, consequentially, that step is followed by a nonlinear, progressively climbing amplitude to an abrupt end rather than a flat plateau. In some prior field test gauges the relative terminal amplitude was more than double that of the initial step (an extremely non-ideal response). Subject to a stress pulse loading of arbitrary form, as in field tests, the recorded pulse shape resembles neither the characteristic step response nor the imposed input pulse. Such a distorted response profoundly displays the dreaded nonlinear response. Critically, neither the recorded pulse shape nor the peak amplitude from the applied field test quartz gauges even approached faithful waveform reproduction.

It should be noted that with the later advent of the AGT pulsed radiation capabilities, quartz gauges were also used to measure stress waves from pulsed electron beams to further understand thermoelastic energy coupling (Graham and Hutchison 1967; Graham et al. 1967b).

Although less directly connected with shock wave phenomena, many other experiments, some of which were designed for posttest recovery, incorporated full-scale or more often subscale reentry vehicle aeroshells. Their purpose was to investigate the buckling of these structures from the pulsed x-ray weapon output. Since the buckling of these shells is characterized by time constants (in milliseconds) several orders of magnitude longer than material shock response (in microseconds), these two response modes can often be decoupled. The shock propagation codes, if properly validated, can calculate the radiation-induced impulse, which is then used as a relatively simple input to the structural response models and codes.

It is also interesting that during these years considerable effort was being devoted to the development of closed-form analytic models for calculating these x-ray-induced impulses. This work was sponsored initially by AFWL and later significantly expanded by Sandia. In the end, these models proved quite successful and over the succeeding years have been applied to many diverse problems, ranging from pulsed power experiments (Lawrence et al. 2002, 2009a, b) and ground-based pulsed lasers for low-Earth-orbit launching (Lawrence et al. 1991) to defense against potentially Earth-impacting meteoroids or asteroids (Lawrence et al. 2012).

Relatively simple models for the impulse-driven buckling of structures and shells were also developed at this time, mainly by scientists at the Stanford Research Institute (SRI). These approaches yielded "pressure-impulse" thresholds (P-I curves) for shell buckling. These curves were all derived from the straightforward properties related to the basic characteristics of the structure and its materials. They provided good estimates for both the vulnerability and survivability of these weapon systems to structural buckling from pulsed x-ray loads (Lindberg and Florence 1983).

## 3.10 People and Places of the 1960s

**Walter Herrmann, promoted to manager of the newly created Mechanical Behavior Department, 1967 (Reprinted with permission of Sandia National Laboratories)**

**Al Chabai, promoted to Division Supervisor of the Underground Physics Division, 1965 (Reprinted with permission of Sandia National Laboratories)**

**Lee Davison, promoted to supervisor of the Shock Wave Physics Research Division, 1968 (Reprinted with permission of Sandia National Laboratories)**

**Luke Vortman, detonating 30,000 lb of explosives at Coyote Canyon (Reprinted with permission of Sandia National Laboratories)**

**Duane Hughes, Dick Traeger, and Orval Jones (*L-R*), discussing Sandia's new Unified Science and Engineering program, 1966 (Reprinted with permission of Sandia National Laboratories)**

**George Samara, promoted to supervisor of the Physics of Solids Research Division, 1967 (Reprinted with permission of Sandia National Laboratories)**

**Phil Stanton received PhD from UT, 1968 (Reprinted with permission of Sandia National Laboratories)**

**Jim Johnson, relaxing after a hard day of theory (circa 1970), private collection of Jim Johnson**

**Glenn Seay, promoted to Physical Research Department Manager, 1964 (Reprinted with permission of Sandia National Laboratories)**

**Bruno Morosin, promoted to supervisor of the newly created Chemical Physics Research Division. 1967 (Reprinted with permission of Sandia National Laboratories)**

**George Ingram holding the new sapphire gauge, 1966 (Reprinted with permission of Sandia National Laboratories)**

**Bob Graham (*left*) and Frank Neilson inspecting a quartz gauge, 1962 (Reprinted with permission of Sandia National Laboratories)**

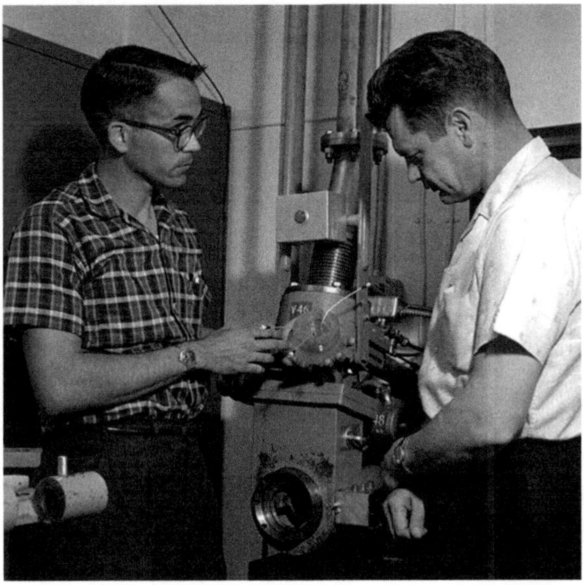

Bob Eckhardt (left) and [illegible] [illegible] National [illegible]

# Chapter 4
# The 1970s: New Opportunities

## 4.1 Background

In 1973 a laboratory-wide reduction in Sandia staff caused major reverberations in shock wave research. As Barry Butcher stated in his recollections:

> This era was one of turmoil at Sandia with budgets slashed to the point that a layoff was necessary. Many had been involved in underground testing, which was now over. Attention turned to keeping components safe from explosive damage. During part of this time a very nervous staff went about their duties with uncertainty about whether a job would be there for them in the morning.

One of the people picked to be laid off was Don Lundergan who, along with Lynn Barker and Red Hollenbach, had originated the shock wave group in the late 1950s focusing on engineering applications. Lundergan had stepped down from supervision around 1968 after Herrmann's promotion to Department Manager and was a staff member at the time. His selection for layoff produced considerable dissatisfaction among the staff, particularly with Lynn Barker, who had worked with him from the beginning of the shock wave program. Barker and Karl Schuler, with input from other staff, immediately wrote a protest letter to the upper management. Although this may or may not have influenced the final decision, the upshot was that Lundergan was taken off the layoff list the following week. However, Lundergan ended up leaving the Solid Dynamics Research Department within a couple of years and joined another group at Sandia that was involved in developing new intelligence tools. He excelled in this new position, achieving the status of Distinguished Member of Technical Staff in 1985. His work in intelligence methods not only generated considerable recognition for himself and the intelligence group but also elevated Sandia's stature in the U.S. intelligence community. One of his major achievements was an innovative surveillance technique for detecting foreign underground nuclear tests. Lundergan's success in another high-risk, high-payoff endeavor attests to his unique talent for identifying and implementing forward-looking technologies.



J.R. Asay et al., *Impactful Times*, Shock Wave and High Pressure Phenomena,
DOI 10.1007/978-3-319-33347-2_4

The 1973 layoff caused some of the technical staff members at Sandia to think seriously about their remaining years before retirement. Jim Johnson had already been recruited from Sandia to work for Terra Tek, Inc. in Salt Lake City. Terra Tek was a small but growing company of about 50 employees that offered attractive stock options to new hires. Lynn Barker joined Jim Johnson at Terra Tek in mid 1974. The departure of these two highly innovative researchers was a major loss to the Sandia shock wave program. However, Barker returned to Sandia in 1981 as a member of Jim Asay's Thermomechanical and Physical Division, which was conducting experimental research in Walt Herrmann's Solid Dynamics Research Department. Barker resumed making major contributions to shock physics research, as he had done earlier, until his retirement in 1990. After leaving Sandia the second time, he commercialized the VISAR with an innovation that involved recording several points on a target with a single interferometer.

In spite of the budget uncertainties in the mid 1970s, the evolution of shock wave instrumentation continued (including a revolutionary development in interferometry). The type and number of materials being studied substantially increased, more innovation occurred in the types of material properties that could be measured, and computer capability substantially expanded, resulting in an enhanced capacity for solving shock wave problems. The major developments in shock wave research in the 1970s that are discussed in this chapter include:

- Velocity Interferometer System for Any Reflector (Sect. 4.2)
- Shock-Induced Phase Transitions (Sect. 4.3)
- Two-Dimensional Computer Codes (Sect. 4.4)
- Composites and Mixtures (Sect. 4.5)
- Damage-Based Spallation Models (Sect. 4.6)
- Geological Materials: Oil Shale (Sect. 4.7)
- Piezoelectric and Ferroelectric Materials (Sect. 4.8)
- Third-Order Elastic Constants (Sect. 4.9)
- Pressure–Shear Loading (Sect. 4.10)
- Shock Rise Time and the Fourth Power Law (Sect. 4.11)
- Acceleration Waves (Sect. 4.12)
- Shock Thermodynamics Applied Research Facility (Sect. 4.13)
- Mass Ejecta from Shocked Surfaces (Sect. 4.14)
- Energetic Materials (Sect. 4.15)

## 4.2 Velocity Interferometer System for Any Reflector

Another game changer in shock wave diagnostics occurred in 1972 that had, and continues to have, a revolutionary impact on shock wave research at Sandia and other institutions throughout the world. The previous interferometer techniques developed by Barker and Hollenbach for shock wave research, i.e., the Michelson displacement interferometer (Barker and Hollenbach 1965) and the Sandia velocity interferometer (Barker 1968), had the serious drawback that the reflecting free

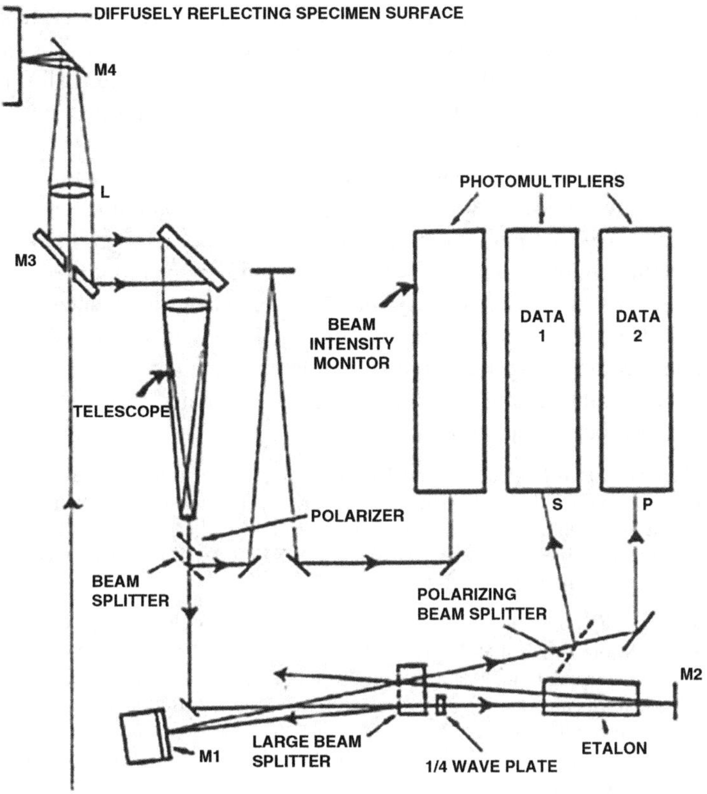


**Fig. 4.1** Schematic of the VISAR for measuring the velocity of any moving surface (Reprinted with permission from Barker 2000a, Copyright 2000, AIP Publishing LLC)

surface of the sample had to remain specular, or mirror-like, during an experiment. This requirement is not maintained in many cases because the reflective surface may become diffuse as a result of shock-induced effects. Some examples are (1) shock-induced phase changes that may occur, (2) shock compression of porous and composite materials that exhibit heterogeneous response on a scale comparable with the recording area of the interferometer, and (3) micro jetting from a shocked free surface. A versatile interferometer technique was badly needed that could be used with any material under any loading conditions.

Barker and Hollenbach (1972) resolved this pressing need with the ingenious development of the VISAR, which is shown schematically in Fig. 4.1. The VISAR is based on optically differentiating the reflected laser signal from a diffused surface, using the two delay legs of the optical paths shown, to determine the particle velocity directly. This is also referred to as the wide-angle Michelson interferometer (Gillard et al. 1968). The approach is not unlike the Sandia velocity interferometer, except the difference in optical delay leg times is small, on the order of 1–2 ns, allowing significant improvement in time resolution, as opposed to the Sandia velocity interferometer. In addition, the reflection from a diffused surface produces a speckle pattern in the reflected light so that when both legs are overlapped, a fringe pattern is produced that is imaged in the interferometer. This image is recorded using photomultiplier tubes, or photodiodes, to obtain the fringe shifting that is related to the surface velocity (Barker and Hollenbach 1972). Two fringe images 90° out of phase, as a result of a quarter wave plate inserted in one leg, are monitored with two photomultiplier tubes and analyzed simultaneously with the beam intensity monitor to determine the particle velocity (Barker 1974). The significant advantage of this approach is that *any* surface can be monitored, in contrast to the Michelson and Sandia velocity interferometers for which specular surfaces are required. The time resolution is also much better than was possible with the previous interferometers, and the precision of the particle velocity measurements is also increased. An interesting anecdote about how the name "VISAR" came about is provided in Barker's recollections:

> I remember when Red Hollenbach and I began to realize what a powerful shock wave instrumentation tool we'd found in the laser velocity interferometer that worked when the specimen surface didn't need to be polished to a mirror surface – a diffusely reflecting surface worked just fine! And unlike the Lockheed interferometer, which preceded ours, ours was at least ten times more accurate!
>
> We decided to look for a name which made a jazzy-sounding acronym such that the acronym came from words that described the interferometer. I think we spent at least an hour one morning trying to include the words "velocity interferometer" and "any reflector" to stress that it works on either diffusely reflecting surfaces or surfaces polished to a mirror finish. When we finally got an "s" in there using the word "system," we had Velocity Interferometer System for Any Reflector, or VISAR. Then we wondered why it had taken us so long to see it. It made sense, it sounded good, and it described the product.

Analysis of the fringes produced in the VISAR is a sophisticated, but straightforward process, as described in early reports (Barker 1971a, 1974). In most experiments at moderate particle velocities, the uncertainty in particle velocity measurement is on the order of 1 % and the time resolution is 1–2 ns (Barker and Hollenbach 1972; Barker and Schuler 1974; Barker 2000a). The time resolution can be substantially improved by using modern instrumentation, such as photodiodes or fast streak cameras, to record the fringe signals, although the accuracy in measuring the amplitude of a single fringe is reduced. An interferometer trace recorded on shocked iron is shown in Fig. 4.2 and a reduced free surface velocity record in Fig. 4.3.

Because of its versatility, the VISAR has been used in a broad range of applications, including:

- measurements of stress loading and unloading (Chhabildas and Asay 1979, 1982; Chhabildas and Grady 1984; Chhabildas and Miller 1985; Wise et al. 1982; Chhabildas and Barker 1988; Davis et al. 2002; Davis 2005; Davis and Hayes 2007; Davis and Foiles 2005),
- shock compaction of porous materials (Asay and Barker 1974; Trott et al. 2000, 2001, 2002, 2006, 2007),
- spallation, polymorphic or melting phase transformations, and shock-induced vaporization (Barker and Hollenbach 1974; Barker 1975; Asay and Hayes 1975; Asay et al. 1988; Chhabildas et al. 2006),

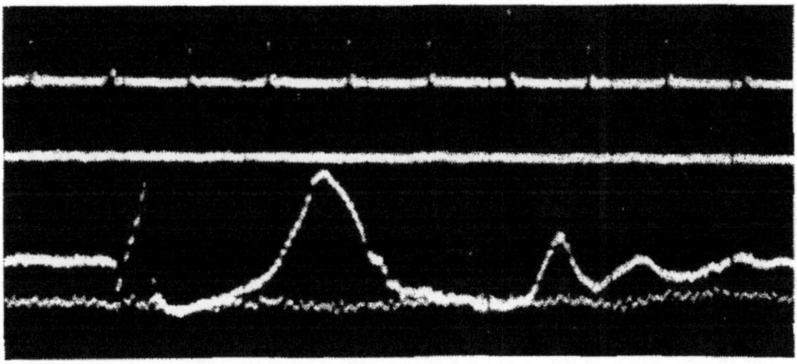

**Fig. 4.2** VISAR record from an iron impact experiment recorded with photomultiplier tubes. Timing marks are shown at the *top*. One of the data traces and the beam intensity monitor (*bottom trace*) are shown (Reprinted with permission from Barker and Hollenbach 1972, Copyright 1972, AIP Publishing LLC)

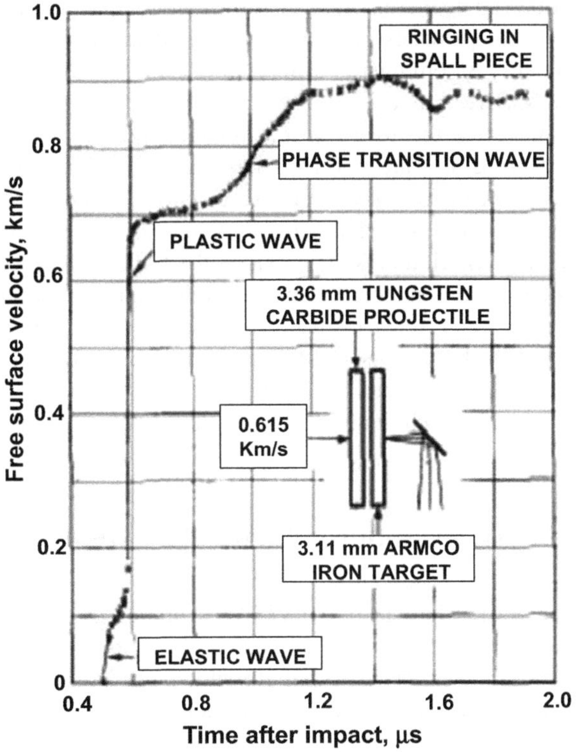


**Fig. 4.3** Reduction of the fringe record to obtain free surface particle velocity of the shocked iron sample. The record clearly shows the elastic wave, the first plastic wave in the α phase, and the second plastic wave in the ε phase. The small increase in velocity near the top of the wave can be used to determine the unloading path and the final ringing can be used to determine spall strength (Reprinted with permission from Barker and Hollenbach 1974, Copyright 1974, AIP Publishing LLC)

- in-bore projectile velocity measurements (Munson and May 1975; Hawke et al. 1991a, b),
- magnetically-accelerated flyer plates to 46 km/s (Knudson et al. 2003d, 2012; Lemke et al. 2011),

- multi-Mbar ramp wave experiments (Chhabildas et al. 1988; Asay and Knudson 2005; Hayes et al. 2004; Davis and Hayes 2007; Davis 2005, 2006, 2008; Davis et al. 2005),
- continuous measurements of the refractive index during ramp loading (Hayes et al. 2003), and
- direct measurement of shock velocity by reflecting from the shock front in a window material that becomes 4.2 Velocity Interferometer System for Any Reflector conductive at high shock pressure (Knudson et al. 2001, 2003a, 2008, 2012).

More recently, the direct shock velocity measurement technique has been used to tens of Mbars to obtain EOS data (Knudson and Desjarlais 2009), as discussed in more detail in Chap. 7.

The growing use of the VISAR was further extended beginning in the 1970s to determine transverse, or shear wave, particle velocity measurements in pressure–shear experiments (Chhabildas et al. 1979; Chhabildas and Swegle 1980) and continues to be used at both the STAR and the Z facilities for these measurements. These applications are also discussed further in Chap. 7. In addition, the time resolution can be increased from a few nanoseconds to picoseconds through a modification that provides optical recording of the fringe data with an electronic streak camera (Bloomquist and Sheffield 1983a, b). Because of its extensive versatility, the VISAR was a *game changer* for shock wave studies and has had a profound influence on this research since its inception.

During his career, Barker has made a number of significant contributions to a wide range of shock wave research, not only in experimental technique and applications but in the development of computer codes and material models. For these pioneering contributions, he received the Shock Compression Science Award in 1999. Bob Graham, another "giant" from the Sandia shock wave program and the Shock Compression Science Award recipient in 1991, paid him the highest compliment[1]: "Lynn Barker was the master experimentalist of the group [Solid Dynamics Research Department] and perhaps the best in the world. His … VISAR has become the most widely used technique for shock studies that stir the pseudo-liquid [shocked solid]."

## 4.3 Shock-Induced Phase Transitions

One of the triumphs of the VISAR stemmed from its ability to study phase transformation in materials. In most cases, this was not possible with the Sandia velocity interferometer because a polymorphic phase transition in a solid causes a mirror surface to lose reflectivity. This problem was resolved with the VISAR, and several phase transition studies were performed early in its development. The most celebrated use was measurement of the $\alpha$–$\varepsilon$ polymorphic transition in iron at 130 kbars. This transition, first discovered at Los Alamos Scientific Laboratory in 1956 with shock wave techniques using pin technology, contradicted the hydrostatic pressure

[1] From Bob Graham's comments commemorating George Samara's work at Sandia at the George Samara Memorial Symposium, Sandia National Laboratories, on May 18, 2007.

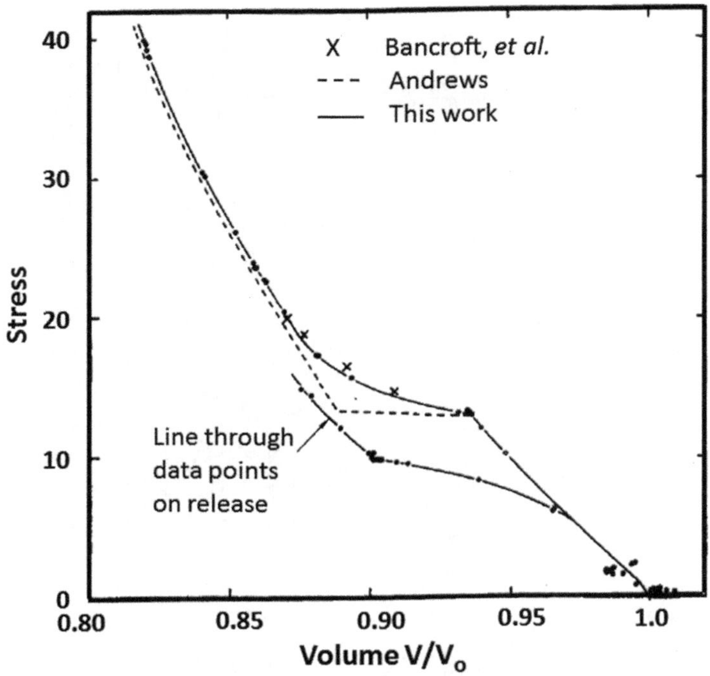


**Fig. 4.4** Shock compression and release of iron above the α–ε phase transition. The transition occurs at about 130 kbars during compression and the reverse transition occurs at around 100 kbars (Reprinted with permission from Barker and Hollenbach 1974, Copyright 1974, AIP Publishing LLC)

measurements conducted at the time by the famous researcher in high-pressure science, P. W. Bridgman. Bridgman did not initially observe this transition in measurements with a hydrostatic pressure vessel, and many did not believe it could happen on the submicrosecond time scales of shock wave experiments. Later, when Bridgman confirmed the Los Alamos results with static methods, the scientific stature and credibility of shock waves for studying phase transitions were forever elevated.

To demonstrate the capabilities of the VISAR in phase transition applications, Barker and Hollenbach performed an extensive set of experiments on Armco iron (Barker and Hollenbach 1974). Using particle velocity profiles shown in Fig. 4.3 at different stress levels, they accurately determined the stress at which the hexagonal close packed ε phase occurred, the time necessary for it to occur, and the stress at which the reverse transition to the initial α (body-centered cubic) phase began. Barker and Hollenbach obtained the first full loading and release paths for the 130 kbar α–ε transition, as shown in Fig. 4.4. This iconic result, as illustrated in Fig. 4.5, is imprinted on the Shock Compression Science Award, which has been awarded every other year beginning in 1987 to researchers who have made significant contributions to shock wave research.[2] The caption on the award reads

[2]As of 2015, Sandia researchers who have received the award include Bob Graham in 1993, Lynn Barker in 1999, Jim Asay in 2003, Dennis Grady in 2007, and James N. Johnson in 2011.

**Fig. 4.5** Icon of the iron phase transition imprinted on the Shock Compression Science Award (Reprinted with permission of the SCCM-1999 Proceedings, ed. by M. D. Furnish et al., Copyright 2000, AIP Publishing, LLC)

"In recognition of his outstanding contributions to the temporal measurement and interpretation of nonlinear physical processes in shock-compressed matter."

The VISAR was also essential for studying melt transitions produced by shock heating at high stresses. These results are obtained by accurate measurement of the unloading wave velocity from shock states near and in the solid–liquid mixed phase region. The transition of this wave velocity from an elastic longitudinal velocity characteristic of solids to a bulk velocity representing fluid response provided a direct measure of the solid–liquid transition. The technique was first demonstrated for shock-induced melting of aluminum by Asay and Hayes in 1975 and later in bismuth (Asay 1977a). Jim Johnson made a significant contribution to the theory of modeling multiphase and melt transitions, in particular, by constructing the first three-phase equation of state, which included the solid I–solid II melting phases for bismuth. This model was incorporated into a wave code developed by Dennis Hayes and used to model Asay's experiments, including the effects of transition kinetics. In Johnson's words:

> Another area in which I was able to learn new aspects of shock wave physics was the subject of phase transformations, working with Jim Asay and Dennis Hayes on bismuth (Johnson, Asay, and Hayes 1974). Bismuth has the very interesting property that single shock

compression can probe not only the solid I-solid II transition, but also the liquid phase. Thus, the presence of a triple point provided a challenge for the theorist. Dennis did some very nice work on how to deal with these questions (Hayes 1975), and I benefited greatly from this interaction.

For solid density aluminum, the shock stress necessary to produce melting is about 1.2 Mbars (Furnish et al. 1999), which requires a two-stage or three-stage light gas gun to achieve this impact stress. At the time of these experiments, this capability was not available, so porous aluminum was used to increase the shock temperature and produce melting at a lower stress. Thermodynamic calculations predicted that melt should occur at shock stresses of about 75 kbars with aluminum of 60 % theoretical density. However, since the VISAR had just become available, it was an ideal choice for use with porous materials in shock wave experiments. Fortunately, Barker's advice on the application of the VISAR to this problem resulted in the quick success of these experiments (Asay and Hayes 1975). Additional experiments on shock-induced melting of bismuth further validated the unloading wave velocity approach for detecting melt (Johnson et al. 1974; Asay 1977a; Furnish et al. 1999; Chhabildas et al. 2000a). Barker was also instrumental in identifying a "fringe benefit" of the VISAR that is the ability to estimate spatial variations in particle velocity by measuring the fringe contrast as the porous aluminum was being compacted to solid density (Asay and Barker 1974).

Several other phase transition studies were carried out for a broad range of materials with the VISAR in subsequent years by Sandians and other researchers. Several papers are listed in the Bibliography by Asay, Barker, Chhabildas, Furnish, Grady, Kennedy, Knudson, Montgomery, Root, Setchell, Trott, Vogler, and Wise, to name a few, which attest to the breadth of studies possible with the VISAR.[3]

## 4.4 Two-Dimensional Computer Codes

In addition to the significant experimental developments in the 1970s, modeling and computational capabilities grew rapidly. This rapid growth was driven by the need to model and simulate a variety of weapon problems. While the shock wave codes of the 1960s were mostly focused on 1-D wave propagation problems, the 1970s saw the implementation of several 2-D hydrocodes. With the development of a host of material models in the 1960s and partial validation using 1-D shock wave experiments, the advent of powerful computers in the 1970s enabled simulations of weapon components and subsystems in two dimensions. However, those simulations

[3]There have been several recent innovations in shock wave research made possible by the VISAR. These include development of multipoint measurements, as pioneered and discussed by L. M. Barker in his recollections; study of heterogeneous deformations, as discussed in Chap. 4; the capability to track shock fronts, as discussed in Chap. 7; detection of metallization of deuterium, as discussed in Chap. 7; and use of streak photography for routine sub-nanosecond resolution, as discussed in Chap. 7.

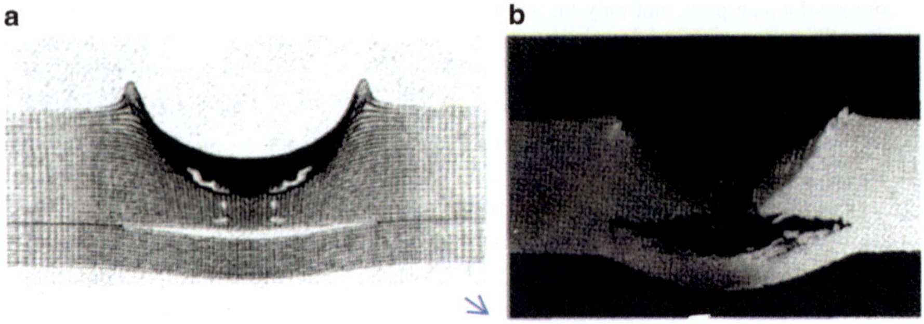

Fig. 4.6 Comparison of a TOODY calculation of a 9.53-mm-diameter nylon sphere impacting a 12.7-mm-thick steel plate at 5.18 km/s. The numerical simulation is shown in (**a**) and the experimental result in (**b**) (Adapted from Fig. 5 of Bertholf et al. 1975 and reprinted with permission Copyright 1975, AIP Publishing LLC)

were often arduous and required considerable time to set up and analyze. As Mike McGlaun put it in his recollections:

> Running a calculation was a laborious process in the 1970s. We punched input on card decks. We walked the decks across the tech area ('sneaker-net'), submitted them across a counter, and then returned a few days later for the output. If we made an input error, we tried again. We ran calculations on Control Data Corporation (CDC) 6600s and 7600s. The process got easier when a remote batch facility was set up in a neighboring building. We could then read in card decks and receive printout in the next building, rather than walking across the tech area.

One of the first big examples of the ability to simulate 2-D experiments was a classic hypervelocity impact experiment that involved a nylon sphere impacting an iron plate at 5.18 km/s. The combined simulation and experiment also demonstrated the importance of the α–ε phase transition occurring during the penetration and damage of the plate, since the impact stress exceeded the stress necessary to initiate the iron α–ε phase transition. Bertholf and colleagues (Bertholf et al. 1975) performed a systematic study of the impact with the 2-D Eulerian hydrocode CSQII (Thompson 1976, 1979) and the 2-D Lagrangian hydrocode TOODY (Thorne and Herrmann 1967). In both cases, they implemented accurate elastic–plastic models for the nylon and steel materials, an accurate value for the spall strength, and finely zoned descriptions of the materials. They also performed simulations of the impact with and without the α–ε phase transition to determine the influence of the transition on numerical solutions and the comparison with the recovered steel samples. Figure 4.6 shows very good agreement between the calculations and the experiment. In particular, inclusion of the phase transition dramatically affects the calculated spall zone. In their words[4] (Bertholf et al. 1975):

> The computational results show that in order to obtain agreement with the experiment, it is necessary to account for the α (bcc)-ε (hcp) polymorphic phase change for iron. Calculations also indicate the need to obtain experimentally the correct spall strength for the steel....

[4] Reprinted with permission from Bertholf et al. 1975, Copyright 1975, AIP Publishing LLC.

> Treatment of elastic-plastic response, a good numerical model for fracture, and a high degree of numerical resolution are also shown to be necessary in order to predict accurately the back-surface spallation observed in the experiment.

Another high profile example of the use of the shock physics codes being developed at Sandia concerned the analysis of the Three Mile Island (TMI) accident, a partial nuclear meltdown that occurred in one of the two nuclear reactors in Pennsylvania on March 28, 1979. This was the worst accident in the history of U.S. commercial nuclear power plants. The partial meltdown resulted in the release of small amounts of radioactive gases and radioactive iodine into the environment. The accident was caused by the failure of a relief valve in the system, which allowed large amounts of nuclear reactor coolant to escape. One of the issues was whether the hydrogen released could cause the concrete structure of the reactor to fail. Dennis Hayes, who was then managing the Fluid Dynamics Department, was asked to aid in the analysis of the accident. He assembled a research team to address this problem. In his words:

> At a place like Sandia, interesting problems can arise unexpectedly. Immediately after the Three Mile Island (TMI) reactor accident in 1979, I got a call asking for help on the hydrogen release problem. The big question: "Would an internal hydrogen detonation breach the concrete containment structure of the reactor?" I assembled a team of Sandia experts. We worked long days and nights and were in constant phone contact with officials at the TMI site. With Sam Thompson and his ANEOS equation of state and CSQ codes, Mike Cieslak on material and chemical issues, Dennis Mitchell on explosive issues, and Walt Murfin, a Sandia expert on the damage shocks cause in concrete, we were able to give definitive guidance to the TMI Emergency Management: there was no danger of containment failure. But the larger question went untested as the large volume of hydrogen produced later had insufficient oxygen to be detonable. I have always believed the good marks we made during this effort helped Sandia establish itself in the nuclear reactor safety arena.

The evolution of the 2-D codes TOODY and CSQII in the seventies and their use in solving complex multi-dimensional problems was a precursor to the next generation of codes and computers that significantly increased computational capabilities in the 1990s and beyond. Some of these problems will be discussed in the following chapters.

## 4.5 Composites and Mixtures

Composite materials, including laminated materials and mixtures, play an important role, for example, in the heat shield for a reentry vehicle. To develop and analyze these components, models of the dynamic response of composites were needed. Lundergan recognized this requirement early and initiated a research program to study composite materials in the late 1960s. The first experiments on laminated layers of aluminum and PMMA, or polymethyl methacrylate, were conducted using the original air gun developed in 1958 and the velocity interferometer later developed by Barker and Hollenbach (Lundergan 1970). Several experiments were performed on these plate composites. When Doug Drumheller was hired in the late 1960s, the group obtained much needed expertise in theoretical mechanics for modeling composites and other complex materials. Drumheller and Lundergan teamed together and with others on several studies of laminated composites, which resulted in a basic understanding of shock wave propagation in these materials. This team published several papers in the early 1970s (Lundergan and Drumheller 1971a, b; Drumheller and Bedford 1973; Drumheller and Lundergan 1975).

In addition to the formal continuum mechanics approach used by Drumheller to describe shock propagation in laminated materials, Barker developed an innovative physical model based on the actual reverberation of waves that occur in individual plate layers and the multiple reflections caused by many layers, resulting in a smoothed out shock wave profile. The collective response is much like a viscous, or Maxwellian, material that spreads waves because of viscous or nonequilibrium forces. This simple but elegant model could be determined a priori from knowledge about the properties of the individual layers. It could also effectively predict a wave risetime from the transit time and reverberation of waves within the layers for use in a Maxwellian model (Barker 1971b; Barker et al. 1974b). The model was first tested by using the characteristics code SWAP (Barker and Young 1974) to calculate the actual steady wave profiles for wave propagation in a layered composite and then comparing the calculations with the predictions of the model. Figure 4.7 shows that

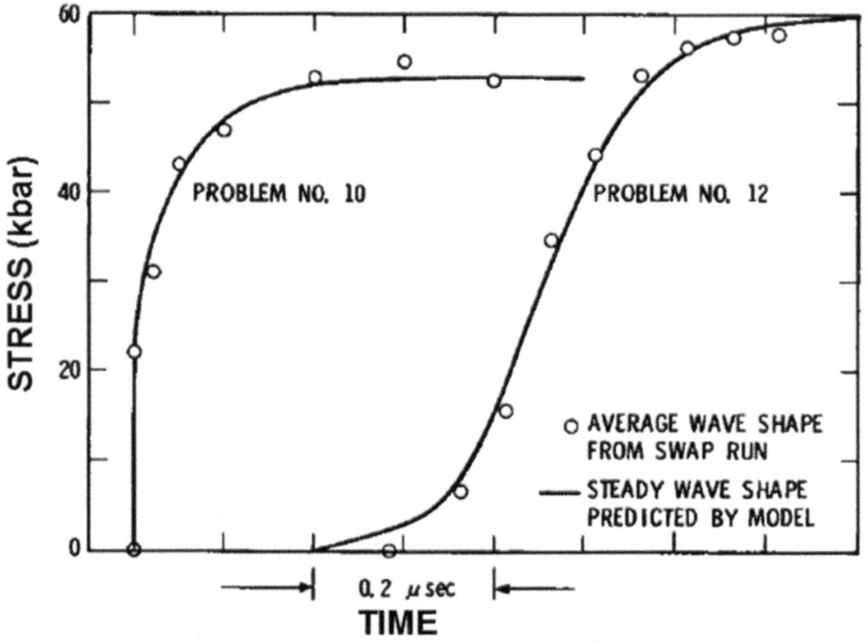


**Fig. 4.7** Comparison of steady wave profiles obtained on an aluminum-PMMA layered composite using the SWAP code (Barker and Young 1974) calculations of the layers and the averaged Barker composite model (From Barker 1971b, Copyright 1971, reprinted with permission of SAGE Publishing)

the agreement is extremely good between the model and the calculated wave profiles in an aluminum-PMMA laminated composite. Barker's composite model was also implemented in the WONDY wave code for more general applications (Lawrence 1970).

To a large extent, the experimental and theoretical work on laminated composites was a prelude to understanding more complex mixtures. Lundergan performed the first experiments on an S-glass-reinforced resin that illustrated the complex nature of shock propagation in such materials. The following year, experimental and theoretical analyses of shock propagation in quartz-phenolic composites were performed (Munson and Schuler 1971). Additional experiments on similar materials were conducted by other investigators (Munson and Schuler 1971; Lee 1972, 1973a; Munson et al. 1977). Drumheller led a theoretical effort to develop continuum models of composite materials and mixtures that could be implemented into the hydrodynamic codes that were then available. These models (Drumheller and Bedford 1973; Drumheller and Lundergan 1975; Drumheller 1978, 1982a, b) were applied to different mixtures and, in particular, to alumina-filled epoxy.

The theoretical work on mixtures continued into the 1980s, although at a lower level, and resulted in material models that could be implemented in codes to simulate the response of various materials (Drumheller 1984; Drumheller et al. 1984). In the late 1990s, interest in alumina-filled epoxy was renewed as part of a research effort to develop predictive capabilities by using massively parallel computer codes.

## 4.6 Damage-Based Spallation Models

Shock-induced spallation occurs in a number of applications involving weapon operation and vulnerability to adverse environments; hence, significant experimental and theoretical research was devoted to this issue. The experimental research on shock-induced spallation in the 1960s and the early modeling by Floyd Tuler and Barry Butcher (1968) led to more sophisticated experimental and theoretical efforts in the 1970s. Based on experimental measurements, Tuler and Butcher developed a simple pressure–time relation that described the temporal evolution of the spallation process, from incipient spall to final fractured surfaces fairly well; however, the model was strictly empirical. A typical spall signal, as evidenced by the pullback in peak free surface velocity $\Delta u$, is illustrated in Fig. 3.11 of Chap. 3. A concerted theoretical effort was initiated in the Solid Dynamics Research Department to connect the observed spallation stresses to physical processes that could be described in terms of the nucleation and growth of voids created in states of rapidly applied uniaxial tension. This approach required knowledge of the physical mechanisms producing the voids under tension and how these grew and coalesced into fracture surfaces.

Most of the shock recovery experiments were centered in the Physics of Solids Research Department managed by George Samara, although these experiments were usually performed on Lundergan's gas gun by the Solid Dynamics Research Department, especially for beryllium studies that required the containment of all debris after the shots. During shock loading and release in the recovery experiments, 2-D effects created at the edge of specimens initially loaded in 1-D strain had to be eliminated or at least minimized. The shock physics group under Lee Davison devoted considerable effort to producing states of tension leading to incipient and full spallation and development of recovery fixtures for post-shock metallurgical examination after an experiment. In pursuit of better recovery fixtures, Stevens and Jones (1972) performed several 2-D simulations of target recovery fixtures to optimize configurations for minimizing radial release effects in recovered samples.

These studies set the stage for a lengthy series of shock recovery experiments on beryllium using the original Lundergan gun that had been moved to Tech Area I (see Fig. 2.8) and had a containment structure built around it. An extensive set of Be recovery experiments were conducted to examine the effects of texture on spallation (Stevens and Pope 1973) and determine residual mechanical properties (Stevens 1974; Pope and Stevens 1973; Pope and Johnson 1975). Another set of recovery experiments was conducted on polycrystalline and monocrystalline aluminum samples to examine void nucleation and growth effects for a face-centered-cubic material (Stevens et al. 1972, 1973).

The extensive shock recovery and continuum wave profile experiments to quantify the spallation process allowed the development of damage-based models that could be incorporated into codes for realistic simulations of shock-induced fracture (Davison and Stevens 1973; Davison et al. 1977; Kipp and Stevens 1976; Kipp and Davison 1982). These advanced fracture models were implemented primarily in the 1-D computer codes and used to solve a variety of problems, as Marlin Kipp notes in his recollections:

> I got involved in ductile fracture modeling with Al Stevens and Lee Davison using WONDY and TOODY. Lee and Al had developed a continuum model that coupled plasticity with void growth representations. The void formation continuously degraded the effective modulus, leading to unloading of release waves as the spall evolved. Al was having problems with instabilities in WONDY using the model, so I was enlisted to figure out a way to make it behave. Differencing the derivatives in the model expressions seemed to be the source of the difficulties, so I resorted to accessing an external numerical ordinary differential equation (ODE) solver (from a Sandia-developed math library), which was called from WONDY and directly solved the differential equations at each cell and time step. I enforced linear variation of the density and artificial viscosity over the timestep; the technique worked very well, giving stable results. We had some VISAR data for planar impact in aluminum and set out to determine a suitable set of parameters for the void growth part of the model. We papered the hallway wall outside Lee's office with results of the runs for each set of parameter variations until we got a satisfactory fit to the data. We would run a simulation in WONDY and wait for a roll of paper with the plots to be delivered by courier; we would then post the plots on the wall and decide on the next parameter variations. We eventually converged on a satisfactory set of parameters that matched the uniaxial strain data from the plate impact experiment. Then I put the model in TOODY, and we applied the same parameter set to a 2-D rod impact configuration and obtained the central void growth seen in the experiments. This was a significant improvement over the minimum stress criterion commonly used to model the spall.

This research effort established Sandia leadership in this area for several years and provided the groundwork for similar work at other institutions.

## 4.7 Geological Materials: Oil Shale

In the mid 1970s, Walt Herrmann reorganized the Solid Dynamics Research Department and established an experimental division focused on geological materials. The motivation for this management decision was to increase the knowledge base for these materials in support of the UGT program in progress and to respond to the energy initiative being established at Sandia to extract oil from oil shale. This experimental division played a leading role in a new energy initiative established by DOE during the energy crisis of the 1970s. Several additional staff were hired about this time, including Dennis Grady, Paul Yarrington, Marlin Kipp, and Bill Brown. Grady recalls:

> I came on board the shock physics group in 1974. I was recruited by Darrell Munson. It was a time of rapid growth for both the shock-physics group and Sandia. Marlin Kipp, Paul Yarrington, and Bill Brown, in addition to me, all joined either the computational group or the experimental group within a few months.
>
> My previous several years were at Stanford Research Institute, where I conducted experimental shock equation of state studies on geological materials for the Defense Nuclear Agency (DNA). In the experimental shock physics group I undertook comparable experimental work in support of Sandia programs concerned with nuclear ground shock and earth penetrating weapon effects.
>
> Two occurrences in this early time contributed to my good fortune. First, the VISAR instrumentation technology, recently developed by Lynn Barker, was made available and I had the opportunity to acquire some of the first-ever high-resolution shock wave structure measurements on rocks and minerals. Second, Lynn Barker had just previously departed for Terra Tek in Salt Lake City, and I had the extreme good fortune to work with Roy (Red) Hollenbach, one of the foremost experts in the application of VISAR technology, during his final years at Sandia Albuquerque.

Grady was the intellectual driver for much of the early experimental shock wave work on geological materials, but several others joined the experimental effort during this period, including Karl Schuler, Richard Schmidt, Pete Lysne, Al Stevens, and later Mike Furnish. Grady further states:

> The energy crisis of the mid 1970s provided the impetus for the three national laboratories to embark on extended efforts to support U.S. energy concerns through technological innovations. Technical issues concerning the effective use and recovery of oil shale, tar sands, deep coal gasification, among other activities, were actively pursued. Methods such as explosive and static hydraulic fracturing (now popularly referred to as fracking) were extensively investigated. A number of shock physics issues were identified and pursued. A key issue was the development of effective explosive fracture and fragmentation methods in support of investigations into the in situ retorting of oil shale deposits that were and still are abundant throughout the Rocky Mountain states. Shock spall and fragmentation studies of oil shale and other rock materials were undertaken in the shock physics group. These efforts provided the basis for theories of fracture and fragmentation that were implemented in computer codes to support the various energy recovery programs.

Figure 4.8 shows Marlin Kipp and Dennis Grady selecting blocks of oil shale for conducting large-scale explosive fragmentation experiments. As part of the energy initiative program and the continuing ground shock program at Sandia, an extremely

**Fig. 4.8** Dennis Grady and Marlin Kipp examining an oil shale sample (Reprinted with permission of Sandia National Laboratories)

large number of reports and publications related to the dynamic compression of geological materials, in general, and oil shale, in particular, were published during this period. These included (Lysne 1970; Stevens et al. 1974; Butcher and Stevens 1975; Grady and Hollenbach 1977; Schuler et al. 1976; Schuler and Schmidt 1977; Grady et al. 1977, 1978; Grady 1979; Grady and Kipp 1979, 1980; Kipp 1979; Boade et al. 1981). The research on the dynamic fracturing of geological materials also laid the groundwork for the recent advances being made in commercial oil recovery using hydrodynamic-fracking methods. In addition to supporting the energy program and studies on geological materials for the Defense Nuclear Agency, the research had spin off to penetration mechanics capabilities being developed at the time in Al Chabai's group (Chabai et al. 1974, 1977; Byers and Chabai 1977; Byers et al. 1978).

## 4.8 Piezoelectric and Ferroelectric Materials

Intense research efforts continued on ferroelectric and piezoelectric materials during this period. Graham, Neilson, and Benedick had perfected X-cut quartz as the stress gauge of choice for most low-pressure shock wave experiments and continued to improve the understanding and accuracy of the quartz gauge (Neilson and Benedick 1960; Graham 1961a, b; Ingram and Graham 1970). The gauge could be used in a variety of applications that were not applicable to the early interferometer techniques, such as in phase transforming materials (Johnson et al. 1974), energetic materials (Kennedy 1970, 1973), and radiation-induced stress pulses. With proper impact control, its accuracy and time resolution were unsurpassed, even for the interferometer techniques prior to VISAR applications with modern recording instrumentation. Depending on impact tilt and the cross-sectional area over which the measurements are made, the time resolution of quartz gauges would be on the order of 1 ns.

A number of further improvements in the development and application of these versatile gauges occurred in the 1970s. Graham performed several studies to determine the nonlinear strain dependence of the piezoelectric, elastic, and dielectric response of quartz (Graham 1972b). Lysne developed a theory of shock-induced polarization with application to quartz (Lysne 1972b). Graham and Jacobson (1973) developed lithium-niobate gauges that were especially useful for pulsed radiation experiments because of their higher current output. The effects of time-dependent electrical breakdown were studied by Graham and Yang (1975), and the effects of shorting or shunting the guard ring used with the gauge was extensively explored (Graham 1975). In particular, Graham and Chen (1975) identified an electromechanical coupling effect for piezoelectric solids and quartz. Several other publications on this prolific shock wave gauge were published in the 1970s (Graham 1974, 1976; Lysne and Bartel 1975; Lysne 1976; Graham and Reed 1978).

In addition to studies of piezoelectricity, several investigations were conducted in the 1970s on ferroelectric materials. These included effects of dielectric breakdown in ferroelectric ceramics (Lysne 1975), shock-induced polarization (Lysne 1977), nonlinear wave propagation in piezoelectric solids (Lawrence and Davison 1977), dielectric relaxation (Lysne 1983), and normal mode and nonlinear response (Chen et al. 1976a, b; Graham and Chen 1975; Chen and Montgomery 1977, 1978, 1980; Chen and Tucker 1981). The 1970s produced vigorous and ongoing studies of piezoelectric and ferroelectric materials, resulting in another area where Sandia excelled within the shock wave community.

## 4.9 Third-Order Elastic Constants

Complementary to electromechanical effects in piezoelectric and ferroelectric materials, Graham initiated a program for accurate evaluation of third-order elastic (TOE) constants of single-crystal brittle materials, including quartz, lithium niobate, and sapphire. By determining the dependence of shock velocity on impact stress under uniaxial strain in the elastic range, it was possible to determine certain second-, third-, and fourth-order elastic constants (Graham 1972a, 1977). These could be supplemented with acoustic velocity measurements versus hydrostatic pressure (Fritz and Graham 1974) to determine the full set of TOE constants for crystalline quartz and other crystals. Later experiments by Root and Asay (2009a, b) using

**Table 4.1** Elastic constants, with units in GPa, determined from shock and shockless compression

| Method | $C_{11}$ | $C_{111}$ | $C_{1111}$ |
|---|---|---|---|
| Shock (Graham 1972b) | 86.8 ± 0.95 | −300 ± 30 | 7500 ± 2500 |
| Shockless (Root and Asay 2009b) | 86.487 ± 0.089 | −319.93 ± 4.76 | 7988 ± 106.0 |

ramp waves determined the same combination of third-order constants by shock loading X-cut quartz. Isentropic loading experiments provided TOEs that were in excellent agreement with Graham's earlier results from shock loading but with considerably reduced error bars. Moreover, from ramp loading experiments, the rate at which the ramp evolves into a shock wave can be used to determine the third- and fourth-order elastic constants. Table 4.1, which is adopted from Table 1 in Root and Asay (2009b), compares the high-order elastic constants from shockless or ramp loading with similar constants from shock compression (Graham 1972a). The good agreement between the two approaches attests to the accuracy of both techniques. For additional discussion of magnetic ramp loading, see Chap. 7, Sect. 7.3. The lower uncertainty for the ramp wave experiments results because the wave velocity is measured continuously along the profile, and deviation from linearity gives the higher-order coefficients. In the experiments, only the shock velocity is determined for each Hugoniot point, and the locus of points is then fit to a high-order polynomial to determine the higher-order elastic constants.

## 4.10 Pressure–Shear Loading

A continuing goal of the shock wave program at Sandia, beginning with Lundergan's initial research on the elastic–plastic behavior of metals, was to determine the compressive shear (yield) strength of materials in the shocked state to develop full constitutive equations of material response. In the late 1970s, Barry Butcher initiated a research program to measure shear stress and longitudinal stresses directly and simultaneously with the longitudinal stress during shock compression. This technique had been demonstrated by Prof. Rod Clifton of Brown University and was often referred to as the "pressure–shear" or "compression–shear" experiment. Previous pressure–shear (P-S) experiments at Brown had shown that these measurements provided a direct determination of high-pressure material strength, which is needed for developing full mechanical models of dynamic material response. Another motivation was that the 1977 powder gun accident disrupted the experimental programs being conducted by Jim Asay and Lalit Chhabildas, who were the major users of the facility at the time. Research on the powder gun was halted as the result of an investigation, and new research efforts were therefore initiated that used the department's gas gun in Tech Area I. (For more information, see the recollections of Asay, Butcher, Chhabildas, and Konrad.)

Another experimentalist in Butcher's group, Herb Sutherland, had been developing a technique to measure longitudinal and transverse velocities simultaneously using two VISARs to determine the fringe shifts at opposite angles from the surface normal (Chhabildas et al. 1979). Combining the two measurements from opposite angles provides the longitudinal and transverse velocities induced at the free surface. Shortly after the powder gun incident, Sutherland transferred to one of the departments involved in fossil energy research and Chhabildas took over leadership of the project. Through a series of experiments, he and his coworkers demonstrated that the two-VISAR approach provided accurate data on the longitudinal and transverse velocities (Chhabildas et al. 1979; Chhabildas and Swegle 1980, 1982; Chhabildas and Hardy 1982), which was a first step to developing experimental pressure–shear loading capabilities at Sandia. The next step was to develop a method for producing the two stress components in shock wave experiments.

The technique developed by Prof. Clifton for pressure-shear measurements (Duprey and Clifton 2000) involved the impact of a plate aligned at an oblique angle on a projectile (up to 30°) against a target inclined at the same angle. At impact, both high-amplitude longitudinal and shear waves are simultaneously produced in the target sample if the interface between the impactor and sample does not slip. If the input shear stress is larger than the critical shear strength of the sample but less than the shear strength of the impactor, the amplitude of the shear wave in the sample is limited to its shear strength. Thus, measurement of the transverse velocity on the sample free surface allows a direct measurement of the shear strength at the longitudinal stress induced by the shock. However, to prevent rotation and maintain alignment of the impact face with the target, the internal bore of the Brown University gun was milled with a slot along its axis that matched a key on the projectile to keep it from rotating.

Since Sandia did not have an oblique impact capability at the time, Chhabildas developed a technique proposed by Jim Johnson (1971) that used an anisotropic Y-cut quartz disk as a generator of high-amplitude shear waves. When Y-cut quartz is impacted in a standard plate experiment, both quasi-longitudinal[5] and quasi-shear waves are produced. If a sample is bonded to the rear surface of the crystal, pressure–shear waves similar to the oblique impact are generated in the sample, and the shear strength can be determined, as with oblique impact. The dual-VISAR technique discussed above can be used to analyze both components of motion. Through a series of experiments on the single-stage gas gun, Chhabildas and Hardy (1982) developed a standardized approach to use the Y-cut pressure–shear generator and the dual-VISAR technique in several applications (Chhabildas and Swegle 1980, 1982).

Application of the technique required numerical simulation of the complete configuration, consisting of the Y-cut quartz shear generator, which generated both longitudinal and shear waves in a sample bonded to its rear surface. These waves could therefore be used to produce shear stress loading at a given longitudinal stress. Jeff

[5] Quasi-longitudinal implies that the major response is longitudinal stress but that there is a coupled shear stress. A similar effect applies to quasi-shear response.

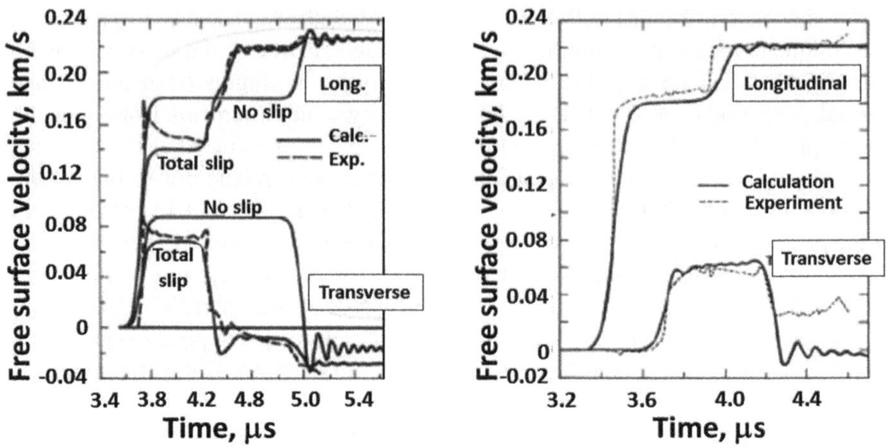


**Fig. 4.9** Pressure–shear experiments using a Y-cut quartz crystal. Surface slippage, as evidenced in the left graph by the difference between the code simulation and experiment when a Y-cut crystal is bonded to another Y-cut crystal. Slippage was eliminated through surface preparation, obtaining agreement between experiment and theory as shown in the right graph (Reprinted with permission from Chhabildas and Swegle 1980, Copyright 1980, AIP Publishing LLC)

Swegle, who joined Sandia in 1975, implemented Johnson's anisotropic material model for crystalline quartz in the 2-D hydrodynamic code TOODY for this purpose. The code simulated several Y-cut quartz experiments on aluminum, resulting in good agreement between the experiment and simulations (Swegle and Chhabildas 1981). As Chhabildas remembers:

> This project gave me an opportunity to team with Bob Hardy and Jeffrey Swegle. Jeff had arrived in 1975, about a year and a half before I did. His expertise was code development, and he was in charge of TOODY. I developed pressure-shear experimental techniques using the output of Y-cut quartz as the generator and gluing the material of interest to the crystal. Jeff did the modeling. We studied the effects on materials such as Al, alumina-filled epoxy, PZT 95/5 and PBX 9404 (with Marlin Kipp.)

A comparison of a pressure–shear experiment using alumina-filled epoxy as a bond is shown in Fig. 4.9. The P-S loading technique was also extended to several weapon materials, including alumina-filled epoxy (Chhabildas and Swegle 1982), PZT 95/5 (Chhabildas and Hardy 1982), and the explosive PBX 9404 (Chhabildas and Kipp 1985).

The Y-cut quartz generator was limited to longitudinal stresses of about 30 kbars because quartz yields at that stress for that crystalline orientation. To extend the stress range, the four-inch air gun originally developed by Lundergan in 1957 was milled with a small slot along the interior bore in the mid 1980s. This modification allowed experiments with slanted impactors, as shown earlier at Brown University, but the gun was not used for this purpose until about 2005 when Vogler and his colleagues initiated a program to study the shear response of granular materials (Vogler et al. 2011). This application is discussed in Chap. 7.

## 4.11 Shock Rise Time and the Fourth Power Law

In the late 1970s, Barry Butcher initiated a project to determine the rise time of high-pressure shock waves. At the time, conflicting data and theories existed in the literature about how viscoplastic properties influenced shock wave rise times and, additionally, how rise times depended on shock amplitude. In 1968, Barker had observed with the Sandia velocity interferometer that the rise time of shocked aluminum decreased rapidly as the shock amplitude increased (see Fig. 3.9). Grady later fit the data and found an inverse fourth power relation of strain rate with shock stress. After joining Sandia in 1976, Lalit Chhabildas was asked to initiate a research program to measure the shock wave rise time of metals to Mbar pressures. At the time the powder gun was operational, but it had limited velocity and pressure capabilities, and the two-stage light gas gun was becoming available, but it had not yet been configured for precision impact shock wave experiments. As Chhabildas states in his recollections:

> To get Mbar pressures, we needed to use the two-stage light gas gun. Unfortunately, at that time the two-stage gun was not set up for well-controlled equation of state studies. Marlin Kipp and Joel Lipkin had used the gun to look at cratering caused by polyethylene (as simulated rain drops) on heat shield materials for space missiles. No standard techniques had been established at Sandia for equation of state studies on the two-stage gun then, let alone for time-resolved techniques such as the VISAR. I talked to Barry about making rise time measurements using VISARs on the powder gun because we would approach Mbar pressures in tantalum and tungsten at impact velocities exceeding 2 km/s. Barry said, 'That would be too easy and there would be no challenges. What we really want to do is get the two-stage gun operational for time-resolved shock profile measurements. Measuring rise times in metals are the means to get us there.'

With the help of Carl Konrad and Bob Hardy, Chhabildas configured the two-stage gun at the STAR facility so that accurate VISAR measurements of shock rise time could be made. Using the VISAR, systematic studies were performed in aluminum, copper, and steel (Chhabildas and Asay 1979). At lower shock stresses (typically, below about 100 kbars), the structure of the shock and thus the rise times could be measured with the existing VISAR instrumentation. Above this level, the VISAR could not resolve the shock rise times in most materials. Combined with the shock wave measurements in beryllium, the collected set of data showed strikingly similar changes in rise time with peak shock stress for all materials studied, as shown in Fig. 4.10.

Dennis Grady analyzed these rise time data in detail and showed that all shock wave data for metals and nonmetals could be fit with a relationship that indicated that the peak strain rate produced during shock loading in homogeneous metals should vary as the inverse fourth power of the peak shock stress[6] (Grady 1981b). This relation has come to be known as the fourth power scaling relation or the "fourth power law." The relation can be explained on the basis that the dissipated energy produced during shock compression times the loading time, or equivalently

[6]The difference between peak stress and the Hugoniot Elastic Limit is actually used in most cases.

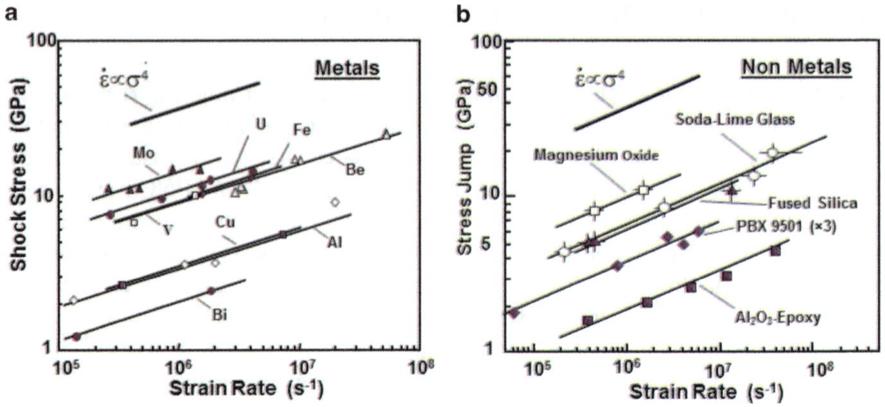


**Fig. 4.10** Shock rise times measured in several metals (**a**) and nonmetals (**b**) versus peak strain rate in the shock (Reprinted with permission from Grady 2010a, Copyright 2010, AIP Publishing LLC)

the action integral, should be constant. Various empirical viscous models were developed to predict a similar rate dependence on stress (Hayes and Grady 1982; Swegle and Grady 1985, 1986a, b), but only recently (Grady 2015) has a firm theoretical foundation been developed to explain this behavior. However, Hayes and Grady's initial 1982 research is noteworthy because shock rise time is computationally determined from an assumed strain rate dependence of plastic viscosity. This allows determination of pressure, temperature, and density in the plastic wave during the compression process and determination of mechanical properties such as yield strength. This application is discussed further in Chap. 5. More recently, Grady (2015) has provided a theoretical foundation for the fourth power law and has identified physical mechanisms that result in this profound effect (Grady 2016). This exciting development may lead to new theoretical and experimental research that clarifies the role of dissipative action as a unifying attribute underlying the dynamic failure of solid materials. Roger Minich of LLNL is also developing a fundamental theory of shock wave dissipative action based on conservation of energy, momentum, and scale invariance of the shock front that results in a fourth power dependence of maximum strain rate on peak shock stress in metals. The physical constants in the theory lead to the observed shock invariant, which have far-reaching implications for shock structure, the entropy produced, the importance of microstructural deformation of mechanisms, and nonequilibrium transport of energy across the shock front.

Recently, the scaling law was verified to peak shock stresses of over 400 kbars in aluminum at a strain rate of about $10^{10}$ s$^{-1}$, as shown in Fig. 4.11 (Crowhurst et al. 2011). Remarkably, these experiments have validated the relation over five orders of magnitude in strain rate. Grady has recently shown that a different, but specific, power dependence of rise time on peak stress exists for other material classes, such as composites, plate laminates, and porous materials. In these cases, the fourth power law does not apply but other power laws do (Grady 2010a; Vogler et al. 2012).

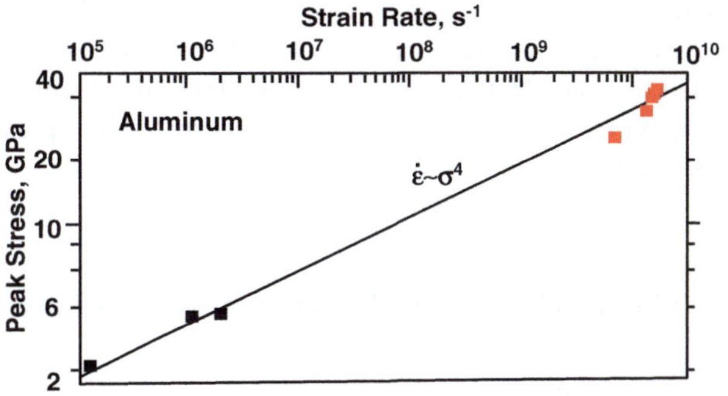


**Fig. 4.11** Use of fourth power law over five orders of magnitude in aluminum. Data points in upper right are from Crowhurst et al. 2011 [Reprinted with permission from Fig. 4 in Crowhurst et al., Invariance of the dissipative action at ultrahigh strain rates above the strong shock threshold, *Phys. Rev. Lett.* **107**, 104322 (2011), Copyright 2011 by the American Physical Society]

## 4.12 Acceleration Waves

In the 1970s considerable attention was directed to understanding the effect of viscous damping mechanisms on the structure of shock waves. A parallel development involved how these effects influenced the evolution of planar ramp waves, which were referred to in the literature as "acceleration waves." Peter Chen had done his PhD thesis on acceleration waves at the University of Washington and joined Sandia in 1965 to provide theoretical support for experimental shock wave work. He immediately initiated a research program, in collaboration with several others at Sandia and elsewhere, to examine the growth and decay of both longitudinal and transverse (shear) acceleration waves with arbitrary structure for a variety of materials. Others soon joined this effort, including Jace Nunziato, who was also a brilliant and prolific theoretician.

Chen and Nunziato were the principal Sandia investigators pursuing acceleration wave research, although many other collaborators were involved. Their collective work over a period of about 10 years included dozens of publications dealing with the issue of planar acceleration waves. Some of the topics included the growth of acceleration waves[7] and thermodynamic effects (Chen 1968a, 1968b, 1973), the propagation of longitudinal and transverse waves, and the effects of precompression on acceleration wave growth (Nunziato 1975; Nunziato et al. 1975, 1978b). However, this extensive research on acceleration waves did not impact Sandia's experimental programs because techniques to produce well-controlled acceleration waves through smooth ramp loading were very limited during that period. Barker and Hollenbach (1970) had demonstrated the ability to produce smooth ramps,

[7] Note that the amplitude of an acceleration wave refers to the magnitude of the acceleration and not to the peak stress attained by the wave.

which are acceleration waves, to about 30 kbars with a fused silica buffer plate, but higher smooth ramp pressures were not possible then. Later, the magnetic drive technique invented by Asay, Hall, and others in 1999 permitted smooth ramp loading or acceleration wave generation of essentially any material to pressures as high as 4 Mbars. This innovative development will be discussed in Chap. 7.

## 4.13 Shock Thermodynamics Applied Research Facility

One of the major gun facilities at Sandia was the powder and two-stage light gas gun facility initially constructed in Area III (Area Y) by Darrell Munson. These guns were the precursors to what became the STAR facility in the 1980s. Munson was one of the four division supervisors in the Solid Dynamics Department managed by Walt Herrmann in the late 1960s. Munson's recollections provide a perspective about how the site originated and later evolved into a world-class research facility:

> Somewhat prior to this time [1962], Bild's directorate had built an explosive facility… in Area Y, a complex surrounding Area III. This consisted of a bomb-proof building (Building 9950) with a firing pad on the roof accessible by a dirt ramp at the rear of the building. The permissible charge was 25 pounds of high explosive. The facility consisted of the bomb-proof laboratory, an explosives storage igloo, and an explosives assembly building. The facility was not being used for its intended explosive tests on materials; instead, it had only a high-voltage exploding foil machine housed inside the building. At this time, the defense needs were to study the vulnerability of missile components for the weapon developers at shock wave stress levels above those produced by the 4-inch compressed gas gun. After some difficult discussions on the need for higher pressure impact material studies, I acquired this facility to add to the capability to produce material responses to very large impact shock waves.

This initial set of buildings became the focus for a large shock wave facility containing several research guns in later years. Carl Konrad was one of the first staff assistants at the site and later became the facility supervisor. Konrad's memories of its early development are captured in his recollections:

> I was hired by Darrell Munson in 1969. Darrell, the Division Supervisor (i.e., level-1 manager), was starting a new gun facility to replace the explosive site at Building 9950 in Area Y. My first assignment was to report to Ray Reed at Building 9956 in Area Y. I was tasked to wire the instrumentation racks in the 9956 control room to the impact chamber, which would be located in the gun bay. I believe I ran 48 rigid 50 Ω (i.e., 50-ohm) cables to the racks and kept each cable in a given rack to the same length within less than one inch. At that time, the gun had not been delivered and 9956 consisted of a 100-foot-long (30.5 meter-long) gun bay and an attached bunker-type control room. Since 9950 was still certified as a 50-pound (22.7-kilogram) high explosive site, a substantial bunker-style room was required for personnel protection. This distinction proved important later in the life of the facility.

Ray Reed (shown in Fig. 4.12) played a prominent role in the acquisition and implementation of a large bore powder gun and was one of the first staff members to use it. Reed recalls the development of this key capability:

**Fig. 4.12** Ray Reed, Dick Lingle, and Bob Hardy (*left to right*) in the control bunker of the powder gun in preparation for a shock wave experiment (circa 1970) (Private collection, C. H. Konrad)

Adjacent to the Area Y site, near Building 9950, Munson's division later built a large Butler building shelter[8] to house a new powder-driven gun to study higher velocity impacts. (Years later, a second staged light gas gun, next to the powder gun, was added in this shelter.) A new adjoining control and instrument room housed an array of recording instruments and sheltered the operators and experimenters from possible harm. Bob May was active in the preliminary architectural set up of the facility. I developed the associated recording facility and instrumentation for the test chamber in preparation for the powder gun, which was then on order.

Sandia's first powder gun and its associated large experimental chamber were designed, fabricated, and first assembled by General Motors (GM) in Santa Barbara, California. From the enclosed muzzle of the gun, a projectile was fired into a walk-in, evacuated cylindrical containment chamber. Experimental targets were set up within the chamber for impact at the muzzle [see Fig. 4.13]. The target chamber was barely large enough for experimenters to stand as they set up the experiments. Transverse ports allowed photographic and x-ray observation during impact, as well as electrical and VISAR monitoring of the tests.

As Sandia's representative, I monitored the single acceptance test firing of a projectile from the powder gun into its target chamber at the GM plant. Projectile impact on instrumented targets occurred at the muzzle. The volume and strength of the evacuated chamber were sufficient to capture and contain the expanded explosive products, remains of the impact-destroyed experiment, and debris from a test. Access to the chamber and closure was at the rear via a heavy hinged circular door that was the full diameter of the chamber.

---

[8] *Editors' note*: A Butler building is a "temporary" corrugated metal structure that has a foundation. With the need for rapid expansion after World War II, Sandia relied on (and still relies on) such structures. Buildings 849 and 851, two Butler buildings installed in 1949, are still in use in Area I. See R.A. Ullrich, C. Martin, and D. Gerdes, *Historic Building Survey and Assessment Sandia National Laboratories New Mexico Site Albuquerque, New Mexico*, Vol. 1: Survey and Assessment, Sandia National Laboratories Report No. SAND2010-6117P (August 2010).

**Fig. 4.13** Bob May adjusting the catcher plates in the impact chamber of the powder gun (Private collection, C. H. Konrad)

This gun facility was initially commissioned with a single 89-mm-bore powder gun in the early 1970s, but was considerably expanded by November 1977. However, the powder gun was designed to be operated with a second-stage barrel with a smaller bore diameter. The projectile in the powder gun would compress hydrogen gas contained in the barrel to a high pressure (about one hundred thousand psi) and accelerate a smaller projectile in the second barrel. This projectile could achieve velocities up to around 7 km/s. Initially, only the powder gun was operational. The combined two-stage gun was made operational in the mid 1970s. A photo of the gun bay is shown in Fig. 4.14, with the powder gun on the left and the two-stage light gas gun on the right. The facility was later increased in square footage to include a machine shop and an extended range for ballistic experiments (not shown). The facility was used continuously from about 1971 to 1977 for a variety of shock wave studies until a major accident occurred on November 1, 1977 that considerably changed the course of future activities, actually for the better. As Carl Konrad, the facility manager at the time, says:

> We had a major accident with the powder gun on November 1, 1977 that destroyed Building 9956. We were shut down for about 1 year. I believe the accident cost about $100,000 to repair 9956; the cost of the accident investigation was 4 or 5 times greater. The guns were not damaged, very little damage occurred to the instruments, and personnel were safe in the reinforced control room. (The control room was originally designed as an explosive bunker because the site was still an explosive site when 9956 was built.) This was a very difficult time and several things came out of the investigation that severely hampered our future operations.

After considerable analysis and approvals, the facility was fully reconstructed and considerably increased in size, resulting in much more efficient operations of the two guns that had been crowded together in the earlier configuration, as shown in Fig. 4.14. Konrad further elaborates:

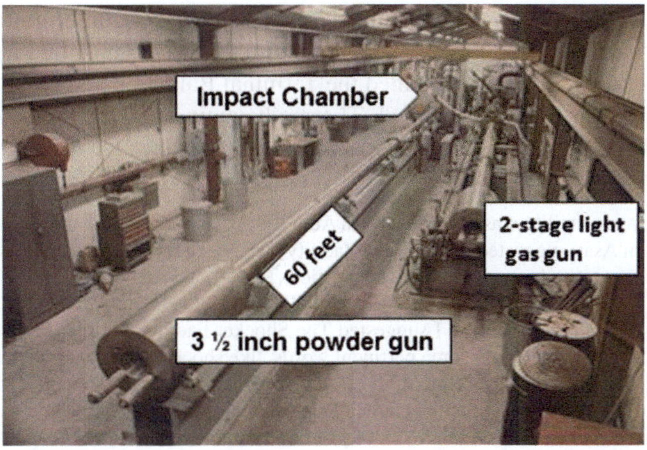


**Fig. 4.14** STAR circa 1976. At that time, the powder gun (*left*) and two-stage gun (*right*) were the only high-velocity launchers at the facility (Private collection, B. M. Butcher)

> The reconstruction following the accident actually started the expansion of the STAR facility into a world-class operation. In quick succession we added a new gun bay east of the original control room for the single stage gas gun(s). A new powder gun bay, a breezeway, control room, optics clean room, target fabrication lab, dark room, and restroom were added. The TBF (Terminal Ballistics Facility) gun was added in the early 1980s.

At the time of the accident, the first-level management of the facility had just transitioned from Darrell Munson to Barry Butcher. Butcher clearly recalled the events leading to the accident:

> Two weeks after I took over responsibility for the powder gun at Site 9950 it blew up. At the culmination of a routine experiment, the door blew off the catcher system venting hot gases into the building. The building was a typical metal building with a steel frame upon which sheets of metal formed the outer wall. The overpressure inside popped many of the rivets holding the outer wall metal sheets and the building looked like someone had gently pushed outward on all the walls. Fortunately no one was hurt. The gun operators, in fact, did not realize anything was unusual about the shot until they exited the control room bunker and saw all the soot and debris. The reason for the accident was traced to the bolts on the catcher tank. Investigation showed that, while there were [48 one-inch-long steel bolts] that needed to be secured with nuts on its rear door, the technicians had started to save some work by not securing all the bolts during a shot [seven were secured on this shot]... in rebuilding the facility, we were able to remedy all the adverse features that had unknowingly been incorporated into it during initial construction. The new version was a world-class facility that thrust Sandia well in front of any other organization in this country.

Later, several other guns were added to the facility, bringing the total to six in the mid 1980s (jokingly referred to as "the six-gun facility" with the "fastest gun in the west"). These included the original "air gun" built by Don Lundergan that was modified with a milled slot in the barrel and used an oblique impact projectile to allow simultaneous longitudinal stress and pure shear motion at impact. The original Lundergan gun is still being used for pressure–shear studies, particularly for studying the dominant effects of pure shear stress in the compaction of granular materials

(Vogler et al. 2011). Other guns included the 4-inch helium gun originally built in Building 855 in Tech Area I, a three-stage railgun that was dismantled in 2003, and a small two-stage light gas gun named the Terminal Ballistic Facility.

The facility was named the Shock Thermodynamics Applied Research facility, or the STAR facility, in the early 1980s. An interesting anecdote of how this name came about is provided in Barker's recollections:

> … when our gun facility was taking shape in Area III, at one of our Monday morning safety meetings Jim Asay suggested that our new facility deserved a name of its own. Asay asked us to think about it and to come up with suggestions for the name at our next safety meeting. I mentioned the naming contest to my wife, Val. She immediately took an interest in it, and we discussed some possibilities. I suggested The Shocker Facility, but that didn't pass the Val test. Then we wrote down some pertinent words, like shock, research, applied research, and finally thermodynamics. After studying the words for a while, Val said 'How about Shock Thermodynamics Applied Research? Does that make sense? That would make it the STAR facility.' I said to Val, 'I think you've just named our new facility!' And that's the way it turned out.

## 4.14 Mass Ejecta from Shocked Surfaces

As the VISAR was used for higher shock stresses and more general loading conditions, several unexpected effects were observed. One of these was loss of light signal when a sample monitored with a VISAR on its rear free surface was shocked to stresses of a few hundred kbars. These shock levels were routinely obtained on the powder and two-stage gas guns. Sample surfaces were usually made extremely flat with a lapping machine, resulting in a slight surface roughness on the order of a few microns. The surface was usually not polished further, since the VISAR was fabricated on the assumption that it could be used on any surface. Since the system was often successfully used at lower shock stresses to measure free surface velocities on samples with similar surface conditions, this was a perplexing result.

The rise time of shock waves at very low stresses was typically a hundred nanoseconds, whereas at stresses above 100 kbars, the shock rise time was not resolvable in most materials; based on the fourth power law, the rise time was probably much less than one nanosecond. This would make the shock thickness on the order of or less than the surface roughness. It was well known that, when a steep shock wave interacted with a surface groove, a jet formed and ejected material from the groove at a velocity much faster than the average surface velocity. Jim Asay thought that this effect was happening on a microscale and developed a VISAR technique to measure the momentum delivered to a thin foil separated a few millimeters from the free surface by the ejecta mass impinging on it. (This technique is now referred to as the Asay foil technique.) With a few basic assumptions, the velocity acquired by the foil determined the amount of mass deposited on the foil and therefore ejected from the surface (Asay 1977b, 1978; Asay and Bertholf 1978). Systematic experiments using shock waves with a controlled rise time (Asay 1977b) and time-resolved holographic experiments (Asay et al. 1976) verified that the effect was caused by micro jetting of material from surface imperfections. This effect had previously been observed in x rays of imploded surfaces and referred to as "fluff," since the x-ray image looked "fluffy." However, this "fluff" had never been quantified before the Sandia experiments. The Sandia technique for measuring ejecta from shocked surfaces was quickly adopted by the weapons community and is still widely used as a standard technique to estimate mass ejection from shocked free surfaces.

## 4.15 Energetic Materials

Another critical aspect of Sandia components and subsystems involves the response of energetic materials at low stress levels where initiation of explosives can be problematic. Technical issues in the 1960s included establishing the shock stress necessary to initiate the reaction, the transition from deflagration (burning reaction) to detonation (prompt explosive reaction), and the effects of 2-D flow on the performance of energetic systems such as detonation cables. A specific application concerned the use of small explosive charges to generate massive amounts of current from ferroelectrics in one-shot power supplies used for neutron generators. In 1962 Glenn Seay, who had previously performed explosive initiation and shock compression research at Los Alamos, was hired as supervisor of the newly formed Dynamic Stress Research Division in the Physical Research Department (which later became the Physics of Solids Research Department). Seay brought expertise in both energetic materials and ferroelectric response to the budding shock wave group. Later in the 1960s, Al Narath (then the Director of Research and later the President of Sandia National Laboratories) and Orval Jones (then a Division Supervisor in the Physical Research Department and later the Executive Vice President) made a strategic decision to initiate research on energetic materials by hiring another experienced researcher, Jim Kennedy, from the Illinois Institute of Technology. Since Los Alamos already had an established program in this area, competition on energetic materials research between the two organizations loomed on the horizon. As noted in Kennedy's recollections:

> I was hired in July 1968 because I had 9 years experience in explosives research and development (R&D) at the Illinois Institute of Technology (IIT) in Chicago. With my hiring, as Al Narath stated to me much later, Sandia intended to get into explosives research and 'compete with Los Alamos.'(!!!) I thought (but didn't say out loud) that this sounded absurd when Narath said it in about 1984, shortly before Narath went to AT&T Bell Labs to become anointed[9] so that he could return to Sandia as its first homegrown president.

[9] *Editors' note*: When Kennedy was hired in 1968, Al Narath was the Director of Solid State Sciences Research. AT&T policy was that Sandia's president had to come from AT&T Bell Laboratories; hence, Narath had to spend time at Bell Labs before becoming the President. Therefore, beginning March 31, 1984, Narath, who was then the Executive Vice President at Sandia, transferred to Bell Labs (*Sandia Lab News*, March 30, 1984, vol. 36, no. 7, p. 1) to serve as the vice president of Government Systems. In that role, he headed up all research and development in support of Bell Labs' work for the Federal government. Narath was the *only* President whose professional career began at Sandia during the nearly 44 years AT&T managed Sandia. He returned in 1989 (see letter addressed to all Sandians, April 17, 1989, in the *Sandia Lab News*, vol. 41, no. 8, p. 1) and served as President from April 1, 1989 to August 15, 1995.

This prescient strategy set the course for a major research effort on explosives and led to a host of fundamental discoveries. Later in the 1970s under Dennis Hayes, this initiative laid the groundwork for the Explosive Components Facility (ECF), which became an integral part of the Sandia weapons program.

Tim Trucano, a student at the University of New Mexico at the time, remembers exciting days as a student intern sitting in the test area where Kennedy was performing experiments:

> My summer of 1971 was a major harbinger of what was to come, but I didn't know it at the time. I was hired as a 'Summer Student Development Associate' and the hiring manager was Lee Davison. The group was called something like the 'Shock Wave Research Division'.... I spent the first two weeks at an explosive firing site. My mentor that summer was Jim Kennedy; he was doing explosive shock initiation experiments out at the site. The site personnel modified the Standard Operating Procedures, while I was there with a line that said something like 'Make sure Tim is in the blockhouse.' I remember explosions from small sheets of Detasheet®, instrumented with fast oscilloscopes, and the resulting enormous quantities of Polaroid photos of oscilloscope traces. I also remember immediately becoming curious about waves, detonations, and so on in those first two weeks.
>
> The culminating event at the site while I sat there was an enormous fuel-air explosive test (outside of Jim's shock initiation research interests) in which about 30 or 40 pounds of TNT/PBX explosive lens launched an aluminum flyer plate into a big gasbag. It was the largest explosion I have ever experienced. After the fact, we stared at the scorched ground and Jim immediately commented that the event had been a deflagration, not a detonation (imagine what a detonation would have been like!). More curiosity generated on my part – how did Jim know that? Deflagration? It turned out that the plastic bag would have been cut into triangular fragments (as I recall) if there had been a detonation because of the cellular instability associated with a propagating det (detonation) wave in the gas.

With guidance from Orval Jones (who had become Department Manager by the time Kennedy arrived in 1969) and Chuck Mader at Los Alamos, Kennedy initiated research on the explosive PBX 9404. In his first project, quartz gauges were used to determine the distance a planar input stress wave propagates in PBX 9404 before the explosive reaction occurs (called "the run to detonation distance"). Through careful work to investigate various aspects of explosive reactions using the newly developed gas guns, he established Sandia's credibility in energetic materials research (Kennedy 1970). Also, around that time, Jace Nunziato joined the effort to provide theoretical support to the growing experimental program. As Kennedy states:

> At that point (1974), Sandia brought a theoretician, Jace Nunziato, from the Solid Dynamics Research Department into the Physics of Solids Research Department to work with me in analyzing the shock wave evolution results. Nunziato had cut his teeth at Sandia working with Walt Herrmann on shocks in porous materials and with Karl Schuler on viscoelastic behavior of PMMA (polymethyl methacrylate) in the shock front. It was widely supposed at that time in the explosives community that, in heterogeneous explosives, shock growth to detonation was due to some prompt energy release at the shock front. Using continuum mechanics-based singular surface analysis (another Sandia strength), Nunziato derived an equation with which the shock-growth behavior could be analyzed to allow an estimate of the energy release in the shock front. Our 1975 data set was on shock waveform evolution at an initial shock pressure that would lead to detonation in 7 mm of shock run distance, a rather strongly driven condition. To our surprise, Nunziato's analysis of the waveform-evolution data showed that the explosive absorbed rather than liberated energy at the shock front.

Kennedy and Nunziato's results bolstered Sandia's research program on energetic materials and also garnered outside attention. Kennedy notes the surprised reaction at LANL when they were made aware of the new results:

> When we shared these results with Bill Davis and Wildon Fickett at Los Alamos soon afterward, LANL was surprised that
>
> 1. Nunziato had independently derived the 'shock-change equation,' which Los Alamos had derived earlier but never published ('[Expletive deleted], Sandia's got the shock-change equation!' – Bill Davis).
> 2. The shock-front behavior was endothermic.
>
> Fickett and Davis seemed motivated by this to get serious about completing a book they were writing, and hence their classic *Detonation*,[10] published in 1979, contains the derivation of the shock-change equation. No further exploitation of the equation has been made, to my knowledge, in shock wave research on explosives before or after the 1976 paper (Nunziato and Kennedy 1976), other than a related one (Chen and Kennedy 1976) on shock evolution in curved waves at the 1976 Detonation Symposium.

Figure 4.15 illustrates the growth-to-detonation response observed for PBX 9404 and the endothermic reaction inferred from the data. Unexpected endothermic behavior at low shock amplitudes is ascribed to shock dispersion by the mixture of ingredients with dissimilar shock impedances. The young shock wave program at Sandia was indeed making an impact and generating competition with the other laboratories in the energetic area; that competition was to continue for years to come. Over the next few years, several significant contributions were made to the understanding of PBX 9404 and other energetic materials that paved the way for additional staff in this area, including Steve Sheffield, Al Schwarz, and Dennis Mitchell.

After Kennedy's 1970s paper, several additional publications addressed a variety of issues, including shock wave evolution in energetic materials, determination of the explosive impulse into a metal plate, determination of the pressure field for initiated explosives, and detonation transfer by a flyer plate, initiation and deflagration of dense explosives, and modes of shock wave growth during explosive initiation. These 1970 to 1978 studies led to several publications by Kennedy and Nunziato (Kennedy 1971, 1972, 1973; Davison et al. 1972; Kennedy and Schwarz 1974; Kennedy, Nunziato, and Hardesty 1976; Kennedy and Nunziato 1976; Nunziato and Kennedy 1976; Nunziato et al. 1977, 1978a, b). This collected work was instrumental in developing a scientific understanding of shock propagation in energetic materials and generated "energetic" competition with the other two laboratories. The research was also promptly integrated into weapon design at Sandia, as mentioned in Dennis Hayes' recollections.

Kennedy's leadership in energetic materials research drew several other researchers to this area. He notes in his recollections:

---

[10] *Editors' note*: Wildon Fickett and William C. Davis, *Detonation* (University of California Press, Berkeley, 1979). The book is now available in a paperback edition, first published in 2000, as a slightly corrected, unabridged republication of the 1979 work, with the title *Detonation: Theory and Experiment* (Dover Publications, Mineola, NY, 2000).

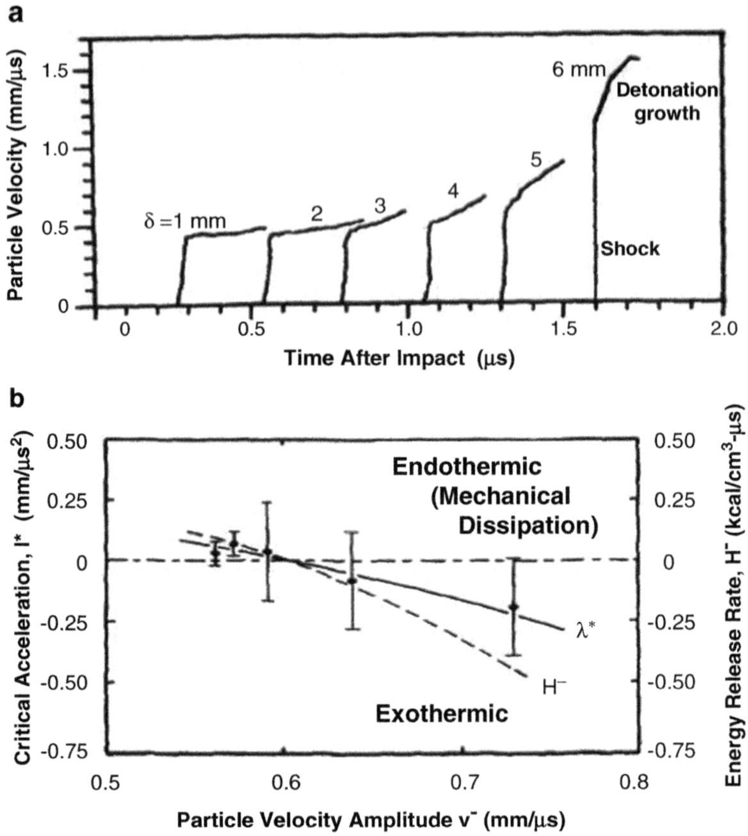


**Fig. 4.15** Growth-to-detonation response in PBX 9404. (**a**) Particle velocity histories for distance $\delta$ that the wave has propagated, showing growth of reaction, which is indicated by the increase in initial shock amplitude with distance. (**b**) Results from a singular surface analysis, where $\lambda^*$ is the critical acceleration for steady wave propagation and $H^-$ is the energy release rate. The variable v- on the x-axis of the figure in part b is singular surface notation used in rational mechanics for the particle velocity amplitude of a shock wave approached from the low pressure side. (Reprinted from Kennedy and Nunziato 1976, Copyright 1976, with permission of Elsevier LTD)

… After I moved from research into component design in 1976, Bob Setchell joined the Physics of Solids Research Department and performed wave-evolution studies on the same explosive with a ramp wave followed by a shock. He found (Setchell 1981) that the ramp wave precompressed the explosive and desensitized it so that the ensuing shock was unable to stimulate any apparent reaction, as judged by the waveform behind the front. After a few millimeters of propagation distance, the ramp wave was overtaken by the shock front and then there was no precompression. At that point, evidence of reaction behind the front was immediately noted…

Don Hardesty joined the Physics of Solids Research Department in 1974 after graduate and post-graduate work in combustion. He worked on explosives until about 1977, when he joined the fledgling Combustion Research Facility activity at Sandia Livermore. While in Albuquerque, he collaborated with Pete Lysne on shock initiation of liquid nitromethane and on the development of a universal Hugoniot for liquids, which is still cited (Hardesty and Lysne 1974). Their work motivated Don Amos, Nunziato, and me to analyze the shock-driven thermal explosion behavior of nitromethane in their experiments by use of a parallel-reactions model (Nunziato, Kennedy, and Amos 1977).

Although he was at Sandia in Albuquerque for just a short time, Don Hardesty made several basic contributions to the understanding of energetic materials (Hardesty and Lysne 1974; Hardesty 1976a, b; Hardesty and Kennedy 1977). He also developed a methodology to determine the change in refractive index of transparent energetic liquids (e.g., nitromethane) from VISAR measurements (Hardesty 1976b). A sampling of energetic materials papers by other Sandia researchers include Lysne and Hardesty (1973), Renlund and Trott (1988, 1990), Renlund et al. (1989), Baer (1988, 1994, 1996, 1997, 2000), Baer and Nunziato (1983, 1986, 1989), Baer and Trott (2002a, b, 2004), and Baer et al. (1986, 1995, 1996a, b, 1998). Mel Baer's brilliant work established a leadership role for modeling inhomogeneous explosives at the mesoscale, thereby emphasizing the role of heterogeneities in the initiation and growth of energetic reactions. This will be discussed further in Chap. 6.

After Kennedy left Sandia in 1980, Bob Setchell became the principal investigator for energetic research. One of his significant contributions was that slow compression of an energetic material by a ramp wave to a stress sufficient to initiate a reaction-suppressed detonation until a shock front was formed. After that, normal detonation behavior was observed (Setchell 1981). Setchell made several other noteworthy findings, including the effects of pulse duration on initiation (Setchell 1982), the effects of precursor waves on initiation (Setchell 1983), grain size effects on shock sensitivity (Setchell 1984), effects of chemical reactivity (Setchell 1986), and microstructural effects of shock initiation in granular explosives (Setchell 1987). These observations were important in developing models for predicting detonation behavior (Nunziato and Kipp 1983; Baer and Nunziato 1983, 1986; Nunziato 1984).

In addition to the fundamental research in Lee Davison's division, in the mid 1970s, Dennis Hayes initiated an energetic materials group that focused on external customers. This effort was started in part because the Physics of Solids Research Department did not want to pursue application-driven work. As Hayes says:

> In 1974 I became Supervisor of the Explosive Components Division. Most of the work was on scheduled components for nuclear weapons. But we did have a rather sizable reimbursable with the Navy to study porous HNS[11] explosives. The Navy was introducing HNS detonators and mild-detonating fuzes on their C-4 missile and desperately needed a more thorough understanding of the behavior of this new explosive. The Explosives Research Department was unwilling to take on this customer-driven work so we built up a capability within the Division. The principals were Dennis Mitchell, who performed front surface impact VISAR measurements on porous HNS that allowed the equation of state and reaction rates to be measured, and Al Schwarz, who built our own VISAR and made innumerable measurements on component performance that could be compared with model results. Jim Kennedy and Steve Sheffield were valuable team members. I did most of the modeling.

[11] *Editors' note*: HNS is hexanitrostilbene, a high explosive with good thermal and vacuum stability that was developed by the Navy in the 1960s.

> The result was a rather complete material description. With the help of Sam Thompson, we introduced reaction kinetics of HNS into CSQ (the two-dimensional progenitor of CTH) and did numerous numerical studies on detonation corner turning and other two-dimensional phenomena important in component design.

Within a few years, applied energetic research became a major Sandia activity and resulted in a new building, the ECF, where this research is focused. A single-stage gas gun was installed in the facility for experimental research and later managed by Lloyd Bonzon, then a division supervisor. Bonzon played a critical role in rescuing experimental shock wave research at Sandia during an extremely trying time in the 1990s that began on "Black Monday" (December 12, 1994). This will be discussed in Chap. 6.

## 4.16 People and Places of the 1970s

**Bob May at the newly installed powder gun, 1970 (Reprinted with permission of Sandia National Laboratories)**

**New Directors Orval Jones (*left*), Solid State Sciences Research, and Al Narath, Physical Sciences Research, 1971 (Reprinted with permission of Sandia National Laboratories)**

**Bob Hardy with target for a powder gun test, 1970 (Reprinted with permission of Sandia National Laboratories)**

**Bob Setchell, adjusting laser, 1975 (Reprinted with permission of Sandia National Laboratories)**

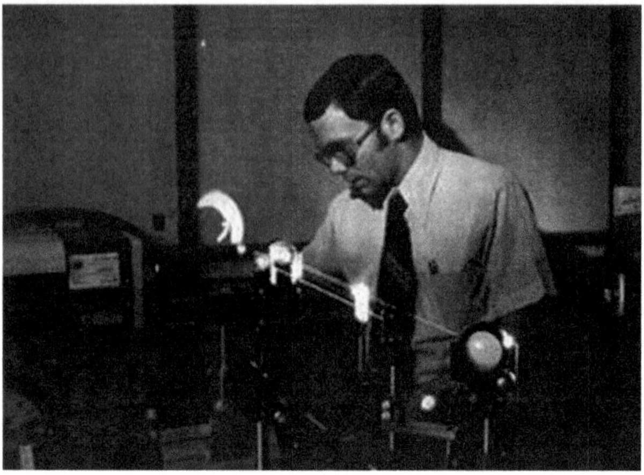

**Dennis Hayes (*left to right*) promoted to Manager of Fluid and Thermal Sciences Department and Jim Asay to Supervisor of Thermomechanical and Physical Research Division. Also shown are Aileen George and Al Chavez, 1978 (Reprinted with permission of Sandia National Laboratories)**

**Dick Rohde, promoted to Supervisor of the Metallurgy I Division, 1970 (Reprinted with permission of Sandia National Laboratories)**

**Al Stevens (*left*) and Chap Chapman, 1976 (Reprinted with permission of Sandia National Laboratories)**

**Larry Bertholf, Jerry Freedman, and Steve Benzley** (*left to right*)**, 1974 (Reprinted with permission of Sandia National Laboratories)**

**Phil Mead and Don Lundergan** (*right*) **editing the "Sandia Laboratories Technical Capabilities," 1976 (Reprinted with permission of Sandia National Laboratories)**

**Charlie Daniels (*left*) and Bill Benedick, inventors of a nonviolent explosive destruct system, 1977 (Reprinted with permission of Sandia National Laboratories)**

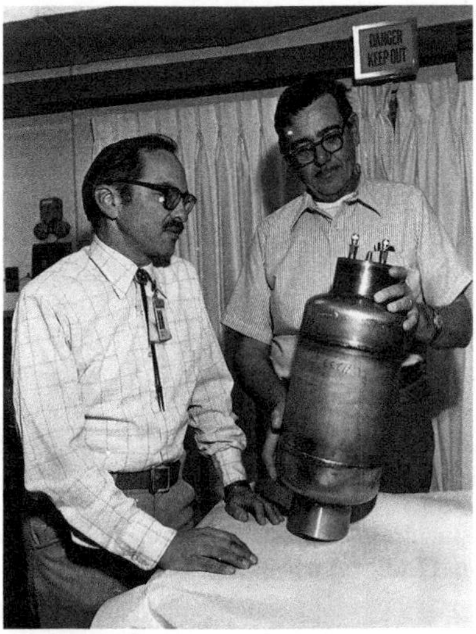


**Lee Davison (*right*) and Bob Henderson examine an explosive line-wave generator, 1972 (Reprinted with permission of Sandia National Laboratories)**

**Jim Mogford, Vic Engel, Gladys Rowe, Bob Esterly, Charlie Winter, Dick Craner, and Jeff Lawrence (*left to right*) at a management staff meeting, 1978 (Reprinted with permission of Sandia National Laboratories)**

**Bob Graham (*left*) and Ron Jacobson examine a lithium-niobate gauge, 1974 (Reprinted with permission of Sandia National Laboratories)**

**Jace Nunziato displays the Pi Tau Sigma Gold Medal awarded to him by ASME in 1974 for "outstanding achievement in mechanical engineering within 10 years following graduation" (Reprinted with permission of Sandia National Laboratories)**

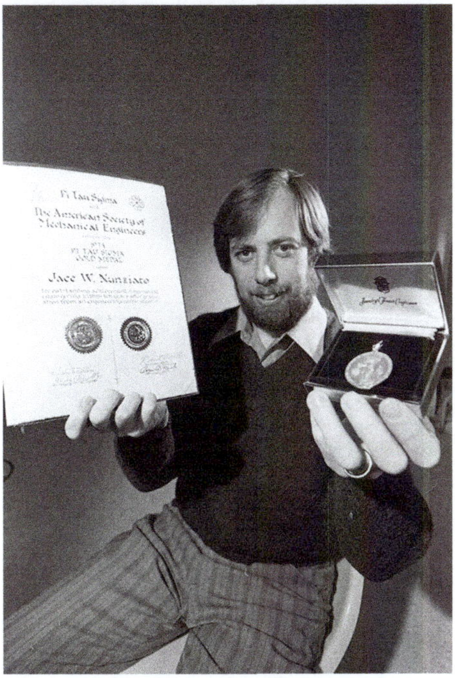


**Jim Kennedy, Shock Physics Research Division 1973, private collection, James E. Kennedy**

**1973**

# Chapter 5
# The 1980s: Heady Times

## 5.1 Background

The previous two decades of shock wave research at Sandia led to (1) advances in experimental techniques, (2) measurements of dynamic material response for a wide range of materials, (3) state-of-the-art material models, and (4) a family of 1-D and 2-D computer codes that could simulate materials used in weapon components and subsystems with considerable accuracy. However, full three-dimensional (3-D) code capabilities were needed for higher fidelity simulations of weapon components and subsystems.

In the 1980s, two major developments had a profound influence on shock wave activities in the following years. First, President Ronald Reagan began the Strategic Defense Initiative (SDI) in 1983. This ambitious program proposed ground- and space-based assets to protect the United States from attack by nuclear missiles. SDI, popularly dubbed "Star Wars," was criticized as being unrealistic, even unscientific. Claims were also made that it would destabilize fragile international agreements based on the concept of mutually assured destruction (MAD) and reignite an offensive arms race. In spite of these concerns, the program proceeded and led to new initiatives in several research organizations.

To accomplish the research goals of SDI, several innovative technologies were proposed by the three DOE national security laboratories (at that time termed the "national weapons laboratories"), many universities, and industrial institutions. By the mid 1980s, considerable funding was flowing from the SDI Innovative Sciences and Technology Office, mostly in the area of applied research. Although none of the proposed concepts were ever fielded in practical applications, much of the basic technology developed through SDI laid the groundwork for future scientific innovations that have continued to influence the Nation's research programs to this day in high energy density physics, supercomputing, advanced materials, and other critical disciplines.



J.R. Asay et al., *Impactful Times*, Shock Wave and High Pressure Phenomena,
DOI 10.1007/978-3-319-33347-2_5

Initially, many scientists were skeptical of a Star Wars-type shield in the near term and thought that several years of research would be needed to establish the feasibility of a practical missile defense technology. In 1993 the research emphasis of the program shifted from national missile defense to theater missile defense, and the scope changed from global to focused regional coverage. However, both the Strategic Defense Initiative and the Theater Missile Defense (TMD) programs had a significant influence on Sandia's shock wave programs.

Sandia began several strategic efforts as part of the original SDI program. One was a new program to develop capabilities for launching gram-sized projectiles to velocities exceeding 15 km/s. Such velocities were about a factor of two higher than possible with existing two-stage light gas gun facilities. The experimental shock wave group, then managed by Jim Asay, a first-level manager in Lee Davison's Solid Dynamics Research Department, participated in this effort and actually exceeded the velocity goal with one of the two approaches that were pursued.

The second major development was the start of massively parallel computing at the national security laboratories. Development of this groundbreaking technology, which employed several thousand parallel processors instead of the single processor used in standard computers, was one of the stated goals of SDI. The intent was to exceed computational speeds by many orders of magnitude over conventional capabilities. Under Director Ed Barsis' leadership, Sandia played an early role in initiating massively parallel computing, and for a short while in the 1990s, Sandia possessed the fastest supercomputer in the world. Massively parallel computing ultimately had a major impact on the ability to perform realistic numerical simulations of dynamic events for both weapons and other scientific applications.

The major advances in shock wave research during this period, as discussed in this chapter, include:

- High-Pressure Material Strength (Sect. 5.2)
- Three-Stage Railgun for the Strategic Defense Initiative (Sect. 5.3)
- Metallization of Hydrogen on the Two-Stage Gun (Sect. 5.4)
- Graded-Density Impactor for "Soft," or Ramp, Impact (Sect. 5.5)
- The Sandia HyperVelocity Launcher (Sect. 5.6)
- Generalized Fragmentation Theories (Sect. 5.7)
- Laser Windows for Mbar Profile Studies (Sect. 5.8)
- Shock-Induced Solid State Chemistry (Sect. 5.9)
- Piezoelectric Polymers for the Bauer Shock Gauge (Sect. 5.10)
- High-Fidelity Ferroelectric Models (Sect. 5.11)
- CTH: A Robust 3-D Hydrodynamic Code (Sect. 5.12)
- Turret Explosion Onboard the USS Iowa (Sect. 5.13)

## 5.2 High-Pressure Material Strength

A major goal of shock wave research is to characterize the mechanical and physical properties of materials fully over a stress range of interest so that continuum models can be developed for accurate simulations. Material strength, which is usually described through the yield strength dependence on strain rate, stress, and temperature, is particularly important. The initial yield strength is easily determined, in most cases, by measuring the elastic–plastic wave structure at low peak shock stresses (typically less than 100 kbars), where elastic response is observable. A number of stress gauges and interferometer techniques have been developed to provide this information. However, the material strength *after* shock compression to high stresses (typically greater than 100 kbars) is difficult to measure with any existing techniques today. Typical approaches such as piezoelectric gauges, interferometer techniques, and pressure–shear techniques have been used at Sandia, whereas lateral-longitudinal gauges have not usually been successful at these stress levels in metals, especially at stresses of several Mbars. Strength information at such ultra-high stresses is needed to address a variety of weapon issues and for other scientific applications.

Asay and Lipkin (1978) first proposed a wave profile technique to extract strength information through a combination of unloading and reloading from a given shock state. The technique, identified as the "self-consistent strength measurement," was demonstrated at shock stresses of about 30 kbars on aluminum, and the measured results agreed well with other techniques, such as compressive yield strength determined under hydrostatic pressure. In the early 1980s, the technique was extended to stresses of a couple hundred kbars in aluminum and several other materials using the powder gun at the STAR facility (Asay and Chhabildas 1981; Chhabildas and Asay 1982; Chhabildas et al. 1982) and later to stresses approaching a Mbar in tantalum and around 2.5 Mbar in tungsten using the two-stage light gas gun (Chhabildas and Barker 1988; Chhabildas et al. 1988).

After demonstrating strength measurements of aluminum to around 200 kbars, Chhabildas and Asay were anxious to extend the capability to the Mbar pressures and to a broad range of materials. Unfortunately, a programmatic need did not exist at Sandia, and discretionary funding for this was extremely limited. However, Dan Steinberg from LLNL had been developing constitutive models for the weapons program, which needed data on material strength for a wide variety of materials to pressures of several hundred kbars. The shock wave group under William J. (Bill) Nellis at LLNL did not have either the time-resolved interferometry or the technical expertise in wave profile analysis to embark on an extensive program and was not interested in pursuing this work. Because of the pressing weapon requirements, Steinberg approached Asay and Chhabildas to perform the experiments. One of the materials was beryllium, which, because of its toxicity, could not be studied without major modifications to the gun facility, so Asay and Chhabildas initially resisted doing this work. However, LLNL persisted and elevated their request to Sandia's upper management. Chhabildas recalls the events at the time:

> Our strength work was especially appreciated by Dan Steinberg of Livermore; he used it to develop strength models now referred to as the Guinan-Steinberg constitutive model for use with hydrodynamic codes. Dan wanted us to measure Be strength in the shocked state! When the initial request came directly to us from Hal Graboske and Dan Steinberg of Livermore, we were not too thrilled and declined the invitation because of all the precautions we would have to take due to the toxic nature of Be. The Livermore folks then approached our upper management directly, and we had no choice but to oblige.

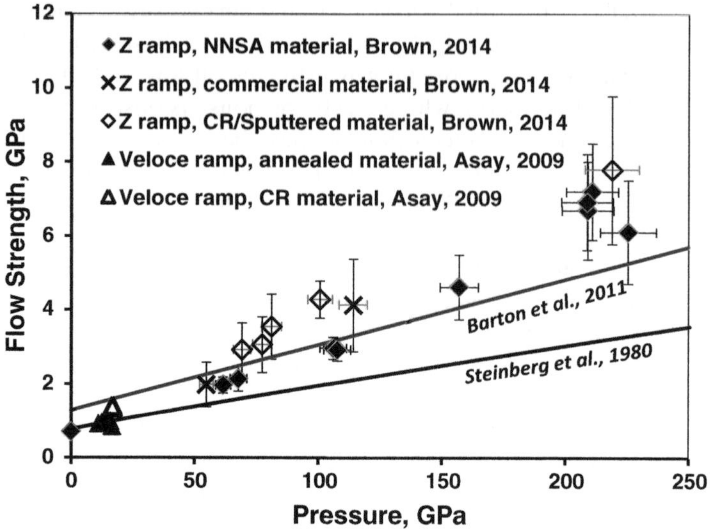


**Fig. 5.1** Flow strength versus shock pressure for quasi-isentropic (ramp) compression of tantalum. The new Z results were obtained for two different Ta preparations. The mean flow strength and the mean pressure are estimated from the measured unloading profiles. The *solid* and *open diamonds* represent the two types of Ta, a standard preparation and a sputtered and cold-rolled (CR) material (Private collection, J. L. Brown et al. 2014a) The curve labeled "Barton et al. 2011" is a dislocation-based multi-scale strength model

The project ended up extending over a 2-year period and provided much needed financial support, which helped maintain the experimental capabilities at the time and also elevated Sandia's stature in the area of high-pressure material strength. Moreover, the strength of several materials was determined with the self-consistent wave profile technique; the results were primarily used by Dan Steinberg and his colleagues at LLNL (Steinberg 1996) in the 1970s and 1980s to develop high-pressure strength models that are still being used (Asay et al. 2009). Research on high-pressure strength continues to be actively pursued at both the STAR and Z facilities. Recently, the self-consistent technique was extended to over 2 Mbars for ramp loading of tantalum on Z by Brown, Alexander, Asay et al. (2013a, 2014a). The new results in Fig. 5.1 represent the most accurate measurements of high-pressure strength made with any technique and are being used to evaluate various material models, including recent multi-scale dislocation-based models being developed at LLNL (Barton et al. 2011). The accuracy of the data over an extended pressure range and the reasonably good agreement, especially since the modeling and experimental studies were independently conducted, with multi-scale dislocation-based model, attest to the increasing capability for predicting continuum behavior from models developed at the microscale and mesoscales.

The ramp wave results obtained on tantalum using Z show generally good agreement with diamond anvil cell (DAC) results (Dewaele and Loubeyre 2005), pressure–shear measurements (Duprey and Clifton 2000), previous quasi-isentropic loading studies (Chhabildas et al. 1990; Chhabildas and Asay 1992; Vogler and

Chhabildas 2006), and Rayleigh–Taylor perturbation growth when different loading rates are considered (Park et al. 2012). The best agreement with strength models occurs with the dislocation-based LLNL multi-scale strength model for predicting material properties, such as dislocation structures, as illustrated in Fig. 5.1. Many empirical strength models (Steinberg et al. 1980; Steinberg and Lund 1989; Preston et al. 2003) tend to be considerably lower than the experimental results and outside most of the data error bars, as shown in Fig. 5.1.

One observation from several strength experiments, particularly for hard, brittle materials, was that the shear stress states inferred from the wave profile measurements did not conform with the expectations of elastic–plastic theory, a commonly used model that had been established for dynamic loading at low stresses. Several explanations have been proposed for this apparent anomaly. One hypothesis by Dennis Grady is especially attractive, although it is still not confirmed experimentally. At extremely high stresses and correspondingly high strain rates, Grady suggested that shock deformation deposits the dissipative energy of the shock wave in small regions that are typically less than a micron in width, but less than the width of the shock. Hence, these localized regions are initially extremely hot during the rise in shock stress, but rapidly cool (by thermal diffusion) after the peak steady state is reached. The outcome is an initially softened state because of thermal trapping and the highly localized temperatures, followed by strength recovery as the localized hot regions cool. The thermal trapping effect is a result of the interplay between the risetime of the shock, where dissipation occurs, and the size of the local plastic region that is being produced. Preliminary estimates of the width of the localized regions are less than a micron at low stresses (Grady and Asay 1982), with the width decreasing rapidly with increasing shock stress. As an extension of earlier work by Hayes and Grady (1982), Swegle and Grady (1986a, b) developed a numerical model based on this idea for use in computer simulations of shock and reshock wave propagation in aluminum. Swegle and Grady's calculations supported the possibility of the shock-induced softening effect. Shock-induced softening also occurs in brittle materials such as alumina (Reinhart and Chhabildas 2003), sapphire (Reinhart et al. 2006), and boron carbide (Vogler et al. 2004), which show major deviations from the elastic–plastic model upon initial shocking followed by reshock from the shock state (Reinhart et al. 2002, 2006). Other mechanisms for the apparent shock softening and recovery include irreversible dislocation motion (Johnson 1993; Winey et al. 2012) and heterogeneous multivalued yield deformation of different grains (Asay and Lipkin 1978).

## 5.3 Three-Stage Railgun for the Strategic Defense Initiative

Of the several research programs begun in response to SDI, the Solid Dynamics Research Department under Lee Davison received funding to participate with LANL on a strategic program to develop a hypervelocity launcher that ultimately could produce sufficiently high velocities to intercept and destroy ballistic missiles aimed at the United States. LANL pursued an explosively driven technique to accomplish this goal. At Sandia, Jim Asay proposed the use of a two-stage gun to inject projectiles at velocities up to 7 km/s into a third-stage electromagnetic launcher, often referred to as a railgun, which would further accelerate the projectile to 15 km/s. Concept feasibility was the first step in the research to produce even higher velocities.

Railgun technology was well established in the United States and elsewhere at the time, although velocities were limited to a little over 5 km/s (Hawke et al. 1991a). This limitation was thought to arise from the ablation of the conducting copper rails used in the bore of the gun that produced the propulsive magnetic field. It was believed that, during initial magnetic acceleration in the breech, the surface of the heated copper rails would vaporize and ablate into the barrel, causing a significant increase in the mass of accelerated material. However, if the projectile is initially inserted into the railgun at a high enough velocity, there would not be sufficient time for ablation to occur. Thermal diffusion calculations indicated that a minimum injection velocity of 5 km/s was necessary to achieve this condition. To obtain that minimum velocity, a two-stage gun was needed. However, the technology for interfacing two quite different concepts, a gas-dynamic two-stage light gas gun and an electromagnetic railgun, was a daunting task. Figure 5.2 shows the concept for the three-stage railgun launcher.

Asay assembled a team of several Sandia staff that included Clint Hall, Carl Konrad, Jim Ang, Lynn Barker, Karl Schuler, Gerry Wellman, and Tim Trucano, all of whom had made significant contributions to shock wave research, with Randy Hickman, Bill Reinhart, Gerry Sauve, Steve Knudsen, Anita Vasey, and John

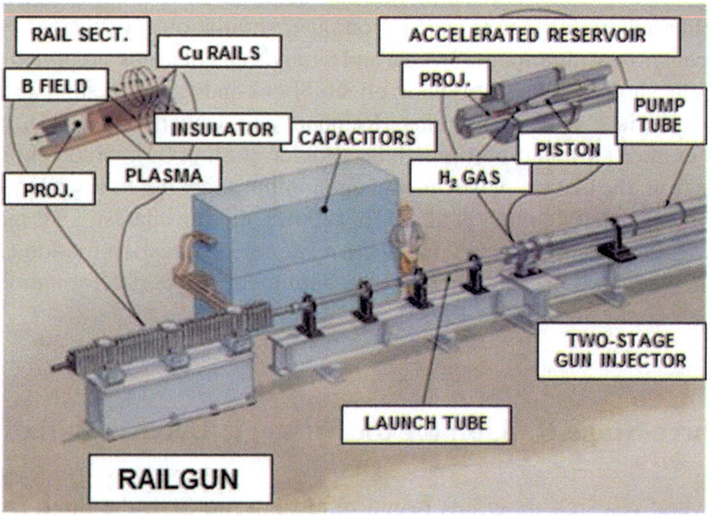


**Fig. 5.2** Three-stage railgun. The two-stage gun on the *right part* of the diagram injects a projectile into the railgun shown on the *left* at about 6 km/s. The railgun further accelerates the projectile to 7.5 km/s (Private collection, J. R. Asay)

Martinez participating as the contractor staff. Professor Mohsen Shahinpoor of the University of New Mexico conducted simulations to optimize the performance of the two-stage gun used to inject projectiles at high velocity into the third-stage railgun. Asay also recruited two LLNL researchers, Ron Hawke and Allan Susoeff, who were well-known experts in this technology from a prior railgun program at LLNL. The project was initiated in the mid 1980s without having any hardware for either the two-stage injector or the railgun on hand. The team designed, purchased, and implemented a two-stage light gas gun injector, the railgun, a 3-MJ power supply for the railgun, and the vast array of diagnostics and instrumentation to monitor and record the overall performance. Many obstacles were overcome, and a three-stage railgun became operational at the STAR facility in about 2 years.

One of the major challenges was how to establish a current-conducting armature between a hydrogen-pressurized projectile entering the railgun section at 6 km/s. The challenge was like striking a match in a hurricane made completely of hydrogen gas (an extremely good insulator) driving the projectile in the second stage. This problem was overcome by inventing a new self-arcing armature in the projectile that automatically initiated a current at the base of the projectile when it entered the railgun section. Clint Hall's recollection provides a perspective of this challenging problem:

> Jim apparently decided I couldn't screw things up too badly and actually gave me a higher profile project…my first REAL one at STAR. I was to be the Mechanical Lead for the two-stage gas gun injected railgun (STARFIRE). Hey, I could handle this right…simply couple a hydrogen-filled gun with a delicate projectile that doesn't like joints and moves at about 6 km/s to a stationary, composite bore hunk that had to be absolutely pristine to avoid pre-arc formation, having a bore that was kinda the same diameter most of the time, needed to have a $10^{-3}$ torr vacuum, no projectile blow-by, oh, and measure the projectile velocity right before the railgun section so you could dump the electrical energy at the right time…. [W]e got the two-stage injector up and operational, modified a tank from the STAR 'boneyard' to work as the impact chamber, and figured out how to make and align joints that actually passed projectiles through them instead of turning them into debris clouds. The problem left was the projectile 'blow-by.' Without solving this problem, in-bore velocity measurements could not be made, and the railgun section would pre-arc on every firing…. I sketched it out and explained [the solution] to Lynn Barker, who said 'Sounds reasonable…' I figured out how to machine it and made one for a test. Well, to make a long story short, it worked the first time out and became one of the enabling technologies for the STARFIRE project. Another projectile innovation that led to a repeatable, self-arcing projectile included the silver-epoxy-loaded fuse (Jim's idea to use the epoxy that would ablate and create a self-timing, initiating arc) with a HIGHLY sophisticated system of Scotch® tape for the mold. All these concepts eventually went into a patent granted in June 1993 entitled 'Hybrid Armature Projectile.'

About 2 years after starting that project, the three-stage railgun was operational and was fired routinely to evaluate its performance. The highest velocity achieved was 7.5 km/s for an injected projectile at 6 km/s.

Several other staff played a major role in making this complex system work. The optimized two-stage gun performance resulting from Mohsen Shahinpoor's calculations minimized acceleration loads on the projectile before entering the railgun section. Gerry Wellman and Karl Schuler performed detailed numerical analysis of the plasma armature used for the projectile acceleration in the railgun (Wellman and Schuler 1988). Tim Trucano did numerical simulations to evaluate and minimize projectile gouging (Barker et al. 1988, 1989), and Jim Ang participated in the projectile design and fabrication (Ang et al. 1993a). Lynn Barker identified mechanisms of pitting caused by sliding metal contact (Barker et al. 1988) that were minimized through the projectile design. Although this unique launcher (Hawke et al. 1991a, b) did not achieve the intended goal of 15 km/s, it did obtain a record velocity for railguns and resulted in Ron Hawke receiving the prestigious IEEE Peter Mark Medal for Outstanding Contribution to Electromagnetic Launcher Technology. The project was discontinued in the late 1980s because of the low incremental velocity attained for the system. By that time, an alternative approach to achieving ultrahigh velocities was showing promise and proved to be successful, resulting in the Sandia HyperVelocity Launcher (HVL, also known as the three-stage light gas gun), which is discussed in Sect. 5.6.

## 5.4 Metallization of Hydrogen on the Two-Stage Gun

The enabling technology that eventually led to the success of the HVL was the outcome of a project to make metallic hydrogen using the two-stage light gas gun. After leaving Sandia in 1974 to join Terra Tek Corporation in Salt Lake City, Lynn Barker returned in 1981 as a research staff member and joined Jim Asay's experimental division. Barker initiated two important projects, both of which required developing techniques to produce shockless (ramp) loading to Mbar stresses. This persistent and personal goal stemmed from Barker's early work in the late 1960s to generate ramp waves with fused silica buffers.

Barker's two projects used vastly different approaches for achieving shockless loading. One was a Laboratory Directed Research and Development (LDRD) project to compress a solid hydrogen sample sandwiched between two tungsten plates, using a high-velocity tungsten impactor in a barrel filled with helium to prevent shock loading. The other was to develop a standard graded-density impactor for use on the two-stage light gas guns, to produce a gradual increase in stress with time up to Mbar pressures upon impact with a target, unlike the abrupt shock obtained by impact loading with a single density impactor. Both projects were extremely challenging and would seed technologies critical not only for developing the HVL but for later research efforts using magnetic loading on Z to achieve projectile velocities much higher than became possible with the HVL.

The goal of the hydrogen project was to metalize hydrogen at Mbar pressures, as predicted by several solid state theories. Previous LLNL experiments using cylindrical magnetic compression suggested hydrogen metallization would occur at about 2 Mbars, but these results were ambiguous. Barker's approach was to start with a solid hydrogen sample at 4.2 K, confined between two tungsten disks in a barrel extension to the two-stage gun. The two-stage gun barrel was filled with helium gas, which gradually compressed the hydrogen to Mbar pressures as the projectile accelerated down the barrel. Both pressure and density were to be measured during compression using a VISAR and four MeV-level x-ray machines.

However, several major obstacles had to be overcome. One was to develop a cryogenic system for producing solid hydrogen at a few degrees Kelvin and then compress it gradually to Mbar pressures. A VISAR monitoring the rear tungsten disk determined the pressure history, and four fast MeV x-ray machines determined the compressed density from the separation between the tungsten disks. The resulting thermodynamic measurements of pressure and density would allow a comparison with theory. Additional conductivity experiments would have followed to determine if the sample became metallic during compression.

The first task, to implement an operational cryogenic system (Barker et al. 1986) and provide a target fixture to hold the sample, required a couple of years and was accomplished. Barker developed a unique target insertion system in which the solid hydrogen sample was prepared in a helium cryostat that Jack Wise had developed. The target fixture containing the hydrogen sample was located immediately outside the gun chamber and was rapidly inserted before the solid sample warmed up. It was then isentropically compressed to high pressure by the projectile in the gas-filled launch tube of the gun. As Asay says in his recollection,

> Lynn conceived a novel technique that prepared the solid sample in a short barrel extension of tungsten that would become part of the two-stage barrel just before firing. The fixture was initially placed in a cryostat immediately outside the impact chamber, cooled to solidify the hydrogen, and then rapidly inserted and aligned to the gun barrel just before impact, kind of like a soft ball pitch. Lynn called this contraption a 'transient insertion mechanism' (TIM), and dubbed it the 'Tiny TIM,' – 'tiny' because it was huge. I don't know if anyone ever told him it looked like a classical Rube Goldberg contraption, although it did. However, it worked.

The second part of the project was simple in concept but difficult to implement. The goal was to compress the hydrogen sample continuously to a few Mbars. This was to be accomplished by filling the barrel of the two-stage gun with helium that was continuously compressed with a high-velocity projectile containing a tungsten facing plate in the launch tube of the two-stage gun. In early experiments, x-ray photographs showed that the front tungsten plate in the sample fixture broke into several fragments as the helium pressure increased. Although several attempts were made to eliminate this problem, they were not successful, and the final objective of making metallic hydrogen on the two-stage gun was not achieved. However, an unexpected, but important, spin-off, which involved the shockless acceleration of thin plates to ultrahigh velocities, was successful and strongly influenced future shock wave programs at Sandia.

## 5.5 Graded-Density Impactor for "Soft," or Ramp, Impact

The other project that Barker started upon his return involved a new approach for producing "shockless," usually referred to as "ramp," loading of samples in a plate impact experiment. This was a continuation of his long-term goal to develop techniques

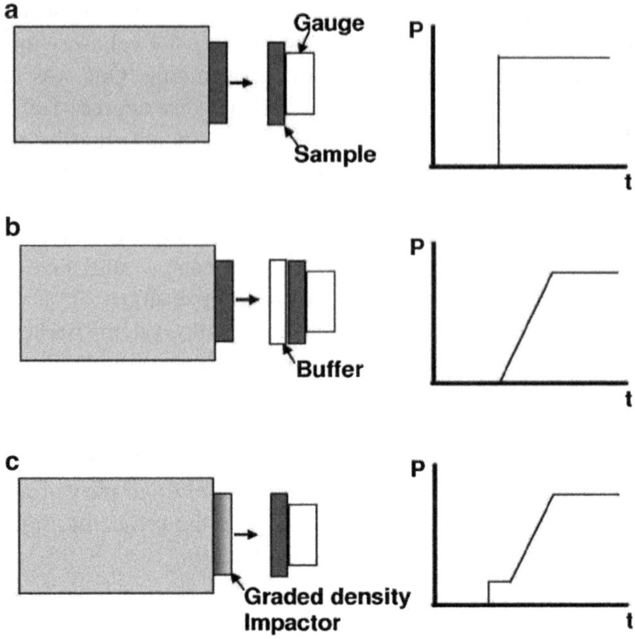


**Fig. 5.3** Impact configurations. The graphs on the *right* illustrate the stress loading history experienced by the left side of each sample. (**a**) Typical impact condition for shock loading of planar samples. (**b**) Fused silica buffer for producing shockless loading of a sample. (**c**) Graded density impactor for direct shock/shockless loading of a sample (Reprinted from Asay and Knudson 2005, Figure 10.2, with permission from Springer Science and Business Media, Copyright 2005)

for shockless compression of materials to high pressure. In 1970, he and Red Hollenbach published a paper, which showed that ramp loading could be accomplished by shock loading of fused silica. This generated a ramp wave that could be used as input to samples bonded to it. However, fused silica was limited to peak stresses of about 35 kbars, and attempts to use this concept with other "glass-type" materials were not entirely successful because of difficulties in reproducibility. The goal of several researchers in the shock wave community was to extend the peak stress achievable with ramp loading to Mbar stresses.

Barker's new approach to shockless, or ramp, loading was to vary the density of the impacting plate continuously from a low initial density at the impact surface to a high density approaching that of tantalum at the back surface. The concept for producing a small initial shock followed by ramp loading is illustrated in part **c** of Fig. 5.3 (Asay and Knudson 2005). Normal planar impact on a flat sample produces an abrupt shock loading and a single shock wave in a fluid or an overdriven shock in a solid, as shown in part **a** of the figure. Impacting a fused silica buffer produces a ramp wave in the fused silica for impact pressures less than 30 kbars, and the resulting ramp loading of a sample attached to the back of the silica. Barker's approach was to fabricate an impactor for the projectile, which had a density varying

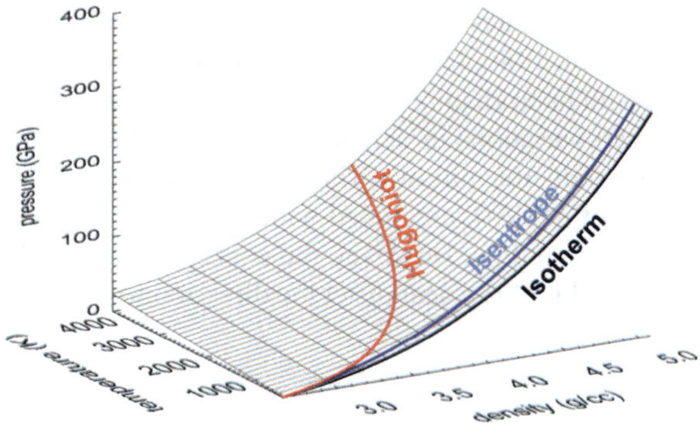


**Fig. 5.4** Temperature states achieved in LiF by shock (Hugoniot) and ramp (near isentropic) compression. Note that temperature states achieved by isentropic loading are similar to isothermal (constant temperature) compression (Private collection, J.-P. Davis)

from a low value typical of a foam on the impact surface to a high density such as tantalum on the back of the impacting plate. This produces a small initial shock in an impacted sample, followed by a gradual increase in pressure as stress waves are reflected from the higher-density regions in the impactor. Calculations showed that peak ramp loading stresses of several Mbars could be achieved.

A major advantage of ramp loading is the resulting low temperatures that can be achieved. Shock loading to Mbar pressures typically produces temperatures of a few thousand degrees Kelvin in metals. In contrast, the temperature states obtained during ramp loading remain very low compared to the shock-induced temperatures. Thermodynamic states produced by ramp waves are nearly *isentropic*, and hence the experiment is referred to as "ICE," for isentropic compression experiment. This was another appropriate acronym conceived by Barker. Figure 5.4 graphically illustrates the huge difference in temperatures obtained by shock versus ramp loading of LiF. Shocking to 2 Mbars produces a shock temperature of about 4000 K, whereas ramp loading to the same pressure results in only a couple hundred degree increase in temperature. Although different materials will exhibit different thermal responses, the major difference between Hugoniot temperature and isentropic temperature at a given density change from ambient will be similar to the results shown in Fig. 5.4.

Barker, Carl Konrad, and later Doug Scott pursued an unconventional approach for making the graded-density impactor, which is described in Barker's papers as a "pillow." He used this term to emphasize that ramp loading produces continuous, or "soft," stress loading, versus the abrupt, or "hard," change in stress obtained through shock loading. Barker and Scott prepared a suspension of several fine micron-sized powders with different densities in an alcohol solution and then deposited the powder on a flat plate through sedimentation. Sedimentation of a fluid suspension follows Stokes' law, creating a difference in sedimentation rates as a function of density, and thus allows higher density particles to settle before lower density ones. The density distribution in the final sediment (Barker 1984) could be pressed and heated into disks to use as impact plates on the two-stage gun. Through painstaking efforts, the process was perfected, and several pillows were tested to demonstrate the capability for producing ramp waves to over 2.5 Mbars. The technique was used to measure the high-pressure quasi-isentropic response of several metals, including the tungsten plates used in the hydrogen metallization experiments (Barker 1984; Barker et al. 1986; Chhabildas and Barker 1988; Chhabildas et al. 1988). Although successful in producing and measuring quasi-isentropic response to Mbar pressures, the small variations in density during manufacturing were not conducive to attain fully reproducible pillows. The difficulty in quality control, combined with Barker's retirement in 1990, precluded widespread use of graded-density impactors for ICE studies at that time. However, the concepts for shockless acceleration of thin plates established in this research set the direction for future approaches that are now used routinely.

## 5.6 The Sandia HyperVelocity Launcher

Near the end of the hydrogen metallization project (discussed in Sect. 5.4), Lalit Chhabildas and Lynn Barker initiated a research effort to understand the basic mechanisms controlling breakup of the tungsten plate in the hydrogen fixture. To determine the effects of material properties such as yield or spall strength systematically, Chhabildas performed well-controlled experiments (Barker and Chhabildas 1990; Barker et al. 1990) on different plate materials with the same mass, using the same peak drive pressures, and measured the "degree of fragmentation" of each (Chhabildas et al. 1990a). The figure of merit used to identify better candidate materials for resisting fragmentation was the average fragment size after breakup. The results were rather surprising. It was generally thought that the yield or spall strength would impede instabilities that cause plate breakup. However, it was found that a single material property did not apply for estimating fragment size or the number of fragments for gas-driven plates. Tungsten possessed the highest yield strength and yet was the least resistant to fragmentation and easily broke up during acceleration. All other materials tested also fragmented during Mbar pressure gas loading and displayed unique fragment distributions. Nevertheless, there was a strong correlation between fragment size and the spall strength divided by plate density, which Chhabildas referred to as the "specific strength" (Chhabildas et al. 1990). That figure of merit identified a very good flyer plate material, namely, a titanium alloy, because of its high spall strength and low density. As Chhabildas recalls this exciting development,

> I observed a correlation between the mean fragment sizes for each material and its specific spall strength – that is, the ratio of spall strength to density (Chhabildas, Barker, Asay, and Trucano 1990). To verify this empirical relation, I selected a titanium alloy, Ti-6Al-4V, as a candidate material to yield the least number of fragments. To my amazement, we launched the alloy intact!

That surprising result was the beginning of the Sandia HyperVelocity Launcher. Shock wave researchers had made persistent attempts, dating back to the 1960s, to launch projectiles to velocities exceeding 10 km/s (the definition of "hypervelocity") for military and scientific applications. Several approaches, including the railgun project, had not been successful. However, the understanding of plate fragmentation gained through the gas acceleration studies was a key breakthrough. The approach taken by Chhabildas and his colleagues involved transferring momentum from a slower moving, heavy two-stage light gas gun projectile to a lighter third-stage flyer plate. The nominal plate sizes and masses required for the flyer plate indicated that Mbar pressures were needed to launch the projectile over less than a microsecond time scale of engagement. Direct impact at these velocities would shock melt and possibly vaporize the flyer plate.

A *shockless* interaction between the projectile and flyer plate was therefore required. The pillow technology developed by Barker was initially used for this purpose, although it was later found by Chhabildas that a carefully designed, multilayered graded-density impactor was easier to manufacture, reproduce, and characterize and it worked extremely well. In addition, a short barrel assembly holding the final flyer plate had to be coupled to the muzzle of the two-stage gun so that the second-stage projectile did not have any free flight to develop "extreme" tilt with the third-stage flyer plate. The third-stage barrel was made of tungsten with an inner bore matching the diameter of the launch barrel of the two-stage gun, similar to that used in the hydrogen metallization project. A thin flyer plate was precisely mounted in the tungsten barrel and separated a few millimeters from the end of the two-stage gun barrel. The approach used the shockless impact of a thin flyer plate by a heavier, multi-layered projectile facing that traveled at up to 7 km/s. Upon impact, the lighter plate was shocklessly accelerated to a higher velocity, much like a heavy ball hitting a lighter one. The modified three-stage gun configuration that was produced is referred to as the Sandia HyperVelocity Launcher (Chhabildas et al. 1991), or HVL, and is schematically illustrated in Fig. 5.5. In this configuration, a graded-density impact plate produces nearly smooth, nearly shockless loading of the flyer plate to maximum pressures of 2 to 3 Mbars and accelerates the flyer plate to velocities as high as 20 km/s (Thornhill et al. 2006).

A myriad of engineering details had to be resolved before launching a flyer plate to hypervelocities. Clint Hall and Bill Reinhart vigorously pursued these issues. Hall comments:

> Before too long, the projectile manufacturing and repeatability, alignment fixturing, plate fabrication techniques, and two-stage performance had all improved. The project was eventually able to launch flat, intact plates capable of reasonably precise equation of state measurements.

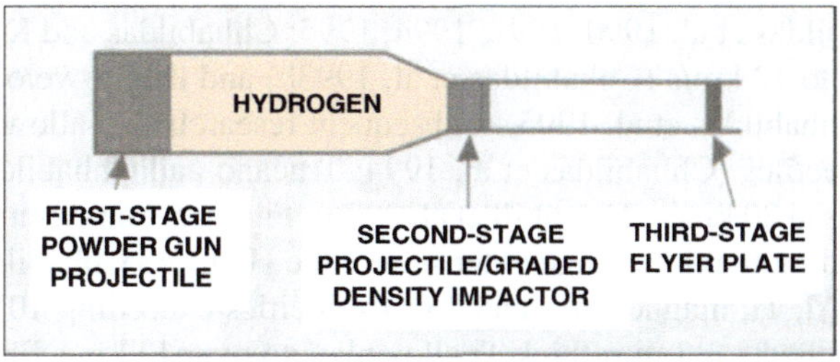


**Fig. 5.5** Three-stage gun concept (HVL) to launch thin flyer plates to hypervelocities. The second-stage projectile is launched to 7 km/s maximum velocity and accelerates the third-stage projectile with a shockless loading pressure to around a Mbar (Private collection, L. C. Chhabildas)

Making the second-stage impactors was also challenging. Originally, a simple, two-layered impactor was used, but it did not produce good shockless loading. The graded-density impactors developed by Barker produced a much smoother acceleration of the third-stage flyer plate. Through extensive modeling and testing, a properly designed multi-layered impactor was easier to fabricate and worked very well. Reinhart remembers:

> These impactors varied from simple PMMA (polymethyl methacrylate) and copper, to the Barker 'pillow' (fabricated by spraying metallurgical powders and using sedimentation techniques so a smooth variation in shock impedance occurred), to the Chhabildas 'plywood' (PMMA/Al/Ti/Cu), to the final configuration still in use, polymethylpentene (TPX)/Mg/Al/Ti/Cu/Ta.

Extensive 2-D simulations were performed to identify the configurations for the multi-layered impactor that would produce minimal shock formation and minimize 2-D effects arising during the impact. This was necessary because, at the time, DNA (the Defense Nuclear Agency, now known as the Defense Threat Reduction Agency) wanted to launch "chunky" projectiles. Understanding these effects was especially important as the launch velocity increased from the initial launch of chunky projectiles at 12 km/s to nearly 20 km/s (Chhabildas et al. 1994; Thornhill et al. 2006) with refinements in the technique. Tim Trucano performed extensive 2-D simulations that were critical to optimizing the chunky design (Trucano and Chhabildas 1993, 1995). He states:

> I feel to this day that my best and most satisfying computational prediction involved my work with Chhabildas (circa 1993–1994) to predict the results of the first experiment on the completed gun. The simulation of the interior ballistics of the performance of the third stage, and the transition of the operation from the second to the third stage, exercised all of the computational capabilities that had been developed in CTH.... Our prediction of the final projectile velocity was within a few percent of the observations, and the shape of the formed projectile was very close to the experiment results.

This remarkable achievement was the epitome of Lundergan's strategic objective of coupling experiment, modeling, and computing to achieve a goal not attainable through each independently.

The development of the HVL was a breakthrough in experimentally accessing new regimes of material response that had profound impact for both weapon and other scientific applications. In relatively short order, flyer plate velocities of 10 km/s (Chhabildas et al. 1990, 1992, 1994, 1995; Chhabildas and Knudson 2005) were increased to 12 km/s (Chhabildas et al. 1993b) and then to velocities exceeding 16 km/s (Chhabildas et al. 1995). Subsequent research also allowed the launch of chunky projectiles (Chhabildas et al. 1994; Trucano and Chhabildas 1995) and particles to around 20 km/s (Thornhill et al. 2006). Figure 5.6 shows a photo of Lalit Chhabildas holding the third-stage extension of the HVL. As Chhabildas mentions, "We launched Al, Ti, magnesium, and Ta to velocities exceeding 10 km/s. In particular, we launched a 0.5-mm-thick Ti alloy plate to over 12 km/s."

For the first time, the HVL allowed scientific study of highly expanded states of matter that had usually been produced by shock loading and unloading. These thermodynamic states are often referred to as "warm dense matter," which is important in a number of astronomical applications and is presently an area of intense scientific

**Fig. 5.6** Lalit Chhabildas at the breech end of the two-stage gun holding the third-stage extension that achieved plate velocities exceeding 10 km/s (Reprinted with permission of Sandia National Laboratories)

interest within the NNSA Stockpile Stewardship Program. Tim Trucano succinctly describes the paradigm shift:

> Because of the conjunction of funding, program drivers, research opportunities, and talent accumulated in the 1980s, the Sandia shock wave effort pursued every aspect of hypervelocity impact. These aspects included how to achieve hypervelocity impact, how to characterize it, how to simulate it both physically (i.e., experimentally) and computationally, and how to assess its consequences. This effort involved developing the experimental capabilities, theoretical understanding, computational capabilities, and specific applications – none of which had really existed prior to the 1980s. As part of the effort, Sandia became closely integrated with a revived national and international community in hypervelocity impact physics.

Early experiments on shock-induced vaporization had been performed with the two-stage gun (Asay et al. 1988; Trucano et al. 1988), but the materials that could be partially vaporized were limited to low melting point materials such as lead and cadmium, and only a small fraction of the material was actually vaporized. The HVL provided an opportunity to extend the suite of engineering and scientific materials for vaporization studies and to probe even full vaporization of zinc impacted at 5 Mbars and released (Brannon and Chhabildas 1995). Additional examples of full vaporization of indium and aluminum are given in other references (Chhabildas et al. 2006).

The HVL also played a role in attracting significant external funding from other government agencies for shock wave research. As Chhabildas says,

> JSC [Johnson Space Center] and MSFC [Marshal Space Flight Center] asked us to evaluate debris shield designs. Our first measurements to evaluate a debris shield design, known as the 'Whipple Bumper Shields,' at impact velocities of 10 km/s (i.e., space-type velocities) could now be reproduced in the lab…. Cmdr. Connell [DNA] even supported a basic research project to investigate the kinetics of shock-induced vaporization in zinc and determine the role of projectile shape on the evolution of debris generated upon impact on a thin plate.

The studies for the National Aeronautics and Space Administration (NASA) provided the first experimental data for evaluating a debris shield design known as the Whipple Bumper Shields at impact velocities greater than 10 km/s. NASA also initiated HVL experiments to simulate the effects of orbital debris known as space "junk" (Lawrence 1990b, 1992a; Lawrence et al. 1995; Chhabildas et al. 1993a; Ang et al. 1991; Boslough et al. 1993, 1994a).

In addition, as a spin-off of the original SDI effort, the Theater Missile Defense program originated a multi-year effort to use the HVL to evaluate projectile lethality at velocities greater than 10 km/s in hit-to-kill missions. Other government organizations that operated conventional two-stage guns also pursued, with assistance from Sandia, upgrades to their facilities to include this capability, as Chhabildas notes in his recollections.

> At the request of Cmdr. John Connell of DNA, the technology was transitioned to NRL and the U.S. Air Force Arnold Engineering Development Center (AEDC) for evaluation of lethality effects on space structures. Bill Reinhart was key in transferring the technology. He is well organized and good at planning and taking care of many details. The impressive thing about those facilities was that we were launching plates for over a millisecond over tens of meters in free flight, taking x-ray pictures, and hitting a target less than a meter from its intended trajectory. At AEDC we were able to launch a 50-mm-thick Ti plate weighing 7.6 g to 9.2 km/s. DNA wanted us to develop a technology to launch 'chunky' projectiles instead of plates (Chhabildas, Trucano, Reinhart, and Hall 1994). We leveraged the effort with an internal Laboratory Directed Research and Development (LDRD) project, which allowed us to achieve 16 km/s.

The launcher also provided a new experimental capability for benchmarking hydrodynamic code predictions of hypervelocity penetration, impact flash, and debris cloud generation (Ang 1990; Ang et al. 1991; Chhabildas et al. 1993a, 2006; Konrad et al. 1994; Lawrence et al. 2006). The HVL continues to be used for a variety of weapon and other scientific applications to the present time (2016).

## 5.7 Generalized Fragmentation Theories

In the late 1970s, Dennis Grady, Marlin Kipp, Karl Schuler, and others developed experimental techniques to characterize the dynamic properties of geological materials. Specific issues concerned fracture strengths under dynamic loading and how such materials, and oil shale in particular, would rubblize under tensile loading. It was noted in explosive experiments, such as shown in Fig. 5.7, that fragment sizes

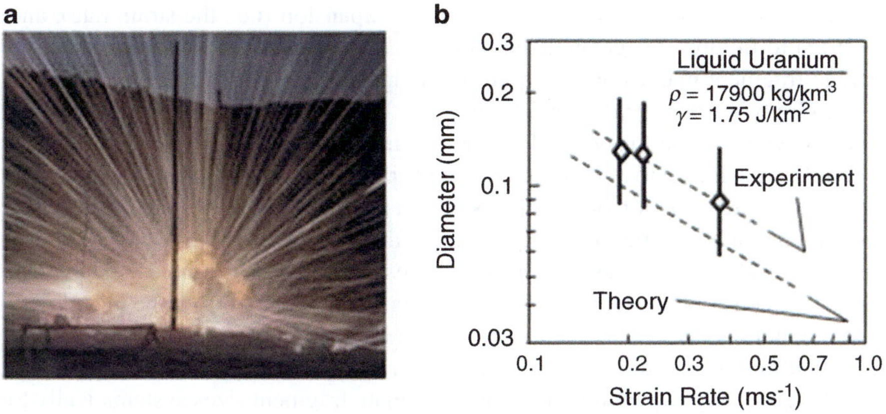


**Fig. 5.7** Explosive fragmentation. (**a**) Photo of explosively driven cylinder fragmentation. (**b**) Comparison of fragment size with the Grady–Kipp fragmentation theory (From Grady 2010b, Figure 2, Copyright 2009, reprinted with permission of Springer Science + Business Media)

produced in a medium around an explosive charge varied with distance from the explosive. As Kipp says,

> In our department, extensive static and dynamic testing of oil shale samples was pursued, and Dennis Grady and I embarked on a dynamic fragmentation study that eventually expanded to many other materials over many years until Dennis' retirement in the 1990s... Not surprisingly, fragments nearest the explosive were very fine and graded out to much larger ones as the distance from the charge increased. Such a size disparity was not going to be good for in-situ mining and retorting, and work by others was pursued to reduce the loading rates and thus the formation of fines...

Since the explosive pressure, and thus the rate of loading (strain rate), depends on peak stress, higher loading strain rates close to the charge meant smaller fragment size, thereby suggesting a strong correlation of fragment size with strain rate.

Analytic models did not exist at that time for describing the relationship between strain rate and fragment size. However, from these and other studies, Grady developed a generalized energy- and momentum-based model that effectively predicted the fragment sizes for specifically applied strain rates (Grady 1981a, 1982a, b, c, 1993, 2010b). The model was based on conserving kinetic energy and momentum in an expanding center of mass system. Since the total energy of the system, such as an expanding shell, is not available for fragmentation, the energy is separated into center of mass energy coordinates and the local kinetic energy of the fragments that are created. The local kinetic energy is available to produce the surface energy of newly created fragments and their local kinetic energy. This approach results in a quantitative expression among the fragment size, the surface energy, and the expansion strain rate. Grady specialized the relation to ductile materials using the yield strength to quantify the surface energy needed in the model for predicting average fragment size and to brittle materials using fracture toughness for determining the energy required to form fragments (Grady 2006). Using fundamental material properties, such as tensile strength, the rate of expansion (i.e., the strain rate), and the strain to failure, this generalized model was remarkably accurate in predicting average fragment masses in several different applications (Grady 1981a, 1982a, b, 1990). Grady's fragmentation model was implemented in 2-D and 3-D hydrodynamic codes (Kipp and Davison 1982; Kipp and Grady 1983; Trucano et al. 1990) by coupling the model to output routines from code simulation events. This approach allowed the determination of fragment size in a large number of system applications through computer simulations that determine the loading conditions for producing fragments, i.e., the peak stress and rate of expansion. An example is shown in Fig. 5.7, which compares the average fragment sizes predicted from the model with experimental results.

In addition to experiments using explosives, Grady developed a new experimental technique using expanding metal rings to relate fragment sizes systematically to the applied tensile loading rate (Forrestal et al. 1978). These experiments established the fundamental relation between fragment size and strain rate and accelerated development of the fragmentation model for metals. As Kipp notes,

> A natural extension of the strain rate dependence of fragmentation was to metals. The first effort was directed towards ring expansion (done electromagnetically by Dennis Grady and Dave Benson) and development of models to represent the fragment formation. I set up an equivalent 1-D expansion of the ring in WONDY with a uniform velocity gradient to simulate the ring conditions and then randomly inserted initiation points with random initiation thresholds. The fracture model was set up to dissipate the proper energy during stretch, and the results agreed well with the data: even with large numbers of initiation points, there was only enough energy to bring a few to completion and obtain the small number of fragments of random sizes seen in experiments. Dennis Grady published a seminal paper on fragmentation that formed the basis of our thinking over many years for ductile, brittle, and liquid fragmentation. With Jeff Swegle's help, the model was implemented in CTH as a monitoring tool (that is, when the fragments were defined to have formed, the material was not agglomerated into actual fragments; instead, the totals could be summarized in bin charts, and the distributions could be obtained by post processing). We found that, if we took the average fragment size from the simulations and imposed an exponential distribution on those averages, the actual distribution could be reasonably represented. It became a powerful tool, and I used it extensively for the remainder of my career on a wide variety of fragmentation–formation problems, ranging from fine liquid droplets to explosive casing expansion. The dependence on strain rate explained clearly why 'pipe-bomb' like explosive devices (characteristic of most fragmenting rounds) broke into strips, since the axial strain rate was typically much smaller than the circumferential strain rate, with large strain rates giving smaller average dimensions than small ones. Spherical devices, on the other hand, with a centrally-initiated spherical explosive charge inside a thin external metal shell, would lead to nominally square fragments. In that case, strain rates would be uniform in all circumferential directions – a result confirmed by radiographs.

Grady's fully developed model was published in 1981 (Grady 1981a). Earlier studies that had been conducted provided substantial data for developing the model (Grady and Hollenbach 1977; Grady and Kipp 1979; Grady et al. 1978). Later papers dealt with numerical aspects (Kipp et al. 1993; Grady and Kipp 1985a) and were used to validate the analytic model.

Surprisingly, the dynamic fragmentation model was found to be of greater applicability than originally intended. In a specific application of scientific interest,

Holian and Grady (1988) used molecular dynamics (MD) computer simulations to study microscopic fragmentation of condensed matter. In the MD simulations, matter expanded adiabatically, leading to fragmentation into distinct clusters of several atoms. A Lennard-Jones pair potential was used to provide an attractive well for atomic interactions. The simulations established that the homogeneous distribution of cluster masses was an exponential distribution as observed and that the average cluster mass could be reasonably well predicted by the Grady fragmentation model. The numerical simulations of how atomic clusters are formed during rapid expansion of the system provided insight into the nature of fragment distributions and also provided reasonably good agreement of average fragment mass with estimates from the fragmentation model.

Grady and his colleagues also used the model to estimate the distribution of galaxies resulting from the "Big Bang." To make these estimates, they made several simplifying assumptions about gravitational forces and uniformity at the instant of the Big Bang to determine how matter fragmented into galaxies later in time. It was assumed that the universe expanded uniformly. The observed luminosity of galaxies in the universe was used to estimate their effective masses and to compare with the calculations. Figure 5.8 shows the agreement between the fragmentation theory and the existing distribution of galaxies, assuming that the galactic luminosity is proportional to absolute mass (Brown, Karpp, Grady 1983). The results illustrate the potential of the theory for estimating fragmentation on a very large scale; however, not knowing the cause of the effective fracture energy (i.e., gravity, dark energy, or dark mass) limits the accuracy of this comparison.

An even more impressive example of the fragmentation model's power was provided at atomic scales. In this case, Grady used the model to estimate the fragments (atoms) created by the collision of 70-MeV nuclei of carbon and silver. The number

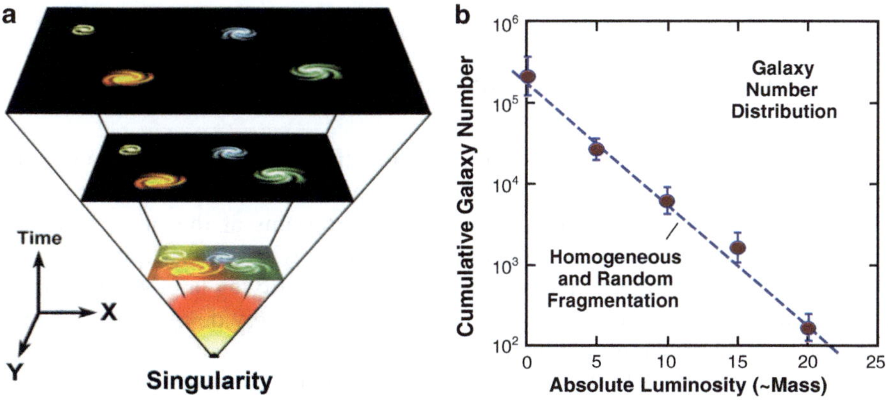


**Fig. 5.8** Fragmentation in the large, i.e., the "Big Bang." (**a**) Graphic representation of expansion (Reprinted with permission from Fredrik, Wikimedia Commons 2004, http://commons.wikimedia.org/wiki/File%3AUniverse_expansion.png, in the public domain). (**b**) Comparison of fragmentation theory and universe expansion data (Reprinted from Brown, Karpp, Grady, 1983, Figure 4, Copyright 1983, with permission from Springer Science+Business Media)

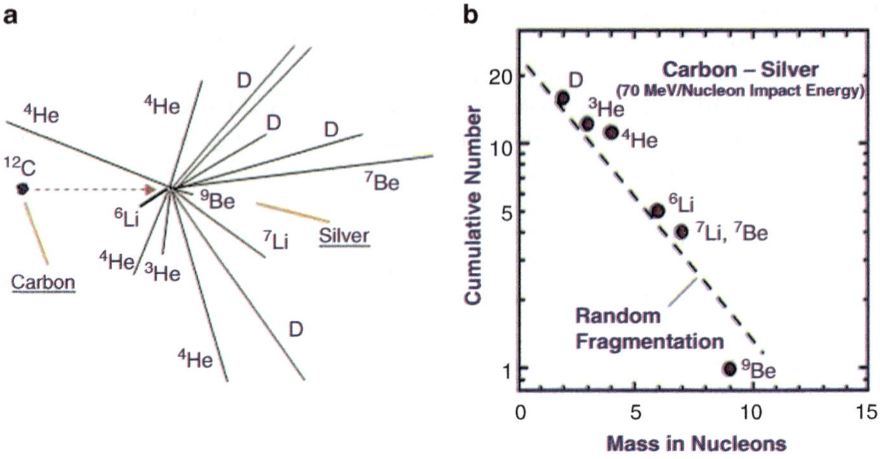


**Fig. 5.9** Fragmentation in the small. (**a**) Collision of carbon and silver nuclei. (**b**) Cumulative isotope number distribution from collision (**a** and **b** are from Figure 4 of Grady 2010b, Copyright 2009, reprinted with permission of Springer Science + Business Media)

of different atoms created in the collision, as illustrated in Fig. 5.9, was measured with a photographic emulsion. A quantitative measure of the atomic fragments was obtained by using the nuclear binding energy as the resisting fracture energy. The figure illustrates the extremely good agreement with the experimental results (Grady 2010b). The above examples illustrate the extreme range in scale for applicability of the fragmentation model. These examples and the theoretical molecular dynamics simulations establish the credence of the model for predicting atomic fission, the breakup of oil shale, the destruction of armor, and the distribution of galaxies.

## 5.8 Laser Windows for Mbar Profile Studies

In the 1980s, a major objective of laser interferometry was to determine near in-situ particle velocity profiles for arbitrary loading at impact stresses of 1 Mbar or more. These pressures were dictated by the weapon programs at the other national security laboratories and by the scientific goals for studying material properties in the vicinity of shock-induced melting. This required identifying laser windows that would remain transparent at these high stress levels and provide a relatively good impedance match to samples of interest.

The seminal work of Barker and Hollenbach in 1970 that determined the optical properties of fused silica, PMMA, and sapphire was the only available information at the time, albeit limited to low pressures (Barker and Hollenbach 1970). In the mid 1970s, Larry Lee performed a scoping study of laser window materials through several experiments on sapphire, LiF, and PMMA to explore the peak stresses for transparency (Lee 1976); the peak stresses were about 0.5 Mbar in sapphire using the powder gun. However, the results were not designed to analyze refractive index changes. In the late 1970s, Chhabildas (Chhabildas and Asay 1979) demonstrated the use of PMMA windows to about 220 kbars. In the early 1980s, Jack Wise investigated several laser window candidates for VISAR use at Mbar pressures. At stresses exceeding several hundred kbars, most candidate window materials are compressed to completely plastic states of stress, making the task of identifying good materials quite challenging. Several materials were explored, including fluids (Wise 1984). Single-crystal sapphire was found to not be a good window because of elastic yielding near 200 kbars. Wise and Chhabildas (1986) discovered that single-crystal LiF (<100> orientation) worked well up to 1.2 Mbar, even though the window yielded elastically at a few kbars. They did a careful optical study of LiF, which illustrated a linear dependence of the refractive index with density to around 1.2 Mbar. Hayes later showed that a linear dependence ensures that the constant correction found for LiF laser windows can also be used for arbitrary waves (Hayes 2001). This relationship is expected to hold for arbitrary loading conditions, providing there is no temperature dependence to the index. Recent experiments by Paulo Rigg and colleagues, however, have demonstrated a temperature dependence on the refractive index in the multi-Mbar range, which must be taken into account in analyzing wave profiles at those stress levels (Rigg et al. 2014).

Furnish et al. (1999) further found that LiF could be used to attain shock wave stresses of about 1.8 Mbars in LiF, although shock-induced absorption of the laser light produced a time-dependent loss of transparency. The transparency loss results from the closure of the electronic band gap caused by shock temperature increases. This assumption is supported in recent studies by Fratanduono et al. (2011), who found that LiF, when compressed by a ramp wave to 8 Mbars, does not lose transparency, apparently because of the lower temperature. Furnish and colleagues also found a linear dependence of refractive index for LiF on density to within their expected uncertainty levels. Setchell further demonstrated the linear dependence of the refractive index with density for Z-cut single-crystal sapphire below its yield point during shock loading and release at low shock temperatures (Setchell 2002). LiF continues to be the most versatile transparent window material for dynamic shock interferometer studies and is usable to pressures of at least 8 Mbars in ramp wave experiments. Similarly, PMMA was also found to be transparent for use as a window material at stresses up to 40 GPa under ramp loading (Anderson, Chhabildas, Reinhart 1998).

The LiF laser window has become a workhorse for high-pressure material studies on Z. A major application has been the use of the window to determine the input pressure history directly for samples in ramp experiments to allow Lagrangian wave analyses. The concept is illustrated in Fig. 5.10. A stripline generator magnetically generates a ramp wave propagating through a LiF window mounted directly on one of the electrodes. A tantalum sample with LiF on its back surface is mounted on the other electrode and compressed by the similar ramp wave propagating through the opposing electrode. Measurement of the particle velocity profile at the back surface of an aluminum electrode used for magnetic loading determines the input in-situ wave profile to a sample mounted on the opposite electrode using a backward analysis technique originally developed by Dennis Hayes (Hayes et al. 2004). Using a transfer function

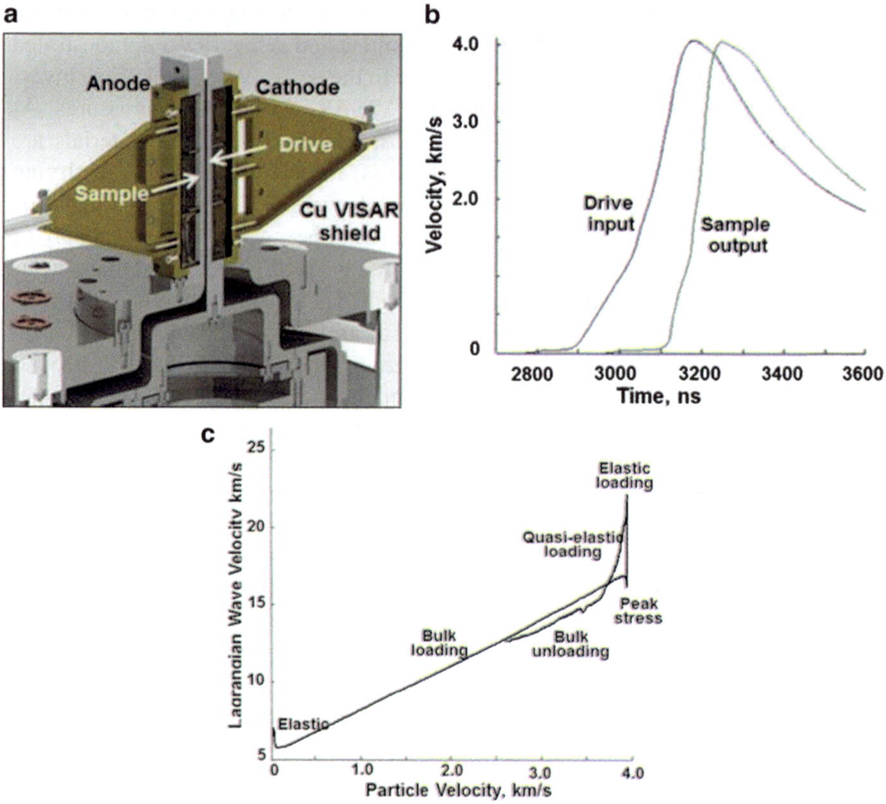


**Fig. 5.10** Ramp wave experiment conducted on Ta at Z using LiF windows. (**a**) Sample configuration. (**b**) Velocity profiles obtained on the drive and Ta sample with LiF windows. (**c**) Wave velocities for loading and unloading (Private collection, J. L. Brown)

convolution developed by Justin Brown (Brown, Alexander, Asay et al. 2013) on the sample-LiF data provides an accurate determination of the in-situ particle velocity profile in the sample. After using the transfer function to estimate the input and output particle velocity profiles for the sample, it is possible to determine wave velocity as a function of particle velocity for both loading and unloading. The two profiles in Fig. 5.10b continuously define the Lagrangian wave speed, or essentially the sound speed after correction for density compression, as a function of particle velocity, as shown in Fig. 5.10c, and hence provide an accurate measurement of the stress-density curve using the equations of motion. The wave velocity for loading and unloading provides an estimate of the bulk sound speed as a function of pressure in the plastic regime, and the triangular portion of the unloading velocity can be used to extract the flow strength of tantalum and other materials at high pressure (Brown, Alexander, Asay et al. 2014a; Brown, Knudson, Alexander, Asay 2014b). LiF windows have been used to pressures as high as 5 Mbars for high impedance materials like gold under isentropic loading and to about 3 Mbars for low impedance materials like deuterium. Several applications will be discussed in the following chapters.

## 5.9 Shock-Induced Solid State Chemistry

Most shock wave research conducted prior to the mid 1980s explicitly addressed mechanical and physical processes occurring during shock compression. In 1979 Lee Davison and Bob Graham published a comprehensive review of the shock wave research being conducted worldwide (Davison and Graham 1979). Graham noted the attempts to understand chemical processes occurring under shock compression being conducted by several laboratories in the Soviet Union. Based on this finding, he initiated a broader evaluation of the literature and, as Bruno Morosin states in his recollections,

> ...[H]e was astonished to find a well-developed scientific effort reporting chemical changes and reactions. This was in contrast to the prevalent opinion of most U.S. experts who agreed that chemistry could not occur on the time scale of shock compression in normally inert solids.

That significant finding led to the formation of a major effort at Sandia focused on the topic of shock-induced Solid State chemistry, which became known as "shock chemistry."

Graham formed a core of researchers with diverse technical backgrounds to address different technical issues related to shock chemistry. This group included himself for shock and explosive physics, Bruno Morosin for x-ray diffraction and general material properties, Gene Venturini for electron spin resonance and other magnetic properties, Marty Carr for electron diffraction, Mark Boslough for geophysical and planetary physics, W.F. Hammetter for phase transformation and chemical effects, J.E. Smugeresky for microstructural analyses, R.T. Cygan for nuclear magnetic resonance (NMR) spectroscopy, Bob Setchell for shock-induced chemical reactivity, I.K. Simonsen for synthesis of intermetallic compounds, Y. Horie for material modeling, Ed Beauchamp for ceramics and sintering effects, and Dave Webb for numerical simulations. Several other Sandians also participated, based on their technical expertise for specific materials.

The initial research efforts were concentrated on several ceramics, including titania, alumina, and zirconia, as well as on iron pyrite, to see whether shock modification produced reproducible chemical effects. The research was later broadened to other materials. A broad range of technical issues was addressed to identify the chemical effects responsible for the many physical and chemical changes observed in these materials during shock compression. Graham led the effort to understand the interplay between shock compression and chemical reactions. Several Sandia research results on shock chemistry were published beginning in 1984 that covered a decade and involved a large number of investigations.

One of the early goals of the shock chemistry program was to carefully design shock recovery fixtures that produced well-defined pressure and temperature states in shocked powder samples to ensure reproducibility of results (Graham and Webb 1984, 1986). After perfecting the design of the recovery fixtures, several powders of various materials and mixtures were studied using explosive loading with different-sized fixtures to control loading conditions. The size and configuration of the sample fixture introduced considerable local interactions and temperature variations in the shocked sample. Following shock loading and recovery, the compacted materials were extensively examined for chemical changes (Graham et al. 1986a, b; Dodson and Boslough 1990). X-ray diffraction was a major diagnostic used to assess the physical changes, as Morosin states:

> The three initial diagnostic tools were conventional x-ray patterns as well as the determination of line broadening of diffraction peaks, the intensity and shape of the electron paramagnetic resonance signatures, and the determination of dislocation densities of the transmission electron microscope images. Consistent and reproducible diagnostic results were obtained from the tiny, but carefully separated samples unloaded from various locations of the capsule as well as between different loading configurations when merged with values given by Webb's 2-D numerical calculations for the sample pressure and temperature.

A large number of experiments were conducted that systematically addressed several thermomechanical and chemical issues. A variety of material powders, including organic, inorganic, and geophysical minerals and particle morphologies, were shock loaded and recovered (Hellmann et al. 1984; Graham et al. 1986a, b; Boslough et al. 1986a; Smugeresky et al. 1988; Graham 1988, 1989). This was followed by different physical and chemical analyses on the post-shocked powders. Specific investigations included x-ray diffraction to determine morphology (Morosin and Graham 1984, 1986; Morosin et al. 1986), changes in optical properties (Boslough et al. 1986b), electron microscopy to determine electronic states (Graham and Carr 1986; Myers et al. 1986), magnetic effects and microwave loss measurements to address dissolution effects (Venturini et al. 1986, 1988; Cygan et al. 1989), nuclear magnetic resonance spectroscopy (Cygan et al. 1990; Cygan and Boslough 1994), and exothermic energy release (Hammetter et al. 1984). These extensive and coordinated experiments and diagnostics were critical and addressed many issues related to shock-induced chemistry, including chemical reactivity during shock loading (Setchell 1986; Morosin et al. 1988a, b); the ability to synthesize metal powders, such as nickel–aluminum alloys (Hammetter et al. 1988); and the synthesis of specific solid phases (Morosin and Graham 1984; Morosin et al. 1988a, 1992; Graham et al. 1986a, 1988a, b; Boslough and Graham 1985; Venturini et al. 1986; Simonsen et al. 1986; Horie et al. 1986a, b; Boslough 1989, 1991, 1992). Numerical simulations were also integrated with the experimental research to define thermodynamic states and mesoscale loading processes responsible for chemical changes. Specific examples include the modeling of Ni–Al powders (Taylor, Boslough, Horie 1988) and the modeling of chemical processes (Horie and Kipp 1988). Both explosive loading and projectile impact were used for these experiments. Figure 5.11 shows a photo of Bob Graham and Bruno Morosin preparing for a gas gun experiment.

In his recollections, Morosin captures the rapid progress made in the program:

> By 1987, this core group had completed and presented eight short papers at the June 21–26 American Physical Society Shock Conference, dominating the shock chemistry area, and these papers were subsequently included in the conference proceedings. These included shock-induced reactions of specific inorganic compounds, defect formation and the resulting effects, such as enhanced catalytic action, the initial evaluation of line broadening of x-ray diffraction patterns, and enhanced sintering characteristics, all of which resulted from shock loading of selected single-phase inorganic compounds as well as a more complicated clay (nontronite), with its implication of evolution for planetary surfaces, namely that of Mars. The major positive step demonstrated that, unlike the Soviet experiments, the Sandia results were very reproducible and consistent with the pressures and temperatures suggested

**Fig. 5.11** Bob Graham (*left*) and Bruno Morosin ready a gas gun experiment (Reprinted with permission of Sandia National Laboratories)

> by the preliminary calculations of Dave Webb. As time went on, these three-dimensional calculations were refined and samples were carefully taken from the recovery capsule as a function of location, subjected to various experiments, and shown to be consistent among various experimental shock configurations. In addition, Bob Graham, with the discussions and modeling results of Yuki Horie, was able to arrive at a conceptual chemical reaction model for shock chemistry, a process involving turbulent local mixing, possible massive defect generation, a short, rapid transit pulsed temperature excursion – all of which were caused by the transit pressure pulse.

The extensive, careful, and well-coordinated set of experiments conducted over a decade through the shock chemistry program contributed significantly to the understanding of thermomechanical, physical, and chemical processes occurring during the shock compression of organic and inorganic powders. Morosin emphasizes that

> One must recall that this was the era of President Reagan, the perceived Soviet threat, and the 'Star Wars' initiative. Out of the shock chemistry effort came the concept of a radically different, long life, inert battery that could be activated upon impact for possible fusing capabilities, that is, essentially a shock wave-activated battery.

## 5.10 Piezoelectric Polymers for the Bauer Shock Gauge

Electrical effects in shocked polymers and elastomers were identified in early shock wave experiments (Graham 1979b, 1980), resulting in the use of polymers as shock wave detectors (Champion and Benedick 1968). Later, in the 1970s, a polymer, polyvinylidene difluoride (PVDF), was found to be piezoelectric, first by Japanese researchers and then by a French scientist, Francois Bauer. As a graduate student, Bauer had conducted careful and systematic research on PVDF at the Franco-Allemand Institut de Recherches de Saint-Louis (ISL) and developed stringent material controls and experimental configurations for its use as a shock wave gauge. This novel work became known to Sandia scientists at an international conference. As Ray Reed recalls:

> At an American Physical Society symposium in Albuquerque, we first met the French scientist, Francois Bauer, who presented a paper on his thesis work with a new Japanese-discovered thin film piezoelectric polymer material, polyvinylidene difluoride, or PVDF. Bauer was familiar with Graham and Benedick's work on X-cut crystalline quartz. Benedick spoke French. Bauer, then a graduate student, needed support for his thesis claim of reproducible dynamic properties from his special processing of PVDF. He convinced a skeptical Graham to test the claimed (but implausible) superior material properties using Sandia's precision gas gun facility. Graham's gas gun tests on Bauer's PVDF were convincing. Graham convinced (equally-skeptical) me, as well. From that encounter, formal symbiotic relations developed among Sandia, Ktech, and the Franco-Allemand (French-German) Institut de Recherches de Saint-Louis (ISL) in France.

This work provided the impetus for accurate shock wave measurements using the PVDF stress gauge and was adopted at Sandia and elsewhere as the standardized "Bauer PVDF shock sensor" (Lee, Williams, Graham, Bauer 1986; Reed and Greenwoll 1989; Bauer et al. 1992).

The PVDF shock sensor is a 25-μm-thick piezoelectric polymer with a well-characterized shock response for loading under uniaxial shock compression to nominal stress levels up to ~250 kbars, although it has been used to higher stresses with less reliance (Anderson, Chhabildas, Reinhart 1998). The charge release during shock loading is accurately measured in the "current mode" with a precision, high-frequency, current-viewing shunting resistor. The time derivative of piezoelectric charge release is measured as a voltage drop across a precisely known resistance, providing a direct measure of the rate of stress loading induced in the gauge. Numerical integration of this measurement over time gives the charge-time profile; the as-recorded stress rate measurement provides unusual sensitivity to shock reflections through the 25-μm-thick sensor and electrical insulation layers. The charge-time profile is related to stress time from the characterized charge–stress relationship; numerical integration then gives velocity–time and displacement–time profiles.

When used in the current mode, the signal produced is proportional to the differential particle velocity across the gauge, or the rate of change of velocity with time (the acceleration). Since particle acceleration is a direct quantity in the differential momentum equation for wave motion, use of the gauge enhances the accuracy of obtaining stress–strain relations for loading by a stress wave. Integration of the acceleration profile provides the particle velocity profile normally measured in interferometer experiments. The ability to measure acceleration directly is a major advance for shock wave instrumentation.

The PVDF gauge found significant use in a variety of applications, including characterization of radiation loading and non-energetic response of energetic materials (Graham et al. 1988a; Lee, William, Graham, Bauer 1986; Lee, Graham, Bauer, Reed 1988; Lee, Hyndam, Reed, Bauer 1990; Lee, Johnson, Bauer et al.

1992; Bauer et al. 1992). Because of phase changes that are prevalent in polymers at high stresses, PVDF is mostly useful for shock stresses less than 200 kbars. It has proven valuable in the design of materials for weapon applications. PVDF-based accelerometers have also been characterized, with the most thorough work believed to be by Forrestal and colleagues (2003). Several workshops were held from 1990 to 1995 to present critical findings on the performance of this sensor.

## 5.11 High-Fidelity Ferroelectric Models

A basic understanding of shock-induced ferroelectric response began with the work of Neilson and Benedick in the 1950s and 1960s and rapidly progressed over the next two decades. This resulted in a detailed fundamental understanding and integration of the results into models for use in simulations. Key researchers involved in this long and extensive experimental effort included Bill Benedick, Bob Graham, Peter Lysne, Frank Neilson, and Bob Setchell, although many others also participated.

To bring the large amount of experimental and theoretical data together into practical models for many components, Steve Montgomery was hired in 1975 by an exploratory firing systems development group to continue work on the analysis of explosively driven ferroelectric current sources. Montgomery quickly learned shock wave physics and identified the key issues to be resolved for component applications. As he recalls,

> While waiting on my security clearance, I studied the pioneering work conducted at Sandia on shock compression of dielectrics. I also became familiar with the basics of shock wave compression by reading review articles and a set of class notes prepared by Dennis Hayes.

Montgomery further recalls some of the early developments at the time:

> Frank Neilson had proposed using the displacement current generated by rapid changes of electric polarization in materials under shock compression as a power source a few years prior to 1960. Research groups at Sandia conducted extensive experiments on a number of different materials, using explosive loading and gas guns, to study this idea in the following decade. For example, the nonlinear piezoelectric response in X-cut quartz crystals was used as a stress gauge for materials loaded in shock compression, and the depolarization of polycrystalline ferroelectric ceramics under shock compression was exploited to develop a variety of shock-wave-activated power sources. Applied research, with the goal of an improved understanding of material polarization changes in ferroelectric ceramics, continued with Pete Lysne by conducting experiments and developing simple models for electrical response and with Pete Chen in developing more detailed coupled electromechanical constitutive models for use in numerical simulations. I was asked to work with Lysne and Chen to gain an understanding of what they were doing and to develop application models.
>
> Chen and Lee Davison had developed a one-dimensional (1-D) response model for coupled mechanical and electrical response when the direction of material polarization and material motion were identical. This special configuration was called axial mode operation. Jeff Lawrence and Davison built a numerical model for this special case in the 1-D wave-propagation code WONDY. My group was interested in what was called the normal-mode

> configuration, for which the polarization direction was perpendicular to the direction of material motion. I converted the axial-mode response model in WONDY to a normal-mode response model and then applied the model to develop a complex device of interest to my group. The device relied on using a previously established explosive loading device to generate the desired shock compression for depolarization of ferroelectric ceramics. The culmination of this work was an 'admiral's test' of the device with our vice president in attendance for that first test. To my great relief and that of the three levels of management above me, the electric output from the device was reasonably close to the result I predicted and development engineers took over the project. After the demonstration of the ferroelectric device, I was asked to take primary responsibility for the firing system to be used in the Pershing II earth penetrator.

However, when the Pershing II project was canceled due to an international arms limitation agreement, Montgomery became involved in other Sandia activities.

Over the next few years, Montgomery was engaged in various shock application problems, particularly through a short stint in Walt Herrmann's Solid Dynamics group. However, he returned to a focus on ferroelectrics in the mid-to-late 1980s. As he further recalls,

> I returned to the study of polarization changes by shock compression in ferroelectric ceramics in 1986. The groups at Sandia developing neutron generators were managed by Dennis Hayes, and he wanted to supplement the prototype testing approach with simulations to understand issues encountered during that development. At the time, a systematic study of the mechanical and dielectric properties of ferroelectric ceramics used in components did not exist. Lalit Chhabildas initiated an experimental study of ceramic response under shock compression. I investigated methods to permit multi-dimensional simulations of the components. As a result, a simulation framework was developed in the SUBWAY code in collaboration with Pat Chavez....
>
> Computational capability was advancing, and three-dimensional (3-D) simulations were now feasible with the availability of the CDC, Cray, DEC, and SUN systems at Sandia. I decided to develop an integrated 3-D finite element framework for SUBWAY and teamed with Lee Taylor to use the explicit finite element code Pronto to replace the functions of TOODY. Early 3-D simulations of the device provided mixed results. Some aspects, particularly the ability to simulate the electric field in the ceramic, were well received by the development group. However, while the electric output from the simulations was similar to the measured output, there was room for improvement.
>
> My interest in the ferroelectric ceramic work and application modeling motivated a transfer to the neutron generator group in summer 1992. Along with supporting several neutron generator development programs, I coordinated work to improve simulation capability by developing a fundamental understanding of ferroelectric ceramics and encapsulation materials under operational conditions.

Until his retirement in 2011, Montgomery improved and refined the ferroelectric model to the point where a variety of complex, but controlled, experiments could be accurately modeled. These included (1) quasi-static experiments by Dave Zeuch (Zeuch et al. 1992, 1993, 1994, 1999, 2000) to understand the dependence of domain switching on crystal orientation; (2) gun experiments by Bob Setchell, Lalit Chhabildas, Mike Furnish, and Mark Anderson to understand coupled electromechanical response; (3) 2-D and 3-D impact experiments with the appropriate instrumentation (Furnish et al. 2000) by Mark Anderson, and Bob Setchell (2003, 2005, 2007; Setchell et al. 2000, 2006, 2007a, b) to understand effects of complex loading; and (4) ramp loading experiments by Jack Wise to further examine loading path dependence.

The results of this extensive experimental work provided the basic data for improvement of the material response models developed in collaborations between Montgomery, Rebecca Brannon, and Joshua Robbins (Montgomery et al. 2002). The result of this all-encompassing program was that a compressive ferroelectric model was generated that reliably predicted component response for use in extremely large computer simulations. The vastly improved understanding of ferroelectrics and PZT 95/5 in particular led to refinements in the computer simulation code, SUBWAY, which is now being used for routine simulations of power supplies used in weapon systems (Montgomery et al. 1995, 1996). As Montgomery notes, "I was routinely conducting simulations…on massively parallel computers with around five million elements, and several component simulations used element numbers ranging in tens of millions." The noteworthy accomplishments of this group of researchers, dating back to the 1950s, emphatically demonstrate the culmination of extensive experimental testing, fundamental model development validated through modern 1-D to 3-D experiments, and the advent of massively parallel computing, which is yet another example of Lundergan's strategic vision coming to fruition.

## 5.12 CTH: A Robust 3-D Hydrodynamic Code

CTH became a principal hydrodynamic code for complex shock wave problems at Sandia and elsewhere in the early 1990s. Development of the code benefitted from continuing management support over several years and the exceptional technical expertise of the staff who worked on it. CTH was the brain child of Sam Thompson who had a long history of developing codes for shock wave applications. Thompson joined Sandia in 1966; his early work involved development of analytic equations of state with Dave McCloskey. Toward the end of the 1960s, he joined Walt Herrmann's growing Solid Dynamics Research Department. The first code for shock wave problems that he developed was the 1-D Lagrangian CHARTD (Thompson 1969, 1970, 1973; Thompson and Lauson 1972). CHARTD was used for a variety of shock wave and radiation problems during this period. In the 1970s, Thompson made several improvements to CHARTD and also implemented the development of 2-D Eulerian hydrodynamic codes, resulting in several versions of the 2-D Eulerian hydrodynamic code, CSQ (Thompson 1976, 1979). In 1980, Thompson was promoted to supervisor of the Thermal Hydraulic Analysis Division. This left a void in the development of multi-dimensional hydrodynamic codes in Herrmann's group, especially for the development of a robust 3-D computer code to handle the increasing demand for realistic simulations of weapon problems and other scientific applications. In addition, support for the 2-D code CSQ waned, and there was no serious effort on 3-D code development. (See Gene Hertel's recollections for an extensive discussion of code development activities at Sandia.)

In 1982, Orval Jones was promoted to Vice President responsible for field engineering, weapons analysis, development testing, and fabrication laboratories. Walt Herrmann was promoted to replace him as Director of Engineering Sciences. In turn, Herrmann promoted Lee Davison, then a supervisor in the Physics of Solids Research Department, to fill his previous position as Manager of the Solid Dynamics Research Department. One of Davison's insightful decisions was to entice Sam Thompson back into the department to manage code development efforts. As Tim Trucano recalls in his recollections,

> Lee convinced Sam Thompson to leave the nuclear reactor safety code development area and return to revive CSQ and begin to develop a serious three-dimensional (3-D) code capability, with a code that would be called CTH. When Sam returned, he brought Mike McGlaun back with him. I ended up working very closely with Sam and Mike during the rest of the decade as well, because it was their computational tools that enabled me to pursue most of the computational work....

Mike McGlaun, who had originally worked on cratering studies for Al Chabai, remembers his new assignment under Thompson: "I came back to the shock physics group in 1986 to design and build CTH. We decided to develop a tightly integrated suite that modeled 3-D, 2-D, and 1-D problems."

In late 1988, Davison changed jobs from the Solid Dynamics Research Department to the Engineering Sciences Department, also under Herrmann. Dennis Hayes became manager of the Solid Dynamics Research Department and, during his 1-year tenure, promoted a robust code development effort. As Hayes recalls,

> In 1988–1989 I moved to the shock physics department. An important change was taking place in the computing world. Parallel computing had arrived, and we staffed that area to implement shock codes and made good progress. Another related effort was transforming the CTH code from a vertically integrated project to a full-fledged software engineering project, with benefits that exist to this day.

Rapid progress was made on 3-D CTH development under Hayes' leadership. In addition to CTH, he implemented a new code development project with James Peery, initially as project leader, to build a 3-D magnetohydrodynamic (MHD) code in partnership with the Pulsed Power Sciences Center. This code was to be the workhorse for analyzing high-temperature plasma states being studied in that Center. The code was initially called RHALE for Robust Hydrodynamic Arbitrary Lagrangian Eulerian, but it was later changed to ALEGRA for Arbitrary Lagrangian Eulerian for General Research Applications (Haill et al. 2003; Robinson et al. 2008, 2011). As it turned out, in the 2000s, this code had a major influence on the high energy density research being conducted on Z and it greatly influenced the transformational effect of shock wave research being developed on that pulsed power machine.

After Hayes left the Solid Dynamics Research Department in 1989 for an assignment in Washington, Jim Asay was promoted to become manager of the department. Phil Stanton was promoted to replace Asay as manager of the experimental division. Sam Thompson also left the department at that time, returning to research on reactor safety, and Mike McGlaun was promoted to supervisor of the code development branch. Paul Yarrington managed the code applications branch in the department, which was primarily engaged in modeling geological materials. At the time, the Defense Advanced Research Projects Agency (DARPA) had just initiated a multi-lab competition to develop a robust 3-D hydrodynamic code intended for use by all Department of Defense (DOD) laboratories and contractors to maintain the best modeling capability for DOD problems and would provide additional external support. The management team of the Solid Dynamics Research Department made a strategic decision to participate in the competition with a team of nine staff members led by Steve Rottler. As McGlaun vividly recalls,

> We started working with… DARPA…. DOD wanted an improved shock modeling capability for their laboratories and contractors. DARPA invested in shock codes at LANL, LLNL, and SNL and funded us to send CTH to key DOD users and add the features they needed. The users were very happy with CTH. In a year, DARPA was heavily investing in CTH. DARPA was a great partner that helped mature CTH and added features important to DOD. Our external user community exploded (no pun intended). By 1995, CTH had over 130 external users. DARPA's support was also politically important inside Sandia because their funding augmented… DOE funding and they had a reputation for developing world-class technology.

Trucano also remembers:

> I believe Sam's first hire upon his return was Steve Rottler, who quickly developed an important role in the LANL-Sandia SDI project; Steve became the first formal PI (principal investigator) for CTH late in the decade, leading CTH to be the code that DARPA selected as the new-generation hydrodynamic code for DOD. Steve eventually became the Chief Engineer (Vice President) for nuclear weapons at Sandia.[1]

As a result of the dedicated and talented efforts of the CTH team, the code was selected to be the code of choice for DOD applications and has enjoyed considerable external funding, and continues. After Rottler's promotion to management in 1990, Gene Hertel took over the project with management skills that have elevated the code to its continuing prominent status in the community. As Hertel states,

> CTH is the most widely used hydrocode in the U.S. and has been the most used code on DOD high performance computers as long as they have been keeping track. CTH is used extensively at Sandia and Los Alamos to simulate a wide variety of phenomena.

Yet another Sandia application involving explosive charges was simulation of a fluted copper liner spinning at different angular velocities, to gain insight into how the spin rate relates to the coherence of the resulting jet. Since the early 1950s, fluted (as opposed to smooth) liners had been used to compensate for the effects of rotation on behavior of shaped-charge jets. However, it was not until the late 1990s that the complete liner collapse and the eventual breakup of the jet from increased angular rotation could be modeled with 3-D numerical simulations. These simulations were conducted with the CTH Eulerian shock wave code, modeling the copper liner as a Mie-Grüneisen solid, with a strain-hardening yield and fracture strength that varied from a spall stress at high strain rates to a uniaxial tensile strength at low strain rates and large strain. The synthetic radiographs generated in the simulations

[1] In 2015, Rottler was again promoted, that time to the Deputy Laboratories Director and the Executive Vice President for Nuclear Security at Sandia.

agreed well with the experimental radiographs from Picatinny Arsenal. Marlin Kipp, Reina Martinez, and Gene Hertel, along with Ernest Baker, Brian Fuchs, and Chuck Chin of the U.S. Army Picatinny Arsenal, received the Neil Griffiths Award at the 18th International Symposium on Ballistics in 1999 for the most significant contribution to a shaped-charge technology for this work (Kipp et al. 1999a).[2] This investigation is another example of the integration and convergence of computational and experimental tools as applied to shaped-charge jet munition studies.

CTH has had numerous other applications in the conventional and nuclear weapon communities (Baer et al. 1996a) as well as studies of fundamental science (Boslough et al. 1995a, b; Boslough and Crawford 1996, 1997, 2008).

## 5.13 Turret Explosion Onboard the USS Iowa[3]

The capabilities developed through the shock wave program aided in the analysis of several problems of national and international importance. One of these in the 1980s concerned the explosion onboard the USS Iowa. The hydrocode CSQ, which was developed as a research effort in the shock wave program, played a major role in identifying the potential key cause of the explosion. The development of models to describe multiphase flow in energetic reactions was also important to the study. In addition, the expertise of people directly involved in shock wave research at Sandia was crucial to using the code and material models to identify the dynamic effects that most likely resulted in an accident.

A large explosion occurred in the middle gun turret of the USS Iowa on April 19, 1989, resulting in major damage to the ship and considerable loss of life (Schwoebel 1999). The USS Iowa was the fourth in the United States Navy to be named for the 29th State and was the only ship of her class that served in the Atlantic Ocean during World War II. The explosion occurred in the center gun room, killing 47 crewmen and severely damaging the gun turret. A photograph showing the aftermath of the explosion was captured by an onboard ship camera in Fig. 5.12.

Two major investigations of the accident were undertaken, one by the Navy and the other by the General Accounting Office, aided by Sandia. These investigations produced conflicting conclusions. The Navy's investigation concluded that one of the gun crew members had deliberately caused it. Sandia's investigation focused on determining a probable cause for the explosion and concluded that it occurred when the gun was being loaded. A team of researchers led by Director Richard Schwoebel

[2] For additional information about this conventional weapon analysis project, see Marlin Kipp's recollection.

[3] Mel Baer, who retired from Sandia in 2010, contributed the material for this section (private communication, 2013). He was one of Sandia's leading shock wave experts in theoretical energetic materials research at that time and a major contributor on the Sandia team that investigated the cause of the accident.

**Fig. 5.12** Turret explosion onboard the USS Iowa. The center gun on the near turret is lying on the deck after the accident (Reprinted with permission from Lt Thomas Jarrell, Public domain, released to the Public ID: DN-SC-90-05388 U.S. Navy, 1989, http://commons.wikimedia.org/wiki/File:USS_Iowa_BB61_Iowa_Explosion_1989.jpg)

worked on this investigation full time for 2 years. In 1999, Schwoebel published a book, titled *Explosion Aboard the Iowa*, on the Sandia investigation (Schwoebel 1999).

Figure 5.13 shows a cross-section schematic of the 16-inch gun that exploded. The breech of the gun contained five powder bags of nitrocellulose propellant that were accidentally over-rammed into its breech. Each powder bag consisted of uniform layers of cylindrical perforated propellant pellets and a "trim" layer that was randomly placed at the end of each bag to equalize the bag's weight. Several computer analyses and small-scale tests were conducted to identify possible causes for the accident. Karl Schuler performed analytic analyses that showed the powder bags had been rammed 24 inches into the breech, 3 inches further than the 21 inches claimed by the Navy. Furthermore, the Sandia study concluded that the propellant train had been subjected to an over-rammed impact load. Ken Gwinn conducted structural analysis of the impacted propellant train, which suggested that high stress localization occurred for the hypothesized loading; this scenario led to breakage of the propellant pellets in a direction transverse to the pellet perforations. To evaluate this possibility, Paul Cooper conducted 450 drop tests using small bags of D-846 powder. The tests demonstrated that the trim layer would often ignite when compressed at high speed. His investigations confirmed that, for over-rammed impact loads, the propellant pellets experienced brittle fracture and energy localization that generated hot fragments sufficient to cause ignition.

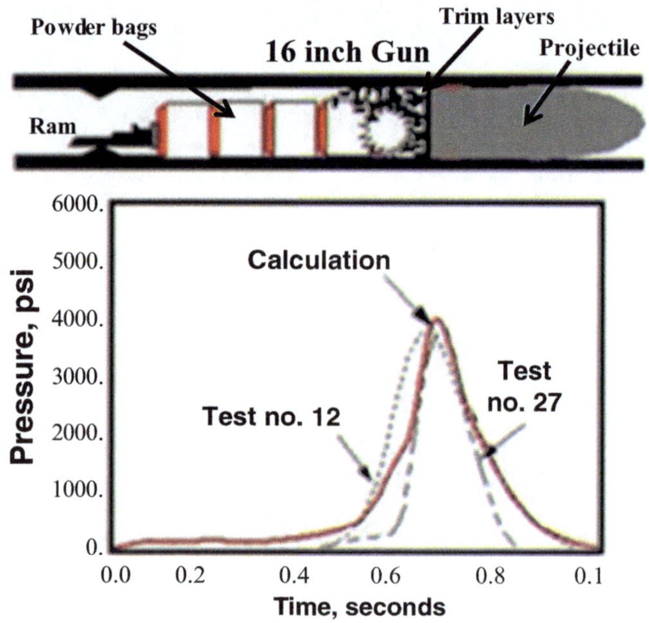


**Fig. 5.13** The top schematic illustrates the location of five powder bags accidentally over-rammed into the breech. Interior ballistic calculations established that the explosion likely initiated at the trim layer in the forward bag (Private collection, M. R. Baer)

Mel Baer performed interior ballistics analyses of the subsequent combustion, which are also illustrated in Fig. 5.12. At the time these calculations were initiated, CTH was not configured to treat multiphase flow—that is, a mixture of propellant gas and unreacted material occurring in the breech during ignition (Baer and Nunziato 1983, 1986, 1989). Baer implemented this capability in CTH to model the combustion process and examined the resulting dynamic loading on individual propellant pellets. His analysis also confirmed that the trim layer was the most vulnerable and established that the explosion most likely occurred in the vicinity of the first (most forward) powder bag. Later tests at full scale in open-breech guns confirmed that, under these conditions, low-velocity ram loading could indeed trigger a violent combustion event.

Based on these initial results, the Navy asked Sandia to participate in full-scale tests with an open-breech gun. These tests established conclusively that the propellant train could be ignited by an over-rammed impact. In July 1991, the team of Sandia investigators presented their final findings to Navy admirals; shortly thereafter, the Navy apologized to the families of the crewmen initially implicated in the Navy investigation. However, some skeptics have still not fully accepted Sandia's conclusions, even though the original Navy scenario has long been disproved. Many are now accepting Sandia's assessment as the most likely cause for the accident.

## 5.14 People and Places of the 1980s

**Tim Trucano, waiting for train in Lake Baikal region near Irkutzk in southern Siberia, private collection, T. G. Trucano; photo by Marvin Ross (LLNL) during a visit to attend a Soviet shock wave physics meeting in August 1990**

**Carl Konrad (*left*) and Dave Cox at powder gun impact chamber (Reprinted with permission of Sandia National Laboratories)**

**Pete Lysne, promotion to Distinguished Member of Technical Staff, 1989 (Reprinted with permission of Sandia National Laboratories)**

**Francois Bauer and Ray Reed (*right*), 1986, private collection, F. Bauer, 1986**

**Walt Herrmann promoted to Director of Engineering Sciences, 1982 (Reprinted with permission of Sandia National Laboratories)**

**Darrell Munson, promoted to Distinguished Member of Technical Staff, 1983 (Reprinted with permission of Sandia National Laboratories)**

**Herb Sutherland, promoted to Distinguished Member of Technical Staff, 1989 (Reprinted with permission of Sandia National Laboratories)**

**Doug Drumheller, promoted to Distinguished Member of Technical Staff, 1989 (Reprinted with permission of Sandia National Laboratories)**

**Paul Yarrington, Archie Farnsworth, Bill Davey (*left to right*) received award of excellence for the use of hydrocodes for weapon applications, 1988 (Reprinted with permission of Sandia National Laboratories)**

**Dennis Grady at STAR facility, elected to APS Fellow, 1990 (Reprinted with permission of Sandia National Laboratories)**

**Jace Nunziato (*left*), and Sam Thompson (*upper right*), supervisory appointments, 1980 (Reprinted with permission of Sandia National Laboratories)**

**Lynn Barker, promoted to Distinguished Member of Technical Staff, 1983 (Reprinted with permission of Sandia National Laboratories)**

**Bob Graham, promoted to Distinguished Member of Technical Staff, 1983 (Reprinted with permission of Sandia National Laboratories)**

**George Samara, elected to the National Academy of Engineering, 1986 (Reprinted with permission of Sandia National Laboratories)**

**Ed Barsis, Director of Computer Science and Mathematics Center, 1988 (Reprinted with permission of Sandia National Laboratories)**

**Karl Schuler, promoted to Distinguished Member of Technical Staff, 1985 (Reprinted with permission of Sandia National Laboratories)**

**Orval Jones (*left*) and Senator John Glenn, 1980 (Reprinted with permission of Sandia National Laboratories)**

**Larry Lee, private collection of F. Bauer**

# Chapter 6
# The 1990s: Black Monday

## 6.1 Background

The 1990s were turbulent times for shock wave research at Sandia because of the near elimination of the experimental program, including experimental facilities. Three management decisions led to this challenging event. The first decision was implementation of a laboratory-wide restructuring of management in the early 1990s. As a consequence, all Sandia departments, including the second-level ones (i.e., the original department management level) that involved shock wave research managed by George Samara and Jim Asay, were dissolved. The first-level divisions (now renamed departments) that had been supervised by Samara and Asay became individual departments under the direct supervision of two different directors. In addition, Walt Herrmann stepped down as the Director of Engineering Sciences; that was the directorate in which Asay's shock wave department had resided. The directorate was then eliminated, and the shock wave divisions that Asay had managed were transferred to Ed Barsis, who was the Director of the Computing Research Center. This resulted in a two-level management structure, with each director supervising the direct-reporting managers, as opposed to the previous situation in which three or four second-level managers reported to each director.

The increased number of direct reports made it challenging for directors to comprehend the day-to-day operations of the individual departments and to coordinate long-term research activities among them. That function was previously the responsibility of the second-level department managers who were no longer in charge. As a consequence, the cohesive, coordinated, and integrated activities in shock wave research that had been a major advantage in competing with external organizations on problems of common interest were severely impacted. At the same time, the resources for shock wave research were decreasing. The new structure therefore encouraged each first-level department to compete individually for the decreasing resources, rather than to focus on integrated research projects that were previously managed under the guidance of a second-level manager. Not only did this management



J.R. Asay et al., *Impactful Times*, Shock Wave and High Pressure Phenomena,
DOI 10.1007/978-3-319-33347-2_6

structure have a major influence on the ability to compete with other laboratories in solving complex scientific and engineering problems; it also had the effect a couple of years later of almost decimating experimental shock wave research at Sandia.

Fortunately, in the mid 2000s, Sandia reinstated the original management structure,[1] so that typically three to four first-level department managers again reported to a second-level manager. This change restored some of the much needed coordinated and integrated approach among departments. In addition, new experimental capabilities were developed as a result of additional resources, such as the internally funded MAVEN program that is discussed in Sect. 6.7 and increased external funding by governmental agencies other than DOE and NNSA; these additional sources of funding were essential to restoring experimental activities.

The second management decision was to bill charges to each department for the space occupied, which put departments with large experimental facilities at a disadvantage. The space charges for the STAR facility alone, over $450k a year (in those-year dollars), made STAR a prime target for closure as funding declined.

The third decision had a more direct effect on experimental shock wave research at Sandia. In the late 1990s, all three weapons laboratories (which are now called national security laboratories) initiated major efforts to acquire massively parallel processing (MPP) computers and to restructure their codes to use the new MPP architecture. Ed Barsis, as the Director of the Computing Research Center, initially took the lead among the laboratories in implementing parallel computing by acquiring several multiprocessing computers, which led to the first MPP computer (the Intel Paragon) at Sandia. To proceed at the fastest pace in reconfiguring the existing codes to run on this new machine, Barsis directed essentially all available resources to that effort. All experimental facilities including the one experimental shock wave department under Barsis, which was then managed by Phil Stanton, had a difficult time competing for resources with other computational departments that were involved in modeling or code development. As Tim Trucano notes,

> Experimental facilities at Sandia found themselves under great stress then because of funding reductions and programmatic uncertainties about future needs. When I was hired in 1980, the shock wave group included theory, computation, and experiment under a single second-level organization. This was a unique strength of our group. By the time we accomplished the tasks that I summarized above, the shock wave group was separating, with an experimental effort in a different organization than the computational effort. The overall computational group had been moved into a computational sciences center that was focused on advancing Sandia's capabilities in parallel computing. The computational shock wave work was an excellent driver for parallel computing. The experimental work was not, not the least because of the increasing funding challenges presented by the laboratory environment.

[1] In the Sandia management structure prior to the early 1990s, the first-level managers were referred to as supervisors and the second level as managers. After Sandia reinstated the original management structure in the mid 2000s, the designation of first-level managers ("supervisors") was changed to managers and that of second-level managers ("managers") was changed to senior managers.

The second-level managers who had previously managed the first-level divisions had to find new jobs. Many became program managers. George Samara became a program manager for the DOE Basic Energy Sciences research at Sandia. Jim Asay was initially a program manager for Ed Barsis and participated in developing multi-lab activities involved in parallel computing. After Dennis Hayes returned from his Washington, D.C., assignment as a director, he managed the new Center in the Defense Programs Office that was established to develop strategic plans and coordinate activities more closely among the research and weapon program organizations within Sandia. Jim Asay joined Hayes' Center as a "crosscut" manager[2] around 1993 and helped to initiate several activities, including the development of research agreements with private industry and formal research agreements with the former Soviet Union.

Phil Stanton's experimental shock physics department under Ed Barsis continued for a couple of years, but he struggled to obtain diminishing internal and external resources. With the end of the Cold War in late 1991, external funding decreased significantly. The experimental research staff decreased to Mark Boslough, Lalit Chhabildas, Dave Crawford, Dennis Grady, and Mike Furnish; as a result of internal transfers and retirements, the number of technicians also decreased. At that time, Ed Barsis decided to eliminate the experimental activities completely. That decision was announced on Monday, December 12, 1994—a day remembered as "Black Monday" by the experimentalists in Stanton's department. Several perspectives of this traumatic event come from those directly impacted.

As Clint Hall recalls,

> At the time, approximately ten Sandia employees and about eight contractors were working at STAR. Jim Asay had been promoted to a level-2 manager and was no longer associated with the site. Our Department Manager, Phil Stanton, called out to STAR and told Carl that he was holding a MANDATORY staff meeting in Area I for all Sandia employees, and there were to be no excuses to miss it. This was in early December, everyone was getting ready for Christmas, and nobody knew what the meeting was about. All the STAR tech folks piled into the van and headed into the area. When we got there, the entire department staff had already assembled in the meeting room, which was now standing room only, and there were plenty of hushed whispers going on. Phil stood up and proceeded to say the following. Effective immediately, the shock physics department was being disbanded due to a decision by the Director Ed Barsis. He told all the STAR techs that we were to do nothing now but concentrate on finding another job at Sandia. The contractors would be given a month to be placed elsewhere, and the staff members were to be reassigned throughout the Center. You could have heard a pin drop.

As Carl Konrad remembers,

> I think we all understood we were in trouble as our external customers were not extending their programs, but I was still caught off guard when Ed Barsis pulled the plug. Black Monday, December 12, 1994, is a chapter at STAR that I prefer not to dwell on. As the guys left the site, I remember I called Dave Cox to ask him if he had applied for the position to operate the 808 gun. He was not aware of the opening and applied on the last day they were

---

[2] In Hayes' new Center, a manager position, referred to as "crosscut manager," was established to review research activities for optimal short- and long-term benefit to the weapons programs.

accepting applications and got the job. In the end Clint Hall, Bill Reinhart, and I were the last three at the site and Lalit Chhabildas was the last staff member....

Lalit Chhabildas recalls that

Dennis, Phil Stanton, and I, together with Tony Chou from the Army Research Laboratory (ARL), visited Ernst Mach Institut in Germany and the Centre d'Etudes de Gramat in France to promote collaboration. After our return, we found out on Monday, December 12, 1994, that Ed was going to shut STAR down in about six months and get rid of all the contractors within a month....

Bill Reinhart remembers that

...1995 appeared to be the end of STAR. The rationale has been stated by others; the last days have been mentioned. What may not have been realized is that, at this time, I was not ready to throw in the towel. 1994 to 1995 had also brought the internet and email to STAR. I cannot remember exactly, but I did send off emails to our secretaries, managers, and director assistants – Lucille Verdugo, Anita Vasey, and others – about the status of STAR and offered up many options for operations continuing at a lower cost (to no avail), but I did manage to attract, probably not desired though, unwanted attention from our director, Ed Barsis. I was in touch with Ktech about the possibility of taking over the facility, as well, and had made contact with management at Phillips Laboratory.... Regardless, the shutdown continued, but still I had not given up.

The last experiment at STAR during that period was on May 5, 1995, and for all practical purposes, the facility was shut down. However, it was resurrected later in the decade under the leadership of Chhabildas, with less technical research staff but more efficient operations.

The experimental effort initially under George Samara's Physics of Solids Research Department, being in a different directorate, did not directly suffer the effects of Barsis' decision, and experimental research within that department continued at a low level. Mark Anderson and Bob Setchell, were the remaining experimentalists with Bob Graham's retirement in 1996. A small experimental effort on energetic materials under Lloyd Bonzon also continued in the Explosive Components Facility, primarily involving Anita Renlund, with Wayne Trott and Mel Baer participating from the Engineering Sciences Center.

In addition to this traumatic event, other losses of shock wave researchers occurred in this decade. Lynn Barker retired in early 1990. Orval Jones, who had joined Sandia in 1961 during the early phase of shock wave research, retired in 1993. However, Jones had not been actively involved in shock wave research since 1973. Lee Davison also retired in 1993 and began editing several chapters on shock wave research in the Springer-Verlag series. Al Chabai retired in 1994 but continued as a consultant to the Nevada programs. Bob Graham, who had made several seminal shock wave contributions during his career, retired in 1996 under the Voluntary Separation Incentive Program (VSIP) that Sandia offered to reduce staffing levels. He continued to participate in shock wave research and wrote several articles concerned with impact events. Dennis Grady also retired under the VSIP in 1996, partly as fallout from Black Monday. He joined a local company, Applied Research Associates, and continued to collaborate with Sandia researchers. Pete Lysne retired in 1996 after spending his first 10 years at Sandia in shock wave research and the rest in geological applications. Carl Konrad retired in 1998 but continued to work with the shock wave group at STAR and Z as a Ktech contractor until he accepted a position in Nevada managing the operation of LLNL's JASPER gas gun to obtain plutonium data. Dennis Hayes left the Defense Programs Office under Roger Hagengruber in 1994, retired from Sandia in 1995, and became President of Lockheed Martin Nevada Technologies in Las Vegas in 1996–1997. He retired from that position in 1997 and became a consultant to the emerging shock wave program on Z that was initiated in 1997. He continued as a consultant through 2010, making significant contributions that helped the resurrected experimental shock wave program survive its early beginnings on Z (see Chap. 7), and he achieved national recognition for research in high energy density physics a few years after 2000.

Needless to say, the modeling and computational shock wave research thrived during this decade. The robust development of the three-dimensional (3-D) code CTH in the 1980s was beginning to pay large dividends through superior capabilities in modeling complex dynamic problems. By the early 1990s, CTH became the workhorse for large computer simulations. It was being used for a broad range of applications, including analysis of the gun turret explosion on the USS Iowa in April 1989 and the effects of planetary impact physics (e.g., the much publicized Shoemaker–Levy impact on Jupiter in 1994). Computational shock wave research continually improved, with new capabilities as defined by the growing user community. In addition, major efforts were being devoted by Allen Robinson to convert CTH to run on the MPP computers and by Dave Crawford to add adaptive mesh refinement to CTH.

Because of the long history of experimental studies, combined with state-of-the-art material model development at Sandia, CTH had emerged as the leading computational tool for shock wave problems, not only at Sandia but nationally as well. As explained by Gene Hertel,

> …numerical simulations can explore material response in pressure and temperature regimes that are beyond currently available experimental or diagnostic capabilities…. Simulations are only useful, however, insofar as there is some reason to have confidence in the accuracy of the solution method and the material response models used in the calculations. Presently, accurate material response models exist for only a small part of the spectrum of materials and loading regimes of interest in shock compression studies.

Hertel's perspective effectively illuminates the benefits of the long history of integrating the three principal research efforts initiated by Lundergan and Herrmann and of the continuing need for a vigorous combined experimental, theoretical, and computational research program.

The major developments in shock wave research in the 1990s are discussed in the following sections:

- Kinetic Energy Kill for Theater Missile Defense (Sect. 6.2)
- Space Debris Impact on the International Space Station (Sect. 6.3)
- The DOD/DOE Memorandum of Understanding (Sect. 6.4)
- Mesoscale Modeling (Sect. 6.5)
- Line VISAR for Mesoscale Studies (Sect. 6.6)

- MAVEN: Model Accreditation via Experimental Sciences for Nuclear Weapons (Sect. 6.7)
- Underground Testing on Subcritical Experiments (Sect. 6.8)
- The Role of the Accelerated Strategic Computing Initiative in Shock Wave Research (Sect. 6.9)
- The Shoemaker–Levy Comet Impact on Jupiter at 60 km/s (Sect. 6.10)
- ALEGRA: The Next-Generation Hydrodynamic Code (Sect. 6.11)

## 6.2 Kinetic Energy Kill for Theater Missile Defense

The HyperVelocity Launcher (HVL) developed in the late 1980s and early 1990s enabled a new horizon of opportunities for research and applications in the following decades. These opportunities included several external programs to investigate phenomena during the impact of a hypervelocity projectile with a missile target. Missile defense engagements comprise a wide range of phenomena, including mechanical response, penetration, material failure, spallation and fragmentation, melting and vaporization, internal debris generation and subsequent interaction with internal parts, emission of light and transport of radiation, external debris, and structural failure. The interactions often occur at velocities exceeding 10 km/s, where thermodynamic effects such as melting and vaporization are produced but are not well understood. Figure 6.1 depicts these phenomena schematically. The

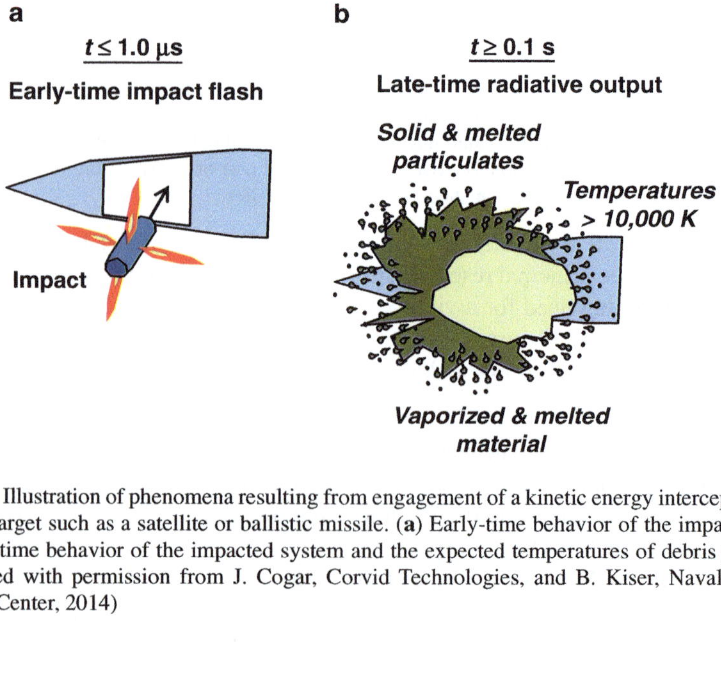


**Fig. 6.1** Illustration of phenomena resulting from engagement of a kinetic energy interceptor with a space target such as a satellite or ballistic missile. (**a**) Early-time behavior of the impact event. (**b**) Late-time behavior of the impacted system and the expected temperatures of debris products (Reprinted with permission from J. Cogar, Corvid Technologies, and B. Kiser, Naval Surface Warfare Center, 2014)

phenomena important to predicting the degree of damage and thus the survivability of an impacted system include (1) shock wave generation and material response, (2) penetration and fragmentation, (3) material failure and structural response, and (4) the effects of internal and external debris. Several effects related to these phenomena were studied at the STAR facility.

At the time, the HVL was the only capability that could produce the appropriate engagement velocities under well-controlled conditions in the laboratory. In addition, several diagnostics at STAR could be used to characterize the different phenomena (e.g., material equations of state and debris propagation resulting from impact). This unique capability created significant new opportunities for experimental shock physics research in the early 1990s. As Lalit Chhabildas recalls,

> NASA and DNA sought the technology. With the external funding, I became manager for these reimbursable projects. This gave me an opportunity to work with a team that included Bill Reinhart, Jeff Miller, Luba Kmetyk, Tim Trucano, Mark Boslough, and Jim Ang. The development of the technology was personally beneficial.

Theater Missile Defense (TMD) was the follow-on research program to the Strategic Defense Initiative (SDI) that began in the mid 1980s under President Reagan. The objective of SDI was to provide defense against missiles aimed at the United States from distances of thousands of miles. The Army's Theater High Altitude Area Defense (THAAD), one element of TMD, was designed to provide extended local defensive coverage by engaging incoming missiles over several hundred miles. The hit-to-kill interceptor being developed was to provide multiple engagement options within this range. THAAD missiles were being designed to collide with the target ballistic missile, rather than destroying it by exploding nearby, as fragmentation warheads do. The Defense Nuclear Agency in Washington, D.C., and the Army Missile Defense Agency (MDA) at Huntsville, AL, were the main government agencies involved in the development of this technology.

Detailed understanding of impact phenomenology, shock physics, and spectroscopy was required to predict the effects, understand the consequences, and analyze the scenarios of kinetic energy-based missile defense engagements. Specifically, combined laboratory-scale experiments, 1-D and multi-dimensional numerical models, and analytic scaling law analyses were needed to develop the engineering models. Figure 6.1 illustrates the various phenomena involved in kinetic engagements. The initial impact produces a flash of light at submicrosecond (less than one millionth of a second) time scales that persists for a very short time. The resulting stress waves from the impact weaken the structure and produce fracture and fragmented debris that can destroy other parts of the target. Finally, structural failure of the ballistic missile target produces external debris with a corresponding optical signature that persists for several hundred milliseconds (thousandths of a second). These multiple effects can prevent the missile from performing its mission. In developing defense strategies using hit-to-kill concepts, each of these different effects and the observable signatures, such as the light flash produced upon impact, had to be understood. This information was essential to assess whether a successful intercept occurred.

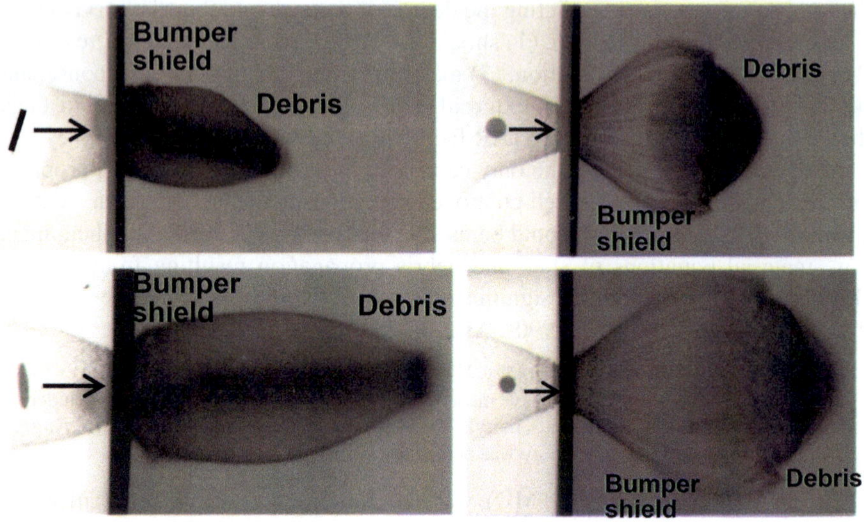


**Fig. 6.2** X-ray image showing the effects of impactor shape and orientation on debris formation from a metal plate after impact (Reprinted with permission from Konrad et al. 1994, Copyright 1994, AIP Publishing LLC)

A variety of fundamental and applied experiments were conducted on the Sandia HVL to understand the impact phenomena. In the early 1990s, DNA supported a basic research program to study vaporization kinetics and evaluate the role of projectile shape on the evolution of debris from impact on a thin plate (Konrad et al. 1994). The typical debris clouds produced in the experiments are illustrated in Fig. 6.2, which shows that the impactor shape and orientation strongly influence the debris cloud shape and therefore the damage induced on other structural components. HVL experiments with tantalum impactors at velocities up to 10 km/s produced full vaporization of zinc targets. The results characterized kinetics effects of multiphase (liquid and vapor) flow on the impact debris (Brannon and Chhabildas 1995).

Understanding the light flash signatures was also crucial for assessing lethality issues. Jim Ang (1990, 1992) performed a large number of experiments to quantify impact flash spectra, as he mentions in his recollections:

> …I had the opportunity to continue my research on impact flash; I was able to work with EG&G Princeton Applied Research Corporation and procure a first-of-a-kind Optical Multi-channel Analyzer (OMA) to capture time-resolved spectra of impact flash signatures. This unique diagnostic led to a project from the U.S. Army Space and Strategic Defense Command to send this OMA to Andy Williams' two-stage gas gun facility at the Naval Research Laboratory. His terminal ballistics catch tank is rated to allow testing on targets that include small amounts of high explosives. There, we were able to capture unique time-resolved spectra that distinguish impact flash optical emissions from the subsequent high-explosive detonation spectral signatures.

In early 2000, Jeff Lawrence initiated a modeling effort to understand and predict the optical signatures produced during hypervelocity impact engagements. The electromagnetic spectrum extends from long wavelengths to gamma rays, resulting in

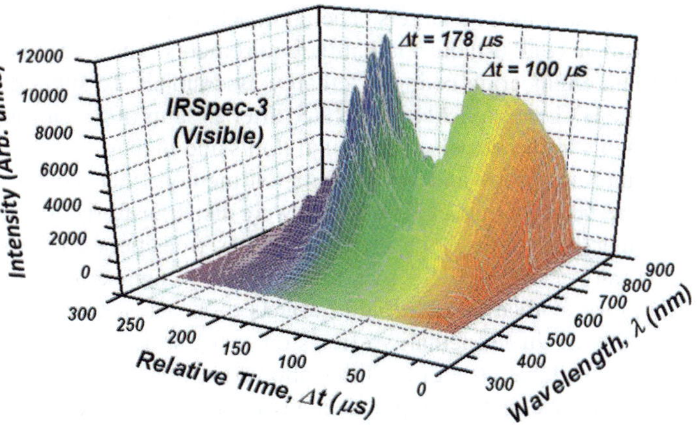


**Fig. 6.3** Impact flash generated by a titanium projectile impacting an Al target at 9.8 km/s. Both the temporal and spectral distributions are illustrated (Reprinted with permission from Reinhart et al. 2008, Copyright 2008, Elsevier LTD)

difficult experimental characterization. Lawrence led a Laboratory Directed Research and Development (LDRD) project to obtain impact flash data over extended spectral regions. The objective was to characterize the impact flash signature for different projectile and target materials, including metal–metal and metal–composite ablators, as well as the impacts on energetic materials (Lawrence et al. 2006; Reinhart et al. 2008). Using the two-stage gun at STAR, the three-stage HVL, and magnetically-launched flyer plates on Z, impact velocities ranging from 6 to 25 km/s were produced. An Optical Multichannel Analyzer was used to characterize the time-resolved spectra from the impact. The spectra from optical to infrared generated during early and late times were measured (Thornhill et al. 2008). Figure 6.3 shows a typical impact flash signature in the visible range for a titanium projectile impacting aluminum. These experiments were useful for developing phenomenological models to understand and predict impact flash signatures for lethality analyses (Lawrence et al. 2006). Specific spectral and temporal features were observed that can be used to compare with theoretical or numerical models or to develop analytic models for predicting optical flash in different engagements. Reasonable agreement of the predicted optical emission and the calculated temperature states with the data was obtained over several orders of magnitude in optical fluence (Lawrence et al. 2006). In addition, the data allowed evaluation of commonly used models, such as the MBBAY,[3] which relates the impact coupling efficiency to the optical fluence produced. To determine the capability of using impact flash as a discriminating measure, the spectral data were compared with CTH calculations of the expected spectral lines and with National Institute of Standards and Technology data.

In 1999, the MDA initiated a 3-year program at Sandia using the HVL to validate hit-to-kill assessment codes being developed for predicting the lethality of kinetic

[3] MBBAY is an acronym for Modified BBAY, which in turn is an acronym for the model developed by (Hans) Bethe, (William) Bade, (John) Averell, and (Jerrold) Yos. The model was modified by Sandia.

energy impacts against missile targets. The approach was to use CTH to simulate the HVL experiments and establish lethality requirements. These results were then integrated into an engineering code for lethality assessment. To support the Army requirements, the HVL was quickly reactivated after a hiatus of 6 years and used for impact experiments in the velocity range of 10 km/s and higher.

The HVL gun technology and instrumentation methods that Sandia developed for studying hypervelocity impact phenomena were also transferred to other organizations. As Chhabildas states, “the technology was transitioned to NRL [Naval Research Laboratory] and the U.S. Air Force Arnold Engineering Development Center (AEDC) for evaluation of lethality effects on space structures.” These are large facilities that evaluated lethality effects on space structures. In addition, the long flight range available at both facilities allowed a more realistic evaluation of hypervelocity lethality effects. The practical advantages of this technology transfer were that realistic projectile and target sizes could be studied because of the much larger bore size of the AEDC two-stage gun.

## 6.3 Space Debris Impact on the International Space Station

The amount of debris in centimeter-sized chunks orbiting in low Earth orbit (LEO) has grown substantially because of the fragmenting debris resulting from the increasing number of satellites launched. The problem was enhanced in 2009 when a Russian communications satellite, Kosmos 2251, struck the solar panel of Iridium 33, a commercial American communications satellite. The panel shattered, Iridium 33 was released, and Kosmos 2251 disintegrated into multiple fragments. The collision created more than 2000 pieces of debris with sizes greater than 10 cm and potentially thousands of smaller fragments. Being hit by a “sugar cube”-sized chunk of space debris at orbital velocities is the equivalent of standing next to an exploding hand grenade because of the high relative velocities and the enormous kinetic energy released upon impact. Centimeter-sized fragments have the potential to destroy satellites completely or to do considerable damage to the International Space Station. Figure 6.4 schematically illustrates the debris problem, based on actual data. The illustration is an artist’s rendition of the actual relative distribution of space debris around the Earth as of early 2010 (Stansbery 2011).

In the early 1990s, the HVL was the only facility capable of studying the effects of debris impact on realistic space station structures at the expected impact velocities. As Chhabildas remembers,

> [Marshall Space Flight Center] asked us to evaluate debris shield designs. Our first measurements to evaluate a debris shield design, known as the ‘Whipple Bumper Shields,’ at impact velocities of 10 km/s (i.e., space-type velocities) could now be reproduced in the lab…!

In collaboration with NASA, JSC, and MSFC, several experiments were performed with the HVL to simulate the debris impact on typical structural materials. The enhanced velocities achievable with the HVL were critically important for this analysis because material phase changes such as melting and vaporization have

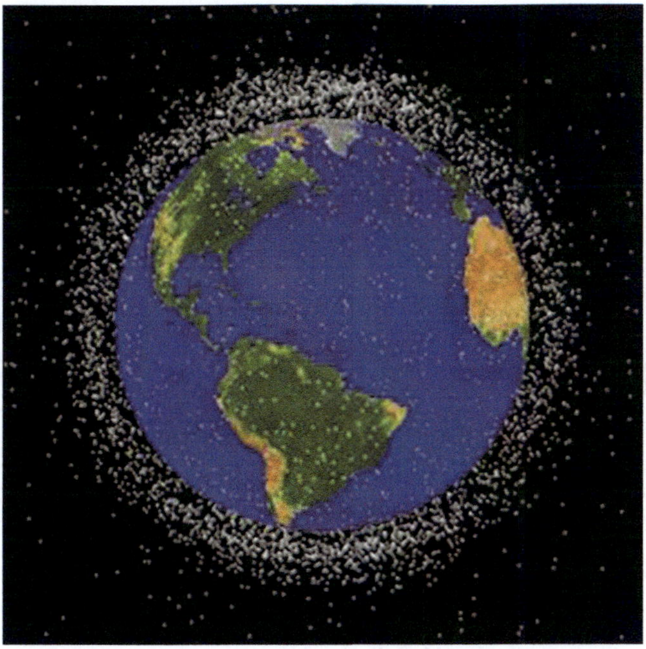

**Fig. 6.4** Conceptual distribution of space debris (sometimes referred to as "the low Earth orbit beehive"). Relative distances and rates are shown to scale, but the orbiting objects themselves are greatly exaggerated in size (Reprinted with permission from the NASA Orbital Debris Program Office, http://orbitaldebris.jsc.nasa.gov/photogallery/beehives/LEO1280.jpg)

pronounced effects on the design of debris shields (often referred to as "bumper shields") to protect underlying structures. Previous debris shield testing was limited to about 7 km/s using standard light gas guns and spherical projectiles, which did not fully address these effects. Space debris, however, is more plate-like, and the resulting debris generated upon impact can be more lethal (Konrad et al. 1994). Based on the HVL experiments, a composite bumper shield of thin metal plates and 3M® Nextel® cloth was designed and tested before being deployed on the Space Station Freedom (Ang et al. 1991; Boslough et al. 1993; Chhabildas et al. 1993a; Lawrence et al. 1995). The follow-on to Freedom, the International Space Station, was even more susceptible to debris impact because of its increased size. Another space asset that required debris shielding was the Hubble Space Telescope. The numerous impacts produced by debris impact on the telescope are illustrated by the drill holes shown in Fig. 6.5, which mark the individual locations for impact by debris particles. The multiple impacts shown on the shield emphatically demonstrate the magnitude of the debris problem and the necessity to characterize the effects of debris impact and the means to mitigate the problem. Chhabildas emphasizes the satisfaction of helping to resolve a national concern in his recollections, "We tested those [bumper shield designs] before they were deployed for use with the space station Freedom. I felt like this was quite a contribution to our national space program and was satisfying and gratifying at the same time."

**Fig. 6.5** Mark Boslough standing next to the Wide Field and Planetary Camera 2 in 2014 that had been used by the Hubble Space Telescope to take the photos of the Shoemaker-Levy 9 impact shown in Fig. 6.12b. Impact micro-craters on the camera structure have been drilled out, leaving the visible holes (Private collection, M. B. Boslough)

## 6.4 The DOD/DOE Memorandum of Understanding

In the 1980s, DOD and DOE signed a memorandum of understanding (MOU) concerning jointly funded shock wave research programs at the national security laboratories on materials of common interest. The MOU program was initiated at Sandia in the mid-to-late 1980s under Max Newsom and managed by Bill Tucker and, later, by Tom Hitchcock. The research included studies of geological materials, metals, ceramics, porous materials, energetic materials, and concrete. A wide range of studies was conducted under the auspices of the MOU, including high-pressure equations of state, material strength, fracture, and fragmentation. This highly successful research initiated under an MOU continued for several years. Dennis Grady managed this project since its inception and was initially the project manager at Sandia for experimental studies conducted within the Solid Dynamics Research Department before the shutdown of STAR. As Grady mentions,

> An extensive research program emerged during the late 1980s that entailed a joint effort between the Department of Defense and the national laboratories on issues of mutual concern. Material property studies through impact shock physics methods were undertaken at Sandia. Extensive experimental studies were performed on low density and high strength ceramics that had potential application for military armor needs. From this effort emerged Hugoniot equation of state and dynamic strength properties for a suite of ceramics that remain, to the present day, the basis for computational analysis of ceramic performance in the terminal ballistic environment.

Grady continued as the project leader until his retirement in 1996, whereupon Lalit Chhabildas assumed that role. Chhabildas initiated new approaches for studying material behavior at intermediate strain rates important to material modeling studies (Chhabildas et al. 1997). As he recalls,

> Upon Dennis Grady's retirement, I inherited…[the MOU funds] to redirect the shock physics activities…Leonard Wilson of the Naval Surface Weapons Center (NSWC) in Dahlgren played a key role in getting us going. He funded studies to characterize PBXN-128, a Navy explosive. We started out gradually in that we used the ECF compressed gas gun first and limited our experiments to gas gun velocities. This is when I developed a technique to achieve uniaxial stress experiments on the [single-stage] gas gun [at Sandia] by using a sleeved cylindrical target material and impacting it with a graded-density, layered impactor…. The surprising outcome was that we could conduct uniaxial-stress experiments at strain rates higher than Hopkinson bar experiments but lower than shock loading experiments! We coined these 'intermediate strain rate' experiments. We investigated ceramics such as alumina, silicon carbide, and boron carbide…

Other new technologies developed include well-controlled fracture fragmentation studies (Vogler et al. 2003) on many metallic tubes and expansion of earlier studies on ceramics to include their strength determination in the shocked state (Reinhart and Chhabildas 2003) to Mbar stresses. In 2004, Tracy Vogler assumed responsibility for the MOU project and initiated additional efforts to study the combined pressure–shear properties of porous metals using the modified gun that was initially constructed by Don Lundergan in the 1950s. This work is identifying the importance of shearing effects in the pulverization of metal powders (Vogler et al. 2011) and continues to the present.

## 6.5 Mesoscale Modeling

A long-standing goal of the shock physics community has been to develop continuum material models that represent the physical effects that dominate material response under shock compression. Examples are heterogeneous explosives that are initiated by hot spot formation occurring during shock loading, premature melting during shock loading of metals and alloys, shock propagation in mixtures and composites, material failure and fragmentation, and the shock response of porous materials. Attempts to include the fine details of material structure in models of dynamic material response date back to the early 1970s with the work of Lundergan (Lundergan 1970; Lundergan and Drumheller 1971a, b, 1972), Drumheller (Drumheller and Bedford 1973; Drumheller and Lundergan 1975; Drumheller 1984; Drumheller et al. 1982), Nunziato (Nunziato et al. 1978a, b), Davison (Davison 1971; Davison et al. 1977), Swegle and Grady (1986a, b), and Barker (1971b). The early work by Barker on shock propagation in plate laminates is especially noteworthy because it connected the measured properties of individual constituents and knowledge of the composite structure to the parameters needed in a continuum scale-based Maxwellian model of dynamic material response in order to predict the average, or continuum, response accurately.

The material scale of a few microns to millimeters, where material features such as grain size, impurities, the sizes of mixture constituents, and shock-induced effects (e.g., micro-shear banding) are important, is referred to as the mesoscale. Finer material features such as dislocation structures are usually referred to as the microscale. A concerted effort to include the effects of the mesoscale in the shock response of materials was led by Mel Baer from about 1975 until his retirement in 2010. His early work involved the development of multiphase models to describe the response of heterogeneous materials without structural details of the material. He and Jace Nunziato published a classic review in 1986 concerning a two-phase mixture theory for the deflagration-to-detonation transition (DDT) in reactive granular materials (Baer and Nunziato 1986). This widely referenced paper presented an overview of a modeling approach that treats fully compressible reactive mixtures. As Baer remembers,

> At that time, multiphase modeling was being developed mainly in the propulsion and gun communities, and it was recognized that these approaches were incompatible with detonation theory. The explosive components organization at Sandia, which sponsored our work, requested modeling assistance to provide a better understanding of DDT as applied to detonator components. A new approach was needed that could couple high-speed deflagration to shock physics and admit detonation states. We developed a continuum mixture theory approach that treated volume fraction as an independent kinematic variable and allowed compressibility of all phases and detonation states. In the appendix of the paper, a brief characteristics analysis suggested the modeling was mathematically well-posed and consistent with classical Riemann theory. This caught the attention of several mathematicians throughout the world, and our continuum mixture approach became the foundation for several subsequent works. Colleagues at Los Alamos National Laboratory (LANL) later took an interest in the continuum mixture model, which they referred to as the B-N (Baer-Nunziato) model. LANL posed some modifications to the model; however, the original B-N model is still regarded as the pioneering work. Unfortunately, Jace Nunziato passed away before the work gained notoriety in the multiphase community.

Baer worked on several additional multiphase continuum mixture theories after that time, including additional studies of explosive materials (Baer et al. 1986; Baer 1994, 1997). A more comprehensive and annotated list of publications on energetic materials is presented in his recollections. Many of the important physical processes in heterogeneous materials during shock compression occur at the crystal or grain level. The multiphase modeling work by Baer and others emphasized the need to include structural details of material morphology in continuum modeling and led to later mesoscale research at Sandia. Baer's first paper on mesoscale modeling explicitly treated wave propagation in a heterogeneous reactive material (Baer 1997). At that time, shock physics modeling was dramatically changing, with several advances and improvements in computational software and hardware. High-order flux limiters and adaptive mesh refinement that were implemented in CTH greatly reduced the effects of artificial viscosity and allowed more realistic calculations of the influence of real material effects on shock wave structures. Furthermore, CTH was revised to run on the new high-performance MPP computers that were being acquired during this period. This allowed full 3-D simulations of material features and led to new modeling ideas. As Baer recalls,

> Bob Graham suggested that greater insight into the shock processes of mixed materials could possibly be revealed by simulations of shock loading of multi-particle ensembles,

such as the studies by Dave Benson at the University of California San Diego that were based on two-dimensional simulations of shock loading of 'circular' particles. However, Benson's approach appeared to be too simplistic and was somewhat removed from the real geometries that typically represent energetic composites. Hence, a new approach was developed using 3-D shock physics analysis. This paper presented the earliest mesoscale simulations of shock loading on ordered arrays of spheres. The mesoscale modeling showed that the classical view of shock waves in heterogeneous materials is not a simple jump state. Furthermore, the simulations showed that mesoscale modeling could possibly provide new insights into energy localization linked to micromechanics, which could greatly improve the existing shock initiation descriptions. This was an exciting discovery, and we continued to improve the representative mesoscale models and motivate new experimental efforts.

Baer constructed a numerical simulation that allowed precise determination of the heterogeneous structure for evaluating the effects of material structure on wave propagation. His simulations of shock propagation in a tin spherical lattice, which included a full description of the EOS and both the melt and the tin (I)–tin (III) polymorphic phase transition, provided the first concrete demonstration that shock loading in heterogeneous materials is not simply a single jump in shock variables, as is usually assumed. These simulations were performed with the 3-D code CTH. In addition, Wayne Trott conducted detailed VISAR experiments on a shocked tin lattice to evaluate the simulations (Baer and Trott 2004). Figure 6.6 shows that

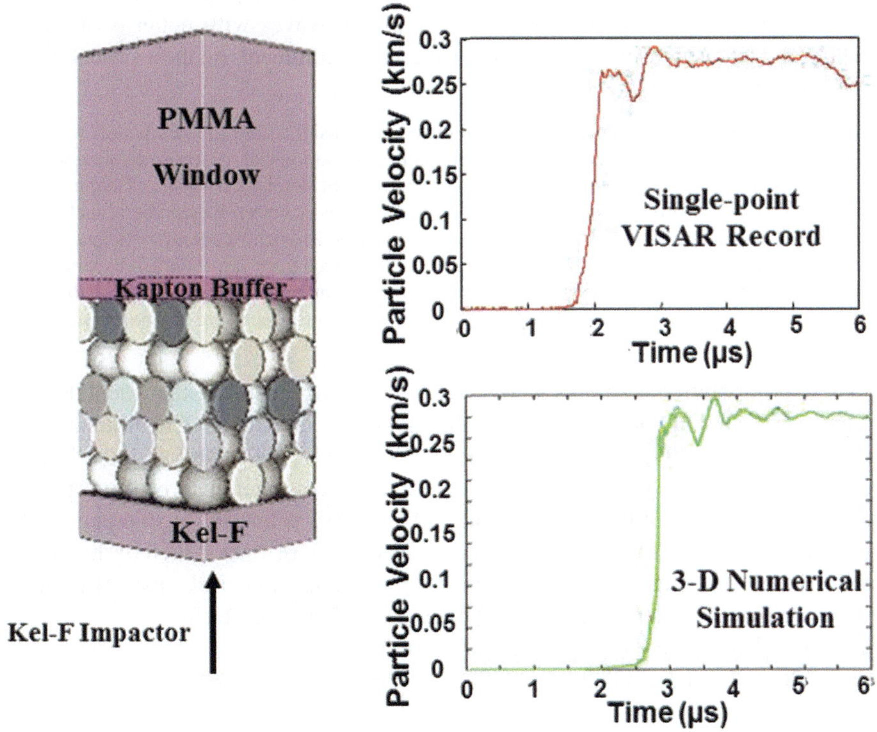


**Fig. 6.6** Numerical simulation of a shock-loaded tin sphere lattice. A complete equation of state was used for tin, including melting and polymorphic transitions. Time scales are arbitrary (Reprinted with permission from Baer and Trott 2004, Copyright 2004, AIP Publishing LLC)

excellent agreement was obtained between Trott's experimental result and the 3-D simulations. With the advent of the Line VISAR instrumentation somewhat after these simulations were performed, more detailed temporal and spatial information allowed the investigation of the effects of mesoscale features on wave propagation. Mel Baer's and Wayne Trott's research on a number of heterogeneous materials with the Line VISAR (see Sect. 6.6) attests to this increased understanding.

As the development of mesoscale modeling improved, better representative microstructures were needed to model shock loading in heterogeneous materials. The work discussed above treated microstructures as ordered spheres, which caused resonant waves to develop in the shock. Furthermore, microphotography indicated that high-density composites usually consist of individual crystals with randomly oriented planar interfaces and internal crystal defects. Randomness and varied particle size distributions were important aspects of real material microstructures. Baer addressed these issues by including more realistic microstructures using a Monte Carlo particle packing approach that produced mesoscale geometries representative of real materials. Later, tomographic techniques became available that could generate and quantify actual 3-D microstructures. Baer demonstrated that randomness enhances homogenization of the wave fields and the effects of energy localization at material interfaces and crystal contacts (Baer et al. 1998). Furthermore, he showed that, when reactive behavior was included in the shock loading simulations, detonation in heterogeneous materials suggested multiple waves with patterns of shock focusing and interactions with rarefaction waves. Baer comments on these observations in his recollections:

> For the first time, mesoscale simulation produced new insights into the detonation wave structure. This was consistent with experimental observations of gaseous detonations, which had long before shown that the classical models of detonation such as Chapman-Jouget (CJ) and Zeldovich-von Neuman-Doring (ZND) are incorrect. Researchers, such as Anatoly Dremin (Institute for Chemical Physics Research, Russian Academy of Sciences), argued that when chemistry interacts with the wave mechanics it produces dispersive behavior at the shock front. This paper presented the first numerical simulation of detonation waves in condensed phase energetic composites that were dispersive and nonplanar at the crystal level.

These computational studies were complemented to include experimental studies under the DOD/DOE MOU that Chhabildas was managing (Chhabildas et al. 2002; Kipp et al. 1998, 2000; Furnish et al. 2000, 2007, 2009) and with the implementation of a new LDRD project (see next section). This later allowed the development of a new DOD/DOE MOU around 2003 to focus primarily on energetics and their reactions at a mesoscale level under the leadership of Baer. Much of the mesoscale studies on wave propagation in heterogeneous materials has also been researched by others, including Tracy Vogler, Mike Furnish, Marlin Kipp, and their colleagues (Borg and Vogler 2008, 2009a, b, 2013; Borg et al. 2009; Vogler and Clayton 2008; Vogler et al. 2004, 2008, 2011, 2012).

## 6.6 Line VISAR for Mesoscale Studies

It has long been known that shock compression of polycrystalline metals and other heterogeneous materials, such as granular materials or composites, will not produce a uniform planar wave. Early unloading and reloading wave experiments (Lipkin and Asay 1977; Asay and Lipkin 1978) on aluminum suggested that the microstructural heterogeneous response at the crystallographic scale appeared to have a major influence on the observed wave profiles. As an explicit example, 3-D numerical simulations of the actual structure of a fine-pored polyurethane foam (see Fig. 6.7) showed substantial deviations from uniform behavior, in both the spatial and temporal variables, and were in agreement with experiments on the modeled material (Kipp et al. 2000). Similarly, experiments on explosive materials dating to the 1970s have suggested that initiation occurs heterogeneously at local "hot spots." Heterogeneous ignition at local hot spots in partially porous explosives was noted by Jim Kennedy:

> After I moved from research into component design in 1976, Bob Setchell joined the Physics of Solids Research Department and performed wave-evolution studies on the same explosive with a ramp wave followed by a shock. He found (Setchell 1981) that the ramp wave precompressed the explosive and desensitized it so that the ensuing shock was unable to stimulate any apparent reaction, as judged by the waveform behind the front. After a few millimeters of propagation distance, the ramp wave was overtaken by the shock front and then there was no precompression. At that point, evidence of reaction behind the front was immediately noted. This indicated that even quite gentle precompression over less than a microsecond is capable of closing pores and eliminating hot spot sites in the explosive where a reaction would otherwise be stimulated by shock compression.

These results helped to spur explicit treatment of hot spot initiation using continuum theories for heterogeneous explosives (Nunziato et al. 1978a, b). Mel Baer extended these results through rigorous modeling of material response at the

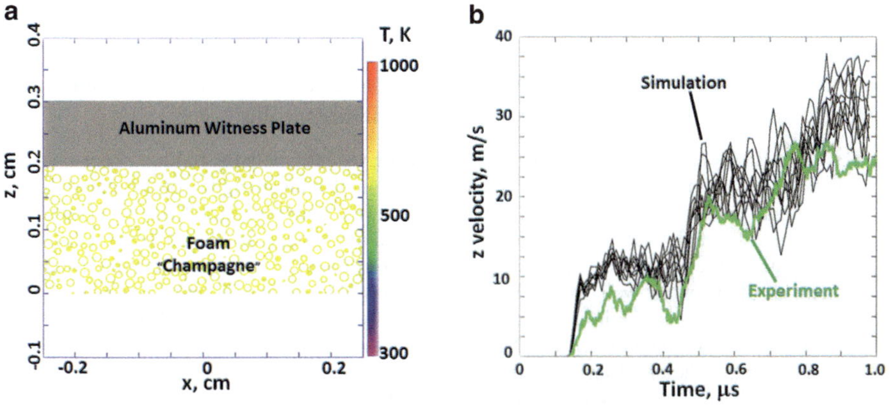


**Fig. 6.7** (**a**) Porous polyurethane foam ("Champagne") plate impacting an Al plate. (**b**) The foam was modeled explicitly in three dimensions with 25,000 hollow spheres, impacting at 565 m/s (Reprinted with permission from Kipp et al. 2000, Copyright 2000, AIP Publishing LLC)

mesoscale level using numerical simulations. He published several papers on mesoscale modeling with coworkers from 1986 to 2004 (see the list of papers by Baer et al. in the Bibliography).

Although measurements of local light emission from hot spots in explosives had previously been measured, direct measurements of the local particle velocity fields had not been quantified. Quantitative data on particle velocity variations in granular and heterogeneous explosives were needed to validate Baer's theoretical predictions. In the late 1990s, Wayne Trott, Lalit Chhabildas, and Jim Asay jointly proposed the use of a modified VISAR to make these measurements. The approach was to image a line varying from about 1 mm to tens of mm on a heterogeneous sample, rather than the few hundred micron-sized spots normally imaged. To accomplish this, a standard VISAR was slightly defocused to monitor several "line" fringes that represent the particle velocity at the projected fringe position on the sample. The fringe information was then recorded with an optical streak camera to provide temporal data about the local particle velocity at that position on the sample. The approach is an adaptation of the Optically Recorded Velocity Interferometer System (ORVIS), which was first developed by Doug Bloomquist and Steve Sheffield for recording the velocity of a single spot on a sample (Bloomquist and Sheffield 1983a, b) and is usually referred to as the Line VISAR.

Trott developed this system (Trott et al. 2000, 2001) into a precision tool to make local particle velocity measurements on several metals, energetic materials, and granular materials (Chhabildas et al. 2002; Trott et al. 2000, 2001, 2006; Furnish et al. 2006, 2007, 2009; Vogler et al. 2008). Spatially-resolved particle velocity data obtained on energetic materials and energetic material surrogates such as sugar (Trott et al. 2007) helped to validate the mesoscale models of material response being developed by Mel Baer and others in the late 1990s and early 2000s (Baer and Trott 2002a, b, 2004; Kipp et al. 1998, 2000). Although the Line VISAR has been extensively used at Sandia to quantify spatial variations in particle velocity on shocked heterogeneous materials, many remaining issues still have to be resolved with respect to its use. Figure 6.8 shows a Line VISAR record and the resulting distribution in particle velocity for granular sugar shocked to a low shock stress. This capability represented a significant advance in the VISAR for material studies at the mesoscale or subgrain scale. However, as Tracy Vogler succinctly states in his recollections,

> The Line VISAR has remained tantalizingly promising yet extremely frustrating. For example, the work on tungsten alloys produced a distribution of spall strengths for the material, but attempts to obtain similar data on silicon carbide were not successful. Such data could have been extremely valuable to the stochastic failure model developed by Rebecca Brannon and her coworkers. The reality is that the instrument is, in general, complex and difficult to get to work well. While Wayne Trott was able to do this, most others have proved unsuccessful at it. Furthermore, the analysis of the streak images obtained from the experiments has proven problematic. Wayne utilized a simple line-out approach to the analysis of the data, whereas Tom Ao developed a Fourier transform approach that tended to smooth out the results. This raised the question of how much of the spatial variability observed in the Line VISAR measurements was due to real differences in behavior from point to point and how much was noise or an artifact of the measurement technique. Questions about repeatability of the measurements have also been raised since a thorough investigation of the Line VISAR for well-controlled phenomena has not been performed.

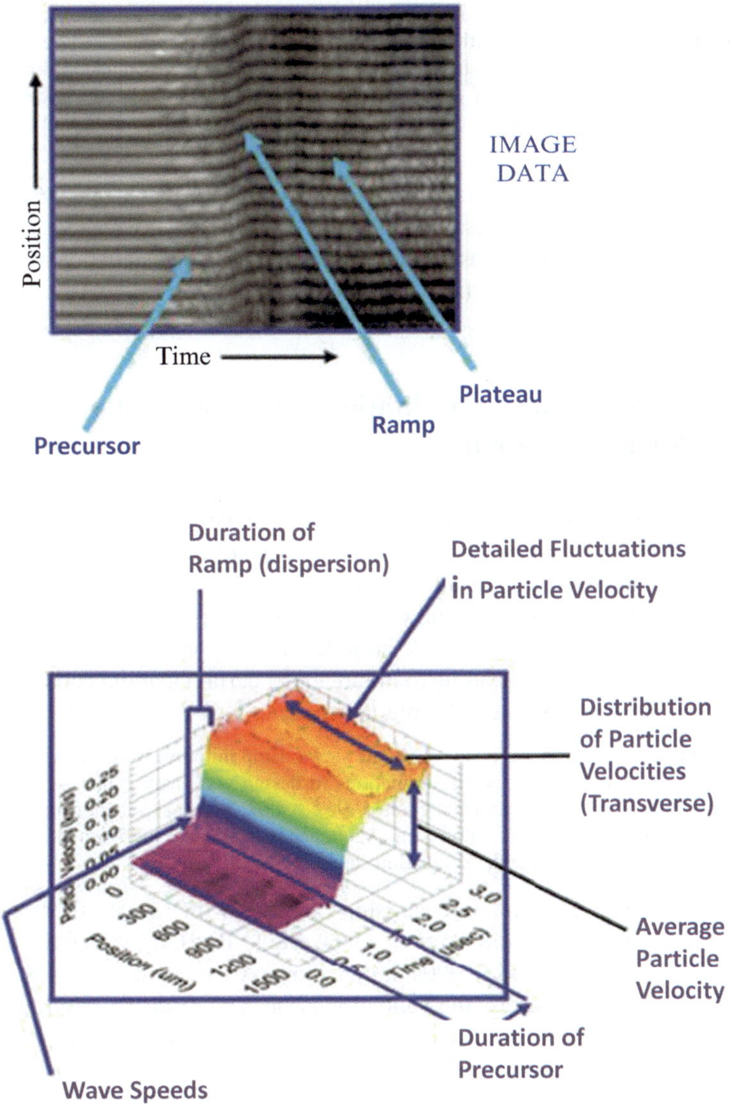


**Fig. 6.8** Line VISAR record obtained during the shock compression of a flat plate sugar specimen. The *top graph* shows the velocity fringe record obtained from a 1-mm-long line on a plastic plate located at the rear side of the sugar sample. The *lower graph* shows the velocity variations measured along the line (Reprinted with permission from Trott et al. 2007, Copyright 2007, AIP Publishing LLC)

Mike Furnish further emphasizes the need to make the Line VISAR data a quantitative measure of heterogeneous material behavior:

> Right now [2013], my focus is back in Albuquerque, mainly on experiments on the DICE (Dynamic Integrated Compression Experimental) and STAR facilities. This includes Line

> VISAR buildup as well as more of the garden-variety shots for spall problems, strength measurements, and such.... With the Line VISAR, the intent remains to get beyond the 'nice image' goal to issues such as instability spacing, effects of material property perturbations, and mechanisms of yield in spallation. We have begun a productive collaboration with AWE [the Atomic Weapons Establishment] on this effort.... Where this will lead, time will tell.

The Line VISAR is actively used by several researchers at present, and hopefully many technical issues will be resolved in future studies. In many respects, this work is still in its infancy; with increased usage and understanding, statistical representation of dynamic material properties is not too distant.

## 6.7 MAVEN: Model Accreditation via Experimental Sciences for Nuclear Weapons

The increased emphasis on simulations, instead of experimental testing, to design and analyze weapon components and subsystems resulted in a concerted effort to ensure that the models and algorithms in those simulations were accurate. Up to the mid 1990s, models based on uniaxial loading or other experiments were used without validating their predictability for complex loading conditions. In particular, dynamic material properties leading to failure under certain loading conditions were still poorly understood. As Steve Montgomery recalls,

> The Department of Energy began the stockpile stewardship initiative to reduce reliance on expensive tests and increase use of simulations to design and analyze nuclear weapons across the nuclear weapons complex. From 1996 to 2002, the stockpile stewardship and component development funds allowed a significant increase in experimental work and code development to improve the simulation capability.

As part of the new Stockpile Stewardship Program, which was authorized by the National Defense Authorization Act of 1994, Steve Rottler, a program manager under Dennis Hayes in the mid 1990s, initiated an experimentally focused research program to specifically address the validation issue. The resulting program, Model Accreditation Via sciences for Nuclear weapons, MAVEN, was begun in late 1997 and managed by Jaime Moya. This effort was initiated, in part, to reinvigorate all experimental research at Sandia. It had a significant influence on revitalizing shock physics research.

MAVEN covered several areas of weapon components and subsystems, including technologies involved in quasi-static evaluation, shock and vibration testing of components, radiation-induced impulse, initiation of explosives, mixed-phase flow in fire science, and pulsed-current supplies for neutron generators. Lalit Chhabildas received funding for two major experimental shock wave efforts under this program. One effort was to extend earlier studies on the shock response of PZT 95/5 and alumina-filled epoxy encapsulants used in power supplies, and the other was to develop experimental techniques for validating computer models and the newly developed Arbitrary Lagrangian–Eulerian (ALE) codes, specifically ALEGRA, which were being used to simulate the power supply performance.

Although significant experimental work had been accomplished on alumina-filled epoxy in the 1970s and early 1980s, a change in the epoxy formulation used in the encapsulants necessitated additional studies. In addition, depoling mechanisms that produced current from shocked ferroelectrics were not fully understood, so additional research also focused on the electromechanical properties of the ferroelectric ceramic PZT 95/5. The objective was to measure *simultaneously* the electrical and mechanical response of the ferroelectric ceramic. The code validation effort, in particular, focused on the ALEGRA code (see Sect. 6.11), which had been initiated in the late 1980s and, as a result of intense development during this period, was in its final stages.

Chhabildas established a team of researchers to accomplish both these tasks. The team included himself, Mark Anderson, Mike Furnish, Carl Konrad, Jeffery Lawrence, Dan Mosher, Bill Reinhart, and Bob Setchell. Steve Montgomery implemented the experimental results into the SUBWAY code, which was used to predict the performance of the power supply. Later, Jeff Lawrence assumed overall responsibility as project manager for the two tasks mentioned above, that is, the better understanding of the PZT 95/5 and the improved capability to model components using ALEGRA. A significant amount of research was performed for these two experimental projects over a 3-year period. As a result, the mechanical responses of the encapsulants and the ferroelectric ceramic and the electromechanical properties of the ferroelectric ceramic were considerably better understood and modeled for component applications. These results were published in a series of papers (Furnish et al. 2000; Setchell 2003, 2005, 2007; Setchell et al. 2000, 2006, 2007a, b) and integrated into material models by Montgomery to simulate the response of ferroelectric power supplies (Montgomery et al. 2002; Montgomery and Zeuch 2004; Montgomery 2008). Several multi-dimensional experiments were also performed at STAR to validate the ALEGRA and 3-D CTH codes under complex loading conditions (Konrad et al. 2000; Chhabildas et al. 2000b, 2003; Furnish et al. 2002).

In addition to the shock wave work, funding was also available to pursue quasi-static testing. This was to have a major impact on the progress in modeling. As Montgomery states,

> During this period, [Jeff] Keck and I developed a number of specialized shock compression test assemblies to help validate models in 3-D loading configurations. Some of these tests were conducted using proton radiography at Los Alamos to visualize component deformation under explosive loading. Additional testing, first by Bob Graham and then by Bob Setchell, used projectile impacts to study component and material response under different 3-D loading histories.... [W]e were able to construct an advanced confinement vessel to allow the ceramic to be studied under complex states of stress and electrical bias across the temperature range of interest. The quasi-static loading experiments showed that the phase transformation in a ferroelectric domain was dependent upon the orientation of its spontaneous electric polarization in the stress field. This significant observation clarified many of the results from the comprehensive shock compression studies on mechanical and electric responses of the ceramic conducted by Bob Setchell, Lalit Chhabildas, Mike Furnish, and Mark Anderson. The shock compression experiments were challenging since, in addition to standard mechanical measurement diagnostics, voltages and currents from the ceramic were also measured. The experimental results provided basic data to improve material response models in the simulation code [SUBWAY] developed in collaboration with Rebecca Brannon and Joshua Robbins.

The significant efforts to understand ceramic response through the Stockpile Stewardship Program (SSP) were effectively completed in 2002. A number of additional studies at reduced funding levels were conducted in collaboration with Setchell, Anderson, and Dave Cox between 2003 and 2008 to improve the understanding of dielectric properties, the effects of initial temperature, and the composition variations on the shock compression response of the ceramic and the encapsulant materials. These studies were effective in achieving the primary motivation of a new initiative to quantify uncertainties in device performance.

## 6.8 Underground Testing on Subcritical Experiments

The SSP focused on shock wave research at the NNSA national security laboratories on the thermodynamic regimes appropriate to the operation of nuclear weapons. This initiative enabled the development of realistic material models for use with the rapidly evolving computational capabilities to simulate weapon performance in the absence of underground nuclear testing. The SSP strongly influenced the development of experimental facilities and scientific research to meet the needs of the Nation's weapons program. It had a similar effect on underground testing of nuclear weapon systems at the Nevada Test Site (NTS), which is now called the Nevada National Security Site (NNSS). As Furnish[4] states in his recollections,

> It was not long after the last nuclear test in September 1992 that the weapons design labs began actively pursuing experiments to address the requirements of the Nation's Stockpile Stewardship Program and make use of NTS. At Los Alamos, the initial subcritical experiment (SCE) slated for NTS, Rebound-I, moved to the stage of developing 'local' tests at the start of 1996. These local tests were conducted at the sponsoring laboratory [Los Alamos, Livermore, and eventually also AWE (the Atomic Weapons Establishment)], in preparation for fielding the experiments at the NTS.

LANL and LLNL had principal responsibility for performing subcritical experiments. Although not a main player, Sandia had research teams in place in Albuquerque and at NTS to specifically support the Los Alamos experiments. There was minimal Sandia participation in LLNL efforts, which began around the same time as the first LANL subcritical experiment[5] (SCE). The United Kingdom's Atomic Weapons Establishment also participated in later SCEs. In his recollections, Furnish captures an amusing anecdote about the British participation. "Vito" (also known as Etna) was a British test conducted in early 2002. The fact that the UK was

---

[4] Most of the material in this section was taken from the personal recollections of Mike Furnish, which provide an illuminating insight into Sandia's involvement in the NNSS subcritical experiments.

[5] A "subcritical experiment" refers to an experiment containing nuclear material that will not produce any nuclear yield during the experiment. A subcritical experiment is sometimes referred to as a subcrit experiment.

involved was secret at the time; in the site lingo, the AWE personnel were known as "Texans."

Sandia's long-standing development and application of shock wave technologies was the foundation for Sandia's participation in research activities in Nevada. David Thompson led the Sandia group at NTS at that time, which included Mike Burke, Rod Shear, and Jerry Chael, among others. In Albuquerque, several departments had inherited test site work from the era of underground testing. However, major changes in the support of underground testing occurred in early 1996, caused by diminishing work at NTS. A Sandia-wide Voluntary Separation Incentive Program (VSIP), instituted around this time, identified the NTS support staff as "impacted" and specified a reduction of four staff members in this area. Five staff involved in NTS work actually retired because of the financial incentives. In the Computer Sciences and Mathematics Center, the VSIP impacted the experimental shock physics research managed by Phil Stanton and also impacted the research of Mike Furnish, who was in a sister computational department managed by Paul Yarrington. Furnish was an experimental shock wave physicist who had been previously engaged in underground testing. He transferred to the NTS support department, which resolved one of the required experimental reductions and simultaneously allowed the fifth of the NTS-related staff members (Carl Smith) to retire under the VSIP. In this way, Furnish and Tom Bergstresser, who had transferred earlier from Yarrington's department, became the core Sandia team responsible for fielding shock wave diagnostics at NTS. Wayne Cook, and later Bill Boyer, managed this effort.

Under the auspices of the SSP, subcritical experiments were to be performed in Area I of the test site. Area I is a 27-square-mile expanse in the Nuclear Test Zone near the center of the Yucca Flat weapon test basin, where four atmospheric nuclear tests had been conducted between 1952 and 1955. Three underground nuclear tests had been detonated in Area I, one in 1971 and two in 1990. Since then, all SCEs have been performed at the U1a complex[6] in Area I, which is a 1000-ft-deep underground facility used for dynamic testing, including subcritical experiments that contain special nuclear material and hydrodynamic tests on hazardous materials that cannot be conducted aboveground. In the 1980s and 1990s, construction commenced to develop the underground complex further so it could be used to perform intentionally designed low-yield tests, including safety tests and other experiments that remain subcritical or produce negligible nuclear energy release. The facility was referred to at the time as Lyner, for "low-yield nuclear experiment research." Lyner was renamed "U1a" during the first subcritical experiment, Rebound-I, to emphasize that SCEs were zero-yield experiments. The schematic of the present U1a complex in Fig. 6.9 shows the location where various subcritical experiments were conducted.

Sandia's participation in the SCE program has involved two major activities: fielding advanced shock wave instrumentation and implementing advanced radiography. Sandia's part of the SCE program was managed by Larry Posey, and later by

[6] "U1a" means Underground Site A in NTS Area I.

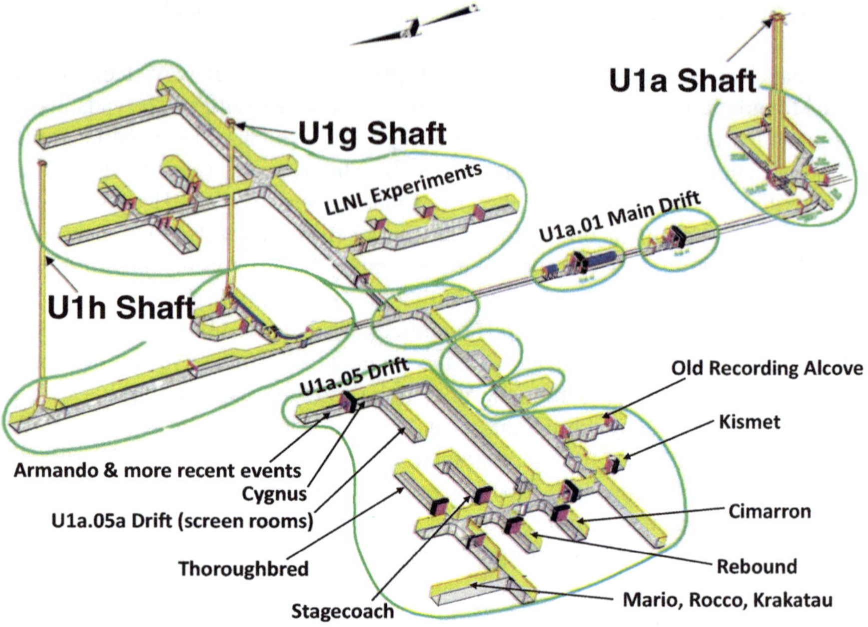


**Fig. 6.9** Schematic layout of the U1a facility at what is now called the Nevada National Security Site. The first subcrit experiment performed at the site was Rebound-I in 1997. U1a.05 was the site for the subsequent Armando, Thermos, Barolo, Bacchus, Gemini, and Leda tests. U1a.05 includes the Cygnus radiography machine, which was jointly developed by Sandia and LANL for Armando and is still in use today (Adapted with permission from "U1a Complex—Status U1a as of 6/30/14"; non-copyright drawing furnished by Bob Caccavale, NSTec)

Paul Raglin, who provided financial support for Sandia's participation and negotiated with LANL and LLNL concerning Sandia's role. Beginning with the first LANL SCE, Sandia's participation influenced the types of shock wave experiments being conducted and also served to direct Bechtel Nevada (BN) support of Sandia projects. BN had responsibility for instrumentation development and for providing personnel support to the two nuclear weapon design labs and to Sandia during subcritical experiments.

Paul Raglin managed the Sandia funding for subcritical experiments when the shock wave program on Z was being formulated in the late 1990s. Discussions between Jim Asay and Paul Raglin identified the need for additional VISAR support at Z, with the possible use of VISARs and BN personnel for Z experiments when they were not needed in Nevada. In addition to helping Z, this would keep the Sandia and the NTS subcritical personnel and instrumentation capabilities in a high state of readiness for the SCEs.

Recognizing this need, Raglin focused financial resources on the acquisition of multipoint VISAR interferometry that could be of dual use and directed the BN operational personnel, mainly Ed Marsh, Greg Mize, and Scott Walker, to use the BN VISAR in shock wave experiments on Z when not needed at NTS. This approach provided extensive experience for the BN operational technicians, who were not initially familiar with the VISAR technique, thereby enhancing their skills on subcritical experiments. This benefitted Sandia's subcritical program by maintaining the ability to participate in an SCE at a moment's notice and simultaneously provided much needed VISAR instrumentation for the fledgling shock wave effort on Z. The dual-use approach continued until the Mario and Rocco subcritical experiments, known as the MaRocco program, were conducted in 2002. Furnish comments on this in his recollections:

> …. This tactic had an advantage for Bechtel and the SCE program as well: the personnel and equipment were kept in a good state of readiness. On several occasions, the NTS schedule allowed a couple of weeks for VISAR setup, but project management was surprised to find the VISAR team ready to record after only two days.

Furthermore, Raglin's strategy considerably accelerated the development and success of the shock wave program on Z, which began in 1997, resulting in publishable results on the equation of state of deuterium and other materials by 1999 (Asay et al. 2000; Asay 2000; Hall et al. 2000; Knudson et al. 2001). This new initiative in shock physics research will be discussed in the next chapter.

Furnish was designated as the diagnostics scientist for Sandia and led the use of VISAR in several preparatory experiments at LANL prior to the actual subcrit experiments at NNSS, beginning with Rebound-I fielded in July 1997. These experiments focused on extending experimental configurations and instrumentation techniques previously developed at Sandia for accurate material measurements on Pu. This was an important contribution to the LANL SCE program. In addition to conventional material property measurements, Furnish also took the lead in implementing mass ejecta techniques previously developed by Jim Asay at Sandia (Asay et al. 1976; Asay 1978) in order to make similar measurements on Pu and other weapon materials. He successfully implemented ejecta experiments on the Cimarron subcrit in December 1998 and the Thoroughbred experiments in mid 2003. Demonstration of the method established its priority for further use in later experiments, although LANL asserted overall leadership of these experiments beginning with the Krakatau SCE in 2004. At that juncture, Sandia consequently reduced its efforts on ejecta measurements in subcritical experiments.

In addition to the dynamic shock wave experiments, Sandia also played a major role in establishing high-resolution radiography as a principal diagnostic for SCEs. John Maenchen initially advocated for and implemented an advanced radiographic technique developed at Sandia, using inductive voltage adder (IVA) technology. This technology was implemented jointly by LANL and Sandia for use on the Cygnus electron beam accelerator located in the U1a complex. Cygnus was initially intended for use only on the Armando subcrit conducted in 2004, but its success in providing high-quality radiographs prompted continued use in subsequent SCEs. As of mid 2015 nearly 3000 shots have been conducted on Cygnus at U1a[7] to prepare

[7] A comprehensive new program for an advanced radiography capability at U1a in SCEs is presently being considered by the NNSA, NNSS, and the three laboratories.

for and then analyze the small number of SCEs that are executed. Although Maenchen was the early driver for IVA technology, Brian Oliver provided continued programmatic and technical leadership for use of Cygnus at NNSS, and Mike Mazarakis was instrumental in perfecting the technology. As of this writing, Bill Stygar, Josh Leckbee, and Randy McKee of Sandia are proposing a variant of the new Russian linear transformer driver (LTD) technology as one of the options for a replacement of the Cygnus radiographic capability.

Sandia has made many significant contributions to the subcritical program, including (1) use of the Sandia-developed VISAR for material studies, (2) application of experimental methods to study material failure and mass ejecta, (3) implementation of advanced radiography, (4) participation by highly qualified personnel in shock physics and radiography, and (5) mentoring and training of many NTS technical personnel participating in SCEs. Furnish's leadership of the Sandia program produced the highly flexible VISAR systems that are still in use and implementation of the ejecta techniques that the other labs are continuing. His recollections provide a more detailed account of these efforts.

## 6.9 The Role of the Accelerated Strategic Computing Initiative in Shock Wave Research

In the 1980s, the largest computers available for numerical simulations were the Cray machines. After the final Cray YMP was procured in 1989 to 1990 and a diversity of individual workstations were purchased, several years elapsed without any new computing capability. The early 1990s marked the beginning of the end of using serial processing computers at Sandia with the advent of MPP. However, by that time all of the shock wave codes had been written for computers with at most a few processors (the workhorse Cray YMP supercomputer at the time had four processors), and a large effort was required to convert the codes to MPP. As Dennis Hayes says about the 1988 to 1989 period in his recollections:

> An important change was taking place in the computing world. Parallel computing had arrived, and we staffed [the Physics of Solids Research Department] to implement shock codes and made good progress. Another related effort was transforming the CTH code from a vertically integrated project to a full-fledged software engineering project, with benefits that exist to this day.

Sandia has a long history in the research and development of technologies to improve the scaling performance of applications on MPP computers. The first significant MPP computer acquired by Ed Barsis' Center at Sandia was the Intel Paragon, which was the progenitor of the first DOE-sponsored Accelerated Strategic Computing Initiative (ASCI) machine acquired by Sandia. The Paragon, purchased in 1993, had 143 Gflops of computing speed (143 billion floating-point operations per second) and 38 Gbytes of memory (38 billion bytes of memory). Prior to the Paragon, Sandia had purchased multiple nCUBE systems as test beds for parallel algorithm development. Researchers Robert Benner, John Gustafson, and Gary

Montry of the Parallel Processing Division of Sandia first won the Karp Prize and then won the first Gordon Bell Prize in 1987 using the nCUBE 10.

The second ASCI machine, Cray Red Storm, was acquired in 2004. It had 497 Tflops (497 trillion floating-point operations per second) and 76.8 Tbytes (76.8 trillion bytes of memory). In addition, two MPP machines were acquired by DOE, one at Los Alamos National Laboratory and one at Lawrence Livermore National Laboratory. The LANL Cray Cielo, acquired in 2011, has 1374 Tflops of computing speed and 286 Tbytes of memory. The LLNL IBM Sequoia was also obtained in 2011 with 1717 Tflops of computing speed and 1500 Tbytes of memory. It is noteworthy that CTH was used in a scaling study on Sequoia employing the entire machine and 1.5 million processors.

With this enormous increase in computational capability, extremely complex shock wave problems could be performed on CTH and other codes using realistic models of material response. As Marlin Kipp mentions in his recollections,

> We had moved from an age of punched card decks to desktops interfacing with machines of all sizes; we were now doing simulations in at least 1-D and 2-D on our desktop machines that had required room-sized machines at the start of my career.... By 2011, 3-D computing with CTH could accommodate an astounding billion cells, compared to what we started with: 6 million cells were all we could manage with CTH in the late 1980s on the Cray. We were addressing problems I never would have believed possible when I first signed on!

An example of the enormous increase in capability is the simulation of an explosive charge next to a thin aluminum plate. The explosive blast shatters the plate into a large number of small pieces, depending on the material strength and the loading rate at a specific location on the plate. With the computing capabilities of the 1980s, the Grady–Kipp fragmentation model had to be used to estimate the fragment distribution by first performing a computer simulation of the part or component to determine the loading rate at various locations and then applying the fragmentation model to estimate the fragment sizes produced at the different loading rates.

In contrast, with present computing capabilities, the individual fragments can be resolved directly and accurately without using the Grady–Kipp fragmentation model if the material properties are properly specified. Figure 6.10 shows an example of this capability, where a structure composed of sheet metal is loaded by an explosive blast. The plot gives the velocities of various components of the structure as they are torn apart. Accurately resolving these thin pieces requires a very fine numerical mesh size, which at present can be achieved using only very large-scale computing platforms (Doerfler and Vigil 2013). The comparison with experimental data provides a critical evaluation of the material properties controlling fragmentation of the plate and thus a systematic approach to optimizing material properties and the structural design to withstand a given blast load. The 3.2-billion-cell CTH simulation depicted in Fig. 6.10 was performed with 4000 processors and 1 Gbyte of memory, which is about one quarter of the Cielo computer's full capacity. These results help designers better understand the response of structures under severe blast loading conditions so that the robustness of the structures can be improved. Nevertheless, depending upon the application, specific details of fragmentation may not be needed in many situations, and Grady's analytic approach could still be a sufficient and efficient way for estimating fragment distributions.

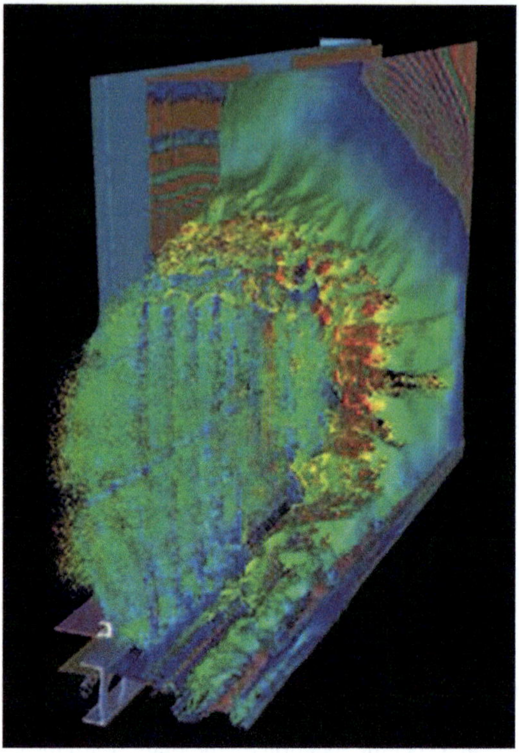

**Fig. 6.10** Sheet metal structure loaded by an explosive blast (Reprinted with permission from Attaway et al. 2011, Sandia National Laboratories)

An additional example of the continuing increase in computing capability, as of 2013, was achieved by Sandia researchers who performed a shock wave calculation using CTH on a massively parallel computer located at LLNL. An announcement was made in the *Sandia Daily News*[8] following this phenomenal success:

> CTH runs one trillion zones on over one million cores: Success! The CTH team announces demonstration of scalability to over one million cores on the LLNL HPC [high performance computer] system Sequoia. The test problem is an interacting blast wave shock physics problem in 3D. The problem consists of one trillion zones, with approximately one million zones per core. CTH is a multi-physics computational tool for simulating multiple materials in high-rate, large deformation and shock physics application[s].

## 6.10 The Shoemaker–Levy Comet Impact on Jupiter at 60 km/s

As the CTH hydrodynamic code was being developed in the early 1990s, a major event occurred that accelerated its development and resulted in international recognition for modeling hypervelocity phenomena at Sandia. This was the impact of the

[8] Bob Schmitt, CTH development team, Sandia National Laboratories, September 2013.

Shoemaker–Levy 9 (SL9) comet on Jupiter that occurred in July 1994. Mark Boslough was the first at Sandia to recognize that this event provided a compelling motivation to validate CTH's capabilities for three-dimensional modeling of hypervelocity impact events. Around a year before the expected impact, he advocated for a research effort to use CTH to accomplish this goal and, after receiving approval, formed a team to model the hypervelocity impact event. Within a year, the team was performing 3-D high-fidelity simulations of the SL9 impact that accurately predicted a vapor plume caused by the impact that would be visible from the Earth. In his recollections, Boslough provides an account of the events leading to the 3-D simulations of the SL9 impact on Jupiter. These simulations resulted in significant external recognition for Sandia and for members of the team, while establishing CTH's capabilities as the premier hydrodynamic code for predicting hypervelocity impact phenomena. A brief summary of this story is provided in this section; a more complete perspective is in Boslough's recollections.

About 20 years ago, an announcement in the International Astronomical Union Circular of March 25, 1993, was the first public description of the comet that became known as Shoemaker–Levy 9:

> It is indeed a unique object, different from any cometary form I have yet witnessed. In general, it has the appearance of a string of nuclear fragments spread out along the orbit with tails extending from the entire nuclear train as well as what looks like a sheet of debris spread out in the orbit plane in both directions.[9]

As Boslough remembers this announcement,

> It was a puzzling find and very exciting to planetary scientists because of its unique character. In those pre-World-Wide-Web days,[10] I first read about it in the July 1993 issue of *Sky & Telescope*. Co-discoverer David Levy, a regular columnist, described the serendipitous events on top of Palomar Mountain that led to Carolyn Shoemaker's first identification of the comet on damaged film, polluted with the glare from Jupiter, from an image taken just before clouds covered the sky for the rest of the night. "I don't know *what* this is," she said. "It looks like... like a squashed comet."[11]

Around that time, Boslough was a staff member in the Experimental Impact Physics Department, working with Lalit Chhabildas and others at the STAR facility

---

[9] These words, according to the book *Shoemaker by David H. Levy: The Man Who Made an Impact* (Princeton University Press, Princeton, 2000), are those of James Scotti, who was observing at the Spacewatch telescope at Kitt Peak the same night and was contacted by the Shoemakers and Levy to see if he could see the object. Scotti sent an email to Brian Marsden, the Director of the Minor Planet Center in Cambridge, MA, on March 26, 1993. A copy of Scotti's email is in David Levy, *Impact Jupiter: The Crash of Comet Shoemaker-Levy 9* (Basic Books, Cambridge, MA 1995), pp. 27–28.

[10] *Editors' note*: Tim Berners-Lee, a software engineer at CERN (the European Organization for Nuclear Research), invented the World Wide Web in 1989. However, it wasn't until April 1993 that CERN announced the technology would be available to use by anyone on a royalty-free basis. In 1994, the World Wide Web Consortium was formed.

[11] Carolyn S. Shoemaker and Eugene M. Shoemaker, "A Comet Like No Other," in *The Great Comet Crash: The Impact of Comet Shoemaker Levy 9 on Jupiter*, edited by John Robert Spencer (Cambridge University Press, Cambridge, 1995). The full quote is "I don't know what this is, but it looks like a squashed comet."

to test the effectiveness of debris shields against space debris. The goal of these experiments was to validate numerical simulations of impacts and explosions for other departments in the Computing Research Center that were focused on computational shock physics. Boslough had just received approval for an LDRD project to model the impact and seismic consequences of the dinosaur-killing impact event 65 million years ago. By modifying a few milestones, the project was redirected to include the SL9 impact on Jupiter. Furthermore, the new Intel Paragon MPP computer had just become available for hydrodynamic calculations, and its availability for planetary applications was timely.

Boslough and his collaborators developed a comprehensive research plan centered around the newly developed CTH hydrocode to simulate the SL9 impact and sought approval from Mike McGlaun, who managed the CTH development project, and from Ed Barsis, the Director of the Computer Sciences and Mathematics Center. As Boslough remembers,

> Ed Barsis gave us the green light [to proceed with the calculation]. Our goals were to test the machine with a big problem, use our results to make recommendations to astronomers for observations the following summer, and then use the observations as a validation opportunity for our models.

With Barsis' concurrence, Boslough, Dave Crawford, and Tim Trucano began utilizing the CTH code to model the impact, first in two dimensions and then in three. Allen Robinson, who was developing a 3-D version of the code referred to as PCTH for use on MPP computers, also joined the effort with the goal of porting PCTH to the Intel Paragon. This was the fastest computer in the world at the time and promised considerably increased resolution of the impact event.

As the Sandia effort got underway, planetary scientists initially hoped the impact would occur on the side of Jupiter facing the Earth. This was not to be. Boslough states in his recollections, "It wasn't long, however, before orbital dynamicists had enough data to calculate the point of impact. The collision was going to be on Jupiter's *back side,* and not directly visible from Earth, as we had all hoped." This made the job of comparing numerical simulations of the event with actual photographic data much more difficult and motivated high-fidelity computer simulations to identify a signature of the impact that would be useful to astronomers.

Boslough further notes in his recollections:

> A convergence of technologies in 1994 enabled the scientific community to make the most out of this event. It was timed perfectly. First, the Intel Paragon allowed us to model the event with sufficient fidelity to make useful predictions. Second, the Hubble Space Telescope, which had just been repaired, was producing exceptionally sharp images. Third, and more easily forgotten, was the role of the Internet that was just beginning to connect various research institutions, allowing rapid dissemination of information.

Development of CTH proceeded at a fast pace from July onward, and the code was beginning to produce interesting results by fall 1993. Boslough commented on the increased understanding being obtained through the Sandia calculations: "By October 1993 Tim Trucano had completed the first two-dimensional simulations showing how the comet fragments would break up as they entered Jupiter's atmosphere."

Trucano's simulations provided an estimate of the fragment sizes that would impact Jupiter's liquid surface and were crucial to understanding the observable effects of fragments traveling at around 60 km/s and impacting the surface. As Boslough noted in a press release for a meeting of the American Astronomical Society's Division for Planetary Sciences, held in Boulder, CO, October 18, 1993:

> Sandia National Laboratories researchers have performed supercomputer simulations to find out what will happen when Comet Shoemaker-Levy 9 collides with Jupiter next summer. Some astronomers have predicted that the impact will be one of the most spectacular celestial events ever witnessed. Telescopes and spacecraft are being rescheduled to observe the resulting display, but what will they see? To help answer that question, the Sandia scientists have made use of a computer code of the type that was originally developed to understand what happens within nuclear weapons.... The computer simulations show that when the comet enters Jupiter's atmosphere, the pressure increase is gradual. The atmosphere is thin at the top, so for the first second or so, the comet slices through almost unhindered. In the next second, however, the pressure builds up rapidly and deforms the comet until it begins to break up.... After the comet begins to break apart, energy is released very rapidly by mechanisms that are not well understood. The Sandia calculations are concerned mainly with the processes that lead to the breakup. One of the surprising results is that, for a collision like the one expected next summer, the comet is torn apart by the large deformations due to atmospheric drag.

In the fall of 1993, Allen Robinson's team was actively pursuing development of the PCTH code that could be run on the Intel Paragon. This was an important advancement because the enormous increase in computing speed and memory available on the Paragon provided considerably better definition of fragment interaction with Jupiter's atmosphere and helped to clarify how the fragments would break up. This was especially important to determine if the size of the impact-generated plume would be observable from the Earth, since the impact was projected to be in Jupiter's shadow but near the boundary (limb) with the observable surface. Early in 1994, the capability to perform 3-D hydrocode calculations on the Intel Paragon was finally available. Boslough describes the very first 3-D simulations of the comet impact obtained by the planetary community:

> In early 1994, Dave had run 3-D simulations for various assumed fragment sizes on the Paragon, showing that the fireballs would be ejected at much higher speeds and to higher altitudes than the other groups had predicted. We also showed that, for an inclined impact (the fragment entry angle would be 45° from the local horizon), the plume would be ejected along the wake – at the same angle but in the opposite direction. Because we had access to the most powerful computer in the world, we had the advantage of higher resolution and three-dimensionality, both of which were required to make these predictions.

This was an important development and provided hope that the resulting plume would be observable from the Hubble Space Telescope and would allow astronomers to infer information about the impact. Mike McGlaun emphasized this groundbreaking development in his recollections:

> Allen's team got a version of PCTH working that demonstrated excellent parallel speedup and good performance.... Mark Boslough, David Crawford, and Tim Trucano were some of the first users. In 1994, they analyzed the Shoemaker-Levy 9 comet impacts on Jupiter using 3-D calculations on the Intel Paragon. PCTH proved spectacularly successful. Astronomers could not observe the comet impacts because they were on the side of Jupiter

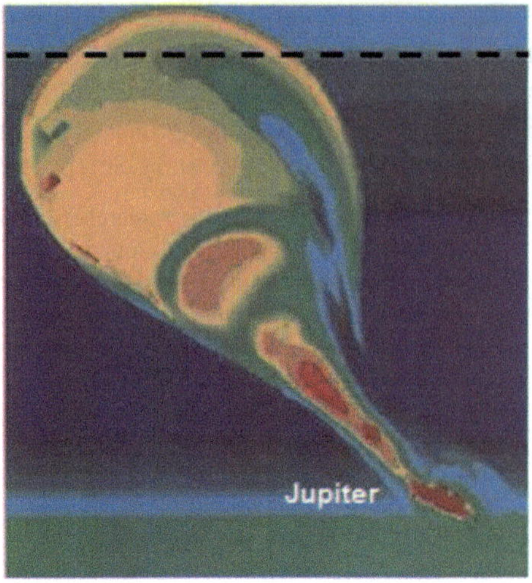


**Fig. 6.11** Simulated impact of a 3-km fragment 55 s after impact on Jupiter. The plume debris above the dashed line would be visible from the Earth. Temperature in the plume is represented by color: blue = 100 K, red = 3300 K (Reprinted from Crawford et al. 1994, Fig. 4, with permission of Springer Science + Business Media)

> facing away from Earth. Our calculations indicated an optical signature should be visible over the apparent visual edge (limb) of Jupiter. Astronomers could infer information about the impacts from that optical signature. Mark, Dave, and Tim received a Sandia Quality Award and published several papers on this work. It was a great demonstration of PCTH's capabilities and the power of massively parallel computers.

Rapid progress was made by the Sandia team in developing the code to the point where it could accurately model the impact and the resulting vapor plume, which they predicted to be visible from Earth. The pioneering research performed by this dedicated team of researchers represented the first 3-D calculations of a large and complex impact event and established the standard of performance for future MPP computer applications.

Figure 6.11 shows a CTH simulation of a 3-km fragment of the comet that consists of water ice traveling at 60 km/s that entered Jupiter's atmosphere about 55 s before impact. The calculation indicates that the fragment deposits most of its six megatons of kinetic energy in the atmosphere in the first 10 s. This accounts for the water ice debris cloud and shock-heated hydrogen and helium atmospheric gases entrained in the plume.

Figure 6.12 shows a comparison of the calculated plume size observable from the Earth for several minutes after impact, along with Hubble Space Telescope (HST) images taken on July 16, 1994, at the corresponding times (Hammel et al. 1995; Boslough et al. 1995a, b; Boslough and Crawford 1996, 1997; Crawford et al. 1994, 1995). In the HST images, the string of fragments making up the comet tail was identified alphabetically. The "G" fragment was about 3 km in diameter. Shown in Fig. 6.12a are 3-D CTH calculations for the plume formed by a 3-km fragment at different times after impact with Jupiter. Shading indicates log (density) with a cut-off at 10-12 g/cm$^3$; times are in minutes after impact (Boslough and Crawford 1997). Since the impact point was not visible from the Earth, the calculated evolution of the fireball was important to astronomers who would be able to view only

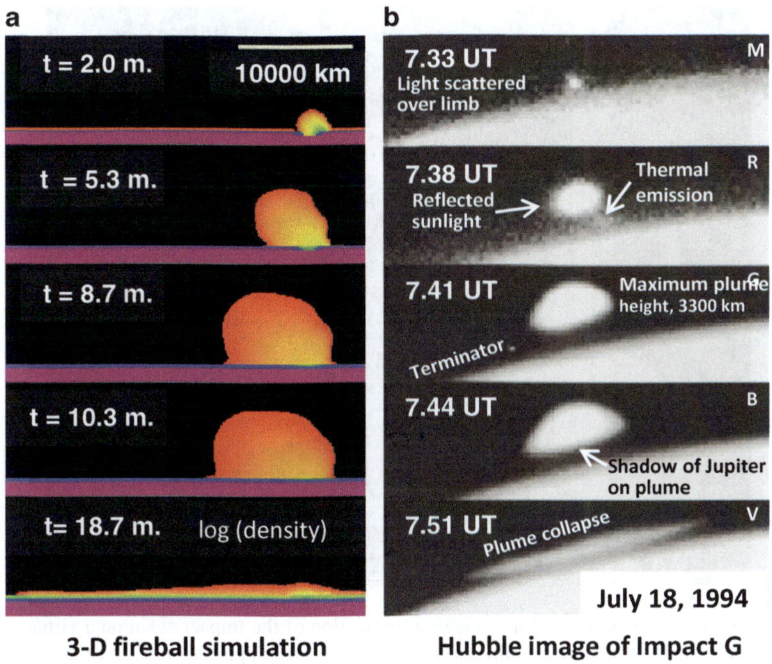


**Fig. 6.12** (**a**) Simulations of 3-D fireball/plume evolution after impact of a 3-km-diameter fragment with times shown in minutes. (**b**) HST images obtained at different times after impact for a similarly sized fragment (Reprinted with permission from Boslough and Crawford 1997, Fig. 6, Copyright 2006, John Wiley and Sons. NASA and the Space Telescope Science Institute (STScI) are acknowledged for use of the Hubble image

part of the plume after impact. HST images of the impact are shown at similar times to the CTH calculations in Fig. 6.12b. In his recollections, Boslough captures a comment from one of the astronomers, Heidi Hammel, upon seeing the photographs of the plume for the first time.[12] In an email communication with Mark in January 2015, Heidi recalled, "We subsequently put together a time lapse set of Hubble images showing the G plume's rise and descent. They were so eerily like Sandia's predicted models that I showed them side by side for years afterwards."

The comet impact on Jupiter, starting on July 16, 1994, was one of the most spectacular events that have been observed from Earth. The Sandia team received considerable international recognition for their groundbreaking and inspiring research of the SL9 impact on Jupiter. The fact that it was possible to accurately predict the evolution of the ejected plume with the computing capabilities and codes that were available in the 1990s is truly remarkable and a tribute to the talents, energy, and dedication of the team. Their pioneering results were enabled by the

---

[12] Heidi Hammel was a Principal Research Scientist at MIT and a member of the Science Observation Team for the Hubble Space Telescope Jupiter Campaign in charge of the Wide Field and Planetary Camera that was used to photograph the impact plume.

**Fig. 6.13** Eugene Shoemaker (sitting) views a replication of the impact event on a virtual reality system at Sandia. Behind him are Jim Asay (*left*), Mark Boslough (*center*), and Craig Peterson (Reprinted with permission of Sandia National Laboratories)

unmatched capabilities of the 3-D PCTH hydrodynamic code and the use of massively parallel computing.

To focus international attention even further on these unique results, Mark organized a special session of the Hypervelocity Impact Symposium (HVIS) that was held after the impact event in Santa Fe, NM, in July 1994. Gene Shoemaker gave the keynote address at the Symposium, which included Carolyn Shoemaker and a large number of international researchers who presented poster board results of their work on the comet impact. The CTH results were especially well received by the computational, planetary science, and shock physics communities. Figure 6.13 shows Gene Shoemaker reviewing the Sandia results a few days before his keynote address at the Hypervelocity Impact Symposium.

## 6.11 ALEGRA: The Next-Generation Hydrodynamic Code

As the second-level manager of the Solid Dynamics Research Department, Dennis Hayes began an initiative in 1989 to develop a robust 3-D hydrodynamic code that included magnetohydrodynamic (MHD) phenomena. The Pulsed Power Sciences Center had recognized the need for such a code to study the high-temperature plasmas and current flows being produced on the Z accelerator (sometimes called the Z machine, or Z), shown conceptually in Fig. 6.14. At that time, Z was used to

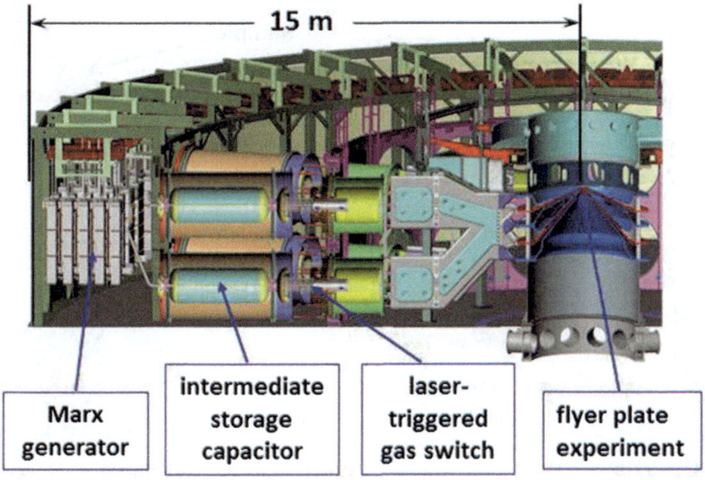


**Fig. 6.14** Schematic of the half-diameter of Z (sometimes referred to as the Z accelerator). Electrical energy is stored initially in the Marx generators. When the machine is fired, the energy is switched through a series of intermediate states that result in energy delivered to a target in the central vacuum chamber in times on the order of 100 ns. In this illustration, a flyer plate experiment is located in the vacuum chamber (Private collection, G. R. McKee, 2014)

produce currents of about 20 million amperes (20 MA) in times of 100–400 ns. (The 2007 refurbished version of Z can now produce over 25 MA with specific current pulse shaping over slightly longer time scales.)

The machine has been used for both experiments using x-ray-generated plasmas for inertial confinement fusion (ICF) applications (Sweeney 2002) and for material property experiments at high pressure using pulsed magnetic fields (Asay and Knudson 2005). In the plasma experiments, the high currents are used to vaporize a cylindrical array of tungsten wires to a plasma state. The top of Fig. 6.15 shows a typical wire array configuration used to create high-temperature plasmas. Typically, very thin tungsten wires in an array containing about 300 individual wires are used to produce a z-pinch plasma. During application of the current pulse in the machine, each wire conducts a portion of the total current and heats to a plasma state. The resulting circumferential magnetic field created around the wire array by current flowing in the plasma sheath interacts through the Lorentz force to accelerate the plasma toward the central axis (the z axis) of the wire array. For this reason, the name "z pinch" is used to describe the final stagnation of the plasma. The cylindrical plasma typically achieves velocities of about 100 km/s before stagnation.

Stagnation of the converging plasma produces temperatures of several million degrees and soft x-ray emission from the z-pinch plasma. The intense radiation can be used to implode a small capsule located inside a hohlraum containing the z pinch. Implosion of the capsule containing a hydrogen-based fuel mixture occurs through energy absorption on its outer surface, causing rapid convergence of the container and subsequent high densities and temperatures in the hydrogen mixture.

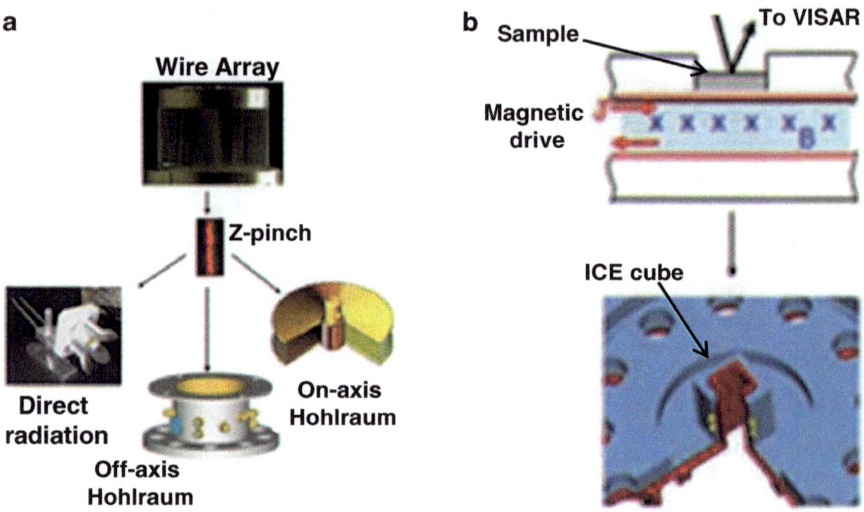


**Fig. 6.15** Test configurations used on the Z machine. (**a**) In plasma experiments, a wire array is used to generate intense z-pinch plasmas. (**b**) Configuration used to produce magnetic loading of a planar sample (Private collection, J. R. Asay)

Spectroscopy and other diagnostics are used to investigate the physical processes important for ICF implosions using pulsed power. The capsule can be either directly irradiated with the z-pinch radiation from an on-axis hohlraum, or indirectly irradiated off axis, as shown in the figure. The current approach to ignition with pulsed power being investigated on Z is magnetized liner inertial fusion (MagLIF). That direct-drive concept uses a laser to preheat the fusion fuel within a cylindrical liner that has been compressed by the magnetic field. Preheating the fuel reduces the compression required to achieve ignition temperatures, and the presence of an applied magnetic field inhibits thermal conduction, thereby enhancing alpha particle deposition within the fuel (Slutz et al. 2010; Sefkow et al. 2014; Gomez et al. 2015).

The high-temperature plasma can also be used to irradiate a thin sample disk and drive a planar shock wave in the disk by ablating the front surface. This approach was first attempted during the early development in 1997 of shock wave capabilities on Z (Hall et al. 2000), but was discontinued because the configuration did not provide a spatially uniform, constant-pressure drive. This major limitation severely affected the accuracy of shock wave velocity measurements using the z-pinch source. Subsequently, as described in detail in Chap. 7, shock wave researchers at Sandia discovered that the optimal configuration for shock wave experiments was not a z-pinch source but instead a short circuit to produce high-pressure magnetic loading of material samples, which is illustrated at the bottom of Fig. 6.15.

A new hydrodynamic code coupled with electromagnetic effects was needed to model the phenomena in z-pinch experiments and later in the magnetically driven experiments also shown in Fig. 6.15. Therefore, a coupled MHD code had to include not only the accurate equation of state and material properties already in conventional hydrodynamic codes, but also magnetic and electrical effects. The framework for the MHD code was based on Sandia's previously developed hydrodynamic codes, as exemplified by CTH and PRONTO, which had been integrated into the new ALE coding structure. The ALE formulation allowed material to flow through a stationary mesh (pure Eulerian), to move with the material (pure Lagrangian), or to move independently from the material motion. This approach permitted a calculation to proceed in Lagrangian fashion until the mesh became too highly distorted, at which time the highly deformed mesh points could be moved to reduce the distortion to acceptable levels. The resulting Sandia MHD code ALEGRA is the finite element-based version of the ALE technique. Tom Haill gives a detailed account of ALEGRA development and application in his recollections.

Dennis Hayes from the Solid Dynamics Research Department and Don Cook of the Pulsed Power Sciences Center initiated a jointly funded program to build ALEGRA in early 1989, and Jim Asay continued the project when he became the Solid Dynamics Research Department Manager in late 1989. James Peery was initially the project leader for this effort until he left the Department in the late 1990s and Allen Robinson took it over.

ALEGRA development began in the early 1990s under the direction of Mike McGlaun. As McGlaun recalls,

> Tim Trucano and I began investigating a beyond-CTH shock physics code in the late 1980s. Sandia's inertial confinement fusion community needed capabilities that CTH did not provide. We believed the best underlying strategy would be to develop an Arbitrary Lagrangian-Eulerian (ALE) code. The mesh in an ALE code could be Eulerian or Lagrangian or have an arbitrary motion.
>
> James Peery, Kent Budge, and Mike Wong began developing a 3-D ALE code named RHALE in 1990. They studied the CTH algorithms as well as the algorithms in Sandia's PRONTO finite-element Lagrangian transient dynamics code. The plan was to use the best of both. They decided to use C++ rather than FORTRAN. Later they renamed the code ALEGRA. This was a very challenging project. ALEGRA demonstrated tremendous capability. It is a specialized code and not appropriate for all applications.

ALEGRA subsequently evolved into a full 3-D MHD code with state-of-the-art EOS capabilities and the basic models for a broad range of materials. It incorporates realistic models of material conductivity and plasma properties and is routinely used in designing and modeling z-pinch experiments. The code has also been used for modeling Sandia weapon components. ALEGRA has had a significant impact on dynamic material properties experiments conducted on Z. This and other theoretical applications will be illustrated in Chap. 7.

## 6.12 People and Places of the 1990s

**Barry Butcher, promoted to Distinguished Member of Technical Staff (DMTS), 1990 (Reprinted with permission of Sandia National Laboratories)**

**Jeff Quintenz, promoted to Senior Manager, 1998 (Reprinted with permission of Sandia National Laboratories)**

**Mike McGlaun, promoted to Supervisor of Computational Physics and Mechanics Division, 1990 (Reprinted with permission of Sandia National Laboratories)**

**Steve Montgomery promoted to Distinguished Member of Technical Staff, 1999 (Reprinted with permission of Sandia National Laboratories)**

**Marlin Kipp, promoted to Distinguished Member of Technical Staff, 1992 (Reprinted with permission of Sandia National Laboratories)**

**Peter Chen, promoted to Distinguished Member of Technical Staff, 1996 (Reprinted with permission of Sandia National Laboratories)**

**Billy Thorne, promoted to Distinguished Member of Technical Staff, 1990 (Reprinted with permission of Sandia National Laboratories)**

**Mike Desjarlais, promoted to Distinguished Member of Technical Staff, 1996, and to Senior Scientist in 2000 (Reprinted with permission of Sandia National Laboratories)**

**Mel Baer, promoted to Senior Scientist, 1998 (Reprinted with permission of Sandia National Laboratories)**

**Walt Herrmann, promoted to Laboratory Fellow in 1992 and elected to the National Academy of Engineering in 1993 (Reprinted with permission of Sandia National Laboratories)**

**Rick Spielman standing above the Z pulse-forming section, 1996 (Reprinted with permission of Sandia National Laboratories)**

**Clint Hall (*left*) and Bill Reinhart loading a projectile in the powder gun (Reprinted with permission of Sandia National Laboratories)**

**Lalit Chhabildas elected American Physical Society Fellow in 2000, private collection, L. C. Chhabildas**

**Bob Graham receives Shock Compression Science Award "in recognition of his research achievements in shock compression of condensed matter and in particular the development of piezoelectric shock wave instrumentation" (Reprinted with permission from the SCCM-1993 Proceedings, ed. by M.D. Furnish et al., Copyright 1994, AIP Publishing LLC)**

# Chapter 7
# The 2000s: A New Millennium

## 7.1 Background

After the turn of the century, dramatic changes occurred in experimental and theoretical shock wave research at Sandia. In the 1950s and 1960s, computational capabilities to design and interpret shock wave experiments were extremely limited. Use of slide rules and small desktop calculators was common. In the 1950s instruments to measure the fine details of shock compression and dynamic material response were limited. In those early decades, the innovation and intuition of experimentalists and modelers were critical in advancing shock wave research in spite of these constraints. By the late 1960s, dynamic phenomenological models to describe the shock compression of complex materials such as composites and porous materials were beginning to be established. The pioneering loading and diagnostic technology developed in the 1960s was instrumental in advancing knowledge in the later decades. This was especially true for high-pressure applications until the mid 1970s, when time-resolved gauges became available for routine use at Sandia. Bob Graham and his team concentrated on developing the piezoelectric gauge, which is also known as the quartz gauge (Graham 1961a, b; Graham 1975; Neilson and Benedick 1960; Neilson et al. 1962; Graham and Ingram 1968; Graham and Reed 1978). Meanwhile, Lynn Barker and his team concentrated on developing optical interferometric gauges and, in particular, the Velocity Interferometer System for Any Reflector (or VISAR), which grew out of a wide-angle version of the Michelson interferometer (Barker and Hollenbach 1965, 1972; Barker 1968, 2000a). This allowed considerable progress in understanding dynamic compression processes. Among early researchers, there was an excitement in developing pioneering new capabilities and solving complex dynamic material problems using the new gauges.

By 1990 many early shock wave researchers who had made seminal contributions had left Sandia or transferred to other organizations not directly involved in that research. Of those remaining, a large number retired in the 1990s and early 2000s. Some of those who left in the 1990s and later include Lynn Barker and Jace



J.R. Asay et al., *Impactful Times*, Shock Wave and High Pressure Phenomena,
DOI 10.1007/978-3-319-33347-2_7

Nunziato, 1990; Peter Chen, Lee Davison, Walt Herrmann, and Orval Jones, 1993; Al Chabai, Charlie Daniel, Bob May, Ray Reed, and Sam Thompson, 1994; Gerry Kerley, 1995; Dennis Grady, Peter Lysne, Bob Graham, Dennis Hayes, Karl Schuler, Carl Smith, Phil Stanton, and Al Stevens, 1996; Barry Butcher and Bruno Morosin, 1997; Dennis Mitchell and Steve Passman, 1999; Dick Rohde, 2000; Jim Asay, 2002; Darrell Munson and Jeff Swegle, 2003; Doug Drumheller and Larry Pope, 2004; Jeff Lawrence and Herb Sutherland, 2005; Bob Hardy and Mike McGlaun, 2006; Lalit Chhabildas, 2007; Bob Setchell, 2008; Mel Baer and Wayne Trott, 2010; and Marlin Kipp, Steve Montgomery, and Paul Yarrington, 2011. The loss of these key researchers left voids in certain technical areas, but Sandia had the foresight to hire extremely talented people in the late 1990s and early 2000s, thereby sustaining its reputation as a premier shock wave research organization.`

Personnel losses also had an adverse effect on maintenance and operation of the shock wave facilities constructed in previous decades. With Bob Graham's retirement in 1996, Bob Setchell was the remaining experimentalist from George Samara's original department involved in shock wave research during the late 1990s. Setchell continued to use the helium gun facility built by Graham in the 1960s for research on a variety of topics, including the refractive index of sapphire under shock compression (Setchell 2002); mechanical, shock-induced depoling properties; and microstructural effects of shock-compressed FEs (Setchell et al. 2000, 2006, 2007a, b; Setchell 2003, 2005, 2007). Setchell's latter results led to significant refinements in the material models for ferroelectric ceramics during this period (Montgomery et al. 2002; Montgomery and Zeuch 2004; Montgomery 2008). After Setchell retired in 2008, the helium gun facility discontinued all operations and was transferred to the STAR facility.

Mel Baer and Wayne Trott had established a coordinated research program to investigate mesoscale mechanisms of shock initiation and growth in energetic materials using a Line VISAR. When they both retired in 2010, this innovative research effort suffered a serious loss, but it is being continued by Marcia Cooper and other researchers. Today, energetic materials research continues at the ECF, and to some extent at STAR, but with an increased focus on weapon applications instead of fundamental research.

A significant change in experimental shock physics occurred in the late 1990s with the development of the magnetic shock wave drive on Z.[1] This fortuitous event happened through a combination of circumstances. These included (1) the NNSA Stockpile Stewardship Program initiated through the National Defense Authorization Act of 1994, which identified a need for focused shock wave research at the *extremes*

[1] Z is a powerful pulsed power accelerator at Sandia that began its life in October 1996 by producing intense x rays from imploding z-pinches following a three-month modification of the then 11-year-old PBFA II (Particle Beam Fusion Accelerator II). Since June 1997, Z has been used for experimental studies in shock physics, weapon effects, radiation transport, inertial confinement fusion, and dynamic material studies. At the end of July 2006, Z was taken offline for a complete refurbishment to replace outdated components, enhance reliability and precision, allow pulse-shape flexibility, increase diagnostic access, and improve high-energy performance. The first shot following this extensive refurbishment was in September 2007.

of pressure and temperature; (2) the breakthrough in z-pinch technology by Tom Sanford in 1995 (Sanford et al. 1996; Sweeney 2002), which increased the power output of plasma sources by several orders of magnitude; (3) the foresight and the risk taken by the Pulsed Power Sciences Center management in exploring new opportunities for materials research using Z; (4) the early strategic decisions to develop shock wave capabilities critical to the SSP; (5) the coalescence of several experienced individuals in shock physics and pulsed power who stuck through early struggles to find a reliable and accurate source for producing shock and ramp waves with Z; and (6) the hiring of highly qualified university graduates who brought new ideas and enthusiasm to the budding shock wave research program in the early 2000s.

Another significant development was the effort to implement MPP at Sandia that began in the 1990s and grew rapidly in the early 2000s through the Accelerated Strategic Computing Initiative (ASCI). The 3-D MHD code ALEGRA (Haill et al. 2003; Robinson et al. 2008, 2011) benefitted directly from this development and initially focused on pulsed power and z-pinch physics applications. The extremely large problems that could be solved with MPP computers also enabled *ab initio* quantum molecular dynamics (QMD) equation of state (EOS) calculations, initially for low-molecular-weight elements and then for a broader class of materials. Highly talented theoreticians, including Mike Desjarlais, produced a focal point for this unique capability in the Pulsed Power Sciences Center that allowed integration of QMD theories with the highly accurate shock wave data being obtained on Z by Marcus Knudson, Clint Hall, Dan Dolan, Seth Root, and others. This combined team of theoretical and experimental researchers made a series of breakthroughs in generating accurate EOS data that are still inaccessible on other experimental platforms. The combination of experiment, theory, and computation under a single (second level in management structure) department represented a return to the unified approach advocated by Don Lundergan and implemented by Walt Herrmann in 1966.

A number of factors were responsible for rejuvenating shock wave research at Sandia: (1) the discovery of Z as a high-quality shock and ramp wave driver, (2) the new researchers joining the shock wave program, and (3) the emphasis on fundamental understanding required through the SSP. The researchers who joined the Sandia shock wave program in the late 1990s and later included Scott Alexander, Justin Brown, Kyle Cochrane, Jean-Paul Davis, Mike Desjarlais, Dan Dolan, Kathleen Holland, Marcus Knudson, Rudy Magyar, Seth Root, Chris Seagle, Thomas Mattsson, Heath Hanshaw, Luke Shulenburger, and Tracy Vogler. Each of these individuals brought new skills and approaches that have resulted in a rich, diverse shock wave program. Moreover, all of the gun launchers previously constructed at the STAR facility became active and are used today for various internal and external programs.

With Asay's retirement in 2002, the materials research was managed under Senior Manager Dillon McDaniel; Z activities in shock wave research continued under the leadership of Chris Deeney. Lalit Chhabildas, who was already managing the STAR research programs, was promoted to Department Manager of those programs in 2003. After Chhabildas' retirement in 2007 and Deeney's transition to a

Washington assignment in 2006, Clint Hall was promoted to Department Manager and the two shock wave activities at Z and STAR were combined into a single department, with Thomas Mehlhorn serving as the Senior Manager. Several innovative and strategic directions continue to influence shock wave research at Sandia today; these included hiring Dawn Flicker to lead the theoretical studies from 2008 to 2011. Until recently, materials research at Z and STAR was under Mark Herrmann, the Senior Manager of the High Energy Density Science Department, with Dawn Flicker managing the experimental dynamic materials programs on Z from 2011 to 2013, Thomas Mattsson managing the theoretical studies from 2011 to present in HEDP[2] Theory, and Gordon Leifeste managing STAR and Veloce in Solid Dynamics Experiments, beginning in 2011. A new HED[3] Material Physics Department was created in June 2013, with Dawn Flicker being promoted to the Senior Manager of that department in July 2013 and Mark Herrmann becoming the Director of the Pulsed Power Sciences Center in September 2013. John Benage (formerly of LANL) became Manager of Dynamic Material Properties in 2013. During the temporary management of the Center by Duane Dimos after Mark left in September 2014[4] and with the return of Keith Matzen as the Center Director (after a "sabbatical" of about 2–1/2 years as the Director of Nuclear Weapons Science and Technology Programs), shock wave research has continued to prosper in the Center.

The general approach to shock wave research in the 2000s at Sandia is significantly different from the previous decades. Early research using explosives and gun launchers was relatively inexpensive (a few thousand dollars per shot in today's dollars), and experiments could be conducted within a few days of being defined. Typically, only a few people, in addition to the researcher, were involved in an experiment. Although this traditional approach continues at ECF, STAR, and Veloce (which will be discussed later), much of the emphasis on basic shock wave research is now performed on the Z. Z experiments are very expensive (a hundred thousand dollars or more for each shot), and the planning and scheduling of experiments must begin at least 3–4 months in advance. In addition, the EOS requirements and technical approach for each experiment are now subjected to intense scrutiny by the national HED Council before approval. This puts considerable emphasis on ensuring that proposed experiments will meet the programmatic needs of the NNSA SSP or other major programs. Once specific experiments are approved, the preparations usually involve detailed theoretical and computational analyses, and a coordinated team effort, including up to 50 members of the operations crew who are involved in executing experiments on Z. As a specific example related to shock wave research, the design of a magnetically-driven flyer plate experiment requires extensive calculations with ALEGRA and circuit codes to define the temporal current profile to prevent melting or vaporization of the flyer plate by either current diffusion or intense shock formation.

[2] HEDP is the acronym for high energy density physics.

[3] HED is the acronym for high energy density.

[4] Mark Herrmann is now Director of the National Ignition Facility at Lawrence Livermore National Laboratory.

Despite these constraints, a large fraction of the Z shot schedule is now dedicated to materials experiments. The unique pressure and temperature regimes accessible in the Z materials research program keep it firmly integrated into the missions of the national SSP. With the magnetically-induced ramp loading capability available on Z, Mbar-level stress can be achieved in samples at low temperatures. Only diamond anvil cells (DACs) can produce isothermal hydrostatic pressure states, but presently the upper pressure of DACs is around 3 Mbars and the samples are typically a few microns in size, as compared to the mm-sized samples for isentropic compression experiments on Z. Magnetically-launched flyer plates on Z produce highly accurate shock pressures to tens of Mbars with correspondingly extreme temperatures. This permits researchers to access the thermodynamic region known as high energy density physics. This is a region of material response that encompasses extremely high pressures, densities, and temperatures in solids, liquids, and plasma states. These states of matter are produced in planets, stars, and nuclear weapons. At the opposite extremes are material states at very high pressures, but low temperatures and at very low densities and pressures but at high temperatures. A combination of shock-ramp (shock loading to an intermediate state, followed by further ramp loading to peak pressure), ramp-shock (ramp loading to high pressure followed by shock loading), or shock loading followed by unloading experiments has the potential to access *many* of the thermodynamic states that lie between these extremes. Although intense lasers like the National Ignition Facility (NIF) can achieve higher shock pressures than Z, laser experiments do not intrinsically provide the accuracy compared to magnetically-driven flyer plates of the Hugoniot, unloading, and quasi-isentropic loading EOS data over the wide range of thermodynamic states accessible by Z (Knudson et al. 2003b, 2004).

The highly accurate EOS data, in turn, provide a critical evaluation of *ab initio* theories and the means to differentiate among various theories. The Z shock physics program has produced a number of firsts. Examples include (1) highly accurate shock Hugoniot data for deuterium to nearly 2 Mbars (Knudson et al. 2004, 2015), as discussed in Sect. 7.4, that resolved the controversial shock Hugoniot data obtained on the Nova laser at LLNL (Collins et al. 1998; Knudson et al. 2003a, b, c); (2) discovery of an ultrahigh-pressure triple point of carbon (Knudson et al. 2008), which has implications for planetary compositions and masses; and (3) determination of the ultrahigh-pressure EOS of water (Knudson et al. 2012), which has inferences for the masses of giant planets, such as Neptune, Uranus, and the exoplanets.

The major developments in shock wave research in the 2000s that are discussed in this chapter include:

- Development of Shock Wave Capabilities on Z (Sect. 7.2)
- Ramp Loading to Multi-Megabars (Sect. 7.3)
- Magnetically-Driven Hypervelocity Flyer Plates (Sect. 7.4)
- *Ab Initio* Equation of State Theories (Sect. 7.5)
- Containment of Toxic Materials (Sect. 7.6)
- Compact Pulser: Veloce (Sect. 7.7)
- Magnetically Applied Pressure Shear (MAPS) (Sect. 7.8)

- STAR in the New Millennium (Sect. 7.9)
- Effects of Shear Stress on Granular Material Compaction (Sect. 7.9.1)
- Reverse Taylor Impact Studies (Sect. 7.9.2)
- Shock-Induced Vaporization: Kinetic Effects (Sect. 7.9.3)

## 7.2 Development of Shock Wave Capabilities on Z

From the mid 1950s to the early 1990s, the principal methods used for shock wave studies involved either plane-wave explosive lens or flat plate impact from gun-driven launchers. In the mid 1990s, high-powered lasers were being developed at many laboratories for ICF applications (Da Silva et al. 1997). A spin-off of this evolving discipline involved deposition of the laser energy onto a small flat disk of a few hundred microns thickness or into a gold-plated hohlraum that releases a soft x-ray pulse, typically a few ns in duration, which is subsequently deposited onto a flat disk. The rapid heating and vaporization of a surface layer produces an intense shock wave that can be used for EOS studies (Da Silva et al. 1997). However, the laser-generated shock wave is not steady (constant pressure amplitude), which can seriously complicate EOS studies. Nonetheless, the laser shock technique has matured over the past 20 years, with the result that EOS properties can now be measured at multi-Mbars with accuracies of a few percent.

At the time, laser technologies were developed for ICF and EOS applications at the other laboratories, fast pulsed power techniques were pursued for both ICF and weapon applications at Sandia. Fast pulsed power is defined as the ability to deliver large amounts of current (tens of megaamperes) to a load in times of a few hundred nanoseconds (ns). This capability was developed progressively through a series of pulsed power systems, extending from SPASTIC and HERMES I in the 1960s to PBFA I and II in the 1990s (Sweeney 2002). In 1996 the research on PBFA II changed directions, moving away from a light ion source to use as a current source for imploding wire-generated plasmas. The converted machine was named the Z accelerator (often called the Z machine or simply Z) and generated about 20 MA current to implode cylindrical plasma magnetically. Z was used for almost a decade for ICF research until the Congress appropriated resources for a major refurbishment in 2004 to increase the output current and operational efficiency. The upgraded Z was completed in 2007. The new machine generates over 25 MA of current and consequently higher pressures for EOS experiments, since the peak pressure scales as the square of the current.

The fast pulsed power capability led to a revolutionary advance in the ability to perform ultrahigh-pressure ramp and shock wave experiments. Z originally built to produce intense x-ray line sources, provided an opportunity for the rebirth of a robust experimental shock wave program at Sandia. The beginnings of this capability started with Don Cook, who served as the Director of the Pulsed Power Sciences Center from 1993 to 1999. In the summer of 1996, Cook requested that Jim Asay form a shock wave program for high-pressure EOS studies on Z. As Asay remembers:

> The transition back to shock wave research began one morning during the early summer of 1996 when Don stopped by my office.... [H]e wanted to develop a shock wave capability using Z.... The motivation for this new thrust arose from reviews of Sandia's pulsed power sciences program that emphasized the need ... to study high energy density physics (HEDP) to support the newly established NNSA (National Nuclear Security Administration) Stockpile Stewardship Program.

Asay formed a small group initially consisting of Lalit Chhabildas, Clint Hall, and himself. Chhabildas focused on reestablishing all the past capabilities at the STAR facility with Bill Reinhart, while Hall and Asay concentrated on establishing a new shock wave capability on Z. Hall and Asay worked together at the beginning of this effort, with Marcus Knudson joining the group in 1998. The three worked closely as a team, with the support of several staff at Z in developing shock wave capabilities over the next 5 years. Carl Konrad also joined the effort shortly after it started and was instrumental in implementing the VISAR techniques on Z. After Asay's retirement in 2002, Hall and Knudson, along with Chris Deeney who replaced Asay as manager, considerably expanded the Sandia shock wave program on Z to an internationally recognized effort.

In October 1996 the principal mode of operation at Z employed a cylindrical array of several hundred thin, high-density wires (about 1/10 of the diameter of a human hair) that are vaporized to form a conductive plasma when large currents are applied. The resulting circumferential magnetic field around the array accelerates this conducting plasma toward the center (the z axis, thus the term "z pinch"). Stagnation of the imploded plasma within a cylindrical enclosure (referred to as a hohlraum) produces an even higher temperature plasma that radiates intense x rays with energies of a few eV (a few tens of thousands of degrees) over a time period of a few ns. The interior of the gold-plated hohlraum acts as a blackbody radiator and reemits x rays at a specific wavelength. This intense x ray burst is used to implode a small capsule containing a hydrogen-based fuel mixture for ICF applications. It could also be used to irradiate a thin cylindrical sample disk placed on the outer diameter of the cylindrical hohlraum. The resulting impulsive energy deposition in the thin disk produces a planar shock wave resulting from ablation of the front surface. Energy deposition calculations indicated that up to 5 Mbar shocks could be produced in many materials such as aluminum, exceeding that possible with a two-stage light gas gun.

Asay, Hall, and later Marcus Knudson investigated the z-pinch mode for about 2 years from 1997 to 1999 with the goal of developing a high-quality shock wave capability on Z. Kathleen Holland, a postdoctoral student from Caltech, also assisted in the experiments. Hall recalls:

> We began putting infrastructure in place for optical shock arrival sensors and VISAR systems, but encountered and subsequently overcame a number of obstacles like precision in timing mandated by small sample sizes, changing of the index of refraction in optical fibers from the inherent radiation, decaying pressure pulses, and non-uniform ablative loading.

A problem with using the VISAR in shock wave experiments on Z was loss of optical signal occurring immediately after the radiation pulse from the z pinch. The intense radiation pulse produced radiation-induced darkening of the fiber optics, resulting in loss of the VISAR signal. Kevin Fleming from the Explosive Components Facility, who provided valuable assistance in developing fiber optic techniques for interferometry on Z, proposed a novel solution to the problem. He showed that shielding fiber optic cables to the VISAR with bicycle cable was effective for minimizing radiation darkening. This was a quick and inexpensive solution and enabled the use of the VISAR on Z for shock wave experiments. These issues were systematically resolved and the first shock wave experiment with a z-pinch source, Z Shot 135, was conducted about a year later on October 6, 1997. As the excerpt from the Z Operations home page mentions, "This will be the first EOS shot. We have three secondaries [cylindrical attachments to the primary hohlraum containing the z-pinch]—shock break out, VISAR, and XRDs.[5]"

Asay and Hall, with insightful recommendations from Professor Yogi Gupta at Washington State University, made the strategic decision immediately after the shock wave effort started in late 1996 to focus on developing the cryogenic capabilities necessary to obtain Hugoniot data on liquid deuterium ($D_2$) as one of the first materials to study. This was a high-risk decision, since a reliable shock wave source had not been demonstrated on Z, and cryogenic capabilities at 20 K for shock wave experiments were not then available at Z. This capability had to be developed from scratch and would take over 2 years to accomplish. Furthermore, several shock wave diagnostics (i.e., multiple VISARs and spectrographic equipment) and considerable resources were needed to implement these precision capabilities. This was difficult without a demonstrated shock wave source. At the time, the funds allocated through the Pulsed Power Sciences Center were not sufficient to achieve these objectives.

Fortunately, rapid progress was made on all fronts. Gerry Yonas, Vice President of Systems, Science, and Technology at Sandia, and Jeff Quintenz, the Deputy Director of the Pulsed Power Sciences Research Center, agreed to share the cost of developing the cryogenic system, which became operational for shock wave physics studies in early 1999 (Hanson et al. 2000). Lalit Chhabildas and Bill Reinhart at STAR provided a new laser source, a discarded VISAR, and loaned another one for use at Z. Mike Furnish and Paul Raglin assisted with financial support through the use of two multi-point VISARs from the subcritical experiments program at the Nevada Test Site (NTS), which is now known as the Nevada National Security Site (NNSS). The hands-on support of three experienced technicians was also leveraged from NTS to field these VISARs on Z. After he left Sandia in 1990, Lynn Barker invented and patented[6] a multi-beam VISAR (Barker 2000b; also, see Barker's recollections in this book). A variation of this concept was developed for use on Z and contributed significantly to the success of experiments by allowing highly accurate high-pressure EOS measurements. John Porter donated several of his dedicated shots on Z to test configurations for shock wave experiments and also helped support the development of the cryogenic system. The Z operations manager at that

[5] XRDs are x-ray diodes for measuring the radiation flux, in this case from the z-pinch source.

[6] L. M. Barker, "Multi-beam VISAR using image coupling from one optical fiber bundle to another through the VISAR interferometer," U.S. Patent 5,870,192, issued February 9, 1999.

time, Johann Seaman, redirected significant operational funds to configure the instrumentation, which allowed VISAR experiments in Z's intense radiation environment. With these parallel efforts, rapid progress was made, and the first shock wave EOS experiments with a z-pinch source were reported by Hall at the Shock Compression of Condensed Matter conference in the summer of 1999 (Hall et al. 2000). A working cryogenic system for shock EOS measurements on Z was also reported at the same conference by Dave Hanson (Hanson et al. 2000).

The z-pinch approach for producing planar shock waves was successful in generating intense shock pressures (Hall et al. 2000), but the EOS accuracy was not to the required standards available on guns. Steady shock waves were not produced by the pulsed radiation source and the spatial uniformity (planarity) was not good. Tailoring the energy deposition process showed promise for producing longer pulse durations, less shock attenuation, and better planarity but was discontinued when a purely magnetic drive was discovered in early 1999 that produced gun-quality ramp loading and, later, shock loading (Asay 2000; Asay et al. 2000). As Knudson comments about this period,

> …the z-pinch load on the Z accelerator proved to be at least one too many stages of pulse compression. We had felt that the resulting x-ray burst was insufficient as a source for dynamic compression. Rather, we desired a source capable of producing steady shock waves for durations of several tens of ns. We therefore spent nearly a year investigating the use of low-density, carbon-based foams (such as TPX®[7]) as a way of 'conditioning' the radiation that reached a sample, such that the resulting radiation and foam tamper would support a significantly longer, steady shock. These efforts, while somewhat promising, ultimately proved to be inefficient and were unsuccessful. We were unable to produce any results that warranted publication.

However, the path to producing quality shock waves on Z, which had been pursued for 2 years with z pinches, was about to take a sudden turn for the better.

## 7.3 Ramp Loading to Multi-Megabars

The discovery of a high-quality magnetic drive on Z was serendipitous. Rick Spielman, the co-developer of z-pinch technology, needed a noninvasive diagnostic for measuring current close to the z-pinch source where the radiation fields are extremely high. His discussions with Jim Asay and Clint Hall early in 1999 led to the conclusion that the VISAR would be a good tool for measuring particle velocity and, thus, unfolding the data to obtain hydrodynamic pressure, magnetic field, and current near the z-pinch source. The first experiments with this approach were successful (Leon et al. 1999; Asay 2000) and led to further developments (Hall et al. 2001a, 2002) that ultimately enabled highly accurate ramp wave experiments to about 4 Mbars. As Clint Hall remembers,

> … Jim and I placed two small disks (one copper and one iron) into the anode plate on a wire array load and then attached shielded VISAR diagnostics to the rear of each to measure

[7] TPX is the brand name for polymethylpentene (PMP), a thermoplastic polymer.

> their respective free surface velocities. As it turned out, the Z shot was fired late on a Friday afternoon, with the raw VISAR data saved and delivered to me to be analyzed on the following Monday morning. I made it until Saturday around lunch time before I had to head into work to see what we recorded. I sat down, opened the files, and reduced them. When I had done so, I got a very nice trace from the iron, but it had an odd feature I couldn't understand. It just so happened that Lalit was in his office working too, so I went down and showed the trace to him. He immediately thought that it might be clearly showing the alpha-epsilon phase transition in the iron, but couldn't remember at what pressure it was supposed to occur, so we jumped into the company truck and headed into Area I hoping to find someone that might be there and remember. The best we could do was to find a viewgraph on a poster outside a door (that we just had to 'borrow') showing the iron transition from a gun experiment. Sure enough, they correlated, and the birth of magnetic loading for ramped wave experiments on Z gained credibility.

The first experiment was not designed to address drive uniformity and theoretical approaches were not yet in place for accurate analysis of the ramp waves. The experiment did, however, demonstrate that a high-quality magnetically-driven ramp wave could be produced and led to the improvements that rapidly followed. In particular, technical discussions with Art Toor and 2-D MHD simulations by David Reisman, both at LLNL, were instrumental in developing an experimental configuration for producing planar, nearly isentropic compression in test samples on Z.

The basic concept for an isentropic compression experiment (ICE) using magnetic drive is shown in Fig. 7.1. Figure 7.1a shows the machine configuration for producing planar loading of cylindrical samples about 12 mm in diameter by magnetic pressure produced between the anode and cathode. The configuration has been called an "ICE Cube." A pulse of electromagnetic energy produces current on the anode and cathode of the target configuration. Current flowing on or near the surface of each conductor produces a magnetic field in the gap between the two (typically about 1 mm) that interacts with the current on each conductor through the Lorentz force, resulting in magnetic pressure on the conductor. The applied magnetic pressure or stress, illustrated on the right side of Fig. 7.1b, loads the conductor and produces a ramped stress wave that propagates into a sample mounted on the conductor's rear surface. If designed properly, the sample does not experience magnetic or thermal loading and is subjected to only the smooth pressure loading shown in Fig. 7.1b.

An insightful innovation that allowed the simultaneous planar loading of several specimens on a single Z experiment was implemented by Clint Hall following discussions with Toor and Reisman. This configuration consisted of four individual panels of aluminum or copper with diamond-machined counter-bored holes for mounting specimens, as illustrated in the upper left of Fig. 7.1. The panels were bolted together, making a hollow square cube (anode) that was mounted over the cathode of the current-carrying center section of the machine. Both parts were shorted at the top as indicated in the left side of Fig. 7.1b. With current flowing on the two surfaces, the resulting magnetic field in the gap (typically about 1 mm), produces nearly uniform magnetic pressure on the inner surface of the sample cube (Reisman et al. 2001a) that initiates a stress wave propagating through the thin metal floor of the cube and into the sample, as illustrated on the left side of Fig. 7.1b. Clint called this fixture an "ICE cube," which appropriately describes its function. Typical dimensions of the ICE cube are about 30 mm on a side with sample diameters on the order

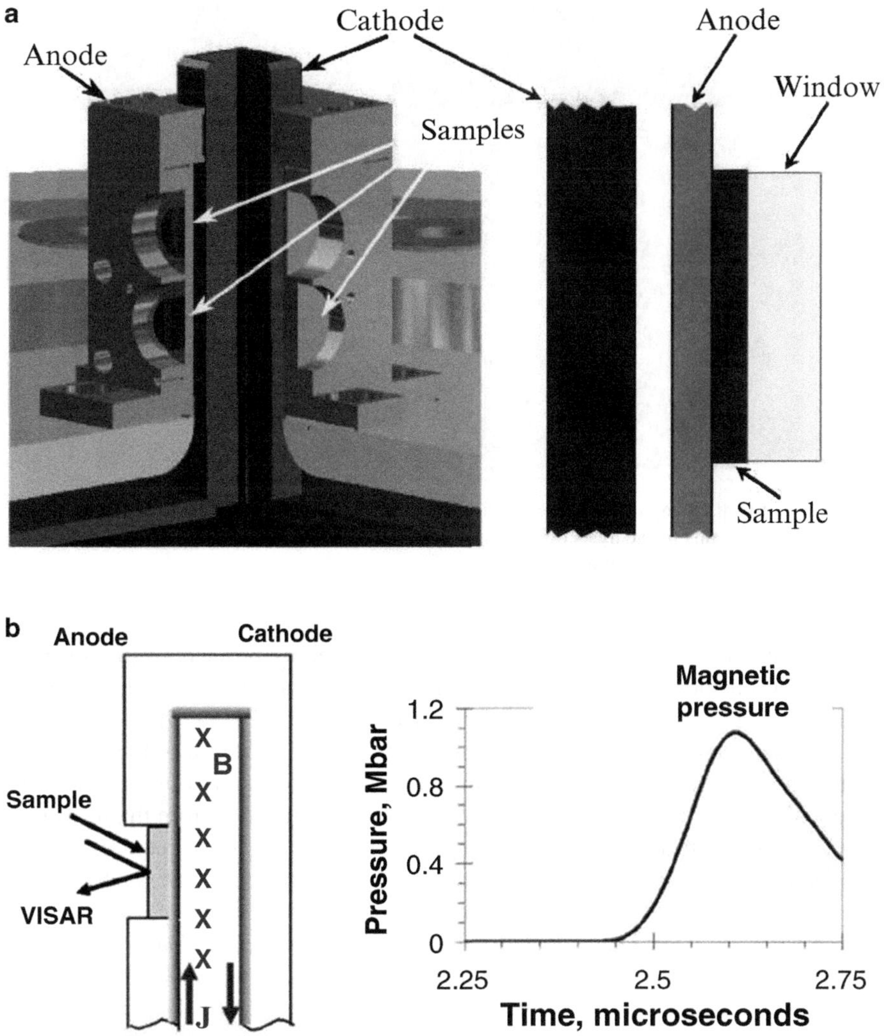


**Fig. 7.1** (**a**) A schematic of the ICE cube hardware. (**b**) On the *left*, a drawing of the basic concept for magnetic drive. J is current density and X identified by B is the induced magnetic flux. On the *right*, the magnetic pressure induced in the sample (From Asay and Knudson 2005, Figs. 10.3 and 10.7, reprinted with permission from Springer Science+Business Media, Copyright 2005)

of 10–15 mm and thicknesses of 1–2 mm. Figure 7.2 shows Hall setting up an experiment with an aluminum ICE cube. Copper was used initially for the ICE cube because of its high conductivity, although most ramp experiments are conducted with an aluminum ICE cube. Typical current rise times range from 200 to 300 ns and are shaped in time to minimize shock formation (Hayes et al. 2004; Davis 2006).

The most important feature of magnetically-induced ramp loading is the smoothly increasing current resulting in a corresponding increase in pressure with time, as shown in Fig. 7.1b. This allows a relatively straightforward method to analyze the

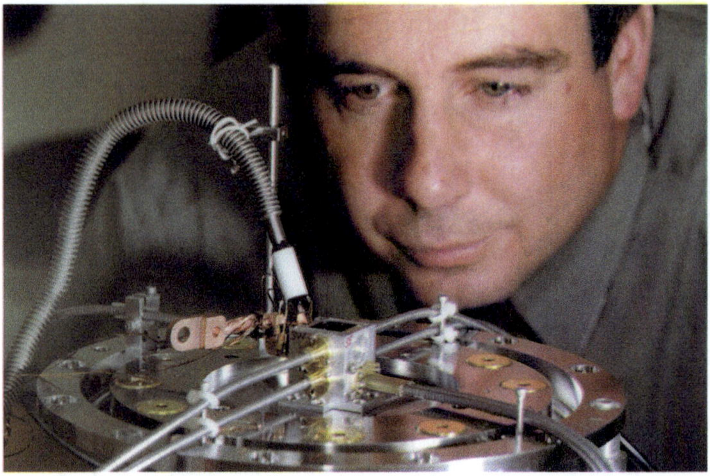

**Fig. 7.2** Clint Hall setting up a Z experiment on an Al ICE cube. Fiber optic cables transmit and receive VISAR signals for analysis of the ramp- or shock-loaded sample (Reprinted with permission of Sandia National Laboratories)

evolution of stress waves produced in the sample to deduce pressure or stress versus density states achieved by ramp compression (Hayes et al. 2004; Hall et al. 2001a, 2002; Davis 2005, 2006; Davis et al. 2005; Reisman et al. 2001a, b). In particular, Hayes developed a technique using the WONDY code that integrated the equations of motion backward in space, rather than forward in time, as is usually done in numerical simulations to determine the input pressure histories from the measured particle velocity profiles on the aluminum panels. This, however, requires a good EOS for aluminum. The technique continues to be used to the present. Smooth compression is important in many applications, including the detection of phase boundaries (Asay et al. 2000; Davis et al. 2002; Davis and Hayes 2007; Davis and Foiles 2005), measurements of dynamic yield strength (Asay et al. 2008, 2009, 2011; Vogler and Chhabildas 2006; Vogler 2009), and acceleration of flyer plates (Hall et al. 2001b; Knudson et al. 2003d). Another critical development that involved the ability to shape the current pulse was accomplished by firing different modules in Z to obtain the optimum pressure drive history to delay formation of a shock wave in the sample (Davis 2006; Hayes et al. 2004; Lemke et al. 2011). This development significantly increased the sample thicknesses that could be studied and mitigated shock formation during launching of the flyer plate.

The use of magnetic loading to produce smoothly increasing Mbar ramp loads was a major breakthrough in the use of Z for material studies. Hopkinson bar uniaxial stress experiments are routinely used to study material response at strain rates up to about $10^3$ $s^{-1}$. Chhabildas developed a sample configuration that allowed uniaxial stress measurements under ramp loading to strain rates of about $10^5$ $s^{-1}$ (Chhabildas et al. 1997) and uniaxial strain conditions at strain rates of $10^6$ $s^{-1}$ to stresses up to 2.5 Mbars (Chhabildas et al. 1988). Magnetically-induced ramp compression is the first accurate method for extending these kinds of material studies to

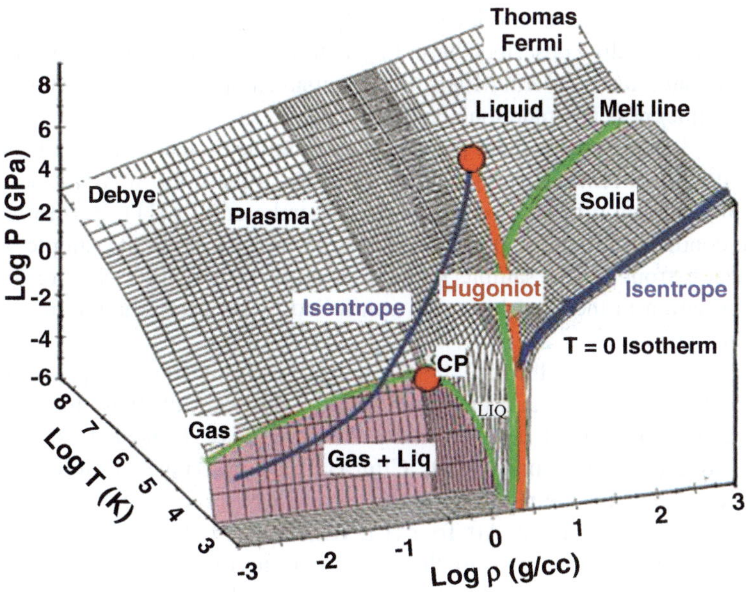


**Fig. 7.3** Approximate pressure, density, and temperature states achieved in aluminum for shock and ramp loading that produces near-isentropic states. The combined capability enables the study of new physical regimes, including cool solid states at high pressure, hot states at high pressure, and vapor and plasma at expanded states. CP represents the critical point for aluminum (From Asay and Knudson 2005, Fig. 10.13, reprinted with permission from Springer Science + Business Media, Copyright 2005)

strain rates of $10^5$–$10^8$ $s^{-1}$ for uniaxial strain loading and to higher pressures. Ramp compression is much less dissipative than shock loading, and is nearly equivalent to thermodynamic states produced by isentropic loading. For liquids that exhibit only bulk response, but even for solids, the plastic dissipation produced by strength effects is relatively small; therefore, the compression process is referred to as "quasi-isentropic." Since the dissipation is small, the thermodynamic states produced by ramp loading permit access to previously inaccessible states of matter at high pressures but much lower temperatures than shocks. The difference in pressure–density–temperature states between shock and ramp loading of aluminum is illustrated in Fig. 7.3 (Asay and Knudson 2005).

Achieving the capability to ramp load materials to Mbar pressures had been an unreachable goal of the shock wave community for several decades, beginning with the first isentropic compression experiments (Barker and Hollenbach 1970). Barker and Hollenbach used fused silica buffers to produce ramp waves that could be used to ramp load samples. Although the graded-density impactors developed later by Barker and Doug Scott in the 1980s (Barker 1984) showed promise for achieving this goal, the impactors had limitations in reproducibility and uniformity that precluded widespread use. The layered-material impactor developed by Chhabildas (Chhabildas and Barker 1988; Chhabildas et al. 1988, 1990, 1992) for the HyperVelocity Launcher also produced nearly shockless loading and was very reproducible, but it did not produce smoothly increasing pressure when compared to the magnetic loading described here. Lasers have also demonstrated the ability to ramp load materials to several Mbars, but on small samples, typically a few tens of microns in thickness (Fratanduono et al. 2011).

Magnetic loading avoids these problems but introduces other issues such as those concerned with current diffusion and the shock up of the stress wave introduced in the sample. Current diffusion occurs rapidly (Lemke et al. 2011). But, fortunately for good conductors such as Al and Cu, the stress wave front in the sample outruns the diffusion front so the sample remains at ambient conditions before loading. Shock formation in the sample can be controlled by properly shaping the current versus time profile.

Since its inception in the late 1990s, magnetically-induced ramp loading has continued to be perfected and improved through the work of many investigators, including (in alphabetical order) Tommy Ao, Jim Asay, Jean-Paul Davis, Chris Deeney, Mike Furnish, Thomas Haill, Clint Hall, Dennis Hayes, Randy Hickman, Marcus Knudson, Ray Lemke, David Reisman (LLNL, now at SNL), Seth Root, Chris Seagle, Rick Spielman, Art Toor (LLNL), Tracy Vogler, and Jack Wise. The technique has been used to measure the EOS properties of a large number of materials with an accuracy approaching that possible with shock loading. These results have been invaluable for developing material models needed for weapon and planetary science applications.

## 7.4 Magnetically-Driven Hypervelocity Flyer Plates

Generating high-quality shock waves on Z involves the use of high-pressure ramp wave loading to launch thin flyer plates to ultrahigh velocities, as for the HVL, but far exceeds all previous capabilities. This groundbreaking advance was enabled by the smooth increase in pressure and the peak pressures produced by ramp loading. This is revolutionary because a "smooth" pressure profile mitigates shock formation and the subsequent heating of the flyer plate that results in possible melting or vaporization, as was first demonstrated with the HVL. The pressure history shown in Fig. 7.1b allows either a sample to be loaded continuously for quasi-isentropic EOS studies or, alternatively, allows the conducting panel itself to be launched as a flyer plate to impact a specimen located behind it (not shown), as in traditional shock experiments. In the flyer plate mode, Marcus Knudson, Clint Hall, Ray Lemke, and others have demonstrated that flyer plates on the order of 17-mm diameter and up to about 1-mm thick can be launched to velocities as high as 46 km/s (Lemke et al. 2003a, b, 2005, 2011). This is about *five times* higher than that possible with conventional two-stage guns and about *four times* higher than the HVL for EOS studies. Using Al flyer plates, impact pressures as high as 40 Mbars can be produced in high-impedance materials. This velocity range considerably extends the pressures available in the laboratory and enables EOS measurements to multi-Mbars with a high degree of accuracy. A typical configuration for launching flyer plates on Z is shown in Fig. 7.4. The diamond-machined base of the countersunk

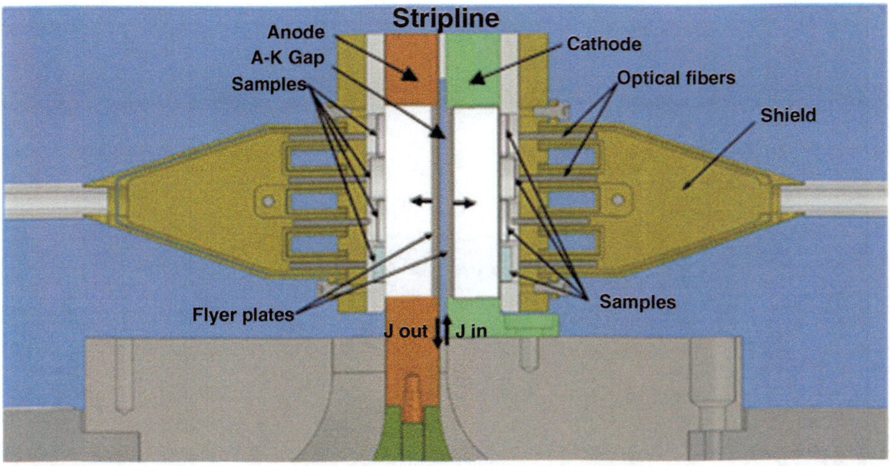


**Fig. 7.4** Stripline configuration to produce magnetically-driven flyer plates. Current flows into the stripline on the cathode and out on the anode. Both are typically aluminum. J is the current density (amps/cm) on the electrodes. The cathode and anode plates are about 1-mm thick and become the flyer plates in the experiment. Magnetic pressure is produced in the A–K gap between the anode and cathode flyer plates through the Lorentz force. Multiple samples are used in each experiment so that several Hugoniot data points are obtained. Optical fibers and samples are shielded from magnetic field effects (Private collection, M. D. Knudson)

anode and cathode stripline conductors is about 1-mm thick and becomes the flyer plates in the experiment. Magnetic pressure is produced in the anode–cathode (A–K) gap between the anode and cathode flyer plates through the Lorentz force. Multiple samples are used in each experiment so that several Hugoniot data points are obtained.

The following quote from Knudson captures the excitement of this development:

> The next significant breakthrough occurred during the 1999 American Physical Society meeting of the Division of Plasma Physics in Seattle, WA. Jim Asay, Clint Hall, and I were at the lobby bar of the Westin in Seattle, discussing the magnitude of the impulse these types of experiments delivered to the electrodes. Back of the envelope calculations suggested that the impulse was sufficient to launch relatively thin anode plates, roughly 1 mm in thickness, to significant velocities, well in excess of 10 km/s. That evening I ran some simple hydrodynamic code simulations and the results were extremely encouraging. Upon returning from that meeting we immediately began planning dedicated experiments on Z to test this new concept. Given that it takes roughly three to four months to design and fabricate hardware for Z experiments, the first experiments would not be performed until the spring of 2000.
>
> In the meantime we presented the idea of magnetically accelerating flyer plates on Z at the 2000 Garwin review, an annual external review of Sandia's pulsed power program. I presented our plans on the first day of the two-day review. I'll never forget the response of David Reisman, who was attending the meeting to present on the progress of the ICE platform, the following day. He had been intrigued with the idea that we had presented, and that evening in his hotel room he had run several MACH2 calculations, including full magnetohydrodynamics. He came to the conclusion that it was a great idea, but unfortunately it would never work. The MACH2 calculations indicated that no matter how thick the initial electrode, the magnetic field diffusion was so severe that the field would burn through the plate before it could reach a terminal velocity, and thus the plate would be vaporized prior to impact.

**Fig. 7.5** Marcus Knudson at the Z machine holding a flyer plate that will be launched to velocities in excess of 10 km/s (Reprinted with permission of Sandia National Laboratories)

> A few weeks later we conducted the planned experiments. Given the fact that we only had a few VISAR channels at the time, we had to rely on shock breakout measurements. These were effectively reflectivity measurements in which a sudden loss of reflectivity occurred upon the arrival of the shock wave at the free surface, to obtain several transit time measurements in a wedged target to infer the shock velocity. I still remember sitting at home on the evening of that first flyer plate experiment analyzing the data, so anxious that my hands were literally shaking as I plotted the breakout time versus distance. Lo and behold, despite the MACH2 simulations, the data suggested that a significant portion of the plate was still at full density upon impact, implying that the diffusion rate in the simulation was too high. This marked the birth of the magnetically-accelerated flyer plate platform on Z.

Marcus Knudson is shown in Fig. 7.5, holding a flyer plate assembly. Clint Hall also remembers this momentous and groundbreaking development:

> Jim, Marcus, and I were sitting at a table between talks, and we began discussing another whole concept. Instead of transmitting the ramp loading directly into the sample of interest, we thought that we could instead launch the bottom of the panel as a flyer plate and impact our samples to obtain high pressure Hugoniot data. We did what any self-respecting physicists would do at that point and drew a sketch out on a cocktail napkin, stuffed it in one of our pockets, and went to our next talk! When we got back to Sandia, we did indeed retrieve the 'highly precise scientific sketch,' did some calculations, performed the experiment, and we exceeded years of HVL development on our first shot with a metallic flyer that was good enough for equation of state experiments moving at about 10 km/s!

Conception and development of the magnetically-launched flyer plate for making accurate shock Hugoniot measurements at ultrahigh shock pressures were truly a game-changing event, as the HVL was earlier using conventional two-stage guns.

Many of the lessons learned for launching flyer plates with high impulsive loads on the HVL were extremely valuable in achieving similar capabilities on Z.

Later *ab initio* calculations by Mike Desjarlais showed that the actual high-pressure electrical conductivity of aluminum would be a factor of two greater near the melting point than the values being used in MACH2,[8] based on Lee and More's electrical conductivity model for dense plasmas (Lee and More 1984); hence, the plate did not melt and vaporize as Reisman had predicted (Desjarlais 2001; Desjarlais et al. 2002). These significant theoretical and experimental breakthroughs had a profound impact in the shock wave community by demonstrating new experimental and theoretical methods for obtaining accurate EOS data at ultrahigh shock pressures. Line VISAR experiments quantified the nonplanarity of the loading for this configuration, which was found to be adequate for EOS studies (Knudson et al. 2003d). A stripline loading configuration, to be discussed later, improved the loading uniformity even more. Since the first experiments in 2000, Knudson has continually refined the flyer plate technique to increase the velocities well beyond ~10 km/s (Hall et al. 2001b; Knudson et al. 2003d) and to improve the EOS accuracy to equal that of gun-launched projectiles (Knudson and Desjarlais 2009).

To explore the range of velocities possible with magnetically-driven flyer plates, Ray Lemke conducted an extensive investigation using ALEGRA MHD simulations, which established an upper limit of about 46 km/s for the present current capability of the refurbished Z (Lemke et al. 2002, 2003a, b, 2004, 2005, 2011). His calculations further show that, with sufficiently high currents, flyer plate velocities as high as 65 km/s may be possible (Lemke et al. 2014). However, this is expected to be an upper limit for any magnetically-driven flyer plate because of the limitations imposed by current diffusion and the requirement that shocks do not form in the flyer plate (Lemke et al. 2011). Generally, the ALEGRA flyer plate predictions have accurately reproduced the VISAR measurements of the final flyer velocity to within 1–2 %, as shown in Fig. 7.6. Because of the complexity of these experiments, ALEGRA calculations are routinely performed to define the experimental parameters for every flyer plate experiment.

Marcus Knudson has championed the use of the magnetic flyer plate technique for studies of several materials of scientific and programmatic interest with unprecedented accuracy and at pressures far exceeding the capabilities of conventional launchers in the laboratory. The method was used to obtain (1) near-absolute Hugoniot data on aluminum, followed by measurement of the release adiabat into the liquid–vapor region (Knudson et al. 2005); (2) shock loading and partial unloading of beryllium to determine the high-pressure EOS and the high-pressure solid and liquid phase boundaries, including an hcp–bcc phase transition near 2 Mbars and determination of its high-pressure yield strength in the vicinity of the polymorphic transition; (3) the Hugoniot of water at extreme pressures, which is revising conventional concepts of the compositions of giant planets and exoplanets[9]

[8] MACH2 is a 2-½-dimensional magnetohydrodynamic simulation code developed by NumerEx. 2-½-D refers to a two-dimensional simulation with symmetries included to approximate a 3-D simulation.

[9] For additional discussion, see Sect. 7.5.

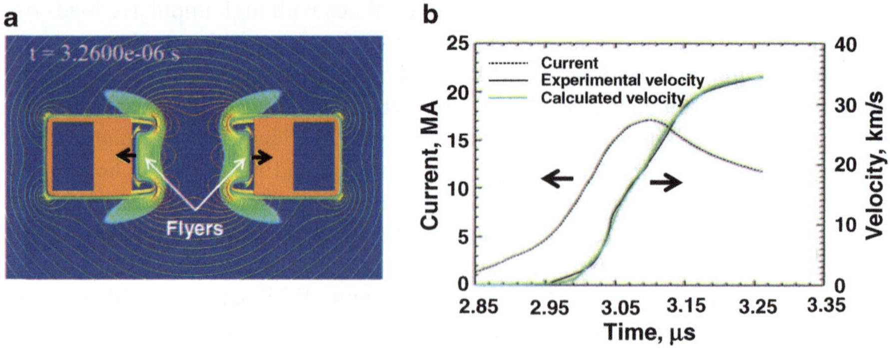


**Fig. 7.6** Stripline configuration for launching aluminum flyer plates. (**a**) MHD simulation of the experiment showing magnetic field symmetry and flyer plate launch. (**b**) Comparison of simulated and experimental flyer plate velocities for a stripline geometry using Z to magnetically launch aluminum flyer plates. The aluminum plate thickness is 0.9 mm, peak current is about 22 MA, and peak velocity is about 34 km/s (Reprinted with permission from Lemke et al. 2011, Copyright 2011, Elsevier LTD)

(Knudson et al. 2012); (4) discovery of a high-pressure triple point and a new solid phase of carbon (Knudson et al. 2008); (5) over a hundred shock Hugoniot/shock wave measurements on quartz to over 16 Mbars that have established quartz as the most accurate standard for high-pressure EOS measurements (Knudson and Desjarlais 2009); and (6) extensive studies of the high-pressure EOS of liquid deuterium that enabled critical comparison with *ab initio* theories and established the Z data as the accepted high-pressure EOS for hydrogen (Knudson et al. 2001, 2003a, b, 2004).

The extensive research on liquid deuterium, beginning in 2000 and continuing to the present, is a noteworthy accomplishment that established the credibility of the Sandia shock wave program on Z and helped secure Sandia's role in the NNSA national program for HEDP. In the late 1990s, laser shock experiments on Nova at LLNL (Collins et al. 1998) indicated that the deuterium shock Hugoniot was about 50 % softer (for a peak compression of about six) at Mbar pressures than predicted by conventional phenomenological models (for a peak compression of about four). With the successful development of the magnetic flyer plate technique in 2000, Knudson obtained shock data on deuterium at similar pressure levels but with substantially increased accuracy (Knudson et al. 2001, 2003a, b, 2004). The Z data indicated a peak compression closer to the classical model for the pressure–density relationship, as indicated in Fig. 7.7. However, investigators involved in the laser effort maintained that their deuterium data were correct; the controversy persisted for several years and involved most of the high-pressure community. To resolve the discrepancy, Knudson's team performed a variety of experiments that included reshock from the shocked state, a reverberating wave in the shocked state, and extending Hugoniot data to higher pressures and temperatures (Bailey et al. 2008);

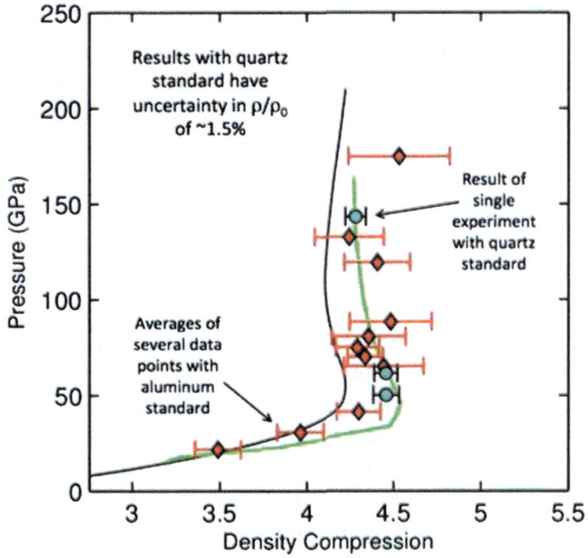


**Fig. 7.7** Experimental Hugoniot data for liquid deuterium obtained on Z in comparison with SESAME EOS tables (*black curve*, Kerley 2003) and QMD calculations (*green curve*, Desjarlais 2003). The *red diamonds* represent averages of several experiments. The *green circles* were obtained with improved quartz shock velocity gauges that significantly reduced the error bars (Knudson 2012). Excellent agreement is obtained with the more accurate Z data and QMD calculations; the Kerley curve is more incompressible than the data at high pressures. The original LLNL data are not shown but attained a density compression of about six-fold (Collins et al. 1998) (Figure from private collection, M. D. Knudson, 2014)

those experiments supported the stiffer response. In addition, the development of the extremely accurate quartz standard for shock wave research (Knudson and Desjarlais 2009) substantially reduced the error bars of the deuterium Hugoniot data, as illustrated by the green data points (Knudson 2012) in Fig. 7.7. The significantly reduced error bars in the deuterium Hugoniot that were obtained with the new quartz standard (Knudson 2012), which allow discrimination of theories such as Kerley's semiempirical model (Kerley 2003) and Desjarlais' *ab initio* theory (Desjarlais 2003). The combined data set, including the most recent data, provides compelling evidence for the Sandia results, which are now generally accepted by the scientific community. Later, laser shock experiments identified earlier systematic errors so that the Z and laser data now agree within the reported errors for each (Hicks, Boehly, Celliers et al. 2009; Knudson and Desjarlais 2009). The improved understanding and the development of accurate models of the hydrogen EOS obtained over the past decade have broad implications for applications to the interiors of the giant planets, astrophysical plasmas, warm dense matter, and inertial confinement fusion.

Recently, Marcus Knudson and a group of researchers on Z have made another landmark discovery. Using the Z facility, they obtained compelling evidence for an *abrupt* transition of dense liquid deuterium from an insulator to a metal at high pressure and relatively low temperatures. Hydrogen is at the top of the alkali metal column in the periodic table, but even as a low-temperature solid, it remains an insulator. In 1935 Eugene Wigner and Hillard Huntington had predicted that, at about 250 kbars and low temperatures, hydrogen should become metallic through a pressure-induced transition caused by closure of the electronic band gap (Wigner and Huntington 1935). Since that prediction, observing such an elusive transition directly has been described as "the holy grail of high-pressure research" (see, e.g., McMahon et al., *Reviews of Modern Physics*, vol. 84, p. 1608 [2012]), with numerous attempts to do so. Theoretical predictions for the phase transition have varied over an astonishing pressure range. Previous experimental attempts by other researchers have included cylindrical magnetic flux compression in the USSR and at LLNL in the 1970s (Grigoryey, Kormer, Mikhailova et al. 1972; Hawke 1974), shockless compression with a two-stage light gas gun at Sandia in the 1980s (as described in Chap. 5, Sect. 5.4), temperature-driven shock experiments on a two-stage gun at LLNL in the 1990s (Weir et al. 1996; Nellis et al. 1999), laser-driven shock experiments at LLNL in the 2000s (Loubeyre, Brygoo, Eggert et al. 2012), and static diamond anvil cell experiments (McMahon et al. 2012). The unique aspect of the experimental effort by Knudson and his colleagues (Knudson, Desjarlais, Becker et al. 2015) was in developing an approach that combined shock and shockless (ramp) compression on a Z experiment, which controlled the temperature state produced by compression, to access the regime where theory predicted deuterium would transition from an insulating molecular liquid to a conducting liquid. As illustrated in Fig. 5.4 for LiF in Chap. 5 of this book and in Fig. 7.3 for aluminum in this chapter, widely different temperatures are achieved by shock and shockless compression at high pressure, which allow specific thermodynamic states to be achieved.

Although Bill Nellis and coworkers (Weir et al. 1996; Nellis et al. 1999; Nellis 2001, 2013) did observe that hydrogen became conductive under shock wave reverberations (when the sample was sandwiched between two sapphire anvils) at around 1.4 Mbars, the metal-like behavior was initially thought to be dominated by the high temperature, about 3000 K, produced by the first shock, and not by a density-driven closure of the electronic band gap as predicted by Wigner and Huntington. In the recent shock-ramp experiments on Z, Knudson and his team used a well-defined, flyer plate-generated shock in order to shock the liquid deuterium initially to relatively low pressure and temperatures states, followed by ramp compression to about 3 Mbars (Knudson et al. 2015) to reach higher pressures while keeping the temperature low. An abrupt onset of reflectivity, characteristic of a shiny metal, was observed and the pressure state was directly measured with an LiF window. The onset of high reflectivity is characteristic of high metallic conductance. By varying the amplitude of the initial shock, Knudson and colleagues were able to trace the pressure–temperature boundary of the metallic transition for the first time, which will aid the identification of the transition in experiments using other techniques, such as isothermal diamond anvil cells. The direct observations of the insulator-to-metal transition are qualitatively different in character and at an order of magnitude higher pressure (at about 2.5 Mbars) than Wigner and Huntington's prediction, but provide compelling evidence of a pressure-induced, density-driven transition as they had predicted.

The hydrogen metallization investigation on Z of the liquid analog to the 80-year-old conjecture about solid hydrogen was inspired by *ab initio* molecular dynamics simulations based on density functional theory performed by the University of Rostock theorists (Lorenzen, Holst, and Redmer 2011). Identification of the metallic phase transition of hydrogen could help explain a long-standing mystery of planetary evolution models that suggest a two-billion-year discrepancy between the ages of Saturn and Jupiter. Hydrogen metallization-driven demixing of hydrogen and helium at high pressure is important to the understanding of the evolution and interior structure of these gas giants. For example, see Salpeter (1973), Hawke (1974), Lorenzen et al. (2011), and McMahon et al. (2012), as well as other references therein. The behavior of dense liquid deuterium is also essential for understanding the simulated performance of nuclear weapons and for improving the performance of inertial confinement fusion targets. Determining the pressure–temperature location for the phase transition provides a new benchmark for theory and advanced simulation codes. Experimental confirmation of this abrupt first-order transition by other researchers will likely invigorate the planetary science and condensed matter communities and open new vistas of high-pressure science on metallization phenomena.

The shock wave group at Z grew rapidly in the decade after the first magnetically-driven flyer plate experiments in 1999, fueled by a number of newly graduated students. In addition, several experimenters from the original group under Walt Herrmann in the 1980s rejoined the shock wave program through internal transfers and reorganizations. Researchers from several different technical areas in the Pulsed Power Sciences Center have strengthened the breadth of the program by adding innovative instrumentation and theoretical capabilities for shock wave studies, notably Jim Bailey, who developed techniques for measuring radiation temperatures in shocked deuterium, and Dave Hanson, who implemented a novel cryogenic capability for performing shock wave experiments on liquids at temperatures of a few degrees Kelvin. Kyle Cochrane, Mike Desjarlais, Heath Hanshaw, Thomas Mattsson, and other recent additions bring strong theoretical skills to the program, particularly in the area of *ab initio* modeling.

The magnetic loading capability available on Z considerably expands the range of EOS data that can be obtained in the laboratory. Through a combination of shock or ramp loading, followed by unloading, it is now possible to study the full EOS surface for a variety of materials, including expanded states, as shown in Fig. 7.3 for aluminum. A research area of intense interest in the national and international communities concerns the behavior of materials at extreme pressures and temperatures referred to as HEDP. The study of hot dense matter covers the range of material response at high compression and several thousand eV, which is pervasive in supernova, accretion disks, and the interiors of stars. These states are also produced in practical applications such as laser plasmas and z pinches. Shock wave experiments on Z have contributed substantially in understanding a variety of HEDP phenomena.

## 7.5 *Ab Initio* Equation of State Theories

Beginning in the late 1990s, the Nova laser experiments on deuterium at LLNL first started producing shock Hugoniot data at higher pressures than possible with gas guns. Later, when a similar set of new data were generated on Z, the deuterium research motivated several theoretical investigators at Sandia to use quantum molecular dynamics (QMD) for predicting Hugoniot curves. QMD is a powerful tool for predicting EOS properties of low-molecular-weight liquids shocked to a few Mbars, but it requires large computational capability. The emerging ultrahigh-pressure shock wave data on deuterium provided a unique opportunity for solving these computationally intensive problems with the fast MPP computers that were also just becoming available. The thermodynamic regime accessed by shock experiments is referred to as "warm dense matter," a region for which the thermal occupation of excited electronic states is non-negligible and ions are strongly coupled.

At Sandia, Mike Desjarlais led the development of QMD formalism for EOS applications early in the 2000 time frame. His first use of the theory concerned the conductivity of aluminum for magnetically-driven flyer plate applications. Existing conductivity models in MHD codes predicted that the high current densities on Z would completely diffuse through and vaporize an aluminum flyer plate before it could be launched to full velocity. However, the initial experiments on magnetically-launched flyer plates by Marcus Knudson indicated that these results were erroneous.

The *Physical Review E* paper (Desjarlais, Kress, Collins 2002) established the power of using QMD simulations to create accurate conductivity models for materials away from ambient. The first application in shock wave research was to the aluminum conductivity, which was needed for modeling the flyer plates. This advance was essential for predictive design simulations and, as an immediate consequence, rapidly advanced the pace of data accumulation by permitting much more accurate pre-shot design.

The QMD simulations established that aluminum was significantly more conductive than predicted by the classical theories, and therefore a significant portion of the flyer plate would remain at ambient conditions upon impact. This conclusion explained the successful flyer plate results observed experimentally and the miscalculation produced in the previous MACH2 simulations by David Reisman (Hall et al. 2001b; Knudson et al. 2003c). Desjarlais' QMD research on the conductivity of aluminum has had a major influence on the design of all flyer plate experiments on Z by defining the operational limits for flyer plate materials, including the thickness and pulse shaping requirements (Desjarlais 2001; Desjarlais et al. 2002). This initial *ab initio* research was the basis for additional QMD predictions of the shock loading and unloading response of several other materials.

A specific example highlights the role that QMD played in resolving the controversy about the high-pressure shock Hugoniot of liquid deuterium. At the time of the controversy over the experimental differences between SNL and LLNL data on the EOS of deuterium, Desjarlais became interested in the problem and initiated a research effort with QMD to predict the Hugoniot of deuterium. Using density functional theory (DFT) and the VASP (for Vienna *Ab Initio* Simulation Program) code developed at the Technical University of Vienna, he developed a set of interaction forces to treat the ions and electrons of deuterium and simulated the prior gas gun data to about 400 kbars. Excellent agreement was obtained with experimental data. Encouraged by this, he extended the calculations to shock pressures of about 1.5 Mbars (Desjarlais 2003), where the Z and Nova laser data significantly disagreed. The QMD predictions showed good agreement with Knudson's Z data (see Fig. 7.7) but serious disagreement with laser data above 400 kbars. To further support the stiffer EOS response, QMD simulations of experimental recompression states produced by shock reflection were performed, again indicating good agreement with experiment and validating the Z results (Desjarlais 2003). Additional QMD calculations also predicted the shock temperature measurements (Bailey et al. 2008) and the Hugoniot data at higher pressures (Knudson et al. 2001, 2003a, b). These highly integrated studies provided compelling evidence for the Sandia results and were pivotal in resolving the deuterium controversy.

Experimental and theoretical collaborations using *ab initio* theories continued on several other materials, including (1) the multi-Mbar shock loading and unloading of aluminum into the vapor region (Knudson et al. 2003d, 2005), (2) the high-pressure Hugoniot of quartz (Knudson and Desjarlais 2009), (3) the identification of new phases of carbon (Knudson et al. 2008), (4) the ultrahigh compression of water (Knudson et al. 2012), (5) the compaction of porous materials (Cochrane et al. 2014), (6) multi-Mbar studies of other fluids and foams (Root et al. 2010, 2013a, b; Mattsson et al. 2010), and (7) most recently the discovery of the metallic transition in liquid deuterium (Knudson et al. 2015). The quartz experiments are especially noteworthy; extensive magnetic flyer plate data to 16 Mbars have established quartz as one of the most accurate standard laser "windows" for high-pressure shock wave measurements (Knudson and Desjarlais 2009). Since most, if not all, laser windows become conductive by shock loading to a few Mbars, the window becomes a *shock velocity* gauge (Knudson and Desjarlais 2009) rather than a *particle velocity* gauge, as originally intended by Barker (Barker and Hollenbach 1970, 1972). Good agreement between the experiments and the QMD calculations was also obtained on quartz and provided physical insight as to its high-pressure response. As Desjarlais commented, "QMD work [on quartz] provided the explanation for the non-linear $U_s$-$u_p$ relation through the observation of an extended dissociation regime and the corresponding consequences for the specific heat."

Knudson's shock wave experiments on carbon (diamond) demonstrated that highly accurate Hugoniot data can be used to identify subtle changes in the phase diagram (Knudson et al. 2008). Knudson was able to deduce that a new phase of carbon, identified as bc8, existed at high pressure and that a triple point occurred among the diamond phase of carbon, the new phase, and liquid carbon at 8.5 Mbars on the Hugoniot. QMD calculations by Desjarlais predicted these results and "the subtle changes in $U_s$-$u_p$ as the Hugoniot traverses the diamond-liquid-bc8 coexistence region were predicted by the QMD calculations and observed in the data,

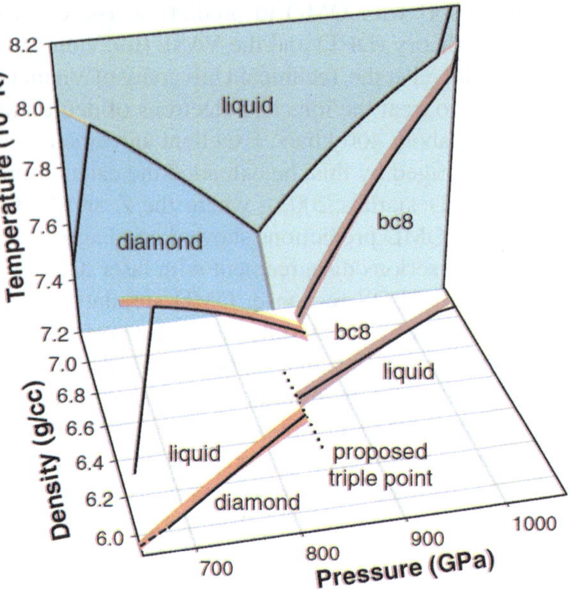


**Fig. 7.8** Coexistence regions in P-ρ-T space for carbon. The Hugoniot (*black line*) enters the diamond phase at ~680 GPa (6.02 g/cc), reaches the diamond-bc8-liquid triple point at ~850 GPa, and exits the bc8-liquid coexistence region (*pink band*) at ~1040 GPa (7.04 g/cc) (From Knudson et al. 2008, reprinted with permission from AAAS, Copyright 2008)

which gave convincing evidence for the diamond-liquid-bc8 triple point and the first experimental evidence for bc8 in carbon." The phase diagram for carbon resulting from these experimental and theoretical studies is shown in Fig. 7.8.

Recent years have witnessed an explosion in the discovery of exoplanets orbiting other stars. Two categories of these planets include "hot Neptunes" and "icy giants." The temperature and pressure states existing within this body of exoplanets establish into which group a newly discovered planet fits. It is important to understand the composition of these planets, as well as the EOS of light elements and compounds like water over wide pressure and temperature ranges, in order to study the composition and evolution of the planets in our solar system and in other stellar-exoplanet systems. In the past, much of this information came from phenomenological equations of state such as ANEOS and SESAME. However, the growing power of *ab initio* EOS calculations was a seascape for determining high-pressure properties of materials important to planetary science. A particular example is water. Freezing conditions for liquid water can be created quickly using isentropic compression techniques, which also allow insight into the timescale for the phase transition (Dolan et al. 2007). Recent flyer plate data on water to 4.5 Mbars by Knudson and QMD calculations by Desjarlais (Knudson et al. 2012) established an accurate and much different EOS for water than has been used in the classical models. A notable consequence is that *ab initio* EOS calculations for water estimate the core temperatures of Neptune and Uranus to be 20 % cooler than those predicted by earlier models (Knudson et al. 2012).

The ICE and shock data obtained on Z and supported with *ab initio* calculations have established the Sandia shock wave program as a major asset for the fundamental science and weapon communities. In his recollections, Jim Asay reproduced a quote from a *Sandia Lab News* article (Vol. 64, No. 4, p. 1, Feb. 24, 2012), which highlights this recognition:

> Both Congress and DOE have recognized the value of the experiments conducted on Z by Sandia and its partner laboratories, Los Alamos ...and Lawrence Livermore ...in support of NNSA's Stockpile Stewardship Program. The Senate's Subcommittee on Energy and Water Development's 2012 appropriation bill commented that 'The Committee understands that (two Z) experiments (on plutonium) yielded fundamentally new and surprising data about the behavior of plutonium at high pressure and this new data [sic: have] been one of the most valuable contributions to the stockpile stewardship program. The Committee continues to strongly support the weapons physics activities at Sandia's Z facility that are critical to sustaining a safe, secure, and effective nuclear stockpile.' Wrote NNSA Deputy Administrator for Defense Programs Don Cook regarding the joint plutonium work on Z carried out by Sandia and LANL, 'These high-quality data will likely provide new insights and challenge our fundamental understanding.... This accomplishment has been one of our most valuable technical contributions to the stockpile stewardship program.'

As of this writing, the Sandia shock wave group maintains a strong theoretical and experimental program in high-pressure EOS research. QMD plays a central role in this extensive effort. Other theoreticians involved include Thomas Mattsson and Kyle Cochrane, who have extended QMD simulations to other fluids, including xenon and carbon dioxide (Root et al. 2013a, b), and to polymers (Mattsson et al. 2010). The combined capabilities include a diverse team of researchers for addressing a broad range of material issues in the NNSA Stockpile Stewardship Program. The successful resolution of the controversy over the high-pressure response of deuterium is a prime example.

## 7.6 Containment of Toxic Materials

The magnetically-driven ramp and shock loading techniques have also been applied to toxic materials, such as beryllium, plutonium, and uranium, which cannot be studied in conventional laboratory environments. Early in the development of the magnetic driver, Jim Asay, Chris Deeney, and Clint Hall proposed that magnetic ramp loading should be a strategic capability of the NNSA SSP for studying certain actinides and other materials. Various approaches for achieving this goal were discussed with colleagues at other national security laboratories. The development of explosively-driven fast pulsed power at NNSS was first discussed with Bob Cauble, Art Toor, and others at LLNL. There was no interest by LLNL in jointly pursuing research on toxic materials using pulsed power techniques at that time (about 2001) and Sandia dropped that option. Another approach pursued by Chris Deeney involved the implementation of a small pulsed power system located in an underground environment at NTS (now called NNSS) and dedicated solely to studying toxic materials. This option was also not acceptable to the other laboratories and, after considerable discussion, it was not pursued.

A third approach described below was given serious consideration. It ultimately led to the successful development of the only experimental NNSA capability for high-pressure ramp loading of toxic materials at that time. Chris Deeney conceived of an innovative containment system that worked very well for this application. In this case, the ICE cube is enclosed within a hardened steel container connected to the current source in the center section of Z. The ramp or shock loading experiment is conducted as usual, but immediately before the data are acquired, 36 detonators ringing the outer part of the containment chamber are triggered to close off the ICE cube from the rest of the machine. The closure action then occurs shortly after the electrical pulse propagates through the target. The chamber and the toxic residue are contained and can then be transported safely to a storage site.

Achieving this capability on Z required a daunting effort over several years in the early 2000s, including (1) engineering the design of an explosively-operated containment system that had never been used in a high-energy pulsed power environment; (2) coordinating several internal and external groups with different expertise in the mechanical, electrical, explosive, and pulsed power disciplines; (3) coordinating safety issues with the Environmental, Safety and Health (ES&H) organizations; (4) partnering with LANL, who prepared the samples and arranged for transportation to and from Sandia; and (5) obtaining the appropriate local, state, and national approvals. Figure 7.9a is a schematic of the containment system, show-

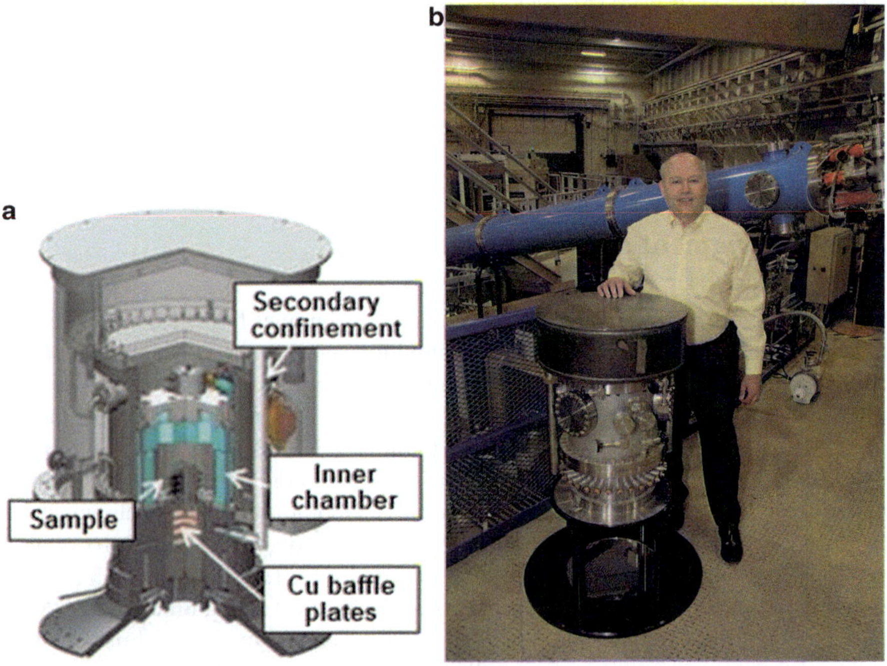


**Fig. 7.9** Containment system to study toxic materials on Z. (**a**) Schematic of explosively contained chamber that isolates the sample after an experiment. (**b**) Randy McKee with a containment chamber to be used on Z. The 36 detonators are visible toward the bottom of the chamber. (Part (**a**) is from the private collection of G. R. McKee; (**b**) is reprinted with permission of Sandia National Laboratories)

ing the complex mechanical–electrical–explosive configuration necessary to ensure containment after a ramp loading experiment on a sample. Several toxic samples are mounted on an ICE cube in the inner (primary) containment chamber. A series of copper baffle plates are placed at the bottom of the sample region to absorb the Mbar pressures that are achieved during generation of ramp loading in the sample. Figure 7.9b shows Randy McKee, Manager of the Pulsed Power Engineering Department in the Pulsed Power Sciences Center, who was instrumental in designing and improving the containment system; he is shown next to an actual containment chamber. The ring of 36 detonators necessary to close off the outer containment chamber is shown near the bottom of the chamber.

Several individuals at Sandia and elsewhere thought this project could not be done, at least not within a short time. One high-level manager at another laboratory was heard to say "they will never shoot Pu on Z." However, Deeney obtained approval for the local management controls and proceeded with the project. He personally organized and managed the various internal and external groups necessary to accomplish the project, since Hall had temporarily left the group. But Deeney was able to entice Clint Hall to rejoin the group, which accelerated development of the project. Hall organized the various working groups into a coordinated team and, over the next 3 years, accomplished the tasks needed for the first test shot with the containment chamber. This was successful as Hall describes in his recollections:

> After several years of work, we were ready to test the first full system on Z. We fired the shot, and all real-time indicators suggested that it functioned properly, but the only way to know for sure was the much anticipated helium leak test. We pulled the system from the machine and hooked it up to the detector. Suffice it to say that I did the 'happy dance' when there were absolutely no leaks! The system worked as well or better than we had hoped, first time out. This turned out to be yet another enabling technology that truly provided a future for Z and the Sandia shock physics effort, even if we didn't fully understand it at the time.

With the successful demonstration of the containment system, additional work was necessary to obtain the internal and external authorizations for experimenting with Pu on Z and to establish an inter-lab team effort with Los Alamos to define the appropriate experiments, prepare the samples, transport them and the waste to and from Sandia, and analyze the results. Deeney and Hall resolved these issues, and the first Pu experiment was conducted with Marcus Knudson as the principal experimenter on May 25, 2006 (Z Shot 1720). Hall recalls this moment:

> Well, to make a long story short, we did shoot three Pu shots on Z over a three-month period with 'the world watching,' they were perfectly contained, and they went off without a hitch. Marcus and Dennis Hayes collaborated to make some nice deductions from the analyzed data, all the logistics went perfectly and lo and behold, we had done the seemingly impossible on a shoestring budget! This experiment was tremendously important as it not only provided some new insights into Pu behavior; it also secured Sandia's future in DMP for many years to come.[10]

[10] DMP is the acronym for a subprogram of NNSA's Science Program. The formal name of that subprogram is Dynamic Materials Properties.

Two more Pu shots were conducted in 2006 with Knudson as the experimenter on June 14 (Z Shot 1731) and July 19 (Z Shot 1757). The purpose of the three experiments was to obtain previously unavailable information related to pit aging.

The combined strategies to choose the liquid deuterium EOS problem as the first objective in the Z shock wave program and the decision to proceed with the development of a shock wave capability on Z for studying toxic materials paid countless dividends. The resolution of the deuterium EOS controversy put Sandia and Z on the map in NNSA's Defense Programs and in the scientific community. The first ramp loading experiments on Pu solidified Sandia's position in the Stockpile Stewardship Program and enhanced the funding for shock wave research at Sandia in the coming years. In addition, the containment system is now opening doors into whole new lines of business, enabling Sandia to bring in an entirely new customer base with highly recognized national needs not connected to weapon physics. Without that system, the problem set for shock wave studies on Z would still be Defense Programs-centric.

Recognition of the role that the Z facility has played in advancing the understanding of nuclear weapons performance is emphatically demonstrated by the FY 2012 Senate Energy and Water Development Appropriations Bill,[11] which recommended additional funding to support nuclear weapons research at Sandia:

> Science Campaign.—... The Committee commends Sandia National Laboratory for successfully and safely performing two plutonium experiments at the refurbished Z facility. The Committee understands that these experiments yielded fundamentally new and surprising data about the behavior of plutonium at high pressure and this new data [sic: have] been one of the most valuable contributions to the stockpile stewardship program. The Committee continues to strongly support the weapons physics activities at Sandia's Z facility that are critical to sustaining a safe, secure, and effective nuclear stockpile.

A lot has happened with respect to experiments with toxic materials on Z, particularly plutonium, since completion of Z's refurbishment at the end of FY 2007 and much will continue to happen. Seventeen more Pu shots have been conducted since the refurbishment through the end of FY 2016. Moreover, the containment system has become a lot more complex than it was for the first three Pu shots in FY 2006, before Z was refurbished. Containment now includes a tent structure around the containment system that maintains negative pressure and contains any toxic materials that are released and other added safety precautions. The Z plutonium experiments continue to yield new, surprising data about the behavior of the material. On May 15, 2013, the current delivered by Z was increased by 45 %, thereby enabling the highest pressure ramp (i.e., shockless) compression experiment ever obtained for plutonium. This accomplishment required very precise pulse shaping at 16 MA currents. The new data will reduce the uncertainties in complex simulation codes that are used to assess the safety and performance of the Nation's weapon stockpile.

[11] Senate Report 112–075, Government Printing Office, 2012), p. 101.

## 7.7 Compact Pulser: Veloce

Although Z experiments provide invaluable data at ultrahigh ramp pressures (up to 4 Mbars) and shocks (to about 40 Mbars) that are difficult to obtain by other means, the experiments are extremely expensive. Scientific and weapon requirements exist for ramp loading data at more modest pressures (less than 500 kbars). Recognizing this need, Clint Hall initiated a program in 2004 to build a pulsed power generator for ramp and shock studies for low-pressure studies that are especially oriented to the Sandia weapons program. The first pulser, built under contract by a French organization and called Veloce, became operational in 2006. A similar compact pulser was built at Washington State University. Both shock and ramp loading techniques have been perfected on the machine (shown in Fig. 7.10), which generates up to 3 MA of current with a risetime of about 450 ns. High-quality ramp loading has been achieved on several metals to peak pressures of about 170 kbars on tantalum with operational costs similar to those for gas gun studies (Ao et al. 2007, 2008; Asay et al. 2008, 2009, 2011). Veloce has also been used to launch flyer plates to velocities of a few km/s, although these velocities are easily obtained with more conventional techniques. Although Veloce is limited to drive pressures of about 200 kbars, efforts are ongoing by David Reisman, now at Sandia, and others to develop a small compact pulser called Thor that will be capable of generating megabar-class ramp pressures.

Veloce has also made significant contributions to the shock wave program on Z by providing a low-cost test bed for new experimental approaches before transferring the concept to Z. An example is the stripline configuration for producing magnetic loading, as illustrated in Fig. 7.11. The anode and cathode plates are separated by a layer of Kapton® 0.35-mm thick, which results in a nearly uniform lateral magnetic field

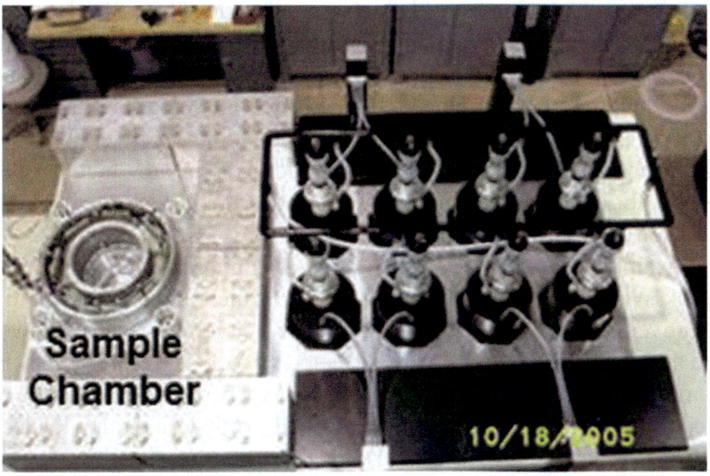


**Fig. 7.10** Photograph of the compact pulser used to produce magnetic ramp loading to peak pressures of about 200 kbars, depending on peak current achieved and the impedance of the sample (Private collection, R. J. Hickman)

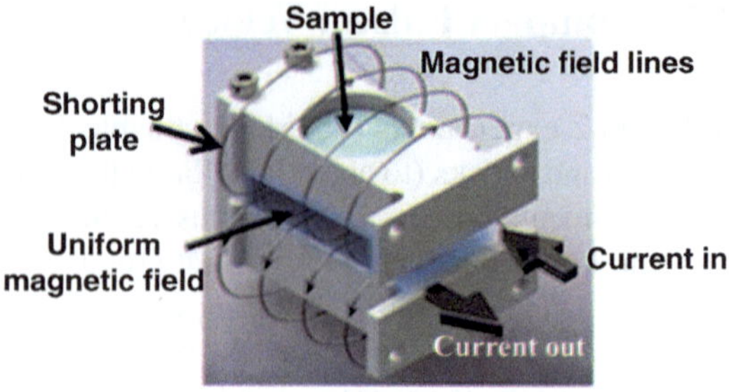


**Fig. 7.11** Stripline technique developed on Veloce for ICE experiments. (Private collection, J. R. Asay)

over the sample dimension in the gap between the two (Chantrenne et al. 2009). Samples are placed on both panels opposite each other and the particle velocity histories on each allow determination of the stress vs. volume response. If the sample response is rate-dependent, one sample is removed and the particle velocity history of the aluminum panel is measured directly. This allows determination of the input particle velocity history to the sample and can be used to determine the material response. This method provides better pressure uniformity across the sample surface in the lateral and longitudinal directions (typically less than 1 % variation from the average particle velocity) compared to the ICE cube configuration. The method also produces more efficient higher pressure loading for a given applied current and allows an accurate analysis of rate-dependent material properties by measuring the panel velocity history directly opposite the sample. Detailed MHD analyses and experimental measurements established uniformity of magnetic loading over the sample diameter to about 1 % (Chantrenne et al. 2009).

Measurement of the input particle velocity on the opposite drive panel allows a direct determination of the input profile to the sample using Hayes' backward analysis (Asay et al. 2009). Standard modeling techniques are then used to simulate the measured sample output profile with any material model, including rate-dependent models. This technique was first developed and perfected on Veloce starting in 2006 and then transferred to the Z facility with minor modifications (Asay et al. 2009, 2011). The stripline configuration is now commonly used on Z for very high pressure measurements and for achieving the highest flyer plate velocities. The configuration, as implemented on Z, is shown in Fig. 7.4 in Sect. 7.4.

Another example of technology transfer to Z is a unique heating technique developed by Jack Wise on Veloce (Wise et al. 2013), which allows samples to be preheated to several hundred degrees before ramp loading. This technique was transferred to the Z facility after it was perfected on Veloce, resulting in substantial savings during development. A high temperature experiment, Z Shot 2189, was successfully conducted on a beryllium sample, heated to 656 C and ramp loaded to high pressure. A soft recovery fixture was also designed by Wise and implemented on Veloce in anticipation of being used on Z (Chantrenne et al. 2009).

## 7.8 Magnetically Applied Pressure Shear (MAPS)

Immediately after ramp loading was demonstrated on Z, there was a concerted effort to develop experimental techniques for obtaining material strength information from unloading wave profiles as is done using shock loading and release wave profiles. Early results were demonstrated on the Veloce machine and then extended to Z (Asay et al. 2008, 2009, 2011; Ao et al. 2009a). Justin Brown significantly extended the technique for use on Z, both in terms of peak pressure and the accuracy of measured strength results (Brown, Alexander, Asay et al. 2013, 2014a; Brown, Knudson, Alexander, Asay 2014b). The results of that work on the flow strength of tantalum to 2.5 Mbars are shown in Fig. 5.1. Nevertheless, an intense effort was launched to develop an alternate technique for use on Z that provides a *direct* measurement of flow strength at high pressure. This effort began in 2007 through a Laboratory Directed Research and Development (LDRD) project under the leadership of Scott Alexander. It is based on the pressure–shear technique developed for oblique impact gas gun studies discussed earlier. The basic concept was applied to magnetic loading techniques, which required a new approach for generation of the shear component of the stress tensor.

The technique was first conceived and implemented on the Veloce pulser. In this approach, a quasi-static magnetic field of up to 10 T (100,000 Gauss) is imposed normal to a sample for several milliseconds before the pulsed current from Veloce is applied on a stripline configuration containing the sample. This external field saturates the conducting stripline and the sample, whether it be a metal or a non-metal. Upon application of the pulsed current at the peak quasi-static magnetic field, a normal longitudinal stress is induced on the sample assembly from the applied current, as described earlier, and a transverse stress is also produced in the driver panel because of the Lorentz force resulting from the external field. The quasi-static external field interacts with the current to produce a transverse motion and a shear stress in the drive plate up to its critical shear strength. The resulting shear wave in the drive plate enters the sample and induces a shear stress in the sample if the shear strength of the driver plate (typically molybdenum) exceeds that of the sample, and the shear stress is transmitted across the interface. The shear stress transmitted through the sample equals its shear strength at the corresponding longitudinal stress. The sample shear strength can then be determined by transmitting the shear wave in the sample into an elastic back plate called an anvil and measuring its free surface velocity. Again, this is dependent upon the shear stress being transmitted across the interface between the two. Presently, several efforts are ongoing to evaluate and develop surface conditions that are most conducive to shear transmission.

The concept for magnetically producing the combined longitudinal stress and pure shear is shown in Fig. 7.12. A dual-VISAR system technique (Chhabildas et al. 1979) is used to determine both the longitudinal and transverse velocities, as in the oblique impact experiment; this will yield the shear strength of the material at the applied longitudinal stress. Although development of the concept on Veloce was extremely complex and required many experiments to resolve the pulsed power, instrumentation, and material issues, these problems were successfully resolved and demonstrated in MAPS

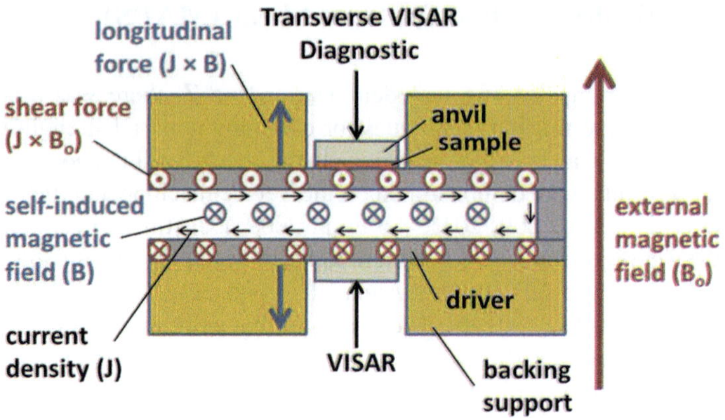


**Fig. 7.12** The magnetically applied pressure shear (MAPS) stripline technique for inducing normal longitudinal stress (pressure) and shear loading on a thin sample. Off-axis VISARs are used to determine the longitudinal and transverse particle velocities, as for the oblique impact technique (Reprinted with permission from Alexander et al. 2010, Copyright 2010, AIP Publishing LLC)

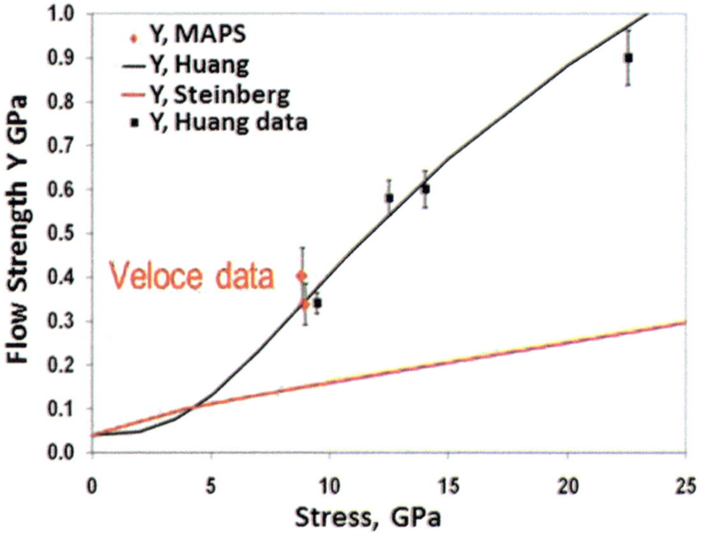


**Fig. 7.13** Comparison of flow strength determined with the MAPS technique and the longitudinal unloading wave profile technique discussed earlier. Al data and model results are from Huang and Asay (2005) (Reprinted with permission from Alexander et al. 2010, Copyright 2010, AIP Publishing LLC)

experiments on aluminum (Alexander et al. 2010). An experimental determination of the flow strength (which is twice the critical shear strength) using the MAPS technique on an aluminum sample is shown in Fig. 7.13 and compared with the flow strength determined from unloading wave profiles (Huang and Asay 2005). The MAPS technique has been transferred to the refurbished Z. Future experiments on Z will investigate the critical shear strength of high Z materials, including tantalum.

## 7.9 STAR in the New Millennium

In spite of the decision to dismantle STAR after a last scheduled experiment in May 1995, the facility was resurrected in the late 1990s through the dedicated efforts of several individuals, mainly Lalit Chhabildas and Bill Reinhart. STAR continues to operate today under Reinhart's supervision, with assistance from John Martinez and Tom Thornhill. Many new techniques and instrumentation have been developed since the facility was reopened. In particular, techniques were developed to launch particles on the three-stage gun to over 19 km/s (Thornhill et al. 2006).

A diverse range of novel techniques were implemented at the facility, including multi-dimensional experiments to validate computer codes (Chhabildas et al. 2000b, 2003, 2006), the effects of microstructure on spallation (Chhabildas et al. 2002; Vogler and Asay 2004; Furnish et al. 2006, 2007; Vogler and Clayton 2008; Vogler et al. 2008), and pressure-shear measurement techniques (Vogler et al. 2011). Other experimental studies at STAR have included the evaluation of 2-D effects of shock loading (Reinhart et al. 2002), well-controlled expanding tube fragmentation (Vogler et al. 2003), hypervelocity penetration (Bessette et al. 2004), dynamic response of concrete (Hall et al. 1999), shock-induced vaporization (Chhabildas et al. 2006), shock compaction of ceramics, mixtures, and powders (Vogler et al. 2007), high-pressure material strength of glass (Alexander et al. 2008), and phase transitions and strength in ceramics, sapphire, and other materials (Reinhart and Chhabildas 2003; Bessette et al. 2004; Reinhart et al. 2006, 2007; Vogler and Asay 2004; Vogler et al. 2004; Vogler and Chhabildas 2006).

Time-resolved spectroscopy was also introduced and expanded over a spectral range from visible to infrared and to the terahertz regime for monitoring hypervelocity impact-generated debris (Reinhart et al. 2008; Thornhill et al. 2008, 2009; Wanke et al. 2007). In addition, Tracy Vogler and his colleagues resurrected the pressure shear compression technique using the oblique impact technique at STAR and have applied it to several investigations of high-pressure shear strength. Vogler has also used it to study the longitudinal shock and shear compaction of granular materials (Vogler et al. 2011).

Management of STAR continued under Lalit Chhabildas until 2007, then under Clint Hall until 2010, and is at present under Gordon Leifeste. As of this writing, the facility is well funded and continues to produce valuable data for a variety of internal and external organizations. Some of the new technology developments are highlighted below.

### *7.9.1 Effects of Shear Stress on Granular Material Compaction*

The study of combined longitudinal and shear loading at Sandia began with the work of Lalit Chhabildas and colleagues in the early 1980s (Chhabildas et al. 1979; Chhabildas and Swegle 1980; Chhabildas and Hardy 1982). The technique used normal impact of a Y-cut quartz crystal to produce combined longitudinal and shear waves that were transmitted to a sample bonded to the crystal rear surface. This technique was effectively used for studying several inert and energetic materials (Chhabildas and Swegle 1982; Chhabildas and Kipp 1985). The main drawback was that the maximum longitudinal stresses in the quartz driver were investigated only to about 30 kbars. Impact techniques were needed to extend the impact stresses to higher values. A slanted, i.e., oblique, impact technique originally developed at Brown University had been used for this purpose (Duprey and Clifton 2000). However, it was not until about 2010 that the slanted impact technique was successfully implemented at STAR (Vogler et al. 2011).

To accomplish this, Lundergan's original projectile launcher, sometimes referred to as the *precision impact launcher* (Barker 1961; Smith and Barker 1962), was transferred to the STAR facility in the 1980s and converted into an oblique impact gun. In this configuration, a straight slot is milled along the interior of the gun barrel and a projectile is fit with a key to ride in the slot during launch. The key prevents the projectile from rotating, thus allowing the impactor, which is mounted at an oblique angle, to strike a sample at the same angle. Upon impact, individual longitudinal and pure shear waves are produced in ratios determined by the oblique angle and the degree of slipping at the surface. This approach allows direct measurement of shear strength at a given longitudinal stress. The technique is referred to as the compression–shear experiment or, more commonly, the pressure–shear (P–S) experiment. The P–S configuration is shown in Fig. 7.14a. Both the normal longitudinal velocity and the transverse velocity, which can be related to shear strength, are measured with two off-axis VISARs, designated as V1 and V2 in Fig. 7.14a. Addition and subtraction of the VISAR signals allow determination of the two velocities (Chhabildas et al. 1979; Chhabildas and Hardy 1982).

Tracy Vogler and his colleagues have implemented this technique for use at STAR with the oblique impact gun and have applied it to several investigations of high pressure shear strength. One of the applications was to study the combined longitudinal shock–shear compaction of granular materials (Vogler et al. 2011). A notable result is that normal shock compaction of tungsten carbide (WC) powder does not change the average granular size of the starting material for shock stresses up to about 45 kbars because of the high strength of WC. This is illustrated in Fig. 7.14b on the left figure. However, if a 2-kbar pure shear stress wave is transmitted through the granular material after longitudinal compressive loading at the same stress, the initial particles are pulverized to a size about ten times smaller than that obtained for longitudinal compression alone (Vogler et al. 2011). The resulting microstructure is shown in Fig. 7.14b on the right side. Note the scales are a factor of five different for the two microstructures shown in the figure. These experiments illustrate that the presence of a small superposed shear stress produces a significant difference in the particle size of the pulverized material, which is caused by grain-to-grain sliding friction. This effect is not observed in the many shock compaction experiments previously performed on granular materials and highlights the need to include grain-to-grain frictional effects during shock compaction.

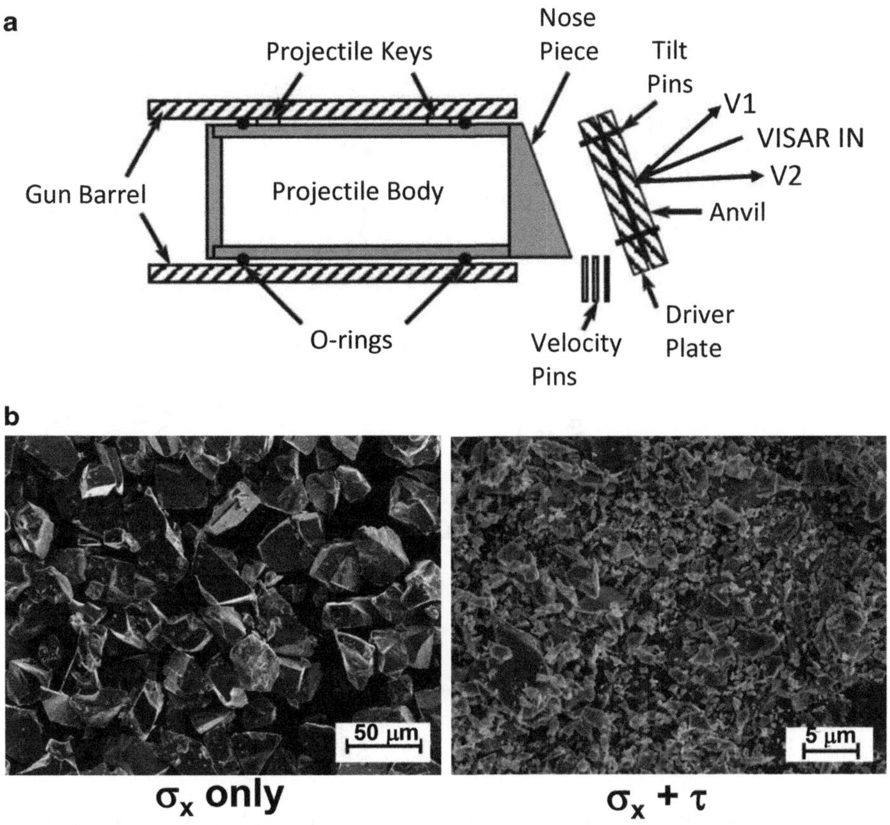


**Fig. 7.14** Slanted impact technique for producing pressure–shear loading of granular materials. (**a**) Experimental technique. A thin granular sample is contained between the driver plate and anvil. A normal laser beam is incident on the anvil, and the off-axis VISAR beams, V1 and V2, are used to determine the longitudinal and shear particle velocities (Reprinted with permission from Vogler et al. 2011, Sandia National Laboratories). (**b**) Results of compaction experiments on a WC granular sample for longitudinal stress loading only and combined longitudinal-shear loading (Private collection, T. J. Vogler)

## *7.9.2 Reverse Taylor Impact Studies*

Hopkinson bar experiments (Kolsky 1949) have been used extensively to determine the failure stress of materials at strain rates from 10 to $10^3$ $s^{-1}$ for uniaxial stress loading, while shock experiments determine the failure stress at strain rates over $10^5$ $s^{-1}$ for uniaxial strain loading. The Hopkinson bar experiment consists of two long, small-diameter rods (loading input and output rods) that sandwich a thin sample between them. Because of the large length-to-diameter ratio of the loading rod, the input stress condition in the sample is essentially uniaxial, or one-dimensional, stress loading. This occurs because the large length-to-diameter rods result in a zero lateral stress load (atmospheric pressure), in contrast to the high lateral stresses

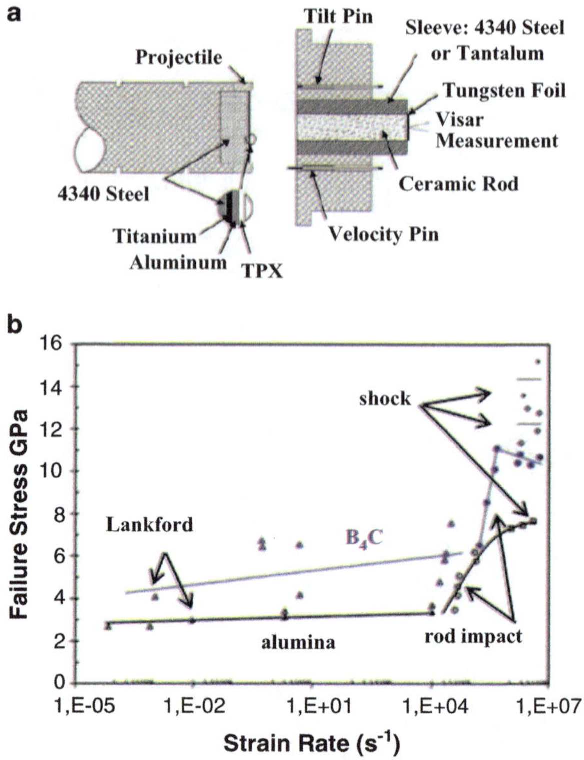


**Fig. 7.15** Sleeved-rod shockless loading technique. (**a**) Experimental configuration. (**b**) Variation of failure stress with strain rate for alumina (*dark line black symbols*) and $B_4C$ (*light line* and *solid circle symbols*) (Private collection, L. C. Chhabildas)

produced by inertial confinement in uniaxial strain shock wave experiments. The failure stress is defined as the dynamic stress at which the material yields (e.g., the HEL in shock wave experiments). Ramp wave experiments discussed earlier provide loading rates of around $10^5$ $s^{-1}$, but are uniaxial strain experiments, so material failure data obtained from these experiments cannot be directly compared with Hopkinson bar results.

Taylor impact experiments (Taylor 1946, 1948), which consist of impacting a small-diameter cylindrical rod onto a rigid barrier, bridge the gap between the two loading rates obtained in Hopkinson bar and shock wave experiments. Post-test observations and high-speed photography are then utilized to determine the plastically deformed contour of the cylinder from which the failure stress can be determined through measurements of deceleration of the cylindrical rod. This technique works very well with ductile specimens because of the ease with which the specimens can be recovered for postmortem analysis. However, the method has not been applied to brittle materials, such as ceramics, because of the inability to recover an intact specimen.

In a novel new test method developed at STAR, a rigid anvil impacted a stationary sleeved rod, and its acceleration profile was used to estimate the dynamic yield stress for uniaxial stress loading. Figure 7.15a illustrates the approach. The use of layered-density materials as an anvil provides a time-dependent loading profile to the sample rather than an abrupt shock, as in normal shock wave experiments

(Chhabildas et al. 1992, 1997). A single-stage compressed gas gun and velocity interferometric techniques experimentally determined the loading behavior of the ceramic rod specimen on its rear surface. The long length-to-diameter aspect of the rod, combined with time-dependent loading, allowed the sample to achieve near-uniaxial stress loading, as obtained in traditional Hopkinson bar tests. For these experiments, time-resolved diagnostics were used for the first time at strain rates of $10^5$ and near-uniaxial stress conditions to determine the failure stress of materials. The rod dimensions were chosen so that the ratio of the length to its diameter is at least four, which is sufficient to obtain uniaxial stress loading. Layered-density materials (Chhabildas et al. 1993b, 1995) impacted both bare and sleeved ceramic rods (Chhabildas et al. 1997, 1998, 1999; Holland et al. 2000), while the VISAR monitored the axial velocity of the free end of the rods. The peak particle velocity determined by the VISAR is a measure of the failure of the material during loading. This test methodology is well suited for brittle ceramics because ceramic rods will invariably fracture during the loading process, especially if they are not sleeved.

Figure 7.15b shows the variation of failure stress vs. strain rate that was obtained from several experiments over $10^{-5}s^{-1}$ to $10^7s^{-1}$ in strain rate. Results of Taylor (rod) impact experiments for Coors AD995 alumina (Chhabildas et al. 1997) and boron carbide ($B_4C$) (Chhabildas and Reinhart 1999) are presented over the strain rate regime of $10^4$ to $10^6$ $s^{-1}$. Experimental results for silicon carbide (Holland et al. 2000) follow the same trend (not shown). The quasi-static compression and Hopkinson bar data (Lankford 1981) are also shown, along with shock Hugoniot results for alumina (squares) and boron carbide (diamonds) at strain rates around $10^6$ $s^{-1}$ (Grady 1992, 1995b). Also included in the figure are the higher ~15 GPa shock Hugoniot values for boron carbide (Brar et al. 1992).

Unique features of this novel test methodology include (Chhabildas et al. 1997) the following: (1) a time-dependent stress pulse generated from graded-density impact allows a smooth and efficient transition from the initial uniaxial strain loading to a uniaxial stress state as the stress pulse propagates through the fairly short rod; (2) the loading rates can be varied from $10^4$ to $10^6$ $s^{-1}$ through a combination of increased impact velocity and different graded- or layered-density impactor designs; (3) intermediate loading rates of $10^4$ $s^{-1}$ obtained in this configuration lie in a region that is not achieved easily by either split Hopkinson bar or shock loading techniques; (4) sleeved rods in combination with graded-density impactors can be used to eliminate the tension damage generated in the sample; and (5) by configuring the specimen during the loading stage by changing both the dimension and the material for the sleeve, the damage induced in the rod can be controlled. This test methodology can also be exercised to evaluate many damage models for both ductile and brittle materials.

With a proper design of the layered-density material and lower impact velocities, strain rates commensurate with split Hopkinson bars can be achieved. This technique would therefore be very valuable for brittle materials. There is considerable challenge in handling brittle materials in Hopkinson bar experiments and, in particular, in trying to prevent these materials from fracturing while the measurements are being made.

### 7.9.3 Shock-Induced Vaporization: Kinetic Effects

The space and defense communities have considerable interest in the effects of shock-induced vaporization for space lethality and vulnerability assessment, as discussed in Sect. 6.2. The technique used to investigate the liquid–vapor debris generated upon impact was first developed in the late 1980s. The early studies of shock-induced vaporization effects examined cadmium (Asay et al. 1990), lead (Asay et al. 1988, 1990), zinc (Wise et al. 1992), and porous aluminum (Kerley and Wise 1988). In these studies, the liquid–vapor products were allowed to propagate across gaps on the order of 10 mm before stagnating against a witness plate. The debris loading of the witness plate measured by a VISAR is then used to infer the thermodynamic state of the debris by correlating the results with computer simulations of the expected debris propagation. The early studies were conducted on the two-stage light gas gun at the STAR facility. Because the impact velocities were limited to around 7 km/s, these experiments did not achieve high levels of vaporization for the shocked and unloaded samples. Later, the three-stage light gas gun (Chhabildas et al. 1993b, 1995) extended these studies to higher impact velocities on zinc (Brannon and Chhabildas 1995) shocked to 550 GPa that did achieve full vaporization upon release from the shocked state. A similar study on aluminum (Chhabildas et al. 2003) suggested difficulty in comparing the experimental results with vaporization predicted by computer simulation for shock stresses lower than 230 GPa, which was predicted by computer simulations, even for gap propagation distances on the order of 100 mm. For aluminum, the shock threshold for vaporization was determined to be around 230 GPa (Chhabildas et al. 2003).

To further probe the shock-induced vaporization process, the kinetics and the transport properties of liquid–vapor mixed debris for aluminum, indium, and zinc, were investigated. For these three materials, the melt, $T_m$, and vaporization, $T_v$, temperatures are vastly different at ambient conditions. Aluminum, zinc, and indium melt at 660, 420, and 156 C, respectively, whereas the boiling temperatures are 2467, 907, and 2270 C, respectively. The melt temperatures, $T_m$, for Al and In are considerably different, but the boiling temperatures, $T_v$, are comparable. The energy required to fully vaporize Al, Zn, and In are 13.9, 2.28, and 2.56 kJ/g, respectively. The vaporization temperatures, $T_v$, for Al and In are comparable, but the energy, $E_v$, required to achieve vaporized states are considerably different. Likewise the energy, $E_v$, required to achieve vaporized states are comparable for In and Zn. All three materials will be totally melted at impact velocities around 10 km/s, but the expansion process leading to shock-induced vaporization will vary, since the boiling temperatures and the vaporization energies are different.

The relative importance of the fundamental thermodynamic parameters that influence the debris transport and propagation characteristics of Al and In in mixed liquid–vapor states produced at impact stresses over 3 Mbars was determined by stagnating the impact-generated debris that expanded both axially and laterally through a gap against a witness plate for gap dimensions over 120 mm. The gap dimension also provides a time duration over which the kinetics of the shock-induced vaporization process occurs before stagnation. The decrease in the loading

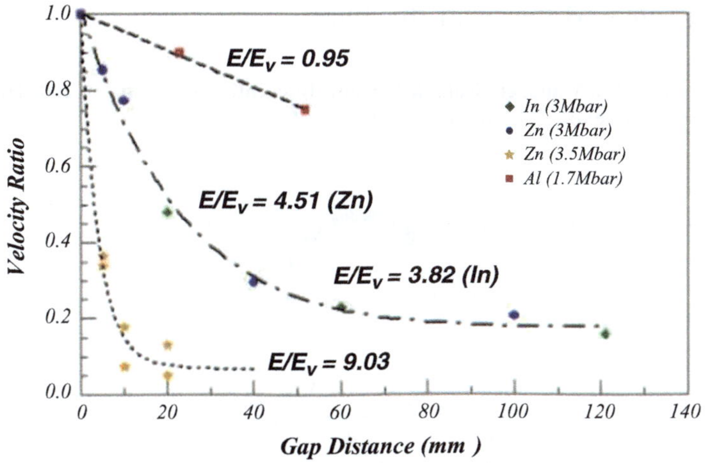


**Fig. 7.16** Variation of the velocity ratio of the peak free surface to impact velocity, i.e., an estimate of peak stress loading on the witness plate by the debris cloud vs. the gap traversed prior to stagnation (Reprinted with permission from Chhabildas et al. 2006, Copyright 2006. Elsevier LTD)

behavior—identifed as the ratio of the free surface velocity to the impact velocity—on the witness plate as a function of gap propagation distance determines the kinetics of the vaporization process.

The velocity ratio in the figure is the ratio of the measured peak free surface (loading) velocity of the witness plate normalized to the impact velocity in the experiment that generates the peak stress. This is shown in Fig. 7.16 for Zn and In, where the impacted shock energy $E$ relative to its vaporization energy ($E/E_v$) is comparable (Chhabildas et al. 2006). Notice the similar propagation behavior for the two materials when $E/E_v$ is similar. Zn is expected to have the highest concentration of vaporized content because both $T_v$ and $E_v$ have the lowest magnitude of the materials investigated in this study. When results of all studies of shock-induced vaporization in Zn ($E/E_v = 9$) and Al ($E/E_v = 0.9$) are combined, a better understanding of the shock-induced vaporization process emerges. In other words, the kinetics of shock-induced vaporization is time-dependent, and the greater $E/E_v$ is, the faster the shock vaporization process. In this study $E/E_v$ is varied from 1 to 9. These first-of-a-kind measurements demonstrate the role of kinetics of vaporization process as a function of impact stress or impact energy induced in the material. A major question that needs to be addressed is whether this represents a *universal behavior*, as the data on Zn and In appear to suggest. Another question concerns the changes in the concentration of liquid to vapor as the debris propagates as a mixed phase flow prior to its stagnation. This information is needed to develop better material models of expanding vapor flows for computer simulations.

## 7.10 People and Places of the 2000s

**Jim Ang, promoted to Manager, Scalable Systems Integration Department, 2004 (Reprinted with permission of Sandia National Laboratories)**

**Dillon McDaniel, Vladimir Fortov, Jeff Lawrence (*L-R*), 2003; Fortov was elected president of the Russian Academy of Sciences in 2013, private collection, R. J. Lawrence**

**Chris Deeney (*left*) and Don Cook visiting the gun facility used for actinide studies at the Nevada Test Site, 2011 (Photo reproduced and unedited from http://nnsa.energy.gov/dr.-chris-deeney-and-dr.-don-cook-standing-front-secondary-confinement-chamber-used-jasper-experiment, according to the Creative Commons License (https://creativecommons.org/licenses/by-nd/2.0/legalcode), Photo was accessed by J. R. Asay, November 2014)**

**Jim Bailey, promoted to Distinguished Member of Technical Staff, 2002 (Reprinted with permission of Sandia National Laboratories)**

**Wayne Trott, promoted to Distinguished Member of Technical Staff, 2000 (Reprinted with permission of Sandia National Laboratories)**

**Mary Ann Sweeney, received an Individual Defense Programs Award of Excellence and promoted to Distinguished Member of Technical Staff, 2013 (Reprinted with permission of Sandia National Laboratories)**

**General Smolen presents Award for Exemplary Civilian Service to Steve Rottler (*right*), Director of New Mexico Weapon Systems Engineering Center, 2004 (Reprinted with permission of Sandia National Laboratories)**

**Marcus Knudson, elected American Physical Society Fellow, 2014 (Reprinted with permission of Sandia National Laboratories)**

**Jack Wise (*left*), promoted to Distinguished Member of Technical Staff, 2016, and Bill Reinhart, 2013, private collection, W. D. Reinhart**

**Lalit Chhabildas, promoted to Manager of the Solid Dynamics and Energetic Materials Department, 2003 (Reprinted with permission of Sandia National Laboratories)**

**Tim Trucano, promoted to Senior Scientist, 2006 (Reprinted with permission of Sandia National Laboratories)**

**Bill Proud, Mike Furnish, Linn Furnish *(left to right)*, private collection, M. Furnish**

**Award dinner for ALEGRA team members *(left to right)* Dan Carroll, Allen Robinson, Kent Budge and Sue Carroll, 2001 (Reprinted with permission of Sandia National Laboratories)**

**Keith Matzen (*back*), Director of Pulsed Power Sciences Center, explaining z-pinch experimental configuration to a Japanese visitor, 2013 (Reprinted with permission of Sandia National Laboratories)**

**Jim Asay elected to National Academy of Engineering, 2003, private collection, J. Asay**

**Jim Johnson receiving the 2011 George E. Duvall Shock Compression Science Award, private collection, James N. Johnson**

**Tracy Vogler co-chairing the 2011 SCCM Conference. He was promoted to Distinguished Member of Technical Staff in 2016, private collection, Tracy Vogler**

**John Benage, Dawn Flicker, Gordon Leifeste, and Thomas Mattsson (*left* to *right*) in front of part of Thor, a compact pulser that will be capable of megabar pressures. Dawn is the Senior Manager of the HED Materials Physics Department. John is Manager of Dynamic Material Properties, Gordon is Manager of Solid Dynamics Experiments, and Thomas is Manager of HEDP Theory (Reprinted with permission of Sandia National Laboratories)**

# Chapter 8
# Looking to the Future

## 8.1 Looking Back

The synopsis of shock wave science presented in this book describes some of the pioneering research conducted at Sandia over the past 60 years. The shock wave program was organized and conducted rather differently from that of similar research programs at other institutions. Two separate shock wave research efforts were established in the 1950s, one focused on scientific understanding of shock compression processes and the other on engineering applications.

On the scientific side, the initial focus of shock wave research under Frank Neilson dealt with understanding the fundamental electromechanical response of ferroelectric and piezoelectric materials subjected to shock loading. This work derived from the need to develop new devices to meet the evolving requirements of the Nation's nuclear weapons program. Along with Bill Benedick and Bob Graham, Neilson performed many experiments that shaped a basic understanding of these materials and a theoretical description of the electromechanical response. Initially, experimental research was conducted with explosive plane wave generators developed at Los Alamos to produce precise loading conditions to correlate the electrical output with shock stress. The Los Alamos research paved the way for the use of these materials in weapon components and was also important for the development of diagnostic sensors for experiments. In particular, it had immediate pay off in the early 1960s, including the first time-resolved stress gauge for shock wave studies and the development of explosively activated power supplies for weapon components. From this early initiative, a research organization, later named the Physics of Solids Research Department, emerged under the guidance of George Samara with a considerably broadened emphasis on understanding the physical and chemical properties of a wide variety of materials subjected to shock compression.

On the engineering side, Don Lundergan emphasized engineering solutions to weapon problems by focusing primarily on the classes of materials used in these systems. The motivation was to understand the response of contact fuzes and other



J.R. Asay et al., *Impactful Times*, Shock Wave and High Pressure Phenomena,
DOI 10.1007/978-3-319-33347-2_8

components fielded at that time. Lundergan recognized that solution of these complex problems would require extensive experimental research, along with the integrated development of new material models and hydrodynamic codes to simulate the components and systems. Accordingly, he initiated a coordinated research effort to couple the results from new experimental capabilities with advanced material models for use in the emerging hydrodynamic computer codes. This objective was achieved by combining the three separately managed activities within an organization under a single manager.

Lundergan organized and led the Physical Properties Division in 1962 to begin implementing this research goal, and he recruited several key individuals who successfully enabled achievement of his vision. In particular, he initially enlisted two innovative experimentalists, Lynn Barker and Red Hollenbach, who developed novel shock wave techniques and diagnostics that are still in use today. Lundergan added several other experimentalists and modelers to the department in the early 1960s. In the mid 1960s, he hired Walt Herrmann, a research associate from MIT, who had proven experience in material modeling and code development. When Lundergan stepped down from management in the late 1960s, Herrmann managed a department involved in shock wave research, naming it the Solid Dynamics Research Department, in 1968. That department and name remained in effect until it was dissolved by Sandia reorganization in the early 1990s. Herrmann was the driving force for enabling Lundergan's vision through substantial staff increases and a cohesive organization of individual departments (at the time, called divisions) with a concerted focus on applying the integrated experimental, modeling, and simulating capabilities to solve applications of nuclear weapons and other problems of national interest.

Research in this area continued over the years on various fuze designs and continues to the present. Now it is possible to simulate fuze performance accurately for virtually any material used in contact fuzes and other weapon components for essentially any impact environment. Furthermore, the research conducted on energetic materials resulted in the development of improved detonators and explosively actuated electrical generators used in firing sets. This research allowed not only the development of complex material models used in these devices, including composites and ferroelectric phase transformations, but also the implementation of the models in hydrodynamic computer codes to certify nuclear weapon components and systems.

A landmark accomplishment of experimental shock wave research activities was the development of planar impact techniques using high velocity plate impact launchers. While Neilson's team initially used plane wave explosive generators for shock wave research, Lundergan initiated a major transition to using smooth bore guns in the late 1950s. In this configuration, flyer plates mounted on a gun-launched projectile produce precise impact loading with better safety and control than high explosive sources, more flexibility in defining the impact states, and better accuracy for measuring the kinematic properties of the shock wave, such as its velocity and either the stress or particle velocity in the shocked state. Lundergan's development of the first precision impact launcher was a game-changing capability that was the start of a family of guns at Sandia and elsewhere that progressively increased the launch velocities and corresponding impact pressures.

Launch capabilities for equation of state studies rapidly increased from an initial velocity of about 0.3 km/s using air as the pressurized gas in the 1950s, to slightly over 1 km/s using helium in the 1960s, to over 2 km/s with a propellant-driven (powder) gun by the early 1970s, to 7 km/s with a two-stage light gas (hydrogen) gun in the early 1980s, and finally to a three-stage light gas gun (the HyperVelocity Launcher) with a maximum velocity of 20 km/s in the 1990s. In the late 1990s, it was discovered that magnetic fields from fast pulsed-current sources could launch intact flyer plates to even higher velocities; 46 km/s was achieved with this technique by 2010. This enormous velocity increase resulted in nearly a thousand-fold increase in the shock pressures that could be precisely produced and measured in the laboratory, from tens of kbars in the early 1960s to a few Mbars in the 1990s on the two-stage gun to a few tens of Mbars in the 2000s with magnetically-driven flyer plates. This range of impact velocities and pressures allows shock studies over the full spectrum of weapon applications, from low pressures appropriate to weapon delivery systems to the operational pressures of modern nuclear weapons.

Another strategic capability evolving from gun technology and progressively developed over several decades was the ability to compress materials smoothly to Mbar pressures, referred to as ramp or quasi-isentropic loading. This additional game-changing technique resulted from a long-term commitment by the Solid Dynamics Research Department to probe cooler thermodynamic regions at high pressure than could be achieved by shock compression. The first experiments to develop ramp loading were conducted in the late 1960s using fused silica driver plates, followed by several graded-density impactor approaches, culminating in 1999 with magnetic loading on the Z facility that subsequently produced the first smooth pressure loading to several Mbars. Furthermore, the magnetic loading technique allowed more complex loading histories to be produced in a sample. Through technical advances to improve loading conditions and data analysis, the accuracy of ramp loading data on Z is approaching that possible with traditional shock wave techniques. As a result, the accurate determination of the full equation of state surface of any material is now possible up to about 4 Mbar pressures. This landmark accomplishment is now being mimicked on other platforms, including laser compression, which can produce higher ramp loading pressures, although with limited accessibility of the equation of state surface and higher uncertainties in equation of state data because of the smaller sample sizes.

By the mid 1970s, experimental diagnostics in the Physics of Solids Research Department and the Solid Dynamics Research Department had evolved to a mature state, resulting in the ability to perform high-resolution measurements of shock wave phenomena to pressures of a few hundred kbars. The quartz gauge developed at Sandia around 1960 was the first time-resolved stress gauge for these measurements and was limited to stresses of around 30 kbars. The VISAR, invented at Sandia by Lynn Barker in the early 1970s, extended these peak stresses to several Mbar pressures by the 1980s and the 1990s on the two-stage gun and to over tens of Mbar by 2010 on the Z platform. The VISAR, with its capabilities enhanced over the years (see, in particular, Barker's recollections), was also a game changer for shock wave instrumentation and continues as the primary diagnostic for shock wave studies in laboratories throughout the world.

Sandia developed several other diagnostics for shock wave studies, including lithium niobate PZT gauges, ferroelectric stress gauges, polyvinylidene difluoride (PVDF) stress gauges, the Line VISAR (also referred to as the Line ORVIS), and the pressure–shear interferometer; these are discussed in the preceding chapters. As a collective set, the temporally- and spatially-resolved diagnostics have allowed in-depth studies of material response, not only over extended pressure, temperature, and density ranges but also at a scale sufficient to observe heterogeneous phenomena, such as that caused by grain orientation or other localized deformation. The enormous increase in the ability to probe detailed dynamic material response over wide pressure, specific volume, and temperature regimes has enabled the development of next-generation material models to describe these responses.

A wide range of material models has also been developed in the Sandia shock wave program. These include models for the dynamic response of porous materials, the viscoelastic response of polymers, the dynamic response and failure of metals, and the electromechanical behavior of piezoelectric and ferroelectric materials. Also developed were analysis methods for the shock behavior of composites, the initiation properties of energetic materials, the kinetics of phase transformations, and *ab initio* equations of state using quantum molecular dynamics. Many of these models have been implemented in hydrodynamic codes to describe material response over large dynamic ranges. These improvements have enabled solutions to a wide variety of applications related to weapons and fundamental science.

Concurrent with the experimental and modeling advances, significant efforts were devoted to developing hydrodynamic codes to simulate dynamic material response. These started with the 1-D Lagrangian code, WAVE, which Walt Herrmann brought from MIT. This code evolved into the 1-D Lagrangian code, WONDY. Another 1-D code, CHARTD, was also developed by Sam Thompson at the time. The computational capabilities progressed to 2-D codes in the 1970s and 1980s, including the Lagrangian code, TOODY, and the Eulerian code, CSQ. These hydrodynamic codes, particularly Thompson's CSQ and later his 2-D CTH, were used to solve several scientific problems in the 1970s and 1980s, including the Three Mile Island nuclear accident and the USS Iowa explosion. Thompson's 3-D hydrodynamic code, CTH, became the workhorse for solving complex problems in the mid 1990s and incorporated the suite of material models developed both at Sandia and at other laboratories in prior decades to address an array of scientific and practical applications. CTH has played a pivotal role in a number of nuclear, DOD munitions, and conventional problems, such as interpreting the vapor plume produced by the Shoemaker-Levy comet impact on Jupiter. The wide use of CTH throughout the shock physics community was managed for many years by Gene Hertel. The electromechanical code SUBWAY, developed by Steve Montgomery, allowed accurate descriptions of shock-induced currents from piezoelectric and ferroelectric devices. ALEGRA, one of the next-generation codes developed at Sandia by Mike McGlaun, James Peery and colleagues, extended these capabilities to include 3-D magnetohydrodynamic phenomena important to pulsed power and other applications.

The widespread application of Sandia shock wave research to more complex scientific and practical problems has demonstrated the progress made over the past six decades. Many of these applications could not be envisioned at the beginning of the Sandia shock wave program. The preceding chapters in Part I and the individual recollections in Part II provide a glimpse of how people became involved in this effort, the struggles they experienced, the paths leading to success or failure that caused a redirection of research efforts in some cases, and the moments of exhilaration when ideas came to fruition. The hard work and dedication of many individuals have integrated shock compression science at Sandia into the mainstream of high-pressure science. The exceptional impact of this work is underscored by the many awards that individual researchers have received, the national problems that have been solved, and the international recognition of the research.

In addition to major contributions in developing weapon components and subsystems, the shock wave program at Sandia made significant contributions to the nuclear weapon testing program. The early tests were carried out right after the end of World War II to determine weapon effectiveness, nuclear yield, and explosive capability. Nuclear tests provided information on how weapons perform under various conditions and how structures responded when subjected to nuclear explosions. LANL and LLNL were primarily responsible for testing nuclear weapons, but Sandia played an important supporting role in providing state-of-the-art diagnostics throughout the testing phase. Nuclear weapon testing was performed aboveground either at the Nevada Test Site or at various sites in the Pacific Ocean until the United States, the United Kingdom, and the Soviet Union signed the Limited Test Ban Treaty of 1963, which allowed only underground nuclear testing. Underground experiments continued at the Nevada Test Site (NTS) in the United States until the last nuclear test was conducted in September 1992.

Sandia was an active participant up to the cessation of atmospheric and underground nuclear testing. In the early phases of underground testing, Sandia researchers, including Luke Vortman, Al Chabai, and Carter Broyles, provided the instrumentation and diagnostic advances used in many of the tests; this led to the characterization of shock propagation, explosive cratering, and shock interaction with structures. The evolution of scaling relations and the use of the SLIFER technique, which is discussed in Chap. 2, were substantial Sandia contributions to quantify nuclear yield. These measurements, in combination with scaling laws, provided the yield of a nuclear device within hours following an event, versus the *weeks to months* required with standard radiochemical analyses. With the development of the time-resolved quartz gauge and interferometry, Sandia made some of the first measurements of radiation-induced stress pulses to quantify the response of components to pulsed nuclear radiation. Later, extensive tests of geological materials by Karl Schuler, Dennis Grady, Lalit Chhabildas, Mike Furnish, and others provided material properties that led to accurate simulations of ground shock propagation and interaction with structures.

Following the cessation of nuclear testing, LANL and LLNL began actively pursuing non-nuclear experiments at NTS to address the requirements of stockpile stewardship. The NNSA Stockpile Stewardship Program, initiated by the National Defense Authorization Act of 1994, focused shock wave research on thermodynamic regimes appropriate to operation of nuclear weapons. This new thrust prompted initiation of subcritical experiments at NTS to subject materials and systems to relevant thermodynamic conditions without nuclear yield; the available data were significant and examined the simulation capabilities of various codes being used to predict and analyze weapon performance. Sandia played another important role in subcritical experiments by using the VISAR and other diagnostics to provide critical data on the response of many actinides subjected to shock loading. Mike Furnish and Tom Bergstresser used these diagnostics to (1) validate equation of state models, (2) determine material failure mechanisms, and (3) study mass ejecta from actinides and other materials. Those experiments attest to the pivotal contributions of Sandia expertise in shock wave research to NNSA's nuclear weapons program.

Through its heritage of leading developments in experimental techniques, instrumentation diagnostics, and theoretical modeling over the last six decades, the Sandia shock wave program is well positioned to continue that leadership. In particular, the High Energy Density Material Physics Department under Dawn Flicker possesses several unique capabilities that underpin its strategic position, which can be summarized as follows.

- Extensive expertise in developing and using the Line VISAR has been developed over several years and is available to perform and analyze mesoscale-level experiments for applying this technology to complex materials.
- The STAR facility is unique for "one-stop experiments" because of its full range of launcher capabilities, which include the HyperVelocity Launcher and an extensive suite of diagnostics for impact lethality and material property studies.
- The combined capabilities of the oblique impact gun and the magnetically applied pressure shear (MAPS) technique are unique in the community for direct shear strength measurements over the full pressure range to Mbar pressures.
- Z is the only facility that can produce smooth ramp compression of millimeter-sized samples with microstructural features appropriate to applied materials. When combined with magnetically-driven flyer plate shock compression and unloading experiments, Z allows the study of the full pressure, volume, and energy surface to several Mbars. The ability to structure the loading history arbitrarily, such as the shock-ramp technique developed on Z, is a paradigm shift for shock compression science that expands the equation of state surface that can be studied.
- Experimental research in the department is closely integrated with theoretical quantum molecular dynamics *ab initio* modeling.
- To offset the high cost of Z experiments, the Veloce pulsed power facility is available for ramp loading experiments to several hundred kbars at a much lower cost (a few thousand dollars per experiment on Veloce, as compared to a few hundred thousand dollars per shot on Z).
- The magnetically-driven cylindrical loading capability that is under development on Z will be available as an alternate technique for nonplanar material studies to several Mbars.
- Novel direct-temperature diagnostics are being actively pursued for combined shock and ramp loading.
- The Sandia shock wave effort is closely integrated with the next-generation Advanced Photon Source at Argonne National Laboratory, which will allow first-of-its-kind material studies at atomic and microscopic levels.

## 8.2 Looking Forward

Sandia's shock wave program has contributed significantly to the nuclear weapons program and to resolving many other national problems, in part due to the development of advanced material models and simulation capabilities. Significant progress has been made in understanding and predicting dynamic material response over the past six decades, but substantial work remains, particularly in characterizing the failure of materials and their subsequent response. This is especially important for developing accurate models to describe the dynamic response of heterogeneous materials both in compression and tension. Several technical challenges remain for present and future generations of shock wave researchers, some of which are summarized below.

- **Resolution of model uniqueness is essential to develop predictive capability**. It has been recognized since the beginning of shock wave research that different models of material response can be used to describe the same experimental observation, typically represented by a time-resolved wave profile. A well-known example is the unloading and reloading response of materials from shocked states, which can be adequately modeled with rate-independent mesoscale-based models or with rate-dependent dislocation-based models (as discussed in more detail by Asay and Lipkin, 1978 and by Johnson, 1993, Winey et al. 2012, and colleagues). Several other examples can be cited that illustrate the nonuniqueness issue arising from measurements at the continuum level. To achieve true predictive capability, novel experimental, diagnostic, and computational techniques are needed to identify deformation processes leading to the distinguishing characteristics of material response. An unambiguous interpretation of the physical process is the necessary first step that can lead to uniqueness in modeling and eventually to a predictive capability.
- **Accurate temperature measurements are needed for thermodynamically consistent descriptions**. A goal of high-pressure modeling is to determine thermodynamically consistent equation of state surfaces from which all other thermodynamic variables can be derived. This requires measurements of pressure, volume, and temperature (P, V, T) to determine the Helmholtz free energy function, as opposed to the pressure, volume, and energy (P, V, E) determined from shock and ramp wave measurements. Measuring the temperature in shocked materials has been a goal of many shock wave programs for several decades, with hundreds of man-years devoted to this effort. Even though accurate temperature measurements have been obtained in cases with transparent materials, such measurements cannot be applied routinely and arbitrarily for any material. Radiance measurements are usually made, but these are not possible in opaque materials such as metals or at transparent window interfaces. Innovative new approaches are needed to determine the temperature states achieved in shock and ramp wave experiments. A recent innovation using the emissivity of a shocked gold foil viewed through a transparent window appears promising (Dolan et al. 2013a) if it can be perfected for general use. Other novel approaches should be rigorously pursued, given the enormous benefit to equation of state modeling.

- **Measurements of material response at the atomic scale are needed for rigorous confirmation of *ab initio* equations of state**. The bulk of shock and ramp wave experiments are conducted at the continuum scale. Material property data at the atomic or microscope level would be useful to remove ambiguity in interpreting material properties and to develop more realistic equation of state models at an atomistic level. An example is the determination of material strength at high pressures. These data are normally obtained through continuum measurements, such as wave profile measurements, and are based on considerable assumptions and interpretation. This problem may be circumvented with measurements of strength at the atomic level. Considerable progress has been made using x-ray diffraction to determine lattice spacing and to infer strength properties during shock loading. By comparing such data with continuum measurements, this approach could remove ambiguities in interpretation (Turneaure and Gupta 2007; Comley et al. 2013).
- **Material models are needed to bridge the length scale from the atomic to the continuum**. The role of mesoscale (grain-scale) effects on continuum response has been long recognized and actively pursued. Most mesoscale models are based on grain-scale dimensions, although recent multi-scale models have the ability to couple dislocation mechanisms effective at the microscale to material strength measurements at the continuum level (Becker 2004; Barton et al. 2011). Although these approaches are promising, adjustable parameters are still used to model the dislocation processes and to describe the continuum response, again raising the specter of nonuniqueness. It is possible that this deficiency could be removed by atomic-level measurements in real time that quantify dislocation distributions and other microscopic deformation mechanisms.
- **New diagnostic tools are needed to probe high compression states**. Wave profile measurements of shock deformation features have been the primary tool for probing dynamic material behavior for half a century and have been an effective source for developing continuum models. However, new approaches are needed to probe material behavior over the full-length scale from atomic to continuum. The recently established Dynamic Compression Sector (DCS) at the Advanced Photon Source[1] (APS) promises to be a viable tool to probe materials at the atomic and microscopic levels. With this unique capability, intense x-ray pulses from the high-energy synchrotron beam line are used to probe shocked states produced by different on-site platforms, including single- and two-stage light gas guns, laser-induced shock loading, and magnetically-driven shock and ramp wave loading. This first-of-its-kind capability will probe real-time material response at the microscopic scale and make time-resolved x-ray diffraction and imaging measurements. Implementing this technique simultaneously with continuum measurements should provide not only a fundamental understanding of how materials compress dynamically at atomic scale lengths but also evaluate

[1]The DCS is a collaborative effort between the Institute of Shock Physics at Washington State University, the Advanced Photon Source in Argonne, Illinois, and the DOE/NNSA national security laboratories.

statistical concepts invoked to infer continuum response. In addition, this will help determine actual deformation mechanisms at multi-scale lengths.

- **Experimental techniques are needed to study material response in more complex geometries.** As discussed throughout this volume, the overwhelming majority of shock wave measurements are made for one-dimensional planar loading. This configuration produces well-defined states of uniaxial strain for developing material models. However, the approach neglects more complex deformation modes, such as shearing or local deformation arising from the relative motion of internal material segments. Oblique planar impact is a step toward directly quantifying shear response but is still 1-D planar loading. Examples of complex geometries include determination of material properties for spherical or cylindrical convergence, which is also one-dimensional loading, but involves the relative motion of material segments. Both these approaches are considerably more difficult to produce and diagnose than planar loading,[2] but significant progress has been made on both fronts with Z (Dolan et al. 2013b; Lemke et al. 2011). Ray Lemke and Dan Dolan have demonstrated the ability to make accurate equation of state measurements to around 10 Mbars for a converging Be cylinder that is 3 mm in outside diameter using magnetic compression. This technique is promising for probing material properties, particularly equation of state and strength, in pressure regimes and deformation configurations not achievable with planar loading (Trunin et al. 2001).

As illustrated by the success of the Solid Dynamics Research Department throughout its history, the close integration of experiment, modeling, and hydrodynamic code development is an essential requirement for weapon science and other applications. The Physics of Solids Research Department, under the guidance of Frank Neilson, Orval Jones, Robert Graham, and George Samara, developed a comprehensive and theoretical understanding of piezoelectric and ferroelectric materials. These philosophies continue to the present day in the shock wave program managed by the Pulsed Power Sciences Center, which is sustaining a substantial effort in high-pressure shock wave research using both the Z and STAR facilities. The experimental research is complemented by a strong theoretical modeling effort and continued enhancement of the ALEGRA magnetohydrodynamic code. Several recent papers based on first principles theory highlight this close interaction, reflecting Lundergan's 1950s vision of an integrated shock wave capability.

Although not directly connected to the pulsed power shock wave program, the departments in the Computational and Mathematical Sciences Center and the Lethality & Threat Department in the Integrated Military Systems Center as well as the energetic research at Explosive Components Facility at Sandia continue to develop state-of-the-art material models. In addition, the latest computing technology is used to simulate a broad range of customer-driven problems, whether they are weapons-related or involve other investigations of national interest. A recent example is analysis of the asteroid breakup near Chelyabinsk, Russia, in 2013 (Brown, Assink, Astiz et al. 2013b).

---

[2] Ray Lemke, private communication in 2014 with Jim Asay, concerning the analysis of material response during cylindrical convergence.

With the Pulsed Power Sciences Center's mix of world-class capabilities, the shock wave program at Sandia is well positioned and likely to continue, at least in the near term. The shock wave research staff at Sandia has always preserved an optimistic outlook, especially in the dark times of the 1990s. This attitude has served the program in good stead over the past six decades and hopefully will continue to do so in the future.

# Part II
# Memories of Shock Wave Research

# Chapter 9
# Memories of Shock Wave Research at Sandia

**Editors' Note** These individual recollections present a window into the personal experiences of people who participated in the shock wave research program at Sandia. We made a strong effort to contact and encourage as many people as possible to participate. Over 80 people were contacted and about 40 provided recollections of their personal experiences. Each contributor was asked to provide a summary of their role in shock wave research at Sandia, bringing out any interesting events or anecdotes that happened along the way. We purposely gave very little guidance on style and format, which is why some recollections are written as a "stream of consciousness," while others are more technical and consist of short annotated summaries of major papers by the submitter. The result is a rich and interesting mix that highlights individual personalities, personal struggles, and technical successes. The years provided for each person represent his or her year of hire at Sandia and the year he or she left Sandia. If this is different from the year the person left the shock wave program, that year is also noted in a footnote. Quotations from the individual recollections are used liberally throughout Part I to bring out the personal aspects of shock wave research at Sandia and provide a pointer to further reading in the recollections. In some cases (e.g., George Samara and Walter Herrmann), recollections by others were used to capture the seminal contributions that these researchers made to the shock wave field. In total, the recollections are a unique glimpse into the shock wave program at Sandia. A short montage of photos is also given at the end of Chaps. 2–7 to associate "faces" with the recollections. These photos often tell a story in themselves.

Because these recollections might be read as individual pieces, we have attempted to define nearly all acronyms within each recollection. The one notable exception is the VISAR, the acronym for Velocity Interferometer System for Any Reflector, because this is used so frequently in both Part I and in Sandia publications relating to shock wave physics. Moreover, in some instances, we have used the "modern" version of an acronym rather than the historical definition of that acronym. A particular example is the Department of Defense, which recently changed its acronym from DoD to DOD.



J.R. Asay et al., *Impactful Times*, Shock Wave and High Pressure Phenomena,
DOI 10.1007/978-3-319-33347-2_9

# James A. Ang

## (1989–Present[1])

My research and development (R&D) career started in combustion and fire research at the University of California at Berkeley. I finished my PhD in the summer of 1986 and accepted a position with General Research Corporation (GRC), a small defense contractor in Santa Barbara, CA. Having gone straight through my academic career with a Bachelor of Physics and Bachelor of Mechanical Engineering and on to graduate school, part of the attraction of this job was the opportunity to live in Santa Barbara. This position provided me with the serendipitous introduction to GRC's small group of shock wave physicists and impact lethality scientists, including Alex Charters and Bill Isbell. Alex Charters was a researcher at the U.S. Army Ballistic Research Laboratories at the Aberdeen Proving Grounds before and during WWII. When I was contemplating a potential move to Sandia's shock physics group, Alex advised me that this was a tremendous opportunity to join research leaders like Walt Herrmann and Sam Thompson. Bill Isbell was the first to commercialize VISAR systems, perhaps an enterprise that motivated Lynn Barker to start Valyn VIP. GRC had a long history of interactions with the Department of Energy (DOE) national security laboratories. GRC commercialized Sandia's clean room technology and built a successful business around this technology. Another line of business for GRC was supporting DOE and the Defense Nuclear Agency with fielding instrumentation and diagnostics for underground nuclear tests.

I spent three and a half years at GRC, where I supported our Systems Engineering and Technical Assistance contract for the Department of Defense's Strategic Defense Initiative Organization (SDIO), now known as the Missile Defense Agency. While at GRC, I developed a theoretical and analytical model to explain optical impact flash emissions. I also led GRC's systems engineering analysis for the Defense Advanced Research Projects Agency (DARPA) Electrical Energy Gun System (EEGS) study, which was chaired by David Hardison. As a part of my work on railguns, Bill Isbell and I attended a workshop on railgun technology in Durango, Colorado. At this workshop I met Jim Asay. I cannot recall the specifics of our conversation, but I remember that I was impressed with Jim and the HELEOS[2] railgun project. Upon completing the DARPA EEGS study, I sensed the time was right to move on. My GRC supervisor advised me that to pursue research involving more than systems engineering, I should go to a national security laboratory. It did not take me long to arrange for an interview trip to Sandia. In retrospect, my career change came at a very good time, since within a couple years of my departure, SDIO funding was cut back, and underground nuclear testing ceased.

[1] Jim Ang left active participation in shock wave work in about 1994. As of this writing, he leads a computational group developing capabilities for solving shock wave problems.

[2] Editors' note: STARFIRE was a joint SNL/LLNL project to develop a Hypervelocity Electromagnetic Launcher for Equation of State (HELEOS) experiments.

I joined Jim Asay's experimental shock wave physics division in December 1989. The other divisions in the group were Mike McGlaun's shock wave code development division and Paul Yarrington's shock wave weapons analysis division. The three divisions made up Dennis Hayes' Solid Dynamics Research Department, and we were under the Engineering Sciences Center led by Walt Herrmann. Three of us started at approximately the same time: James Peery with McGlaun, Elliot Fang with Yarrington, and I with Asay. I am proud to be associated with this group of successful Sandians.

When I first joined Sandia, I was assigned to the HELEOS railgun project. The concept was simple—launch a projectile with a two-stage light gas gun into the breech of a railgun, which provides the third stage of acceleration. There were a number of engineering challenges associated with the integrity of the projectile and its armature, precise timing for the discharge of the capacitor banks, etc. The project was successful in achieving railgun acceleration of an injected projectile. This was testament to the engineering development skills of the team. I recall being impressed that Jim Asay had assembled a team for the HELEOS project that included Ron Hawke and Allan Susoeff from LLNL—an experience that prepared me well for many subsequent collaborations with counterparts at our sister national laboratories. Ultimately, the HELEOS project was superseded by Lynn Barker, Lalit Chhabildas, and Bill Reinhart's successful development of the graded-density impactor to launch flyer plates. I understand that Z is now launching flyer plates at velocities up to about 45 km/s. It is interesting and very impressive to read how far the electromagnetic launch technology has progressed.

I was grateful that I had the opportunity to continue my research on impact flash; I was able to work with EG&G Princeton Applied Research Corporation and procure a first-of-a-kind Optical Multichannel Analyzer (OMA) to capture time-resolved spectra of impact flash signatures. This unique diagnostic led to a project from the U.S. Army Space and Strategic Defense Command to send this OMA to Andy Williams' two-stage gas gun facility at the Naval Research Laboratory. His terminal ballistics catch tank is rated to allow testing on targets that include small amounts of high explosives. There, we were able to capture unique time-resolved spectra that distinguish impact flash optical emissions from the subsequent high-explosive detonation spectral signatures.

Jim Asay steered me to the ballistic holography work that Gary Hough developed at the Missiles and Electronics Group of LTV Corporation in the late 1980s. On Jim's recommendation, I also visited Charlie McMillan and his facility at LLNL in the early 1990s to learn about the holographic system he had developed to measure the size distribution of particles ejected from shocked metal surfaces. This background investigation helped me create a small project to develop the concept that was first proven by Gary Hough to use pulsed lasers to capture holographic images of terminal ballistic impact-generated debris bubbles. Gary's key insight was that, while impact-generated debris bubbles may move too fast to image the surfaces of those fragments directly, the holographic images of the "shadowgraphs" of the debris fragments against a stationary background do record three-dimensional (3-D) information. With very capable support from Scott Gosling, my technologist, we were quickly able to integrate and demonstrate our own version of this 3-D diagnostic capability. Key differences and advances over Gary's original work (Hough et al. 1990) included the use of a Q-switched, frequency-doubled Nd-YAG laser to capture the images. This allowed the use of a continuous-wave, frequency-doubled Nd-YAG laser for image reconstruction. Scott Gosling also designed and developed a computer-controlled, three-axis, stage-mounted microscope to analyze and capture quantitative 3-D information of impact debris, including position and size data for each individual fragment shadowgraph.

With additional optical engineering analysis from Bill Sweatt and Bruce Hansche in Sandia's photometrics organization, we were able to quantify the resolution of our holographic diagnostics system to less than 5 μm. This resolution was well beyond the precision of the 3-D simulation capability at the time. Ultimately, when Ed Barsis, our Center Director, decided to shut down the STAR facility, our pulsed laser holographic diagnostic capability was also mothballed. The focus was then on the development of massively parallel computing capabilities and applications, not on experimental capabilities. After leaving the shock wave physics community to work in the high performance computing (HPC) field, I am reminded from time to time that we may eventually need high-resolution 3-D experimental data to validate our advanced 3-D modeling and simulation capabilities.

Sandia's shock wave community built a broad and deep foundation of experimental, computational, and theoretical capabilities that encompass many decades of contributions to our nuclear weapons community and beyond. In the context of the many Sandians who spent a majority of their careers in shock physics, I only had a brief intersection with the field, but I am proud of the small part I played.

## Key Sandia Contributions in the History of Massively Parallel Processing

Twenty-five years ago, Sandia (Gustafson, Montry, and Benner 1988) developed a seminal analysis of "weak scaling" that helped establish explicit message passing on distributed memory, massively parallel processing (MPPs) as the way to move high-performance technical computing beyond the ubiquitous vector-based supercomputers of the 1980s.[3] At the time, conventional wisdom was that Amdahl's Law[4] would limit the practical number of multiple processors to only a handful—e.g., only four processors were available in a Cray YMP supercomputer. The key insight was that, by using larger-scale parallelism to solve larger problem sizes, the serial portion of an application could be a smaller fraction of the total computational time.

[3] Editors' note: Benner, Gustafson, and Montry received the 1987 Gordon Bell Prize in its inaugural year for this achievement as well as the Allan Karp Challenge Award. The three formally received these awards at a March 2, 1988, computing conference in San Francisco.

[4] The law is named after Gene Amdahl, the IBM computer architect. The speedup of an application that uses multiple processors in parallel computing is limited by the time needed for the serial portion of the application.

Sandia introduced this "weak scaling" concept to the HPC community and helped usher in the era of MPP supercomputing.

Sandia has a long history in the R&D of technologies that improve the scaling performance of applications on MPP. One key technology that Sandia pioneered was lightweight kernel operating system software for large-scale MPP supercomputers. This research was motivated by the ability of lightweight kernel operating systems to minimize the overhead effects of operating system noise on application performance. Lightweight kernel operating systems also maximize the amount of memory available for the application. The lightweight kernel system software has been used in four different generations of MPP supercomputers at Sandia.

A second key enabling technology was Sandia's R&D of graphical partitioning algorithms. Early in the development of MPP supercomputers, researchers recognized that the bulk synchronous programming model with explicit message passing did not automatically provide good scaling performance. It was important to make sure that (1) the application was mapped to the compute nodes in a balanced manner so every node had an equal amount of computing work to do for a given phase of computing and (2) the application was also mapped in a manner that minimized the amount of internode communication that was required once all the nodes finished their phase of computing. To solve this partitioning problem, Sandia (Hendrickson and Leland 1993) developed Chaco: a multi-level algorithm for partitioning graphs. Chaco was useful for efficient domain decomposition to map a scientific or engineering application onto a distributed memory MPP system.

## Seminal Work

1. J.L. Gustafson, G.R. Montry, and R.E. Benner, "Development of parallel methods for a 1024-processor hypercube," SIAM Journal on Scientific and Statistical Computing 9, pp. 609–638 (1988).
2. B.A. Hendrickson and R.W. Leland, "A multi-level algorithm for partitioning graphs," in Supercomputing 1995 Proceedings of the ACM/IEEE Conference on Supercomputing, San Diego, December 1995 (ACM Digital Library, Association for Computing Machinery, Inc., New York, 1995) [on CD-ROM as article no. 28 and on http://dl.acm.org/citation.cfm?doid=224170.224228].
3. G.M. Amdahl, "Validity of the single-processor approach to achieving large scale computing capabilities," in *The American Federation of Information Processing Societies* (AFIPS), Vol. 30 (AFIPS Press, Reston, VA, 1967), pp. 483–485.

# James R. Asay
# (1971–2002)

My Sandia career in shock compression science got its start during the 6 years prior to joining Sandia. In 1964, I enlisted in the Air Force and spent 4 years at Kirtland Air Force Base, Albuquerque, NM, performing research on the high-pressure response of materials using ultrasonic techniques under Captain Donald Lamberson and science advisor Arthur Guenther. After leaving the Air Force in the fall of 1968, I joined the Physics Department at Washington State University (WSU) under Professor George Duvall, who had started a shock wave program a few years earlier. For my PhD thesis project, I studied the effect of divalent impurities, specifically $Mg^{++}$, on the dynamic yielding of <100> LiF single crystal using the quartz gauge developed at Sandia National Laboratories. I received a PhD degree in physics from WSU in December 1970.

I interviewed for a research position at Sandia in September 1970, which included discussions with several staff members and management from Walt Herrmann's and Orval Jones' departments. Both departments were engaged in shock wave research; Herrmann's had more of an engineering approach and Jones' focused more on science issues. Although I found the research in both groups exciting, I leaned more toward Herrmann's department because of the strong ties between experiment, modeling, and computing. Walt wanted to build a stronger experimental effort for high-pressure research, and I was offered a position in the experimental group under Darrell Munson. I was also expecting an offer from Art Guenther at Kirtland, so I was delaying the acceptance of Sandia's offer. Sandia was experiencing a hiring freeze or limited hiring at that time, and Darrell called me about every week to see if I had made a decision. I was also getting phone calls from staff at Kirtland cautioning me against going to Sandia because of the lack of long-term employment security. After a month or so, I had not heard from the Air Force and I accepted the Sandia offer, which was one of the smartest decisions I ever made.

**Arrival at Sandia.** I reported to Sandia in early 1971. My initial responsibilities included developing research programs on the powder gun in Area Y, later referred to as the Shock Thermodynamics Applied Research (STAR) facility. At first I was in Darrell Munson's group, but shortly afterward Barry Butcher became my Division Supervisor through reorganization. In support of higher-pressure material studies, I contributed to projectile designs and experimental configurations for accurate wave profile measurements under precise impact conditions. Walt Herrmann also encouraged me to continue my research on LiF to resolve a controversy about elastic precursor decay. The dislocation theory of elastic yielding was well established and had been extended to shock wave experiments by John Gilman of the University of California Los Angeles, John Taylor of Los Alamos National Laboratory (LANL), and Jim Johnson and others at Sandia and LANL. For this theory, prediction of elastic precursor decay required only knowledge of the initial dislocation density and the stress dependence of dislocation velocity. However, when observed initial densities were used in the theory, numerical simulations predicted very little decay of the elastic wave with sample thicknesses. Much higher dislocation densities, typically by a factor of $10^3$, were necessary to match the experimental results.

Walt and others speculated that a very thin elastic peak of a few microns was washed out by the impact tilt and the area-averaging effect of quartz gauges, which had a typical recording area of a few mm in diameter. I conducted interferometer experiments on ultrapure LiF that should have been able to resolve the issue, since the 0.1-mm-diameter recording area of the interferometer effectively eliminated the tilt issues. However, when the experiments were conducted, the precursor amplitudes reproduced the data obtained with quartz gauges at WSU. The dislocation-based model that I developed to describe elastic yielding by dislocation nucleation at the elastic front and used in a characteristics code being developed by Darrell Hicks and Diane Holdridge gave excellent agreement with the experiment. The upshot of the modeling was that fast creation of dislocations had to be occurring at the elastic shock front to account for the rapid decay.

During my career, I have usually chosen research areas that had programmatic impact, were not overly populated with researchers, and were challenging. I would typically stay on a given problem for about 4 years, which was long enough to make a few major advances and publish several papers, and then move on to another area. The research on high-pressure strength and development of high-pressure isentropic loading are exceptions, since my involvement in these areas has continued to the present.

**Shock-induced melt.** One of my early goals was to develop techniques to detect and characterize shock-induced melt in metals. Several investigators had previously shown that standard approaches, such as shock velocity measurements, were insensitive to the onset of melt during shock loading. I reviewed Russian work, particularly by Al'tshuler, which discussed techniques for measuring sound velocities at high shock pressures. I decided to tackle the melt problem by measuring both shock and unloading wave profiles for shock states near and in the solid–liquid mixed-phase region. Unloading wave profiles would theoretically allow measurement of the initial unloading wave velocity as it transitions from longitudinal to bulk velocity when melt occurs. Aluminum and bismuth were chosen for the study. David Baldwin was my staff assistant on this project.

The peak velocities, and therefore the stresses, obtainable with impact launchers were extremely limited at the time. The powder gun was the highest velocity gun at Sandia, with an upper limit of about 2.2 km/s. The impact stresses at this velocity were not sufficient to shock melt metals like aluminum (the required shock stress is about 1.2 Mbars). For this reason, I decided to use porous (distended) Al to increase the temperature in the shock state to a point where melt would occur at much lower shock stresses. The approach was to first shock the porous sample to solid density and high temperature, allow sufficient time for thermal equilibrium, and then introduce a controlled unloading wave to determine the sound velocity at that stress. Porous Al of about 60 % solid density was needed to induce shock melt near 75 kbars, which could be accessed with the powder gun. Shock melt of bismuth could be achieved on the gas gun using either preheated solid Bi or porous Bi.

To make the porous material technique work, a high degree of porosity was necessary to obtain a large temperature increase after shock compression as well as small pore sizes to achieve thermal equilibrium quickly. Making these materials was not easy for suppliers: there were several pitfalls in obtaining high-quality porous samples. Finally, I obtained porous Al of 60 % solid density with 1-μm-diameter pore sizes from the Y-12 Division at Argonne National Laboratory. I was not able to obtain similar quality porous Bi. The first melt experiments were therefore conducted by preheating solid Bi to near melt and shocking it into the melt phase. This resulted in partial melting at shock stresses over the range of a few kbars, depending on the initial temperature, up to the triple point of 17 kbars between the solid I-solid II-liquid phases. Thermodynamic calculations indicated that a discontinuity in the shock wave profile should occur when the shock wave first entered the solid–liquid mixed-phase region.

A series of shock wave profile experiments on Bi were conducted to cover the melt region, resulting in the first profiles ever measured for shock waves entering a solid–liquid mixed-phase region. At about the time these experiments were performed, Jim Johnson developed a three-phase thermodynamic model for Bi that included transformation kinetics, and Dennis Hayes implemented the model in an existing one-dimensional (1-D) wave code. This capability allowed the detailed evaluation of phase transformation kinetics occurring in the experiments by comparing the simulations with the wave profiles. In contrast to expectations, the comparisons clearly showed that elastic–plastic effects from strength loss at melt and possibly kinetics dominated the features predicted by the thermodynamic calculations. Melt could therefore not be detected from the shock wave profiles alone.

However, I persisted in efforts to identify shock-induced melt unambiguously by taking another approach. The samples were first preheated to states near the melt boundary, shocked into the solid–liquid mixed phase, and then unloaded from that state. Time-resolved measurements of the unloading waves were used as a diagnostic to measure the sound velocity at that state. It was thought that melt could be identified as a transition from the measured elastic longitudinal velocities in the solid phase to bulk wave velocities in the melt region. This assumption was borne out by the observation that the measured initial unloading velocity rapidly decreased toward the bulk velocity characteristic of liquid as melt proceeded. These results were the first use of wave profile methods to detect shock-induced melt and showed that the technique could be applied with a high degree of precision.

When the first Bi shock compression experiments were nearing completion, high-quality porous Al was finally obtained, and shock loading and unloading experiments with Al proceeded on the powder gun. The newly developed VISAR was critical for measuring the unloading wave velocities in shock-compacted porous materials. The results of several experiments near and in the melt phase demonstrated a dramatic drop in the initial unloading velocity from elastic longitudinal to bulk at the expected melt pressure, as was observed for Bi. In addition, the experimentally determined melt transition pressure agreed extremely well with equation-of-state calculations by Dennis Hayes. The experiments that used unloading waves to identify shock-induced melt in Al and Bi provided strong confirmation of the approach, and several investigators subsequently used the technique to detect shock-induced melt in other materials at high pressure. David Cox was my staff assistant in the powder gun experiments.

**Particle velocity dispersion as an indicator of heterogeneous response.** A novel spin-off of the Al experiments was the strong dispersion in the peak particle velocity observed immediately after shock compaction. The magnitude of the velocity variation decreased as the material approached a uniform shock state. A complete loss of fringe contrast occurred in the VISAR when the shock first arrived at the recording surface, followed by a gradual return to full contrast after about 50 ns. I was Lynn Barker's office mate at the time, and we puzzled over this for a while. I think Lynn came up with the notion of particle velocity dispersion, which would produce different fringe shifts at different locations on the surface, resulting in an overall loss of fringe contrast. I turned this idea into a model for fringe contrast and calculated the variation in particle velocity necessary to match the observations. A "fringe benefit" of the experiment was that a potentially serious limitation of the VISAR ended up giving estimates for the nonhomogeneous response of a porous material when shock loaded and also provided an estimate of the time necessary for mechanical equilibrium—a case of "making lemons into lemonade." The technique is still used, particularly by Y. Mescherykov and coworkers in Russia, who have developed mesoscale models of material behavior using the approach.

**Mass ejecta studies.** As we pushed the VISAR measurements to higher pressures, we found that high-pressure shock waves reflected from a free surface would cause an immediate extinction of the return light signal. Since the VISAR was developed for use on any surface and worked on regular lapped surfaces at low shock pressure, the source of the light loss was perplexing. A hypothesis was that it resulted from ejection of a fine spray of sample material, referred to as "fluff" by researchers at LANL, when an intense shock wave reflected from a free surface. However, the source of the ejecta was unknown and the amount could not be predicted for a given surface or material. Several investigators speculated that a variety of sources, including surface oxide, surface impurities, grain boundaries, impurities or inclusions near the surface, or surface roughness, could cause the effect; however, no data supported these hypotheses.

I decided to investigate the physical mechanisms controlling mass ejection, using discretionary funding from our department and matching funds from the Pulsed Power Sciences Center. Researchers in that Center had a direct interest in the effect because of their program for controlled fusion from shock-imploded capsules containing hydrogen isotopes. Since the amount of ejected material was presumed to be extremely small, a highly sensitive method was needed to detect and quantify it. At the time, no diagnostics could make such a measurement. I developed an approach that involved the transfer of momentum from the high-velocity, low-density ejecta to a thin foil separated by a known distance from the surface. Simultaneous time-resolved measurements of the ejecta and the particle velocities, combined with an assumption of how much mass would stick on the foil (the coefficient of restitution), could be used to quantify the mass–velocity distribution of the ejected mass.

I started with very thin optical beam-splitting pellicles that allowed detection of ejecta mass with a sensitivity of about 50 ng per square centimeter. In the first experiments, the collection foils were separated from lapped surfaces of flat Al plates shocked to 250 kbars. This shock amplitude produced sufficient ejected mass to record time-resolved VISAR data on the foil. Over the next few years, additional diagnostic techniques to characterize the ejecta effect were explored. For one of these, VISAR foil measurements were combined with simultaneous four-pulse differential holography developed by Paul Mix and Frank Perry to verify that minute amounts of mass were ejected at discrete locations and, in fact, correlated with surface pits and scratches. Research over about 4 years confirmed this mechanism and led to a predictive model of ejected mass. Development of the model was enabled by Larry Bertholf's two-dimensional (2-D) simulations of different defect geometries using the TOODY hydrodynamic code, which quantified the amount of mass ejected for candidate defect geometries. In addition, the effects of melt and vaporization were investigated in later experiments on the two-stage light gas gun and modeled by Tim Trucano with 1-D Lagrangian and Eulerian wave codes.

**Material strength studies under shock loading.** One of my interests dating to the late 1970s involved the compressive strength of materials at high shock pressure—an area of intense interest to the scientific and weapons communities. At that time, essentially no data and very little research existed on the topic. The importance of material strength was becoming increasingly apparent in computer simulations of weapon systems; however, because of the lack of knowledge of the pressure dependence of strength, most calculations assumed constant strength. The major problem then and now is that sensitive techniques are not available to determine the material strength accurately after shock loading to high pressures. Techniques at low pressures, such as lateral longitudinal stress or compression–shear measurements, are not reliable at high shock pressures.

I developed a wave profile technique to estimate shear strength in the shocked state based on a 1961 paper by Dick Fowles (Fowles 1961), who established the relations to describe compressive strength for uniaxial strain in terms of the quasi-static yield strength for uniaxial stress. His model describes the elastic–plastic structures of both compressive shock waves and unloading waves from the shock state. It was known from low-pressure strain rate data that pressure and plastic work hardening should increase the compressive strength, but quantitative data were not available at high shock pressures. My approach was to first shock a material to high pressure and to then measure both the unloading and reloading wave profiles from the shocked state in two separate experiments. Based on Fowles' model, the elastic portion of both waves can be combined to estimate the compressive strength in the shocked state. To test the idea, I conducted wave profile experiments on Al at an impact stress of about 30 kbars. The compression profiles showed the characteristic elastic–plastic wave structure, in agreement with the model previously observed by others, but the unloading and reloading profiles differed considerably from the predictions.

Based on these observations, Joel Lipkin and I published two papers (Lipkin and Asay 1977; Asay and Lipkin 1978) that provided a tentative explanation of the results and set the stage for a sustained research program to investigate high-pressure material strength. The first paper dealt with potential material mechanisms responsible for deviations from elastic–plastic predictions. This paper proposed that the shear stress states produced by shock compression were not uniform throughout different grains and that the variation in shear stress could be described by a distribution function in a continuum model. When this approach was implemented in the WONDY code, the results for both unloading and reloading were reasonably modeled. This approach explicitly assumed that the shocked state is heterogeneous at a small scale compared to macroscopic dimensions and that the phenomena influenced physical phenomena at the macroscopic, or continuum, scale. I remember discussing this issue with Walt Herrmann at least once. I am not sure he ever believed this level of detail was necessary to describe continuum response. He thought it was like describing "fleas on fleas." However, the modern approach to predictive, physically based material models is based on the interaction of microscale (dislocation dimension), mesoscale (grain dimension), and macroscale effects.

The second paper used the basic elastic–plastic assumptions to show that the unloading and reloading results could be combined to determine the shear stress state and the shear strength independently in the shocked state without knowledge of the hydrostatic pressure. This model was referred to as the "self-consistent model" and it is still used by several investigators. Lalit Chhabildas and I extended the model to determine material strength properties at high pressure on the powder gun and later on the two-stage light gas gun. Determination of strength properties from wave profile measurements at high pressures crucially depended on the calibration by Jack Wise and Lalit of single-crystal LiF as a laser window to about 1.8 Mbars. Their groundbreaking work is still being used for wave profile studies in laboratories throughout the world.

Our shear strength measurement techniques received considerable attention in the shock wave community and the other weapons laboratories. Dan Steinberg at Lawrence Livermore National Laboratory (LLNL) was interested in extending these studies to several weapons-relevant materials. He had tried to persuade Bill Nellis, group leader for experimental shock physics at LLNL, to develop wave profile capabilities for similar experiments at LLNL, but the group was not interested. Because of this, Dan funded us to do the work, and, over a couple of years, we performed strength experiments on several materials of interest to the LLNL weapons program.

One of the requested materials was beryllium, and, in addition to strength properties, Hal Graboske at LLNL wanted us to investigate a possible phase transition in this material and determine the Grüneisen ratio. Lalit, Carl Konrad, and I discussed this but decided it would be too risky and costly to develop a containment structure on the powder gun, so we decided against it. However, Graboske pursued his request with Sandia's upper management, resulting in a directive from our vice president, John Galt, to do the work. We constructed a simple containment structure from wood and plastic sheet and performed a series of shock loading and unloading experiments. This work provided unique data on the high-pressure strength of Be and confirmed that a phase transition did not occur over this stress range.

Our experiments for LLNL had several benefits: Steinberg used the data to develop a generalized strength model that is still being used at the weapons labs, the additional funding we received helped to sustain our experimental efforts, and the shock risetime measurements on several metals helped Dennis Grady validate the "fourth power" relation between strain rate and peak stress. That relation has become a general "law" for predicting shock wave risetimes.

**Ramp wave generation.** In 1970 Lynn Barker and Red Hollenbach discovered that the anomalous compression response (negative curvature in stress–strain) of fused silica produced ramp waves when shocked to about 30 kbars. Because of this unusual response, test samples could be bonded to the back surface of a fused silica buffer and loaded with a ramp, or shockless stress, wave with predictable risetime. The major limitation was a peak stress of 30 kbars, since fused silica resumed normal material response above that. This major accomplishment for shock wave physics allowed compression data to be acquired off the principal Hugoniot. We had an interest in extending the 30-kbar pressure limit by using different materials or approaches. In late 1979, I talked to Bill Benedick about the problem. Bill knew a lot about a lot of things and also had a lot of things stockpiled at his firing site. He mentioned that a commercial product called Pyroceram®, a ceramic used as kitchen counter tops, exhibited similar behavior over a larger-pressure range. I acquired samples of this material from Bill and performed a series of experiments in which Al samples were mounted on Pyroceram® buffers shocked to about 200 kbars. This produced controlled ramp waves in Al samples of different thicknesses. Analysis of the multiple wave profiles allowed determination of the quasi-isentrope of Al over this range. Pyroceram® had considerable promise as a higher-pressure ramp generator, and several people expressed interest in using it. However, further study showed that density variations within the material strongly limited its reproducibility, so that interest gradually faded away.

We tried other methods in the continuing quest to develop high-pressure ramp wave generators; Lalit constructed an impactor consisting of a series of thin disks that produced a quasi-smooth increase in pressure at the impact surface through sequential small steps in loading stress. Lynn Barker developed a graded-density impactor through particle sedimentation techniques that produced a small initial shock upon impact and a smooth increase in pressure to peak stress. Although a promising technique, the graded-density disks were hard to make and suffered from small variations in local density, which produced slightly nonplanar impact. Jack Wise and Clint Hall pursued plasma spraying techniques to create a graded-density impactor that also produced a smooth increase in impact pressure but had minor variations in density through the plate. These techniques are not routinely used at present. The magnetic loading technique avoids many of these problems and is now used routinely to generate high-quality ramp waves to pressures as high as 4 Mbars.

**Promotion to supervisor.** In December 1978 I was promoted by Walt Herrmann to supervisor of the Thermomechanical and Physical Division previously managed by Barry Butcher. This was a first-level management position now referred to as department manager. Although I was reluctant to leave technical work, I viewed the opportunity as a way to have broader impact on shock wave research at Sandia. In this position I did little direct research but participated in several efforts through continuing collaborations. My objectives were to build a stronger experimental program, continue development of advanced instrumentation, expand STAR into a unique world-class capability available to the shock wave community, and focus on developing ultrahigh launcher velocities for equation of state studies in the multi-Mbar pressure region. Ultrahigh velocities (>10 km/s) were outside the scope of most existing Sandia research but, through discussions with several scientists, were identified as a critical need for accessing high-pressure states to model nuclear weapons. This hypervelocity capability would also open the regime of shock-induced vaporization, which is important to x-ray vulnerability problems at Sandia. At that time, no facilities could produce accurate shock wave data in this material response region.

The division was comprised mostly of experimentalists, although Doug Drumheller, a highly respected continuum mechanics theorist, was also a member. Experimentalists included Dennis Grady and Lalit Chhabildas; Jack Wise and Mike Furnish were hired within a few years. Dennis Grady graduated from WSU somewhat after I did and had strong theoretical and experimental abilities in a range of areas, including fracture and fragmentation, shock wave structure, and multi-dimensional shock wave propagation. Lalit was from Cornell and extremely knowledgeable in high-pressure physics; Jack had a background in aerodynamics and was an extremely careful, detailed experimentalist. Mike attended Stanford and Cornell and had a wide knowledge of geophysics. Toward the end of the 1980s, we hired Jim Ang from General Research Corporation, which strengthened our capabilities in hypervelocity impact phenomena. In addition, Lynn Barker returned to Sandia and rejoined the group in 1980, which considerably strengthened our capability for instrumentation development. Altogether, this was an extremely strong and multi-talented group of experimentalists. Tim Trucano also became a member of the division in the early 1980s, which resulted in a powerful conjunction of theory, modeling, and experimental and computational expertise in a single group. This turned out to be important for solving complex problems quickly and allowed us to tackle a wide range of issues without having to negotiate often with other groups for computational or theoretical support. The group had many significant achievements, including development of general fragmentation relations, understanding high-pressure response of geological materials, discovery of a general "law" for shock wave structure, achieving hypervelocity launch capabilities using modified guns and later magnetic drive, and the successful quest for a multi-Mbar ramp wave generator using pulsed power techniques. Because of these achievements and others, we garnered significant recognition in the weapons and scientific communities.

**Line VISAR development.** Barker and Hollenbach's pioneering development of interferometry represented a profound advance in the ability to study wave structures and resulted in a significantly improved understanding of high-pressure material phenomena. The VISAR is the epitome of this development and has had a far-reaching impact on shock wave programs worldwide. Arguably, no other single diagnostic has had such broad impact. However, several new diagnostics developed at Sandia since then have significantly increased our understanding of material response.

In the mid 1980s, Doug Bloomquist had shown that the return light signal from a recorded laser spot on a shocked surface could be imaged and recorded by a streak camera instead of by the photomultiplier tubes in the standard VISAR. To accomplish this, the internal optical components in a VISAR are slightly misaligned so the combination of the delayed and prompt optical return signals results in a series of fringes recorded with a streak camera. Displacement of the fringes as the surface moves provides a direct measure of particle velocity. Doug called this modified VISAR an Optical Recording Velocity Interferometer System, or ORVIS.

In the late 1990s, based on prior work by Kurt Baumung of Karlsruhe, Germany, we realized that a line rather than a spot on the recording surface could be imaged and recorded with the VISAR. In principle, this allows measurement of the spatial variations in particle velocity along the line and quantifies the effects of heterogeneous material response. This information is needed to develop multiscale models of dynamic response. Wayne Trott, who was leading the development of advanced instrumentation for energetic materials research, had been using ORVIS to record the surface velocity at a point. I encouraged him to submit an LDRD (Laboratory Directed Research and Development) proposal to expand the technique to record time-resolved line motion for studies of heterogeneous material response. Wayne followed up on this suggestion and developed a robust line-imaging VISAR (or ORVIS) system and a data analysis method that effectively characterizes irregular surface velocities. To date, the capability has been used to study heterogeneous shock response of both inert and energetic materials by Wayne, Lalit Chhabildas, Mike Furnish, and others. Mel Baer, one of the principal modelers of energetic materials at Sandia, used the Line VISAR data to develop multiscale material models.

**VISAR for pressure–shear measurements**. Another technique we developed extended the ability of the VISAR to measure longitudinal and pure shear response of shocked materials. In this approach, longitudinal and transverse motions are simultaneously produced in a sample by oblique impact or other means. Two VISARs viewing the surface at two off-axis angles, sometimes referred to as the Transverse VISAR, are used to determine both particle velocities directly. To test the idea initially, a Y-cut quartz crystal was used to generate a coupled longitudinal shear wave during shock loading, as originally proposed by Jim Johnson. By bonding a sample to the crystal's back surface, both a longitudinal and a pure shear stress wave are produced in the sample. Lalit and Jeff Swegle used this approach on Al and alumina-filled epoxy to determine the pressure–volume and shear stress–strain relations simultaneously. The major limitation of the technique is a 30-kbar stress limit at the onset of dynamic yielding in Y-cut quartz. However, the Transverse VISAR worked with other pressure–shear techniques, such as oblique impact and the magnetically applied pressure shear (MAPS) technique. The additional knowledge of both stress components allows the development of better constitutive models for computer simulations.

**Differential holography.** We also explored the use of both dynamic and differential holography to characterize the discrete motions of particles or surfaces during shock loading. Differential holography works in much the same way as interferometry, except that the fringe patterns produced by the interfering photographic holograms at two different times describe the 2-D displacement. Paul Mix and Frank Perry developed differential holography. It was used in the studies of surface ejecta discussed earlier and in crystallographic surface motions on lithium niobate shocked beyond its elastic limit. In the latter experiments, a hologram of the free surface before a shot partially exposed a photographic plate at four viewing angles. During the shock event, four additional holograms, each separated by 15 ns, were superposed at the same angles, producing a topographic fringe pattern of the surface at four different times. The four snapshots showed localized deformation during yielding, corresponding to the increasing differential displacement of fractured regions as time increased.

Later, Jim Ang in my group used a similar holographic approach to produce movies of particle distributions in debris clouds created during hypervelocity impact events. Previous work had shown that shock-induced debris and vaporization were highly nonuniform processes occurring in most hypervelocity impact events. There was considerable interest in characterizing the size distribution of debris for analyzing damage to spacecraft structures. Jim developed a dynamic holographic technique that provided this information and was used in several lethality studies at STAR.

**Three-stage railgun.** In the mid 1980s, a new initiative for missile defense, which is sometimes referred to as "Star Wars" or the Strategic Defense Initiative (SDI), began under President Reagan. Star Wars involved Sandia and the other two Defense Programs laboratories. One of these efforts required high-velocity projectile launchers of at least 15 km/s. I proposed to my Department Manager at the time, Lee Davison, and to the Sandia SDI program manager that we pursue the development of a high-velocity railgun, later dubbed STARFIRE. In addition to the SDI objectives, this would significantly extend the equation of state region accessible for shock compression studies. The proposed gun, our first attempt at an ultrahigh-velocity launcher, used a hybrid approach consisting of a two-stage light gas gun as an injector to an attached third-stage electromagnetic railgun that would accelerate the projectile magnetically over 3 m. Many in the railgun community thought this approach would prevent erosion of the conducting copper rails in the breech of the railgun during initial acceleration that had limited the previous railgun performance to about 6 km/s. Japanese investigators had tried the hybrid concept, but their gun suffered a catastrophic failure and the project was discontinued.

To initiate the project, we bought a 12-mm bore, two-stage light gas gun from Physics Applications, Inc. and installed it in the main gun bay at STAR. (The gun has been modified and is presently operational at the Institute of Shock Physics at WSU.) I recruited the leading high-velocity railgun expert in the United States, Ron Hawke of LLNL, and hired him half time as the overall project leader. He and his staff associate, Allan Susoeff, worked directly on the project for about 3 years. Clint Hall was the Sandia mechanical leader, and several others were hired as contract employees, including Randy Hickman, Gerry Sauve, and Bill Reinhart, to assist in the construction and operation. Carl Konrad integrated the massive amount of diagnostics used to assess the gun performance into a new, highly efficient control room and participated in testing the gun. Toward the end of the project, Jim Ang participated in perfecting the instrumentation and projectile launching techniques. Karl Schuler also made significant contributions to understanding the plasma armature performance of the projectile through numerical studies.

The VISAR was extremely useful in determining whether the railgun increased the velocity to the projectile injected from the two-stage light gas gun. This was determined by gluing a thin piece of Scotch-Brite® tape to the front surface of the projectile. The tape consisted of microscopic pores that look like miniature cubical reflectors and returned most of the laser light back on axis to the input beam. This light was inserted into the VISAR and used to determine the projectile velocity accurately and continuously during the full launch through the second and third stages of the gun. With this approach, it was straightforward to measure the velocity gain in the railgun.

At first we were not successful in initiating current flow in the thin conducting fuse attached to the rear of the projectile when it exited the two-stage gun and entered the railgun section. Establishing current conduction with the resulting magnetic pressure is necessary for magnetic propulsion. Achieving a current of tens of kiloamps between two conducting rails, behind a projectile traveling at 6 km/s in a hydrogen environment in less than 1 μs, is a daunting challenge. Various techniques were tried, but nothing worked. Clint Hall solved the problem through a novel projectile design, which resulted in automatic initiation of current in the fuse upon first entry into the railgun. With this fix, the projectile was boosted magnetically with an additional velocity increase of up to 1.5 km/s in 3 m, depending on the charge voltage. A U.S. patent was received for the railgun projectile design. Tim Trucano, in collaboration with Lynn Barker, helped mitigate a major problem via numerical simulations that involved high-velocity metal-on-metal sliding contact. We finally achieved a record experimental velocity of about 7.5 km/s. However, the velocity was an upper limit because the current-carrying plasma formed in the fuse was not functioning as expected. With no theoretical guidance to fix this problem, the project was discontinued in 1989.

**Building capability in high-velocity guns.** In the 1970s about four or so staff assistants were at STAR, with Carl Konrad as the facility supervisor and Bob Hardy as the alternate. Carl was supervisor until STAR was shut down in the early 1990s. Bill Reinhart became supervisor after STAR was resurrected in 1996 and has continued in that role. The facility initially comprised around 3000 $ft^2$ for operating the powder gun, which could also be converted into a two-stage gun for higher velocity experiments. Staff offices and sample preparation rooms were in other buildings at the site. After the November 1, 1977 accident with the powder gun, the sheet metal building that housed the gun was completely destroyed. Upon completion of safety reviews the following year, the gun bay was rebuilt and both a powder gun and a two-stage gun assembled from spare parts were located side by side in the new building. To relieve the congestion of this arrangement, another building was constructed somewhat later to house each gun separately.

An x-ray processing lab and other gun bays were added later to house the air gun originally in Tech Area I as well as a railgun, with the gun bay for the railgun converted to Dennis Grady's Terminal Ballistics Facility (TBF) for ballistic impact studies. In addition, the original "impact machine" built by Don Lundergan in 1958 was moved to the facility and modified by milling a small groove along the interior length of the barrel. A projectile with a slanted impact surface, inclined typically about 30° from normal, could then be launched and impacted against a similarly slanted target, producing simultaneous longitudinal and pure shear stress states. This gun, now known as the "oblique impact gun," is actively used for shear strength studies on metallic powders and other materials. With these extensive additions, the total laboratory area at STAR grew under Carl Konrad's able management from about 3000 $ft^2$ in 1981 to 15,000 $ft^2$ in 1991, making it one of the largest and most flexible facilities available for material property studies. Its capabilities spanned a range of impact velocities from a few tens of m/s to nearly 20 km/s, which is unique at a single facility. For a while in the 1980s until the railgun project was discontinued, STAR was a "six-gun" facility, with the "fastest gun in the west."

STAR has been a focal point for much of the shock wave research at Sandia both in the early days and even more at present. Its combined capabilities are unmatched anywhere; it has an international reputation and has produced a great number of major advances in understanding high-pressure science.

**Hydrogen metallization project on the two-stage gun.** When Lynn Barker returned to Sandia in 1980, he took on two major challenges. One was an LDRD project to perform shockless, or isentropic, compression of hydrogen on the two-stage gun to Mbar pressures to test high-pressure metallization theories. The project continued his goal to develop techniques for ramp, or quasi-isentropic, loading of materials, beginning in 1970 with the fused silica ramp generator, and required a cryogenic system to prepare solid hydrogen. Lynn conceived a novel technique that prepared the solid sample in a short barrel extension of tungsten that would become part of the two-stage barrel just before firing. The fixture was initially placed in a cryostat immediately outside the impact chamber, cooled to solidify the hydrogen, and then rapidly inserted and aligned to the gun barrel just before impact, kind of like a soft ball pitch. This had to be done before the hydrogen, initially cooled to a few degrees Kelvin, was vaporized. Lynn called this contraption a "transient insertion mechanism" (TIM) and dubbed it the "Tiny TIM,"—"tiny" because it was huge. I don't know if anyone ever told him it looked like a classical Rube Goldberg contraption, although it did. However, it worked.

The hydrogen sample was confined between two tungsten plates in the tungsten barrel extension. Helium gas placed in the main two-stage barrel launch tube beforehand was gradually compressed by the high-velocity projectile during launch to create a smooth increase in pressure to a few Mbars on the first tungsten plate. A VISAR monitored the rear surface of the second plate to determine the pressure in the sample. Four 1-ns flash x-ray machines developed by Jim Chang in the Pulsed Power Sciences Center measured the separation of the two tungsten plates through small sapphire-filled slots and thus the density of the solid hydrogen during compression. Tim Trucano and Gerry Kerley played significant roles in designing the system through numerical analysis and theoretical modeling of hydrogen.

This complicated project required modification of the gun to compress solid hydrogen isentropically, installation of four fast x-ray machines, and development of the unique cryogenic cell and insertion mechanism. Although Lynn was successful in getting the elements of this complex configuration to work and he conducted several experiments, flash x-rays showed that the first tungsten plate was susceptible to shock-induced instabilities that developed during the acceleration of the heavy plates by gases. Even though the strength of tungsten is very high, it was insufficient to prevent formation of these instabilities, and the plate fractured immediately into fragments. It was not possible to determine the high-pressure isentrope of hydrogen, and the project was discontinued. However, an important spin-off led to the development of the ultrahigh launch capability that we had been pursuing all along.

**Development of the HyperVelocity Launcher.** At about that time, Lalit was investigating the breakup of different plate materials under high-pressure acceleration using layered flyer plates to produce nearly continuous loading. He made the significant discovery that the tendency of a plate to break up under these conditions was controlled by the compressive strength of the plate divided by its density, or what he called the normalized strength. Plates with higher normalized strength resisted breakup and could be accelerated to higher velocities before fragmenting. Titanium alloy was one of the best choices because of its high strength and low density.

Using results from the hydrogen experiment, Lalit, Bill, and Clint Hall developed a modification to the two-stage gun that substantially increased the launch velocities from a typical upper limit of about 7 km/s. They built a tungsten extension to the gun barrel that contained the plate to be accelerated. Numerical simulations of the configuration by Tim Trucano were necessary to optimize the design. The approach adopted was to launch a layered projectile impactor with the two-stage gun and impact a titanium plate in the barrel extension. Quasi-smooth acceleration of the plate launched it to a higher velocity than the original projectile through momentum transfer. This unique capability, referred to as the HyperVelocity Launcher (HVL), achieved impact velocities of 11 km/s for equation of state studies and about 20 km/s for lethality studies.

**Development of impact "pillows" for high-pressure ramp loading.** As a major step in the continuing effort to develop a high-pressure isentropic compression experiment (ICE), Lynn Barker undertook another innovative project upon his return to Sandia in the early 1980s. He developed a specialized projectile impactor for producing high-pressure ramp waves. The approach involved the fabrication of a flat disk with a controlled gradation in density using sedimentation of a fine powder suspension in a fluid. The resulting disk had a low-density layer on the impact surface with a gradual increase in density, ending with a high-density layer such as tungsten on the rear surface. This configuration resulted in an input loading history consisting of an initial small step in pressure, followed by a smooth increase to a peak pressure determined by the final density of the impactor and the projectile velocity. Lynn described the graded-density impactor as a "pillow" because of the initial "soft" impact, followed by the smooth increase in pressure.

High-quality pillows were produced after many months of painstaking development, but small variations in density resulted in slight nonuniformity of the input stress, thus limiting the accuracy for determining the quasi-isentropic compression response of material in ramp wave measurements. Furthermore, the apparatus used to produce the controlled sedimentation of powders was complicated and difficult to use. The pillow sedimentation technique was not continued when Lynn retired from Sandia in 1990. However, the project was an important step in continuing our goal to develop a high-quality Mbar ramp generator. The magnetic loading technique finally achieved this objective.

**Promotion to Department Manager.** In late 1989, I was offered the position of Department Manager under Walt Herrmann, who was then the Director of Engineering Sciences. In accepting this position, my goals were to build even stronger ties between the code development and experimental programs and to expand modeling efforts to include the integration of material response at multiple length scales. The department consisted of the experimental division I had vacated, a code development division headed by Mike McGlaun and a code applications division, which was heavily involved in shock wave work at the Nevada Test Site under Paul Yarrington. My experimental position was soon filled by Phil Stanton from the Explosive Component Department.

**Development of the 3-D CTH code.** A computational goal of my department was to continue the development of advanced computer codes already in progress, including a three-dimensional (3-D) magnetohydrodynamic code, ALEGRA, for pulsed power applications and the 3-D hydrodynamic code, CTH. ALEGRA, under the initial direction of Mike McGlaun and later supported by James Peery, Allen Robinson, and Tim Trucano, was to play a crucial role in a new shock wave program in the Pulsed Power Sciences Center that was initiated about 8 years later. At that time, however, the Defense Advanced Research Projects Agency (DARPA) had started a program to support a general-purpose hydrodynamic code for weapon applications by the Department of Defense (DOD). After an evaluation period of a couple of years, during which users were to judge the technical capabilities of candidate codes at the Department of Energy (DOE) weapons laboratories, the agency was to provide ongoing support of the selected code for use by all DOD contractors. LANL and LLNL had existing hydrodynamic code and were vigorously competing for this lucrative award. Many thought that the LANL code, MESA, would be chosen because of its advanced state of development. Sam Thompson and Mike McGlaun had developed a similar code, CTH, at Sandia, but there had not been an earlier attempt to involve the code in the competition.

Mike McGlaun, Paul Yarrington, and I decided that we would assemble a research team and seriously compete for the DARPA support, since this would have long-lasting financial rewards. Steve Rottler was recognized as one of the most experienced analysts in the department and was appointed as team leader. He assembled nine staff members and spent the next few years improving the capabilities and user interfaces of CTH for DOD applications. Several technical and customer evaluations were held during the development phase. CTH was on par with the other two codes in technical capabilities, but far excelled in customer satisfaction. Through Steve's leadership and the team's hard work, CTH was finally chosen to support DOD contractor applications and is still used for that purpose. When Steve was promoted in another department, Gene Hertel became project leader and further perfected CTH for use by the DOD community. CTH is still the "code of choice" for many modelers of impact phenomena and had influence on several Sandia and national problems.

**Shutdown of STAR.** In 1991–1992 Sandia upper management, under President Al Narath, reorganized and streamlined the overall management structure. The upshot was that the second level of line management, the department managers, was eliminated, and those line managers became various kinds of program managers. Walt Herrmann also stepped down as the Director of Engineering Sciences about this time and Ed Barsis replaced him. I became a program manager, mainly in the computing area, under Ed and was initially involved in various activities, including the development of Sandia programs in parallel computing. Ed decided to disinvest in the experimental shock wave program in Engineering Sciences, even though STAR was leading other world-class organizations, and its staff was internationally recognized for their expertise. Ed therefore redirected the experimental research funds to the parallel computing program. As a consequence of this upheaval, most experimentalists either left or were redirected to support computing efforts in Engineering Sciences. This unfortunate turn of events involved Lalit, who was instructed to learn CTH for customer applications.

Ed also made the decision to dismantle STAR completely. Fortunately, Lalit was able to secure sufficient external funds to maintain a small shock wave effort at the facility that persisted for several years until things got better for experimental shock physics. Some of the key instrumentation devices and operational information on the guns were cleverly pigeonholed by the remaining STAR personnel to prevent them from being salvaged. In addition, Lloyd Bonzon, a first-level manager in the Energetics Department, took over management of STAR. Lloyd's recognized skills for efficiently operating facilities were crucial for maintaining facilities during those tough times.

In December 1995 the experimentalists were put on an impacted list under Ed's mandate to eliminate all experimental shock wave staff. Most facility staff and experimentalists left or were placed elsewhere. All contractors involved in maintenance and operation of STAR, including the gun cleaners, were terminated. Dennis Grady retired from Sandia under the Voluntary Separation Incentive Program (VSIP). Mike Furnish joined the geoscience group involved in testing at the Nevada Test Site. Carl, the facility manager, took an Environmental Safety and Health position at Sandia. Clint Hall, the alternate manager, took a temporary job in the drafting department of the Pulsed Power Sciences Center. Bill stayed at STAR temporarily, doing everything that everyone else had previously done, including cleaning the guns, preparing experiments for Dennis Grady's DOD/

DOE Memo of Understanding (MOU) program, conducting the experiments, and analyzing the data.

In early 1996 Lalit moved to the Pulsed Power Sciences Center, inheriting Dennis Grady's DOD/DOE MOU program. He and Bill performed rod impact experiments on the TBF gun and strategized on how to reopen STAR. Under Don Cook, the Pulsed Power Sciences Center helped by providing a small amount of seed money. Lloyd agreed to take over management responsibility and operation of STAR, which ensured efficient operation. STAR was back in operation by April 1996, with John Martinez, a Ktech contractor who previously worked at the site, helping in gun cleaning and operation, as needed. Clint, Bill, and Lalit were the only staff from the original group that had access to the guns and the experience to operate them.

**A stint in the Defense Programs Office.** Around 1992, Dennis Hayes returned from a Washington, D.C. assignment and established a new group in the Defense Programs Office under Vice President Roger Hagengruber. One purpose of the group was to develop stronger ties between Sandia's research organizations and the weapons programs. Dennis Hayes asked if I would be interested in a position in the new group. After several weeks of deliberation, I accepted, much to Ed's dismay. My initial assignment was as a "crosscut" manager[5] for research, which meant I was to help focus the different research efforts at Sandia to support the intermediate- and long-term needs of the weapons programs. I served in this position, working on various Defense Programs problems, until summer 1996, when I had an opportunity to return to my real interest—shock wave research.

**The start of pulsed power shock wave research.** The development of fast pulsed power systems at Sandia led to a revolutionary new approach for performing high-precision shock physics experiments at ultrahigh pressures. Fast pulsed power is defined as the delivery of large amounts of electrical current—that is, tens of mega-amperes (MA)—to an electrical load in a few hundred nanoseconds. This capability is available at the Z facility, which is operated by the Pulsed Power Sciences Center. My role in developing shock wave programs with this technology occupied the last 5 years of my Sandia career before I retired in 2002, and I have continued in a consulting role in that Center.

Important factors influencing my reentry into the shock compression field and my specific contributions to shock physics on Z involved three major events: changes in Sandia's management structure and divestment of experimental shock physics, my involvement in Defense Programs activities from 1992 to 1996, and Don Cook's visionary goal of developing a shock wave effort in the Pulsed Power Sciences Center. If the management structure had not changed and I had stayed in the Engineering Sciences Center, I would probably have been satisfied with that position and not been inclined to make a change. Although my stay in the Defense

[5] The role of crosscut managers was to encourage interactions between the applications groups and the research organizations in order to focus short- and long-term research that directly benefitted the weapons programs.

Programs Office broadened my perspective, I didn't have the same passion for that work as for shock wave research and was anxious to get back to research.

The transition back to shock wave research began one morning during the early summer of 1996 when Don stopped by my office after a meeting in the Defense Programs building. He said he wanted to develop a shock wave capability using Z and asked if I would be interested. The motivation for this new thrust arose from reviews of Sandia's Pulsed Power Sciences Program that emphasized the need to use Z and laser facilities to study high energy density physics (HEDP) to support the newly established NNSA (National Nuclear Security Administration) Stockpile Stewardship Program. Z had recently achieved a major milestone for producing an intense x-ray source that had significant potential for inertial confinement fusion (ICF) applications. Don didn't have much financial support for this new initiative, but he wanted to start it at the earliest opportunity.

I was eager to get back into shock compression science and jumped at the opportunity to take on this challenge. It provided a perfect match with my approach to problems: it was a novel unexplored area of research, there were strong program needs, no one else was developing a pulsed power shock wave effort, and the major challenge was getting the right people, the right equipment, and sufficient funding. Although not obvious at the time, each of these obstacles could be overcome. After a couple of months, the details of the transition were worked out and I officially transferred to Don's Center. The pulsed power shock wave program began in earnest in fiscal year 1997.

In addition to a lack of financial resources, the research staff and shock wave diagnostics were not immediately available. Fortunately, Clint Hall had previously taken a position in the Engineering Department in the Center, although he had bid on a position in the Engineering Sciences Center and his transfer was eminent. Lalit Chhabildas was working on shock wave programs at STAR as a member of the Z Pinch Physics Department under Tom Mehlhorn in the Center. After a few months, I got both Clint and Lalit reporting to me and we had the beginnings of a shock wave group. As was the case for Don Lundergan's initial shock wave efforts, this was a bottoms-up approach, and fortunately it had a "rags to riches" outcome, as Red Hollenbach had characterized the shock wave program in 1958.

Clint focused on developing the infrastructure for shock wave experiments on Z. His innovative talents for implementing new mechanical systems and his ability to make things work were a huge factor in developing a shock wave capability quickly. Lalit focused on bringing STAR back to its previous stature as a world-class facility. Because of his extensive external contacts and knack for identifying the right problems for customers, he was able to develop several new hypervelocity lethality programs that utilized its unique capabilities and led to stable funding. STAR had existing experimental methods and instrumentation available. The synergism of shock wave diagnostics at Z and STAR was crucial to develop shock wave capabilities on Z, even though we initially had to use discarded (pigeonholed) interferometer systems and recording diagnostics. We were also able to get Carl Konrad back from his stint in robotics to work on transferring the VISAR instrumentation techniques from STAR, which helped immensely in implementing the VISAR quickly at Z.

**Z-pinch-driven shock waves.** Z was developed initially through the ICF program to study the physics of high-temperature plasmas. This very large facility (>100 ft in diameter) can deliver over 25 MA of current in 100–300 ns to a low impedance load containing a shock wave experiment. The electrical energy produced can be applied in different modes of operation. The principal operational mode, and the original one developed for ICF applications, employs a cylindrical array of very thin (less than 1/10 of a human hair in diameter), high-density wires (typically tungsten) that conduct a combined current up to 25 MA during the pulse duration. Vaporization of the wires by resistive heating and formation of a cylindrical plasma sheath creates a circumferential magnetic field, which accelerates the plasma toward the center of the array at velocities approaching 100 km/s. Stagnation of the plasma at the center (the z-axis, hence the origin of the name "z pinch" and thus "Z machine") produces temperatures of a few electron volts in a gold-plated hollow cylinder, referred to as a "hohlraum." This intense radiation source can be used to implode a small capsule ablatively that contains a hydrogen-based fuel mixture and is located in the hohlraum, which up to that point had been the focus of most Z research. An interesting aside is that the "z pinch" was touted as a major destructive device in the movie "Ocean's Eleven."

The high-temperature plasma can also be used to irradiate a cylindrical disk that is attached to the hohlraum and drive a planar shock wave ablatively through it. In the first shock wave efforts on Z, we pursued the radiation approach for producing shock waves. However, the technique was discontinued after about a year, first, because it did not provide the spatially uniform and constant pressure drives necessary for high-quality equation of state experiments and, second, because a much better shock wave drive was discovered. For the ablatively driven experiments, Bob Cauble (LLNL) helped design configurations to optimize the soft x-ray spectrum produced in the hohlraum, and Tim Trucano performed simulations to design the best shock wave drive configuration. Several staff in the Center and at other facilities made significant contributions to our program; many were voluntarily "donated," since we didn't have much funding: Clint Hall led the engineering design of the shock wave assemblies, Kevin Fleming of the Energetic Components Department assisted in transferring a previously developed fiber optic system for the VISAR, John Porter allowed us to "ride along" on several of his Z experiments and also dedicated several shots to evaluate hohlraum configurations for optimizing shock waves, and Johann Seaman, who coordinated instrumentation for Z shots, dipped into his funding pool to purchase diagnostics and later constructed a dedicated screen room adjacent to Z's center exclusively for VISAR use. John Porter's motivation in supporting the effort was to develop a shock wave diagnostic to calibrate the peak radiation temperature in the hohlraum as an aid in modeling the efficiency of z pinch sources.

For equation of state applications, several problems had to be solved, including the x-ray darkening of fiber optic cables used to transfer the laser signal to and from the target area. These issues were resolved and shock wave Hugoniot data were obtained on Al to about 1 Mbar. However, the error bars for the equation of state data were not acceptable, and the technique was discontinued in early 2000 when the magnetically driven flyer plate technique was developed. Nevertheless, we did acquire reasonable shock wave data with the radiation drive that were reported at several technical conferences.

Although we had not even established the concept for making accurate shock wave measurements when we started the program in early 1997, I had conversations with several people, including Professor Yogi Gupta at WSU, about the first material on which we should focus. Our initial experiments with the radiation drive were on Al, but we felt that accurate Hugoniot data on Al would not generate much interest in the scientific or weapons communities. Our driving goal was to establish Z as an essential facility for HEDP research as early as possible.

**Cryogenic capabilities on Z.** At that time, the laser group at LLNL led by Gilbert ("Rip") Collins had made good progress in obtaining shock wave data at pressures inaccessible with conventional methods and had already generated high-pressure Hugoniot data on deuterium. Their results were significantly different from most theoretical predictions and the 50 % softer response for deuterium indicated by the data was favorable for producing ignition of hydrogen-based fuel with the National Ignition Facility being developed at LLNL. (Those results were later proven to be in significant error through careful shock wave experiments on Z by Marcus Knudson, who had been hired about a year after the shock effort on Z was started.) We decided that, even though much more difficult, accurate equation of state data on liquid deuterium would put Z on the map and would help to establish it as a better platform for high-pressure material studies or at least put it on par with the laser techniques.

To perform the deuterium equation of state experiments, we needed cryogenic capabilities on Z. Although we did not have a good shock wave driver, I advocated for and found funding to start a cryogenic effort at the beginning of the project. Dave Hanson and his staff assistant, Bob Johnston, were in the Pulsed Power Sciences Center and available to develop the system. Dave was an expert in building practical cryogenic systems and was also proficient in pulsed power systems. My Center funding was miniscule and not sufficient to support the effort, although Dave's manager, John Porter, was willing to share the cost of development, since it would also be needed for experiments with the z pinch source.

I needed an additional $600 k for this effort, so early one morning, I went to see our vice president, Gerry Yonas, who was managing LDRD funds for the Center. Gerry said pulsed power should be funding this, not LDRD. But he said "if pulsed power will put in half, I'll put in the other half," probably thinking they would not. I took this as an opportunity and immediately went to the Deputy Director for the Center, Jeff Quintenz, who agreed to provide equal funds. Later that morning at the Director's meeting in Gerry's office, Jeff mentioned that pulsed power would do its part. Gerry gulped but had committed to matching the funding so we were on track for a cryogenic system. Through Dave's and Bob's talents, we had a working cryogenic system on Z within 2 years. A unique feature of the capability was that the cryogenic cell was protected during the energy released at the target by Z (equivalent to a few pounds of TNT) and could be used several times, significantly cutting the cost of doing cryogenic equation of state experiments.

**Discovery of magnetic drive on Z.** The other mode of using Z to create stress waves is better suited for producing highly precise planar compression. In this direct pressure-producing method, current is applied to two flat, parallel Al plates placed about 1 mm apart and shorted at one end. This produces a nearly uniform magnetic field over the inner surface of the plates. The increasing magnetic pressure creates a propagating ramp wave in the plate and therefore in the sample attached to its rear surface. Typical plate dimensions are about 5 cm in length by 2 cm wide containing a sample about 1.5-cm diameter and 1–2 mm thick. The noteworthy feature is that a smooth increase in input pressure over about 400 ns is applied to the sample, up to around 4 Mbars with Z's present capabilities. The smooth pressure increase produces a propagating ramp wave through the plate and into a sample attached to its back surface, steepening as it goes and eventually forming a shock wave. The sample is therefore loaded to high pressure under nearly isentropic conditions. Although this thermodynamic path is not at constant temperature, it is much closer to isothermal loading than that for shock compression, which typically produces several thousand degrees of temperature increase. The data resulting from ramp compression experiments are therefore easier to compare with data from isothermal loading experiments, such as diamond anvil cells (DAC).

In early 1999, Clint and I tested the direct magnetic loading technique through Rick Spielman's project to develop a new current diagnostic. Rick wanted to use a nonelectrical method to measure the current near the center of Z, where the magnetic and electric fields are intense. Existing electrical diagnostics did not give reliable results in this region. Rick, Clint, and I discussed the possibility of using VISAR measurements to determine the peak pressure on a thin plate located near the center of the steel convolute plate that conducts the current applied to the load. We thought that by measuring the peak pressure produced in the plate, we could determine the magnetic field and hence the current density.

In achieving Rick's objective, we also wanted to demonstrate the feasibility of using Z to perform high pressure ramp loading for equation of state measurements. To perform the first magnetic drive experiment, small holes were drilled in the steel convolute plate. A copper disk was placed in one hole and an iron disk in the other, with VISAR interferometers measuring the response of both. The disks were flush with the plate and conducted current on their front surfaces. This configuration replicated the magnetic field that the convolute plate experienced and thus the current at that location. In analyzing the results, Clint and Lalit found that the wave profile for iron showed the classic two-wave structure characteristic of the $\alpha$–$\varepsilon$ phase transition first observed with shock waves. The unique two-wave structure produced in the iron transition was one of the reasons we had chosen a sample of iron for the test; observing the transition would ensure the measurement was reliable. This was a very exciting result and strongly implied that the magnetic approach could be used for material studies. In light of this breakthrough, other experiments were conducted, including a shot with two thicknesses of iron at the same radius. The experiment produced high-quality VISAR data that allowed analysis of ramp wave evolution in iron and, consequently, the stress–volume response of iron and the peak stress needed to estimate the current density at that location.

**Mbar ramp wave studies on Z.** After we demonstrated these first ramp wave experiments on Z, Clint and I began an intense effort to improve the configuration for accurate experiments to Mbar pressures for any material. We had ongoing discussions with Art Toor and David Reisman of LLNL about optimal configurations. During these discussions, a "squared anode" configuration was suggested for mounting the samples to produce more planar loading. However, it was not apparent that this shape would produce planar 1-D loading of thin disks. A large number of MHD simulations by David Reisman established that this arrangement did produce stress waves that were planar to within a few percent. Clint engineered the concept into a robust design for ramp wave experiments that is still being used. The design involved four equal-size plates of Al or Cu screwed together to form flat panels to attach the samples and was referred to as an "ICE cube" (for an isentropic compression experiment cube). The initial design has undergone several refinements and improvements through 2-D and 3-D MHD simulations by Ray Lemke, Jean-Paul Davis, Tom Haill, and Sophie Chantrenne (a Ktech contractor at the time) and through materials experiments by Marcus Knudson and others. Paul Raglin, who managed Sandia's subcritical experiments program at the Nevada Test Site (NTS), made an enormous contribution to the shock wave program that significantly accelerated our progress by making VISAR instrumentation developed for NTS available at Z, providing funding for Z experiments, and strongly arguing and negotiating with LANL and LLNL for support of the program.

With the magnetic loading concept as the focus for high-pressure equation of state studies, we made significant progress in a short time. Four papers on the use of Z for shock and ramp wave experiments were given at the Topical Group Conference on Shock Compression of Condensed Matter at Snow Bird, Utah, in summer 1999 (Asay 2000; Asay, Hall, Holland et al. 2000; Hall, Asay, Trott et al. 2000; Hanson, Asay, Hall et al. 2000). The contributed and invited presentations described the radiation-drive approach, the magnetic loading approach, and our progress in developing capabilities to study cryogenic fluids.

New approaches were needed to analyze the high-pressure data in the ramp wave experiments. Dennis Hayes developed the first accurate approach using a backward integration technique to define the ramp input to a sample. This approach estimates the input pressure history and was used, along with the measured output profile, to determine the continuous stress–volume response of a sample. Jean-Paul Davis (Sandia), Steve Rothman (Atomic Weapons Establishment), and Jon Eggert (LLNL) developed other methods that had advantages over the backward integration technique. Most recently, a new Sandia employee, Justin Brown, has developed a fast Fourier transform approach that promises to provide even more accurate analysis. This range of analysis techniques has proven valuable for cross checking the stated accuracy of each, identifying potential systematic errors, and establishing confidence in the results.

**New staff joined the Z shock wave program.** In the following years, the shock wave effort in the Pulsed Power Sciences Center grew rapidly by attracting several talented graduate students. In the late 1990s, when the program was just getting started, we were fortunate to hire Kathleen Holland and Jean-Paul Davis from the

California Institute of Technology and Marcus Knudson from WSU to help establish the shock wave effort. Other graduates joined later, including Dan Dolan, Scott Alexander, Tracy Vogler, Tommy Ao, Seth Root, Chris Seagle and Justin Brown. In addition, several members of the original shock wave group from the 1980s and 1990s, including Mike Furnish, Randy Hickman, and Jack Wise, returned to work on various aspects of shock physics. Several other researchers in the Center also made significant contributions by adding state-of-the-art instrumentation techniques—i.e., Jim Bailey, who developed spectroscopic techniques for measuring the temperature of shocked liquids, and Dave Hanson, who developed the novel cryogenic capability for performing shock wave experiments at temperatures as low as 4 K. Furthermore, the *ab initio* theory developed by Mike Desjarlais, Thomas Mattsson, and others added a critical component to the overall shock wave program. In particular, Mike's *ab initio* calculation of the high-pressure conductivity of aluminum was a breakthrough in the ability to launch thin flyer plates magnetically. Jeff Quintenz and Keith Matzen, the Center Directors following Don Cook, also played vital management roles to ensure the success of the shock wave effort on Z.

That shock wave effort had grown sufficiently large by 1999 to justify a separate group under a second-level manager. Since I was already a second-level manager, a position was created called weapon physics in a group referred to as Shock and Z-Pinch Physics. I was also assigned management responsibility for the NTS group under Bill Boyer, and the STAR facility staff still reported directly to me. I had to replace myself in the Shock and Z-Pinch Physics group and advertised for that position. Several people applied, including Chris Deeney, who was then working on z-pinch plasma physics in the Center. Chris was clearly the most qualified, but was on assignment at DOE/NNSA. That was intended to be a 2-year assignment but he agreed to leave after 1 year, so I waited 8 months for him to take over daily responsibilities as manager of the shock group. This was one of my better management decisions and has had lasting consequences for the shock wave program at Z. When I retired in 2002, my level-2 position was dissolved but was reestablished a couple of years later. At that time, Chris was promoted to the level-2 position and he promoted Clint as manager of the shock wave group. Clint served in this capacity until he left for another Sandia assignment in 2010, and Dawn Flicker was promoted to that position. Chris left Sandia in 2007 to accept a position at NNSA, where he was responsible for all materials research at the Defense Programs laboratories.

**Magnetically driven flyer plates.** Several groundbreaking achievements have occurred using ramp waves on Z to study material response at Mbar pressures. Jean-Paul Davis and his colleagues obtained near-isentropic pressure measurements of tantalum to about 4 Mbars. Tracy Vogler and his colleagues used the unloading wave profile technique developed in 1978 to determine the compressive strengths of Al to about 500 kbars and Ta to about 1 Mbar. Tommy Ao, Tracy Vogler, Marcus Knudson, Justin Brown, and I made strength measurements in Al, Ta, and LiF to pressures from 170 kbars to about 2.5 Mbars (for Ta). Justin is presently developing experimental and computational approaches to improve the accuracy of the compressive strength data at several Mbars on Z.

In fall 1999, Marcus and Clint proposed using magnetic ramp loading to launch flyer plates to high velocities. The three of us discussed this during a meeting of the American Physical Society Division of Plasma Physics in Seattle and agreed to proceed. In the flyer plate mode, the Al conducting plate is accelerated by the magnetic field produced during ramp loading. If the input current history is properly designed, the plate is accelerated continuously to a high velocity. Velocities far exceeding conventional techniques can be easily achieved, considerably extending the pressure range for equation of state studies. A potential problem is that the current pulse time history must be precisely controlled to ensure that shock waves will not form in the Al plate before the final velocity is attained. Otherwise, the flyer plate may melt or vaporize and disintegrate during launching.

Determining the actual thermodynamic state in the flyer plate during launch requires accurate MHD simulations of the experimental configuration. Both an accurate equation of state of the flyer plate and its conductivity at high pressure and temperature are needed. Mike Desjarlais developed a quantum molecular dynamics model for the conductivity of liquid Al at high pressure, which differs significantly from the model previously developed by Dick Lee and Richard More (LLNL). His model predicted that mm-sized plates can be accelerated intact to high velocities. Calculations with the Lee-More model indicated that Al flyer plates could not be launched to high velocity. However, based on Mike's theory, Marcus, Ray Lemke, and others demonstrated that Al flyer plates of 15-mm diameter and up to 1 mm thick can be launched to about 46 km/s—that is, about five times higher than possible with conventional techniques using two-stage light gas guns—thereby allowing impact pressures as high as 40 Mbars. This capability has far exceeded our original goal to develop an ultrahigh velocity of 15 km/s on the railgun.

**Shaping current drives for improved ramp generation.** One of my contributions to high-pressure ramp waves and magnetically driven flyer plates involved the realization that the current history had to be precisely shaped in time to prevent formation of shock waves in the sample or flyer plate. One evening I was analyzing how a ramp wave would steepen with propagation distance and realized that the problem could be solved in Lagrangian coordinates using shock velocity and particle velocity data and an analytic characteristic approach. A simple linear ramp in input pressure will form a shock quickly in a sample. However, by shaping the current with a specific nonlinear temporal history, shock formation is deferred to larger sample thicknesses. For the idealized current shape, a single shock will form at a specific sample location, and this is the maximum thickness that can be studied using ramp waves of a specified peak pressure. Achieving the capability to shape the output current arbitrarily required quite an effort by modifying the laser-triggered system but has paid huge dividends to the shock wave program. Several people have examined the current shape requirements in detail, including David Reisman (formerly of LLNL and now at Sandia), Dennis Hayes, and Jean-Paul Davis. Jean-Paul's extensive work, in particular, has allowed studies of ramp compression to multi-Mbar pressures and launching of flyer plates to velocities exceeding 40 km/s. This work had a major influence in modifying the laser triggering circuits that fire the 36 capacitors used to produce current in Z.

**Deuterium equation of state.** By 2000, Marcus and the Z accelerator team had begun acquiring data on deuterium with the magnetic flyer plate technique. Over the next few years, they produced a highly accurate Hugoniot to 1.8 Mbars, which resulted in the revision of earlier Hugoniot data obtained with laser techniques. The new data also agreed to within experimental uncertainty with Mike Desjarlais' *ab initio* calculations, providing significant confidence in both experiments and theory. Several subsequent publications on deuterium by Marcus and the shock wave team have extended the Hugoniot data, which has become the standardized data set and is now accepted by most researchers. In addition, the team has obtained equation of state data on water and diamond to 18 Mbars, xenon to 8 Mbars, unloading of shocked Al at 5 Mbars into vapor states, and beryllium to 5 Mbars. These pioneering results significantly exceeded the initial goal in 1997 of obtaining worldwide recognition for Z and firmly established the facility as one of the premier organizations for ultrahigh-pressure equation of state research. The tie among experiment, theory, and computation established through such experiments also achieved Don Lundergan's vision in the early 1960s of bringing together these three aspects of shock wave research.

**Toxic material studies on Z.** The equation of state capability on Z has been extended to the study of toxic materials, such as beryllium and plutonium, which cannot be handled in most laboratories. This capability was developed through a clever containment design conceived by Chris Deeney and then implemented by Marcus and Clint Hall for the first three plutonium experiments on Z. Those first experiments were conducted in 2006 before Z was shut down for a complete refurbishment. The present containment system is significantly different than the one used in 2006 because of increased concern about the safety of workers and the viability of the refurbished Z. John Porter, Bill Stygar, Randy McKee, and their staff contributed significantly by enhancing the containment infrastructure and performing detailed safety analyses to obtain Sandia and DOE approval. As part of the containment system, a toxic sample is located within a steel container, and an ultrafast closure valve prevents the escape of the vaporized toxic material. Upon firing Z, an explosive charge is initiated that closes the gap between the anode and cathode and isolates the containment chamber from the rest of the accelerator. Before closure, the sample configuration receives the current from Z and is subjected to the magnetic load, and VISAR data are collected and relayed to recording instruments. The containment system has been thoroughly validated for use with toxic materials, and Z is presently the only dynamic high-pressure capability aboveground available for high-pressure equation of state studies on these materials. The team that manages the shock physics program on Z is to be commended for integrating the shock wave program into this national effort to support the NNSA Stockpile Stewardship Program.

**Development of Veloce.** In 2004, Clint initiated a project to build a compact pulsed power generator, named Veloce, for low-pressure material studies. The goal was to off-load the high demand for ramp wave experiments on Z by providing similar data to about 170 kbars. Another purpose was to develop new techniques at a faster turnaround rate—such as different source configurations, soft recovery, magnetic pressure–shear generation, and preheating that could then be transferred to Z. In addition, the pulser was constructed so it is inexpensive to operate and can perform a large number of experiments compared to Z. Veloce was built by a French company and is capable of delivering about 3 MA of current to a 15-mm-wide stripline load containing the sample. Tommy Ao and Sophie Chantrenne performed a careful study of stripline geometries to identify a planar magnetic drive that is uniform to within about 1 % for material studies. Randy Hickman assumed management of Veloce and has kept it busy on a variety of projects. A large number of material studies have been performed with the system.

**Magnetically produced pressure shear experiments.** Around 2007, Scott Alexander, Tom Haill, and I developed an innovative technique on Veloce, referred to as MAPS, for magnetically applied pressure–shear, which uses crossed magnetic fields to ramp load a sample simultaneously to high pressure under longitudinal stress and then apply a pure shear stress ramp wave to determine shear strength directly at the longitudinal stress state. A Transverse VISAR measures the transverse velocity produced by the shear stress. This novel concept provides a direct measure of compressive strength at high stress, a long-term goal of the community, and is being pursued by Scott. A patent disclosure was submitted for the concept but we did not apply for an actual patent.

**Closing remarks.** The breadth of high-pressure equation of state capabilities available has established Z as a critical component in the international investigation of HEDP. Material phenomena in this region occur in astrophysical environments. At the high-temperature end, hot dense matter covers hundred fold compressions and temperatures of several thousand electron volts, conditions that are pervasive in astrophysical environments (e.g., supernovae, accretion disks, and the interiors of stars). These states are also produced in practical applications, such as laser plasmas and z-pinches. Understanding the phenomena is critical to achieving ICF in a laboratory setting. Similar phenomena occur during the operation of nuclear weapons, this has added to the significant role that Z plays in our Nation's nuclear weapons program.

The major impact Z has had on shock wave research was succinctly captured in a 2012 article in the *Sandia Lab News* (Vol. 64, No. 4, p. 1, February 24, 2012):

> Both Congress and DOE recognized the value of the experiments conducted on Z by Sandia and its partner laboratories, Los Alamos (LANL) and Lawrence Livermore (LLNL) in support of NNSA's Stockpile Stewardship Program. The Senate's Subcommittee on Energy and Water [D]evelopment's 2012 appropriation bill commented that "The Committee understands that (two Z) experiments (on plutonium) yielded fundamentally new and surprising data about the behavior of plutonium at high pressure and this new data [have] been one of the most valuable contributions to the stockpile stewardship program. The Committee continues to strongly support the weapons physics activities at Sandia's Z facility that are critical to sustaining a safe, secure, and effective nuclear stockpile." Wrote NNSA Deputy Administrator for Defense Programs Don Cook regarding the joint plutonium work on Z carried out by Sandia and LANL, "These high-quality data will likely provide new insights and challenge our fundamental understanding of this extraordinarily complex material… This accomplishment has been one of our most valuable technical contributions to the stockpile stewardship program."

After retiring from Sandia in 2002, I spent 4 years at the Institute of Shock Physics (ISP) at WSU, working with Professor Yogi Gupta and several graduate students and postdocs. After retiring from WSU in 2006 as professor emeritus at ISP, I returned as a consultant to the shock wave program at Z until 2012.

The overall theoretical and experimental shock wave programs on Z and STAR are presently strong and robust under senior manager Dawn Flicker and managers Thomas Mattsson, John Benage, and Gordon Liefeste (who directs shock wave research at STAR and the Nevada National Security Site). In addition, the group has integrated the capabilities of experiment, theory, and computation in a single second-level department, similar to the initial structure developed by Walt Herrmann (no relation to Mark Herrmann, the Director of the Pulsed Power Sciences Center from October 2013–September 2014, who became the head of the National Ignition Facility in September 2014).[6] The broad talent of the existing shock wave group matches that of the early Sandia investigators, and the group is every bit as strong and innovative as the shock wave groups of the 1950s to 1990s. Notably, the new generation has made a "sea change" in shock physics at Sandia, in the national program, and worldwide, carving their own niche in the twenty-first century shock physics to best of "world class."

I have been fortunate to work with many talented people during my career, starting with Captain Don Lamberson and Dr. Arthur Guenther at Kirtland Air Force Base, who pointed me in the direction of high-pressure research, and Professor George Duvall at WSU, who gave me the freedom to work on a problem of my choosing as a PhD thesis; to Lynn Barker, who provided initial guidance in using the VISAR and the inspiration to continue many of his initial research projects; to Clint Hall, who has a knack for developing complex systems; to Marcus Knudson, whose expertise and intuition for shock wave research have enabled the recognition of the Z magnetically driven pressure technique in the international community; to Lalit Chhabildas, who has partnered with me in most of my research projects and who really made a lot of things work and directly initiated several of my recognitions; to my family; and to the many other talented colleagues with whom I have been fortunate to collaborate. Paul Raglin deserves special recognition for his programmatic skills in initiating the shock wave work in the late 1990s by negotiating with the other Defense Programs laboratories to make resources available to the initially struggling Z shock physics program. In addition, Professor Yogi Gupta, of the ISP at WSU, was extremely supportive in advocating for shock wave research to Sandia and DOE management, which substantially accelerated the Z research program. Finally, I have benefited from a large helping of serendipity.

[6] Editors' note: Keith Matzen, the Director of the Pulsed Power Science Center from January 2005 to July 2013, began serving as the Director again in March 2015, following almost two years as Director of the Nuclear Weapons Science and Technology Programs at Sandia. Duane Dimos served as the temporary Center Director from September 2014 to March 2015.

# Melvin R. Baer
# (1975–2010)

I retired as a Senior Scientist in Engineering Sciences in 2010, having been at Sandia National Laboratories for 35 years. I received a BS (1970), MS (1972), and PhD (1975) in Mechanical Engineering from Colorado State University. My doctoral studies concerned combustion instability in liquid phase rockets. I joined Sandia in the aerodynamics organization as a Member of Technical Staff in 1975. In 1978 I transferred to the Engineering Sciences Center and was promoted to a Distinguished Member of Technical Staff in 1989. I became a Senior Scientist in 1997. I have conducted fundamental and basic research in the field of energetic materials, including explosive and pyrotechnic research, mesoscale studies, detonation behavior, and in equation of state studies, reactive processes, hazards assessments, and computational shock physics. I was a Member of the Sandia team that investigated the USS Iowa accident and a team member for the National Transportation Safety Board investigation of the TWA 800 accident. I also served on the technical team for the Department of Defense (DOD) Advanced Energetics Initiative, the Basic Science Research Assessment panel for the Defense Threat Reduction Agency, and as a board member for the Standing Committee on Operational Science and Technology of the Joint Improvised Explosive Device Defeat Organization. Currently, I am a panel member for the National Academy of Sciences Board for Arms and Armament, and I have a staff augmentation position with Sandia Staffing Alliance, mentoring new staff in energetic materials research.

## Key References with Comments about Research and Application Activities

1. M.R. Baer and J.W. Nunziato, "A two-phase mixture theory for the deflagration-to-detonation transition (DDT) in reactive granular materials," *International Journal of Multiphase Flow* **12**, no. 6, pp. 861–889 (1986).

This widely referenced paper presented an overview of a modeling approach that treats fully compressible reactive mixtures. At that time, multiphase modeling was being developed mainly in the propulsion and gun communities, and it was recognized that these approaches were incompatible with detonation theory. The explosive components organization at Sandia, which sponsored our work, requested modeling assistance to provide a better understanding of DDT as applied to detonator components. A new approach was needed that could couple high-speed deflagration to shock physics and admit detonation states. We developed a continuum mixture theory approach that treated volume fraction as an independent kinematic variable and allowed compressibility of all phases and detonation states. In the appendix of the paper, a brief characteristics analysis suggested the modeling was mathematically well posed and consistent with classical Riemann theory. This caught the attention of several mathematicians throughout the world, and our continuum mixture approach became the foundation for several subsequent works. Colleagues at Los Alamos National Laboratory (LANL) later took an interest in the continuum mixture model, which they referred to as the B-N (Baer–Nunziato) model. LANL posed some modifications to the model; however, the original B-N model is still regarded as the pioneering work. Unfortunately, Jace Nunziato passed away before the work gained notoriety in the multiphase community.

Several other multiphase shock physics efforts at Sandia, such as the works of Steve Passman and Doug Drumheller, contributed much toward the development of the original B-N model, which was later adopted in the shock physics code, CTH.

2. M.R. Baer, R.J. Gross, J.W. Nunziato, and E.A. Igel, "An experimental and theoretical study of deflagration-to-detonation transition (DDT) in the granular explosive, CP," *Combustion and Flame* **65**, pp. 15–30 (1986).

This paper described the application of the continuum mixture B-N model to the explosive CP, 2-(5-cyanotetrazolato) pentaammine cobalt (III) perchlorate. CP was of much interest to explosive components researchers; it was the detonator explosive of choice at Sandia. The paper reported on a combined theoretical and experimental study of DDT in granular CP. Most importantly, we demonstrated that compaction had a dominant effect leading up to the transition from high-speed deflagration to detonation. We investigated the effect of initial density, showing that low and very high loading densities impeded the transition process and that a "sweet spot" for DDT to occur existed. Subsequent work in the UK by Professor Field and colleagues further evaluated the DDT behavior in CP. After production of CP ended, the explosive became unavailable as a detonator material. Sandia research in detonators then shifted to developing semiconductor bridges, optical initiators, and miniaturized devices.

3. M.R. Baer and J.W. Nunziato, "Compressive combustion of granular materials induced by low-velocity impact," *9th International Symposium on Detonation*, edited by J.M. Short, E.L. Lee, OCNR 113291-7, pp. 293–305 (1989).

The DOD HARP (Hazards Assessment of Rocket Propellants) project largely sponsored this effort. At that time, the propulsion community began to incorporate a large fraction of explosives (primarily the nitramines, RDX and HMX) in their propellant formulations. A concern arose as to whether the new additives would greatly alter the potential for DDT in high energy propellants. For this effort, we conducted collaborative modeling and experiments to investigate the low impact thresholds leading to reactive flow behavior. Our paper applied the B-N model to several granular explosives, which replicated the experimental studies by Harold Sandusky of the Naval Surface Warfare Center. The work demonstrated that the continuum mixture approach could describe DDT behavior in a broad class of energetic materials. Based on this paper, subsequent development of the model was done in collaboration with Erik Matheson of Lockheed Martin to extend the approach to treat impact-induced explosive detonation transition (XDT) behavior in propellants. The work was sponsored by the Navy under the Propellant Hazards Technical Analysis Program project. This project sponsored incorporation of the B-N model into the CTH shock physics analysis. The coupling of mechanics to the multiphase approach was termed the CDAR (Coupled Damage And Reaction) model. Much XDT work has remained classified and thus very little documentation is available in the open literature.

4. M.R. Baer, "A mixture model for shock compression of porous multi-component reactive mixtures," in *High-Pressure Science and Technology 1993*, edited by S.C. Schmidt, J.W. Shaner, G.A. Samara, and M. Ross (American Institute of Physics, Melville, NY, 1994), AIP Conference Proceedings **309**, pp. 1247–1250.

This reference extended the B-N model to multicomponent mixtures with particular application to ballotechnics and pyrotechnic mixtures consisting of metallic and organic reactive materials. At that time, many shock physics analyses of these classes of energetic materials were modeled using effective single material models, which had limitations in predicting shock response. Thus, the formalism of mixture theory was extended to include an arbitrary number of constituents to determine admissible phase interaction relationships that could describe shock response as well as the effects of momentum and energy interphase. An application of the modeling was discussed, which treated shock loading on a porous mixture of aluminum and iron oxide powder.

5. M.R. Baer, E.S. Hertel, and R.L. Bell, "Multidimensional DDT modeling of energetic materials," in *Shock Compression of Condensed Matter 1995*, edited by S.C. Schmidt and W.C. Tao (American Institute of Physics, Melville, NY, 1996), AIP Conference Proceedings **370**, pp. 433–436.

This paper implemented the multiphase B-N model into the shock physics code, CTH. Prior modeling was implemented in a one-dimensional (1-D) method of lines[7] shock physics code that was specifically developed to resolve conservation equations that had stiff terms such as interphase momentum and energy effects, including reaction effects. Since much prior experimental work investigated uniaxial impact loading, the 1-D modeling was adequate. However, experiments at LANL by J.M. McAfee and B.W. Asay et al. (1991) suggested that many modes of DDT in granular explosives may be linked to the formation of multiple compaction wave behavior. The B-N model in CTH was applied to the LANL experiments to demonstrate that lateral release effects associated with weak confinement greatly changed the development of compaction and inhibited the reactive growth processes leading to DDT. It was shown that the DDT behavior observed by LANL was actually caused by the effects of the release of weak confinement, producing rarefaction waves that interacted with compaction rather than the weak reaction of the energetic materials producing new modes of compression and wave behavior.

6. M.R. Baer, R.A. Graham, M.U. Anderson, S.A. Sheffield, and R.L. Gustavsen, "Experimental and theoretical investigations of shock-induced flow of reactive porous media," in *Proceedings of the 1996 Joint Army Navy Air Force Combustion*

[7] Editors' note: The method of lines is a general method for solving time-dependent partial differential equations by using finite differencing for the spatial derivatives and ordinary differential equations for the time derivative.

*Subcommittee and Propulsion System Hazards Subcommittee Joint Meeting* (Chemical Propulsion Information Analysis Center, Johns Hopkins University, 1996), pp. 123–132.

This work presented a joint theoretical and experimental effort to resolve the shock wave response of porous explosives and simulant energetics. Polyvinylidene fluoride and magnetic particle gauges provided detailed wave structures to determine the dispersive wave fronts in granular materials. All of these experimental observations were replicated in the B-N model.

7. M.R. Baer, "Numerical studies of dynamic compaction of inert and energetic granular materials," *Journal of Applied Mechanics* **55**, pp. 36–43 (1988).

This paper applied the B-N model to a wide set of impact experiments for the specific investigation of compaction behavior. Steady wave analysis was also presented that admitted finite wave structure.

8. P. Embid and M.R. Baer, "Mathematical analysis of a two-phase continuum mixture theory," *Continuum Mechanics and Thermodynamics* **4**, pp. 279–312 (1992).

This work began as a Sandia-sponsored project with the University of New Mexico under Professor Embid. It extended the mathematical structure of the B-N model and defined the field equations in characteristics form. At that time, new ideas were appearing in the literature to incorporate Riemann solvers, which minimized the effects of artificial viscosity in numerical simulations. Key to this effort was determining the characteristics and the associated eigenvectors of the multiphase conservation equations. This paper provided the foundation for several numerical methods for multiphase materials such as those developed by R. Saurel and his colleagues in France.

9. M.R. Baer, "Continuum mixture modeling of reactive porous media," in *High-Pressure Shock Compression of Solids IV: Response of Highly Porous Solids to Shock Loading*, edited by L. Davison, Y. Horie, and M. Shahinpoor, Chapter 3 (Springer-Verlag, New York, NY, 1996).

This was a chapter in a book of collected works that provided a comprehensive review of the multiphase continuum mixture model development and application to several experimental studies.

10. M.L. Hobbs and M.R. Baer, "Nonideal thermoequilibrium calculations using a large product species database," *Shock Waves* **2**, pp. 177–187 (1992).

Prior to 1990, much of the thermochemical assessment of explosives and associated energetic materials was conducted using the TIGER code, which was developed by M. Cowperthwaite and W. Zwisler at the Stanford Research Institute. There was a need to generate a complete species library that could be used to assess pyrotechnic and ballotechnic materials. This paper was an overview of the implementation of the BKW (Becker–Kistiakowsky–Wilson) equation of state using the complete JANAF (Joint Army Navy Air Force) library consisting of 900 gaseous and 600 condensed species. With this extensive species library, product phase boundaries and composition maps were generated for several explosive and pyrotechnic mixtures. Following implementation in TIGER, Lawrence Livermore National Laboratory (LLNL) took ownership of the code and revised the numerical implementation, creating a new code called CHEETAH. The DOD labs, which sponsored the CHEETAH development as a Memorandum of Understanding (or MOU) project, requested that the Sandia species library be included in CHEETAH as a user option denoted as the BKWS database (where the "S" in "BKWS" stands for "Sandia library"). Much of this library became the foundation for the molecular potential equations of state that are now the preferred ones in CHEETAH.

11. M.L. Hobbs, M.R. Baer, and B.C. McGee, "JCZS: An intermolecular potential database for performing accurate detonation and expansion calculations," in *Propellants, Explosives and Pyrotechnics* **24**, no. 5, pp. 269–279 (1999).

Although the prior extensive thermochemical library was based on the BKW equation of state for the TIGER and CHEETAH codes, we extended the library to include intermolecular potential interactions for thermochemical analysis of non-ideal explosives. Existing codes such as PANDA (G. Kerley at Sandia) and CHEQ[8] (M. Ross and others at LLNL) used molecular potential-based equations of state; however, their species library was limited to CHNO explosives rather than to general energetic formulations. The original TIGER code included the Jacobs–Cowperthwaite–Zwisler (JCZ) equation of state based on Lennard-Jones potentials. The original JCZ equation of state database option in TIGER had relatively few species; we expanded the library to include a generalized species set. This paper documented that effort and demonstrated that molecular potential-based equations of state provided better predictions of detonation and expansion states. As a result, thermochemical analysis could be used to determine the Jones Wilkins Lee (JWL) equation of state (often used in shock physics analysis) even when cylinder expansion data was unavailable. This option was included in CHEETAH and denoted as JCZS. Currently, CTH includes a capability to generate product equation of state during execution that also uses this JCZS library.

12. M.R. Baer, "A numerical study of shock wave reflections on low density foam," *Shock Waves* **2**, pp. 121–124 (1992).

This work applied the B-N model to low-amplitude shock loading of low-density foams. At that time, several experimental studies used low-density foams to protect blast gauges that produced anomalous measurements. Shock tube studies by B. Skews (University of Witwatersrand, South Africa) showed that the foams amplified wave response rather than attenuated shocks. The multiphase B-N model replicated Skew's shock tube results and showed the amplification was caused by material stagnation effects. This work was the first to demonstrate that a continuum mixture theory model can be applied to describe shock loading of low density materials.

---

[8] Editors' note: For a description of the CHEQ code, see F.H. Ree, "A statistical mechanical theory of chemically reacting multiphase mixtures: Application to the detonation properties of PETN," *Journal of Chemical Physics* **81**, 1251 (1984).

13. M.R. Baer, "A multiphase model for shock-induced flow in low density foam," in *Shock Waves @Marseille III: Shock Waves in Condensed Matter and Heterogeneous Media*, edited by R. Brun and L.Z. Dumitrescu (Springer-Verlag, Berlin Heidelberg, 1995), pp. 169–174.

This paper extended the modeling of shock loading in low density materials used in shock tube experiments, including multiple and unsupported foams. Details of the constitutive relationships were provided, and the various multiple wave features were resolved, including compaction waves and gas penetration in foams, which were consistent with experimental observations.

14. M.L. Hobbs, M.R. Baer, and R.J. Gross, "A constitutive mechanical model for energetic materials," Twentieth International Pyrotechnics Seminar, Colorado Springs, CO, edited by J. Austing and A. Tulis, pp. 423–436 (1994).

Early in the 1990s, DOD supported an MOU project with Sandia focusing on the "cookoff" of energetic materials. The Sandia cookoff program was headed by Art Ratzel, a first-level manager in Engineering Sciences who managed the team of researchers that included Engineering Sciences staff, staff from the Explosive Components Facility, and researchers from the Combustion Research Facility at Sandia Livermore. This was a follow-on project to the DDT program described in references 3 to 8 above. At that time, it was recognized that thermally initiated combustion in condensed phase energetic materials was closely linked to DDT phenomena, and much of what had been developed in modeling could be directly applied to this problem. The Department of Energy (DOE) also recognized the importance of investing in a better understanding of cookoff as applied to nuclear weapons. Thus, the DOD/DOE MOU also funded separate cookoff programs at LLNL and LANL. DOD and DOE were greatly interested in modeling the level of violence following thermal explosion of energetic materials. To address this problem, the effort focused on coupling quasi-static thermal and mechanics analysis with shock physics. The key to the effort was determining an appropriate constitutive model for thermally degraded energetic material. This seminar presented a preliminary model that treated thermal damage as a collection of pores, similar to the classical pore collapse model developed by Carroll and Holt[9] of LLNL. Most importantly, this work motivated subsequent Sandia experiments, whereby the measurement of internal pressures and temperatures could help guide modeling.

15. M.R. Baer, R.J. Gross, D.K. Gartling, and M.L. Hobbs, "Multidimensional thermal-chemical cookoff modeling," in *Proceedings of the Joint Army Navy Air Force Propulsion Systems Hazards Subcommittee Meeting* (Chemical Propulsion Information Analysis Center, Johns Hopkins University, 1994), CPIA Publication **615**, p. 323.

[9] Editors' note: M.M. Carroll and A.C. Holt, "Static and dynamic pore collapse relations for ductile porous materials," *Journal of Applied Physics* **43**, 1626 (1972).

This paper presented an overview of the methodology of coupled finite-element heat transfer and quasi-static mechanics with CTH shock physics analysis. In the years that followed, coupled system-level analysis motivated the development of a new computational effort, called SIERRA, which was funded by the Accelerated Strategic Computing Initiative. Much of the computational framework of SIERRA has roots in this earlier approach.

16. A.M. Renlund, J.C. Miller, M.L. Hobbs, and M.R. Baer, "Experimental and analytical characterization of thermally degraded energetic materials," in *Proceedings of the 32nd Joint Army Navy Air Force Propulsion Systems Hazards Subcommittee Meeting* (Chemical Propulsion Information Analysis Center, Johns Hopkins University, 1995), pp. 35–39.

When the MOU cookoff program was formed, a collaborative experimental program was also established. Anita Renlund led the experimental program, which developed a hot-cell device that could simultaneously measure the transient pressures and thermal fields leading up to thermal runaway. Much of this foundational study was supported by modeling that ultimately led to an improved experimental apparatus called SITI (Sandia Instrumented Thermal Ignition).

17. M.R. Baer, M.L. Hobbs, R.J. Gross, and R.G. Schmitt, "Cookoff of energetic materials," in *Proceedings of the 11th International Symposium on Detonation*, edited by J. M. Short, J. E. Kennedy, Office of Naval Research Report ONR 33300–5, pp. 852–862 (1998).

This reference provided an overview of the Sandia modeling that was applied to a series of cookoff experiments. LANL had a separate experimental program and developed an apparatus very similar to SITI, which incorporated a thermal trigger hot wire to fix the location of the thermal runaway. After sufficient heating, the energetic material was thermally initiated, and high-speed photography revealed that large cracks formed prior to the thermal event. LANL researchers claimed this was the essential feature that leads to a violent response. That claim was disproved using the B-N model in CTH, which showed that the specific area associated with the large cracks was insufficient to cause rapid pressure buildup; instead, it was the thermal damage at the crystal level that was responsible for rapid combustion. Later, experimental work by Rich Behrens at Sandia Livermore revealed that thermal damage does occur at the crystal level as decomposition bubbles nucleate and grow during thermal decomposition, producing a very high specific surface area. This confirmed the conclusions of the Sandia modeling.

18. M.R. Baer, "Shock wave structure in heterogeneous reactive media," *Proceedings of the 21st International Symposium on Shock Waves*, Great Keppel Island, Australia, pp. 923–927 (1997).

This was the first paper that motivated much of Sandia's later mesoscale work. At that time, shock physics modeling changed dramatically, with several improvements in computational software and hardware. High-order flux limiters and adaptive mesh refinement were implemented in CTH, which greatly reduced the effects of artificial viscosity. Furthermore, CTH was revised to run on the new high-performance parallel computers. Full three-dimensional (3-D) simulations were then possible, which led to new ideas in modeling. Based upon prior work on multiphase continuum mixture theory, much of the important shock physics in heterogeneous materials was associated with processes that occurred at the crystal or grain level (known as the "mesoscale"). Bob Graham suggested that greater insight into the shock processes of mixed materials could possibly be revealed by simulations of shock loading of multi-particle ensembles, such as the studies by Dave Benson at the University of California San Diego that were based on two-dimensional simulations of shock loading of "circular" particles. However, Benson's approach appeared to be too simplistic and was somewhat removed from the real geometries that typically represent energetic composites. Hence, a new approach was developed using 3-D shock physics analysis. This paper presented the earliest mesoscale simulations of shock loading on ordered arrays of spheres. The mesoscale modeling showed that the classical view of shock waves in heterogeneous materials is not a simple jump state. Furthermore, the simulations showed that mesoscale modeling could possibly provide new insights into energy localization linked to micromechanics, which could greatly improve the existing shock initiation descriptions. This was an exciting discovery, and we continued to improve the representative mesoscale models and motivate new experimental efforts.

19. M.R. Baer, M.E. Kipp, and F. van Swol, "Micromechanical modeling of heterogeneous energetic materials," in *Proceedings of the 11th International Symposium on Detonation*, edited by J. M. Short, J. E. Kennedy, Office of Naval Research Report ONR 33300-5, pp. 788–797 (1998).

As we improved the development of mesoscale modeling, better representative microstructures were needed to model the shock loading in heterogeneous materials. Our earlier work treated microstructures as ordered spheres, and the shock loading caused resonant waves to develop. Furthermore, microphotography indicated that high-density composites actually consist of an ensemble of bimodal crystals with randomly oriented planar interfaces and internal crystal defects. Most importantly, randomness and varied particle size distributions were aspects of the microstructures. Better microstructures were generated using a Monte Carlo particle packing approach that produced mesoscale geometries closer to realistic configurations. (Later, tomographic techniques were developed that could generate actual 3-D microstructures.) This 11th Symposium on Detonation paper showed that randomness enhances the homogenizing of the wave fields, and the effects of energy localization were seen to be located at material interfaces and crystal contacts. Furthermore, when reactive behavior was included in the shock loading simulation, a view of detonation in heterogeneous materials is suggested that consists of multiple waves with patterns of shock focusing and interactions with rarefaction waves. For the first time, mesoscale simulation produced new insights into the detonation wave structure. This was consistent with experimental observations of gaseous detonations, which had long before shown that the classical models of detonation such as Chapman-Jouguet (CJ) and Zeldovich-von Neuman-Doring (ZND) are incorrect.

Researchers, such as Anatoly Dremin (Institute for Chemical Physics Research, Russian Academy of Sciences), argued that when chemistry interacts with the wave mechanics, it produces dispersive behavior at the shock front. This paper presented the first numerical simulation of detonation waves in condensed phase energetic composites that were dispersive and nonplanar at the crystal level.

20. M.R. Baer, "Computational modeling of heterogeneous reactive materials at the mesoscale," in *Shock Compression of Condensed Matter 1999*, edited by M.D. Furnish, L.C. Chhabildas, and R.S. Hixson (American Institute of Physics, Melville, NY, 2000), AIP Conference Proceedings **505**, pp. 27–33.

This plenary lecture paper introduced mesoscale modeling as a potentially new area of study in shock physics. It reviewed much of our mesoscale modeling and suggested different ways to incorporate porosity and internal crystal defects in microstructures as a framework to determine energy localization effects. New ideas for including microstructure in shock initiation models were suggested; however, homogenization methods linked to the continuum macroscale still needed to be developed. In the years that followed, mesoscale studies were included as a separate topical area at the biennial Shock Compression of Condensed Matter meetings of the American Physical Society and at the detonation symposia.

21. M.R. Baer and W.M. Trott, "Mesoscale descriptions of shock-loaded heterogeneous porous materials," in *Shock Compression of Condensed Matter—2001*, edited by M. D. Furnish, N. N. Thadhani, Y. Horie (American Institute of Physics, Melville, NY, 2002), AIP Conference Proceedings **620**, pp. 713–716.

The earlier work in mesoscale modeling suggested that shock loading in heterogeneous materials caused space-time fluctuations of the thermal, mechanical, and chemical wave fields, which should be observable provided that measurements could be made at the crystal or grain level. This served as a challenge to the shock physics experimentalists, and a Laboratory Directed Research and Development (LDRD) project was funded to determine if such measurements were possible. Wayne Trott was the project leader; he developed an improved line-imaging interferometer to measure the spatial variations of particle velocities at the crystal level. Localized particle velocities were measured in planar impacts on sugar (to simulate HMX without reaction effects). The data provided details of the dispersion waves at a window interface. These mesoscale experiments verified that multiple waves evolve, as was suggested by the mesoscale modeling. Mesoscale modeling was applied to these experimental conditions and statistical methods were investigated, which could possibly link measurements at the window interface to regions containing shocked agglomeration of crystals. Despite the nondeterministic random nature of the microstructure, various wave fields (i.e., temperature) seemed to possess well-defined probability density functions, or PDFs. This was the first indication that statistical PDF methods, similar to those developed in turbulence theory, could possibly be applied to shock loading in heterogeneous materials. Much of this paper was devoted to showing that many aspects of the wave fields had well-defined PDFs and that the characteristics of these distributions could be directly linked to certain aspects of the microstructure.

22. M.R. Baer and W.M. Trott, "Theoretical and experimental mesoscale studies of impact-loaded granular explosives and simulant materials," in *Proceedings of the 12th International Detonation Symposium,* San Diego, CA, edited by J.M. Short, J.L. Maienschein, Office of Naval Research Report ONR 333-05-2, pp. 939–950 (2002).

This paper presented extended studies of impact experiments that probed mesoscale shock behavior in sugar simulants and granular layers of HMX using the Line VISAR technique. The paper introduced statistical methods, which could be applied to both modeling and experimental measurements. The effects of reaction in HMX were investigated near the threshold of initiation; the reactive behavior clearly had a stochastic nature. Although the shocks traversed random microstructures, experimental observations indicated that organized reactive waves ultimately evolved. This work suggested the growth of reaction was directly linked to the dispersive nature of the shock fields. (Underlying this conclusion was an indication that shock initiation behavior is likely dictated by the energetic material microstructure.)

23. M.R. Baer, "Modeling heterogeneous energetic materials at the mesoscale," *Thermochimica Acta* **384**, pp. 351–367 (2002).

This review paper covered the Sandia mesoscale modeling effort. The methodology is outlined, and methods are suggested to link the mesoscale effect to the continuum level using statistical methods.

24. M.R. Baer and W.M. Trott, "Mesoscale studies of shock loaded tin sphere lattices," in *Shock Compression of Condensed Matter—2003*, edited by M.D. Furnish, Y.M. Gupta, J.W. Forbes (American Institute of Physics, Melville, NY, 2004), AIP Conference Proceedings **706**, pp. 517–520.

Although much of our LDRD work has been centered on modeling microstructures that had random configurations, realistic geometries were actually nondeterministic because they are stochastic by nature. Direct experimental measurements in real heterogeneous materials would always be faced with uncertainty as to the geometrical configuration at the mesoscale. To circumvent this problem, a series of experiments was conducted on tin sphere lattices. Wayne Trott developed a technique to pack 500-μm-diameter tin spheres hexagonally in a lattice that could then be impacted and probed with a Line VISAR. This paper presented collaborative experiments and modeling. The Line VISAR measurements showed that resonant wave behavior evolved as predicted by the mesoscale modeling. This work demonstrated the importance of determining the appropriate constitutive and equation of state models for the constituents forming the microstructure.

25. W.M. Trott, M.R. Baer, J.N. Castañeda, L.C. Chhabildas, and J.R. Asay, "Investigation of the mesoscopic scale response of low-density pressings of granular sugar under impact," *Journal of Applied Physics* **101**, 024917 (2007).

This paper summarizes our experimental and numerical studies funded by the Sandia LDRD program. It focused on impact experiments of low-density pressings of sugar—a simulant of the explosive HMX. Line VISAR measurements of spatial and temporal variations at the crystal scale provided unique insights into the structure of the dispersive compaction waves and guidance in numerical simulation, including the development of statistical methods for numerical and experimental data interrogation.

26. M.R. Baer, C.A. Hall, R.L. Gustavsen, D.E. Hooks, and S.A. Sheffield, "Isentropic compression experiments (ICE) for mesoscale studies of energetic composites," in *Shock Compression of Condensed Matter—2005,* edited by M. D. Furnish, M. Elert, T. P. Russell, C. T. White (American Institute of Physics, Melville, NY, 2006), AIP Conference Proceedings **845**, pp. 1307–1310.

In 2003, a new MOU project began with a focus on mesoscale studies of energetic materials. At that time, the Z machine was being explored as a new way of determining equations of state. This project became a multi-laboratory effort with a focus on explosive material. The new technique probed near-detonation stress states without inducing reactive behavior. Furthermore, the ICE configurations allowed simultaneous loading of several materials on a single shot. This was ideal for studying energetic materials at the mesoscale because individual constituents and mixture composites could be simultaneously investigated. Much of our prior mesoscale studies funded by the LDRD program and the Engineering Sciences Research Foundation showed that mesoscale modeling required accurate equations of state and constitutive properties at high pressures. This paper reported on a specific test (Z Shot 1251) done on PBX 9501 and its constituents. An ICE configuration was designed that had multiple samples of the energetic composite PBX 9501, monolithic HMX crystals, Estane, and "dirty-binder" consisting of fine HMX crystallites mixed with Estane binder materials. Isentropic compression to 50 kbars was produced simultaneously in these materials. A forward method for data reduction to determine the equation of state for all constituents was described in the paper.

27. C.A. Hall, M.R. Baer, R.L. Gustavsen, D.E. Hooks, E.B. Orler, D.M. Dattelbaum, S.A. Sheffield, and G.T. Sutherland, "A study of polymer materials subjected to isentropic compression loading," in *Shock Compression of Condensed Matter—2005,* edited by M.D. Furnish, M. Elert, T.P. Russell, C.T. White (American Institute of Physics, Melville, NY, 2006), AIP Conference Proceedings **845**, pp. 1311–1314 (2006).

In ICE, time-varying magnetic pressure is applied on drive plates containing very thin layers of material. Collaborators from LANL developed a method to create precise layers of energetic composites and monolithic crystals of explosive mounted on the drive plates. In separating the response of the various constituents of the composites, polymer binder layers were also created and mounted. Binder materials are typically soft and difficult to process. Nonetheless, an ICE configuration was created that contained many polymer binders typically used in explosive composites. This paper presented results from that Z test.

28. M.R. Baer, M.L. Hobbs, C.A. Hall, D.E. Hooks, R.L. Gustavsen, D. Dattelbaum, and S.A. Sheffield, "Isentropic compression studies of energetic composites and

constituents," in *Shock Compression of Condensed Matter–2007,* edited by M. Elert, M.D. Furnish, R. Chau, N.C. Holmes, J. Nguyen (American Institute of Physics, Melville, NY, 2007), AIP Conference Proceedings **955,** pp. 1165–1168.

As our isentropic compression studies of explosives advanced, forward analysis became linked to optimization software (DAKOTA) that, in turn, provided a framework for assessing sensitivity analysis. This paper provides details of that new technique.

29. M.R. Baer, C.A. Hall, R.L. Gustavsen, D.E. Hooks, and S.A. Sheffield, "Isentropic loading experiments of a plastic bonded explosive and constituents," *Journal of Applied Physics* **101**, 034906 (2007).

Having developed a more precise method for data analysis, the ICE studies on PBX 9501 were revised to better define the equation of state and constitutive model parameters. With these inputs, a mesoscale model of the composite microstructure was created and ramp loading was simulated. Of most interest was the determination of the particle velocity fluctuations that would accompany the isentropic loading. Numerical results indicated that particle velocity variations up to 10 % should occur, which motivated continued development of the Line VISAR for our magnetic compression studies.

30. M.R. Baer, S. Root, D. Dattelbaum, D.E. Hooks, R.L. Gustavsen, B. Orler, T. Pierce, F. Garcia, K. Vandersall, S. DeFisher, and B. Travers, "Shockless compression studies of HMX-based explosives," *Shock Compression of Condensed Matter—2009*, M.L. Elert, W.T. Buttler, M.D. Furnish, W. W. Anderson, W.G. Proud (American Institute of Physics, Melville, NY, 2009), AIP Conference Proceedings **1195**, pp. 699–702.

Much of our prior work involved collaboration with LANL researchers. As this project matured, the team of researchers expanded to include LLNL and the DOD labs. Magnetic compression experiments had then transitioned to using the Veloce pulser, which was partially funded by the mesoscale MOU project and provided quicker turnaround than Z in experimental testing. This paper discussed tests containing samples of HMX-based composites from each laboratory. The HMX content of the composites was similar; however, different binders were used. The paper showed that subtle changes in the binder produced observable differences in the response.

31. M.R. Baer, S. Root, R.L. Gustavsen, T. Pierce, S. DeFisher, and B. Travers, "Temperature dependent equation of state for HMX-based composites," in *Shock Compression of Condensed Matter—2011: Proceedings of the Conference of the American Physical Society Topical Group on Shock Compression of Condensed Matter*, edited by M.L. Elert, W.T. Buttler, J.P. Borg, J.L. Jordan, T.J. Vogler (American Institute of Physics, Melville, NY, 2012), AIP Conference Proceedings **1426**, pp. 163–166.

This paper examined a series of Veloce tests whereby heated samples of HMX-based composites were subjected to ramp loads surpassing detonation-like pressures. The intent was to observe the effects of β to δ phase transition in HMX during ramp loading. However, the data analysis revealed that an undeterminable degree of phase transition occurred prior to ramp loading.

32. M.R. Baer, R.G. Schmitt, E.S. Hertel, and P.E. DesJardin, "Modeling enhanced blast explosives using a multiphase mixture approach," in *Fluid Structure Interaction and Moving Boundary Problems,* edited by S. Chakrabarti, S. Hernández, and C.A. Brebbia (Wessex Institute of Technology, UK, 2005) **84**, pp. 393–401.

In 2001, the Office of the Secretary of Defense (OSD) established the advanced energetics program, which funded research in new concepts and formulations in energetic materials. Volumetric explosives, particularly thermobaric explosives, were assigned as a major technical area of study.

Sandia studies of thermobaric explosives began in 1988 and were funded by the intelligence agencies to provide an assessment of Russian research that could be potentially exploited in enhanced blast technology. OSD asked Sandia to provide experimental guidance and modeling support based on our prior research in this area. Thermobaric materials are single-cycle explosives that have a long-duration thermal pulse, which accompanies and supports shock and air blast output. The Russians first developed these explosives as fuel-rich slurries of metal and explosive liquids. These materials were chosen because they do not have strength effects; hence, post detonation products are dispersed in a large volume of ambient air to induce secondary combustion effects.

This paper provided an overview of extending multiphase shock physics to include enhanced blast. The methodology for linking multiphase combustion to structural analysis was discussed, and mesoscale modeling was conducted to indicate that the detonation of metal-loaded energetics likely causes fragmentation of the dispersed powders.

33. M.A. Cooper, M.R. Baer, R.G. Schmitt, M.J. Kaneshige, R.J. Pahl, and P.E. DesJardin, "Understanding enhanced blast explosives: a multi-scale challenge," in *Proceedings of the 19th International Symposium on Military Aspects of Blast and Shock (MABS—19)*, edited by E. Benggeli, Calgary, Canada, October 1–6, 2006.

This paper provided an overview of the various aspects of shock physics associated with enhanced blast (i.e., thermobaric) explosives. To develop an appropriate model of these materials, nonideal detonation and secondary combustion effects had to be incorporated in the shock physics analysis.

The Defense Threat Reduction Agency took ownership of the advanced energetics program and immediately sought to speed up the research and testing of thermobaric explosives toward developing a weapons development program. Although it was known that thermobaric explosives were slurry systems, the military services were reluctant and resistant in fielding these new materials because that meant developing new methods for assessing performance and certifying safety requirements. It was felt that these constraints could not accommodate a rapid weapons development effort. Hence, thermobaric materials were "redefined" as aluminized explosives similar to those used in Navy underwater weapons to circumvent the assessment requirements. Thus, much of the prior research on thermobaric explosives was disregarded (including the Russian researchers that invented thermobaric concepts). It was decided to conduct an extensive testing effort to "brute force" develop a thermobaric formulation using existing aluminized explosives formulations. This was an unfortunate decision because the discovery of a new thermobaric formulation was not achieved and advanced energetic funding was diverted to build expensive test facilities rather than investing in fundamental studies of thermobaric behavior. Sandia interest in this program faded and our work in this area ceased.

34. M.R. Baer, D.K. Gartling, and P.E. DesJardin, "Probabilistic models for reactive behaviour in heterogeneous condensed phase media," *Combustion Theory and Modeling* **16**, no. 1, pp. 75–106 (2012).

This paper presented new statistical-based models to bridge the mesoscale to continuum levels. Much of this approach follows the PDF methods that have been successfully applied to turbulent combustion. Similar to the turbulence problem, the distribution of states contains a spectrum of scales that arise and evolve over space and time. A PDF transport relationship is derived that describes the evolution of a joint distribution including temperature, strain, and strain rate as random variables. Moments of the PDF are defined that lead to a statistically based reactive flow description, which readily adapts to existing finite element or finite volume analyses. As an example application, the modeling approach considers the impact of a fragment, which embeds in an explosive and does not lead to shock initiation but does lead to delayed reaction due to cookoff effects. Impact causes energy localization at a mesoscale "hot spot," and, with the PDF approach, these states at the tails of a temperature distribution interact with the surrounding states, sustaining a reaction until an ignition state is attained.

This new modeling approach potentially can be applied to a broad class of shock and thermal initiation problem. Currently, this PDF mesoscale approach is being investigated to model slapper initiators as funded by a new LDRD project at Sandia. Although I have since retired from Sandia, I have returned on a part-time basis to help mentor new staff in this area of study.

# Lynn M. Barker
# (1955–1974, 1980–1990[10])

**Editors' comments**: In a letter to Jim Asay, dated September 11, 2011, Lynn Barker provided a list of papers written over a period beginning in the early 1960s that highlighted the development of techniques for precision shock wave measurements. Several of these have become classic articles that continue to be used in modern research programs. The collected papers provide a good recollection of the efforts by Lynn and his colleagues to evolve, from a fledgling shock wave capability in the 1950s, to a premier research program in the 1960s and beyond that is now a first-in-class shock wave research program at Sandia. The following sequence of papers by Lynn and his colleagues identifies the major developments in time-resolved instrumentation and the application to several scientific problems.

## Early Papers and Reports by Lynn Barker on Instrumentation and Data Reduction[11]

The following is a list of my 17 early papers on shock wave research, including instrumentation and data reduction, by our fledgling Sandia shock wave science group. These papers were published during 1962 to 1975. My 1999 award paper (Shock Compression Science Award) on the development of the VISAR appears before this list because it documents the background of most of the instrumentation described in the rest of the papers, which are arranged in chronological order.

L.M. Barker, "The development of the VISAR, and its use in shock compression science," in *Shock Compression of Condensed Matter 1999*, edited by M.D. Furnish, L.C. Chhabildas, and R.S. Hixson (American Institute of Physics, Melville, NY, 2000), AIP Conference Proceedings **505**, pp. 11–17.

This paper is a record of Barker's acceptance speech during the 1999 George E. Duvall Shock Compression Science Award at the 1999 Topical Conference on Shock Compression of Condensed Matter in Snow Bird, Utah. The paper provides a good overview of the development of interferometry, starting with the displacement interferometer through the development of the VISAR, which can be used in essentially any dynamic material experiment. Lynn presents several applications and variations of the technique in the paper:

1. J.H. Smith and L.M. Barker, "Measurement of tilt, impact velocity, and impact time between two plane surfaces," Sandia Corporation Report SC-4728 (RR), December 1962.

[10] Lynn retired from Sandia in 1990 and formed his own company to make VISARs. He continued to interact in the shock wave community through 2002.

[11] Editors' note: The annotations about Barker's American Physical Society Shock Compression Science Award paper and these 17 early papers were provided by Jim Asay.

This early paper discusses the technique developed for accurate measurement of the impact time and misorientation, or tilt, of a flat projectile upon impact with a flat target. Rigorous attention to these details ensured that accurate shock wave data could be obtained with plate impact techniques being developed in the late 1950s and early 1960s.

2. B.M. Butcher, L.M. Barker, D.E. Munson, and C.D. Lundergan, "Influence of stress history on time dependent spall in metals," *AIAA Journal* **2**, No. 6, pp. 977–990 (1964).

This is one of the first papers to contain a careful discussion of the shape and time dependence of stress wave interactions that result in dynamic tensile loading and subsequent spall of a flat specimen. Extensive studies were conducted on several materials, including aluminum, copper, brass, steel, and silver. The spall thresholds are reported for each.

3. L.M. Barker and R.E. Hollenbach, "System for measuring the dynamic properties of materials," *Review of Scientific Instruments* **35**, pp. 742–746 (1964).

The use of a slanted resistor wire on the free surface of an impacted planar sample for accurate determination of the free surface particle velocity and shock velocity was first discussed by Barker in two separate Sandia Corporation reports (Barker 1961, 1962). This journal paper describes the air gun developed for plate impact studies, the details of the slanted wire resistor technique, and the resulting stress–strain loading path of aluminum to 20 kbars that was obtained with the technique.

4. L.M. Barker and R.E. Hollenbach, "Interferometer technique for measuring the dynamic mechanical properties of materials," *Review of Scientific Instruments* **36**, pp. 1617–1620 (1965).

According to Charles Karnes' recollections, the earliest experiment to test the feasibility of using Michelson, or displacement, interferometry in shock wave experiments was made by Lynn Barker and Red Hollenbach in late 1963 using a vacuum-driven "projectile" in a short aluminum tube to impact a flat sample monitored with a Michelson interferometer. The development of the technique with a real gas gun and improved instrumentation on soft aluminum specimens began afterward and resulted in this 1965 paper, which describes the refined Michelson interferometer technique in detail. Although preliminary, the interferometer results revealed an unexpected overshoot elastic wave, which is discussed in a 1966 paper on 1060 aluminum (See Barker's ref. no. 6 below).

5. D.E. Munson and L.M. Barker, "Dynamically determined pressure-volume relationships for aluminum, copper, and lead," *Journal of Applied Physics* **3** (4), pp. 1652–1660 (1966).

This paper reports the use of both the slanted resistance wire technique and the Michelson interferometer to measure the free surface velocities of impacted aluminum, copper, and lead. Good agreement is observed between dynamic shock Hugoniot data and ultrasonic data, and the importance of shear strength was clearly evident in the shock wave experiments.

6. L.M. Barker, B.M. Butcher, and C.H. Karnes, "Yield point phenomenon in impact-loaded 1060 aluminum," *Journal of Applied Physics* **37** (5), pp. 1989–1991 (1966).

This study used a Michelson interferometer technique to measure the elastic yielding in shocked 1060 aluminum. An unexpected overshoot in elastic stress and relaxation from the peak elastic response were observed and discussed in terms of dislocation theory.

7. L.M. Barker, "Fine structure of compressive and release wave shapes in aluminum measured by the velocity interferometer technique," in *Proceedings of International Symposium on the Behavior of Dense Media under High Dynamic Pressures* (Gordon & Breach, New York, 1968), pp. 483–504.

The newly developed velocity interferometer technique is described in this paper. This major advance in interferometry for shock wave experiments used the superposition of undelayed and optically delayed laser beams to differentiate Michelson displacement data optically and to obtain a direct measure of particle velocity. Results are presented for the compression and release of aluminum shocked to 90 kbars. The risetimes of the shock waves produced to 90 kbars were observed to vary approximately as the fourth power of the peak shock stress; this shock wave structure data suggested a fundamental relationship between shock strain rate and peak Hugoniot stress. This relationship was formalized theoretically in later papers by Dennis Grady.

8. J.N. Johnson and L.M. Barker, "Dislocation dynamics and steady plastic wave profiles in 6061-T6 aluminum," *Journal of Applied Physics* **40** (11), pp. 4321–4334 (1969).

Advanced material models and computational techniques were rapidly developed in the 1960s, along with these new instrumentation techniques, which permitted the critical examination of physical processes occurring during shock compression. Johnson and Barker's paper is the first at Sandia connecting shock wave structure and the evolution of steady shock waves to dislocation mechanisms.

9. L.M. Barker and E.G. Young, "SWAP-9: An improved stress wave analyzing program," Sandia Corporation Report SLA-74-0009, August 1974.

The first Stress Wave Analysis Program, called SWAP-7, was introduced in April 1967. SWAP-7 is a characteristics computer code for rapid solution of one-dimensional strain rate-independent shock propagation problems. After its introduction, SWAP-7 was widely used in a variety of applications because of its ability to model hydrodynamics and elastic–plastic materials, radiant energy deposition, spallation, and a variety of other phenomena. Solving many of these problems was impractical or impossible with the more common finite-difference programs used at the time because of the large computer run times required. A number of improvements were made to SWAP-7, resulting in a new version, SWAP-9, which had an increased ability to solve problems involving gases, detonations, and vaporization resulting from energy deposition.

10. L.M. Barker and R.E. Hollenbach, "Shock wave studies of PMMA, fused silica, and sapphire," *Journal of Applied Physics* **41** (10), pp. 4208–4226 (1970).

This is the first paper that relates changes in the refractive index of shocked materials to corrections in the refractive index under shock loading. These data are needed to obtain accurate particle velocity information in interferometer applications that use transparent materials as laser windows.

11. L.M. Barker, "Velocity interferometer data reduction," *Review of Scientific Instruments* **42** (2), pp. 276–278 (1971).

This paper discusses the analysis approach used on velocity interferometry data to improve the time resolution of the instrumentation. The method described in the paper can substantially improve particle velocity information over the delay time of a few ns, which is the typical time delay of the velocity interferometer.

12. L.M. Barker and R.E. Hollenbach, "A laser interferometer for measuring high velocities of any reflecting surface," *Journal of Applied Physics* **43** (11), pp. 4669–4675 (1972).

The previous developments in displacement and velocity interferometry by Barker and Hollenbach provided the groundwork for development of the VISAR, which expanded the use of velocity interferometry to any dynamically loaded surface. The VISAR has better time resolution and accuracy than previous techniques and, most importantly, can be used on any reflecting surface. After more than 40 years, the VISAR is still the standard diagnostic in most shock wave experiments and has been successfully used to shock pressures over 20 Mbars. In these ultrahigh-pressure experiments, the shock velocity in a window, such as single-crystal quartz mounted on the back surface of a specimen, is directly measured since the window becomes conductive and thus reflective at these shock pressures and temperatures.

13. L.M. Barker, "VISAR data reduction," Sandia Corporation Report SLA-73-1038, February 1974.

This is the first report on analysis of "push-pull" VISAR data. It provides useful hints about how various effects, such as changes in the phase difference of the two measured VISAR signals or optical contrast of the signal, affect the accuracy of the velocity data. More recent VISAR analysis programs are a variation of the methods reported in this document. Since the initial preparation of the VISAR data reduction program by Barker, Sandia has maintained a modern version that is available through a license agreement.

14. J.R. Asay and L.M. Barker, "Interferometric measurement of shock-induced internal particle velocity and spatial variations of particle velocity," *Journal of Applied Physics* **45** (6), pp. 2540–2546 (1974).

This paper discusses how the loss of fringe contrast in an interferometer experiment can be used to estimate the variation in measured particle velocity at the focal area (typically about 200 μm in diameter) of the laser signal. The effect was first observed in VISAR experiments on shocked porous aluminum, which exhibits spatial variations in particle velocity produced during compaction of the pores.

15. L.M. Barker and K.W. Schuler, "Correction to the velocity-per-fringe relationship for the VISAR interferometer," *Journal of Applied Physics* **45** (8), pp. 3692–3693 (1974).

As more data were obtained with the VISAR, it was noted that measured particle velocities tended to be about 3 % higher than measurements made with other techniques. After careful fringe analysis used for data reduction and corroborating projectile velocity experiments, Barker and Schuler showed that the frequency dispersion of laser light in the glass plates (*etalons*) of the interferometer itself was responsible for the discrepancy. When this effect was accounted for, near-perfect agreement between the VISAR data and results from other experiments was observed.

16. L.M. Barker and R.E. Hollenbach, "Shock wave study of the α-ε phase transition in iron," *Journal of Applied Physics* **45** (11), pp. 4872–4887 (1974).

This paper reports use of the VISAR to study the forward α-ε phase transition in Armco iron that is shock loaded above 130 kbars and the reverse transformation ε-α that occurs during the unloading from peak shock stress of about 400 kbars. The free surface velocities measured with the VISAR provided information on the kinetics of the forward transition, sound speeds, Poisson's ratio after shock compression, and the effect of a strong magnetic field on the transition stress.

17. L.M. Barker, "α-phase Hugoniot of iron," *Journal of Applied Physics* **46** (6), pp. 2544–2547 (1975).

Earlier studies of the α-phase of iron indicated that softening of the Hugoniot state occurred as the 130 kbar α-ε transition stress was approached. This unexpected effect was interpreted as the phase transition beginning to occur at a stress level below 130 kbars. Barker reanalyzed previous data and combined these with data from the earlier paper to show that the premature phase transition does not occur and that the Hugoniot behavior of α-iron is normal to stresses approaching the phase transition.

## Early Papers and Reports by Lynn Barker on a Model for Stress Wave Propagation

The following papers concern my theoretical model for calculating stress wave propagation in composite materials:

1. L.M. Barker, "A model for stress wave propagation in composite materials," *Journal of Composite Materials* **5** (2), pp. 140–162 (1971).
2. L.M. Barker, C.D. Lundergan, P.J. Chen, M.E. Gurtin, "Nonlinear viscoelasticity and the evolution of stress waves in laminated composites: A comparison of theory and experiment," *Journal of Applied Mechanics* **41**, pp. 1025–1030 (December 1974).

## Contributions by Non-Sandians to VISAR Technology

The basic VISAR was developed at Sandia by Lynn M. Barker and R.E. Hollenbach in 1972. It worked well, but was difficult to learn to use; the hardware was quite large, and it was not at all mobile. Below are three later papers by B.T. Amery, Willard Hemsing, and Lynn Barker, all non-Sandians (Lynn having retired in March 1990), which added significantly to the VISAR technology. Lynn formed his own company, Valyn International, in September 1990 and invented a safer fiber optic probe and improvements to the VISAR that greatly improved its user friendliness. On February 13, 1993, Lynn was granted U.S. Patent 5,202,558 for the flexible fiber optic probe. On January 2, 1996, Lynn was granted U.S. Patent 5,481,359 for the ease-of-use features of the Valyn VISAR (see below). On February 9, 1999, Lynn was granted U.S. Patent 5,870,192 for the Valyn Multi-beam VISAR (see below). On January 13, 2004, Lynn and his son, Zane B. Barker, were granted U.S. Patent 6,678,447 B1 for the adjustable multi-beam splitter and fiber coupler. These improvements have contributed to the widespread use of VISARs as the ultimate in velocity vs. time measurements of shock wave events:


1. B.T. Amery, "Wide range velocity interferometer," in *Proceedings of the 6th International Detonation Symposium*, Coronado, CA, edited by S. J. Jacobs, D. J. Edwards (Office of Naval Research, Arlington, VA, 1976), ONR ACR-221, pp. 673–681.
2. W.F. Hemsing, "Velocity sensing interferometer (VISAR) modification," *Review of Scientific Instruments* **50** (1), pp. 73–78 (1979).
3. L.M. Barker, "The accuracy of VISAR instrumentation," in *Shock Compression of Condensed Matter 1997*, edited by S.C. Schmidt, D.P. Dandekar, and J.W. Forbes (American Institute of Physics, Melville, NY, 1998), AIP Conference Proceedings **429**, pp. 833–836.
4. L.M. Barker, "Multi-beam VISARs for simultaneous velocity vs. time measurements," in *Shock Compression of Condensed Matter 1999*, edited by M.D. Furnish, L.C. Chhabildas, and R.S. Hixson (American Institute of Physics, Melville, NY, 2000), AIP Conference Proceedings **505**, pp. 999–1002.
5. Lynn M. Barker, Inventor, "Multi-etalon VISAR interferometer having an interferometer frame of high stiffness with a linear elongated slide bar," U.S. Patent 5,481,359, January 2, 1996.
6. Lynn M. Barker, Inventor, "Multibeam VISAR using image coupling from one optical fiber bundle to another through the VISAR interferometer," U.S. Patent 5,870,192, February 9, 1999.


## Papers by Lynn Barker on Gouging of Gun Barrels and Sled Tracks

These three papers from the 1980s have to do with the study of gouging of gun barrels and sled tracks due to high-velocity sliding contact between two metals:


1. L.M. Barker, T.G. Trucano, and J.W. Munford, "Metal surface gouging by hypervelocity sliding contact," in *Shock Waves in Condensed Matter 1987*, edited by

S.C. Schmidt and N.C. Holmes (Elsevier Science Publishers B.V., Amsterdam, The Netherlands, 1988), pp. 753–756.
2. L.M. Barker, T.G. Trucano, and A.R. Susoeff, "Railgun rail gouging by hypervelocity sliding contact," *IEEE Transactions on Magnetics* **25** (1), pp. 83–87 (1989).
3. L.M. Barker, T.G. Trucano, and A.R. Susoeff, "Gun-barrel gouging by sliding metal contact at very high velocities," in *Proceedings of the 39th Aeroballistic Range Association Symposium*, Albuquerque, NM, October 10–14, 1988.[12]

## Papers by Lynn Barker on Isentropic Loading of Solid Hydrogen

These five papers from the 1980s are concerned with dynamic isentropic loading of solid hydrogen:

1. L.M. Barker, "High pressure quasi-isentropic impact experiments" (invited paper), in *Shock Waves in Condensed Matter 1983,* edited by J.R. Asay, R.A. Graham, G.K. Straub (Elsevier Science Publishers B.V., Amsterdam, The Netherlands, 1984), pp. 217–223.
2. L.M. Barker, T.G. Trucano, J.L. Wise, and J.R. Asay, "Experimental technique for measuring the isentrope of hydrogen to several megabars," in *Shock Waves in Condensed Matter 1985*, edited by Y.M. Gupta (Plenum Press, New York, NY, 1986), pp. 455–459.
3. T.G. Trucano, L.M. Barker, J.R. Asay, and G.R. Kerley, "Numerical studies of the dynamic isentropic loading of solid molecular hydrogen," in *Shock Waves in Condensed Matter 1985*, edited by Y.M. Gupta (Plenum Press, New York, NY, 1986), pp. 461–465.
4. L.C. Chhabildas and L.M. Barker, "Dynamic quasi-isentropic compression of tungsten," in *Shock Waves in Condensed Matter 1987*, edited by S.C. Schmidt and N.C. Holmes (Elsevier Science Publishers B.V., Amsterdam, The Netherlands, 1988), pp. 111–114.
5. L.M. Barker, L.C. Chhabildas, T.G. Trucano, and J.R. Asay, "Gas-accelerated plate stability study," in *Shock Compression of Condensed Matter 1989*, edited by S.C. Schmidt, J.N. Johnson, and L.W. Davison (Elsevier Science Publishers B.V., Amsterdam, The Netherlands, 1990), pp. 989–991.

[12] If a paper was prepared on this topic for the Aeroballistic Range Association, ARA members should be able to obtain a CD-ROM containing the collected proceedings.

## An Early Chronology of Lynn Barker's Contributions to Shock Wave Research

**Editors' comment**: In addition to the papers above, which reflect the development and application of advanced instrumentation techniques, the following text was extracted from a letter written on March 14, 1999, by Barker to Don Lundergan. It provides additional insight into the significant contributions that Barker has made to the shock wave field:

| | |
|---|---|
| 1959 | I stepped down from section supervisor and transferred into your [Don Lundergan's] budding shock wave group. Shock waves were completely foreign to me at the time. I remember visiting Dick Fowles at SRI[13] with you and trying to understand Dick's explanation of the elastic limit in 1-D strain and the hysteresis in the shock loading-unloading path. |
| 1960 | We had the gun in Area Y. It was outside the instrumentation building and had a sawdust pile as a catcher. We put a heavy wire mesh around the pile to catch pieces that glanced off the sawdust, but they went right through the mesh! You asked Sandia Safety to take a look at our operation. They recommended a method such that we just stepped over the gun barrel!—That's all! |
| 1960 | We used pins to measure displacement time and differentiated to get free surface velocity vs. time. But we learned how to minimize and to measure tilt. |
| 1961–1962 | I spent a year at Columbia University working toward a PhD in physics. Because of my wife's poor health, I returned to Sandia in 1962 after finishing all of my course work, but lacking a research project and a dissertation for a PhD. |
| 1962–1963 | Red and I developed the slant wire instrumentation technique and did a pretty good study of 6061-T6 Al with it. We could see an elastic precursor on resumed loading after a pause. |
| 1964 | We began thinking about laser interferometry for shock wave instrumentation. We bought a Michelson interferometer for about $200 just to learn interferometer techniques. |
| 1965 | We developed the displacement interferometer technique. We got unprecedented accuracy, but were limited to low surface velocities and specimens that could be polished to a mirror finish. |
| 1966 | We developed the velocity interferometer. It measured velocity directly—with no loss in resolution due to differentiation. Much higher velocities were okay, but mirror specimen surfaces were still required. |
| 1967 | At the IUTAM[14] Conference in Paris, I presented a paper on the velocity interferometer and its use in a study of the wave profiles (particularly plastic wave risetimes) of 6061-T6 Al. It was an important paper—the data are still quoted and used. The conference limit on paper size was 10 pages, but mine was 20 pages long. The conference organizer asked me to make two papers out of it—one on the instrumentation and one on the Al study. I felt it all belonged in one paper and offered to publish it elsewhere. I got a two-word reply: "Extension accepted." |

[13] Editors' note: SRI, a nonprofit corporation, was founded by Stanford University in 1946 as the Stanford Research Institute. It became independent of Stanford in 1970 and several years later changed its name to SRI International.

[14] Editors' note: IUTAM is the International Union of Theoretical and Applied Mechanics.

| | |
|---|---|
| 1968–1970 | We fielded velocity interferometer experiments on three NTS tests to demonstrate whether laser interferometry could be adapted to the underground test environment. The first two tests didn't attempt to actually make a measurement, but rather were to determine whether calculations of shielding, etc., were okay. We got very good data on the third test, including an excellent measure of the Grüneisen constant for fused quartz. |
| 1970–1972 | Developed and published the first papers on the VISAR. |

## On the Origin of Some Shock Wave Acronyms

**Editors' comment**: In a letter to Jim Asay, dated January 13, 2013, Barker provided additional insight into how some of our commonly used shock wave acronyms at Sandia originated.

I remember when Red Hollenbach and I began to realize what a powerful shock wave instrumentation tool we'd found in the laser velocity interferometer that worked when the specimen surface didn't need to be polished to a mirror surface—a diffusely reflecting surface worked just fine! And unlike the Lockheed interferometer, which preceded ours, ours was at least ten times more accurate!

We decided to look for a name which made a jazzy-sounding acronym such that the acronym came from words that described the interferometer. I think we spent at least an hour one morning trying to include the words "velocity interferometer" and "any reflector" to stress that it works on either diffusely reflecting surfaces or surfaces polished to a mirror finish. When we finally got an "s" in there using the word "system," we had Velocity Interferometer System for Any Reflector, or VISAR. Then we wondered why it had taken us so long to see it. It made sense, it sounded good, and it described the product.

Then, later on when our gun facility was taking shape in Area III, at one of our Monday morning safety meetings, Jim Asay suggested that our new facility deserved a name of its own. Asay asked us to think about it and to come up with suggestions for the name at our next safety meeting. I mentioned the naming contest to my wife, Val. She immediately took an interest in it, and we discussed some possibilities. I suggested the Shocker facility, but that didn't pass the Val test. Then we wrote down some pertinent words, like shock, research, applied research, and finally thermodynamics. After studying the words for a while, Val said "How about Shock Thermodynamics Applied Research? Does that make sense? That would make it the STAR facility." I said to Val, "I think you've just named our new facility!" And that's the way it turned out.

I joined Sandia in 1955, but in 1961 left to go to Columbia University in New York to study for my PhD. I returned to Sandia in 1962. In 1974, I was recruited to work at Terra Tek in Salt Lake City, UT. I returned again to Sandia in early 1981, where I stayed until my retirement in March 1990. Later that year, my wife Val and I formed our own company, Valyn International. I persuaded Val to take on the job of chief executive officer of our company, while my job was mainly Research and Development. We spent the next dozen years successfully inventing, improving, and marketing products for the shock compression science community. Our company still operates (2014) as Valyn VIP, under the leadership of our son Zane Barker and his wife, Wendy.

# Mark B. Boslough

## (1983–Present[15])

**Editors' comments**: Mark Boslough graduated in Applied Physics from the Geological and Planetary Sciences Division at the California Institute of Technology (Caltech). He joined the Shock Wave and Explosives Physics Division led by Bruno Morosin. The early part of his career focused on a variety of experimental projects, including a new research effort on shock-induced chemistry initiated by Bob Graham and research on hypervelocity impacts of debris shields with Lalit Chhabildas. In the early 1990s, his research turned to more emphasis on the numerical simulation of hypervelocity impact events and planetary physics. He has become internationally known for his wide-ranging investigations of cometary and asteroid impact events. In his recollections, Mark describes how he became involved in investigating planetary airbursts. He provides a fascinating account of efforts to use the rapidly evolving computing capability at Sandia to model the impact of Comet Shoemaker–Levy 9 on Jupiter.

### Opportunity Knocks into Jupiter at 60 km/s!

> It is indeed a unique object, different from any cometary form I have yet witnessed. In general, it has the appearance of a string of nuclear fragments spread out along the orbit with tails extending from the entire nuclear train as well as what looks like a sheet of debris spread out in the orbit plane in both directions.[16]

This was the first public description of the comet that came to be known as Shoemaker–Levy 9 (SL9), as the discovery announcement in International Astronomical Union Circular No. 5725 of March 25, 1993. The words were those of James V. Scotti of the Spacewatch program at the University of Arizona, who confirmed the object's existence. It was a puzzling find and very exciting to planetary scientists because of its unique character. In those pre-World Wide Web days,[17] I first read about it in the July 1993 issue of *Sky & Telescope*. Co-discoverer David Levy, a regular columnist, described the serendipitous events on top of Palomar Mountain that led to Carolyn Shoemaker's first identification of the comet on damaged film, polluted with the glare from Jupiter, from an image taken just before

[15] Mark worked in experimental shock wave science until the mid 1990s, followed by a couple of years in simulations of shock wave problems. He was then involved in other activities at Sandia until around 2006 when he became active in planetary applications that continue to the present.

[16] Private communication on March 26, 1993 from James V. Scotti to Brian G. Marsden of the Central Bureau for Astronomical Telegraphs. The description that Scotti sent then appeared in IAU Circular No. 5725 that same day.

[17] Editors' note: Tim Berners-Lee, a software engineer at CERN (the European Organization for Nuclear Research), invented the World Wide Web in 1989. However, it wasn't until April 1993 that CERN announced the technology would be available to use by anyone on a royalty free basis. In 1994 the World Wide Web Consortium was formed.

clouds covered the sky for the rest of the night. "I don't know *what* this is," she said. "It looks like… like a squashed comet."[18]

David went on to speculate about how the comet might break apart, and what might happen to it:

> As Comet Shoemaker-Levy continues to evolve, we may see some of the fragments fade to nothing. Maybe a few will last a year or more. Right now, all we can do is wait, watch, and speculate. No matter what the outcome, all of Comet Shoemaker-Levy's offspring will live long in our memories.[19]

He was right about one thing. We have not forgotten. But he underestimated the magnitude of the memories because he didn't know at the time what really would happen to the comet. His article was already in the newsstands when I happened to notice a small, 5-column-inch story in the *Albuquerque Journal* on June 12, 1993: "Jupiter, Comet Could Collide in Huge Crash." New calculations by comet co-discoverer (and Carolyn's husband) Gene Shoemaker showed that the remaining chunks of the comet might collide with Jupiter on July 20, 1994. It was thought that the collision could unleash the energy equivalent to a billion megatons of TNT, more powerful than the one that wiped out the dinosaurs 65 million years ago. One scientist speculated that the impacts would cause Jupiter to flare up to 25 times its normal brightness for a couple minutes.

I had always been interested in planetary impacts and, as a graduate student at Caltech, had taken a course about impact craters from Gene, which included a field trip along with Carolyn to Meteor Crater, Arizona. That was where I first learned how legendary Gene already was to crater *aficionados*. When I told the cashier at the Meteor Crater visitor center that our guide would be Gene Shoemaker, she was thrilled he would be there and exclaimed, "That's like getting a field trip to heaven with God!"

In 1993, I was a staff member in the Experimental Impact Physics Department, making use of the STAR facility to test the ability of bumper shields to protect spacecraft from space debris, among other things. Part of our department's role was to validate numerical simulations of impacts and explosions for our sister departments, which were focused on computational shock physics. What could be better for validation than the ultimate natural impacts and explosions? So I brought the newspaper clipping in to work and showed it to Mike McGlaun, manager of one of the other groups. It didn't take long to get him interested, and he said he'd talk to Ed Barsis, our Center Director.

Fortuitously, I had just gotten my first LDRD (Laboratory Directed Research and Development) project funded. It was a computational project to model the impact and seismic consequences of the dinosaur-killing cometary impact. With a few modified milestones, we had some funding to pay for our time. All we needed was

---

[18] From D.H. Levy, "Pearls on a string," *Sky & Telescope* **86**, no. 1, p. 39 (1993), reprinted with permission from *Sky & Telescope*. The original quotation was from C.S. Shoemaker and E.M. Shoemaker, "A Comet Like No Other," Chapter 2, p. 7 in *The Great Comet Crash: The Impact of Comet Shoemaker-Levy on Jupiter*, edited by J.R. Spencer and J. Mitton (Cambridge University Press, Cambridge, UK, 1995), reprinted with permission from Cambridge University Press.

[19] D.H. Levy, "Pearls on a string," *Sky & Telescope* **86**, no. 1, p. 39 (1993), reprinted with permission from *Sky & Telescope*.

permission to use the brand-new Intel Paragon, the most powerful machine in the world. And we needed people who knew how to run the code or could learn how. Dave Crawford, a new postdoc, was immediately in: his background in planetary geology and experimental impact physics was perfect. Best of all (unlike me), he was a computer whiz. Tim Trucano agreed to start running simulations and teach us how to use CTH, Sandia's three-dimensional (3-D) multi-material shock physics code. Allen Robinson, who had written a parallel version of the code that he was porting to the Paragon, would also help. A big unclassified problem would be a perfect exercise to test the new hardware and software. With additional help from Marlin Kipp and Gerry Kerley, we would quickly be running simulations.

On July 13, I wrote a memo to Ed Barsis, Mike McGlaun, and members of the team. Subject: "Apojove Meeting: What should we do about the impact of comet Shoemaker-Levy 1993e?"

> Comet 1993e was discovered last March, and is predicted to impact Jupiter. Its orbit has been refined over the last month, and it is now widely believed the impact will take place in July, 1994. The orbital period is about two years - it was captured by Jupiter last July. It will reach apojove, the high point in its highly eccentric orbit tomorrow, July 14, and will spend the next year accelerating up to its impact velocity of 60 km/s. At that velocity, the largest fragment will have a kinetic energy equivalent of about one billion megatons of TNT, releasing an order of magnitude more energy than the K/T dinosaur-killer. The resulting flash will overwhelm all objects in the sky except the sun, and should be visible in the daytime sky.
>
> We are well-positioned to make a major contribution toward both computational predictions and observations of the event. It is important that we get an early start and that we come up with a unified plan. Let's have an 'apojove' meeting tomorrow, so we can begin moving in the right direction the same day that the comet does. Time: 9:00 am, Wednesday, July 14, 1430 conference room (880/C-35C).

This was not an exaggeration, based on what we knew at the time. Planetary scientist Clark Chapman had written in the News and Views section of *Nature*, on June 10[20]:

> John Lewis estimates that a fireball formed perhaps 1,000 km below Jupiter's cloud deck might ascend through the clouds in an awesome display. If the comet fragments were to hit the front side of Jupiter, the explosions would be visible from Earth in broad daylight as the separate objects followed each other down into Jupiter's clouds.

Ed Barsis gave us the green light. Our goals were to test the machine with a big problem, use our results to make recommendations to astronomers for observations the following summer, and then use the observations as a validation opportunity for our models.

It wasn't long, however, before orbital dynamicists had enough data to calculate the point of impact. The collision was going to be on Jupiter's *back side* and not directly visible from Earth, as we had all hoped. At a comet "Pre-Crash Bash" on August 23–24, 1993, at the Lunar and Planetary Laboratory in Tucson, 120 scientists met for a brainstorming workshop. Despite the disappointment about no direct observations from Earth, researchers expressed the hope that it would be observed

---

[20] C.R. Chapman, "Comet on Target for Jupiter," *Nature* **363**, pp. 492–493 (1993), reprinted with permission from Macmillan Publishers Ltd.

by the Galileo spacecraft, on its way to Jupiter at the time of impact. Maybe, if we were lucky, we'd see evidence of the impact flash reflected from Jupiter's moons.

Three groups presented preliminary computational simulations, including Dave Crawford. According to a summary write-up, dated September 20, 1993, by workshop organizer Jay Melosh,[21]

> Kevin Zahnle described his work on energy deposition of entering comets, following up on his model originally developed to model airblasts from meteorites entering Venus' atmosphere and the 1908 Tunguska event on Earth. He argued that the projectile would break up as it entered, thus increasing its drag and depositing its energy in a relatively small altitude interval. The energy deposited would then heat the surrounding gas, which would expand upward in a classic fireball. For km size projectiles, most of this energy deposition takes place about 200 km below the upper cloud decks, and so is not directly visible, but the fireball should become visible when it rises back up into the upper atmosphere, about 1 minute later. He argued that ablation is not important for large projectiles, although what 'large' meant was debated by the participants, and he reported that a check by Chris Chyba showed that his results changed somewhat when a different ablation model is used. Temperatures in the shock wave ahead of the entering projectile were estimated to reach 30,000° during the high speed part of the entry. Zahnle closed his talk by showing a video of a numerical computation of the expansion of a fireball in a stratified atmosphere modeling that of Jupiter.
>
> Several participants showed results from numerical computations currently in progress. Tom Ahrens presented a computation using an SPH hydrocode result that followed the entry of a 10 km diameter projectile. He and T. Takata found that the projectile penetrated relatively deeply, depositing most of its energy 500 km below the 1-bar level. The impact generated a 6 km/sec plume and partitioned about 70 % of its energy to atmospheric heat. David Crawford described similar computations that use the Sandia hydrocode CTH, although his work mainly focused on the projectile. It was clear that several groups are ready to do hydrocode computations of the comet entry, but due to the need for lengthy supercomputer runs none has yet encompassed the entire impact process. A major point of interest was whether any atmospheric gasses would be ejected at high enough speed to enter space around Jupiter and perhaps affect the radiation belts or magnetosphere. None of the computations presented showed such high speed jets, but it was not clear that any of them have been carried out with sufficient accuracy to be sure of their absence. Other issues were the importance of thermal radiation (not included in any of the hydrocodes) and the need for an accurate equation of state.

By October 1993 Tim Trucano had completed the first two-dimensional simulations, showing how the comet fragments would break up as they entered Jupiter's atmosphere. Dave and I got permission to attend and present Sandia's work at the October 18 meeting of the American Astronomical Society's Division of Planetary Science meeting in Boulder. At the request of the AAS (American Astronomical Society) press officer, I drafted a press release:

> Sandia National Laboratories researchers have performed supercomputer simulations to find out what will happen when Comet Shoemaker-Levy 9 collides with Jupiter next summer. Some astronomers have predicted that the impact will be one of the most spectacular celestial events ever witnessed. Telescopes and spacecraft are being rescheduled to observe the resulting display, but what will they see? To help answer that question, the Sandia scientists have made use of a computer code of the type that was originally developed to under-

[21] Reprinted with permission from Jay Melosh; see summary link, http://www.surveyor.in-berlin.de/himmel/SL-9/Jupiter-SL9.txt.

> stand what happens within nuclear weapons. The computer program can simulate the extreme states of matter that result from energetic events such as explosions and hypervelocity impacts. In recent years, it has been used to help design shields to protect satellites from collisions with orbiting space junk. Now it has been used to study the collision of the ultimate piece of space junk: a 3 km diameter body of ice moving at 60 kilometers per second.
>
> The impact will be similar in size to the one 65 million years ago that left a 180-km diameter crater in Mexico and is thought to have wiped out the dinosaurs. For that asteroid-earth impact, the atmosphere had almost no effect on the asteroid. When the object slammed into the ground the pressure increased instantaneously, and shock waves converted the kinetic energy to mechanical and thermal energy, resulting in an explosion a million times as big as the biggest nuclear test. However, Jupiter does not have a solid surface, so next year's event will be very different in character. Jupiter's hydrogen/helium atmosphere increases in pressure gradually over a few hundred kilometers from the vacuum of space to the equivalent of hundreds of earth atmospheres.
>
> The computer simulations show that when the comet enters Jupiter's atmosphere, the pressure increase is gradual. The atmosphere is thin at the top, so for the first second or so, the comet slices through almost unhindered. In the next second, however, the pressure builds up rapidly and deforms the comet until it begins to break up. At the time the comet starts to disintegrate, it has lost less than two percent of its kinetic energy. This means that as much as 98 percent of the comet's energy is carried to a depth of more than 120 km, where it is catastrophically released.
>
> After the comet begins to break apart, energy is released very rapidly by mechanisms that are not well understood. The Sandia calculations are concerned mainly with the processes that lead to the breakup. One of the surprising results is that, for a collision like the one expected next summer, the comet is torn apart by the large deformations due to atmospheric drag. Other models have previously suggested mechanical wave interactions within the comet body lead to states of tension that exceed the strength of the comet, causing it to break apart.

Only two other competing groups were doing supercomputer simulations. One was led by my PhD advisor, Tom Ahrens, and the other consisted of partners Kevin Zahnle and Mordecai-Mark MacLow. They also issued press releases and were more explicit in their efforts to predict something that would be observable to astronomers. According to Ahrens, the plume from a 10-km-diameter cometary impact would be ten thousand times as bright as Jupiter and last several minutes. Unfortunately, it would cool down by the time it rotated into the view of Earth-based telescopes and only be visible in the infrared. Zahnle and MacLow also predicted a bright fireball, as hot as the Sun, rising above the cloud tops on Jupiter's far side. They pointed out that it would be visible to the NASA probes Galileo and Voyager 2 and that it might even brighten Jupiter's moons enough to be detected from Earth.

At the end of the year, we were still focused on the entry problem. How deeply would the fragments penetrate before they exploded? In December, we told *Sandia Science News* that as much as 98 % of the energy of the largest pieces of the original comet would be carried beneath Jupiter's clouds, where it would be explosively released. This claim eventually led to the major point of contention among the modeling groups, with Zahnle and MacLow claiming that the fragments would explode much higher in the atmosphere than we said they would. However, they were using a code with very limited equation of state options, so their comet fragments expanded during the descent. Their fragments were therefore larger and had more drag than ours. Because we had a better equation of state, our comet fragments remained mostly in the solid phase until they vaporized beneath the clouds.

The *Sandia Science News* article went on to say:

> The current results will give atmospheric scientists a starting point from which to determine whether the collisions will make giant mushroom clouds, create a new red spot on the planet, or be swallowed up without a trace. It will also help astronomers know what to expect to see.[22]

But unlike the other groups, we still hadn't made any specific predictions of what might be observed. Our disagreement with the other group about the depth of penetration was not going to be settled because no measurement could unambiguously make that determination.

In early 1994, Dave had run 3-D simulations for various assumed fragment sizes on the Paragon, showing that the fireballs would be ejected at much higher speeds and to higher altitudes than the other groups had predicted. We also showed that, for an inclined impact (the fragment entry angle would be 45° from the local horizon), the plume would be ejected along the wake—at the same angle but in the opposite direction. Because we had access to the most powerful computer in the world, we had the advantage of higher resolution and three dimensionality, both of which were required to make these predictions.

Unfortunately, like penetration depth, there was no apparent way to determine plume height or angle for something that was to happen on the back side of Jupiter. That is, until the predicted impact location started to change.

A convergence of technologies in 1994 enabled the scientific community to make the most out of this event. It was timed perfectly. First, the Intel Paragon allowed us to model the event with sufficient fidelity to make useful predictions. Second, the Hubble Space Telescope, which had just been repaired, was producing exceptionally sharp images. Third, and more easily forgotten, was the role of the Internet that was just beginning to connect various research institutions, allowing rapid dissemination of information.

The University of Maryland had set up an electronic bulletin board where I was able to keep up with the latest developments. That's how I found out quickly that on February 1, Jet Propulsion Laboratory had revised their orbit and impact predictions. Here's what I emailed to the Sandia modeling team:

> Attached are the latest impact predictions for the comet train, posted to the bulletin board last week. According to the table a few of the main fragments are predicted to hit about 4 ½ degrees past the limb. With the 1.7 degree uncertainty given, they could possibly hit within 3 degrees, making a direct observation of the fireball possible (and it will be rotating toward us). Even better, at Gene Shoemaker's talk in Houston on Saturday, he showed images from the repaired Hubble Space Telescope. There are several smaller fragments that are OFF the axis of the train, i.e.[,] in somewhat different orbits. Gene said orbits had not been calculated for these off-axis fragments, and [he] would not speculate about whether they might hit on this side of the limb, but in my opinion there is still hope for directly viewing a smaller hit.

From that moment on, our primary focus was on the 2 % of the impact energy that would contribute to the potentially observable fireballs. The projected impact points continued to migrate closer to the limb. We worked hard to get the word out,

[22] *Sandia Science News*, vol. 28 (December 1993), Sandia National Laboratories, second page of unnumbered four-page article, reprinted with permission from Sandia National Laboratories.

but we were still considered outsiders in the planetary astronomy community and were not yet on the "A" list for national science journalists. On March 19, 1994, local reporter Lawrence Spohn of the now-defunct *Albuquerque Tribune* wrote an article entitled "Sandia computer plots Jupiter impact," in which Dave was quoted:[23]

> Depending on how close they are to Jupiter's horizon — as viewed from the Earth — the fireballs, or portions of them, actually might be visible through special telescopes on Earth. 'In fact some of the fragments will not disappear behind the planet until they have already entered the thin upper atmosphere (about 250 miles) above the cloud,' Crawford said. 'Less than a minute later, the top of the hot fireball will rise back into view[,]' he predicted, although the Sandia team cautioned that still may not be bright enough to be seen from Earth.

This was the first publication of our prediction. It's always nice to get local press, but we quickly realized that the *Albuquerque Tribune* was not going to help to (1) get our prediction to the astronomers who were finalizing plans for their observational campaign and (2) document our prediction in a way that we would get full credit for it as a *scientific* prediction.

To help take care of the first issue, Dave and I prepared a poster for presentation at the spring meeting of the American Geophysical Union (AGU) and I traveled to Baltimore to present it. I was delighted when Heidi Hammel stopped to talk to me. She was a principal research scientist at the Massachusetts Institute of Technology who was the member of the Science Observation Team for the Hubble Space Telescope Jupiter Campaign in charge of the Wide Field Planetary Camera. She recently described her memory of the event in a telephone conversation with Mark:[24]

> I remember going to the AGU meeting, which had a special session about predictions for SL9. Mark and Dave had a poster predicting these giant impact plumes. Mark said we had to be sure to image the limb of Jupiter with Hubble, because those plumes would be so high that the tops of them would be visible from Earth, even though the impacts would happen just out of view. I was pretty skeptical, since my expectation was that these comet fragments would vanish without a trace into Jupiter. But, nevertheless, who knew? So we went ahead and set up an imaging sequence for Jupiter's limb for the A impact (the first one expected) as well as limb-imaging sequences for the impacts of the brighter (and thus probably larger) fragments.

Now we had real skin in the game, but we wanted to take it further and go on record in the scientific literature. We knew we had to get a rapid publication out *before* the impact. We submitted a paper to AGU's *Geophysical Research Letters* (GRL) on June 1, in which we said in the abstract, "For sufficiently large fragments, impact-generated fireballs will rise into line-of-sight over the Jovian limb (less than 1 min after impact for a 3-km diameter fragment)."[25] Our paper was accepted 10 days later and published in the July 1 issue, an unprecedented turnaround time for that journal, according to the editor.

---

[23] L. Spohn, "Sandia computer plots Jupiter impact," *Albuquerque Tribune*, March 19, 1994, p. A-5. Quote is reprinted with permission of David A. Crawford, Sandia National laboratories, in email to J. Asay, March 7, 2015.

[24] Private communication from Heidi Hammel to Mark Boslough, 2014.

[25] M.B. Boslough, D.A. Crawford, A.C. Robinson, and T.G. Trucano, "Mass and penetration depth of Shoemaker-Levy 9 fragments from time-resolved photometry," *Geophysical Research Letters* **21**, 1555 (1994).

We weren't the only ones publishing specific predictions. Tom Ahrens and his team made essentially the same prediction in the same July 1 issue of GRL, saying in the abstract, "Earth-based observers can detect these plumes as these expand over the SW limb of Jupiter and come into Earth view some minutes after impact!"[26] Exclamation points are unusual in scientific literature!

Not all scientists were as confident as we were that there would be fireworks. On the same day as the GRL issue, science writer Dick Kerr published a news article in the July 1 issue of *Science* on pages 31–32 entitled, "Bets Range from Boom to Bust for Jovian Impacts." He wrote:[27]

> ... astronomers may get a view of fireballs rising into view from the impact sites on the planet's backside, waves rippling across the Jovian clouds, and internal gases spewing out. Alternatively, the collision could simply fizzle—in full view of every telescope in the world from modest amateur rigs to the Hubble Space Telescope.

But Kerr also cited a study suggesting that the largest fragment would only be about a half km in diameter, much smaller than the ones we modeled:

> One-kilometer impacters, according to most studies, would be enough to produce fireballs rising into view from Jupiter's far side, dredge up bountiful amounts of Jupiter's interior, and send easily observed ripples around the planet. At half a kilometer, no one is making any promises.[28]

Fortunately, we had included a size caveat in our prediction.

Kerr closed by reminding readers that predictions based on computer models depend on assumptions: "Computer simulations have yet to agree on how deeply a 1-km sphere of ice would penetrate, whether the impacter would ultimately explode, and how high the fireball might rise."[29]

We had selected GRL, a subscription-only journal, for our formal prediction paper, but we decided that we wanted our predictions to be more widely known. So we wrote it up for *Eos*, the newspaper-like Transactions of the American Geophysical Union that gets mailed to every member of the organization. Wanting to get attention, we needed a catchier title. We called our paper "Watching for Fireballs on Jupiter" and timed it perfectly. It appeared in the July 5 issue and showed up in members' mailboxes only a few days before the first impact, fragment A, which collided with Jupiter on July 16.

Only one prediction paper followed our last one. It was in the high-profile journal *Nature*, written by comet expert Paul Weissman. It was called "The big fizzle is coming." Predictions are always more meaningful when there are disagreements.

---

[26] T.J. Ahrens, T. Takata, J.D. O'Keefe, G.S. Orton, "Radiative signatures from impact of comet Shoemaker-Levy 9 on Jupiter," *Geophysical Research Letters* **21**, 1551 (1994).

[27] R. Kerr, "Bets Range from Boom to Bust for Jovian Impacts," *Science* **65**, p. 31 (1994), reprinted with permission from American Association for the Advancement of Science.

[28] *Ibid.*, p. 31, reprinted with permission from American Association for the Advancement of Science.

[29] *Ibid.*, p. 32, reprinted with permission from American Association for the Advancement of Science.

In the final days before impact week, Dave and I were asked to do many local television interviews. It was trial by fire, as neither one of us had any prior TV experience. We also decided to put our money where our mouths were. We both bought amateur 8-inch telescopes and plane tickets to Hawaii, where Jupiter would be high in the night sky for many of the impacts. We would be watching the planet at night and reading the University of Maryland bulletin board in the daytime.

My most memorable experience was watching Jupiter from the top of Haleakala as one of the largest fragments collided. We didn't see our predicted plume. It just wasn't bright enough to see with our little telescopes and our eyes that weren't sensitive to infrared radiation. But we saw the remarkable aftermath firsthand, as the dark spot from the collapsed plume rotated into our view.

Heidi Hammel recalls the excitement at the Space Telescope Science Institute:[30]

> When those first images came in, which would show the A plume, we saw a bright spot right on the limb. But none of us really believed it at first. 'It's got to be a moon,' I said. 'Probably Io, since it is so bright. Where's the astronomical almanac? Someone check for the location of Io.' But when we found that no moons were predicted to be at that location, we started to get excited. All of us crowded in around the screen to wait for the next images, which would be when the impact site itself rotated into view (about an hour later). The iconic image of astronomers crowded around the screen — and the video of champagne bottles — followed, as we got our first glimpse of big black impact sites on Jupiter. We subsequently put together a time lapse set of Hubble images showing the G plume's rise and descent. They were so eerily like Sandia's predicted models that I showed them side by side for years afterwards.

When we returned from Hawaii, our work wasn't over. Many of the observed phenomena had not been anticipated by us or by anyone else. We now had a wealth of data that still needed to be explained and the answers weren't immediately obvious.

Much to our annoyance, planetary scientist Clark Chapman (now a close colleague and friend) downplayed our predictions in his News and Views piece for *Nature* magazine, called "Dazzling demise of a comet," in which he said:[31]

> As the crash approached, comet researchers hesitated to predict more than a modest display — in Paul Weissman's phrase, a 'cosmic fizzle.'... The spectacular show during the subsequent week exceeded even the most optimistic predictions about the event's potential visibility from the Earth, nearly 800 million kilometres away. No one doubted that significant events would occur at the impact sites around Jupiter's limb, hidden from our sight. But the prominent visibility of high plumes, hotspots and enormous long-lasting dark splotches on the face of Jupiter was wholly unexpected. What does it mean?
>
> One of the most impressive and unexpected aspects of the comet impacts was the dramatic plumes that reached more than 2000 km above Jupiter's cloud tops, clearly depicted in pictures by the Hubble Space Telescope. Some numerical modellers had tentatively hoped that one or two of the later impacts, which occurred much closer to the Earth-facing side of Jupiter, might manage to peek above Jupiter's limb before thinning into invisibility... Indeed the effects are so large, even for the largest estimated sizes of the comet fragments, that it seems likely that our understanding of atmospheric impact physics will have to be revised to account for the amazing phenomena that have been seen....

---

[30] Heidi Hammel email communication to Mark Boslough, Jan. 25, 2015.

[31] C.R. Chapman, "Dazzling demise of a comet," *Nature* **370**, pp. 245–246 (28 July 1994), reprinted with permission from Macmillan Publishers Ltd.

The scientific community's work was now cut out for it to put together a comprehensive model. At Lalit Chhablildas' suggestion, I organized the first post-impact conference session at the Hypervelocity Impact Symposium in Santa Fe later that summer. We called it "Comet Day," and it was attended by Gene and Carolyn Shoemaker, as well as by many of the observational astronomers and computational modelers who had made predictions. Miles O'Brien of CNN attended and interviewed me—my first time on national television.

Over the next year, we ran more simulations and published our paper, "Numerical modeling of Shoemaker-Levy 9 impacts as a framework for interpreting observations," in the July 1, 1995 issue of GRL. Our principal conclusions were stated in the abstract:[32]

> Among the observations that have already been at least partially explained in a way that appears to be consistent with the impact models are: the sources and timings of multiple flashes observed from Earth, the temperatures and durations of the single flashes observed from the Galileo spacecraft, and the asymmetry of the plumes and ejecta patterns observed by the Hubble Space Telescope. Further modeling subsequent to the impacts has shown that (contrary to our pre-impact expectations) fireball trajectory data do not provide strong constraints on either fragment mass or maximum penetration depth.... After more data become available and correlated, and more simulations are performed, we expect that fragment size estimates will become more precise.

Dave Crawford began working on a definitive estimate of the size of the comet, which he published in a paper[33] "Comet Shoemaker-Levy 9 fragment size estimates: How big was the parent body?" He concluded that the biggest fragments were greater than 1 km in diameter, but of low density. The original comet, before it broke up, was probably about 1.4 km in diameter. The fact that our plume sizes and velocities were underestimated had a simple explanation. Even with the most powerful computer in the world at the time, our simulations lacked sufficient resolution.

While Dave was working on size estimates, I turned my attention back to the Earth. A week after the impact, Congress had changed NASA's authorization bill to require the agency to conduct surveys for threatening asteroids and comets. Regarding planetary defense, David Levy put it this way, "Shoemaker-Levy 9 killed off the giggle factor."[34] There was also increasing interest within the national labs, and, thanks in part to the efforts of Sandian Dick Spalding, the Department of Defense began releasing satellite data on large "superbolides" in the Earth's atmosphere including a massive explosion over the south Pacific on Feb. 1 of that year.

Even though our simulations were not fully resolved, we noticed something about them that had implications for Earth impacts. In our animations, the eye is drawn to the ballistic fireballs rocketing upward into space, but in terms of energy, most of the

---

[32] M.B. Boslough, D.A. Crawford, T.G. Trucano, A.C. Robinson, "Numerical modeling of Shoemaker-Levy 9 impacts as a framework for interpreting observations," *Geophysical Research Letters* **22**, pp. 1821–1824 (1995), reprinted with permission of John Wiley and Sons, copyright 2012.

[33] See *Annals of the New York Academy of Sciences*, Vol. 822, pp. 155–173 (May 1997).

[34] C.R. Chapman, in *The Great Comet Crash: The Impact of Comet Shoemaker-Levy 9 on Jupiter*, ed. by John R. Spencer and Jacqueline Mitton, p. 105, Cambridge University Press, Cambridge, UK, 1995), reprinted with permission of Cambridge University Press.

action is at the bottom, where the comet exploded. Even after the explosion, the vaporized comet continued to move downward. Was that also true for explosions in the Earth's atmosphere?

I started looking at the history of airbursts in the Earth's atmosphere, the most famous being the 1908 Tunguska explosion in Siberia. Another one was at Sikhote-Alin, also in Siberia, in 1947. It had been witnessed by a painter, who rendered it on canvas. It ended up on a 1957 Soviet postage stamp that had been reprinted in various publications on meteorites. It showed an unusual cloud that looked remarkably like our plume model.

I started running simulations of small impacts into the Earth's atmosphere on my desktop workstation, but for lack of computational horsepower, they were limited to two-dimensional axial symmetry. Nevertheless, I convinced myself of two things: small impacts like Tunguska (1) can produce plumes that rise hundreds of kilometers into space, potentially endangering satellites in low-Earth orbit, and (2) can produce a jet of vapor that continues to move downward under its own momentum, causing more damage at the surface than a nuclear explosion of the same yield.

Interest in the impact threat continued to grow after 1994. In April 1995, Dave and I were invited to present papers at the United Nations International Conference on Near-Earth Objects, organized by John Remo of Harvard University and Sandian Bill Tedeschi. The following month, Lawrence Livermore National Laboratory held a meeting on planetary defense, where I first presented my conclusions about plume-forming impacts on the Earth. Edward Teller was in the front row with his eyes closed, but Gene Shoemaker was attentive (but skeptical).

The following year I extended my Tunguska model in an effort to explain the mysterious yellow-green silica glass found only in the Libyan Desert of western Egypt, publishing only an abstract associated with a meeting on that subject in Bologna, Italy. I tried to tie everything together in a single, final paper for the proceedings of the United Nations conference, called "Shoemaker-Levy 9 and Plume-forming Collisions on Earth."[35] John Remo was the editor, and he arranged to have our simulation, along with the 1957 postage stamp showing the apparent Sikhote-Alin plume, on the cover of the book. That would be my last word on the subject for the next 8 years as I turned my attention to other work.

## Postscript

In the fall of 2005, I got a call from a British filmmaker. She wanted to know if I could do some simulations in support of a documentary she was producing for the BBC (British Broadcasting Corporation) and *National Geographic* about the Libyan Desert Glass. During my 8 years away from the field, a lot had happened. Both the code and Sandia's computers had become much more powerful. Dave had added

[35] See *Near-Earth Objects, The United Nations International Conference*, edited by John L. Remo (Annals of the New York Academy of Sciences, New York, NY), Vol. 822, pp. 236–282 (1997).

adaptive mesh refinement, and we had a new supercomputer called Red Storm. There would be no issues with resolution.

I was ultimately invited to join an expedition to the Great Sand Sea. With new LDRD funding, I was able to run simulations that resurrected interest in planetary airbursts (perhaps too much uncritical interest as well, I later discovered). My documentary appearance led to more invitations, including a 2008 expedition to ground zero of the Tunguska explosion on the centennial anniversary. In 2009, I was invited to be the "token skeptic" for an episode of NOVA focused on a group that, in my view, had taken my airburst models too far. Inspired by an incorrect animation of one of my simulations for the Libyan Desert Glass documentary, they had proposed that a giant comet explosion had changed the climate and wiped out the North American megabeasts and Clovis Culture 12,900 years ago.

I presented my airburst models at Planetary Defense Conferences every 2 years, arguing that the risk was greater than previously acknowledged. Then, on Valentine's Day evening in 2013 (in our time zone), a half-megaton airburst occurred in Chelyabinsk, Russia. Within a few days, I got a call from the producers of NOVA: Did I want to go? It turned out to not be a plume-forming event, because the asteroid entered the atmosphere at a very shallow angle. But because of the huge quantity of data from mobile phones and dashboard videos, my colleagues were able to determine the trajectory and energy deposition rate precisely in a way that I could use to initialize and validate a high-fidelity model. I was also able to bring back some meteorites for analysis.

A couple months later, we held our biannual Planetary Defense Conference in Flagstaff, Arizona, including a hastily organized technical session about Chelyabinsk. In attendance was Carolyn Shoemaker, whom we honored for her lifetime achievements at our banquet. We presented her with a pendant made from one of the meteorites I had brought back from Russia.

I also gave a presentation about an idea I had been considering for years. If airbursts generate upward plumes, an equal but opposite force must be pushing downward against a planet. I had concluded (in contradiction to opinions that have prevailed for nearly two decades) that the observed wave in Jupiter's atmosphere was generated by this reaction force. And if that happened on Jupiter, then it could be argued that such a plume-forming airburst could create a similar wave at a fluid interface on Earth: a tsunami!

While I was preparing my presentation, I thought I'd run the idea by Heidi Hammel, who had been my friend and pen pal ever since we first met that spring in Baltimore. She was pleased that I was still interested in her 1995 paper that showed images of the observed waves, and she made an observation about our favorite comet: "Shoemaker-Levy 9 is the gift that keeps on giving."[36]

---

[36] Private communication by email from Heidi Hammel to Mark Boslough, 2014.

# Barry M. Butcher[37]

# (1962–1997)

My first experience with Sandia probably goes back to my interview trip. Professor Ham Maybe, a former professor at Cornell, roamed the halls of Thurston Hall at Cornell recruiting for Sandia. Dick Claassen was also involved. They finally trapped me, and an interview trip was planned for November 1961 when I started the long trip to Albuquerque from Ithaca. I remember taking the Electra plane, probably from Rochester to Chicago. It was one of the smoothest prop planes in the sky, but also one of the noisiest. From Chicago, I went to Kansas City on the first TWA jet I had experienced and then way out into the unknown to Albuquerque. I was housed at the old Sundowner Motel on Central. The interview was uneventful; I remember the names George Anderson, Reid Holland, Frank Neilson, and others.

Returning to Cornell, I got a telephone call several weeks later from Don Lundergan. Don explained that they had somehow missed me during my visit to Albuquerque and would like to talk to me about a new group at Sandia involving shock wave research. Could I possibly come to New York City and meet with them before any decisions were made? I agreed to this and must have flown to New York on Mohawk, which was an experience in itself in those days. We met in a room at a hotel, I believe, on 10th Avenue and probably 42nd Street up on the 9th or 10th floor. Don was there, and a gentleman named Lynn Barker, who was currently a student at Columbia, sponsored by Sandia. Don explained that the group was just being formed. Later, I received an offer and agreed to go to Sandia, to get "industrial experience." We would plan to stay only a few years and then move on to a teaching position. The culture at Cornell at that time was extremely academic, and any deviation into industry was considered of lesser value. I was supported, however, in my choice by Dale Corson, dean of the Engineering School, who had been at Sandia during the very early days. He called me into his office and told me he was pleased with my choice.

### 1962–1968

- Staff member specializing in the mechanics and physics of shock and stress wave propagation
- Experimental stress and shock wave materials testing (Gas Gun Technology)
- Modeling expert in the areas of dynamic fracturing (spallation) and porous material compaction
- Development of a finite-difference stress and shock wave propagation code

**Hooray!** The clearance had arrived. Initially my office cubicle was in Building 806 on the second floor. The cubicle contained a government gray desk and had AT&T (Western Electric) light brown partitions. I shared it with an office mate, Jack Cannon. Others in the group were Bill Hartman and, eventually, Lynn Barker when

[37] Editors' note: Barry's active involvement with shock wave research ended around 1979.

he returned from Columbia. Charles Karnes joined the group sometime later. In the lab were Red Hollenbach (in charge), Tom Looby, Wayne Brookshire, Bob Newman, and Bob Simpson. The managers were Charlie Bild (Vice President), Lew Berry (Director), Bob Sowell (Department Manager), and Don Lundergan (Division Supervisor). Darrell Munson was the Section Leader. (The Section Leader position was abolished some time later.) Charlie Bild was the visionary in the organization, which was concerned with the materials part of Sandia's mission.

Our initial assignment was to investigate stress wave propagation in materials in order to provide the material property information to determine how damage would occur. Sandia's interest was that weapon systems were becoming more and more complex as the trend toward miniaturization happened. The initial systems had been thrown together in haste without concern about size. Operation was sequential, with each component finishing its job before the next one started. However, size and timing were very important with regard to weapon delivery on airplanes and even by artillery shells, and some components would have to function at the same time. Now the challenge was to make things manageable.

Coupled with miniaturization, another factor was the energy used to power weapon components. These components consisted of all the gyros, triggers, radar, etc. needed to make the system work. All required power, either in electrical form or by some sort of mechanical actuation. At the same time, the systems would be stored for long periods of time, essentially in a dormant state. This requirement made energy produced by explosives very attractive. Explosives could be stored for long periods and then detonated to produce plenty of power. But, at the same time, if a "firecracker" went off next to an important component, the last thing that should happen would be that the "minor explosion" would destroy the component before it had a chance to perform its function. On the other hand, detonated explosives cause damage by stress (pressure) waves that propagate out from the bang. Hence, our mission was to support knowledge of exactly how stress waves propagate through various media, specifically solids. Stress wave propagation in liquids (waves) was well understood, but the weapon systems were made of solids, rather than liquids. Small elastic stress wave motion in solids was also well understood (sound waves), but when an explosion went "barroom" and threw things long distances, causing extreme deformation and fracture, that was something else. Solid material properties under extreme conditions were needed to predict such phenomena.

The research procedure was to cause impacts and observe their consequences. An experiment had a simple one-dimensional configuration, with one flat plate, usually metal, impelled at a high velocity against another flat metal plate, hitting it as simultaneously as possible with the back surface of the target instrumented in some manner to observe when and how it started to move. If the impact was sufficiently slow, the result would be simply elastic, as with two billiard balls colliding, with the initially moving one reduced in velocity, while the impacted one flew off at a velocity dependent on their relative masses. The impact was much greater than for two billiard balls colliding, however, with permanent deformation and fracture of the plates after impact. The mechanism for such experiments was an air gun, approximately 4 inches in internal bore and 12–14 ft long. The impacting plate was the front of a cylindrical plastic projectile, approximately 6–10 inches in length, which was drawn from the front of the barrel back into a position next to the air cylinder, which was the breech of the gun. The experimental plate was then mounted on the other end of the barrel. To fire the gun, the breech chamber was pressurized with air at up to 1000 psi, depending on the projectile velocity desired. The breech air was then dumped into the barrel through a valve and the projectile was shot down the barrel, impacting the target. Projectile velocities were limited by the speed of sound at our altitude, but could be as much as 1000 ft/s.

The consequence of the impact on the target plate was observed by several ingenious methods. First, several electrically isolated thin wires were located at a measured separation. As the projectile came down the barrel, the wires (at different voltage levels) were shorted out. Measurement of the time change of voltage with an oscilloscope during impact could then be used to determine the velocity of the projectile as it shorted out the pins. The simultaneity of impact was measured in a similar manner. Four pins, their tips mounted flush with the impact surface and at different voltages, were set at the quadrant points around the periphery of the impact area. As impact occurred, the pins were shorted out and the results recorded on the oscilloscope. If the impact were perfect (i.e., all points of the impact surface struck simultaneously), all "flush" pins would short out at the same time. This was never the case.

The real consequence of the impact, the motion of the free surface of the plate opposite to the surface that was impacted, was the information we needed. Initially, this motion was measured by a resistive "slant wire" with one end touching the rear surface of the target and at a very slight angle with the surface. As the surface started to move outward because of the reflection of the shock wave from it, the wire would be shorted out. The angle of the wire was such that the effect of the surface closure would occur *before* a disturbance propagated along the wire. By recording on an oscilloscope the change in the voltage across the high-resistance wire by the part of the surface touching the wire, the motion of the surface could be photographed. Later, Lynn Barker devised much more sophisticated means of recording surface motion using interferometer techniques. In these techniques, a laser beam was shone against a shiny spot on the surface and the motion was recorded with an interferometer. None of this would have been possible were it not for the invention of the laser with its single wavelength.

Initially, tests were done at a facility in Area III amidst the spiders and rattlesnakes. Guns had previously been used for impact experiments, but with the exception of Bob Graham's facility, these were rudimentary in terms of our air gun. The control room was full of state-of-the-art oscilloscopes and counters. The data were recorded with Polaroid® film, by taking pictures of the single sweeps of the scopes triggered by impact. The film was then digitized with a rather costly digitizer and the results, as I remember, were printed out on IBM punch cards. We shot the experimental sample into a sawdust pile and then dug to recover it.

Later, I guess management took pity on us and found an old Quonset hut, Building 855, inside Tech Area I, next to the shop building. It was a relief to only have to walk across the area from Building 806 to get to the lab. There, a new facility was constructed, which eventually consisted of two guns and catcher tanks, rather than the sawdust pile. One of the guns was used for materials that were likely to cause contamination, such as beryllium. Cliff Witten did many experiments on this gun, which involved suiting up in full protective gear during recovery of the samples. Much time was spent with the Health Physics organization to make sure no contamination escaped.

The Physical Research Department within Organization 5100[38] was our competitor in this work. Bob Graham had built one of the first compressed gas guns, and he was heavily involved in the Sandia Quartz Gauge, a quartz crystal device that produced a measurable signal when shocked. The charter for this group was much different than ours, however. Under Frank Neilson (the visionary) and George Anderson, with the experimental work conducted by Bill Benedick, their mission involved research on explosively-driven electrical signal technology that could be used for weapon devices. This was considered to be legitimate "physics" rather than the "mechanical" work we were doing in the Materials and Process Directorate. Nevertheless, some overlap existed with our efforts, which promoted competition. Likewise, scientists in GMX-6 at Los Alamos, including Bob McQueen and others, were interested in material response under very high pressure shock-type loadings. Their interest, however, was focused on the ultrahigh pressures at which materials essentially behaved as fluids with no strength, principally to characterize elements of the periodic table. This information was used to estimate compressibility, assuming simple mixture theory, for nuclear weapons calculations.

For my part, I was soon asked to specialize. A phenomenon known as spall was of concern. Spall occurs when a compressive shock wave is reflected from a free surface. As the wave travels back into the material and interacts with a rarefaction wave, it grows in tension until it is sufficiently strong to fracture the material. The consequence is that the material between the surface and the fracture flies off at a high velocity. This is the same phenomenon as when a marble hits two inline marbles at rest. The impacting marble stops and the marble farthest away flies off at the impact velocity. The possibility of spall in systems activated by explosive energy is rather unsettling to designers because fast-moving debris could fly through space and impact special parts before they have a chance to function.

Not too long after my arrival, the decision was made to develop a computational capability. Walt Herrmann, on the faculty at the Massachusetts Institute of Technology (MIT) at that time, had developed a computer code for stress wave calculations. Our mission was to interest him in coming to Sandia. As I recall, he came to visit in the winter, because he wanted to do some skiing. Having done so, he sustained a leg injury during his time on the slopes. (I seem to recall that something was broken.) Hence, he had to lengthen his visit to Albuquerque and rest in a hotel room until he was able to travel. What a recruiting opportunity! We had a captive audience who could not go anywhere. I guess we did not do anything terribly wrong, for he decided to come to Sandia and head up a computational division.

[38] Editors' note: The title of Organization 5100 in the mid 1960s was Physical Research.

Energy absorption by porous materials of metal and plastic (i.e., foam) was another area I was asked to investigate. Energy is dissipated as a fragment moves through the foam and the foam is severely deformed. Actually, in an explosive environment, an awful lot of foam is needed to slow down the rapidly moving fragments, but the presence of the foam is certainly better than no foam at all since it helps keep things in place. Foam is also useful when very high energy from x rays is deposited within it. With foam, the vaporized material has room to expand rather than being pushed like a piston. Some of my work concerned this phenomenon. With underground testing in progress during this time, I evaluated the effectiveness of this material response on several underground nuclear tests to see if we could duplicate the results mathematically. Those tests were indeed big physics, and I am still overwhelmed by them.

## Underground Testing Is a Story in Itself

During this period, our Division was asked to move into testing somewhat exotic materials, whose purpose was to protect heat shields on missiles designed to counteract a Russian nuclear attack. Tommy Guess and Larry Lee were hired for this work. The materials were developed to minimize radiation damage in the form of stresses induced by radiation deposition, in the event of a nuclear detonation. They were low absorbers such as graphite and beryllium. The purpose of the second gun in Building 855 was specifically to test these materials. Much of the information we obtained was exchanged with the British at the time.

This brings us to the subject of computers. When I left Cornell, computers were the furthest from anyone's minds. The slide rule was still king. So I naturally assumed that the computer state of the art was a curiosity. When I arrived at Sandia, I was introduced to the CDC 1604[39] computer. A computer center in Building 805 was on the first floor and accessible to everyone. You simply submitted a deck of IBM punch cards for your program to a lady inside the controlled temperature room, and, lo and behold, several hours or days later, a printed paper output appeared in your mail slot. Often you got nothing because you had fouled up the input deck or mispunched it or sometimes for no reason at all. Sometimes things worked and sometimes they did not. The best advice I received from a helpful lady programmer was that if your program did not work and everything was submitted correctly, you should wait a day and then resubmit the job because the systems people were probably dinking with the system. Later, as computers changed to the CDC 3600 version and ultimately to a Cray, we used terminals to access them, but there were still times when it was best to wait a while if something did not work because "the systems people were probably dinking with the system."

---

[39] Editors' note: The CDC 1604, released in 1959, was the Control Data Corporation's first fully transistorized computer. Seymour Cray, the lead designer, later left (in 1972) to form his own company, Cray Research Inc.

Little did I realize that the computing power at the labs was truly awesome. Even calculations at our desks were state of the art. The big innovation was when we got terminals (out in the hall close to our offices) that linked to a WANG[40] system in Phoenix, then desk terminals that linked together to let us talk to each other, and finally linked personal computers. We had DEC[41] computers for a while, but then an AT&T executive came by and convinced the laboratory president that we should have AT&T computers because Sandia was being managed by AT&T. The most mind-boggling innovation in my mind was the optical loop that was created around the main technical area. This was a high-speed transmission fiber optic loop that (at that time) could transmit unbelievable amounts of information at unbelievable speeds, sort of like a subway with stations at various buildings. Transmission centers (stations) inside each office building linked all the desktop terminals in the building to the loop, sort of like the brass rings that used to be so much fun at the old merry-go-round in Forest Park[42] near where we lived in New York City. (You got on one of the painted horses on the outside of the ride; the merry-go-round started; you went round and round and reached out to the stationary arm to try and grab the brass rings while you were swept by. When the ride slowed down, you tossed the rings into a basket and it all began again.) Information around the optic loop was sort of the same. It was fed into the loop from a station; it went around and was deposited at the proper recipient's station and on to his terminal.

The initial reason for such massive computing power at Sandia was to transmit and analyze field data from nuclear tests. As computer programs evolved, computations were used to simulate experiments and nuclear tests, especially after the Comprehensive Test Ban Treaty, when testing was no longer permitted. The computations were strictly reserved for scientific purposes. For example, we used computers to predict the rate of encapsulation of waste in the Waste Isolation Pilot Plant (WIPP) using finite-element structural studies. I was sort of taken aback one time, when the computing organization at Sandia called up and asked for authorization to conduct a WIPP calculation over a weekend. I was told that we had already used up 6 hours of Cray time, and did I want to continue? The Cray at that time was one of the fastest computers in the world. Not to worry! The Cray was not fully utilized during weekends and they were glad for some business.

---

[40] Editors' note: Wang Laboratories was founded in 1951 by Dr. An Wang as a one-man electrical fixtures store atop a garage in Boston's South End. He had a PhD from Harvard in applied physics. In 1955 Dr. Wang received U.S. Patent 2,708,722 for the magnetic pulse controlling device, a doughnut-shaped iron ring that is the principle upon which magnetic core memory was based before the introduction of the microchip. Shortly afterward Wang, with Dr. Ge-Yao Chu (a former schoolmate at Shanghai Jiao Tong University), incorporated Wang Laboratories.

[41] Editors' note: These were Digital Equipment Corporation VAX computers.

[42] Editors' note: Forest Park's original carousel burned to the ground in a fire in 1966. The present Forest Park Carousel is a century-old wooden carousel carved in 1903 by Daniel Carl Mueller. From 1903 to 1971, Mueller's carousel was located at Lakeview Park in Dracut, Massachusetts. It is now located at Woodhaven Blvd and Forest Park Drive in Queens and operates from late March to October. See http://www.forestparkcarousel.com/.

I was amused to note that administrative data were not allowed in the loop system. Maybe it is now. For example, suppose we needed the Graphic Arts Department to prepare graphs for viewgraphs or illustrations. It would have been so easy to just send the data to them via the optical loop. However, this was not possible, and instead we had to walk across the Technical Area[43] and hand carry the material to the graphics experts. They then used their own little computer to prepare the material. Of course, this is all changed now and networks are available for all functions.

The upshot of all this was that, in going through this computer evolution, I failed to realize that Sandia was truly on the cutting edge of technology. Other laboratories such as Los Alamos, Livermore, and MIT had equal or better capability, but many technical organizations outside the national laboratories did not. For example, Cornell did not have a computer center until probably 5–8 years after I left graduate school.

## 1968–1978

- Division (Department) Supervisor, Experimental Impact Physics Division, Sandia National Laboratories
- Staff direction, counseling, program development and research related to shock wave applications
- Director of the Hypervelocity Impact Facility
- Modeling of shock response, vulnerability, and hardenability materials
- Examination of thermal shock physics using electron-beam technology
- Experimental participation in Nevada underground tests
- Use of finite-difference shock wave codes
- One of the principal investigators in the Light Water Reactor Molten Core-Concrete Interaction Program

After several years of research, I was encouraged to go into supervision. My initial responsibility was a group doing research with the air guns. Later, however, I was asked to take over a group that was working with a much larger powder gun, with plans for a light gas gun.

Sometime during this period, I was also in charge of the Building 9930 explosive site, involved in explosives tests. Carl Konrad and Bob Hardy operated this facility. The powder gun was a modified cannon, with a breech that took a charge essentially of black powder. When the powder was ignited, the charge detonated, causing a much higher gas pressure than could be achieved with gas compressed by a mechanical compressor. The tests were similar to those for gas gun experiments, with the debris from the impact contained in a catcher tank (impact chamber) that was about 6 ft in diameter and 8 ft long. Access to the catcher tank was from the back through a swinging door secured with large nuts on bolts spaced around the periphery of the outside diameter of the tank. When an experiment was completed, the tank was depressurized, releasing the residual gas of the detonation, the nuts (about 50 of about 3-quarter-inch thread diameter in size) were removed, and the tank was

[43] Editors' note: Tech Area I

cleaned for the next shot. During an experiment, the gun operators and experimenter were safe in a bunker. All data were collected electronically on counters and oscilloscopes and the pictures from these were recorded on Polaroid® film, which was later analyzed digitally.

I don't remember when the light gas gun became operational, but it consisted of a two-stage arrangement, with the first stage a powder gun, which compressed a light gas in the second barrel, causing the projectile to go much faster. For example, with a one-stage air gun, the velocity was limited to about 960 ft/s because that was the velocity of sound at our altitude. The air simply could not push a projectile any faster. With a gun cavity containing compressed light gas, the sound velocity was much faster, and much higher projectile speeds caused much higher pressures. I was always amused that the guns were positioned so that they pointed directly at an explosive site some distance away where our competitors in the physics group performed explosive tests.

Two weeks after I took over responsibility for the powder gun at Building 9950, it blew up. At the culmination of a routine experiment, the door blew off the catcher system, venting hot gases into the building. The building was a typical metal building with a steel frame upon which sheets of metal formed the outer wall. The overpressure inside popped many of the rivets holding the outer wall metal sheets and the building looked like someone had gently pushed outward on all the walls. Fortunately no one was hurt. The gun operators, in fact, did not realize anything was unusual about the shot until they exited the control room bunker and saw all the soot and debris.

The reason for the accident was traced to the bolts on the catcher tank. Investigation showed that, while there were some 50 or 60 bolts that needed to be secured with nuts on its rear door, the technicians had started to save on some work by not securing all the bolts during a shot. First, they secured only 40. Then it dropped to 30. Finally, during the accident, I think it was determined that the nuts on only something like 11 bolts had been tightened, and these were insufficient to hold the expanded gases produced by detonation of the black powder. I was not faulted in the investigation that followed because I did not have enough time to determine what procedures were being followed, but I paid a stiff penalty in that almost a year of my time was spent responding to questions, writing new safety procedures, and defending my staff. Over $100,000 of damage was caused, all of which was required to fix the building. At that time, whenever damage of that magnitude occurred to government property, the cost of the ensuing investigation probably amounted to several times more.

The amusing thing was that the gun was fine, the staff was safe, and none of the electronic equipment was harmed. They all functioned as they were supposed to. The gun simply belched a huge charge of sooty smoke, and that is what caused the damage. An observation about the entire event was, however, that, in rebuilding the facility, we were able to remedy all the adverse features that had unknowingly been incorporated during the initial construction. The new version was a world-class facility that thrust Sandia well in front of any other organization in this country—

and it served Sandia well for many years. In addition, I felt completely vindicated several years ago when newspaper accounts reported that exactly the same accident had occurred at Los Alamos with a similar gun that they owned: they blew the back off the catcher tank.

This era was one of turmoil at Sandia with budgets slashed to the point that a layoff was necessary. Many had been involved in underground testing, which was now over. Attention turned to keeping components safe from explosive damage. During part of this time, a very nervous staff went about their duties with uncertainty about whether a job would be there for them in the morning. Later, as things loosened up, we were permitted to examine other types of projects to help with the cause.

### 1978–1986

- Division (Department) Supervisor, Geomechanics Division, Sandia National Laboratories
- Staff direction, counseling, program development and research related to Geomechanics application
- Direction and improvement of capabilities at the Rock Mechanics Laboratory
- Coordination of the fossil energy recovery (oil shale), Nevada and WIPP waste storage, BES[44] research initiative, and SPR[45] oil storage rock mechanics laboratory programs
- Experimental rock mechanics and material response modeling for Nevada nuclear test containment, Nevada, and WIPP waste storage
- Finite-difference and finite-element code use and modification
- Budget preparation, merit review, property acquisition and records

My lateral transition to supervisor of the Geomechanics Division was a welcome change because it was closer to my original interest in Civil Engineering. Soil mechanics studies at Columbia and the materials research at Cornell complemented this direction as well. The Rock Mechanics Laboratory at Sandia (Building 849) contained a number of mechanical testing machines similar to those I had been in charge of at Cornell. These were in the form of hydraulic presses that would apply loads as large as a million pounds to rock samples, measuring load vs. deformation and load at fracture.

[44] Editors' note: BES is the acronym for Basic Energy Sciences. This Department of Energy program is one of the Nation's largest sponsors of fundamental research in the natural sciences.

[45] Editors' note: SPR is the acronym for the Strategic Petroleum Reserve, which was established in 1975, following the 1973–1974 oil embargo. The Department of Energy operates SPR to store crude oil in various underground salt caverns in the event of an energy emergency.

## Albert J. Chabai
## (1958–1994)

Over the years Sandia's capability in shock wave physics, both in research and in field applications,[46] has been notable. My involvement has been principally with field applications and, more particularly, with measurement of shock waves produced in air and Earth materials by nuclear explosions.

Some examples of Sandia's significant contributions in field applications to the national defense posture of our country are as follows. In the early 1960s, Bob Bass and I adopted the SLIFER (Shorted Length Indicator by Frequency of Electrical Resonance) technique of Lawrence Livermore National Laboratory to make continuous shock position vs. time (r vs. t) measurements in nuclear tests providing shock wave attenuation data. After measurements in volcanic tuff, granite, desert alluvium, and salt media, we discovered an essentially universal relationship for shock position vs. shock arrival time as a function of yield that was independent of the medium. This power law relationship was noted by Bob Brownlee of Los Alamos National Laboratory (LANL), who engaged Sandia in having SLIFER cable measurements made on many LANL weapon development events at the Nevada Test Site[47] (NTS). Adjacent to the device emplacement hole, one to three vertical holes (at about 1 million dollars per hole) were drilled, into which we placed the SLIFER cables. Accurate placement surveys were made of cable locations relative to the detonation center.

This technique allowed us to estimate nuclear yield within an hour after detonation, which was extremely valuable to the nuclear device engineers, allowing them to obtain data immediately on the performance of their weapon designs. On most of these tests, a post-shot hole was drilled down to the camouflet[48] produced by the explosion, and a radiochemistry sample was obtained to determine the yield and performance of the device. But these results were not available for months after the test, and this is why our SLIFER measurements were of such value to LANL. This universal relationship that we discovered was later refined by Don Eilers of LANL using more data and has been adopted as the standard for possible use in on-site yield measurements of nuclear detonation tests conducted by foreign nations.

After the Russians conducted exoatmospheric nuclear detonations, our Nation became concerned about the vulnerability of our ICBM (intercontinental ballistic missile) defense to x rays. In the mid 1960s, the United States initiated a program to harden our ballistic missiles to x rays. A series of tests was conducted at NTS in tunnels with evacuated horizontal pipes to simulate exoatmospheric conditions. Different nuclear sources were utilized to provide x rays with various spectral distributions, and experiments were conducted within the pipes to measure the

---

[46] Editors' note: A "field application" is one in which a nuclear test is conducted underground in Nevada or, formerly, aboveground.

[47] Editors' note: Now called the Nevada National Security Site.

[48] Editors' note: An underground cavity formed by a nuclear explosion.

effects of x-ray deposition on materials. Often the level of x-ray deposition was high enough that the surface of exposed material samples was heated above the vaporization level, producing "blowoff" from the sample and generation of a consequent shock wave in the remaining solid material. The quartz gauge developed by Frank Neilson at Sandia was a very significant instrument for measuring this shock wave. This measurement technique was utilized extensively on all x-ray vulnerability tests at NTS, not only by Sandia, but by many other experimenters. The quartz gauge provided, *with high accuracy*, the wave shape and peak shock wave amplitude produced in the sample by blowoff from x-ray exposure. Among the many measurements made (e.g., impulse, temperature, shock pressure, displacement), it is fair to say that the quartz gauge provided the bulk of the most important quantitative data, allowing the United States to develop warheads for the Navy and the Air Force that were hardened to x rays, thereby providing a credible deterrent.

From the mid 1970s to the mid 1980s, the Defense Nuclear Agency (DNA) discovered a severe problem related to the kill and destruction of enemy silos. This problem had to do with energy partitioning into the ground and air from large yield, near-surface burst explosions of radiating nuclear weapons. The only data from large radiating devices were from tests in the Pacific Proving Ground (PPG). The data indicated that the primary kill mechanism was the crater produced by these explosions. In the mid 1970s, computer codes and computer power were sufficiently developed to enable calculations of the craters produced by these explosions. The calculations could not duplicate the large crater sizes in geological materials that were predicted by the empiricism of the PPG data. DNA sponsored an extensive computational effort involving Sandia and four other agencies. Hydrocodes with radiation transport included were applied to predict the effects of a near-surface detonation of a radiating nuclear weapon. After 2 years of improvement and development of the codes, all five agencies were able to attain agreement on the physics and the results for energy partitioning (ground shock, air blast, and radiation transport) for a surface or a near-surface burst detonation of a radiating nuclear weapon.

At the same time, DNA was planning a series of nuclear tests at NTS to increase our understanding of near-surface bursts and validate the hydrocode results. The first test, MINI JADE, was conducted in 1983 in an underground hemispherical cavity 11 meters in radius. A grout medium (called MINI JADE grout) served as the silo medium simulant. Both static and dynamic properties of this medium were extensively measured. Dennis Grady at Sandia provided the Hugoniot and shock dynamic properties of the grout. For this test a non-radiating device was employed so as to produce a ground shock in the medium without a contribution from radiation and to allow verification of the code predictions in a grout medium whose properties were very well established. This test was well instrumented by various agencies including Sandia, and excellent data were obtained on the ground shock produced by the near-surface burst explosion. The hydrocode predictions for the MINI JADE ground shock were confirmed by the data.

MINI JADE was a precursor test in which the ground shock instrumentation was developed and tested and the code predictions without radiation transfer were validated. In 1988 DNA conducted the MISTY ECHO event, also in an 11-m hemispherical cavity. The near-surface burst produced a radiating detonation, and the medium below the detonation was again MINI JADE grout. Very extensive ground shock instrumentation was fielded in the grout medium. Sandia had SLIFER cables to measure ground shock position vs. shock arrival time and also fielded impedance mismatch gauges and PVDF (polyvinylidene difluoride) gauges to measure the shock pressure from a few megabars down to about 100 kbars. Excellent data were obtained on the ground shock produced by the radiating detonation, and the theoretical calculations were verified. The four impedance mismatch measurements were made close to the surface and the pressures were in excess of 1 Mbar.

After nearly a decade of DNA effort, hydrocode computations with radiation transfer were corroborated, and confidence was established in our nuclear deterrence and its ability to destroy enemy silos. Consideration of this problem continued for a number of years after MISTY ECHO, particularly at Sandia, where decisions had to be made on the most appropriate and reliable fusing for near-surface bursts. Paul Yarrington and his group investigated this question in detail.

# Lalit C. Chhabildas

# (1976–2007)

**Pre-Sandia Years: Physics Department, Rensselaer Polytechnic Institute, Troy, New York**

Little did I know when I came to the United States in September 1966 from Bombay, India, that I would someday be writing my recollections of Sandia National Laboratories' history in shock physics. In many respects, the decision to go to Sandia was rooted in what I did at Rensselaer Polytechnic Institute (RPI) for my thesis project in high-pressure research. When I joined RPI as a graduate student, I was hoping to become a theoretical nuclear physicist. Instead, I became an experimental high-pressure physicist. My thesis advisor, Dr. Howard Michael Gilder, had come to RPI as an assistant professor after getting his PhD just 6 months before I arrived. I came as a teaching assistant as one of the laboratory instructors and worked for Professor Harry F. Meiners. Professor Meiners recommended me to Professor Gilder, who was looking for a research assistant while building a brand new high-pressure laboratory. Professor Gilder left a note in my mailbox asking if I was interested in a research assistantship and enticing me with a summer assistantship starting in 1967. Needless to say, the opportunity of full-time employment in the summer was very attractive indeed. I still remember to this day talking to Dr. H.B. Huntington's postdoctoral researchers, who were in labs adjacent to Mike Gilder's new laboratory. I was gathering their personal impressions and experiences of him since there wasn't a past track record of his research accomplishments. They said he was a nice guy and encouraged me to join his team. Incidentally, I was the first member of that team.

To make a long story short, I helped Mike Gilder set up the static high-pressure and high-temperature laboratory, and my thesis topic was self-diffusion measurements in single-crystal, c- and a-axis zinc, at high pressures. Coming from India, in general, you do not have a lot of hands-on experience. I was exposed to learning how to make single crystals and how to design a high-pressure (about 1 GPa) vessel and a high-temperature molten tin bath with controllers to provide a uniform temperature environment (of over 300–400 C) for the high-pressure vessel. The temperature set up allowed self-diffusion measurements to an extremely high degree of precision at high pressures for the first time and allowed us to examine the temperature dependence of the activation volumes unambiguously, but it was in conflict with other measurements being done at the University of Arizona. I defended my PhD thesis in August 1971. My fiancée at the time, Annette Winslow, typed the thesis at least three or four times on a typewriter with a carbon copy. To accommodate any recommended changes, she had to retype the entire thesis. Fortunately, it was only 45 pages because my advisor's philosophy was to pattern the thesis as a journal paper so it could be sent directly to a journal. The thesis committee at my defense objected to that philosophy. Mike Gilder quickly proposed a compromise of including two of our prepublications as appendices, which increased the thesis to just about 72 pages. That was a quick relief on my part. My thesis was published as a journal article (Chhabildas and Gilder 1972).

There was a recession in 1971, and I was not a naturalized citizen. These factors made my job search difficult. Mike Gilder was having some family issues he needed to resolve. He hired me as a postdoc and handed me his entire laboratory including three students (Michael Current, Neil Shea, and Alan Baker). I stayed two more years as their advisor, with Dr. Hilliard B. Huntington acting as their official advisor. Michael Current and Neil Shea graduated with PhDs and Alan Baker graduated with a master's degree. My search for full-time employment was still not fruitful. Dr. Huntington recommended me to Professor Arthur L. Ruoff in the materials science and engineering department at Cornell University in Ithaca, New York. Incidentally, many of Dr. Huntington's PhD graduates worked with Professor Ruoff prior to landing full-time employment. He offered me a research associate position to work with one of his PhD graduate students. His comment to me was "I want him to graduate and I want you to get him out of here within a year." While I was interviewing with Professor Ruoff, I got a postdoc offer from Professor Dan Decker at Brigham Young University in Provo, Utah, with the research topic of using sodium chloride as a high-pressure gauge. The start date was September 1 at Cornell and we were expecting our first child in September. I requested that the start date be October 1, to which Professor Ruoff said okay—you will still start September 1 anyway except that you will be on vacation for the first month. I can never imagine starting a new position anywhere for which you are on vacation the first month.

## Pre-Sandia Years: Materials Science and Engineering Department, Cornell University

We decided to go to Cornell because we were expecting our first child; we wanted to be close to Annette's family and did not want to move thousands of miles away from our roots. We moved just a week after our first daughter was born. At Cornell, we were conducting research on ionic crystals in an attempt to determine the pressure–volume behavior to an extremely high precision. That precision was necessary because we wanted to determine experimentally the second derivative of bulk modulus B with pressure P. A linear relation of B vs. P gives rise to the Birch–Murnaghan relationship and the corresponding equation of state. A nonlinear relation with accurate determination of the second-order quadratic coefficient would allow us to investigate the departure from the traditional Birch–Murnaghan relationship and to evaluate other equations of state in the literature at the same time. Our first experiments were on a meter-long single-crystal lithium-fluoride cylindrical rod. Using a meter-long crystal would allow measurements of the change in length $\delta l/l$ to a very high precision. The longer the rod length l, the larger the magnitude of $\delta l/l$ with increasing pressure. The experiments were conducted to pressures up to 1 GPa. We measured $\delta l/l$ to one part in $10^6$ with a He–Ne laser. However, the pressure measurement limited the accuracy. Nevertheless, these LiF studies indicated that the material was too stiff to get reasonable and accurate values for the second derivative of the bulk modulus relation with pressure. This was essentially our conclusion, which was sufficient for Kwang Yul Kim to finish his thesis and publish a journal paper (Kim et al. 1976).

Our next step was to use an ionic crystal that was more compressible than LiF. We chose a meter-long sodium chloride single crystal. Its initial bulk modulus $B_0$ is about a third that of LiF, and, hence, we would be making $\delta l/l$ measurements that were nominally three times as large as observed in LiF. We were optimistic and believed this would allow measurement of the second derivative of the bulk modulus. We were definitely in the right direction. Even though the change in length was larger compared to LiF, it was still not sufficient. The good news was that the measurements were extremely precise. This allowed accurate estimates for $B_0$ and the first derivative $B_0'$, and we were able to estimate the second derivative of the pressure, $B_0''$. A nonlinear least squares fit indicated that $B_0''$ was negative but that its magnitude was comparable to the uncertainty. At least for ionic crystals, it appeared that the second derivative is negative. This significant result suggested the equation of state of a material is not as stiff as what the Birch–Murnaghan relationship would suggest, especially at high-volume compressions. This meant that pressure estimates in Bridgman anvil cells were probably on the high side. You have to keep in mind that, in those days, extrapolation of pressure–volume relations from experiments at pressures up to 1 GPa was primarily used as a standard to determine higher pressures in the Bridgman anvil apparatus. Even though I did not go to Brigham Young to work on the NaCl pressure gauge, my high-precision research at Cornell (Ruoff and Chhabildas 1976) was useful in discussing its merit as a pressure gauge to estimate high pressures in the Bridgman anvil apparatus.

At the time, tungsten carbide, due to its high dynamic yield strength, was the material of choice for high-pressure anvils. The use of diamond anvils to generate static high pressures was catching on and was first demonstrated by Dr. G. Piermarini, who worked at the National Bureau of Standards. He was able to use gasketed diamond anvils and was pushing the static high-pressure envelope both in magnitude (over 50 GPa) and for high accuracy. Professor Ruoff was quite impatient, ambitious, and excited and got himself a set of diamond anvils. I was the first to use these at his lab. We investigated sulfur (Chhabildas and Ruoff 1977) and detected a conducting phase transition at around 23 GPa. As a new researcher, using diamond anvils to go to stresses exceeding 30 GPa in a very small volume in the laboratory was quite exciting to say the least. Believe it or not I broke quite a few diamonds! They were probably misaligned or perhaps had defects that limited the yield strength. Coming home and telling your wife that you broke large diamonds was quite a treat!

This was early 1976, and I was once again actively looking for a permanent position, except I was now a naturalized citizen. Early in 1974, about 3 months after I had joined Ruoff, he had said, "When you get your citizenship I have just the right place for you to go." He named Sandia Laboratories as where he thought I belonged. I don't know what he saw in me, but he was quite convinced I should be at Sandia. He had spent a summer at Sandia in 1966 and gotten exposed to shock physics, i.e., dynamic high pressures. Barry Butcher was Ruoff's first PhD student and had joined Sandia's shock physics group. Close collaboration existed between the shock and static high-pressure researchers in those days, motivated by a need to "calibrate and determine" static pressures in anvils using the shock Hugoniot. I requested employment applications from Los Alamos, Livermore, and Sandia. Sandia and Livermore never sent applications, but I did get an interview with Jerry Wackerle at Los Alamos in April 1975. I had never seen a powder gun, a two-stage light gas gun, i.e., smooth bore guns, nor a VISAR. Richard Warnes was putting together the VISAR. I was totally awestruck, and my lack of knowledge of those tools came through loud and clear. Jerry did not select me for the position; I was disappointed but not surprised. Ruoff told me to call him up and find out why he did not select me. I was not too thrilled about calling him up and asking why I had been rejected. Ruoff's convincing argument to me was "If you find out what you did wrong or what they did not like, you would be much better prepared for the next interview." I took Ruoff's advice and, against my better judgment, called Jerry Wackerle and asked him what he found lacking in my application. Was it because he was not able to get in touch with my thesis advisor, Mike Gilder, from RPI? Mike never came back from France and was difficult to reach. I think Jerry Wackerle was just as uncomfortable as I was. He told me he had seven candidates and I was the least experienced in shock physics. Apparently, Jerry Wackerle selected someone who had the most shock physics experience. However, looking back, the interview trip was quite educational and informative; I was exposed to some of the tools used extensively in shock physics at the time.

Soon after that, and to my surprise, I found a Sandia employment application waiting in my mailbox. The cover letter said the application was sent at Professor Ruoff's request as he had recommended me for a position. I mailed my application to Sandia and also went on a campus career interview at Cornell with Richard Schwoebel and Venkatesh Narayanamurti, representing Sandia and Western Electric, respectively. (I never heard back from Venkatesh Narayanamurti, who later came to Sandia as Vice President of Research.) I asked Richard Schwoebel if he knew Barry Butcher and, to my good fortune, he did. I got a call from Barry for an interview. Later, I found out that Schwoebel had discussions with Ruoff while at Cornell and he had sent my resume to Barry in the shock physics group. When Barry called Ruoff asking for a reference, Ruoff told Barry he needed a response in the next 4 weeks to help him decide if he should look for a postdoc replacement ASAP. Ruoff valued the research we were doing on sulfur and did not want to be upstaged by other researchers in the field. This worked in my favor because he really valued my research and did not want me to leave until the research was done unless he got a good replacement. That is the best recommendation one can give to a prospective employer! I went for an interview in mid August 1976. My early job interview at Los Alamos became very valuable because I was exposed again to smooth bore guns—a compressed gas gun, a powder gun, and the two-stage light gas gun as well as the VISAR. I was no longer ignorant of the technology and was able to make some meaningful contributions to the discussions. I got the job offer in early September and joined Sandia later that year in December. Little did I know then that I would be there for over 30 years!

## Sandia National Laboratories

Barry Butcher was my immediate Supervisor; Walt Herrmann was our Department Manager and John Galt was our Director. I bring up John Galt for a few reasons. First, at my interview with John Galt, he was quite delighted to learn that Professor

Hilliard Huntington was my default advisor at RPI. As mentioned above, even though my thesis advisor was Mike Gilder, I reported to Professor Huntington when Gilder was on sabbatical in France. I was pleased to learn that John Galt and Professor Huntington were college classmates. The whole interview was about him and Huntington during their college days; Galt was pleased to find out more about his college buddy. That definitely did not hurt my chances of getting hired. Second, I think Galt was very focused. When he found out Livermore had an urgent need to characterize beryllium, a toxic substance, he overruled all of us and asked us to get involved. (He was the Vice President at the time.) Galt was a risk taker; he set up a high-visibility laboratory research program to metallize hydrogen using isentropic loading techniques that Lynn Barker was developing. I am not aware of his active involvement in other high-visibility projects but I am sure there were many.

When I joined Sandia in 1976, Jim Asay, Herb Sutherland, and I were the only shock experimentalists in Barry's group, and Barry was responsible for the compressed gas gun in Building 855. Dennis Grady was the other shock experimentalist in Darrell Munson's group; Darrell was responsible for the powder gun and the two-stage light gas gun in Building 9950. Carl Konrad and Robert Hardy were at the Building 9956 gun facility, which is now called the STAR facility. Dave Cox came about 6 months before I did to assist Jim Asay with all experimental activities. We were never heavily staffed in experimental shock physics in the late 1970s. This was also true in the department headed by George Samara, where Lee Davison was the supervisor of the experimental division. Bob Graham, Pete Lysne, Jim Kennedy, and Bill Benedick were the only experimental staff in that group.

**Rise time measurements.** My first assignment was to measure shock wave rise times to estimate the plastic viscosities of metals at Mbar pressures. There was conflicting evidence by orders of magnitude in the Russian literature on what the magnitude of viscosity was in metals. I did not know enough about shock physics—let alone what viscosity was all about. Jim Asay was very helpful, to say the least, in getting me going. He mentored me in shock physics, gave me appropriate reading material, answered a lot of questions, and tutored me whenever necessary. I probably talked to him every day during my first year, if not longer. Richard Kinchen was assigned as my technical assistant to help with my experimental setups. To get Mbar pressures, we needed to use the two-stage light gas gun. Unfortunately, at that time the two-stage gun was not set up for well controlled equation of state studies. Marlin Kipp and Joel Lipkin had used the gun to look at cratering caused by polyethylene (as simulated rain drops) on heat shield materials for space missiles. No standard techniques had been established at Sandia for equation of state studies on the two-stage gun then, let alone for time-resolved techniques such as the VISAR. I talked to Barry about making rise time measurements using VISARs on the powder gun because we would approach Mbar pressures in tantalum and tungsten at impact velocities exceeding 2 km/s. Barry said, "That would be too easy and there would be no challenges. What we really want to do is get the two-stage gun operational for time-resolved shock profile measurements. Measuring rise times in metals is the means to get us there."

Realizing that the main objective was to get the two-stage gun started, Carl, Bob Hardy, and I designed the two-stage gun projectile as a projectile velocity measurement system using fiber optics in a barrel extension at the muzzle end of the barrel, and we used coaxial pins to trigger the oscilloscopes as we had done on the powder gun. In those days, the projectile sabot[49] was in two pieces, consisting of a polycarbonate body referred to as Lexan® with a high-density polyethylene boot to prevent blow by. The design has evolved over the years; the current design is merely a Lexan® body with the impactor facing inserted and glued at one end. To measure projectile velocity, we were hoping to use laser "break screens" at three locations to estimate the velocity. However, the fibers picked up luminous signals from the blow by gas traversing much faster than the projectile, making the measurement unreliable. Instead of breaking the laser signal, we decided to monitor the *reflected* signal as the projectile passed. This worked beautifully and we now had a reliable projectile velocity measurement system. In those days, we had analog systems. The time resolution depended on the time base used on the oscilloscopes to measure the velocity of the projectile as it traversed more than 120 mm. This limited our accuracy to 1 %. Other laboratories made measurements when the projectile was in free flight over meter-long dimensions and got an accuracy of 0.1 % even though this was at the expense of getting greater tilt in the experiments. The tilt in our experiments when impacted at the muzzle end was only about 4–8 mrad. With current digital recording systems and nanosecond resolution, measurements can achieve an accuracy of 0.1 %. This technique is still used on our two-stage gun. We made the first time-resolved measurements in aluminum, copper, and steel on the two-stage gun and realized the measurements were limited by the time resolution of the recording devices. We got a good publication even though we did not obtain risetime measurements (Chhabildas and Asay 1979); this was my first Sandia shock physics publication.

These results led to further research. I took Lynn Barker's risetime data at lower pressures up to 9 GPa and plotted stress vs. strain rate on a log-log plot to speculate what the rise time should be at Mbar pressures. Dennis Grady took the same results, fitted them to find the exponent, and came up with the fourth power law that is now found to be universal for all metals and metal oxides. We also observed that polymethyl methacrylate (PMMA) would not be a good transparent window at pressures above 22 GPa. This limited the measurements of shock loading and release profile in aluminum only up to 40 GPa. Nevertheless, we were able to determine the Al yield strength in the shocked state at 40 GPa (Asay and Chhabildas 1980). We were now on a search for a new transparent window material.

**Powder gun incident.** Our risetime measurements in metals were almost complete by summer 1978, and I got quite excited about extending the strength measurements to copper after my first VISAR shock loading and release profile measurement in Al on the two-stage gun. I wanted this to be a systematic and comprehensive study over a wide range of the impact velocities available from the compressed gas gun and the

[49] Editors' note: This is a device to ensure the correct positioning of the projectile in the barrel of the gun.

two-stage gun. During the last week of October 1977, I had conducted shock loading and release experiments in Cu on the powder gun at 2.2-km/s impact velocities. The gun crew noticed that fragmented Cu was "damaging" the catcher tank at its midsection. The following week, on November 1, 1977, an experiment was scheduled for Bob Graham to characterize the polyvinylidene fluoride (PVDF) gauge sandwiched between two Cu anvils at an impact velocity of 2.2 km/s. From the week before, the crew remembered the pit marks caused in the catcher tank by fragmented Cu. They decided to move the catcher assembly normally butted against the tank door toward the middle of the tank to reduce the pit marks and tank damage. This small change, however, turned out to be quite detrimental; the catcher assembly, which is pivoted by rails on top of the gun, swung and struck the catcher door twice—first by the projectile and debris impact and second by the blast resulting from propellant burn. Also, only 8 bolts were used instead of the 48 for which the door was designed. The door swung open, and the resulting overpressure in the powder gun bay launched some of the sheet metal quite a distance from the building and the building was damaged. Fortunately, no one was hurt. The concrete bunker, which was the control room, did its job. It protected the staff (Carl, Bob Hardy, Bob Graham, and maybe G.T. Holman). Carl had the presence of mind to save all the acquired data, which was necessary to conclude that no one had overloaded the propellant in the breech. The tank door ended up resting against the rails of the two-stage gun. Since both guns were in the building, for all practical purposes, both were shut down until an investigation was completed.

Right after the accident, Rich Kinchen, my technical assistant, decided to move on. I am not sure if the accident was the main cause. We were both learning shock techniques at the same time, and I guess I was not much of an inspiration. I was assisted by Bob Hardy, while he still had his main assignment to maintain and care for the powder gun and the two-stage gun, even though they were not operational. I believe Bob and Carl provided a lot of support to the committee responsible for the accident investigation.

**Pressure–shear loading.** Jim and I were now starting to depend upon using the single-stage compressed gas gun in Building 855. We developed a program to investigate Rayleigh–Taylor (R–T) instabilities. Sandia had an excellent microminiature machine shop that machined grooves, wavy surfaces, etc. onto metallic samples. Jim took advantage of that capability to show that mass ejection resulted when the main shock interacted with the well controlled defects machined on the free surface of a target. We also had wavy surfaces machined onto the surfaces. Jim was looking at R–T instabilities in which the wavy surface was the free surface, and he was impacting the target with a low impedance flyer. He wanted to look at the growth of instabilities. I had done viscosity measurements using risetimes as the main measurement. Since we had read up on the Soviet literature, I decided to use the wavy surface, which was now the impact side, to estimate viscosity. We did our few experiments on the Building 855 air gun, which is now the compressed gas gun at STAR. Jim got pulled away to make internal pressure measurements on the powder gun to estimate the peak pressure loading inside the new catcher tank chamber at various locations to quantify safe loading on components (doors, ports, etc.). Herb Sutherland, who had been tasked to do pressure–shear measurements, decided to change assignments and I inherited the pressure–shear project. Each of us got sidetracked by these new projects, and the R–T and viscosity projects never got traction. The accident investigation lasted for over a year.

During that time I developed diagnostic techniques using VISARs to measure particle velocity changes accompanied by propagation of a large-amplitude shear wave in a pre-compressed material. Earlier, a team of researchers led by Professor Rod Clifton at Brown University had developed a pressure–shear loading technique using slotted barrels. This was later complemented by a novel interferometric technique those researchers had developed to measure shear displacements. Basically, a diffraction grating is deposited onto the surface normal to the direction of wave propagation with the resulting diffraction pattern in the same plane as the direction of shear motion. When a laser beam is incident on the diffraction grating, a diffracted beam pattern obeying $n\lambda = d\sin\theta$ will be reflected, where n is the order of the diffraction beam reflected at an angle $\theta$, $\lambda$ is the wavelength of the laser, and d is the diffraction width. By combining two beams at $\pm\theta$, we got a fringe frequency proportional to the shear displacement. At Sandia we decided to monitor the two diffracted beams directly at $\pm\theta$ using two independent VISARs. In this instance, the VISAR yields a velocity that is a linear combination of both the longitudinal and shear components. By adding the two beams, we determined the normal velocity component and, by subtracting them, we estimated the shear velocity component. We checked the technique by impacting anisotropic crystal Y-cut quartz. Upon normal impact you get quasi-longitudinal and quasi-transverse waves. We did not need a slotted barrel. At about this time, I believe we were reorganized; John Galt was our Vice President and Orval Jones became our Director. I remember quite well when Bob Hardy and I had a "show and tell" with Orval on our experimental setup. I commented this was quite an art. Although he was quite impressed, Orval retorted that "we should now be doing science."

This project gave me an opportunity to team with Bob Hardy and Jeffrey Swegle. Jeff had arrived in 1975, about a year and a half before I did. His expertise was code development, and he was in charge of TOODY. I developed pressure–shear experimental techniques using the output of Y-cut quartz as the generator and gluing the material of interest to the crystal. Jeff did the modeling. We studied the effects on materials such as Al, alumina-filled epoxy, PZT 95/5, and PBX 9404 (with Marlin Kipp). Jeff put an anisotropic model for Y-cut quartz into TOODY to simulate the one-dimensional (1-D) experiments. In the experiments, all changes in stress or particle velocity occurred in the propagation direction, even though we were making pressure–shear measurements. There was talk of developing a standalone 1-D pressure–shear code that, for some reason, never came to be. Since the pressure–shear states generated by Y-cut quartz are very limited, we sent out a 4-inch barrel compressed gun to get a slot milled into the barrel to provide us with the needed variability in shear stress states. This was the gun system Don Lundergan had designed and the first gas gun impact system Sandia had developed. I never got the project going rapidly, perhaps because of my lack of experience in project management and also because I did not have the luxury of a full-time technician assigned to me. In addition, I was involved in building a dual-air delay leg VISAR (with a 15-ns delay) to give the fringe sensitivity needed for the pressure–shear measurements. More significantly, my interest lay in understanding material deformation kinetics under shock compression as opposed to building the infrastructure. The example we developed was a technique to transmit 0.65-GPa shear stresses across interfaces, which was quite impressive. Two publications resulted from our efforts (Chhabildas and Swegle 1980; Swegle and Chhabildas 1981).

**Strength of materials, aka reshock and release experiments.** In late 1978, Jim Asay was promoted to my supervisor and the powder gun was once again commissioned for standard operations. This had required a thorough review of operations and an extensive rewrite of safe operating procedures, and, I must add, it became a comprehensive document for the better. *The investigation committee also recommended that we join the Aeroballistics Range Association (ARA) to familiarize ourselves with range operations, including safety issues.* At the time, I got more and more enthralled with the elastic precursors Jim Asay had observed in Al upon reshock at low pressures of around 2 GPa. I observed similar behavior in OFHC copper[50] at 10 GPa. I did my first experiments on reshock and release behavior in Al to over 20 GPa on the powder gun and once again observed elastic precursors upon reshock (Asay and Chhabildas 1981). PMMA was the most convenient transparent window at those stresses at the time. The poorly mismatched impedance perturbed the release structure considerably. Larry Lee, formerly at Sandia but then at Ktech, had evaluated sapphire and LiF windows for transparency losses and found that sapphire lost transparency above its elastic limit of 22 GPa, while LiF remained transparent at powder gun impact velocities up to 2.2 km/s, which is around 20 GPa. I did one of the first experiments to determine the refractive index correction for LiF on the powder gun at about 20 GPa. The results were very encouraging, and when Jim hired Jack Wise in 1979 to replace himself and investigate LiF further, Jack was given the task of optically characterizing and evaluating LiF as a window material over the range of the powder gun and two-stage gun velocities. We found that the window was still transparent at up to 1.2 Mbar, which was quite astounding considering the material is approaching 50 % volume compression (Wise and Chhabildas 1986). Both projects—strength measurements in Al and optical characterization in LiF—turned out to be quite significant. First, Sandia became a leader in strength measurements at high shock pressures; second, LiF windows are now a standard used over a wide pressure range—not just to Mbar pressures—and have provided many opportunities to extend time-resolved VISAR studies on the two-stage gun. The shock and reshock studies have continued and been extended to Mbar stresses in many ceramics and Al by Bill Reinhart and Tracy Vogler.

In the meantime, our Vice President John Galt tasked us to determine the strength of beryllium as a national need. How did this happen? Our strength work was especially appreciated by Dan Steinberg of Livermore; he used it to develop strength

[50] OFHC copper is oxygen-free, high-conductivity copper.

models now referred to as the Guinan–Steinberg constitutive model for use with hydrodynamic codes. Dan wanted us to measure Be strength in the shocked state! When the initial request came directly to us from Hal Graboske and Dan Steinberg of Livermore, we were not too thrilled and declined the invitation because of all the precautions we would have to take due to the toxic nature of Be. The Livermore folks then approached our upper management directly, and we had no choice but to oblige. Jack Wise got sidetracked as soon as he joined Sandia because he was actually hired to continue the R–T studies Jim had started; instead, he ended up working with me on the Be studies. Jack and I used many Be experiments to characterize the refractive index corrections for LiF as a transparent window under shock loading. Even in Be, we noticed an elastic precursor upon reshock at stresses up to 25 GPa.

The Be program was personally good for me. I became well known in the shock physics community, especially at Los Alamos and Livermore. At Bob Graham's recommendation to the International Association for the Advancement of High Pressure Science and Technology (AIRAPT) committee, I gave my first invited talk at the 1981 meeting in Sweden. However, being the youngest new hire, I was ranked low on the totem pole. Orval Jones (the Director) and John Galt (the Vice-President) did not fund my trip, and I paid my own way. That was quite a rude awakening to say the least. Since the AIRAPT meeting coincided with the ARA meeting in Karlsborg, Sweden, I went to Karlsborg to represent Sandia's case to become members of ARA. There was some hesitancy on ARA's part because our investigations focused on shock physics plate impact studies, as opposed to aeroballistics operations. Later, I found out that Professor Akira Sawaoka of the Tokyo Institute of Technology, Dr. Alois Stilp of the Ernst Mach Institut in Germany, and Dr. Henri Bernier of the *Commissariat à l'Energie Atomique et aux Énergies Alternatives* in Paris were some of the members who recommended us highly.

When Jim Asay was promoted around December 1978 to lead the group, he hired Jack Wise to replace him in 1979. Around 1980–1981 Lynn Barker returned to Sandia and joined the group. Jim also convinced Dennis Grady to be part of the group, since he was the sole shock physics researcher in what was then Darrell Munson's division. He also hired David Ek from Brigham Young University and beefed up the technician support, with Ron Moody to assist Dennis, Jeff Miller to help me, and John Nevers for the gun and facility operations. I am not sure if this was Jim's or the management's decision but all the guns were consolidated under one roof. I do know that Orval supported it! Orval once took Department of Energy (DOE) auditors from Washington, D.C. to visit the gun facility. Jim had taken the day off and gone fishing! We had the powder gun and two-stage gun side by side with very little room separating them. The powder gun catcher tank door would lean against the rails of the two-stage gun in the open position. Navigating between those two guns, especially with the powder gun door opened, was quite a challenge. When Orval and the auditors, dressed in suits, got them stained and dirty with the soot from the powder gun catcher tank, he was not pleased! Orval immediately asked Jim to put in a request for a new building to separate the two guns. Not only did we get separate buildings for the powder gun and the two-stage gun, we also got a building for the compressed gas gun and the oblique impact gun. I don't know if Jim knew it or not, but he was often referred by many as "an empire builder." With Jeff assisting, we assembled a low-temperature capability to −65° F, for use with the 100-mm single-stage gas gun and used it to investigate pressure–shear loading in ferroelectrics with Y-cut quartz crystals (Miller and Chhabildas 1985). We also assembled the 4-in. oblique impact gun and got the air delay leg interferometer completed.

**Shock loading quartz to Mbar pressures.** With Dennis in the group, we teamed to investigate both fused quartz and single-crystal X-cut quartz. This was partly motivated by geological applications, with considerable interest expressed by Binky Lee, who at the time was at the Defense Nuclear Agency (DNA). Dennis investigated single-crystal calcite. Binky Lee was interested in determining the release states of quartz from the shocked state to as low a stress as possible. We decided to use a thin plate reverberation technique with tantalum plates to investigate quartz. In the past, sapphire was the only material used as a reverberation plate up to its elastic limit. We were pushing the limit and extended the reverberation technique to almost 30 GPa, and we determined the release states of quartz behavior in the mixed-phase regime (Chhabildas and Grady 1984; Chhabildas and Miller 1985). The single-crystal quartz study also pushed the limit of LiF for use as a transparent window to 1.2 Mbar when we investigated shock loading and release wave profiles in quartz to over an Mbar.

**Isentropic loading experiments.** I submitted an abstract to the Hypervelocity Impact Symposium (HVIS), summarizing the investigation of quartz up to Mbar pressures in 1986. They rejected my abstract! HVIS was being revitalized because the United States had started a major defense program, the Strategic Defense Initiative (SDI), and the goal of the symposium was to educate researchers on the state of art as it existed in 1986. Although the symposium rejected my abstract, the organizers asked me to present a survey of time-resolved diagnostic tools that had been developed since the last symposium in the 1960s. The invitation came in July, and they wanted a manuscript by late August. I had no choice but to turn down the invitation. The organizers called back repeatedly. Later I found out Walt Herrmann, my Director, wanted me to take on this task and I guess I had no choice. Jim gave me time to review the loading and diagnostic technology in detail for the next few months. This is where I learned that the Russians were using layered materials to generate isentropic loading techniques to synthesize boron nitride as a super-hard material.

Personally, I was starting to work more and more independently. Jim gave me the opportunity to be the project manager of the strength studies in metals for Livermore. In particular, we studied vanadium, molybdenum, and tantalum–molybdenum alloys as part of the project. Livermore was interested in such materials for armor applications under the Defense Advanced Research Projects Agency (DARPA) program. SDI was also becoming a visible program, and proposals were floated to use tungsten as "high-velocity pebbles" for defense applications. Lynn Barker was developing "pillows" using particle sedimentation techniques for isentropic loading. This meant we could use pillows to characterize tungsten strength properties under shock and quasi-isentropic loading. Jim and I went to Los Alamos to discuss this specific need with Frank Harlow and his colleagues. They were quite excited that we had the potential to determine material properties under shock and isentropic loading! Lynn Barker's "pillows" were just what we needed for the research. I was quite "nervous" because no equation of state studies had been done to Mbar pressures with isentropic compression experiment (ICE) loading, let alone to determine strength properties! In the process of developing the technology, Lynn had used it to generate a loading isentrope for aluminum up to 20 GPa. I used Lynn's pillow to characterize tungsten under ICE loading to 80 GPa on the powder gun. In principle, at least two sample thicknesses were needed since we did not know the time-dependent stress input at the impact interface. Measurements at two locations allowed us to implement Lagrangian analysis techniques to estimate the stress, particle velocity, and volume behind the stress front as the material underwent isentropic compression. This required a considerable impedance mismatch correction since the impedance of the LiF window is much lower than W. The research was extremely challenging and exciting since none of this had been tackled before.

We got some interesting results when we applied the pillow technique to achieve 2.5 Mbars with ICE loading on the two-stage gun in tungsten. We thought we saw a phase transition at around 2 Mbars. The question was as follows: Is it real, or is it an artifact of our new techniques? Is the pillow spalling? We were hoping it was a true material phase transition since it had not been observed before. Jim recommended we repeat the experiment. If we used pillows again and saw the same profiles and phenomena, we would still not have isolated the cause "conclusively." This is when I decided to introduce the technique of using layered materials to introduce isentropic loading into W. The knowledge I gained from writing the HVIS review paper was beginning to pay off. Lo and behold, to our dismay we could not repeat the experiment. We were a bit disappointed but in the process developed a new technique to construct a layered, graded-density impactor for isentropic compression. For the first time in the shock physics literature, we observed the strength of materials under ICE loading was much higher compared to shock loading. There are many reasons, but a dominant one is that temperature effects are suppressed in ICE loading. For more information see two papers (Chhabildas and Barker 1988; Chhabildas, Asay, and Barker 1988).

The ICE studies were later extended to tantalum and molybdenum. The Ta results were quite interesting because we observed shear strength softening under shock loading but an increase in strength under ICE. We speculated that something interesting was occurring, such as a phase change in Ta around 50–60 GPa, but we never followed up on it. Our Mo results were never published but the Ta results were published in an HVIS publication (Chhabildas, Barker, Asay, and Trucano 1990) and proceedings of the Dynamic Behavior of Materials and its Applications meeting (Chhabildas and Asay 1992).

**HyperVelocity Launcher.** While I was completing my investigation of W strength under isentropic compression, Lynn Barker was trying to launch a 1-mm-thick W disk by compressing low-pressure helium in a two-stage light gas gun using a standard projectile with a Ta facing launched at around 6.5 km/s, as a task for SDI. Part of the intent was to launch plates to hypervelocities. Upon acceleration, however, the plates would fragment. The belief was that the low density He driving gas pushing on the high density W caused R–T instabilities, which led to plate fragmentation. Jim asked me to team with Lynn to promote an understanding of the fragmentation. I didn't think it would work and told Jim we should *cut our losses and work on something else*. Jim was optimistic because, in private discussions, he learned the Russians were launching Mo plates to 16 km/s using explosive techniques. The details of the techniques, however, were quite sketchy. Jim would not take "no" for an answer and convinced me it was worth pursuing. So we decided to add more materials (Ta, Mo, Al) to make sure both the mass and the driving conditions for each were the same. Bill Reinhart, a Ktech employee at the time, assisted Lynn with the experiments. All our experienced staff, led by Carl Konrad, were assigned to work on the railgun. The results were fascinating: we fractured all the materials, but the fragmentation statistics were quite different for each. I observed a correlation between the mean fragment sizes for each material and its specific spall strength—that is, the ratio of spall strength to density (Chhabildas, Barker, Asay, and Trucano 1990). To verify this empirical relation, I selected a titanium alloy, Ti-6Al-4V, as a candidate material to yield the least number of fragments. To my amazement, we launched the alloy intact! Science does work sometimes even though it is empirical.

The results suggested R–T instabilities were not the main cause for fragmentation; we did not observe a correlation between mean fragment size and yield strength. The primary cause for fragmentation was nonuniform loading over the plate diameter from edge effects, viscous drag effects at the barrel wall, and turbulence. We took the gas out, used graded-density pillows, impacted the Ti alloy plate directly around 6.5 km/s, and launched the Ti alloy plate to 9.5 km/s and a 6061-T6 Al alloy plate to 10.4 km/s (Chhabildas, Barker, Asay et al. 1991; Chhabildas, Barker, Asay et al. 1992).

These results were quite exciting, and we got a lot of attention, including many invited talks nationally and internationally (in the UK, India, and Australia) and articles in newspapers and magazines. In 1991 I was gone 6 weeks on a lecture circuit internationally! (While in Australia, I found out Sandia had eliminated all department manager positions.) The technology opened lots of opportunities externally. I remember going with Walt Herrmann and Steve Rottler to Boeing, the National Aeronautics and Space Administration (NASA) Marshall Space Flight Center, and the Missile Defense Agency (MDA), where we saw Bob Becker in Huntsville to inform them about our new HyperVelocity Launcher (HVL) and our CTH capabilities. This was the start, I believe, of a new way of doing business at Sandia. With the fall of the Berlin wall and the end of the Cold War, there was no focus on a long-range strategic vision and the need for the DOE labs. Although we did not succeed right away, both CTH and our HVL became some of the best technologies needed and was used extensively by the Department of Defense (DOD).

Later, we concentrated on using layered, graded-density materials because we could tailor them more conveniently to get the specific pulse shapes to launch a variety of materials. We launched Al, Ti, magnesium, and Ta to velocities exceeding

10 km/s. In particular, we launched a 0.5-mm-thick Ti alloy plate to over 12 km/s (Chhabildas, Dunn, Reinhart, and Miller 1993b).

**HyperVelocity Launcher applications.** NASA and DNA sought the technology. With the external funding, I became the manager for these reimbursable projects. This gave me an opportunity to work with a team that included Bill Reinhart, Jeff Miller, Luba Kmetyk, Tim Trucano, Mark Boslough, and Jim Ang. The development of the technology was personally beneficial. It allowed me to grow and mature professionally, network with the outside community, and have an opportunity to look at the bigger picture, while researching many relevant technical issues. Actually, by leading a team with diverse talent, I also learned how to manage staff and take advantage of modeling and simulation (M&S) capabilities to assist with experimentation. M&S tools using CTH were crucial to our success that led to the development of the HVL, which is now referred to as the *three-stage* light gas gun. My NASA counterparts were Jeanne Crews at Johnson Space Center (JSC) and Scott Hill at Marshall Space Flight Center (MSFC) as well as Cmdr. John Connell at DNA and Andrew Williams at the Naval Research Laboratory (NRL).

**Debris shield designs.** JSC and MSFC asked us to evaluate debris shield designs. Our first measurements to evaluate a debris shield design, known as the "Whipple Bumper Shields," at impact velocities of 10 km/s (i.e., space-type velocities) could now be reproduced in the lab (Ang, Chhabildas, Cour-Palais et al. 1991; Boslough, Ang, Chhabildas et al. 1993)! I got to know Gene Hertel well because he had just joined the group and did CTH simulations that mimicked our experiments (Chhabildas, Hertel, and Hill 1993a). He, along with Luba Kmetyk and Sam Thompson, had transferred to the Solid Dynamics Research Department from the nuclear reactor safety group. The debris generated upon impact by a plate on a thin bumper plate was more collimated and focused compared to those generated by a sphere, making it more lethal to a substructure. This was crucial because space debris is more platelike, and most NASA experimental studies at the time were with a sphere and limited to around 7 km/s. NASA, JSC, and MSFC came up with a "stuffed" bumper shield design consisting of plates and Nextel® cloth material to absorb the debris products. *We tested those before they were deployed for use with the space station Freedom. I felt like this was quite a contribution to our national space program and was satisfying and gratifying at the same time.*

**Technology transfer.** At the request of Cmdr. John Connell of DNA, the technology was transitioned to NRL and the U.S. Air Force Arnold Engineering Development Center (AEDC) for the evaluation of lethality effects on space structures. Bill Reinhart was key in transferring the technology. He is well organized and good at planning and taking care of many details. The impressive thing about those facilities was that we were launching plates for over a millisecond over tens of meters in free flight, taking x-ray pictures, and hitting a target less than a meter from its intended trajectory. At AEDC we were able to launch a 5-mm-thick Ti plate weighing 7.6 g to 9.2 km/s. DNA wanted us to develop a technology to launch "chunky" projectiles instead of plates (Chhabildas, Trucano, Reinhart, and Hall

1994). We leveraged the effort with an internal Laboratory Directed Research and Development (LDRD) project, which allowed us to achieve 16 km/s (Chhabildas, Kmetyk, Reinhart, and Hall 1995).

Cmdr. Connell even supported a basic research project to investigate the kinetics of shock-induced vaporization in zinc and determine the role of projectile shape on the evolution of debris generated upon impact on a thin plate. Rebecca Brannon had joined Sandia in late 1991 or early 1992, and her manager, Paul Yarrington, was looking for a project to get her started. She provided the modeling and simulation (M&S) expertise to complement the experiments with shock-induced vaporization in zinc. Using Ta impactors at 7, 8, 9, and 10 km/s, we fully vaporized zinc at shock stresses up to 5.9 Mbars and determined the time-dependent kinetics effects of multiphase flow (liquid and vapor) resulting from shock-induced vaporization (Brannon and Chhabildas 1995). We also conducted the first shock loading and release experiments in Al and Ti-6Al-4V at 10 km/s impact velocities. We didn't write up our results until later (for the 1998 HVIS meeting). DNA also supported research to investigate the effect of projectile shapes (plate, sphere, and cylinder) on the debris generated upon impact of a thin plate and its effect on a substructure (Konrad, Chhabildas, Boslough et al. 1994). Because we were busy with many HVL research activities, we teamed with Andrew Piekutowski at the University of Dayton Research Institute and Scott Mullin at the Southwest Research Institute (Mullin, Littlefield, Chhabildas, and Piekutowski 1994). Bill Reinhart, working with Andy Piekutowski, learned how to optimize the x-ray radiography setup and implemented many of those techniques at STAR.

**STAR facility shut down.** Jim was now assigned to the program office as a crosscut manager. Sandia had gotten rid of all department manager slots. This prevented integrated and coordinated activities among different branches. Phil Stanton, our Department Manager, as well as Mike McGlaun and Paul Yarrington, the managers of our sister divisions, were now reporting directly to Ed Barsis. DOE had developed an Accelerated Strategic Computing Initiative program and were channeling more resources into computing. Ed had a passion to implement parallel processing and wanted Sandia to be the first DOE lab to come up with teraflop computing speeds. The passion for experimentation was waning because everyone was sold on the idea that we would be able to solve many defense problems using high-speed computing. Sandia management introduced a new budgetary scheme, in which both offices and facilities would start paying for floor space. The annual cost of floor space for STAR was estimated to be over $600 k. Moreover, at STAR we had over half a dozen contractors who were hired "temporarily" to implement the railgun activities. Even though the program ended around 1991, we never got rid of the extra staff. This made our department, with its facilities, expensive to operate. We were quite productive and successful in making many new contacts and increasing our budget under reimbursable programs, since we were the only folks that could mimic space impact conditions in the laboratories at speeds over 10 km/s. However, that was not enough.

Dennis, Phil Stanton, and I, together with Tony Chou from the Army Research Laboratory (ARL), visited Ernst Mach Institut in Germany and the Centre d'Etudes de Gramat in France to promote collaboration. After our return, we found out on Monday, December 12, 1994, that Ed was going to shut STAR down in about six months and get rid of all the contractors within a month. David Cox went to the Physical and Chemical Sciences Center to assist Bob Setchell, Jeff Miller moved to purchasing and contracts, and Ron Moody was assigned to the geological group. We were allowed to keep only one contractor to clean the guns. Bill Reinhart volunteered as the gun cleaner and experimenter so we would be efficient. I tried to talk Bill out of that because I wanted him to look for better opportunities elsewhere. In hindsight, perhaps this was the best decision he made since he became a crucial member of my team at Sandia. The last experiment was on May 5, 1995, and, for all practical purposes, we were shut down. Carl, Clint Hall, Bill, and I were the only folks left operating STAR through May 19: Clint decided to enroll at the University of New Mexico and got placed with Art Ratzel; Carl took an environmental, safety, and health position with the robotics group; and Bill helped Dennis with experiments at the Explosive Components Facility (ECF) on his DOD/DOE Memorandum of Understanding (MOU). We had to make sure no "vultures" came and we prevented Ed and others from removing guns and accessories from STAR.

Ed redirected over $1 M toward computing resources. Mike McGlaun became our Department Manager and Phil Stanton was assigned to a computing department that needed a manager. I went on to do parallel CTH simulations using the PCTH code. I also assisted with the ALEGRA code, which was known as RHALE at the time, and worked with James Peery to validate the code using 1-D experiments. Needless to say, this was my least pleasurable time at Sandia. All this time, I was in touch with Don Cook, the Director of Pulsed Power Sciences, who was sympathetic to what was going on. In September 1995, at the senior leadership off-site meeting, Don had discussions with Ed and offered to take over all shock physics and STAR activities in his directorate. Don shared this concept with me; when I discussed this with Dennis, he was lukewarm about the proposal since he was not sure how we would fit in.

Don's proposal never had a chance because, in December 1995, the entire shock computational and experimental group was put on an impacted list; I have often wondered if Ed did this intentionally so he would not have to transfer the "financial" resources to pulsed power. Sandia was downsizing and offered a Voluntary Separation Incentive Program (VSIP). As an impacted group, we had to get rid of three staff members. Dennis took the VSIP and retired in April 1996, Mike Furnish went to a geophysics group to replace Al Smith, and I moved to pulsed power in March 1996, inheriting Dennis's MOU and Bill Reinhart. I thought the whole situation was ironic in that Ed also retired around January 1996, once Sandia had set a record for teraflop computing speeds.

Tom Mehlhorn was my manager in the Pulsed Power Sciences Center, while Bill was still a member of Ktech staff and was assigned to Lloyd Bonzon at ECF. Lloyd took over control of STAR and offered to operate it in a campaign mode. This allowed us to operate at a reduced floor space rate since we were not using STAR full time and not getting any services from the facilities directorate. We did rod impact studies at ECF and started planning to reopen STAR. Jim Asay came to pulsed power later that year. Don Cook provided seed money of around $20 k to get STAR restarted. We reopened STAR in April 1996 with John Martinez helping part time as needed. The first experiment was a fragmentation study using the Terminal Ballistics Facility (TBF), a small two-stage light gas gun with Bill Reinhart and Clint Hall doing the setup. Don was present for the first experiment. Jim and I were in France and Germany attending the ARA and HVIS conferences. We were, once again, on our way to exploring unchartered territory. Jim would develop and implement shock physics programs for use on the Z facility, and I developed programs to reinvigorate the activities at STAR using traditional gun technology. Little did I know then that the isentropic loading techniques we developed on the three-stage light gas gun to launch projectiles to 16 km/s would form the basis for continuing and extending these studies on the Z machine and propelling flyer plates to yet higher velocities!

**DOD/DOE MOU.** After Dennis Grady's retirement, I inherited $600 k in DOD/DOE MOU funds to redirect the shock physics activities. Fortunately, we had enough supplies and parts (pins, projectiles, target plates, impactors, windows, etc.) to last us a couple of years. We never got rid of those supplies. This was crucial since we used most of the $600 k to conduct experiments and pay salaries. Leonard Wilson of the Naval Surface Weapons Center (NSWC) in Dahlgren played a key role in getting us going. He funded studies to characterize PBX N128, a Navy explosive. We started out gradually in that we used the ECF compressed gas gun first and limited our experiments to gas gun velocities. This is when I developed a technique to achieve uniaxial stress experiments on the gas gun by using a sleeved cylindrical target material and impacting it with a graded-density, layered impactor. My CTH "lessons" were paying off. The experiments were designed with CTH, which I used to understand the complex multi-dimensional wave interactions that occur in cylindrical geometry. *The surprising outcome was that we could conduct uniaxial stress experiments at strain rates higher than Hopkinson bar experiments but lower than shock loading experiments!* We coined these "intermediate strain rate" experiments. We investigated ceramics such as alumina, silicon carbide, and boron carbide (Chhabildas, Furnish, and Grady 1997). I was also spending half time doing ALEGRA code validation. With project development, planning for experiments, and ALEGRA validation studies, I had little time for data analysis. I spent time teaching Bill Reinhart shock physics so he could start doing the data analysis. I was fortunate that Bill was motivated and started to plan shock physics experiments based on what he was learning.

Leonard Wilson was also interested in evaluating the response of concrete, based on large or small aggregate size. This was a funded program for a couple of years, and Waterways Experimental Station sent us the concrete samples for free. We leveraged their resources and our DOD/DOE MOU project stretched out quite a bit. I had no choice but to get both Clint Hall and Bill Reinhart involved in conducting data analysis. These were both busy and interesting times; I helped Clint Hall use the concrete studies as a thesis project to get his master's degree. With no active shock physicist on the staff, Mark Anderson from the Physical and Chemical Sciences Center also did a thesis project with me by investigating the PVDF gauge response embedded in PMMA under isentropic loading to 40 GPa (Anderson, Chhabildas, and Reinhart 1998). The PVDF gauge did not perform well due to PMMA polarization effects but we observed that PMMA remained transparent to 40 GPa since we were able to measure the VISAR interferometric signals through PMMA buffers. PMMA is opaque under shock loading at about 22 GPa. I believe that, due to lower temperatures with the ICE technique, we suppressed the bond breaking that occurs under shock loading around 22 GPa. Clint and Mark got their master's degrees in 1998. Clint also got the best paper award for the concrete study at the ARA meeting that year (Hall, Chhabildas, and Reinhart 1999). These projects helped us get the powder gun and the TBF guns operational at STAR, and we were still using the ECF compressed gas gun. At that time, the control and operations of STAR were still managed by Lloyd Bonzon.

**MAVEN program.** Our DOD/DOE MOU was doing quite well with complementary funding from DOD agencies such as the ARL, the Air Force Research Laboratory (AFRL), and the NSWC. In addition, many customers provided finished samples such as ceramics (from ARL), AERMET[51] steel (from NSWC), and tantalum (from AFRL). This kept our operating costs down and made our MOU funding stretch further. In the meantime, in 1996 Sandia realized the error in shutting down facilities and introduced a program called MAVEN. The objective was to provide data for modeling, analysis, and validation. We got two MAVEN projects, one to characterize PZT 95/5 ceramics and the other to validate the ALEGRA code. Jaime Moya was the project manager for the entire MAVEN program. Although I don't remember the details, I believe each of our projects was around $600 k. Now I had over a million dollars and no staff to execute it. We formed a team that included Robert Setchell, Mark Anderson, Dave Cox, and Mike Furnish for the PZT ferroelectric studies using Bob Graham's compressed gas gun in the Physical and Chemical Sciences Center; Carl Konrad, Clint Hall, Bill Reinhart, and Mike Wong, as well as Dan Mosher from Livermore, helped with the ALEGRA validation studies on the TBF gun at STAR. Carl and Clint also helped Jim Asay with the early Z experiments. I know they spent more time on Z than at STAR—which made me realize that rank did have its privileges. These were important projects; *the former provided the first simultaneous mechanical and electrical characterization of PZT 95/5 for use with the SUBWAY code developed by Steve Montgomery, and the latter provided the first experimental data for use with ALE[52] codes, which transitioned hydrodynamic high strain rate deformation behavior to long-term structural response*. We noticed that the highly porous polystyrene foam was not modeled correctly and had to extend the study to characterize the foam to determine its compaction behavior. When Paul Hommert became the Director of Engineering Sciences, he redirected many of these projects to complement programs that originated in his directorate. We lost our DOE

---

[51] Editors' note: AERMET is a steel alloy that contains cobalt, nickel, chromium, molybdenum, and carbon.

[52] Editors' note: ALE stands for Arbitrary Lagrangian–Eulerian.

funding again! Nevertheless, the MAVEN program helped us for a couple of years to bring STAR back by providing the funds to upgrade the oscilloscopes and laser systems. Digital oscilloscopes were just coming online; it was difficult to get capital equipment to upgrade many of our recording systems. Some lasers and interferometers at STAR went to Z to help out with its diagnostic needs.

**MDA program on the three-stage light gas gun.** The Missile Defense Agency needed to validate engineering codes used for lethality at impact velocities related to space engagement. One approach was to use hydrodynamic codes to simulate the lethality requirements and use the computational data to develop the engineering code. Bob Becker from MDA and Dan Kelly from Sandia approached us in early 1999 to provide the equations of state data for materials of interest, using the three-stage light gas gun to validate system-level simulations. This required that we get our three-stage gun operational! We had not shot the gun since 1993, and here in late 1999, we had to get it ready and get the data within 6 months! On top of that, Carl Konrad announced in early January that he was retiring from Ktech and would move to Nevada to work on programs relevant to the JASPER two-stage gun at NTS. Since we needed a Sandia operator to run the STAR facility, we made Bill Reinhart a permanent Sandia employee, Tom Thornhill replaced Carl under the Ktech contract, and we got help from Kathleen Holland, a postdoc working on pulsed power projects who got assigned to the MDA project. In a very short time, we got the three-stage gun up and running. Most of the credit goes to Bill for efficiency and timeliness. Gerry Kerley provided theoretical support and Dan Carroll much needed M&S support. Nancy Winfree from ARA also joined the team; that was fortunate and timely because Kathleen left for California once she got her PhD. (Even though we had hired her as a postdoc, she had not completed her thesis defense.)

So now we were operating the TBF gun, two-stage gun, and three-stage gun with a huge supporting staff to implement the project. One million dollars a year for 3 years from MDA expanded our capabilities quite a bit. Silica phenolic and graphite epoxy were materials of interest, and we determined the shock loading and release behavior at engagement (impact) velocities up to 12 km/s on the three-stage gun. We also determined the equation of state of Ti-6Al-4V on the three-stage gun, observed a kink in the shock velocity vs. particle velocity history at around 10 km/s impact velocity, and were concerned about systematic errors that were not taken into account. Later, Gerry Kerley independently predicted melt in Ti-6Al-4V, and our three-stage gun experiments just happened to confirm his melt predictions. (When Gerry predicted melt, he was retired and had moved to Blacksburg, VA, and started his own consulting company.) During the third year of the project, we did code validation studies to determine the effect of debris propagation on substructures at impact velocities up to 12 km/s. Gregg Bessette, who had recently joined Sandia, and Jeff Lawrence were tasked to provide the M&S support. At the end of the 3-year project when I asked Bob Becker how else we could help him, he said, "I thought it would take you five years to accomplish what you did and I had budgeted accordingly. Unfortunately, you delivered what I wanted in three years and it was your fault that you finished it early." This has been one of our problems with getting involved with a lot of reimbursable projects or "Work for Others."[53] *Once you deliver, your customer has no need to continue funding and you have to plan ahead and develop new technology to engage new customers.* Even though I was concerned that my funding was not going to be as plentiful, I took Bob Becker's comment as a compliment. At this point, we started planning to get involved with spectroscopy measurements as a tool in target typing to assist MDA in the future.

**MDA/CLP Program.** Around the same time, John Cogar of NSWC at Dahlgren and also funded by the MDA/Corporate Lethality Program (CLP) was interested in investigating tributylphosphate liquid as a surrogate for certain chemical agents. This allowed us to team with Nancy Winfree and Dennis Grady in ARA for M&S. This led to a joint program in which we transferred our technology to the Defence Science and Technology Laboratory at Porton Down, UK, and assessed real chemical agents. We assisted them in getting their shock and ballistics technology going. Our counterparts were Major Sylvia Ferry at MDA/CLP and Nicholas Robinson in the UK.

**LDRD on dispersive particle velocity measurements in heterogeneous materials.** Wayne Trott, Jim Asay, and I independently submitted LDRDs in 1998 to develop the line-imaging VISAR, now known as ORVIS, to investigate spatial inhomogeneity in heterogeneous materials relevant to the DOD/DOE missions such as concrete, foams, alumina-filled epoxy, etc. Because it was submitted by the three of us independently, Paul Hommert, at that time the Director of Engineering Sciences, funded the project with a lot of behind the scene convincing by Jim. Wayne was using ORVIS to investigate and measure the velocity of laser-driven flyers. We noted that the fringe record indicated the curved nature of the propelled flyer. It dawned on us that, with some additional research and development, we could use the technology to determine the particle velocity dispersion in shock-loaded heterogeneous materials. This was my first LDRD since we reopened STAR, and Wayne took the lead in developing the capability. We used the LDRD to determine (a) the heterogeneous loading response of concrete, Ta, and foam; (b) the multi-dimensional loading of alumina-filled epoxy, Ta, and sapphire (i.e., edge effects); (c) the heterogeneous spallation behavior of Ta at incipient spall conditions; and (d) the dynamic yield response of boron carbide and spallation behavior of silicon carbide (Trott, Castañeda, O'Hare et al. 2001). Marlin Kipp did the 3-D mesoscale modeling of foam and concrete to simulate these experiments. *I guess we had just developed an experimental tool to investigate mesoscale response of shocked materials.*

In addition to being funded by different organizations within MDA from 1998 to 2006, we were quite proactive and also submitted LDRDs. We were quite aggressive and developed new capabilities even though we were supposed to be shut down and understaffed. I recruited Jean-Paul Davis and Tracy Vogler. Jean-Paul decided to participate on Z-related projects, while Tracy decided to join our team in 2001. Jim retired in September 2002 and moved to Washington State University and I was

[53] The term "Work for Others" was formerly used to refer to projects funded by organizations other than the Department of Energy.

promoted, becoming the Department Manager in March 2003, in part taking Jim's acting level I manager position. Dillon McDaniel became my level II (senior manager) boss.

**Shock kill of biological agents.** Larry Larsen, who has a medical degree, had his office right across from mine in Building 962. After a couple of years saying hello in the hallways, we finally made acquaintance as we were walking into the building together and found out about each other's background and current research area. I inquired if his medical background could help us investigate shock kill of biological agents. This turned out to be a complementary LDRD project in which his team provided the microbiological aspects pertaining to research, while my team provided the shock research capabilities. Jerry McDowell funded the project because of interest in defeating biological systems. This was a highly successful and fun project: I learned new things about microbiology and colony-forming units, a field that of which I had no knowledge. Both teams worked very well and even won an award of excellence for the research from Sandia.

**LDRD on impact flash.** In FY[54] 2004 Jeffery Lawrence was awarded an LDRD titled "Impact Flash." The objective was to develop a technology and pursue impact flash signatures to identify the target by spectroscopy. This 2-year effort gave us an opportunity to develop techniques to determine the spectral output as a function of time and wavelength in both the infrared and visible regimes. We developed the technique to investigate Al, indium, and Comp B,[55] and we also applied the technique to materials of interest to MDA, such as silica phenolic. This project also allowed us to investigate rate-dependent, shock-induced vaporization kinetics systematically in Al and indium, while we were simultaneously determining the spectral signatures from these systems (Chhabildas, Reinhart, Thornhill, and Brown 2006; Lawrence, Reinhart, Chhabildas and Thornhill 2006). The most significant aspect of the project was that it kept us engaged with the MDA/CLP program through FY 2006. I understand that the program has since expanded to include modeling and simulation to predict the spectra observed by combining CTH results with what I call the spectral code.

**LDRD on advanced diagnostics for impact flash spectroscopy on light gas guns.** When Michael Wanke from the Physical and Chemical Sciences Center saw our results on impact flash spectroscopy, he got interested in extending the studies to the terahertz or mm wavelength regime and submitted an LDRD. That LDRD was approved in FY 2005 and allowed us to extend our capabilities to the terahertz regime and to develop advanced diagnostics such as semiconductor detectors. Terahertz detectors extended the impact flash spectroscopy to late time regimes after an impact engagement when the debris products would be at a much cooler temperature (Wanke, Grine, Mangan et al. 2007). This is quite unique in that we

[54] Editors' note: FY is the acronym for fiscal year. The government fiscal year is from October 1 to September 30.

[55] Editors' note: Comp B is an explosive used in artillery projectiles, rockets, hand grenades, and other munitions.

were now investigating the results of an impact engagement from early times at very high temperatures with visible spectroscopy to late times with terahertz spectroscopy. As indicated above, many of these technologies made us unique and allowed us to be engaged with MDA through 2006.

**DOD/DOE MOU on dynamic properties of warhead materials.** As indicated above, I inherited the DOD/DOE MOU from Dennis Grady after his retirement. We kept Dennis on as a consultant and that was perhaps the best investment we made. Dennis, even though he was not a Sandian anymore, remained a Sandian at heart and was instrumental in making this MOU a success. We collaborated on many ideas after his retirement. In many respects, he never left Sandia. *Even though the DOD/DOE MOU was only $600 k, we were able to leverage many DOD agencies since this was a good vehicle to network with the DOD staff.* After Tracy Vogler joined the staff, I handed over the project to him in 2004 to gain experience with managing technical projects and give him the opportunity to network with others in the technical community. We also developed many new experimental techniques and explored new frontiers. Examples have *included developing an intermediate strain rate loading technique* that allows strain rate regimes higher than Kolsky bar but lower than those under shock.

**Dynamic fragmentation.** We also implemented techniques to measure the dynamic fragmentation of a cylindrical tube on the TBF gun. Impacting a stationary plastic/Lexan® cylinder located at the bottom half of a cylinder with an identical moving Lexan® cylinder produces reaction products due to dissociation and expands the cylinder radially. The process is very reproducible and provides a well controlled environment to use state-of-the-art multiple high-speed diagnostics simultaneously for fragmentation studies in the laboratory. This was a collaborative program with NSWC at Dahlgren and with Livermore. Justin Brown, a college student at the time, assisted with the data analysis and presented these results at an American Society of Mechanical Engineers meeting and also at an ARA meeting, where he won the best paper award. Tracy Vogler extended these studies to include interferometry and complemented them with CTH simulations. His paper received the best paper award at the 2002 HVIS conference in the Netherlands (Vogler, Thornhill, Reinhart et al. 2003). Even though Sandia is no longer pursuing the dynamic fragmentation project, the research continued and is flourishing at other DOE labs.

**Strength of ceramics.** Around this time, the Army's vision shifted to developing future combat systems. A part of their goal was to develop lighter but stronger armor systems, which necessitated determining the strength of ceramics in the shocked state. We applied shock-release and reshock experiments to ceramics to estimate the strength of alumina, sapphire (with Bill Reinhart), boron carbide, and silicon carbide (with Tracy) to megabar pressures using both the powder gun and the two-stage light gas gun. By this time, all the guns at the STAR were operational. This was a leveraged program with Datta Dandekar of ARL; ARL provided the ceramic samples for the experiments. In addition, we used the Line VISAR to estimate the variation in particle velocity distributions in boron carbide and silicon carbide to determine the heterogeneous loading and spall response. These experiments, for the first time, established that shock and reshock experiments always display an elastic precursor upon reshock even at megabar pressures. The experiments formed the basis of the Johnson–Holmquist models for ceramics and were used to design advanced armors and human vests and future combat systems.

**Reactive materials.** The DOD community was interested in developing materials to enhance lethality. One candidate material was aluminum Teflon®. A million dollar effort per year was funded for 2 years by NSWC, for which we developed an equation of state model up to almost 80 GPa. The spall signatures were different from those of most materials and appeared to be stretching out very gradually. Reshock and release waves also displayed the leading edge traversing at the same velocity. Leonard Wilson of NSWC was our point of contact for this effort.

**Glass.** Our contributions to develop equation of state models and the strength of ceramics in the shocked state got us engaged with the Tank Automotive Research Development and Engineering Center (TARDEC). They wanted to investigate transparent armors; the candidate materials were borosilicate and starfire glass. The project was timely in that I had just hired Scott Alexander and we used the project to expose him to the fundamentals of shock physics. Guess what? When he wrote the paper for HVIS in 2007, he got the best paper award. This project lasted 3 years at approximately $500 k per year and continued for a year after my retirement in 2007. My point of contact at TARDEC was Doug Templeton.

**Porous sand.** Once Tracy Vogler took over the DOD/DOE MOU, he got interested in porous and granular materials. He developed techniques to investigate shock propagation using multi-stepped targets. I believed they coined the phrase "pyramid" target. Justin Brown got involved in analyzing the data and also using CTH to fit the experiments to P-α and P-λ[56] models. This prepared him well for his graduate studies, and he went to the California Institute of Technology for his PhD. Tracy had extended these studies to other granular materials, and John Borg of Marquette University spent a summer with us to investigate porous materials at a mesoscale level. I had known John when he was at the NSWC and had worked with him on the chemical agent program for tributylphosphate. Some of these projects represented our initial exposure to mesoscale modeling. Even though we did not have a formal program on mesoscale modeling and experiments, as a team we accomplished quite a bit.

**Acknowledgments.** My Sandia years were productive, fulfilling, and extremely gratifying. This is what I conclude as I go back and recollect my career. This occurred in spite of a huge cultural shift in going from an abundant entitlement-type research environment in the early years to justifying your existence in the latter years since the Berlin wall came down. Although I was quite concerned at first, I adapted rather well. I attribute that success to my father, Chhabildas Chhaganlal Mandalaywala, who was in business all his life. I knew he had good years and not

[56] Editors' note: The P-λ model is a generalization of the P-α model to account for the dynamic compaction of multiphase mixtures of porous materials.

very good years as business picked up or went through a lull during unforeseen political events over which he had no control. These uncertain events convinced me that getting an education and getting involved in research was one way of bringing stability into your life. “Business is not for me; it is too stressful” is what I said as I went into research to lead a comfortable and relaxed life. The irony in many respects is that since the early 1990s, I became a businessman, in particular, once we developed the three-stage light gas gun. It is quite apparent that I picked up considerable business attributes subconsciously from my Dad.

As I reflect on things, I think we have probably averaged a million dollar per year in reimbursable projects since 1992. That is a minimum of $15–20 M return into which many of our research projects blossomed. This taught me quite a few things: (a) it takes a minimum of 3 years to develop a new project internally or externally, (b) you have to establish uniqueness as you develop the technology, (c) you have to develop new technology and new projects continuously to remain relevant, and (d) you have to have a good team that supports you and is willing to contribute their best for *all to succeed.*

I would like to acknowledge Barry Butcher and Walt Herrmann for taking the risk in hiring me. Ever since I walked into Sandia, Jim Asay has been an excellent mentor and taught me not only the basics of shock physics, but also how to think and to plan programs and projects strategically. Jim is unique: he is an extremely good manager and very good technically. It is this combination that has led to many of the successes we have had as a group. As we were muddling through the lean years in the mid 1990s during the STAR shutdown, my friendship with Dennis Grady as a colleague and research partner was quite important and crucial. I would like to thank Carl Konrad and Robert Hardy who, during my early years, taught me about the operation of the powder gun and two-stage light gas gun. We are indebted to Don Cook for having the vision to make shock physics a core competency for pulsed power research. This gave us a fresh life and outlook on many programs. Even though the gun technology and pulsed power research never got integrated as well as I would have liked, it laid the foundation from what was developed earlier as it transitioned to Z. More significantly, we have continued to develop many new research opportunities. Examples of these opportunities include studies of fragmentation, mesoscale techniques, spectroscopy, and optical pyrometry—just to name a few.

I am very fortunate to have had the opportunity to team with Bill Reinhart, who is well organized and ran STAR as a business venture when we really needed it. His planning, scheduling, and desire to learn shock physics were important attributes that made our STAR operations cost effective and were instrumental in meeting our Work for Others customer deliverables. I was also very fortunate to hire Tracy Vogler, whom I could entrust with many of my projects when I became the manager and who was dependable for both technical accomplishments and new project developments. As I mentioned earlier, I had an excellent team, both as a project manager and later when I became the Department Manager, and I would like to acknowledge their dedicated contributions that allowed all of us to shine. Even though I did not really want to be a manager, I realized that being a manager of a research group had its privileges—you can control your own destiny much better.

# Michael P. Desjarlais

# (1986–Present[57])

**Editors' comments**: Mike Desjarlais led the theoretical development of *ab initio* quantum molecular dynamics (QMD) theories to describe the experimental Hugoniot and ramp loading data obtained at ultrahigh shock pressures on the Z accelerator. Desjarlais made a major contribution to the experimental program by using these theories to show that the conductivity of aluminum at high pressures and temperatures was considerably higher than predicted by existing theories. This groundbreaking discovery showed that near-ambient temperature, magnetically-driven flyer plates are obtained with the current-generating capability of Z and have established the credibility of ultrahigh shock wave measurements on Z. In a note to Jim Asay, Desjarlais provided the following annotated discussion of five papers involving the progress of applying QMD analysis to experimental high-pressure equation of state studies.

Five key papers summarize the development of *ab initio* theories for describing high-pressure equation of state data obtained on the Z accelerator:

1. M.P. Desjarlais, J.D. Kress, and L.A. Collins, "Electrical conductivity for warm, dense aluminum plasmas and liquids," *Physical Review E* **66**, 025401 (2002).

The *Physical Review E* paper established the power of using QMD simulations to create accurate conductivity models for materials away from ambient. The first application in shock wave research was to the aluminum conductivity, needed for modeling the flyer plates. This advance was essential to doing predictive design simulations and, as an immediate consequence, rapidly advanced the pace of data accumulation by permitting much more accurate pre-shot design.

2. M.P. Desjarlais, "Density-functional calculations of the liquid deuterium Hugoniot, reshock, and reverberation timing," *Physical Review B* **68**, 064204 (2003).

The deuterium *Physical Review B* paper was the first application in our shock physics research to equation of state properties. While others had performed QMD simulations previous to my work, they had failed to establish agreement with the gas gun Hugoniot data that were widely accepted as accurate. In this paper, I demonstrated that the convergence of both energy and stress was essential to getting good Hugoniot results and that prior work had not met the stricter requirements for converging the stress. I also included zero point corrections for the reference state,

[57] Editors' note: Upon joining Sandia, Mike was initially involved in the ion beam program. When that ended in 1996, he began development of better conductivity models in support of the z-pinch program, which led to the development of *ab initio* tools and better conductivity models. As of this writing, he is continuing *ab initio* studies of high-pressure equation of state properties. Mike became a Distinguished Member of Technical Staff in 1996 and was promoted to Senior Scientist in 2011.

another contribution that earlier work had neglected. This research gave good agreement with both the gas gun data and the data from Z and did not agree with the high compression ratio results from the Nova laser at Lawrence Livermore National Laboratory.

3. M.D. Knudson, M.P. Desjarlais, and D.H. Dolan, "Shock-wave exploration of the high-pressure phases of carbon," *Science* **322**, pp. 1822–1825 (2008).

The diamond *Science* paper demonstrated just how powerful the QMD work could be in helping to interpret data, provided the data quality was high, as it is in Marcus Knudson's Z work. The subtle changes in $U_s$-$u_p$ as the Hugoniot traverses the diamond-liquid-bc8 coexistence region were predicted by the QMD calculations, observed in the data, and gave convincing evidence for the diamond-liquid-bc8 triple point and the first experimental evidence for bc8 in carbon. The QMD work also allowed us to construct that now famous plot (thanks to Prof. Yogi Gupta of Washington State University) of the 3-D Hugoniot coexistence landscape.

4. M.D. Knudson and M.P. Desjarlais, "Shock compression of quartz to 1.6 TPa: Redefining a pressure standard," *Physical Review Letters* **103**, 225501 (2009).

The quartz *Physical Review Letters* paper shows once again the very good agreement between the experiments and the QMD calculations. The QMD work provided the explanation for the nonlinear $U_s$-$u_p$ relation through the observation of an extended dissociation regime and the corresponding consequences for the specific heat. This, in turn, helped motivate the Hugoniot fit.

5. M.D. Knudson, M.P. Desjarlais, R.W. Lemke, and T.R. Mattsson, "Probing the interiors of the ice giants: Shock compression of water to 700 GPa and 3.8 g/cc," *Physical Review Letters* **108**, 091102 (2012).

The water *Physical Review Letters* paper made extensive use of our ability to calculate accurate release isentropes with the QMD, which is essential to getting accurate impedance matching results for low impedance materials. Note that the equation of state data from Z compares well with predictions of the University of Rostock (Germany) group that has, through collaboration with Sandia, adopted our QMD methods.

# George E. Duvall[58]
## A Tribute by the Editors

Several universities were instrumental in making major contributions to the shock wave program at Sandia National Laboratories, but without a doubt the shock physics program at Washington State University (WSU) under Prof. George E. Duvall's leadership is unsurpassed. Prof. Duvall trained and mentored a large number of graduate students in shock wave research, and many went on to achieve stellar careers in this field. Prof. Duvall was a pioneer and many regard him as the father of shock compression science in the United States.

Originally from Louisiana, Prof. Duvall entered Oregon State University (OSU) to major in physics, but his studies were disrupted in 1941 because of World War II. He spent 4 years studying underwater acoustics at the University of California in support of the war effort. In 1945 he returned to OSU to finish his bachelor's degree in Physics and then joined the Physics Department at the Massachusetts Institute of Technology, obtaining a PhD in 1948. Upon graduation, he was hired by the General Electric Company at Richland, Washington, to address nuclear reactor problems. In 1953, he joined the Poulter Laboratory at the Stanford Research Institute (SRI) in Palo Alto, California, becoming the Director of the Laboratory in 1962. His leadership established the Laboratory's excellence in theoretical understanding of shock and detonation wave propagation.

During his tenure at Poulter Lab, he became acquainted with Prof. William Band, who was the chairman of the Physics Department at WSU. Thus began a lasting professional and personal friendship that resulted in Prof. Duvall joining the Physics Department faculty at WSU in 1964. He initiated theoretical studies of shock wave

---

[58] Photograph of George E. Duvall taken when he received the Shock Compression Science Award. Photo was reproduced with permission from *Shock Compression of Condensed Matter 1989*, edited by S.C. Schmidt, J.N. Johnson, and L.W. Davison, North-Holland Publishing, p. viii (1990). Photo copyrighted 1990, Elsevier LTD. The Physics Department at WSU graciously provided the background information on Duvall's early history that was used in this tribute.

problems at the university in collaboration with Prof. Band and shortly thereafter expanded to include experimental research by recruiting G. Richard (Dick) Fowles from SRI. This was a period of rapid growth for shock wave research. At the time, the experimental facilities and corresponding diagnostics being developed by Don Lundergan, Lynn Barker, Bob Graham, and others at Sandia were making ground-breaking strides in the field. Professors Duvall and Fowles had contact with Don Lundergan, who had developed the first Sandia 100-mm-bore gas gun launcher for experimental shock wave research. They instigated plans for a similar facility in the Physics Department at WSU. By 1968, that facility, named the Shock Dynamics Laboratory, became fully operational with a variety of instrumentation capabilities. It served as a principal tool for experimental shock wave research by many graduate students, including one of the editors and several of the contributors to the recollections in this book. The facility is still operational with considerably expanded capabilities.

As a professor in the WSU Physics Department, Duvall focused his intellectual energy on a variety of scientific problems that included equations of state for condensed matter, thermodynamic processes during shock wave transitions, physical mechanisms and kinetic effects associated with shock-induced phase transitions, and dynamic yielding under shock compression. His deep understanding of a broad spectrum of shock wave phenomena attracted a large number of students, resulting in his supervision of over 25 doctoral dissertations during his 24-year tenure at the university. He had an enthusiasm and passion for shock wave physics and a simple but elegant theoretical approach, which helped shape scientific appreciation of the discipline in his students. He instilled in them a passion and devotion to the subject, as is apparent by their subsequent careers.

Because of his seminal contributions to shock compression science, Prof. Duvall received the Shock Compression Science Award in 1989 "In recognition of his outstanding contributions to shock wave physics and his educational and organizational leadership in the shock physics community." The award was first established in 1987 by the friends of the Topical Group on Shock Compression of Condensed Matter and is given every 2 years. In 2007, the award was renamed the George E. Duvall Shock Compression Science Award in his honor. Six of his former students have received this prestigious award, including three who joined Sandia and contributed to the shock wave program.

A large number of universities have participated in elevating shock wave research at Sandia to a world-class level, but Prof. Duvall at WSU has been a dominant leader and one of the principal forces for Sandia's excellence in the field. Over a dozen of his graduate students joined the Sandia shock wave program. These researchers distinguished themselves in a variety of specialized areas, including:

1. development and application of multiphase equations of state in hydrodynamic codes to determine kinetic effects of phase transitions,
2. development of generalized fragmentation theories for a variety of military and scientific applications,

3. development of a general relationship to describe high-pressure shock wave structure,
4. development of new methods to determine the flow strength of shocked materials at Mbar pressure levels,
5. development of an unprecedented time-resolved optical recording interferometer that enabled studies of shock wave structure and extended to studies of heterogeneous material response,
6. development of magnetically produced ramp waves to Mbar pressures that had been a goal of the shock wave community for several decades, and
7. extension of magnetically-driven flyer plates to velocities of several tens of km/s, which enabled the first precise comparison of Hugoniot data with *ab initio* equation of state theories.

These accomplishments have had a profound impact on the high-pressure science community. In addition, several of Prof. Duvall's students achieved middle-level or high-level management positions at Sandia, and one became the president of a large research organization.

Prof. Duvall died January 3, 2003, in Vancouver, Washington, at nearly 83 years of age, leaving an impressive legacy of accomplishments. His pioneering research will be long remembered and his dedicated approach to shock wave problems will hopefully expand and grow through the research of his former students.

A number of other university professors have also had a major influence on Sandia's shock wave research. Although it is not possible to include all of these professors because of space limitations, we specifically mention one here. Prof. Y. M. (Yogi) Gupta has continued Prof. Duvall's leadership by expanding the shock compression science program at WSU and establishing stronger ties with the DOE national laboratories, especially at LANL, LLNL, and SNL. He has played a leading role in supporting the overall shock wave program at Sandia and the development of the Z pulsed power accelerator for shock wave studies. Other professors who have contributed to the Sandia shock wave program are identified in the individual recollections and through joint publications that are listed in the Bibliography. These extensive collaborations have served to make Sandia's shock wave research the premier program that it is today.

## Michael D. Furnish

## (1987–Present[59])

First, here is a word about how I came to the field of shock physics. In 1982, in the midst of squeezing olivine in diamond cells at Cornell (actually, finishing classes), I found myself at home in Los Alamos for Christmas break. I'd been hearing about "shock wave standards" for pressure measurement and thought it would be a good idea to get to know them better. So I called up Bob McQueen in the Shock Wave Physics group (M-6), and he said "Come on down" to Ancho Canyon. He escorted me in at 9 in the morning, and I naively thought we would talk for an hour or so. Four o'clock that afternoon, I walked out with my belly still empty and my head feeling like it was going to pop with everything he, John Shaner, Rob Hixson, and others had stuffed in there.

This led me to apply to work at Los Alamos National Laboratory (LANL) for the summer of 1984, collaborating with Mike Brown, newly in the Geophysics Program at University of Washington, and later that summer with Dennis Grady at Sandia. I ended up with a set of single-crystal olivine shots on the two-stage gun in Ancho Canyon and another set on the STAR powder gun at Sandia. These wound up in my thesis, as well as in the *Journal of Geophysical Research* and in *Shock Waves in Condensed Matter*, the proceedings of the fourth American Physical Society (APS) Topical Conference on Shock Waves in Condensed Matter. For the latter, I remember walking into my first APS conference in a strange city (Spokane) in 1985 to be greeted by someone who told me my parents were not disciplining their dog very well. Bob McQueen, the Los Alamos shock physics pioneer, was a dog trainer, too.

From late 1985 until mid 1987, I was at the university squeezing various materials in diamond cells, as well as analyzing shock data with Mike Brown and his collaborators as a postdoc.

In mid 1987, I joined Jim Asay's group at Sandia and came under Dennis Grady's wing for a little while in a study of Mini Jade Two[60] using the powder and two-stage guns at STAR. At that time, the Thermomechanical and Physical Division was the experimental shock physics group, as a sister division to Paul Yarrington's computational applications group and Sam Thompson's computer modeling group. Sam's group was just starting the CTH project and dabbling in parallel processing. That was a really nice setup, like a tripod, under Lee Davison's overall lead (and, in turn, under Walt Herrmann as director). And, in some ways, it resembled what Tom Mehlhorn and Mark Herrmann recently built up in the Dynamic Material Properties Department of the Pulsed Power Sciences Center.[61]

Over the next 9 years, my work on geological materials evolved. With the gas gun at STAR, we studied a wide range of grouts, tuffs, limestones, granites, and similar

---

[59] As of this writing, Mike Furnish continues shock wave experimentation at Sandia using a variety of facilities including the STAR facility and Veloce.

[60] Mini Jade Two is a high-silica, high-water content grout. See Grady and Furnish (1988, 1990).

[61] Editors' note: In 2013, Mark Herrmann was promoted to the Director of Pulsed Power Sciences and Dawn Flicker was promoted to his previous position as the Senior Manager of the department.

materials. I provided quite a bit of support to the Defense Nuclear Agency (DNA), working with the Field Command Defense Nuclear Agency and many contractor shops supporting weapon effects and ground shock studies—i.e., RDA Logicon Corporation, Science Applications International Corporation, Systems, Science and Software, and so forth. Then we had a task to perform gas gun studies of natural snow and "snow simulants" for use in underground tests. What happens when you set off a device in the air above a deep snow blanket? Jim Asay referred to this as the "snow job," and I was the principal investigator. One day I found a stack of boxes waiting for me in the department office. The top one had foam (a snow simulant). The second had insulation. The third had something of similarly low density. So I was lightly grabbing the boxes off of the table. The fourth box had a thick slab of tantalum. I didn't mention that I was also doing shock studies of bcc (body-centered cubic) metals with Dan Steinberg and Lalit Chhabildas, looking at strength under loading and release. I recently returned to this topic (and to steels as well as bcc metals), more aware of the nature of nonsteady waves and rate dependence.

I have notes from about 1990 at which we met with PBFA II (Particle Beam Fusion Accelerator II) scientists to discuss the possibility of using that Sandia accelerator for dynamic material research. We could not figure a way to use the very high energy density to get where we wanted to go, so we went back to what we were doing. Seeds were planted, however, and in recent years PBFA II (after 1997 renamed and reconfigured as Z) has become a vital facility for dynamic material research to extremely high pressures.

A particularly challenging project in 1993 required measurement of the high-pressure equation of state of silica, beginning with synthetic stishovite. Eichi Ito of the University of Tokyo supplied us with small pieces of that material, which we shocked in a reverse-ballistic fashion (sample into projectile) on the two-stage light gas gun. The difficulty here was ensuring that the VISAR diagnostic was looking at the correct spot on the target. The gun systems at the time were not really set up to handle such small (3-mm-diameter) samples. Our uncertainties were therefore fairly large. Such a project would be so much easier now with the experimental configuration developed on Z than with the two-stage light gas gun. Recently, I have conducted several such Z tests in collaboration with Yingwei Fei of the Carnegie Institute of Washington, D.C.

Another interesting project on ferroelectric materials led to a number of papers in collaboration with the Physical Chemistry Center and the neutron generator staff. This was a chance to become familiar with the Line VISAR, as well as to learn that electromagnetic fields can drastically affect the mechanical response of materials—something that may not be intuitive to many but that scientists in the Pulsed Power Sciences Center have come to appreciate.

Two things came together about 1996. First, my work on grouts and rocks brought me closer and closer to problems involving the Nevada Test Site[62] (NTS), and, second, the gun work at Sandia started to dry up. Ed Barsis, our center director, decided he would rather buy computers than pay the space charges to keep STAR

[62] Editors' note: Recently the name was changed to the Nevada National Security Site.

open. So he closed STAR. There is a long story here—STAR eventually reopened (as discussed in the 1990s chapter of this book). So, I'll discuss it in more detail momentarily; when STAR closed, I changed groups to begin supporting the Sandia efforts at NTS.

Let me back up to introduce the status of the Nevada experiments at that time. It was not long after the last nuclear test in September 1992 that the weapons design labs began actively pursuing experiments to address the requirements of the Nation's Stockpile Stewardship Program and make use of NTS. At Los Alamos, the initial subcritical experiment (SCE) slated for NTS, Rebound-I, moved to the stage of developing "local" tests at the start of 1996. These local tests were conducted at the sponsoring laboratory [Los Alamos, Livermore, and eventually also AWE (the Atomic Weapons Establishment)], in preparation for fielding the experiments at the NTS.

Sandia had teams in Albuquerque and NTS to support the Los Alamos efforts. We had much more limited participation in the Livermore efforts, which also began around the same time. David Thompson led the Sandia group at NTS, which also included Mike Burke, Rod Shear, and Jerry Chael (among others), with Nira McCoy of Bechtel Nevada as the secretary. (Nira continues in that position to this day.) In Albuquerque, several departments had inherited test site work from the underground test era. However, major rearrangements were in process at the beginning of 1996 because of the diminishing work. A Sandia-wide Voluntary Separation Incentive Program identified that effort as "impacted," specifying that the staff should be reduced from 20 to 16. As it turned out, five staff wanted to retire with the incentive. Meanwhile, the Experimental Impact Physics area (which had suffered the loss of STAR) was also "impacted" and expected to shed two of its ten staff. I transferred from Paul Yarrington's department to the NTS support department. This resolved one of the Experimental Impact Physics area reductions and simultaneously allowed the fifth NTS-related staff member (Carl Smith) to retire with the incentives. In this way, Tom Bergstresser (who had earlier transferred to the team) and I became the core of the Sandia team concerned with fielding velocimetry and pyrometry diagnostics on SCEs under the management of Wayne Cook and later Bill Boyer. Doug Garbin was the Sandia seismologist (and was later joined by Rob Abbott).

Rebound-I was a set of three experiments detonated simultaneously in a single, mined zero room[63] with coupon samples of plutonium impacted by a metal flyer plate driven by explosive plane wave lenses and boosters. Nine samples were included on each experiment, with seven being used for transit time measurements (i.e., Hugoniot measurements) and the other two for diagnostics of opportunity to give information on release properties, sound speeds, and spall strength. The experiments were labeled "L," "C," and "M," which were abbreviations for "Low," "Center," and "More" (references to the relative Hugoniot pressures achieved). Informal communications early in the project, however, used the names of The Three Stooges: Larry, Curly, and Moe.

---

[63] Editors' note: The zero room is the site of the test, where the explosive (conventional or nuclear) is sited.

Tom Bergstresser began setting up a pyrometry system for Rebound-I, while I assembled a VISAR system around a 1391-nm (infrared) VISAR originally built for Kevin Fleming to use on the Distant Zenith underground test. Gene Herrera and Gordon Hansen supported this effort, primarily in the data transmission area. The effort to assemble the diagnostics also included several Bechtel Nevada people, initially Steve Becker on the pyrometer and Dave Oschwald on the VISAR, with Ed Marsh working on both diagnostics. The development proceeded on two fronts simultaneously: conducting several local shots at LANL and setting up at the Lyner facility at the Nevada Test Site.

The two developmental fronts were very different. The local test shots were set up rapidly, with plywood shot assembly holders connected by a tangle of cables to the Chamber 8 instrumentation bunker under the clear New Mexico sky. The first trial (L) was conducted in late March 1996, with surrogate metals in the assembly. That one included the VISAR diagnostics (a Sandia VISAR on one sample fielded by my team, and a LANL VISAR on another fielded by Will Hemsing and his team). A second L trial was conducted a short while later, with trials of C and M also conducted at various LANL firing sites. Shot execution timing was determined by when the experiment was ready, subject to safety constraints such as forest fires and associated air traffic.

The Nevada shot development was much more formal, following the habits of nuclear underground tests (UGTs). The shot execution timing and conduct were approved by Hazel O'Leary, the Secretary of Energy. The rigmarole of securing approvals occupied several years of Rob Hixson's life. Regulations regarding the package shipments to Nevada from LANL constrained many aspects of the experimental design. Of particular concern for the "diagnosticians" was a requirement that no diagnostics penetrate an acrylic anti-contamination sheet in the packages; this ruled out the use of rear-surface shorting pins as time-of-arrival gauges. The cabling requirements were negotiated many months ahead; the cables ran from the experiment drift to the diagnostics drift "downstairs" (960 ft underground), and then separate read-out and control cables ran to the recording trailers "upstairs." For Rebound-I, the Sandia recording was done in an aboveground trailer. Between the VISAR (which produced electrical signals from the photomultiplier tubes) and the digitizers (which expected to receive the signals) were fiber optic transmitters, about 3000 ft of fiber optics, and fiber optic receivers. The timing of the signals relative to the shot was determined by injecting the output of the "load ring" into the VISAR output via a delay generator. All underground equipment needed to be operated remotely; U1a was unoccupied at shot time.

The Lyner (Low Yield Nuclear Experiment Research) facility was renamed U1a during the Rebound-I project to emphasize that these were *zero-yield* experiments. That the yield was zero was established by means of neutron measurements and seismics; if a yield were nonzero, it would have been a treaty violation (even though the Comprehensive Test Ban Treaty has never been ratified by the U.S. Senate). The original Lyner facility name came from the 1990 "Ledoux" nuclear test conducted in a drift near the U1a shaft. Here, "U1a" means underground site A in NTS Area 1. The name U1a as applied to the overall facility has persisted over the years, even though the primary access is now through the U1h shaft at the north end of the main underground corridor.

Approvals held the final shot schedule at bay for some time. These included approvals to transport the package from Los Alamos, assemble it to the explosives, deliver it to U1a, and conduct the shot. Sandia and LANL tested the VISAR systems with electrical slapper equipment (exploding foils) furnished by Will Hemsing. A dry run named Monarch occurred in August 1996. However, the actual Rebound (the "I" was dropped) event was not executed until a cold, rainy July 2 the next summer. Virtually all the data were considered to be of good to excellent quality. In a last-minute addition, Hemsing had secured permission for a parallel underground recording of the Sandia and LANL data. Although the results of the underground and aboveground recordings were similar, the long fiber run and the fiber optic transmitter and receiver pairs did introduce noticeable distortion of the waveforms.

Holog, the first Lawrence Livermore National Laboratory (LLNL) SCE, was 10 weeks after Rebound. Sandia was not asked to participate. LLNL used primarily Fabry–Perot velocimetry on this and its following SCEs.

The next SCE began life with the name Boomerang, but that eventually changed to Stagecoach and began an Old West theme in shot names. Sandia moved all its high-bandwidth recording (on-package diagnostics such as velocimetry and pyrometry) underground. Low-bandwidth recording, such as containment instrumentation, remained in the recording trailer "upstairs" in the U1g trailer park. This marked an important step away from UGT traditions, capitalizing on the fact that SCEs used at most tens of pounds of explosive vs. kilotons for UGTs. Other aspects of UGT protocols persisted. Both Rebound and Stagecoach were conducted in single-use alcoves, and the experiments were fired behind about 40 ft of stemming (grout fill).

Stagecoach was similar to Rebound, except with five packages instead of three and slightly different target pressure levels. The majority of the packages and samples were used for wave speed measurements. The lowest pressure package (CJ, or "Creative Juices") included five VISAR channels looking at samples configured to give shock-release and spall information. I had purchased a number of channels of VISAR manufactured by Valyn International, the company Lynn Barker had formed after he retired from Sandia. These instruments required frequency-doubled Nd-YAG lasers, and new lasers from Coherent Inc. were now available that did not require external chillers and ran on standard 110 VAC 15-amp circuits. The Sandia setup in the detection alcove included the Valyn VISARs, coherent lasers, and legacy LeCroy 7200 1 GHz digitizers (together with the standard Burleigh modulators, Stanford delay generators, and various modules to allow remote operations). The Sandia classified operations upstairs were now located in a Porta-Kamp building, leaving the B72 trailer for containment recording.

Sandia participated in the local confirmatory for CJ at the R306 firing site at LANL. Although the confirmatory itself went well, the setup was interrupted by a heavy rainstorm two nights before the shot. With the contouring of the soil atop the bunker, many gallons of muddy water dumped through an access pipe directly onto one of the VISAR units. In consultation with Lynn Barker, Dave Oschwald was able to get the instrument functioning again in a single day.

The Stagecoach shot in March 1998 began inauspiciously in the morning with no electricity in the Porta-Kamp. At T-3 hours, the electricians were hard at work trying to find the fault. We were finally up and running with two hours to spare. When the shot fired, the shock from the high-explosive load was sufficiently large that we could feel it through our feet in the noisy Porta-Kamp a half mile away and 962 ft up. The Sandia diagnostics worked very well. I had written a set of scripts to allow rapid first-cut analysis of the VISAR data, so at the T+4 hours meeting, we were able to lay out much more information than simply "We recorded data on all channels" (the UGT standard at these quick-look briefings). With Hemsing from LANL providing similar information, the bar began to be raised as to what the VIPs expected to hear 4 hours after a shot.

So what to do between shots? Paul Raglin held the purse strings for Sandia's funding to participate in these experiments and had significant influence on the Bechtel Nevada fiscal directions involving Sandia efforts. He saw the sense in capitalizing on the equipment obtained for the SCEs and the personnel in place to operate it, between experiments. So this equipment began to see action on the Z facility during lulls in the SCE schedule. This tactic had an advantage for Bechtel and the SCE program as well: the personnel and equipment were kept in a good state of readiness. On several occasions, the NTS schedule allowed a couple of weeks for VISAR setup, but project management was surprised to find the VISAR team ready to record after only 2 days.

All of the SCEs overlapped substantially in their schedules. Stagecoach planning for Sandia could be traced back to a meeting in early October 1996, 9 months before Rebound fired. Similarly, the very different "Icebound" shot (later renamed Cimarron) was in the planning stage prior to the beginning of 1997, even though it was not actually executed until December 1998.

Cimarron occurred in December 1998; that shot was designed to look at multidimensional shock interactions and ejecta production. The task of making one-dimensional measurements (Hugoniot, release, spall, strength) was assumed by the guns at the Plutonium Facility in Los Alamos and by JASPER[64] at NTS. Mary Hockaday at LANL was the diagnostic coordinator for Cimarron. Because the physics goals were different from those of Rebound and Stagecoach, we fired a number of local tests at LANL before the design was finalized and fielded. These highly instrumented tests developed a reputation for always firing after dark.

In one of the two 1998 local tests at the Eenie firing site, we were delayed 2 days by a vacuum leak, but the location of the leak was finally solved. Because the Eenie bunker was so small, only Ed Marsh and Dave Oschwald were allowed to stay inside as the VISAR crew; Dennis Barker and I were among the folks exiled to the canyon divide a quarter mile away. The explosive load on the Bullwinkle package was only a few pounds, but that is still a safety issue. When the shot finally fired, it looked like a fireworks display. As we returned to the bunker 20 min later, the dust was still heavy and a large chunk of the temporary building that had housed the shot

[64] Editors' note: The Joint Actinide Shock Physics Experimental Research (JASPER) facility contains a two-stage gas gun.

had been thrown clear over the bunker into the fortunately empty parking lot. Our data looked good, though!

During the Cimarron event, Sandia moved into a new recording alcove at the end of the U1a.03 drift nicknamed the "Taj Mahal" because it was relatively spacious. The LANL VISAR diagnostic also moved into new quarters next door. This substantially shortened the cable runs required and set the stage for later events in U1a.03g, a new drift just to the north of the Rebound site. The underground working environment felt more like a laboratory than before. That we could remove our helmets and self-rescue gear when working in the room was also nice.

Thoroughbred followed in mid 2003 in the wake of Cimarron. That event involved several multi-dimensional packages and lots of diagnostics. A pop-up camera was even included in the alcove to allow post-shot visual assessments. The Sandia VISAR and pyrometry diagnostics were on the "Boris" package ("Natasha" was the other package). Laser multi-moding nearly "ate our lunch" until we figured out that we needed to operate the lasers at a large fraction of full power, attenuating the beam as needed. This multi-moding was visible when the Lissajous figures collapsed into a tangle that looked like a star, so we called this the "star phenomenon."

The next family of SCEs to emerge in the LANL U1a program, Vito and the "MaRocco" series, adopted a test setup reminiscent of vertical nuclear underground tests. A large drill, known as a Davy Drill, was brought into U1a.03g and used to create a shaft 5 ft in diameter and 35.5 ft deep for each of these events, as well as for the later "Krakatau" event. Then the package, mounted onto a "raclette" containing the diagnostic probe heads, fibers, and patch panels, was lowered into the hole and grouted in prior to being fired. This was modeled after the full-size racks used for vertical UGTs. I did not have a direct role in the Vito event.

The "MaRocco" program was comprised of the Mario and Rocco SCEs, on-site confirmatories, and earlier local tests at LANL. "MaRocco" was designed to elucidate spall properties as well as information about ejecta. The early concepts for the diagnostics centered on surface velocimetry and Asay foils,[65] but it became clear that these were not going to get at the needed physics. I remember asking myself one morning about "Asay windows"[66] and making up a vugraph about this, emailing it to a couple of the Los Alamos folks, then hopping in a car to drive up to meet with them. When I got to Los Alamos, the offices were empty, and everyone was in a

[65] Editors' note: "Asay foils" are metallic foils used to detect ejecta from a shocked surface. A detailed description is in J.R. Asay, L.P. Mix, and F.C. Perry, *Applied Physics Letters* **29**, 284–287 (1976). See also A.V. Fedorov, A.L. Mikhailov, and D.V. Nazarov, Chapter 9 in *Material Properties under Intensive Dynamic Loading*, edited by M.V. Zhernokletov and B.L. Gluskak in collaboration with W.W. Anderson, F.J. Cherne, M.A. Zocher (Springer, Berlin, 2006), pp. 393–418.

[66] Editors' note: An "Asay window" is a non-radiographic spall and damage diagnostic that uses a transparent window in front of a spalled surface. As successive spall layers collide in "domino fashion" with the window, a VISAR measures the change in velocity of the window/spall surface interface. See D.B. Holtkamp et al., in *Shock Compression of Condensed Matter—2003*, edited by M.D. Furnish, Y.M. Gupta, and J.W. Forbes (American Institute of Physics, New York, NY, 2004), AIP Conference Proceedings **706**, pp. 473–476; also, see C.W. McCluskey, M.D. Wilke, W.W. Anderson, M.E. Byers, D.B. Holtkamp, P.A. Rigg, M.D. Furnish, and V.T. Romero, *Review of Scientific Instruments* **77**, 113902 (2006).

conference room with copies of my vugraphs being passed around. That ended up being a key diagnostic on the series.

The Rocco local test was conducted at the PHERMEX[67] site at LANL in December 2001. As with many confirmatories,[68] a disposable steel building housed the test itself, while personnel, interferometers, and recorders were in the spacious bunker below. Ed Marsh, Greg Mize, and Scott Walker from Bechtel Nevada fielded the VISAR diagnostic for Sandia, with me as the diagnostic scientist. While we continued to use VISAR systems with up to seven points per interferometer cavity and digitizer recording, LANL had manufactured units with 36 points per cavity using streak recording. This setup involved a certain amount of tradeoff of quality for quantity, but it was used for the next several SCEs as a "Line" VISAR. After several days of setup, the rapidly dropping temperatures of the winter evening nearly cost us our data. We could watch the returned intensities dropping in front of our eyes. Finally the shot was fired, and I was left with the task of reducing rather marginal quality VISAR data.

"MaRocco" at NTS, with four shots, was conducted during an ambitious 4-month period in 2002. The two confirmatories were actual explosive events conducted inside large steel balls. The package was mounted in a "racklito," a term introduced by Mark Wilke, the LANL diagnostic coordinator. It was important that the confirmatories be successful, both to validate the diagnostic readiness and as a source of comparison data for the surrogate material vs. the plutonium used on the actual shots. The first confirmatory nearly failed for the VISAR and other key diagnostics because of a spurious pre-trigger signal. I remember the feeling of disbelief as I watched the fringe records rolling off of the printer in the B72 trailer—they were all flatlines. The laser was modulated by a Pockels cell built to provide a 60-μs pulse of light to the target and, in this way, avoid burning out the fiber optics and provide an additional factor of safety. However, the cell fired early due to the pre-trigger; so, at the actual shot time, only a small amount of residual light illuminated the target. As it turned out, there was *just enough* light to elucidate the data needed from the confirmatory. Mark Wilke climbed into the Sandia recording trailer, inspected the results, and then thanked the Sandia crew profusely, stating that we had "just saved Los Alamos twenty three million dollars." The remainder of "MaRocco" went well and created lots of fodder for John Budzinski to digest at LANL.

The use of a pulsed laser was new for "MaRocco." On previous events we had used continuous laser light and verified the proper VISAR contrast with a low-bandwidth scope showing D1 vs. D2 (i.e., 90° phase-shifted fringe data), which should be a circle if the VISAR tuning is good. We also began recording all the VISAR signals in full quadrature, removing differential amplifiers from our systems.

There was a longer lead-up to the next set of SCEs. Krakatau was a U1a experiment that was joint with the Atomic Weapons Establishment (AWE) and LANL. The nearly

---

[67] Editors' note: PHERMEX, the Pulsed High-Energy Radiographic Machine Emitting X-rays, was LANL's hydrodynamic test workhorse for more than 40 years until it was replaced by DARHT, the Dual-Axis Radiographic Hydrodynamic Test facility.

[68] Editors' note: Confirmatories, in this usage, are preparatory tests conducted prior to the real experiments or events in order to corroborate that all systems, equipment, and diagnostic instrumentation are operating properly.

simultaneous Unicorn was set for U6c, a separate 600-ft bore a few miles away. Sandia elected to participate in Krakatau with the VISAR and pyrometry diagnostics. The first part of this SCE was participating in the "local" shot "Leuser" at AWE. This involved fielding diagnostics in England. Greg Mize, Ed Marsh, and Scott Walker of Bechtel Nevada took the lead in setting up the Sandia VISAR effort; I was again the diagnostic lead. This gradually stretched into a 2-year project, in part because of AWE safety requirements, and Greg, Ed, and Scott each ended up logging nearly 6 months in the UK. Eventually, the shot was fired in April 2005, at which point the roles were reversed and Andy Rimmer's AWE crew became the "away" team at the Nevada Test Site.

Diagnostic instrumentation continued to progress. During the definition of the Krakatau SCE, I convinced the LANL planners that fiber optics and probes for photon Doppler velocimetry (PDV) should be included in the experiment as a trial diagnostic, in addition to the standard VISAR and pyrometry. The PDV had recently been developed and showed distinct promise for free-surface environments where ejecta might be present. The LANL weapon physics (P-23) team, led by Dave Holtkamp, eagerly accepted the offer to field the PDV. In the end, both the PDV and the VISAR performed well on foils and free surfaces, although the PDV showed an attractive ability to see ejecta simultaneously. Both the "Kerinei" surrogate experiment and the Krakatau subcritical event proceeded well, with the latter being fired in February 2006. Because of authorization basis[69] holdups, Unicorn suffered long delays as well and was finally fired in August 2006.

The next two major evolutions in LANL-related SCE design came about elsewhere in U1a while Krakatau was in progress. The U1a.05 drift was being mined and finished for the Armando SCE, which included the Sandia-designed Cygnus radiography machine, as well as the more traditional velocimetry and shadowgraphy diagnostics. The construction of Cygnus was a significant Sandia effort, with installation and validation at U1a as a joint task with LANL. The Armando zero room was built as a bulkhead fitted with ports for x-ray entry and with a separate large steel box containing the receiving charge-coupled devices for later readout of the radiography images. During the design of Armando, LANL took a page from LLNL's successful use of zero rooms to conduct multiple shots (e.g., the "Oboe" and "Piano" series of SCEs) and changed plans to confine the events to a 6-ft-diameter steel ball. Armando was executed with radiography and limited velocimetry in May 2004.

In addition to the new radiography capability, a whole new set of six diagnostic screen rooms was built in the adjacent U1a.05 drift.

To capitalize on the new radiography capability in the "Armando" zero room, the "Thermos" series was fielded in 2007. This was the first LANL series to include specimen recovery. During these tests, the screen rooms were manned (it was not necessary to control the data acquisition process from the "uphole" trailers). Twelve aluminum containers the size of coffee thermoses were fitted with small Pu disks and explosive drivers. LANL and Sandia worked together to design optical probes

[69] Editors' note: 'Authorization basis' refers to documents used to authorize that activities to be conducted at a facility or on a test are in compliance with Department of Energy orders and policies with respect to environment, safety, and health.

for the VISAR and PDV to monitor the sample motion, while Cygnus radiography was used to gauge deformation. The samples were captured in foam, and the containers returned to LANL for recovery and metallographic examination of the plutonium. Typically, two experiments could be done in a week, gradually inundating the LANL teams with post-shot analysis. At the end of 2007, an additional five Thermos shots were performed at pRad, the proton radiography facility at LANL, and included Sandia velocimetry.

Following these experiments, another lull occurred before the "Bacchus/Barolo" series was fielded in front of Cygnus in the U1a.05 drift. In the lead-up to this series, the Sandia VISAR team replaced the 19-beam Bechtel Nevada (now National Security Technologies, or NSTec) VISAR with a set of four fixed-cavity VISARs. One development that was simultaneously good and bad was the rigorous imposition of comprehensive checklists. Although Sandia, Los Alamos, and NSTec (which had taken the place of Bechtel Nevada) had been using checklists for years, the process of change control and documentation became substantially more formal.

Diagnostic timing and verification were done with a large number of electrical slapper tests (now with a Sandia/NSTec slapper) and also of "impulse tests," which used a closely tied pair of fast pulse lasers to determine the relative timing of the various data channels.

The Bacchus shot, which was led by AWE, had a very different physics design than the LANL Barolo-A and Barolo-B shots. However, the diagnostic suites were nearly identical. Therefore, it made sense to perform all three experiments (plus the confirmatories) in a single fielding effort.

The "Barolo" confirmatory was in September 2009. However, for a variety of reasons, the "Bacchus" event actually proceeded first (in September 2010), and Barolo-A and Barolo-B were executed in December 2010 and February 2011, respectively.

Subsequent LANL-led SCEs (Pollux and Castor of the "Gemini" series, and the current "Leda/Lyra" series) have involved Cygnus radiography and massive PDV deployment. Sandia, under the field (Nevada) lead of Dan Bozman and the New Mexico lead of Mark Kiefer, supports the Cygnus effort and is expected to play a key role in developing and deploying the next-generation radiography in U1a. An ongoing effort is also planned to field a large-bore powder gun in U1a; Sandia may participate with velocimetry in this effort. These projects at the Nevada National Security Site (previously named the Nevada Test Site) continue to provide vital information on the behavior of plutonium under dynamic conditions.

These years working on the NTS projects had the drawback of travel, but certainly included nice aspects as well. The team was terrific, the place is beautiful, and there is plenty of wildlife (no … I really mean the zoo kind). And the nights are clear and dark—the tail from Comet Hyakutake stretched a third of the way across the sky as viewed from just outside Mercury. But they really roll up the sidewalks at night, and it is a long commute from Vegas.

While all this was going on, my Albuquerque work continued as well. For one set of experiments on Z, collaborating with Jeff Lawrence and others, we irradiated a "partitioned pipe" to study the motions of the individual partitions under x-ray drive. For this research, we wanted a bunch of VISAR channels. Because of the

SCEs work, we had VISAR channels available, so I had a bundle of fiber optics laid to the center section. As far as I know, that was the first multi-VISAR experiment on Z. Also, being continually pushed by John Remo of Harvard, we arranged a number of ride-along experiments to look at the mechanical response of meteorites and other materials to x-ray flux, trying to measure the momentum coupling coefficients. Some of these materials were really quite junky! The experimental results have now been published (Furnish 1993; Furnish, Boslough, Gray, and Remo 1995; Furnish and Remo 1997).

I also worked with Bill Reinhart and Lalit Chhabildas in extracting physics from the HyperVelocity Launcher at STAR to analyze the very high-pressure strength behavior of aluminum across the melt boundary and also the limits of transparency of LiF under shock loading. This resulted in a Best Paper Award at the 1998 Hypervelocity Impact Symposium (Furnish, Chhabildas, and Reinhart 1999). My memory of this was arriving in a foggy Huntsville by bus at 2:30 am, deciding to skip the early morning session for a little extra sleep, then getting blasted out of bed by Clint Hall at 6:30 am for a "strategy planning session" at breakfast. Then at the opening session, I was trying to stay awake and then saw the award announcement appear on the screen and felt my eyes opening wide.

As another sidebar, I was persuaded to edit the APS Shock Compression of Condensed Matter (SCCM) conference proceedings beginning in 1999 and continuing through 2007 (recruiting Mark Elert of the U.S. Naval Academy to help in 2005 and 2007). In 2009 I co-chaired the Nashville APS SCCM conference, so I recruited Billy Buttler of LANL to relieve me as editor. I also served a term as chair of the SCCM Topical Group and 3 years as secretary/treasurer. This Topical Group is another great team with which I have enjoyed working.

Right now, my focus is back in Albuquerque, mainly on experiments on the DICE (Dynamic Integrated Compression Experimental) and STAR facilities. This includes Line VISAR buildup as well as more of the garden-variety shots for spall problems, strength measurements, and such. The CTH and WONDY wave codes continue to be useful in experimental design and error analysis. With the Line VISAR, the intent remains to get beyond the "nice image" goal to issues such as instability spacing, effects of material property perturbations, and mechanisms of yield in spallation. We have begun a productive collaboration with AWE on this effort. I wrote a program in Matlab to reduce the Line VISAR data based on fringe position evolution. Matlab is a challenging language for programming, but creates a framework to allow me to finally leave behind the old Fortran language. Where this will lead, time will tell.

Seth Root has persuaded me to come on board a project with insensitive high explosives (also requiring the Line VISAR) in support of the DOD/DOE Memorandum of Understanding under the Joint Munitions Program; this is now a major focus of my effort. I am also tackling an interferometry problem associated with electrical charge accumulation that shares analysis issues with the Line VISAR. As with the Line VISAR routine, the most time-consuming part is getting the computer to talk to the scientist.

I count as key among my many blessings the teams I'm working with here at Sandia, including the LANL, NSTec, and Ktech folks and others. Thanks, all!

# Dennis E. Grady

# (1974–1996)

I came on board the shock physics group in 1974. I was recruited by Darrell Munson. It was a time of rapid growth for both the shock physics group and Sandia. Marlin Kipp, Paul Yarrington, and Bill Brown, in addition to me, all joined either the computational group or the experimental group within a few months.

My previous several years were at Stanford Research Institute, where I conducted experimental shock equation of state studies on geological materials for the Defense Nuclear Agency (DNA). In the experimental shock physics group, I undertook comparable experimental work in support of Sandia programs concerned with nuclear ground shock and Earth-penetrating weapon effects.

Two occurrences in this early time contributed to my good fortune. First, the VISAR instrumentation technology, recently developed by Lynn Barker, was made available and I had the opportunity to acquire some of the first-ever high-resolution shock wave structure measurements on rocks and minerals. Second, Lynn Barker had just previously departed for Terra Tek in Salt Lake City, and I had the extreme good fortune to work with Roy (Red) Hollenbach, one of the foremost experts in the application of VISAR technology, during his final years at Sandia Albuquerque.

Shock equation-of-state measurements on geological materials in support of DNA and the Department of Energy (DOE) continued until concerns with ground shock effects began to wane in the early 1980s. I joined John Shaner from Los Alamos Laboratory for some of the first technical talks with Soviet counterparts on nuclear test ban verification at the Nevada Test Site.[70] From these 1970s structured shock data arose some of the early ideas on the universality of shock structure and the underlying importance of heterogeneity in the shock failure and flow of solids. Joined with concurrent work on the shock wave structure in metals in the shock physics group, the fourth-power rule governing the shock-structuring viscosity of solids emerged.

The energy crisis of the mid 1970s provided the impetus for the three national security laboratories to embark on extended efforts to support U.S. energy concerns through technological innovations. Technical issues concerning the effective use and recovery of oil shale, tar sands, and deep coal gasification, among other activities, were actively pursued. Methods such as explosive and static hydraulic fracturing (now popularly referred to as fracking) were extensively investigated. A number of shock physics issues were identified and pursued. A key issue was the development of effective explosive fracture and fragmentation methods in support of investigations into the in situ retorting of oil shale deposits that were and still are abundant throughout the Rocky Mountain states. Shock spall and fragmentation studies of oil shale and other rock materials were undertaken in the shock physics group. These

[70] Editors' note: Now known as the Nevada National Security Site.

efforts provided the basis for theories of fracture and fragmentation that were implemented in computer codes to support the various energy recovery programs.

In the late 1970s, several mishaps in the transportation of nuclear weapons led to concerns with the accidental detonation of nuclear weapons. Although a nuclear reaction is unlikely under most accident scenarios, the explosive dispersal of highly toxic materials is a very probable outcome. This program eventually evolved into an assessment of the consequences of both accidents and terrorist threats involving nuclear weapons. The program also broadened to include concerns with nuclear reactor safety. It was an advantage that theories of fracture and fragmentation that arose in the earlier shock physics support of energy programs played a central role in assessing the effects of nuclear weapon explosive dispersal. The shock physics group provided the material shock equation-of-state and dynamic strength studies necessary to the computational assessment of myriad scenarios. The needed experimental studies on uranium were conducted on the Sandia experimental gun facilities. For the comparable material studies on plutonium, access to a dedicated experimental facility at Los Alamos was kindly provided.

In the mid 1980s, Earth-penetrating weapons evolved into a major thrust for Sandia, and shock physics provided a substantial supporting role. Using the gun facility and VISAR instrumentation, small-scale penetration experiments into rock materials in a reverse geometry allowed high-resolution measurement of impact-induced acceleration shock spectra during the penetration event. Having some familiarity with the geology of New Mexico, I aided in locating a local geology for which material for penetration experiments could be performed on all size scales. With several trips to San Ysidro, New Mexico, we identified a material that would become known as "San Ysidro sandstone." Cubic meter-size sandstone blocks were carted back to Albuquerque for sled track penetration tests. Several Davis gun penetration tests were conducted on site near San Ysidro. Scaled penetration testing was conducted on the same sandstone at the shock physics gun facility in support of the program.

Comparable quality shock physics research continued in the Soviet Union through the 1970s and 1980s, and we stayed on top of it mostly through the open literature. In 1990, in the interest of *glasnost*, Soviet scientists in the shock physics field decided on short notice that they would open to international participation a conference that previously had been for Soviet scientists only. They contacted a scientist at LLNL asking if he could make it happen. He, in turn, contacted Sandia and Los Alamos and several other organizations. The upshot was that in a short period, about a dozen scientists, mostly from the three national laboratories, were on their way to Russia to attend this first-ever international meeting. Tim Trucano and I were the Sandia participants. The journey entailed a flight to Moscow and then an extended flight on a Russian airline, an adventure in its own right, to Irkutsk, Russia, for the shock physics meeting on the shores of Lake Baikal. There we met Russian scientists that we had known for decades only through the literature.

A concern during that transition brought about by *glasnost* was the stability of the Russian scientists during the period when Russian efforts in nuclear weapons research were rapidly winding down. The U.S. national security laboratories played an important role through financial support of non-nuclear research during this period. The Sandia shock physics group played an active role in this effort. One effort led to an extensive study of the spall and shear strength of metals at elevated temperatures. At that time, modest funding to the Russian scientists produced quite significant scientific research return, and such efforts effectively supported ongoing laboratory programs.

An extensive research program emerged during the late 1980s that entailed a joint effort between the Department of Defense and the national laboratories on issues of mutual concern. Material property studies through impact shock physics methods were undertaken at Sandia. Extensive experimental studies were performed on low-density and high-strength ceramics that had potential application for military armor needs. From this effort emerged Hugoniot equation of state and dynamic strength properties for a suite of ceramics that remain, to the present day, the basis for computational analysis of ceramic performance in the terminal ballistic environment.

I retired from Sandia in April 1996 after almost 21 years of fascinating and challenging work in the Sandia shock physics group. I found that one never entirely leaves this field of endeavor. I continue to find opportunities to interact with the group as it has continued to evolve and flourish over the subsequent years.

## Robert A. Graham

## (1958–1996)

## Recollections of William (Bill) J. Nellis, Harvard University[71]

My first recollection of meeting Bob was at the 1979 AIRAPT (Association Internationale pour Avancement de la Recherche et de la aux Hautes Pressions) conference in France, though it might have been at organized shock sessions (probably organized by Bob) at the 1976 APS (American Physical Society) Winter Meeting at Stanford. In either case, I met Bob soon after I had arrived at LLNL (Lawrence Livermore National Laboratory). In reflecting back on those days, I realize he was a great mentor to me in this new field that I had recently joined and a source of valuable advice. Bob was also my own personal historian about the early days of shock research at Los Alamos and Sandia. In particular, I learned a lot from him about early work on dynamic strength at Sandia.

In those days there was a lot of activity in organizing the shock community and Bob was heavily involved. I remember attending the first meeting in 1979 in Pullman, Washington of what eventually came to be called the APS Topical Group on Shock Compression of Condensed Matter (SCCM). Somehow I came away from that conference as co-chair with Lynn Seaman of the 1981 shock compression meeting at SRI (Stanford Research Institute) in Menlo Park, CA. Three Russian scientists actually attended that conference. Up to that time, it was unimaginable to think of scientists from the Russian defense community attending a conference with their American counterparts. I sent the invitations to the Russians from the former Soviet Union to attend, but I do not remember who suggested the names. A month after the SRI meeting, I received a phone call from Bob in which he said it would be most appropriate if the Soviet scientists reciprocated for their visit to the United States by inviting American scientists to a similar conference in the Soviet Union. I agreed and communicated that sentiment to the Soviets. In September of 1982, Bob and I

[71] Bill Nellis spent most of his career in shock wave research at Lawrence Livermore National Laboratory. After his retirement, he joined Harvard University.

visited the Soviet Union, which was the beginning of a most productive and enjoyable 30-year interaction with our Russian colleagues. This is one of the many good things Bob has done for our community. Thanks, Bob, for taking that initiative. I've continued this idea with other countries with similar results ever since.

The things I remember most are Bob's enthusiasm for excellence in shock compression research, his drive to establish professional relations with scientists around the world, and his determination and persistence in organizing the shock community into the APS Topical Group on SCCM, which Bob most certainly did and from which the members now benefit so much.

## A Tribute by the Editors[72]

Few people have had the transformational impact on a scientific field that Robert (Bob) A. Graham has. Throughout his career at Sandia National Laboratories, Bob made pioneering contributions to the shock wave field that even today continue to guide the scientific community. Early in his career, he led the development of new experimental methods for precision measurements of dynamic material properties, which in turn enabled a fundamental understanding of ferroelectric and piezoelectric materials. These studies resulted in a myriad of applications. His groundbreaking development of the Sandia Quartz Gauge was a game-changing advance that established a new paradigm for probing the fine structure of shock waves and enabled a deeper understanding of mechanical, physical, and chemical properties under shock compression. His personal leadership in establishing shock compression science as a credible scientific discipline has persevered through the professional organizations he founded and the many collaborations he established. His legacy in shock wave research endures with a new generation of shock wave scientists by setting the standard of excellence for aspiring researchers in the field.

Bob graduated from the University of Texas (UT) at Austin in 1954 with a BS degree in civil engineering. He went on to receive an MS in Mechanical Engineering at UT in 1958. His research for the master's degree was performed under the tutelage of Professor E. A. Ripperger (Ripp) in the Engineering Mechanics and Aerospace Department. Ripperger was well acquainted with Sandia through several summers working on a number of technical problems, including the design of a tall inclined drop tower for impact testing of weapon delivery systems. He was instrumental in sending several of his students, and notably Bob Graham, to Sandia in the 1950s and 1960s.

After joining Sandia in 1958, Bob was involved with shock wave research from nearly its beginning. When he joined the budding shock wave effort in the newly

---

[72] These recollections would be incomplete without including those of Robert A. Graham. His leadership and vision have made shock compression science into a premier scientific discipline. Since he was unable to provide his personal recollections, the editors were compelled to summarize his overwhelming accomplishments. Much of the specific information on Bob Graham's contributions was extracted from J.W. Forbes, "The history of the APS Shock Compression of Condensed Matter Topical Group," in *Shock Compression of Condensed Matter – 2001*, edited by M.D. Furnish, N.N. Thadhani, Y. Horie (American Institute of Physics, College Park, MD, 2002), AIP Conference Proceedings **620**, pp. 11–19.

established research directorate, he had the good fortune to work with Frank Neilson, a creative genius in his own right, and with Bill Benedick, a superb experimentalist; this trio energetically moved the shock wave field forward in those early days. The three of them were successful in unraveling the mysteries of the "explosive-to-electric" response of ferroelectric crystals under shock loading. Bob actively participated in explosive testing of ferroelectric (FE) and piezoelectric (PE) materials and developed a keen sense for demanding precision and rigor in shock wave experiments—a standard of excellence he maintained throughout his career.

Through this collaboration, a fundamental understanding of FE and PE materials emerged within a few years and led to a patent in 1961 for the Sandia Quartz Gauge, the first time-resolved stress wave gauge. The gauge was a game changer for probing the fine details of shock wave structure. The in-depth understanding elucidated by the quartz gauge allowed advanced material models of mechanical, physical, chemical, and energetic material response; examples of these applications are dislocation mechanisms of elastic yielding, kinetic properties of structural phase transitions, detection of shock-induced melting, reaction rates in energetic materials, pulsed radiation-induced stress waves, and many more.

After gun launcher technology for precision material studies was first demonstrated in 1958, Bob played a leadership role in elevating that capability to the highest level of precision and elegance. The helium-driven gas gun he developed remains the epitome for precision shock wave experiments, even today. A host of fundamental studies were made with the facility, including development and perfection of the quartz gauge; moreover, many newly hired experimentalists learned shock wave techniques on the helium gun. With the exquisite impact planarity of this gun and the newly developed quartz gauge, it was possible to study dynamic material response with a time resolution of about 1 ns. This was an unprecedented and truly groundbreaking advance, especially in the early 1960s, and the gauge continued to lead shock wave instrumentation until around the mid 1970s. The laser interferometer, invented in the late 1960s and perfected in the early 1970s, became a major competitor to the gauge and significantly expanded the operational range for time-resolved wave profile studies. But the quartz gauge broke ground and paved the way for precision studies of shock wave structure. It also established the value of using highly time-resolved instrumentation to unravel the intricacies of dynamic material response.

Bob's personal research and technical leadership illuminated dynamic material properties and spanned a wide range of subjects. In addition to pioneering research on FE and PE materials that laid the groundwork for their widespread use in scientific applications, he performed research on the pressure derivatives and Curie points of several magnetic alloys subjected to shock compression that has remained unique in the high-pressure community. Along with Frank Neilson and Bill Benedick, he provided the basis for the development of first-of-a-kind shock-actuated ferromagnetic pulsed power sources. His exquisite research on the response of semiconductors and dielectrics to shock compression remains the standard source for understanding the dynamic response of these materials. His work on the band structure of germanium provided the first large strain measurement of the shear deformation potential of the band gap. He advocated for, organized, and led a large research project to understand the fundamental properties and broad implications of shock-induced solid state chemistry, usually referred to as "shock chemistry." His groundbreaking research in this area provided important insights into the nature of chemical reactions in highly transient environments and opened new vistas for microchemistry science and technology in the shocked state. A radically different concept that emerged from this research was a long-life battery that could be activated upon impact and used in specialized applications.

One of Bob's preeminent contributions was his leadership in establishing shock compression science as a credible discipline through his role in advocating for and in instructing others in the unique capabilities of this technology. He was a mentor to many aspiring shock wave researchers, including some of the authors of this volume, and went to great lengths to ensure that young researchers in the field had the opportunities to interact with and present their results to the rest of the community. He accomplished this through nominations for invited and plenary talks, participation on select national review committees, and coauthorship of technical and major review papers on shock wave technology. He has been a prolific writer of comprehensive reviews of shock wave technology, beginning with a major review (Graham 1958) that defined the status of dynamic loading techniques at the time and presented a path for future developments in the field.

Throughout his career, Bob pursued his passion for advancing the science of shock wave research and integrating this new field with the high-pressure scientific community. He was the prime mover and worked hard and relentlessly to establish the APS Topical Group on Shock Compression of Condensed Matter. The formation of that Topical Group had its roots in shock wave studies performed as part of the Manhattan Project. In the late 1940s, shock wave researchers participated in the annual APS Meetings. The Washington, D.C. APS Meeting in May 1947 was the first professional meeting with the presentation of a significant number (21) of shock wave papers (Forbes 2002). In the late 1960s, the APS Executive Secretary established focus sessions on shock wave research to accommodate the growing interest in the field. In 1967, Glenn Seay, Orval Jones, and Lee Davison held the first focus session on shock compression science at the Pasadena, CA APS Meeting. These gatherings continued through the 1970s. Toward the end of that decade, topical conferences were established by the APS to provide additional official focus on specific technical areas. The first topical conference on shock compression science was organized and co-chaired by Dennis Hayes of Sandia and Professor George Duvall of Washington State University. That meeting was held during the summer of 1979 in Pullman, WA.

Based on the success of the topical conferences, the APS formally organized Topical Groups in selected physical disciplines in the early 1980s (Forbes 2002). Bob Graham was asked by the APS to organize a Topical Group on shock compression science. Bob petitioned key participants from a broad base in the scientific community to support this effort, obtaining signatures from 10 fellows and 27 regular members of the American Physical Society.[73] His petition was accepted by the APS Executive Committee, and the Topical Group on Shock Compression of Condensed Matter was officially formed in 1985, with its first official meeting held in Spokane, WA, in the summer of 1985. Subsequent meetings have been held biannually since then. At the Spokane meeting, Bob was elected as founding chair of the Topical Group on SCCM, and his dedication in that leadership role paid significant dividends in promoting and integrating shock compression science into the broader scientific community. As an example, the Topical Group has been able to advance a number of members to APS Fellows, which would have been difficult without a focus on shock wave science.

As another example of his professional leadership, Bob was persistent in convincing the APS to establish an official award for excellence in shock wave research. Although the APS was initially resistant because of their belief that the shock compression field was too narrow to warrant a long-standing award (Forbes 2002), Bob, along with Bill Nellis of LLNL and Professor Neil Ashcroft of Cornell University, was successful in overcoming that resistance. The Shock Compression Science Award, which later became the George E. Duvall Shock Compression Science Award, was formally approved by the APS in 1987. The first award was presented jointly to Robert McQueen, Melvin Rice, and John Walsh for their pioneering research on the high-pressure equation of state of many materials. The award has been given biannually since then.[74] Bob Graham was a prime mover in making this happen and also in designing the iron phase transition icon for the award.

Besides having over 200 professional publications, Bob founded a new book series (Shock Wave and High Pressure Phenomena) in 1993 with the publication of his book, *Solids under High-Pressure Shock Compression*, by Springer-Verlag. Thirty books have since been published in this subject area, covering a wide range of topics including the fundamentals of shock wave propagation, energetic materials, dynamic fracture and fragmentation, heterogeneous materials, ultrahigh-pressure techniques, and high energy density physics. This achievement is yet another example of Bob's leadership in establishing shock compression science as a premier scientific discipline and in integrating diverse technical disciplines into the field. Bob served as editor in chief of this series for several years, during which he was also a managing editor of the international journal, *Shock Waves*.

Another example of Bob's technical accomplishments, scientific inventions, and professional leadership in the shock wave community is evidenced by the awards he has received. Of his numerous awards are the following: the IEEE C.B. Sawyer Award for outstanding contributions to the study of piezoelectric and ferroelectric materials in 1984, a precedence-setting honorary Doctor of Science degree in Materials Science and Technology awarded by the Tokyo Institute of Technology in 1991, and the Shock Compression Science Award in 1993.

Bob Graham's scientific genius, innovative talent, and professional leadership elevated shock wave research to international stature in the scientific community. His pioneering contributions establish him as an icon for shock compression science.

---

[73] Five Sandians signed this petition, including James Asay, Lee Davison, Bob Graham, Bruno Morosin, and Gerold Yonas.

[74] The Shock Compression Science Award was established in 1987 by the friends of the American Physical Society (APS) Topical Group on Shock Compression of Condensed Matter Physics. Three other Sandians have received the award: Lynn Barker in 1999, James Asay in 2003, and Dennis Grady in 2007. James N. Johnson, a Sandian from 1967 to 1973 who later joined Los Alamos National Laboratory, was the 2011 recipient. Beginning with the 2009 APS Topical Conference on Shock Compression of Condensed Matter, the award name was changed to the George E. Duvall Shock Compression Science Award, in memory of George Duvall, a professor at Washington State University for over 24 years. Duvall established a center of excellence in shock wave research by organizing WSU's Shock Dynamics Laboratory in 1968. George was the recipient in 1989 of the second award.

# Thomas A. Haill

# (1984–2016[75])

# Development of the ALEGRA Radiation Magnetohydrodynamic Code and its Role in Modeling Shock Physics Experiments

Unlike many of the contributors to these recollections, my background is not in shock physics but in mathematics and plasma physics. My involvement in shock physics spans less than a decade. I have played mainly a supporting role through computer code development; only recently have I been directly involved in application modeling to support shock physics experiments. The ALEGRA radiation magnetohydrodynamic (MHD) code is a work horse for modeling flyer plate and isentropic compression experiments at the Z pulsed power facility; therefore, I will devote this recollection to the origin, development, and applications of ALEGRA.

## Historical Origins

The ALEGRA shock wave physics code project (Brunner, Garasi, Haill et al. 2005; Robinson, Brunner, Carroll et al. 2008) began as the result of two internal Sandia memos in 1988. In the first, in April 1988, George Allshouse stated the need for a three-dimensional (3-D) radiation hydrodynamic code primarily to design light ion fusion capsules. In the second, in May 1988, Pace VanDevender, then Director of the Pulsed Power Sciences Center, called for initiating a project to develop a 3-D radiation hydrodynamic capability to support Sandia's inertial confinement fusion (ICF) program. This latter memo initiated a planning process for what would ultimately become ALEGRA. The planning process involved a small group: Allshouse (who detailed the requirements for the code), Mike McGlaun, and Tim Trucano (who jointly prepared a detailed, unpublished proposal in October 1989 for its development), as well as Ahmed Badruzzaman and Gary Montry. The formal written proposal was preceded by a February 1989 internal memo from VanDevender to Venkatesh Narayanamurti, who was then a Sandia vice president. The major content in the October 1989 proposal had been presented orally for funding approval to a group of managers earlier in the year. Allshouse played the role of a true visionary, and McGlaun was the major figure who translated that vision into a specific proposal.

During the planning process in 1989, McGlaun and Trucano interviewed Keith Matzen. Matzen specifically requested that the project be flexible enough to allow for possible inclusion of MHD physics in order to model fast z-pinch physics.

[75] Editors' note: Tom's initial research at Sandia involved x-ray radiation transport modeling in support of radiation effects testing of components exposed to underground nuclear and above-ground tests. Later, he joined the ALEGRA team to develop state-of-the-art tools for modeling z-pinch and dynamic materials experiments. Tom Haill retired in 2016.

As stated on p. 172 of T.A. Brunner et al. (Brunner, Garasi, Haill et al. 2005), "The remarkable prescience of this request, before a single line of code had been written, is now clear since the light ion fusion program gave way to the Z pinch program in the 1996 time frame."

It is remarkable how much the current architecture and capabilities of ALEGRA can be traced back to the ideas expressed in McGlaun and Trucano's formal proposal, as summarized below:

- The project was viewed as expensive and extensive (lasting at least 5 years).
- The project would start with an existing code.
- The project would develop an advanced code for Sandia's shock wave physics and ICF efforts.
- The code should be able to model the large changes in physical scales in capsule implosions as well as the complicated coupling between the hydrodynamics and other ICF-relevant phenomena. The code should also be capable of handling complex 3-D geometries.
- The strategy to develop the code and implement the software should be flexible and open-ended. This assumption was ultimately a major factor in the 1992 decision to break new ground with the use of the C++ programming language.
- The code would be developed for massively parallel computing platforms, even though that machine architecture was not obvious in 1989.

McGlaun and Trucano concluded that the best way to achieve the required objectives was with a block-structured, arbitrary-mesh-connectivity, multi-material, arbitrary Lagrangian–Eulerian (MMALE) code. They proposed to start the code development with Sandia's existing PRONTO code, which had two-dimensional (2-D) and three-dimensional (3-D) versions (Taylor and Flanagan 1989). PRONTO 3D was a finite-element, block-structured Lagrangian code with arbitrary (but fixed) mesh connectivity based on the Fortran programming language. A conforming mesh would allow the Lagrangian mode to accommodate large changes in length scale. An Eulerian mode, to be added to the new code, would permit modeling of the complex shearing and fluid distortion that characterize turbulent flows.

The code would also be a next-generation extension of Sandia's 3-D Eulerian CTH code (McGlaun, Thompson, and Elrick 1990), which was in wide use by that time. However, a variety of new physics would be added that was not in PRONTO 3D or CTH: radiation transport, two-temperature (electron-ion) fluid approximations for plasma flows, electron thermal conduction, fusion burn physics, charged particle deposition, and MHD. An additional novelty would be the application of "standard" Eulerian shock hydrodynamic algorithms on an arbitrary connectivity, finite-element mesh.

The code project began formally in March 1990. A 2-D Fortran incarnation, RHALE, became available later that year. In late fall 1990, the decision was made to recast RHALE into an object-oriented structure using C++. This change resulted in enhanced data structures and memory management, object-oriented programming constructs, improvements in debugging and code development, and an enhanced capability to use massively parallel computing hardware (Budge and

Peery 1993). The first formal release of the ALEGRA code was in the late 1990s, and an article on the application of ALEGRA to ballistic impacts appeared in 1997 (Summers, Peery, Wong et al. 1997).

In 1995 to 1996 Sandia's pulsed power fusion program transitioned from light ion beam ICF to z-pinch-driven ICF (Spielman, Long, Martin et al. 1995; Spielman, Deeney, Chandler et al. 1997). At that time, I joined the ALEGRA code development team, and, together with Allen Robinson, we developed the transient magnetic field algorithms that allowed ALEGRA to model z-pinch implosions and, later, magnetically-driven flyer plate experiments. Traditional nodal finite-element methods work very well in solving the magnetic equations in the 2-D versions of ALEGRA; however, the solution in 3-D required a more formal treatment. Pavel Bochev,[76] Allen Robinson, and others (Bochev, Garasi, Hu et al. 2003a; Bochev, Hu, Robinson et al. 2003b), pioneered the application of a hierarchy of volume, face, edge, and node-centered finite elements to model z pinches. The MHD version of ALEGRA has been routinely applied to problems of interest since the late 1990s, and the first formal release (Haill, Garasi, Robinson 2003) of that version was in 2003.

Radiation transport first appeared in ALEGRA by including the package for the SPARTAN simplified-$P_n$ (SPN) transport method, which was supplied by Jim Morel of Los Alamos National Laboratory.[77] Later, Kent Budge coded this method directly into ALEGRA, along with a flux-limited diffusion model. However, deficiencies were uncovered with the SPN transport method (Trucano, Budge, Lawrence et al. 1999), and, in the 2000 time frame, Tom Brunner was hired to implement a multigroup flux-limited linearized radiation transport method (Brunner 2003), as well as a multigroup implicit Monte Carlo radiation transport method. The first formal release of the high energy density physics (HEDP) version was in 2005 (Brunner, Garasi, Haill et al. 2005).

Today the ALEGRA code (Robinson, Brunner, Carroll et al. 2008; Robinson, Niederhaus, Weirs, Love 2011) models geometrical domains that may be in 2-D Cartesian (xy) or cylindrical (rz) coordinates or in 3-D Cartesian (xyz) coordinates. The meshes are unstructured and finite element based and may be Lagrangian, Eulerian, or arbitrary Lagrangian-Eulerian (ALE). The massively parallel code is predominantly in C++, which enables an object-oriented approach to implement the physics algorithms. It runs on a large variety of platforms, from desktop workstations to tens of thousands of cores on large computing platforms such as Cielo at Los Alamos. Object-oriented programming allows the coupled physics to be readily

[76] Editors' note: Pavel Bochev, a computational mathematician, received a 2014 Ernest Orlando Lawrence Award from DOE for his pioneering theoretical and practical advances in numerical methods for partial differential equations. J. Pace VanDevender, a previous Director of Sandia's Pulsed Power Sciences Center, received the E.O. Lawrence Award in 1991 for his contributions to generating pulsed power and demonstrating new concepts and designs for pulsed power such as magnetically insulated transmission lines.

[77] J.A. Josef and J.E. Morel, "Simplified spherical harmonic method for coupled electron-photon transport calculations," *Physical Review E* **57**, 6161–6171 (1998).

operator split. The transient magnetics algorithm that is part of the coupled physics uses the MHD approximation, specifically that materials are charge neutral, so the displacement current term in Ampere's Law is neglected and the current density is divergence free. The MHD equations are closed using a basic form of Ohm's Law. The radiation transport algorithms include either flux-limited multigroup diffusion or multigroup implicit Monte Carlo. A limited thermonuclear fusion burn capability computes the total neutron and gamma yields, as well as the neutron time of flight; however, there is no charged particle transport or alpha particle heating capability.

We are now more than 25 years after the initial requests by Allshouse and VanDevender to develop a 3-D radiation hydrodynamic code, and there is no end in sight to the ALEGRA project. Two key elements of our original vision remain quite firm in the current identity of ALEGRA. First, as described in the 2008 proceedings paper (Robinson, Brunner, Carroll et al. 2008), the code continues to serve as a platform for many research and development advances in computational shock wave physics at Sandia. Second, the code continues to progress toward the goal of providing a primary computational tool for our Pulsed Power Sciences Program.

The remainder of this recollection provides a summary overview of ALEGRA applications pertinent to shock physics. I had a role in nearly all of these, either through direct participation, consultation, new features, capability enhancements, or bug fixes.

#### Radiation-Driven Shocks and Z-Pinch Modeling

Some of the earliest applications of ALEGRA were to radiation-driven shocks and to z pinches. Radiation sources from z-pinch implosions contained within primary hohlraums were employed to drive equation-of-state studies in aluminum (Asay, Hall, Konrad et al. 1999; Hall, Asay, Trott et al. 2000). These experiments produced the first time-resolved particle velocity profiles obtained with pulsed radiation sources in the Mbar regime. The experiments were designed and analyzed via one-dimensional radiation hydrodynamic simulations using ALEGRA with the SPARTAN radiation model (Asay, Hall, Konrad et al. 1999; Trucano, Budge, Lawrence et al. 1999; Lawrence, Asay, Trucano, Hall 2000). Good agreement was obtained between experiment and simulation, with late-time differences attributed to radiation effects that had not been corrected for in the experiments and were not modeled in the simulations. The experiments therefore provided important benchmark data for the code.

Radiation-driven jetting experiments on the Nova laser were modeled with ALEGRA, and the results scaled to larger experiments proposed for the Z pulsed power facility (Lawrence, Furnish, Haill et al. 2001; Lawrence, Mehlhorn, Haill et al. 2002). The radiation drive ablated a cylindrical aluminum plug held in a gold washer. Differences in the material properties caused the plug to form a jet along the symmetry axis. ALEGRA was validated against the Nova data and then used to predict the results of scaled-up experiments on Z. The ALEGRA results were post-processed with SPECT3D,[78] a spectral analysis code, to produce synthetic radiographs to demonstrate the expected capabilities of the Z-Beamlet backlighter.

Experiments to measure shock and debris propagation in 5- and 10-cm-long partitioned pipes were modeled on ALEGRA (Furnish, Lawrence, Hall et al. 2001b; Lawrence, Furnish, Haill et al. 2001). These experiments were designed to measure the possibility of failure caused by debris generation in complex experimental designs. The pipes were heavily instrumented with VISAR probes and strain gauges. The simulations agreed qualitatively with the experiments for shock waves in the pipe walls and debris propagation, but overestimated the debris velocity, possibly because the radiation-drive fluence was overestimated. The simulations led to a better understanding of the diverse phenomena present in the experiments.

Hydrodynamic and magnetohydrodynamic instabilities are inherent in many physical systems. ALEGRA code developers participated in an alpha-group study that compared the growth rates for the nonlinear 3-D Rayleigh–Taylor hydrodynamic instability (Dimonte, Youngs, Dimits et al. 2004). Partial verification of ALEGRA MHD was accomplished through detailed numerical simulations of the $m=0$ instability (Stolz and Oliver 2001) and of the magneto-Rayleigh–Taylor instability (Oliver 1999). Moreover, an experimental and computational investigation of nanosecond electrical explosion of thin Al wires in vacuum demonstrated the separation of each wire into a high-density core and a low-density hot corona, as well as the fast rejection of the current from the core to the corona during voltage collapse (Sarkisov, Rosenthal, Cochrane et al. 2005). The mass ablation phase of a wire-array z pinch was studied via steady-state r-θ ALEGRA simulations to identify the dominant physical mechanisms governing the ablation process (Yu, Oliver, Sinars et al. 2007). Eventually, sophisticated wire-array z-pinch simulations using inflow boundary conditions illuminated the 3-D nature of the trailing mass, instabilities during the implosion, and stagnation of z-pinch plasmas (Jones, Garasi, Ampleford et al. 2006; Yu, Cuneo, Desjarlais et al. 2008; Lemke, Sinars, Waisman et al. 2009).

## Flyer Plates, ICE, and MAPS

ALEGRA MHD has been used for many years to model a variety of applications of direct relevance to shock compression of materials. The code has modeled magnetically-accelerated flyer plates for nearly a decade. Early experiments launched metal flyer plates such as Ti, Al, and Cu to velocities of about 10 km/s (Hall, Knudson, Asay et al. 2001b). A concerted effort has elucidated the detailed physics of launching flyers, thereby enabling aluminum flyers to be accelerated to 20 km/s (Lemke, Knudson, Hall et al. 2003a; Lemke, Knudson, Robinson et al. 2003b;

[78] J.J. MacFarlane, I.E. Golovkin, P. Wang, P.R. Woodruff, N.A. Pereira, "SPECT3D – A multi-dimensional collisional-radiative code for generating diagnostic signatures based on hydrodynamic and PIC simulation output," *High Energy Density Physics* **3**, 181–190 (2007).

Knudson, Lemke, Hayes et al. 2003c). Modeling has helped to confirm Hugoniot measurements on Al in the stress range of 100–500 GPa (Knudson, Lemke, Hayes et al. 2003c). The measurements were compared to models for a variety of Al equations of state (Cochrane, Desjarlais, Haill et al. 2006). Further optimization with MHD simulations to shape the current pulse eventually allowed shockless acceleration of Al flyers to 30–34 km/s (Lemke, Knudson, Bliss et al. 2005c). Most recently, through the use of a stripline configuration and shockless acceleration, the hypervelocity launch of flyer plates to 46 km/s was predicted with ALEGRA and achieved experimentally on Z (Lemke, Knudson, Davis 2011).

Current pulse shaping and magnetically-driven shockless flyer plate acceleration enable another category of experiments known as isentropic compression experiments (ICE) (Davis, Deeney et al. 2005). Such an experiment provides data to map an equation of state along an entire isentrope simultaneously. This is in addition to the Hugoniot data generated by the flyer plate technique. MHD simulations are used routinely to model and design isentropic compression experiments. The technique has been used to measure the principal isentrope of Al (Davis 2006) and the strength of lithium fluoride (Ao, Knudson, Asay, Davis 2009a), among other materials. The ICE method has been extended to magnetically-driven radial compression of cylindrical liners, with ALEGRA simulations used to design experiments to determine the principal isentrope of beryllium to a peak pressure of 2.4 Mbars (Lemke, Martin, McBride et al. 2012; Martin, Lemke, McBride et al. 2012).

ICE shots are also conducted at Sandia on a compact stripline pulsed power generator known as Veloce. ALEGRA has been used extensively to model the uniformity of loading in the Veloce samples (Ao, Asay, Davis et al. 2007; Ao, Asay, Chantrenne et al. 2008). One particular class of experiments has demonstrated a novel concept to produce shear strength measurements in materials. Known as MAPS (magnetically applied pressure shear), an applied external magnetic field provides the Lorentz force that, together with the self-generated magnetic field, produces both transverse and longitudinal velocity and pressure waves in a material (Alexander, Asay, Haill 2010; Haill, Alexander, Asay 2011). Strength is probed by passing these waves through a material, where the transmissible shear stress is limited to the strength of the material sample. Careful timing of the passage of the longitudinal and shear waves allows the shear to be probed at peak material pressures. Another class of experiments, conducted on a pulser similar to Veloce at Washington State University and modeled with ALEGRA, involves the intentional fragmentation of flyer plates upon magnetic acceleration (Lawrence Asay, Gupta et al. 2009a; Lawrence, Haill, Freeman, Gupta 2009b; Haill, Mehlhorn, Asay, Gupta et al. 2007). These experiments demonstrated the feasibility of a concept known as a fragment-producing chemical-electrical launcher. The full concept would employ a flux compression generator to drive an electromagnetic launcher that would propel fragments for various applications.

The shock compression of pure and high-Z doped foams over tens to hundreds of GPa is of interest for ICF and HEDP applications. We have studied the compression of polyethylene and polymethylpentene polymers and foams in experiments (Root, Haill, Lane et al. 2013b), quantum density functional theory with the Vienna *ab initio* Simulation Package (Kresse and Hafner 1993, 1994; Kresse and Furthmüller 1996), from the Technical University of Vienna (Mattsson, Lane, Cochrane et al. 2010), classical molecular dynamics simulations using Sandia's Large-scale Atomic/Molecular Massively Parallel Simulator, or LAMMPS, program (Mattsson, Lane, Cochrane et al. 2010; Lane, Grest, Thompson et al. 2012), and mesoscale hydrodynamic simulation using ALEGRA (Haill, Mattsson, Root et al. 2012). This experimental and modeling effort demonstrated the consistency of results in spatial scales ranging from a few angstroms to tens of nanometers, to tens of microns, and to a few millimeters. The experimental and mesoscale hydrodynamic results with pure foams have been extended to platinum-doped foams of equal density in order to test the validity of equation-of-state mixing rules in hydrodynamic codes such as ALEGRA (Haill, Mattsson, Root et al. 2013). These mixing rules are also being examined for gaseous mixtures such as xenon and ethane (Magyar, Root, Haill et al. 2012).

Finally, equation-of-state and electrical conductivity models are critical to high-fidelity simulations. Subtle differences between experiments and simulations have led to efforts to improve these models (Lee and More 1984; Desjarlais 2001). Revised parameter fits to data soon gave way to detailed *ab initio* quantum molecular dynamics simulations based on density functional theory (Rosenthal and Desjarlais 2001; Desjarlais, Kress, Collins 2002; Clerouin, Renaudin, Recoules et al. 2003; Mazevet, Desjarlais, Collins et al. 2005; Clerouin, Renaudin, Laudernet et al. 2005). Equation-of-state and conductivity models used in ALEGRA MHD are constantly being revised and improved because of these efforts.

## Summary

My recollection presented a broad overview of ALEGRA history and applications. ALEGRA is used in many additional applications. These include exploding bridge wires, gas-puff z pinches, dense plasma focus, lightning discharges, railguns and electromagnetic mortars, and Earth and armor penetrators. As mentioned previously, we are now more than 25 years after Allshouse's initial request to develop a 3-D radiation hydrodynamic code, and there is no end in sight to the ALEGRA project.

# Clint A. Hall

# (1984–Present)

# Clint Hall and Shock Physics: Who Would Have Thought It??

I arrived at Sandia fresh out of New Mexico State University with a Bachelor of Science in Mechanical Engineering Technology. Having interviewed a few months earlier with several organizations at Sandia, I had narrowed my choices to two, pulsed power in Area IV and the STAR facility. My choice was made when Jim Asay (Department Manager) and Carl Konrad (STAR facility supervisor) convinced me that I could become the "go-to person" for all things mechanical after a "small learning curve." Suffice it to say that "small" is a relative term. When I arrived on site in June 1984, the first thing I learned was that I actually knew NOTHING at all about what a shock wave was, how they were made, how they were used, or why anyone even cared so much about them. I did know, however, that STAR had a good machine shop and a welder! What more could a mechanical guy ask for?

## Learning the Ropes

The first thing that "glazed my eyes over" was Carl trying to explain a Velocity Interferometer System for Any Reflector (VISAR). I got velocity and then was lost from interferometer on. He started talking about delay legs and interference and etalons, and everything just started to go gray in my mind. Similar things happened when Carl and all the other electronics types out there started talking about digitizers, terminators, signal generators, time marks … good grief, I had to get over to the launchers with their fluid flow, their gas expansions, their debris mitigation, and their energy absorbing tanks before I felt like Jim actually hadn't made a HUGE tactical error in hiring me. Before long, I actually understood the guns and their operations and could kind of set up an experiment on them as long as I had some "help" from the other STAR techs. Maybe it was because I was starting to catch on or maybe because Jim and Carl wanted someone that thought in terms of steel rather than electrons for diversity sake, but I was eventually named second alternate supervisor and pressure advisor at STAR. Felt pretty good about myself right about then because my career goal of becoming the STAR facility supervisor when Carl retired was on its way to becoming reality!

After taking over responsibility for the single-stage compressed gas gun ("air gun") that nobody really cared about and making some improvements that actually made it more user friendly, more reliable, and look more professional, Jim apparently decided I couldn't screw things up too badly and actually gave me a higher profile project … my first REAL one at STAR. I was to be the mechanical lead for the two-stage gas gun injected railgun (STARFIRE). Hey, I could handle this right … simply couple a hydrogen-filled gun with a delicate projectile that doesn't like joints and moves at about 6 km/s to a stationary, composite bore hunk that had to be absolutely pristine to avoid pre-arc formation, having a bore that was kinda the same diameter most of the time, needed to have a $10^{-3}$-torr vacuum, no projectile blow-by, oh, and measure the projectile velocity right before the railgun section so you could dump the electrical energy at the right time. See what I mean … piece of cake! To make matters even more interesting, collaborate with two guys from Lawrence Livermore National Laboratory (LLNL), who were railgun experts, while doing all this. Well, we got the two-stage injector up and operational, modified a tank from the STAR "boneyard" to work as the impact chamber, and figured out how to make and align joints that actually passed projectiles through them instead of turning them into debris clouds. The problem left was the projectile "blow-by." Without solving this problem, in-bore velocity measurements could not be made, and the railgun section would pre-arc on every firing. I remember Sandia was hosting the Aeroballistic Range Association (ARA) Meeting in Albuquerque around 1988, and I was one of the highly technical guys that got to help put up and take down the chairs each day. As my reward, I got to stay and listen to the talks. I also remember being "intently engaged" in them to the point of my mind wandering to the projectile blow-by problem. And that was when the seal design came to my mind. I sketched it out and explained it to Lynn Barker, who said "Sounds reasonable…." I figured out how to machine it and made one for a test. Well, to make a long story short, it worked the first time out and became one of the enabling technologies for the STARFIRE project. Another projectile innovation that led to a repeatable, self-arcing projectile included the silver-epoxy-loaded fuse (Jim's idea to use the epoxy that would ablate and create a self-timing, initiating arc) with a HIGHLY sophisticated system of Scotch® tape for the mold. All these concepts eventually went into a patent granted in June 1993 entitled "Hybrid Armature Projectile." After several years of effort, however, the best velocity the launcher system was able to achieve was about 7.5 km/s with an injection velocity of about 6 km/s, well below the project goal of 15 km/s. After reviewing the rate of major successes vs. incremental ones, the amount of dollars invested, and prospects for major advances, Jim Asay made the executive decision to shut the project down.

An alternative HyperVelocity Launcher (HVL) technology was being explored at the time utilizing momentum transfer to impart the energy from a 6-km/s, two-stage projectile with a variable density impactor to a lighter stationary plate mounted in an extension at the end of the launch tube. Lalit Chhabildas and Bill Reinhart were working on this, it looked interesting, and the railgun program had just been shut down, so I decided "that project might be fun to work on" since they were already asking for my help anyway. Before too long, the projectile manufacturing and repeatability, alignment fixturing, plate fabrication techniques, and two-stage performance had all improved. The project was eventually able to launch flat, intact plates capable of reasonably precise equation of state measurements. One of the things I was most proud of was a method to eliminate the use of Glad® sandwich baggies as an integral part of the HVL plate installation into the barrel extension. By actually machining the part to the correct dimension, you could avoid the use of a "highly precise and repeatable" flat disk cut from a "sandwich baggie" that was wrapped around the impact side of the flyer plate assembly to provide the interference fit required to keep the assembly in place. Yeah, they needed help!

## Going Back to School and Rebuilding the Experimental Shock Physics Program

Two things occurred in 1994 that had a profound effect on the direction of my Sandia career. Discussion was taking place at the national level around closing one of the three nuclear weapons laboratories. Having a Mechanical Engineering Technology degree which, though useful, was less marketable than a standard Mechanical Engineering degree caused me to rethink going back to school part time, in case Sandia was the lab that would be closed. I had decided to do so under the Sandia University Part-Time Program when the second event occurred, hereafter referred to as "Black Monday." At the time, approximately ten Sandia employees and about eight contractors were working at STAR. Jim Asay had been promoted to a level-2 manager and was no longer associated with the site. Our Department Manager, Phil Stanton, called out to STAR and told Carl that he was holding a MANDATORY staff meeting in Area I for all Sandia employees, and there were to be no excuses to miss it. This was in early December, everyone was getting ready for Christmas, and nobody knew what the meeting was about. All the STAR tech folks piled into the van and headed into the area. When we got there, the entire department staff had already assembled in the meeting room, which was now standing room only, and there were plenty of hushed whispers going on. Phil stood up and proceeded to say the following. Effective immediately, the shock physics department was being disbanded due to a decision by the Director, Ed Barsis. He told all the STAR techs that we were to do nothing now but concentrate on finding another job at Sandia. The contractors would be given a month to be placed elsewhere, and the staff members were to be reassigned throughout the Center. You could have heard a pin drop. I have never seen such shock on a group of people's faces in my life. As people began filing out of the room, Phil told Carl and me to stay. He told us that we were to finish up all the work that STAR was committed to and had been paid to do, then officially close and mothball the site. It was a dark day indeed. Carl and I reminded Phil that a minimum of three people were required by approved Safe Operating Procedures to operate the site. He allowed us to keep Bill Reinhart on as a contractor until the work was completed. As people began leaving for other jobs, Bill, Carl, and I decided not to throw in the towel just yet. We kept executing the experiments, but we also began to "hide" the equipment that would be required to restart the facility and didn't let people take anything from that "stash" that were backing up with trucks hauling equipment and supplies away for other efforts at Sandia. We also realized that the ONLY way we were going to be able to save the facility was to get Jim back involved. We met with him several times and he agreed to help us find a new Sandia sponsor. The Explosive Components Facility (ECF) stepped up to the plate and took ownership using a new "campaign mode" of operation that reduced facility costs. The STAR was saved, although hanging on by a thread. Carl and I had found other jobs, so Bill and his contract were transferred to the ECF.

Jim continued looking for a more permanent solution and began discussions with Don Cook, the Director of Pulsed Power Sciences and owner of the recently reconfigured Z machine at Sandia. Don had recently commissioned an independent review of his inertial confinement fusion (ICF) program, which funded Z and its operation. In that review, he had been told to develop additional uses for Z other than just capsule implosions. Interestingly, work at LLNL on the Nova laser had been showing some potential application for using high energy light to drive high-pressure shock waves, even though the pulse lengths were very short. Don felt that Sandia could do something similar with a hohlraum on Z for measuring the dynamic response of materials. Don knew, however, that for a serious program to develop in his Center, Jim would have to head it up, so he was finally able to arm twist Jim into becoming a manager in his Center and begin to build a dynamic material characterization effort.

Now Jim is a smart man, but so was Don. Don gave Jim just enough money to get him to come over, but not enough to support a department, so Jim had to look for someone that was cheap (he ruled out all the staff members with that one), had some knowledge of conducting material response measurements, and would be willing to take a chance on shock physics at Sandia after having just gone through Black Monday recently. Who might Jim have found that had half his time already paid for by the University Part-Time Program, thought he (Jim) walked on water, and would be dumb enough to consider this line of work again? Well, it still took a call from Don Cook to assure me that he was serious about this effort before I decided to come back! Anyway, Jim and I formed the new shock physics effort at Sandia, and we were having to do it with pulsed power instead of conventional launchers.

At the time all this "department building" was beginning, Z was in its infancy and going through growing pains. Everyone who worked on Z felt the machine had tremendous potential, there was a teaming attitude, and that atmosphere posed opportunities for Jim's and my shock physics efforts. We began putting infrastructure in place for optical shock arrival sensors and VISAR systems, but encountered and subsequently overcame a number of obstacles like precision in timing mandated by small sample sizes, changing of the index of refraction in optical fibers from the inherent radiation, decaying pressure pulses, and nonuniform ablative loading. As we were solving these issues and becoming recognized in the community, Jim and I began adding folks to help establish this effort and also began strategizing for the future. I was beginning to understand the research and science of shock physics because Jim and Lalit were teaching it to me as I worked on my master's degree at the University of New Mexico, so we both realized the difficulties inherent with ablative loading and knew it was of limited use, but we had no other options at the time for creating those pressures on Z. As Jim and I pondered solutions to the ablative loading problem, I continued working on my thesis, which looked at understanding the dynamic response differences in concrete as a function of aggregate size under Lalit's direction. The resulting publication won the Aeroballistics Range Association prize for best student paper in 1998 and allowed me to complete the requirements for an MS in Mechanical Engineering from the University of New Mexico in 1999 and, with the degree, to be promoted to a Member of the Technical Staff.

## The Birth and Growth of Magnetic Loading on Z

Being the "good citizens" that we were, Jim and I decided to help Rick Spielman (a pulsed power researcher) make a pressure measurement on Z that he felt confident he could accurately convert to current (the measurement he was really after). To do so, Jim and I placed two small disks (one copper and one iron) into the anode plate on a wire-array load and then attached shielded VISAR diagnostics to the rear of each to measure their respective free surface velocities. As it turned out, the Z shot was fired late on a Friday afternoon, with the raw VISAR data saved and delivered to me to be analyzed on the following Monday morning. I made it until Saturday around lunch time before I had to head into work to see what we recorded. I sat down, opened the files, and reduced them. When I had done so, I got a very nice trace from the iron, but it had an odd feature I couldn't understand. It just so happened that Lalit was in his office working too, so I went down and showed the trace to him. He immediately thought that it might be clearly showing the alpha-epsilon phase transition in the iron, but couldn't remember at what pressure it was supposed to occur, so we jumped into the company truck and headed into Area I hoping to find someone that might be there and remember. The best we could do was to find a viewgraph on a poster outside a door (that we just had to "borrow") showing the iron transition from a gun experiment. Sure enough, they correlated, and the birth of magnetic loading for ramped wave experiments on Z gained credibility.

Jim and I recognized that there were still many issues to be resolved, but this promised to be a gold mine! We began presenting results at conferences in invited talks, contributed talks, and poster sessions to get the word out about what we were doing and where we felt we were headed. We also began taking a careful look at pressure gradients, diffusion, and analytical techniques. With a technology that had so much promise, but so many details to understand before it could become accepted as a routine method for gathering quasi-isentropic material response data, we knew we needed more help. Jim had taken back the STAR facility from the ECF and, with it, Lalit. Because both of them needed help, I was being forced to choose between STAR projects and those on Z by these two task masters. The choice had to be made between working on proven technologies at STAR with a reasonably steady funding stream under Lalit and unproven, though possibly revolutionary, technologies on Z under Jim that did not have steady funding and might never amount to anything. Once again, I cast my lot, career, future, and ability to feed my family with Jim on Z, a choice that fortunately proved to be the right one many, many times from that point on.

As word about the ramp loading began getting out, our funding began to improve and with it the opportunity to begin adding staff. Two additions were pivotal to the future of shock physics at Sandia. The first was adding Marcus Knudson, a very promising staff member fresh out of Washington State University (WSU), and the other being Chris Deeney. Chris was a plasma physicist on detail at Headquarters[79]

[79] Editors' note: Headquarters refers to NNSA (National Nuclear Security Administration) in Washington, D.C.

(HQ) when Jim began talking to him about coming back to Sandia as a manager of the newly reacquired STAR group. Even though Chris did not have a materials background, he was very bright, young, politically correct, and looked like he had all the skills to be a good manager and leader. Chris did indeed join our dynamic material team, and he and I very quickly became rather co-dependent.

Jim knew we needed some theoretical help to understand the magnetohydrodynamics of the ramp loading process and got it through Art Toor and David Reisman from LLNL. We got that help at a Sandia meeting with Jim, Art, David, myself, and a few others from our Z pulsed power groups where we were discussing solutions to the nonuniform current loading from radial convergence of the current as it flowed over our samples mounted flush to the lower surface of the anode plate. Two "eurekas" occurred in this meeting that made ICE loading what it has become today. The first was that the samples could be mounted vertically on the "current return can," and the second was that the can could be square and constructed out of panels (my idea). I worked with the target designers to come up with a panel arrangement that would allow diamond-turned surfaces and self-indexing corners for construction. We tried it out and, voilà, it worked the first time out! At last, the birth of another enabling technology that had an enormous impact on shock physics at Sandia … my baby, the isentropic compression experiment (ICE) cube: C.A. Hall, J.R. Asay, M.D. Knudson et al., "Experimental configuration for isentropic compression of solids using pulsed magnetic loading," *Review of Scientific Instruments* **72**, 3587–3595 (2001a).

Soon after Marcus had arrived, I was giving an invited talk at the American Physical Society Division of Plasma Physics meeting on the ICE cube and our ramp loading technique. Jim, Marcus, and I were sitting at a table between talks, and we began discussing another whole concept. Instead of transmitting the ramp loading directly into the sample of interest, we thought that we could instead launch the bottom of the panel as a flyer plate and impact our samples to obtain high-pressure Hugoniot data. We did what any self-respecting physicists would do at that point and drew a sketch out on a cocktail napkin, stuffed it in one of our pockets, and went to our next talk! When we got back to Sandia, we did indeed retrieve the "highly precise scientific sketch," did some calculations, performed the experiment, and exceeded years of HVL development on our first shot with a metallic flyer that was good enough for equation of state experiments moving at about 10 km/s! My choice to cast my lot with the Z technology when I was forced to choose between Jim and Lalit was quickly being validated. The first paper on the magnetic flyer plate technique was given in C.A. Hall, M.D. Knudson, J.R. Asay et al., "High velocity flyer plate launch capability on the Sandia Z accelerator," *International Journal of Impact Engineering* **26**, 275–287 (2001b).

## Becoming a Department Manager and Plutonium Containment Experiments on Z

With the department, personnel, technology, funding, and future taking shape nicely, the one and only real disagreement that Jim and I ever had took place. After announcing his impending retirement, I asked Jim about his views on my taking over the department as manager. He felt that without a PhD, he could not support my doing so. Obviously, I felt otherwise and decided I had to find some other group that would allow me to continue advancing. I had to make a choice between two offers. One was at Sandia in the Radiation Effects Department, and the other was at the Nevada Test Site (NTS) running the Atlas Project, which was to disassemble a Z-class pulsed power machine that had been built at Los Alamos National Laboratory (LANL), take it out to NTS, rebuild it, run it, and make it successful. Ultimately, I decided to stay at Sandia and actually ended up working in the same building as when I was in Jim's group, just three floors up! It was a very difficult time for both Jim and myself. I could not make myself move out of my old office completely, and just kind of kept it "as is" for several months to maintain contact. I think I was having separation anxiety from Jim, who had been such a fixture in my career for so many years!

After several months during which Jim and I didn't talk much, Chris Deeney came to me and offered me a job in his materials group, promising me that I would be considered if his job opened up after Jim retired. I agreed and came back to the group after only about 6 months. Chris's number one priority for me when I returned was to take the struggling containment chamber project and make it successful with minimal funding and a lot of cards stacked against us from both the other labs and our own internal management. This project proved to be a perfect way for me to demonstrate and improve the meager management skills I had at the time. There were three groups designing separate parts of the system, little communication, significant technical challenges, an unfocused testing plan, a requirement that it work perfectly each time it was used, and a lot of skeptics everywhere we looked. Chris, being the good guy that he was, washed his hands of it and threw it squarely into my lap. The sad thing is that I loved it, and I suspect he knew I would feel that way when he handed me the project. The design groups and I had several "come to Jesus" meetings where I took control of each design effort, forced communication, and made some hard and unpopular decisions about the testing plan, and we finally started making significant progress. After several years of work, we were ready to test the first full system on Z. We fired the shot, and all real-time indicators suggested that it functioned properly, but the only way to know for sure was the much anticipated helium leak test. We pulled the system from the machine and hooked it up to the detector. Suffice it to say that I did the "happy dance" when there were absolutely no leaks! The system worked as well or better than we had hoped, first time out. This turned out to be yet another enabling technology that truly provided a future for Z and the Sandia shock physics effort, even if we didn't fully understand it at the time.

By now, Jim and I were "back to normal" … thank god … and he introduced me to John Taylor, someone that would end up having a HUGE impact on my life and career at Sandia. Interestingly, he was one of Jim's neighbors and they rode to work together most mornings in a van pool. Well, it turns out John managed a large program at Sandia that was highly classified, and Jim became one of his principal investigators for a piece of it having to do with material response. As Jim was getting closer and closer to retirement, he began to pass control of some of his projects off to others. The "Taylor work" was given to me. All of a sudden, I was forced to get all kinds of security checks and clearances, sign away my first born, and learn how NOT to talk to anyone about what I did on this project. Quickly though, John liked how I managed the part I was given and expanded my role and control to be a much bigger part of his program. I really enjoyed the work and was becoming exposed to another whole customer set and way of thinking and conducting research. It turned out to be very fast-paced, result oriented, and of immediate impact to the Nation.

The dreaded day had finally arrived: Jim actually retired and moved off to work with Yogi Gupta at WSU. I don't think there really was much of a decision to be made at that time, but thankfully they gave his senior manager's job to Chris, who had taken over as manager of the "Z department" and Lalit backfilled Chris's job as manager of the "STAR Department." All this movement left the manager of the Z department open, and Chris put it out to bid after several months. Once again, it was decision time for me. Did I feel qualified to run this fast-growing technical effort that had so much promise, so many new young staff, and so little Center management support in reality? What if I couldn't do it…? What if I failed to provide vision, keep the staff, and/or maintain the funding? This wasn't just characterizing materials using tried and true technologies and analysis methods like would be done at STAR … this was WAY outside my comfort zone. Back in the day, I KNEW I could run STAR, and I felt pretty strongly that I could run the STAR Department, but this … this was way more "researchy" with much more at stake. I finally made my decision based on two things. One, that Chris must have felt I could do it because he highly encouraged me to bid on it and, two, that he would be there to "help me" if I looked like I would fail. I finally did bid, and Chris gave me the job.

You have heard the old adage, "When it rains it pours." Within 3 months of my having accepted the Department Manager position for Chris, John Taylor and the HQ Program Technical Advisor Bruce Boughton wanted to meet with me. Turns out that John was taking a new job working in a vice president's office and didn't feel he would have time to continue running his program at Sandia, so he asked me if I would consider taking it over. I truly enjoyed the work that we were doing on this program, the successes we were having at Sandia, but I wasn't sure I had time to make both it and my new department successful. I explained all that to them, but they were insistent that "I was the guy" they wanted to take it over. How do you turn something like that down … you don't … so I accepted the offer. How hard could it be, right? I was about to find out.

The good thing about working for someone who is a REALLY good boss (Jim and Chris both were, but I am referring to Chris in this instance) is that he takes you along to all the meetings, you learn management skills, you have a security blanket, and you have someone to encourage you. There is, however, a down side as well. You aren't the only one that recognizes that. Yes, you guessed it; Chris got a job offer back at HQ to run the Science Campaign, which had become the major funding source for not only dynamic material work at Sandia but also much of the pulsed power mission.

Fortunately for me, the Shock Physics on Z Department and the activities in the dynamic material properties (DMP) subprogram were in pretty good shape when

Chris left for HQ. As a team, he and I had started down several strategic roads including the ability to do contained hazardous materials on Z, the authorization basis for conducting plutonium characterization experiments using that system, solid DMP base funding, the concept for a compact ICE driver, and a reasonably good, diverse staff. Of course, having grown up in this group, I did know about some of the "issues" that were less apparent, and I actually did have some strategic thoughts of my own (imagine that!). Anyway, with a good base, and Chris positioned in the Defense Programs Science office back at HQ, things were looking pretty good for our future in Department 1646 (Dynamic Material Properties).

Several things still needed some work before they were actual mature capabilities that we could use for problem solving though, so we began working them in parallel as a group. The first of these was shooting Pu on Z. The authorization basis had to be completed before we could do the work, the targets had to be designed, the transportation issues had to be addressed, we had to get LANL buy-in, the waste stream, etc. The list was long. Designing and shooting a material response experiment like we were trained to do was the easy part; it was all this other legal stuff that was the headache! I assigned Jeff Gluth to do nothing but work the ES&H (Environment, Safety, and Health) and transportation issues, Marcus to do the experimental design work with LANL, and I got to do the coordination and the "dog and pony show" presentations that went on ad nauseam after Chris had gone to HQ. To make matters worse, we were going to exercise the truly unique quasi-isentropic loading capability on this material instead of the relatively straightforward Hugoniot approach. So not only were we fighting the ES&H issues, another lab associate director saying, and I quote, "They will never shoot Pu on Z," all the folks in the Pulsed Power Sciences Center that didn't want to do these experiments (of which there were many … and vocal ones at that); we were also trying to pull off a home run technically on an incredibly challenging problem in a very visible experiment! Like I had said before, easy, right? Well, to make a long story short, we did shoot three Pu shots on Z over a three-month period with "the world watching," they were perfectly contained, and they went off without a hitch. Marcus and Dennis Hayes collaborated to make some nice deductions from the analyzed data, all the logistics went perfectly, and, lo and behold, we had done the seemingly impossible on a shoestring budget! This experiment was tremendously important as it not only provided some new insights into Pu behavior, it also secured Sandia's future in DMP for many years to come.

### Other Shock Physics-Related Activities

Chugging along in the background while all this was going on were two other significant projects. Having worked with Randy Hickman in the past on the railgun project, I knew he would be the right guy to get the mini-pulser facility, Dynamic Integrated Compression Experimental (DICE), online. I was able to talk Area IV out of a building that was in dire need of a facelift but was perfect for this application. We were able to convince folks that the million or so dollars that it would take to design and construct this "mini-pulser" would be well worth the cost when it became useful as a test bed for developing diagnostics and experimental techniques that would eventually be transitioned onto Z. We leveraged the Special Technologies Laboratories in Santa Barbara (funded through some indirect DMP dollars) to design and build us a state-of-the-art, air-delay VISAR system for the low-pressure work we planned to accomplish in this facility, used hand-me-down digitizers from Z, invested in some streak cameras, and attached an amazingly capable instrumentation room to the pulser. We also designed and built a small, cheap, and cheap to operate single-stage gas gun for doing complementary shock work to the pulser's ramped compressions.

I needed a programmatic driver to justify the space and investments we were making in DICE and began targeting the Sandia mission space of neutron generator (NG) design and production. The ties to this group had faded away over the years, but I felt we needed them so that the vice presidents in their line organizations would support our materials work. I also needed a very detail-oriented researcher that already had credibility with the NG community, documented EVERYTHING to the nth degree, and was looking to become the cornerstone for such work in this group. I knew what I had to do … go get Jack Wise and convince him to leave the Geological Mechanics group and come work for me. Well, it took a little "fancy talking" on my part, but as it turned out he was looking for a new challenge and decided it might not be too bad to work for me. He moved into the group and this fledgling relationship that we as a program so strategically needed began to blossom. Jack was the perfect fit, but the Sandia work was not quite enough to keep him busy. As it turned out, one of the problems that we really needed to work out for DMP was the ability to map phase lines using the isentropic compression capability. To do this, the obvious approach is to elevate the initial temperature of the sample to several states, compress the sample, and cross the phase line from these various initial states to get a description of its placement and slope in pressure–temperature space. Easy, right? Well, technically not so difficult for ramp compressions … but VERY hard for ramp compressions to about 600 C on nasty materials inside a containment chamber with micron-thick glue bonds that maintained perfect integrity at these temperatures. What's life without the occasional difficult challenge….

Everything was going along quite well for the group. We had an amazing team of researchers that all knew their roles, worked very well together, and bought into the strategic department vision. We had developed our own target fabrication and design team (both of which contributed mightily to our successes on Z and DICE), and significant technical advances were coming at a record pace. Our biggest challenge had become getting enough time on Z to try the new ideas Marcus, Dan Dolan, Jack, and Seth Root were conceiving, while still characterizing the materials we were being asked to do for direct programmatic drivers. Z was getting very difficult to access for our increased needs, and DICE didn't have the pulse control we so desperately needed, nor could it access the pressures we wanted for much of the work, so the logical step was to build a "mini-Z" at DICE. Such a machine would bridge the gap between the worlds and let us continue being innovative. We began development of such a machine and named it Genesis. It was nearly fully designed, and the infrastructure ready to accept it at DICE when I left the group in 2009.

### Moving to the Nuclear Counterterrorism Program

One of the visions that Jim had, but was never able to realize fully at Sandia, was to have all the shock physics activities housed within one department. At this time, there were two separate departments in the Pulsed Power Sciences Center related to the Z efforts and to the STAR group, a dwindling group that operated the "Building 808 air gun" and the air gun at the ECF. Lalit was retiring, so I had the opportunity to combine those groups into one, which I convinced management to do. By doing so, we had a "supergroup" capable of kbar through multi-megabar pressure regimes and both Hugoniot and quasi-isentropic compressions that were rivaled by no others. In addition, I was able to hire Mark Anderson away from the 808 gun facility to work on my Nuclear Counterterrorism (NCT) Program in another group, so his old group was officially decommissioned. In effect, for the first time that I am aware of, all the shock physics at Sandia was under one group, with me as the manager.

They say that success is a two-edged sword, and I have to firmly agree. As it turns out, both the materials group and the NCT Program were growing, becoming more diverse, and each was demanding more of my time. In reality, each was a full-time job. To make matters worse, Sandia began to demand more and more responsibility, accountability, and formality from its level-1 managers. In short, the two jobs were becoming too much for me to do alone. The writing was on the wall, and I knew I would soon have to choose between the jobs, just as I had been forced to do years earlier with the STAR and Z career paths. It took a lot of soul searching, I can assure you. How do you walk away from such a functional group where all your strategic objectives over the previous 5–6 years were becoming reality, I was becoming knowledgeable enough to provide realistic technical input to such distinguished researchers as Marcus Knudson, Dennis Grady, and others from Los Alamos National Laboratory and LLNL, and the future looked so bright? It was one of the most difficult decisions of my life, but I knew that the NCT mission was the one that would most benefit from my expertise, whose future was looking bright as well, and was of greater immediate importance to the Nation. Therefore, I chose to maintain leadership of that effort and stepped down as manager of the shock physics group in March 2010.

Although it is at a much different level, and I can't talk about it or publish much anymore, I still maintain an active role in the shock physics community as the national subject matter expert in dynamic material characterization for the NCT Program.

### Final Thoughts

When I came to Sandia in 1984, I had no idea what a shock wave was, much less that I would have any significant impact on such a field of study during my career. When I look back at the past 25 years in which I was fully engaged in this research, I feel both fortunate and blessed to have had a role in several developments that eventually helped shape the Sandia program. My contributions culminated in the opportunity to lead the effort for 6 years, building upon the strong foundation left to me by both Jim Asay and Chris Deeney, yet putting my own mark upon it as well. Just as Jim and Chris must have felt before me, I knew the strength of the group was in the people and that this effort would survive long after my leaving. I was right.

**Special Thanks**

There are many people that have had positive impacts on my shock physics career at Sandia, but I am most indebted to three people. Jim Asay stepped out on a limb by hiring me and then believed in me as a technician and staff member, providing me with many opportunities to succeed that were not only challenging, but stretched my abilities as well. He also taught me most of what I know about dynamic material response and helped form my initial management style through his leadership and example. Chris Deeney believed in my managerial abilities and took risks to his own reputation by providing me with high-profile opportunities that could either "make or break" his department and possibly both our careers in this field—the most significant one being his decision to promote me to manager of this highly technical group. Chris also constantly took me to meetings with him where I learned another management style that would eventually become a significant part of my own, including how to mentor and advance others' careers. Carl Konrad was a patient mentor and great friend, teaching me much of what I learned as a technician, helping me learn how to "just get it done." He was supportive of many of my technical ideas and approaches, while making it fun to work at STAR. I am very pleased to count all as not only mentors, but personal friends as well. I would also like to thank Dennis Grady, Dennis Hayes, and Lalit Chhabildas as well for helping me to learn shock physics through very patient instruction.

## Dennis B. Hayes
## (1957–1995[80])

I just stumbled into this shock wave thing. In late 1959, I was working swing shift in the print shop at Sandia and that allowed attending the University of New Mexico as a freshman during the day. But they closed down the swing shift, so I looked at the weekly job postings for union members and found a data reduction clerk opening in Byron Murphy's division. This division was involved in explosive cratering as part of the Plowshare Program (Al Chabai, Ron Carlson, Hunter DeVault) and in ground motion studies near underground explosions (Bob Bass, Luke Vortman, Bill Perrett, Dorris Hankins). Although this group was not directly tied to the emerging shock effort at Sandia, these application-oriented missions were two of the drivers for the growth of shock programs.

Members of this group were certainly well versed in shock basics: jump conditions, the Mie–Grüneisen equation of state, the equations of motion, thermodynamics, and so on, and my exposure to those was a daily occurrence. And the staff stretched me out in those disciplines. I was becoming familiar with the "finger exercises" of shock wave physics with the added benefit of gaining proficiency with the content of several math, mechanics, and thermodynamics courses before reaching them in the classroom.

Al Chabai was a skilled user and proponent of dimensional analysis as a tool to organize measurements that did not lend themselves to modeling. For instance, his work on scaling of explosive cratering data gained international recognition and is still in use today. Being a good student of Al's work led me to try some similar analysis of my own circa 1964. This Sandia group, along with their Los Alamos National Laboratory (LANL) colleagues, had amassed hundreds of time-of-arrival measurements made very close-in to several dozen underground nuclear explosions. But the data remained largely unanalyzed owing to a lack of sophisticated modeling capability. It turns out that each underground test was assigned a yield based on radioactive isotope ratios gathered in post-shot gases—i.e., a radiochemical analysis. Scaling both distance and time of arrival of the shock by the cube root of the radiochemical yield produced a remarkable, low-scatter universal curve time-of-arrival graph that I proudly tacked above my desk. I was just a technician and too naïve to understand the importance of publication. This result soon found its way into the test community's "public domain" and, I was told, with some improvement, ultimately became one of the United States' tools for monitoring compliance of the 1974 Threshold Test Ban Treaty.

---

[80] Dennis Hayes had a varied career at Sandia involving both technical and management positions. He had a 2-year external assignment as scientific advisor to the Deputy Assistant Secretary of Military Applications at DOE Headquarters in Washington, D.C., 1990–1991. He retired from Sandia in late 1995 to become the president of Lockheed Martin Nevada Technologies in Las Vegas, 1996–1997, after which he continued as a consultant in shock wave research (1998–2011) at Sandia, Los Alamos, and Lawrence Livermore and as a visiting professor at Washington State University.

In 1965, with my BS completed, I joined Al Chabai's newly formed division concerned with x-ray vulnerability of reentry vehicles that were part of the Nation's nuclear deterrent. I developed the VANDAL x-ray deposition code that found wide use at the Lab by component, system, and material organizations and that led to extensive contact with a wide swath of Sandia programs and people. Walt Herrmann had just arrived at Sandia, and I worked with him and Pat Holtzhauser to implement an x-ray deposition capability (with rapid heating leading to pressure buildup and thus to shock generation) into his WONDY code. Much of the other work during this time entailed designing experiments for underground tests and analyzing the results.

In the fall of 1969, I was selected for entry into Sandia's Doctoral Study Program and spent 3 years under the tutelage of Professor George Duvall at Washington State University (WSU). My topic led to the discovery of metastability during the shock-induced phase change in potassium chloride and also measuring the rate at which the 2-GPa phase transformation from b1 to the b2 crystal structure occurs. I had the good fortune to work with fellow student Joe Andrews, who had a very clever and exact way to treat phase transformations in shock codes, which I adopted, expanded, and used in a variety of different ways at Sandia and Los Alamos for the next three decades.

Graduation from WSU in 1972 was followed by a 1½ year stint in Walt Herrmann's shock wave department, where I worked with Jim Asay on shock melting in aluminum (Asay and Hayes 1975), with Jim Johnson and Jim Asay on the multiphase behavior in bismuth (Johnson, Asay, and Hayes 1974; Hayes 1975), and with several other people on other projects. My role was to model and interpret the experiments. I twice taught a full-semester course on shock wave physics. The students were mid-career Sandians caught in the transition from a one-mission to a multi-mission laboratory. The course notes, and to a lesser extent the lecture videos, turned out to be quite popular, and their use is still widespread today at universities and other institutions internationally. I attribute the popularity to the fact that the material is taught at the most elementary level possible. The students can learn the concepts first and add sophistication as needed in their own specialty. The overall experience in Walt's group was a first for me: actual fundamental research at Sandia.

In 1974, I became Supervisor of the Explosive Components Division. Most of the work was on scheduled components for nuclear weapons. But we did have a rather sizable reimbursable with the Navy to study porous HNS[81] explosives. The Navy was introducing HNS detonators and mild-detonating fuzes on their C-4 missile and desperately needed a more thorough understanding of the behavior of this new explosive. The Explosives Research Department was unwilling to take on this customer-driven work so we built up a capability within the Division. The principals were Dennis Mitchell, who performed front surface impact VISAR measurements on porous HNS that allowed the equation of state and reaction rates to be measured, and Al Schwarz, who built our own VISAR and made innumerable measurements

[81] Editors' note: HNS is hexanitrostilbene, a high explosive with good thermal and vacuum stability that was developed by the Navy in the 1960s.

on component performance that could be compared with model results. Jim Kennedy and Steve Sheffield were valuable team members. I did most of the modeling (Hayes and Mitchell 1978). The result was a rather complete material description. With the help of Sam Thompson, we introduced reaction kinetics of HNS into CSQ (the two-dimensional progenitor of CTH) and did numerous numerical studies on detonation corner turning and other two-dimensional phenomena important in component design.

In 1978, I became the manager of the Fluid Mechanics Department. We did almost no shock work. The exception was Jace Nunziato, whom I had promoted to supervisor of one of the divisions. He spun his magic by applying his mixture theory to shock initiation of composite explosives and to deflagration-to-detonation transition. I helped some but mostly watched. I did work on one problem. Heterogeneous deformation in the shock front was in vogue, and so I developed a simple model that accounted for the competition between heat generation from plasticity in shear bands with thermal diffusion occurring during the finite risetime (Hayes and Grady 1982). The result was that the shear bands in a shock front should number just a few, regardless of the strength and the risetime of the shock. However, the holy grail of the fourth power law between stress and strain rate was elusive and not rendered by this simple analysis.

At a place like Sandia, interesting problems can arise unexpectedly. Immediately after the Three Mile Island (TMI) reactor accident in 1979, I got a call asking for help on the hydrogen release problem. The big question: "Would an internal hydrogen detonation breach the concrete containment structure of the reactor?" I assembled a team of Sandia experts. We worked long days and nights and were in constant phone contact with officials at the TMI site. With Sam Thompson and his ANEOS equation of state and CSQ codes, Mike Cieslak on material and chemical issues, Dennis Mitchell on explosive issues, and Walt Murfin, a Sandia expert on the damage shocks cause in concrete, we were able to give definitive guidance to the TMI Emergency Management: there was little danger of containment failure. But the larger question went untested as the large volume of hydrogen produced later had insufficient oxygen to be detonable. I have always believed the good marks we made during this effort helped Sandia establish itself in the nuclear reactor safety arena.

In 1982 I became the manager of the Ceramic Components Department, whose mission had little to do with shock physics. The sole exception was the responsibility for the design of ferroelectric power supplies that are used in neutron generators. However, adequate models were sorely lacking so our work was strictly empirical in the tradition of much of Sandia's component development up until that time. This mission was the driver for Steve Montgomery to later do superb modeling of shock depolarization of ferroelectrics, which has changed the design world in this area.

In 1988 to 1989 I moved to the Solid Dynamics Research Department. An important change was taking place in the computing world. Parallel computing had arrived, and we staffed that area to implement shock codes and made good progress. Another related effort was transforming the CTH code from a vertically integrated project to a full-fledged software engineering project, with benefits that exist to this day.

I spent 1990 to 1991 as the scientific advisor to Admiral Mike Barr, the Deputy Assistant Secretary for Military Applications at DOE Headquarters in Washington, D.C. I spent 1992 to 1995 as the program manager, under Roger Hagengruber, of Sandia's technical capabilities for Defense Programs, which included a broad portfolio ranging from research and advanced development to capital equipment budgets and other activities related to weapons programs. During this time as a Sandia Director, I was essentially disconnected from the details of the research in shock physics.

I retired in 1995 and served from 1996 to 1997 as the president of Lockheed Martin Nevada Technologies in Las Vegas, NV. Upon my retirement in mid 1997 from Lockheed Martin Nevada Technologies, I entered into consulting contracts with Sandia and Los Alamos. At Los Alamos my main project was to develop a multiphase equation of state with phase change kinetics for plutonium. The results are classified and therefore not published in the open literature. Another interesting project was to measure twinning[82] kinetics in the uranium-niobium alloy U-6wt%Nb and to confirm that was the reason for some of the rather bizarre behavior that had been observed in shock experiments. That study was done jointly with Rob Hixson and Rusty Gray of LANL.

Most of my post-retirement effort at Sandia was designing and analyzing ramp wave experiments conducted on Z. See, for example, Hayes, Hall, Asay, and Knudson (2004). This required developing a method to integrate the equations of motion backward in space. The technique maps VISAR records into stress–strain isentropes and has proved quite useful for hyperbolic systems.

Early retirement was almost like a full-time job but it tapered off. I billed 1200 hours the first year and essentially zero during 2011 with the drop off roughly linear over that period. Amazingly, the amount of shock wave-based research during retirement exceeded the aggregate conducted during my regular employment time.

One closing note is that only a small slice of Sandia's mission space involves applications that have contact with the shock dynamics field. Over the years, this has resulted in constraints on program support and growth within the laboratory. In spite of this, the shock dynamics effort has maintained strong management support. It is quite remarkable that Sandia has excelled and, in many respects, has been the preeminent organization in this field for more than three decades.

Personally, I am quite grateful for the opportunity to have worked with so many very capable people on a whole variety of interesting and important problems.

[82] Twinning is a metallurgical term that refers to an inelastic deformation of the crystal lattice.

## Walter Herrmann

## (1963–1993)

"For technical and managerial leadership in material models and tools for numerical simulation of nonlinear dynamic phenomena."

**in Memoriam**[83]

**by Orval E. Jones, Sandia retired**

Walter Herrmann was born on May 2, 1930, in Johannesburg, South Africa. He was the only child born to Gottlob Fredrich Herrmann, a hotel proprietor, and Gertrud Louise Herrmann (nee Ratzlaff). Herrmann received both his undergraduate (1953) and PhD (1955) degrees in Mechanical Engineering from Witwatersrand University in South Africa. Dr. Herrmann then taught Mechanical Engineering at the University of Cape Town for 2 years until 1957, when he was invited by the Massachusetts Institute of Technology (MIT) to work as a researcher in the Cold War effort.

He married Betty Lorraine Allard in 1955. They raised two children, a daughter, Inga, and a son, Peter, before divorcing in 1983. In 1988, Dr. Herrmann married Ednarae Bligh Gross, who preceded him in death in 1996. Dr. Herrmann was an accomplished skier, a mountaineer (he volunteered for many years for mountain rescues in New Mexico), and a cyclist. He also enjoyed river rafting and scuba diving. After retiring from Sandia National Laboratories, he was a volunteer for the New Mexico Museum of Natural History and served as a docent at archeological sites in southern Colorado. Throughout his life Dr. Herrmann was an avid traveler, and in his later years, he developed a passion for tall ships, sailing, and archeology.

[83] Prepared by Orval E. Jones and submitted by the National Academy of Engineering Home Secretary. See *Memorial Tributes: National Academy of Engineering* (The National Academies Press, Washington, D.C., 2011), Vol. 15, pp. 180–183. (Reprinted with permission from the National Academies Press.)

Walter Herrmann came to Sandia National Laboratories[84] via the MIT Division of Sponsored Research, where he was developing his WAVE 1 finite-difference computer program for numerically calculating high-amplitude, nonlinear wave propagation in one-dimensional structures. Sandians C. Donald Lundergan and Charles F. Bild, then the Director of Materials and Process Development, along with others, recognized that computers were becoming powerful enough to realistically simulate shock wave deformation of structures. Such deformation results from the detonation of explosives or high-velocity impact. Indeed, they were proven right; Herrmann was successful in building this capability at Sandia. During the Cold War between the United States and the Former Soviet Union, such capability was urgently needed for the development of new nuclear weapons.

Herrmann joined Sandia in early 1964 as Supervisor of the Deformation of Structures Division. Initially, he had no staff, but he had management support to build a computer code development team. WAVE 1 became the basis for the Sandia WONDY finite-difference computer program, which was soon used in nuclear weapon component development. Next, the Sandia TOODY computer program was developed, which enabled two-dimensional simulation and analysis of structures. Some of the important code developers included Herrmann (a formidable mathematician and computer scientist), Larry D. Bertholf, Darrell L. Hicks, R. Jeffery[85] Lawrence, Samuel L. Thompson, Billy J. Thorne, and Robert J. Walsh. Thompson went on to develop a powerful three-dimensional code that included radiation transport.

These computer programs all required knowledge of nonlinear constitutive behavior of materials at the extreme stresses and strain rates induced by shock wave loading. Building on Lundergan's program of projectile plate impacts, along with data from other Sandia organizations and other laboratories, the experimental data for developing constitutive models of many types of materials—metals, ceramics, polymers, composites, and foams, for example—were obtained. Names of some of the key staff members engaged in this activity during the next 25 years included James R. Asay, Lynn M. Barker (inventor of the VISAR), Barry M. Butcher, Lalit C. Chhabildas, Robert A. Graham, Charles H. Karnes, Darrell E. Munson, and Karl W. Schuler. The experimental data then had to be incorporated into theoretical constitutive models in order to be used in the computer programs that were under development. Some of the key researchers in this activity included Albert J. Chabai, Peter J. Chen, Lee W. Davison, Douglas S. Drumheller, Dennis B. Hayes, Herrmann, James N. Johnson, Orval E. Jones, and Jace W. Nunziato. Many of these individuals and others were also involved in code simulations applied to pressing national problems. A particularly notable example is the work of Paul Yarrington on missile impact and cratering and warhead contact fuzing. Other such contributors included Marlin E. Kipp and Timothy G. Trucano.[86]

[84] Editors' note: At the time Walt Herrmann arrived, the Laboratory was called Sandia Laboratories.

[85] Editors' note: The spelling of Lawrence's name in the original has been corrected here to Jeffery.

[86] Editors' note: Trucano's middle initial in the original (J) has been corrected here to G.

Over a period of more than 25 years, Herrmann recruited, managed, directed, and led by example an extremely productive and internationally known organization of some 45–50 scientists and technicians. In 1982 he was appointed Director of Engineering Sciences—an organization of some 140 Technical Staff that—in addition to the aforementioned solid dynamics activities, general engineering structural analysis and fluid and thermal sciences were included. In 1992, prior to his retirement in 1993, he was named a Sandia Senior Fellow.

Herrmann died on June 4, 2000, at the age of 70.[87]

Survivors include his children, Inga and Peter, and stepsons, Allan and Jeffrey Gross, all of Albuquerque; a stepdaughter, Janis Gross of Longmont, Colorado; grandchildren Rishar, Mariah, and Joshua; and his former wife, Betty Lorraine Allard Herrmann, of Albuquerque.

**Editors' comment**: Walter Herrmann was also elected to the National Academy of Engineering in 1993 in recognition of his leadership in modeling the response of materials at high loading rates.

[87] Editors' note: Herrmann's age at his death in the original (71) has been corrected here to 70.

# Eugene S. Hertel, Jr.

# (1986–Present[88])

## Shock Physics Codes (aka Hydrocodes[89]) at Sandia National Laboratories

### Personal History

My personal history at Sandia spans the period from late 1978 thru 2013 (and should continue for a few more years). I had just finished my PhD dissertation research at the University of New Mexico in physics. My dissertation was completed over the next several months and defended on July 14, 1979. From March 1979 thru September 1986, I was an on-site contractor and worked in several organizations, mostly focused on high-level radioactive waste disposal. During those early days, my path crossed those of two individuals with whom I would later work in shock physics, Sam Thompson and Mike McGlaun. At the time, Sam and Mike were working on the CSQ hydrocode and a number of reactor safety packages. They were heavy users of Sandia's early computational resources (the CDC 6600 and 7600) as was I, but not at their level.

When I became a Sandia staff member, I worked in a computer support department with Larry Bertholf as my second-level manager (known as a department manager then). I didn't realize it at the time, but Larry was involved in the early days of shock physics at Sandia. I continued to interact with Sam Thompson during my early career by providing computer support for CSQ and CTH (more on these codes in a bit). After a short stay back in waste disposal, I went to work for Sam in late 1989. At that time, Sam was a first-level manager (supervisor) in the shock physics department (managed by Dennis Hayes), and Walt Herrmann was the Director of Engineering Sciences. The shock physics department consisted of a code development organization (under Sam), an analysis organization (with Paul Yarrington as the first-level manager), and an experimental organization (with Jim Asay as the first-level manager). I worked on CTH development and some shock physics analysis. Not long after I arrived in the group, Sam was moved to another position; Dennis went to Washington, D.C., for a stint at the Department of Energy (DOE); Mike was promoted to my first-level manager; Jim was promoted to the second-level management position; Walt was moved to another position; and Dennis came back as a new director! And none of this was my fault!

I stayed in the same organization for the next 10 years through a major reorganization (shock physics moved to a new center focused on high-performance parallel computing) and a few management changes. For most of that time, I was the project lead for CTH and eventually for a project under the Joint Munitions Program (JMP) in support of CTH and other technologies. CTH went from being a niche code at Sandia to the most widely used hydrocode in the United States. In late 1999, I accepted a management position back in the Engineering Sciences Center. I left the project lead-

[88] At the time of this writing, Gene Hertel manages a group involved with hypervelocity lethality and continues oversight of the CTH maintenance and distribution.

[89] Editors' note: The formal name for "hydrocodes" is "hydrodynamic codes."

ership of CTH with staff in the high-performance parallel computing center, but continued as JMP project lead and merged that with another project focused on energetic materials. Over the next few years, shock physics staff and expertise moved into my organization, such that CTH effectively lived there. CTH continued to grow in importance, mostly outside of Sandia in the Department of Defense (DOD) and their contractors. During that time, my management roles changed a few times. Initially I was responsible for a mix of energetic material response and fire modeling. That transitioned into energetic materials and thermal analysis. That department is where we added a key CTH developer, Dave Crawford, and effectively owned CTH within Sandia. I also spent a short time doing program development for our core DOE code development effort. During this time, the two key CTH developers (Crawford and Bob Schmitt) were moved into the core code development organization with the hope that they would help that team understand external interactions and code support. Unfortunately, that attempt didn't succeed for several reasons. I then took over a department that supported the Missile Defense Agency with lethality assessments, a technical area that I had supported as a staff member in the 1990s.

As 2015 starts, I am still managing lethality assessments with CTH and Zapotec (a coupling of CTH and the Lagrangian code PRONTO). CTH is the most widely used hydrocode in the United States and has been the most used code on DOD high-performance computers as long as they have been keeping track. CTH is used extensively at Sandia and Los Alamos to simulate a wide variety of phenomena. Just recently, Bob Schmitt completed a scaling study with CTH on the newest Advanced Simulation and Computing supercomputer (Sequoia). CTH was exercised on the entire machine, a little over 1,500,000 cores! CTH was found to scale very well from a few thousand cores up to over a million. It wasn't that long ago when running on a few thousand cores was considered state of the art.

## Code Development History at Sandia

Numerical simulations can offer a significant complement to experimental methods for studying dynamic material behavior. For example, simulations allow non-intrusive investigation of material response at interior points of a sample. No gauges, wires, or other instrumentation are required to extract the information on the state of the material. The response at any of the discrete points in a numerical simulation can be monitored throughout simply by recording the material state at each time step. Arbitrarily fine resolution in space and time is possible, limited only by the availability of computer memory and time. A somewhat pithy statement has been used for this concept: "An experiment is the full truth partially revealed, a simulation is the partial truth fully revealed."

If we assume validated material models, numerical simulations can explore material response in pressure and temperature regimes that are beyond currently available experimental or diagnostic capabilities. Hypervelocity impacts can be studied in velocity regimes that exceed present projectile launcher capabilities, and detailed investigation of material response can be made at temperatures for which the use of diagnostic apparatus in any experimental study would be precluded, e.g., in materials subjected to intense x-ray energy deposition, in expanding products of chemical explosives, etc.

Furthermore, in many cases, numerical studies can be done in less time and at lower cost than experimental studies. In general, small changes in problem parameters—such as material thicknesses, impact conditions, or material properties—involve minor changes to the problem specifications. These changes can often be made in a matter of minutes, whereas corresponding changes in an experimental configuration may require additional machining and assembly of hardware and can therefore be quite time consuming.

Simulations are only useful, however, insofar as there is some reason to have confidence in the accuracy of the solution method and the material response models used in the calculations. Presently, accurate material response models exist for only a small part of the spectrum of materials and loading regimes of interest in shock compression studies. One notable shortcoming of present techniques is the lack of capability to predict failure and post-failure response for most materials of practical interest. Also, the response of many materials is strongly influenced by micro-mechanical factors, which cannot be modeled explicitly and are not sufficiently well understood to be incorporated in macroscopic models. Composites, which are of great and increasing importance, are an example of a class of materials for which fidelity is lacking in numerical simulations of dynamic response in shock compression processes.

Numerical simulations are designed to solve, for the material body in question, the system of equations expressing the fundamental laws of physics to which the dynamic response of the body must conform. The detail provided by such "first-principle" solutions can often be used to develop simplified methods for predicting the outcome of physical processes. These simplified analytic techniques have the virtue of numerical efficiency and can be preferable to numerical simulations for parameter sensitivity studies. Typically, however, rather restrictive assumptions are made on the bounds of material response in order to simplify the problem and make it tractable to analytic methods of solution. Thus, analytic methods lack the generality of numerical simulations, and care must be taken to apply them only to problems where the assumptions on which they are based will be valid. This can apply also to hydrocodes, particularly with respect to material models and their validity.

The problems solved by hydrocodes are quite challenging for several reasons. The geometry can be very complex. The phenomena can be highly nonlinear, including, for example, nonlinear material models, fracture, or buckling. The codes must be minimally capable of resolving the gross behavior of the discontinuous fields without undue expense. The material models may be very complex, encompassing strain history-dependent plasticity, internal voids, or anisotropic response.

## Overview and Historical Perspective

Hydrocodes in the United States (and at Sandia) generally fall into two categories, loosely referred to as Lagrangian and Eulerian. Lagrangian codes solve the conservation equations in the material frame, so that the computational mesh deforms with the material as a function of time. Eulerian codes solve the conservation equations in the spatial frame, so that the computational mesh is fixed in space as the deformation of the material occurs as a function of time. The solution technique for contemporary Lagrangian codes is generally a finite-element or finite-difference method, and the solution technique for Eulerian codes is generally a finite volume or finite-difference method. Gridless techniques, such as smooth particle hydrodynamics (SPH), are Lagrangian in nature. Hybrid methods are also possible using features of both Lagrangian and Eulerian codes and are commonly referred to as arbitrary Lagrangian–Eulerian (ALE) methods. Over the last 40 years, Eulerian, Lagrangian, and ALE codes have been developed at Sandia.

All hydrocodes today solve the conservation equations (mass, momentum, and energy) with either Lagrangian or Eulerian techniques, and almost all hydrocodes use explicit time steps. All methods have quirks and pluses and minuses depending on the specific code. Several general comments can be made for the major classes of codes.

Because Eulerian techniques solve the conservation equations on a mesh fixed in space, the material moves through the computational mesh. The conservation equations are solved with either finite-difference or finite volume numerical techniques. The Eulerian techniques have several pluses: they have a simple mechanism for creating free surface in fracture and fragmentation, they are computationally robust for large deformation material mechanics, and they have a very simple scheme for mesh development and material insertion. They also have several minuses: they can be computationally intensive in CPU[90] and memory requirements, they have problems with surface recognition (i.e., material interfaces), they have a hard time modeling velocity discontinuity (e.g., sliding), and poor numerical techniques can lead to excessive shock diffusion.

Lagrangian techniques solve the conservation equations on a mesh that moves with the material. The conservation equations are generally solved using finite-difference or finite-element techniques, with finite-element techniques becoming increasingly popular. More novel approaches include Free Lagrangian and SPH. Straightforward Lagrangian techniques have several pluses: they can be computationally less intensive in CPU and memory requirements, and they can accurately recognize material interfaces or surfaces. They also have several minuses: free surface creation in fracture and fragmentation is difficult and relies on ad hoc algorithms, large deformation frequently leads to mesh tangling or inversion, mesh development can be extremely difficult in three dimensions and may be the limiting factor in very large-scale computing, and contact surface algorithms can be difficult to implement even in two dimensions and certainly in three dimensions. The difficulties noted here are the driving factors behind the development of meshless techniques such as SPH.

Many hydrocodes have been written and used over the last 40 years, but only a few of those have gained a reputation or widespread use. I will focus on the Sandia codes in the latter category. The 1960s and early 1970s were times of rapid development in computational shock physics. Herrmann at the Massachusetts Institute of Technology (and later at Sandia) developed the early one-dimensional (1-D) Lagrangian code WONDY. He also developed the initial version of the two-dimensional (2-D) Lagrangian code TOODY. Herrmann went into management by

[90] Editors' note: CPU is the acronym for central processing unit.

the mid 1970s, so he didn't have enough time to develop codes. Marlin Kipp was the point of contact for WONDY. Several people worked on TOODY, including Larry Bertholf, Jeff Lawrence, Darrell Hicks, and Billy Thorne. Jeff Swegle maintained TOODY in the 1970s. Lynn Barker developed an early method of characteristics codes named SWAP in the same time frame.

The Sandia "family tree" for Eulerian codes starts with CHARTD and progresses to CSQ and finally CTH. CHARTD (Coupled Hydrodynamics And Radiation Transport Diffusion) was developed by Sam Thompson (in 1969) and was a 1-D solution (planar, cylindrical, and spherical geometries) of the conservation equations in the shock domain. We normally consider CHARTD as the first in the lineage of CTH, but it actually used a Lagrangian mesh and solved the equations with a finite-difference technique. CHARTD was written in FORTRAN[91] 66 and, typical of codes of that era, it was completely self-contained. CHARTD ran on CDC mainframe-style computers and was maintained using a software configuration management program named UPDATE. Sam actually wrote a FORTRAN version of UPDATE and a graphics post-processing software package known as RSCORS (Revised Stromberg Carlson Optical Recoding System, which was one of the first computer graphics terminals). The mindset of having all the necessary software to build, maintain, and display results carried through the subsequent code development efforts that Sam shepherded. The first version of CSQ (CHARTD Squared) had limited applicability because it allowed only two materials in a calculation. The second version, CSQII, followed shortly afterward (in 1973) and allowed up to ten materials. CSQII modeled both 2-D cylindrical and rectangular geometries using a more standard finite-difference, fixed mesh numerical technique. Eulerian codes actually integrated a Lagrangian form of the conservation equations through the time step as a part of the overall solution sequence. The Lagrangian step is followed by a remap step that returns the physical variables in the distorted mesh to the original mesh. This two-step process is the only known way to solve the conservation equations with general equations of state and material strength. Development of CTH (CSQ to the Three Halves) started in 1986, with Mike McGlaun, Sam Thompson, and Millie Elrick (a Kaman Sciences Corporation contractor) making the bulk of the contributions to the early code. In October 1987, McGlaun ran the first meaningful three-dimensional (3-D) simulation with CTH. We generally consider this to be the birth date of that code. By 1990, CTH was used outside Sandia under Export Control restrictions from the Department of State as a defense article. Over the next 25 years, a large team of individuals (in excess of 30) made regular contributions to CTH development.

At this time, only two individuals are still at Sandia (myself and Steve Rottler), who were involved in the early 1990s in the development of CTH. In the early 1990s, a totally separate, parallel version of CTH (3-D only) was developed, known as PCTH. Allen Robinson was the principal developer. PCTH was developed in C++ for massively parallel computing platforms, such as the Intel Paragon. It used a domain decomposition concept with explicit message passing to communicate

[91] Editors' note: The name FORTRAN is a contraction of FORmula TRANslation. The first FORTRAN compiler was developed by IBM in 1954–1957. After 1977 it was referred to as Fortran.

between computational nodes (also known as multiple instruction, multiple data, or MIMD). As massively parallel computers became more common, external users familiar with CTH started to request access to PCTH. Rather than adding CTH features (2-D capabilities, material models, code heuristics) to PCTH, parallel capabilities were added to CTH. By the mid 1990s, CTH was running on massively parallel computers. These changes kept CTH in the forefront of shock physics technology during the 1990s and into the new millennium. Prior to data parallel capabilities, 2-D problems were typical and used a few hundred thousand cells; large 3-D problems used several hundred thousand to one million cells. Afterward, 3-D was typical and large problems used hundreds of millions of cells. This time frame coincided with the first Accelerated Strategic Computing Initiative (ASCI) supercomputers. One of Sandia's first ASCI computers ran a one billion cell simulation with CTH!

The next major advance with CTH numerics was integration of a fully parallel adaptive mesh refinement scheme. At a DOE code development conference in the mid 1990s, I was asked if we had considered adaptivity in CTH. At the time, CTH was just beginning to be a productive parallel code so it was a question outside my comfort zone. I believe my answer was that there were no technical reasons why it couldn't be done but we had no plans to do so at that time. It took a couple of years and some innovative thinking by Dave Crawford to develop a concept for the integration of adaptivity in CTH. Dave made a conceptually simple extension of the data parallel concept to break the connection of a data patch with a processor. His model uses the concept of a patch of cells that contain all CTH capabilities. This patch is fairly small ($10^3$) and uses the standard messaging passing techniques to communicate with neighboring patches. In this case, the patches are not constrained to living on one processor but multiple patches can (and do) live on a single processor. Communication is not by an explicit message but by a pointer to memory. In addition to a more complicated communication network, infrastructure for changing the resolution and making decisions on when to change the resolution was added. A constraint of 2:1 resolution between neighboring patches is enforced. In reality, all parts of the simulation where one cares about accuracy run at the highest resolution. This technique can be very effective for simulations that are sparse geometrically, which are a large percentage of typical problems. This addition has been a very effective tool for the last several years and has continued to keep CTH in the forefront of numerical shock physics. Problems using hundreds of millions of cells with CTH in relatively short time periods are commonplace.

The Lagrangian code work of Herrmann and Sam Key was continued in much the same way as in the Eulerian community. The Sandia "family tree" for Lagrangian codes has two main branches. Hermann's branch has WONDY and TOODY; Key's has HONDO, PRONTO, and PRESTO. Recent versions of these codes are similar in architecture in that they rely on a finite-element formulation to solve the conservation equations, but differ significantly in their capabilities and features. I have very little experience or knowledge of the Herrmann-inspired codes but can say that versions of WONDY still exist and are used to help interpret experimental data. The 2-D solid dynamics code HONDO was used in the 1970s. The development of PRONTO (in both two and three dimensions) started in the mid 1980s and has replaced HONDO.

HONDO, PRONTO, and PRESTO are fundamentally not shock physics codes: they don't normally solve an energy equation nor do they have a shock equation of state. Both PRONTO and PRESTO focus on relatively low strain rates where volume changes can be ignored. But both codes can solve the energy equation with simple equations of state. Neither would be the first choice for shock analysis; however, they form a part of the Sandia code suite for large deformation analysis.

The Eulerian and Lagrangian branches are intertwined with the ALE code now known as ALEGRA. ALE techniques generally solve the conservation equations on an unstructured mesh with the additional complexity of allowing material convection with respect to the mesh. This formulation allows the user to designate whether material should flow through a stationary mesh (pure Eulerian), whether the mesh should move with the material (pure Lagrangian), or whether the mesh should move independently from the material motion (arbitrary). This latter capability permits a calculation to proceed in Lagrangian fashion until the mesh becomes too highly distorted. At that time, mesh points in the most deformed portion of the mesh can be moved to reduce the distortion to acceptable levels. The advantage is that numerical dissipation associated with remapping is avoided until large deformations occur and then is limited to only those regions where there are severe mesh distortions and the mesh must be moved. ALEGRA is the Sandia version of the technique and is finite element based. ALEGRA development started in the early 1990s under Mike McGlaun and was based on PRONTO. ALEGRA's first name was RHALE, which stood for Robust Hydrodynamics Arbitrary Lagrangian Eulerian. ALE techniques were always hoped to be the best of both worlds, with more accurate surface definition of unstructured meshes and with the more robust functionality of Eulerian codes. Unfortunately, the opposite may be true. ALE codes have proven to be complex, slow, and not very robust. Circumstantial evidence indicates that these codes are used as either pure Lagrangian or Eulerian more than as "arbitrary." During early development of RHALE and ALEGRA, it was widely expected that this code would surpass CTH in usage. Fortunately for the broader community, this has not been the case. ALEGRA development continues but at a fairly low level. Usage is limited to a few (but very important) applications.

Coupling of the techniques described above is also possible. In particular, the penetration of a rigid rod into deforming material is very difficult to model with any of these individual techniques. In this case, the best technique is to treat the rod as a Lagrangian body and the target as an Eulerian body. A direct coupling of Lagrangian and Eulerian codes is necessary to simulate such phenomena accurately. CTH and PRONTO (and also EPIC, a DOD code) have been coupled in a timestep-by-timestep fashion to solve the same class of problems. Zapotec is a coupling algorithm that allows CTH and a finite-element code to solve problems involving both shock response and structural response. Zapotec usage is more limited than CTH but it has unique features that make it an important application in itself. There is a current effort to replace PRONTO with our modern ASCI code called SIERRA/SM. As this is being written, the development of Zapotec 3.0 is essentially complete and has been exercised on several applications.

# Roy (Red) E. Hollenbach (1950–1982[92])

**Editors' comments**: Don Lundergan intended to prepare a brief history of shock wave developments at Sandia and asked several people for input. The following response is from Red Hollenbach.

## In a Letter to Don Lundergan

Don, you know you are requesting information from the worst memory in town! Trying to remember things that happened over 40 years ago and the specific people associated with those events would be very difficult. Somebody would be overlooked. Maybe your scheme of combining memories will work out best. At any rate, I'll just jot down a few ideas and see what you think.

I'm wondering, Don, if you could write up a sort of "rags to riches" saga about the shock wave work. I think we were both in Don Cotter's group when you started pushing shock waves. I don't know how you got saddled with me—I guess I was just plain lucky. It was the start of the bulk of my career at Sandia. We started with nothing—a cast-off shock tube, no instrumentation, two people—and I'm not sure we were on the project full time. We progressed from firing the gun out across the mesa to a sawdust-filled bunker that challenged us to find the test piece after each shot. Early test results were obtained from postmortem visual (eyeball and microscope) examination of the "target." Somewhere about this time our group seemed to really grow—more people, better facilities, lots of instrumentation and much more sophisticated techniques (pins, slanted resistor) for obtaining and reducing the data from each shot.

Now here is the "riches" part. A whole department of people, two fully-instrumented test facilities, and a laser velocity interferometer technique that I think far exceeded your "one-shake resolution." (Someday you can refresh my memory: What is a "shake"? [93])

This same velocity interferometer technique that Sandia decided was not patentable has provided and is still producing extremely accurate data on the behavior of materials that have been subjected to shock waves. I don't know if Sandia's history would be interested in our retired entrepreneur friend (Lynn Barker) marketing these devices around the world.

[92] Roy (Red) Hollenbach was hired at Sandia in 1950 and retired in 1982. His involvement in shock wave work covered the period from the late 1950s to the late 1970s.

[93] Editors' note: A shake is 10 ns.

# James N. Johnson

## (1967–1973[94])

My family and I arrived in Albuquerque in the late summer of 1967. I had just finished a postdoc with George Duvall at Washington State University (WSU). I was hired at Sandia by Don Lundergan of the Deformation of Materials Division. The Mechanical Behavior Department Manager was Walter Herrmann. During the interview process, I had also talked to Glenn Seay and Orval Jones of the Applied Physics Research Department, but chose to join Lundergan's group, probably because of my familiarity with the *Journal of Applied Physics* paper by Barker, Lundergan, and Herrmann (1964) on strain rate dependence in shocked aluminum. My PhD research at WSU had been on elastic precursor decay in iron and its relationship to dislocation mechanics (Johnson and Band 1967).

When I arrived, the Mechanical Behavior Department had a very active research thrust in shock compression of porous materials. My first Sandia report (Johnson 1968a) was on steady wave propagation in porous materials, based on Herrmann's P-α model. I also became familiar with the experimental work of Lynn Barker on time-resolved measurement of plastic wave profiles in aluminum; the standard alloy of aluminum was 6061-T6 with rolled texture rather than extruded (maybe the other way around). This difference caused some head scratching along the way when an unusual response was measured in a few shots. Lynn was very quick to find the reason for the anomaly. Lynn and I published a paper on steady plastic waves in aluminum (Johnson and Barker 1969).

The partnership of Lynn Barker and Red Hollenbach in the "Serendipity Lab" sets a standard for experimental shock wave research that has never been equaled in my opinion and experience. I have never forgotten the fact that the speed of light is very nearly one foot per nanosecond after Lynn explained to me that cable lengths were important in the new nanosecond regime in which shock wave experiments were being conducted at Sandia.

I would be derelict in my perceived duty to provide a thoughtful and complete recollection of important events and personalities involved in shock wave research at Sandia without mentioning Lynn Barker's musical talents. Lynn was raised on an Arizona ranch and learned his cowboy music in the best place possible: the bunkhouse. I was so impressed with these songs in 1967 that I went down to K&B Music on Central Avenue (in Albuquerque) and bought a $35 classical guitar. Lessons on TV from Laura Weber, and with local musician Ed Clarke, did little for me to attain the soul of Lynn's playing and singing. I still try to do justice to two songs I learned from Lynn: Lewie Wickham's "A Little Too Late" and Texas Jim Lewis's "Seven Beers with the Wrong Woman."

Sometime in 1968 (my best recollection), Sandia management detected possible duplication of effort or even slight hostility between the Mechanical Behavior and the Applied Physics Research Departments, and as a result they were combined under a single directorate with Al Narath as the referee. Narath decided which

[94] Upon leaving Sandia in 1973, Jim Johnson joined Terra Tek in Salt Lake City, UT, for a few years and then joined LANL, continuing research in the shock wave area.

departments were going to have prime responsibilities in specific areas of shock wave research. Personnel were transferred accordingly. As Lee Davison described it, "Jim was dragged, kicking and screaming, into the Shock Wave Physics Research Division." If this were true, it was only due to my professional immaturity. The Mechanical Behavior Department had been slanted toward Mechanical Engineering, and the Shock Wave Physics Research Division had more of a physics pedigree, although overlap blurred this distinction considerably. My new Division Supervisor was Lee Davison, a Caltech PhD in modern rational mechanics, and our Department Manager, Orval Jones, was a Caltech PhD in engineering mechanics.

In my new organization, I shared an office with Pete Lysne, an experimentalist studying porous materials and piezoelectric effects under impact conditions. The "deans" of experimentation in the Physical Research Department were clearly Bob Graham and Bill Benedick, and the measurement technique was the quartz gauge. Using X-cut quartz as a dynamic stress gauge had a long history in this department, as described more thoroughly elsewhere in these recollections. The idea I want to emphasize here is that in my first 3 years at Sandia, I had crash courses in many new subfields: porous materials, rational continuum mechanics, piezoelectric properties of solids, explosives, solid–solid phase transformations, and (most importantly) the elastic–plastic impact response of single crystals.

it was a policy of Orval Jones to have a materials scientist on the staff who is also fully trained in metallurgy in the Physical Research Department. I believe the first was J.R. (Reid) Holland. (I did not know him.) The second was Dick Rohde. I was able to work with Dick on dynamic deformation twinning[95] in iron. Following Rohde's transition to management, Lee Davison hired Larry Pope. Larry and I worked on the dynamic elastic–plastic response of single-crystal beryllium. The attention to shock wave response of single crystals, in general, had begun sometime before the work on beryllium.

I had published a paper in 1969 on polycrystalline metals (Johnson 1969), summing the contributions from a *statistical distribution* of random glide planes and slip directions. It occurred to me that, as data were being generated on various single crystals, the same mathematics could be used directly by summing the contributions from *individual* glide planes and directions in face-centered cubic (fcc), body-centered cubic (bcc), rock salt, and hexagonal close packed (hcp) single crystals. Orval Jones became very enthusiastic about this possibility, and we, along with Tom Michaels (a WSU PhD student), published a paper in 1970 on dislocation dynamics and single-crystal constitutive relations (Johnson, Jones, and Michaels 1970). In this work, planar shock waves were limited to specific propagation directions that produced pure longitudinal motion.

At some point in this study, we asked what happens for *nonspecific* crystallographic propagation directions. Fortunately, as often happens in large, multidisciplinary institutions, you ask around and soon find someone who can tell you something very helpful. This was the case, as I discovered in talking to Sam Key in the old Deformation of Structures Division. Sam gave me some notes on linear-elastic, anisotropic, plane wave

[95] Twinning is a metallurgical term that refers to an inelastic deformation of the crystal lattice.

propagation. This included general propagation directions and coupled quasi-longitudinal and quasi-transverse plane waves. I applied these ideas to Y-cut quartz and off-axis hcp crystals, first in the linear-elastic range and later in the elastic–plastic regime. Sam had evidently moved on to other research areas, because he volunteered that I was now to be the local expert in anisotropic wave propagation.

Rational continuum mechanics and names like "Clifford Truesdell" were unknown to me in 1967. That is, until I was suddenly confronted by Peter Chen, a Sandia mathematician. Up until that moment, I thought we (physicists) had been rigorous enough in our formulation of constitutive laws, thermodynamics, strain measures, etc. I had never heard of "the second Piola-Kirchhoff stress tensor" or "the right Cauchy–Green deformation tensor." My previous education at WSU *had* involved a seminar by E.H. Lee (Stanford University) on the polar decomposition theorem and how the deformation gradient might be broken down into elastic and plastic components: did you take the elastic part first and then the plastic part or the opposite order? An interesting mathematical question, but I didn't know the world of dynamic material behavior hinged on such things. As time went on, kinder, gentler minds educated me on this subject. Walt Herrmann's Sandia report on one-dimensional continuum mechanics was very helpful, as were Karl Schuler's and Lee Davison's balanced approaches.[96] I suspect this paragraph will cement my reputation as a scientific Neanderthal and Philistine.

Another area in which I was able to learn new aspects of shock wave physics was the subject of phase transformations, working with Jim Asay and Dennis Hayes on bismuth (Johnson, Asay, and Hayes 1974). Bismuth has the very interesting property that single-shock compression can probe not only the solid I–solid II transition, but also the liquid phase. Thus, the presence of a triple point provided a challenge for the theorist. Dennis did some very nice work on how to deal with these questions (Hayes 1975), and I benefitted greatly from this interaction.

On a final note, I recall an important event (for me) that took place at an American Physical Society (APS) Winter Meeting in the west, the traditional annual meeting place for shock physics prior to the creation of the biennial APS Topical Conferences on Shock Compression in Condensed Matter. I believe the year was 1968 and the venue was San Diego, CA. Glenn Seay was present, and I assume the time was after he had moved to General Atomics from Sandia. Someone was giving a talk on the subject of release from the shocked state. Glen asked the following question: why does the compressive elastic precursor almost always show a sharp yield point, while the release wave, in the same material, almost always shows a very diffuse reverse transition that continues to spread with propagation distance? The speaker had no answer, and that question remained in my mind until the 1990s, when I thought I might have an answer. This has to do with the presence of dislocation pileups and pinned loops that will always provide some reverse plastic flow immediately upon release from the shocked state, and I published numerous papers based on these ideas (Johnson 1993; Winey, Johnson, and Gupta 2012). I include this remembrance as but one more example of the lasting intellectual effect my tenure at Sandia had on me and for which I remain grateful as I write this in 2013.

[96] For example, see the following references in the Bibliography: Davison and Johnson (1970), Davison and Stevens (1973), Davison (1984), Davison (2008).

# Orval E. Jones[97]
# (1961–1993)
# A Tribute by the Editors

Through most of the central 30-some years described in this book, Orval E. Jones was an important player in that shock physics history. He arrived at Sandia in 1961, after earning his PhD at the California Institute of Technology. His initial assignment was as a Member of the Technical Staff working on the electrical and mechanical response of solids to shock waves. His first office mate was Bob Graham, another preeminent figure in our history. Three years later he became the Supervisor of the Dynamic Stress Research group, and 4 years after that, he was promoted to manager of the Physical Research Department. At the time of his first promotion, he called it a journey of "self-exploration" and made the change with some trepidation. He still wonders what his life would have been like had he remained with a career path as a researcher.

Very early on, in 1962, he was the principal author of a study (Jones, Neilson, and Benedick 1962) that showed how quartz gauges could be used to measure shock wave profiles and the dynamic yielding behavior of metals. That paper was considered one of the early classic studies coming out of Sandia shock physics research. Even though he was in management for most of the remainder of his Sandia career, he continued to make many important contributions to shock physics over the years. These included the development of the quartz gauge, the first measurements of time-resolved dynamic yielding, the initiative to include dislocation models for

[97] This brief tribute to Orval Jones was adapted by the editors from several articles appearing in the *Sandia Lab News*, in particular, in the October 1, 1993 issue, on the occasion of his retirement from the Labs, and also in the November 27, 1978 issue, in a description of the Engineering Sciences Directorate, shortly after he became head of that organization. The picture is from the June 11, 1982 issue, when he was first promoted to Sandia Vice President. These materials are reprinted with the permission of Sandia National Laboratories.

understanding dynamic yielding, and, later, the increased emphasis on numerical modeling to understand shock physics and structural response phenomena. Orval relates another often-repeated story regarding some advice he received from one of his first managers: "Around here money is just like oil. You squirt it on to make things move faster." Things have certainly changed since those early days!

In the 1970s, Orval had a number of different responsibilities at the director level; the first, starting in 1971, was as the Director of Solid State Sciences Research. Three years later he became the Director of Nuclear Security Systems, and in another 3 years, he became the Director of Nuclear Waste and Environmental Programs. It was only a year after that, when he became the Director of Engineering Sciences in 1978, that he returned to a more direct involvement with shock physics research.

Shortly after first becoming a director, Orval had to deal with one of his, and probably Sandia's, largest crises: the 1973 layoff of hundreds of Sandians. It was Orval's recurring worry that this suggested a very limited future for the Labs and we might never regain the same level of research activity as in the past. However, it was not too long after that when then Sandia President Morgan Sparks, in the middle of the first oil and energy crisis, asked all Sandians to come up with new ideas to address these issues. Many of the concepts arising from this challenge led to new programs in energy research and development, a number of which involved practical applications for shock physics research; Orval's worries for the Labs' future then abated.

Of major significance to shock physics was his last director-level assignment as head of Engineering Sciences, which started in 1978. As Director, he implemented a major increase in the emphasis on the advanced development of computational capabilities in the areas of multi-dimensional wave propagation, structural response, fluid flow, and heat transfer. Some felt that this new focus was being promoted at the expense of experimental efforts; however, the importance of verification and validation (often referred to today as simply "V&V") for the new computer codes—in particular, validation—required extensive justification generated through accurate, high-quality experiments. This was especially important when the different numerical models were extended well beyond conditions that were readily accessible in laboratory-based experiments.

The last phase of Orval's tenure at Sandia started in 1982, when he was asked to be the Vice President for Technical Support. This was followed, just a year later, by his assignment as VP for Defense Programs. In 1986, he assumed his final role at Sandia, one which he held until his retirement 7 years later in 1993, as Executive VP for Programs. After first becoming Vice President in 1982, he made the decision not to close the Kauai Test Facility permanently, even though there was little work for the installation at that time. It was only a few months later that President Reagan announced the Star Wars program, which, in the following years, breathed new life into Kauai as a major missile launch facility. In fact, it is still active today, over 30 years later. This illustrates the difference between careers in research and engineering compared to those in management. Orval says that technical work generally stands on its own merits, while appreciation for management often comes only later, after your decisions shake down through the chain of command. Both paths can lead to great satisfaction.

Orval recalls several outside events that had major impacts on him as well as on the Labs. In the 1970s the first world oil and energy crisis helped to broaden the Labs' role to include energy research and other work-for-others areas, leading to its position as a multi-program laboratory. In the 1990s, the management transition from AT&T to Martin Marietta, and shortly thereafter to Lockheed Martin, raised the possibility of significant changes. But because Sandia Corporation had always maintained a somewhat independent and effective management style, those changes were not disruptive. Also, at the outset of the 1990s, the fall of the Soviet Union led to new programs for the Labs that will likely influence our activities far into the future. As testimony to his long role in Sandia management, in 1993 Orval was recognized by the Secretary of Energy, Hazel O'Leary, with the DOE Distinguished Associate Award for his "…distinguished and unique contributions to nuclear weapon safety and security and for his outstanding leadership in establishing programs of utmost importance to the security and economic well-being of the United States."

Throughout his many years in Sandia management, Orval evolved a set of principles that we could all do well to follow. First, and maybe most important, maintain your credibility and integrity. In dealing with colleagues, develop an ability to "put yourself into the other person's shoes" to appreciate his or her reactions, understand who is "losing face" in personal or group interactions and consider the consequences, and discuss your complaint about a person first with that person. Also, spend more time talking to a person if the impact of the conversation on him or her is either negative or large. And finally, treat other people, regardless of their position, higher or lower, as you would want to be treated yourself.

As quoted in the October 1, 1993 issue (page 8) of the *Lab News*, Orval tells the story of a conversation he had with the deputy scientific Director of the Russian weapons laboratory, Arzamus 16, Yuri Trutnev. In a visit to Sandia in 1992, Trutnev asked him, "When you look back on your career, do you have any regrets about dedicating it to nuclear weapons?" Orval answered, "No, … I have always believed peace is established through strength, a fully worthy cause to invest one's career in." That response brought a rare smile from the former adversary. Orval smiles too, when he tells this story; it is among the fondest recollections of his 32-year Sandia career.

Orval Jones died on February 25, 2017 with his family at his side.

# Charles H. Karnes

## (1963–1997[98])

**Editors' comment**: The following comments by Charlie Karnes were conveyed in an email message to Jim Asay on October 1, 2011.

Ann and I had lunch with Don and Elaine Lundergan today. I mentioned to Don something that you should know about. Two or three years ago I was talking to Orval Jones about the development of shock wave work at Sandia. He said that Don Lundergan deserves a lot of credit for Sandia's contributions in the field of shock waves because Don recognized the importance of shock wave computations and hired Walt Herrmann, and the rest is history.

**Editors' comment:** email to Jim Asay on October 2, 2011

Here is what I remember about Lynn Barker's first interferometer experiment. I reported to Sandia in June 1963, and I believe it was in September 1963 that Lynn had the idea of using a Michelson interferometer to measure the free surface displacement history of an impacted target. He and Red Hollenbach set up an experiment in our lab on the second floor of Building 805. It was a simple, but elegant experiment.

They used an aluminum tube about 3–4 ft long and 4-inch inside diameter having a wall thickness of 0.5 inch. A lapped aluminum impact plate was bonded to a polyurethane foam cylinder and placed in one end of the tube, the breech end. Another lapped aluminum plate, the target, was put on the other end of the cylinder. The mating surface of the target end of the tube was precision ground and lapped square. The plastic cylinder with the impact plate was held at one end of the aluminum tube, the barrel, with a thin wire, which was attached to something fixed in the lab. The target plate with the interferometer components was at the other end of the barrel. The barrel was soft evacuated, creating an atmospheric-pressure gas gun, a "vacuum gun" if you will, using atmospheric pressure as the propellant. The vacuum held the target plate firmly against the end of the barrel. A photomultiplier was set up to monitor the fringes, with the output being displayed on an oscilloscope. A low-powered He-Ne laser was used to illuminate the interferometer.

Another thin wire was attached to the target plate with the interferometer setup. Lynn held onto that "target wire" to keep the components from hitting the floor. To fire the "gun," Red cut the thin wire holding the projectile in place, and the projectile flew down the tube and impacted the target plate. That was the first shock wave interferometer experiment. I was in the lab watching it.

Lynn had to count the fringes, feed the results into a computer program, and differentiate them numerically to get the free surface velocity history. It was later, after many more experiments using a real gas gun, that he had the idea of using an optical delay leg to optically differentiate the displacement to get a direct measurement of the free surface velocity so that each fringe represented a change in surface velocity rather than a change in displacement.

---

[98] Charlie retired from Sandia in 1997, but left the shock wave program in the late 1970s for another position at Sandia.

# James E. Kennedy

# (1968–1986[99])

# Contribution to Shock Wave Research History at Sandia

I was hired in July 1968 because I had 9 year's experience in explosives research and development (R&D) at the Illinois Institute of Technology (IIT) in Chicago. With my hiring, as Al Narath stated to me much later, Sandia intended to get into explosives research and "compete with Los Alamos."(!!!) I thought (but didn't say out loud) that this sounded absurd when Narath said it in about 1984, shortly before Narath went to AT&T Bell Labs to become anointed[100] so that he could return to Sandia as its first homegrown president.

Yet several significant contributions to the energetic materials field have indeed been made at Sandia. As the first staff member concentrating specifically on research related to explosives, I offer this overview of those contributions over the period 1968 to 1986, until I left Sandia.

## Shock Initiation of Detonation: Shock Wave Evolution in a Heterogeneous Explosive

Orval Jones was a first-level supervisor when I was interviewed in early 1967 and offered a position at Sandia. He was the Department Manager of the Physics of Solids Research Department when I arrived in mid 1968. Orval said they kept the position open that long awaiting my arrival because a hiring freeze was on and they couldn't hire anyone else anyway. Orval recommended that I apply a Sandia strength, gas gun impact studies, to the field of explosives. To identify an appropriate subject for study, Orval referred me to Charles Mader of Los Alamos, who was Orval's personal friend; both owned mountain cabins at Taos Ski Valley. Mader told me that he was glad to help out Orval, who had helped him out in a time of career

---

[99] Jim Kennedy left Sandia in 1986 and joined LANL, where he continued research on energetic materials.

[100] Editors' note: When Kennedy was hired in 1968, Al Narath was the Director of Solid State Sciences Research. AT&T policy was that Sandia's President had to come from AT&T Bell Laboratories; hence, Narath had to spend time at Bell Labs before becoming the President. Therefore, beginning March 31, 1984, Narath, who was then the Executive Vice President at Sandia, transferred to Bell Labs (*Sandia Lab News*, March 30, 1984, vol. 36, no. 7, p. 1) to serve as the Executive Vice President of Government Systems. In that role, he headed up all research and development in support of Bell Labs' work for the Federal Government. Narath was the *only* President whose professional career began at Sandia during the nearly 44 years AT&T managed Sandia. He returned in 1989 (see letter addressed to all Sandians, April 17, 1989, in the *Sandia Lab News*, vol. 41, no. 8, p. 1) and served as President from April 1, 1989 to August 15, 1995.

jeopardy. Mader recommended shock initiation studies of the HMX-based explosive, PBX 9404 (referred to as 9404 in the following discussion), with Lagrangian gauges and arranged to get samples prepared for Sandia through Al Popolato of LANL. Impact studies of shock initiation of 9404 began in 1969 with O.B. Crump as my technician and Lee Davison as our first-level supervisor. The first experiments utilized quartz stress gauges on either the impact face of the projectile or behind the sample, so that observations could be made of shock wave evolution during shock growth toward detonation. The general observation was that energy was released behind the shock front, thereby leading to a pressure excursion behind the front and ultimately to a "pressure hump" well behind the front. My paper for the 1970 Detonation Symposium (Kennedy 1970) represented the first U.S. publication on shock wave evolution toward detonation and was similar to results found earlier by Dremin of the USSR for shock wave evolution in TNT. A second paper at the 1972 Combustion Institute meeting (Kennedy 1973) showed that the evolution to detonation did not follow the "single-curve buildup" notion, which means the shock front did not build up in a manner corresponding to the remaining run distance to detonation according to the Pop plot.[101]

At that point (1974), Sandia brought a theoretician, Jace Nunziato, from the Solid Dynamics Research Department into the Physics of Solids Research Department to work with me in analyzing the shock wave evolution results. Nunziato had cut his teeth at Sandia working with Walt Herrmann on shocks in porous materials and with Karl Schuler on viscoelastic behavior of PMMA (polymethyl methacrylate) in the shock front. It was widely supposed at that time in the explosives community that, in heterogeneous explosives, shock growth to detonation was due to some prompt energy release at the shock front. Using continuum mechanics-based singular surface analysis (another Sandia strength), Nunziato derived an equation with which the shock-growth behavior could be analyzed to allow an estimate of the energy release in the shock front. Our 1975 data set was on shock waveform evolution at an initial shock pressure that would lead to detonation in 7 mm of shock run distance, a rather strongly driven condition. To our surprise, Nunziato's analysis of the waveform-evolution data showed that the explosive *absorbed* rather than liberated energy at the shock front. When we shared these results with Bill Davis and Wildon Fickett at Los Alamos soon afterward, LANL was surprised that:

1. Nunziato had independently derived the "shock-change equation," which Los Alamos had derived earlier but never published (*"[Expletive deleted], Sandia's got the shock-change equation!"*—Bill Davis).
2. The shock-front behavior was endothermic.

[101] The Pop plot is named after Al Popolato of LANL.

[102] Editors' note: Wildon Fickett and William C. Davis, *Detonation* (University of California Press, Berkeley, CA, 1979). The book is now available in a paperback edition, first published in 2000, as a slightly corrected, unabridged republication of the 1979 work, with the title *Detonation: Theory and Experiment* (Dover Publications, Mineola, NY, 2000).

Fickett and Davis seemed motivated by this to get serious about completing a book they were writing, and hence their classic *Detonation*,[102] published in 1979, contains the derivation of the shock-change equation. No further exploitation of the equation has been made, to my knowledge, in shock wave research on explosives before or after the 1976 paper (Nunziato and Kennedy 1976), other than a related one (Chen and Kennedy 1976) on shock evolution in curved waves at the 1976 Detonation Symposium.

In the course of these studies, Crump and I developed the first dual-delay-leg VISAR because of the uncertainty in shock jump amplitude that occurred during rapid shock evolution toward detonation in explosives (never published). We did experiments at five different Sandia gas guns in those years and seemed to develop a "black-sheep" reputation because O.B. was so good at gaming the system and getting any resources and facility access that he needed.

Nunziato and I postulated that the endothermic behavior at the shock front was due to dissipation of mechanical energy into heat, associated with a myriad of shock interactions at interfaces between the stiff explosive crystallites and the softer binder material. We sought to explore mechanical behavior in the absence of reaction in the explosive material by conducting acceleration wave experiments at low compression amplitude. This was done in gas guns by impacting a fused silica plate with another fused silica plate at a low enough velocity that a ramp wave (acceleration wave) would emerge from the target plate. This ramp wave then passed through an explosive sample affixed to the target plate. We measured the ramp waveform behind the explosive sample as a function of thickness of the 9404 explosive and found that acceleration at the wave front decayed with distance in the 9404, indicating the explosive was not shocking up at that loading condition (about 8 kbars peak pressure in 9404), but the acceleration wave front was dying away. These results were published. Subsequent unpublished measurements of the shock Hugoniot at low shock amplitude revealed a region of reverse curvature in the U-u plane, indicating that 9404 would not support a shock weaker than about 12 kbars. This behavior again was attributed to dissipation due to shock reverberations between stiff explosive crystallites and softer binder materials and was consistent with highly lossy behavior also noted in acoustic measurements in 9404 (Sutherland and Kennedy 1975; Sutherland, Kennedy, and Nunziato 1977).

I also conducted short-pulse shock initiation experiments with 9404 (unpublished) that showed the now familiar growth of the shock wave due to reaction, followed by shock amplitude decay caused by the rarefaction wave, and then continued reaction growing to detonation, associated with short-pulse initiation. After I moved from research into component design in 1976, Bob Setchell joined the Physics of Solids Research Department and performed wave-evolution studies on the same explosive with a ramp wave followed by a shock. He found (Setchell 1981) that the ramp wave precompressed the explosive and desensitized it so that the ensuing shock was unable to stimulate any apparent reaction, as judged by the waveform behind the front. After a few millimeters of propagation distance, the ramp wave was overtaken by the shock front and then there was no precompression. At that point, the evidence of reaction behind the front was immediately noted. This indicated that even quite gentle precompression over less than a microsecond is capable of closing pores and eliminating hot spot sites in the explosive where a reaction would otherwise be stimulated by shock compression.

## Gurney Modeling

As a side interest stemming from my PhD dissertation, I wrote a widely distributed Sandia report on the Gurney model that describes how detonating explosives drive metal (Kennedy 1971). It was based on a simple model developed by the solid state physicist (!!) Ronald Gurney at Aberdeen during World War II to predict metal fragment velocities as a function of the ratio of the mass of metal to the mass of explosive and did not involve shock behavior in the analysis. I extended the application of the model by pointing out how to evaluate efficiency in the use of explosives. I described a specific impulse model for detonating explosives that drive heavy loads (which Sandia was already doing at the Light-Initiated High Explosives Facility) and how the Gurney model could be combined with other simple physics to provide, in some cases, closed-form analysis of explosive system performance that could allow optimization of a design or analysis objective. Published examples of the use of this model include:

1. implosion or explosion of strain-hardening tubes based on plastic work of deformation (Herlach and Kennedy 1973)
2. detonation transfer by flying plate impact (Kennedy and Schwarz 1974)
3. estimation of Gurney velocity of an explosive from chemical composition and density, based on Kamlet's parameter $\Phi$ (Hardesty and Kennedy 1977)
4. indirect drive of material based on inelastic-collision analysis (Kennedy 1972)

I recognized the opportunity to use the Gurney model to help in the design of electrical slapper detonators, in which an electrically vaporized metal foil drives a dielectric film to high velocities to shock initiate an explosive pellet. This insight was based upon discussions with Ron Lee and Dick Weingart of LLNL regarding which metals are the most effective in driving the flyer in a slapper. The LLNL results fit the Gurney idea that some of the supplied energy must be used to accelerate the metal vapor that drives the flyer; thus, low metal density is favorable for that application. Tillman (T.J.) Tucker and Phil Stanton developed this idea into a Sandia report in 1975, correlating an "electrical Gurney energy" delivered by a firing circuit with the current density at the time of burst of the bridge foil. Using those results, I modeled the performance of slappers and found that thin flyers are sufficient for initiation of the explosive. I then performed experiments that reduced the firing energy of slapper detonators by a factor of 25, based on predictive guidance from this model (while receiving flak along the way from some Livermore folks).

The Gurney model was not always received kindly at Sandia and the Department of Energy. When I first spoke about it at a meeting, I was derided by the Atomic Weapons Research Establishment (the predecessor of the AWE, which was formed in 1987) for promoting such a simplistic model "when we have computer codes that can solve those problems much more accurately." Lee Davison urged me to pay less attention to Gurney and more to my shock wave research. I asked whether my Gurney report would become a negative rather than positive factor in job performance evaluation, and Davison replied that, having been raised on a ranch, he knew that every steer produces some steak and some hamburger, and you just accept the hamburger along with the steak. So the Gurney model was clearly the hamburger.

### Don Hardesty's Work on Explosives

Don Hardesty joined the Physics of Solids Research Department in 1974 after graduate and postgraduate work in combustion. He worked on explosives until about 1977, when he joined the fledgling Combustion Research Facility activity at Sandia Livermore. While in Albuquerque, he collaborated with Pete Lysne on shock initiation of liquid nitromethane and on the development of a universal Hugoniot for liquids, which is still cited (Hardesty and Lysne 1974). Their work motivated Don Amos, Nunziato, and me to analyze the shock-driven thermal explosion behavior of nitromethane in their experiments by the use of a parallel-reactions model (Nunziato, Kennedy, and Amos 1977).

### Spark Detonator Studies

T.J. Tucker, Dwight Allensworth, and I collaborated in 1968 to 1971 on the development and characterization of high-power spark detonators (Tucker, Kennedy, and Allensworth 1971) that employed low-density pentaerythritol tetranitrate (PETN) in contact with the detonator electrodes, which is similar to the construction used in exploding bridgewire (EBW) detonators. (Orval Jones worried that I would exercise my industrial experience and steal all of Dwight's time away from T.J. He told T.J., "Don't let him do that!" Tucker and I laughed together about that.) We used low-inductance circuits triggered by the breakdown of a spark gap within the PETN powder. Measurements of voltage and current histories were analyzed to yield histories of power and energy delivered into the powder. The findings indicated close matches between the power and energy required to produce detonation in the PETN in spark and EBW detonators and in the time required to establish detonation. The data are valuable for consideration of electrostatic spark hazards as well as for intentional detonation of low-density PETN. These studies are more fully discussed in Chapter 11 of *Non-shock Initiation of Explosives*, Vol. 5 of Springer-Verlag's Shock Wave Science and Technology Reference Library, edited by B.W. Asay (2010). That volume also shows that the initiation of low-density PETN by pulsed lasers operates in a way that is quite similar to that of spark and EBW detonators.

### 1972 New Mexico Section of ASME Conference and Workshop

The NM Section of the American Society of Mechanical Engineers (ASME) in those years conducted a conference on a different subject each year and published the results. Lee Davison and I were asked to organize a conference on engineering applications of explosives. In the New Mexico spirit, the invited speakers, who were largely from Sandia and Los Alamos, presented a set of talks that described the fundamentals and many of the applications of explosives. [See Jones (1972) and Kennedy (1972).] Contributed papers by folks from other institutions and industries rounded out quite an instructive set of information that was published by the local ASME section as *Behavior and Utilization of Explosives in Engineering Design* (1972), and more than 1000 copies of these proceedings were distributed.

## Splinter Gate

The catcher section of the gas gun in Building 808 at Sandia was a steel structure that clamped around an assembly of several 4×4 pine beams that the projectile would hit after the experiment was over. The catcher section stopped the projectile and tore up the beams. The smashed wooden pieces were sent off to Reclamation[103] at Sandia. At some point someone recognized that they could simply take the broken beams home and use them as kindling, and the calls to Reclamation stopped. The Reclamation police figured out that something like that had happened and started an investigation into who was taking the government property in an unauthorized way. Some staffers were suspended briefly as punishment for that.

## Explosion at the STAR Facility

I was on three accident investigation teams in my 17 years at Sandia, and all those accidents happened in 1977. One accident was a case in which the door of the large catcher tank at the STAR powder gun blew open after a shot. Venting from the catcher tank literally tore strips off the roof of the metal building. The entire staff was in the fortified control room so nobody was in danger.

The catcher tank was about 6 ft in diameter, so a person could walk in to set up the target assembly. The tank door had a bolt circle of 48 large bolts, which held the door against a flange sealed with an O-ring, so that a vacuum could be pulled on the tank and barrel before a shot. The door began to warp with use, so not all the bolts could be threaded into position. The gun operators didn't think the number of bolts mattered as long as they could still pull a vacuum, since the tank was always under negative pressure after a shot, when they came out of the control room to clean up. The number of engaged bolts dropped to 20 and then 8 and was down to 7 by the time the accident occurred.

Analysis by the investigation team determined that there was actually a considerable positive pressure in the tank right after a shot, especially since the final expan-

[103] Editors' note: The Reclamation organization at Sandia was responsible for acquiring equipment, experimental hardware, office furniture, etc. no longer used or needed by an organization and reapplying this to other uses.

sion of propellant gases into the tank space was a Joule expansion (i.e., at constant energy) rather than an isentropic expansion. The operators never knew about the positive pressure because they delayed their approach to the tank while they secured data records from the shot (Polaroid® film records from the oscilloscopes), and that gave the gases time to cool enough for the pressure to drop below atmospheric. Vice President of Research Narath's concise statement about the accident findings was "You guys can't operate that gun again until you understand how it works!"

**Bill Benedick**

I knew of a fellow named Benedick at Sandia long before arriving, because we shared a temperamental technician named Wally Mandleco. Wally was a retired marine who was a good technician, especially in emergencies. He could shine in resolving emergencies because he was often the person who had created the emergency, so he knew more than anyone about it. Wally went back and forth two or three times between IIT in Chicago, where he worked for me, and Sandia, where he worked for Bill. Wally suited Bill fine because both of them were really good at "just getting things done." Now that's almost a lost art.

Bill worked on projects that had both scale and technical substance. One that I recall was a time he and Charlie Daniel built a concrete tunnel at Building 9926 to study hydrogen steam–air reactions in order to simulate the events at the 1979 Three Mile Island accident. I thought he was working at large scale in the early 1970s when he studied fuel–air detonations in a 50-ft-long pipe assembly covered with plastic sheet, which Bill had also constructed. But that was nothing compared to the instrumented concrete tunnel. Bill also led the work to build a walk-in detonation chamber alongside Building 9926.

# Marlin E. Kipp

# (1974–2011)

# Pre-Sandia Years at Lehigh University

I was hired at Sandia in May 1974. I had finished my PhD studies in Applied Mechanics at Lehigh University in 1973 under the tutelage of Professor George Sih, with a focus on the theory of quasi-static fracture mechanics in elastic materials. (I had tried only one set of experiments and concluded that I was dangerous in the laboratory: reports indicated that stable radial outward growth of cracks in glass could be formed under a flat circular punch. I replicated that in 2-inch cubes of glass, but when I tried the same configuration in Plexiglas®, no crack formed. I increased the load until the block burst into several pieces ejected around the room—the cracks grew inward onto the axis—completely different behavior than the glass.) I stayed on at Lehigh as an assistant professor, while looking for a position elsewhere. In fall 1973, I got an unexpected call from Larry Bertholf at Sandia, asking me to interview in his shock physics group—Doug Drumheller had recommended that he call me. (Doug had finished at Lehigh prior to me and had already gone to Sandia.) I came for an interview in October 1973 and found that everyone I talked to was working on interesting problems. Larry offered me a position, which I accepted.

## Sandia National Laboratories

Walt Herrmann was the manager of the divisions comprising the shock physics efforts, and Orval Jones was the director. The department included code developers, code theorists, material modelers, and experimentalists. Much activity supported the underground test program, including ground shock and crater formation as well as radiation damage and shocks from radiation deposition. The collaboration was close, with a marvelous atmosphere of learning from one another—a very refreshing environment. And for me, coming from a statics world, having the opportunity to work in this entirely new area of shock physics, was great—I had never heard of a microsecond or a kilobar before I showed up at Sandia.

**Learning to use the hydrodynamics codes.** Considerable effort had been invested in the one-dimensional (1-D) Lagrangian code WONDY, including a rezone option; Jeff Lawrence was the keeper of that code and tutored me in its use. I eventually became the code's keeper when Jeff moved on within the Laboratory. WONDY was popular because of the ease of adding and evaluating new material models. I used it extensively for that purpose and sent out many copies over the years. The two-dimensional (2-D) Lagrangian code in Larry's department was TOODY. Rupe Byers was a key player in that effort; rezoning was being used to make TOODY more versatile. When Jeff Swegle joined the division a few years later, he assumed ownership of the TOODY development. CSQ was the Eulerian code in the department. CSQ was built and maintained by Sam Thompson, with Hank Lauson supporting it as well. Darrell Hicks was working on various concepts of interfaces, multi-materials, and mathematical proofs for code difference solutions. As I interacted with the developers and analysts, it was easy for me, being newly exposed to these large codes, to develop a starry-eyed view that the codes could reveal marvelous infallible truths. However, that view was tempered as I watched what the experimentalists were extracting from their shots at the gun facility, and I learned to have a healthy skepticism of the extent to which the codes could reproduce reality. There was, in fact, no doubt that these extremely powerful tools provided (and continue to provide) invaluable insight into and support for the experimental efforts. Nevertheless, some physical phenomena had not yet been included in the material models, and care in the interpretation of the code results was always required. (Of course, one also had to make sure the context of the experimental data was understood – e.g., how and when multi-dimensional effects might creep into the supposed 1-D experiments.)

Dennis Hayes had recently returned from Washington State University and had taught a shock wave physics course at Sandia that was recorded on a series of videotapes. So, over the course of a few weeks, I watched each tape, did the homework problem sets, and was tutored by Dennis on various related aspects of shock phenomenology. For me, it was a great introduction to the field. (And Dennis' notes have continued to aid several generations of staff to gain insight into the physics of shock waves.)

Over my 37 years within the shock groups at Sandia, I focused on several main areas of modeling and simulation work: hypervelocity impact, dynamic fracture, spall, fragmentation, explosive reaction kinetics, neutron generator standoff, radiation deposition, conventional weapon analyses, and source terms for dispersal from explosive and hazardous material events. In addition, a wide variety of "walk-in" problems came my way to provide support to folks across the Labs.

**Hypervelocity impact.** When I arrived, one of the active experimental research areas was hypervelocity impact at velocities upward of 4 km/s. Recent tests with polymeric spheres impacting steel targets had revealed some unusual spall behavior; it turned out that the $\alpha-\varepsilon$ phase transition in iron at 13 GPa led to a release wave structure that created peculiar fracture behavior (i.e., a spall plane with a connecting fractured plug). Complementary plane wave impact experiments with VISAR instrumentation extracted a three-wave structure that revealed the existence of the phase transition under shock conditions. Sam Thompson had included the transition in the ANEOS tabular equation of state for iron, and his CSQ simulations demonstrated how the shock and release from the impacting sphere formed the wave structure in the steel, leading to the peculiar spall behavior. Over the ensuing years, I would occasionally include the transition in my simulations, but I rarely found that it had a significant effect on the deformation and fracture. (That was after discovering that, sometime along the way, the phase transition had been turned off in the code, and I was in fact running two identical simulations!) I worked with Joel Lipkin on similar impact conditions in aluminum (Lipkin and Kipp 1976), using CSQ to model the particle velocity histories at the back surface he was obtaining (with the monitoring point being typically slightly off-center, but exhibiting a clearly defined spall signal). The fracture models were limited to minimum stress criterion and a Tuler–Butcher integral to failure—both leading to fairly precipitous fracture although we had obtained some good agreement between the simulations and data. We presented our work at a planetary science conference in Flagstaff. (When I gave my talk, I mentioned in passing that I had set up the impact configuration in CSQ, turned it loose on the machine, and had run it to completion. There were laughs in the audience, and when I later asked Al Chabai what precipitated the chuckles, he said it was virtually unknown to be able to run problems in Eulerian codes without manual intervention and restarts.) It was at that conference that Gene Shoemaker took us on a field trip to the Meteor Crater and showed us what a large impact crater looks like. We walked into the crater, with Gene showing us details of the deformation, the shock-induced material remnants, and the fault lines the meteor hit leading to the nonsymmetric crater shape.

**Ductile fracture modeling.** I got involved in ductile fracture modeling with Al Stevens and Lee Davison using WONDY and TOODY (Kipp and Stevens 1976; Davison, Stevens, and Kipp 1977; Davison and Kipp 1978). Lee and Al had developed a continuum model that coupled plasticity with void growth representations. The void formation continuously degraded the effective modulus, leading to unloading of release waves as the spall evolved. Al was having problems with instabilities in WONDY using the model, so I was enlisted to figure out a way to make it behave. Differencing the derivatives in the model expressions seemed to be the source of the difficulties, so I resorted to accessing an external numerical ordinary differential equation (ODE) solver (from a Sandia-developed math library), which was called from WONDY and directly solved the differential equations at each cell and time step. I enforced linear variation of the density and artificial viscosity over the time step; the technique worked very well, giving stable results. We had some VISAR data for planar impact in aluminum and set out to determine a suitable set of parameters for the void growth part of the model. We papered the hallway wall outside Lee's office with results of the runs for each set of parameter variations until we got a satisfactory fit to the data. We would run a simulation in WONDY and wait for a roll of paper with the plots to be delivered by courier; we would then post the plots on the wall and decide on the next parameter variations. We eventually converged on a satisfactory set of parameters that matched the uniaxial strain data from the plate impact experiment. Then I put the model in TOODY, and we applied the same parameter set to a 2-D rod impact configuration and obtained the central void growth seen in the experiments. This was a significant improvement over the minimum stress criterion commonly used to model the spall.

**Rod impact of layered oblique targets.** Larry Bertholf had a contract with the Ballistic Research Laboratory (BRL) in Aberdeen, Maryland, to study long-rod impact into oblique layered targets. Bill Brown and I were doing some of the simulations, trying both CSQ and TOODY in 2-D planar geometry because no three-dimensional (3-D) codes were available in the department yet. After one busy weekend computing, Larry called Bill and me into his office to inform us we had managed to consume nearly all the contract money in computer charges—he was not happy. At that time, computer charges were upward of $600 per hour. I would be involved with staff at BRL intermittently in subsequent years, doing simulations of other conventional warheads (with Allen Robinson and Fred Zeigler) and also providing independent assessments of contractor competition rounds.

**Modeling high-explosive behavior.** In the late 1970s, I began working with Jace Nunziato on modeling homogenous explosives, with nitromethane as the focus, as considerable shock-to-detonation uniaxial strain gas gun data were available for that explosive (Kipp and Nunziato 1981; Nunziato and Kipp 1983). Using the same ODE solver technique in WONDY as for the ductile fracture model, we solved the evolutionary equations of mass and energy balance for the explosive reactant and product gases and the reaction kinetics rate laws. We used a two-stage reaction to capture the initial shock heating at the impact interface, the growth into a supra-detonation wave that overtook the initial shock, and the subsequent steady detonation wave. The model agreed quite well with the 1-D data, but we had to use a few inert buffer cells at the impact interface, since the artificial viscosity tended to create a slightly elevated temperature and an under dense reactant in those first few cells, precipitating an early start to the reaction. Given our 1-D success, I put the model in TOODY to look at boundary effects on the detonation that influence the critical diameter behavior (i.e., shock reflections or release that would affect the reaction rates). The trends were captured, but we found a bifurcation occurred in the explosive response that depended on spatial resolution and time step—a situation we were unable to resolve satisfactorily. Darrell Hicks worked on the numerical theory of having disparate characteristic times for the reaction kinetics and the Courant-based resolution value.

Another modeling effort was mounted to address the shock-induced reaction in granular explosives. Jace derived the basic equations—now three phases (instead of the two for nitromethane): reactant, hot spots, and product gases. Our primary application was for PBX 9404, a relatively insensitive high explosive, using a series of wave profile experiments from Jim Kennedy that showed the evolution of the detonation wave behind the shock front eventually growing into a steady wave. (The evolution from a low-amplitude shock was quite distinct between homogeneous and heterogeneous explosives.) After extensive parameter studies and modification of the reaction rate law, we could reproduce the wave profiles with the model. The Pop plot curve (the distance to detonation as a function of initial shock amplitude) was also reproduced fairly well. Then Bob Setchell did some carefully designed thin pulse shots into 9404, in which the shock amplitude was the same, but the durations of the release waves were tailored so one shot ramped up to detonation and the other failed to initiate. We could not get the model to capture that behavior. I varied the hot spot formation, using artificial viscosity as the energy dumped into hot spots (via an analogy to shear deformation) and produced some very nice results. Nevertheless, we remained "defeated" by those two thin pulse shots. As our computing capability evolved, we could do some explicit modeling of granular explosives, watching hot spots form during cavity collapse (with Mel Baer).

**Modeling ground shocks from nuclear bursts.** A significant effort in our department focused on ground shock modeling from nuclear bursts, led by Bill Davey, Paul Yarrington, and Al Chabai. I was not involved until many years later, when the Defense Threat Reduction Agency (DTRA) would occasionally request (often through Paul) simulations that ranged from the consequences of multiple conventional or nuclear explosions to code evaluation checks. In those later years, no ongoing support existed for ground shock modeling, and simulations tended to be scoping in nature rather than retaining a capability to obtain detailed results. The one exception was a DTRA effort in which the local geological structure was specified in considerable detail, and some conventional explosive shots were fired. Another large-charge conventional explosive shot of 700 tons was to be fired at the Nevada Test Site to look at tunnel collapse, and extensive work had been invested in preparing the tunnel instrumentation, transferring the geology into the codes, and running simulations; unfortunately, the project was shut down prior to the shot. I did find that, in granite, a fairly good match could be obtained for the codes to the peak stress and particle velocity vs. scaled distance with the extensive nuclear data that was reported. But I was unable to match the pulse shape from gauge records—my wave shapes were far too narrow; I never did figure out why the peak stress could be satisfactory and not the duration, since the erosion of a basically triangular wave should be a function of its width. The nuclear yield was modeled with an equivalent energy deposited into a "pill" and provided a reasonable means to model fully contained devices (a technique passed on to me by Paul Yarrington). Extensive simulations were made of the underground structures, such as tunnels, that are subjected to various forms of attack.

**Dynamic fragmentation and oil shale research.** During the Carter administration, the national security laboratories mounted a major effort on various energy thrusts. One area was oil shale—how to extract the kerogen from the shale efficiently. (Old-timers in the mining business have noted the roughly 20-year cycle that interest focused on oil shale when crude oil prices rose; but the more labor-intensive work required for oil shale meant that, as labor costs escalated, the shale would again retreat in feasibility.) In our department, extensive static and dynamic testing of oil shale samples was pursued, and Dennis Grady and I embarked on a dynamic fragmentation study that eventually expanded to many other materials over many years until Dennis' retirement in the 1990s. In addition to oil shale, we studied Arkansas Novaculite, a fine-grained homogeneous rock that gave excellent reproducibility. The major observation from this brittle material was that the apparent fracture stress depended on the strain rate (e.g., data from quasi-static, split Hopkinson bar, and spall experiments), and the average fragment size was inversely dependent on the strain rate. From these observations, Dennis constructed a basic model to capture the characteristics, I implemented it into WONDY, and we made considerable progress in modeling the behavior. The damage variable in the model directly degraded the material modulus, and this continuous coupling was very effective in representing the rock response. When we included the model in TOODY with parameters appropriate for oil shale, we were able to capture the consequences of borehole explosions and simulate multiple borehole shots to estimate the fragment size distribution. Not surprisingly, fragments nearest the explosive were very fine and graded out to much larger ones as the distance from the charge increased. Such a size disparity was not going to be good for in-situ mining and retorting, and work by others was pursued to reduce the loading rates and thus the formation of fines. As part of those studies, Los Alamos National Laboratory (LANL) examined how to represent various diameters of ANFO (a mining explosive composed of ammonium nitrate and fuel oil), and I implemented the Becker–Kistiakowsky–Wilson (BKW) release isentrope in WONDY's suite of explosive models. Communications with Chuck Mader of LANL included the parameters they had derived for ANFO. Later, Jones Wilkins Lee (JWL) parameters were determined at Lawrence Livermore National Laboratory (LLNL), and we also used those; the output of the ANFO charges is diameter and confinement dependent, so care had to be taken to use the appropriate parameter sets.

**Fragmentation in metals.** A natural extension of the strain rate dependence of fragmentation was to metals. The first effort was directed toward ring expansion (done electromagnetically by Dennis Grady and Dave Benson) and development of models to represent the fragment formation. I set up an equivalent 1-D expansion of the ring in WONDY with a uniform velocity gradient to simulate the ring conditions and then randomly inserted initiation points with random initiation thresholds. The fracture model was set up to dissipate the proper energy during stretch, and the results agreed well with the data: even with large numbers of initiation points, there was only enough energy to bring a few to completion and obtain the small number of fragments of random sizes seen in experiments. Dennis Grady published a seminal paper (Grady 1982a; also see Grady 2010b) on fragmentation that formed the basis of our thinking over many years for ductile, brittle, and liquid fragmentation. With Jeff Swegle's help, the model was implemented in CTH as a monitoring tool (i.e., when the fragments were defined to have formed, the material was not agglomerated into actual fragments; instead, the totals could be summarized in bin charts, and the distributions could be obtained by post-processing). We found that, if we took the average fragment size from the simulations and imposed an exponential distribution on those averages, the actual distribution could be reasonably represented. It became a powerful tool, and I used it extensively for the remainder of my career on a wide variety of fragmentation–formation problems, ranging from fine liquid droplets to explosive casing expansion. The dependence on strain rate explained clearly why "pipe-bomb" like explosive devices (characteristic of most fragmenting rounds) broke into strips, since the axial strain rate was typically much smaller than the circumferential strain rate, with large strain rates giving smaller average dimensions than small ones. Spherical devices, on the other hand, with a centrally initiated spherical explosive charge inside a thin external metal shell, would lead to nominally square fragments. In that case, strain rates would be uniform in all circumferential directions—a result confirmed by radiographs.

**Effects of fragmentation of metal spheres during impact**. Dennis Grady embarked on a series of steel sphere fragmentation experiments in which the spheres were launched into thin polymethyl methacrylate (PMMA) targets, using flash x rays to image the steel fragments—the PMMA being transparent to the x rays. The radiographs captured the breakup of the spheres, and a wide range of target thicknesses and impact velocities demonstrated clear trends in the breakup characteristics. I used the codes to look at deformation and internal strain rates in the spheres during impact, and we could correlate the strain rates to the debris sizes. The one aspect I could not reproduce was the shape of the hole in a PMMA target—a thin washer off the front and rear surfaces, with a hole diameter basically that of the original sphere diameter. The code always allowed the size of the hole to grow to about twice the sphere diameter, and variations in equation-of-state models, yield models, and fracture parameters simply would not provide a satisfactory solution; I remained defeated on that score. Dennis used the hypervelocity guns in subsequent experiments to study the fragmentation of other materials, including copper spheres against steel targets (to capture the fraction of copper liquid ejecta).

When it became clear that our computers would support 3-D computing and no in-house capability had been developed, the HULL family of codes was brought in, with Sydney Westmoreland as the support staff. In the 1980s Sam Thompson and Mike McGlaun commenced the development of CTH—a multi-dimensional shock physics code (with 1-D rectangular, cylindrical, and spherical, 2-D rectangular and cylindrical, and 3-D rectangular coordinates). The equation-of-state package was initially the ANEOS package. Gerry Kerley later substantially extended the equation-of-state capabilities, including his multiphase models, reaction kinetics explosive models, and LANL's extensive collection of equation-of-state tables. One of the first problems I ran was with 6 million computational cells on the Cray using all the memory—a copper ball obliquely impacting a steel plate at about 4 km/s—to simulate one of Dennis' experiments. The cavity formation, copper and steel ejecta streams, and target rear-surface breakout were all in good agreement with the radiographs. This led to an exercise to model explicitly the steel sphere debris that impacted an aluminum witness plate in the steel sphere PMMA experiments. I took the expansion rates and average fragment sizes from the impact and randomly inserted fragments into a disk-shaped space of the right size when it would impact the aluminum; the scattered and overlapping craters formed by the debris pieces in the simulation had excellent correspondence to the real witness plate.

**Impact of a protoplanet on the Earth.** During a summer visit to our department, Jay Melosh (Department of Planetary Sciences, University of Arizona) and I worked on 3-D CTH simulations of a proto-Mars-like object impacting the Earth at glancing angles, looking at the ejecta for evidence that it could eventually form our Moon (Kipp and Melosh 1987). CTH had just recently come on line with its 3-D capabilities. We built a spherical model of the Earth with its iron core and mineral mantle (eventually including crude initial densities, temperatures, and pressures). At the time, we had only central gravity in CTH. (Dave Crawford added self-gravitation later for much better physical representation.) We could only run the simulation to about 2000 seconds in elapsed time—enough to see the Earth's distortion and initial plume of material jetting outward from the region of impact, but not far enough to see coalescence into an orbiting body. Jay reported these large Earth impact results at planetary conferences. They generated widespread interest by popular science magazines, which led to broad public dissemination of the slides of our simulation.

Another astrophysical problem of interest was how asteroids might be deflected or dispersed by large conventional or nuclear charges. Simulations of how to use such techniques effectively were pursued with Mark Boslough.

**Running jobs on early parallel computers**. The Paragon was our first production parallel computer; a version of CTH had been written and installed on the Paragon. Users had to run jobs on a scheduled basis, working at the console in the computer annex. It was common to start a job in the afternoon; someone in the computer department would occasionally monitor the machine activity during the night, as indicated by panels of blinking lights. When it was clear the machine had stalled, the person whose job was running would get a call in the middle of the night to come in and restart their job; I made a lot of midnight runs to reboot the Paragon and restart my jobs. On our next large parallel production machine, Janus, we also initially had to run from a console and move our data on and off the machine with compact tapes. Eventually, we could do the transfers electronically and remotely. Extensive weapon-related simulations were done on that platform—a workhorse for CTH applications.

**Simulations of the respirable fraction of a fragmented metal.** I started respirable fraction studies with explosively-driven silver devices at the request of Matt Sagartz. The spherical devices had a thin interior shell of silver (comprised of two hemishells) and an external shell of explosive. I was looking for a correlation with the respirable fraction of metal obtained from experiments in a large hemispherical air-supported tent. Typically, 25–30 % of the silver melted or vaporized and then recondensed, based on post-test filter analyses. With a full three-phase (solid–liquid–vapor) tabular equation-of-state model, I was able to obtain about the correct total fraction of residual material in liquid–vapor states, but the proportions of melt-vapor tended to be incorrect. CTH was providing a prompt estimate at the end of the shock and release—that is, in the first 25–30 μs; the filters were sampling in the minutes time frame. The fragmentation model provided reasonably accurate estimates of the solid debris sizes, which consisted of flakes of silver found after the experiment. The liquid droplet sizes estimated from the release strain rates were also representative of the filter data. We had found that the kinetic energy was sufficient in the expanding melt material, as it would stagnate in the air and fireball explosive products, leading to secondary breakup and perhaps vaporization physics processes that were not in the code. We modeled variations in dimensions, explosive mass, and metal type.

Fred Harper had the lead role in these studies, and I worked closely with him over the next 15 years until I retired. In my 2-D axisymmetric simulations, a jet would always form, but there was no evidence for such a jet in the experiments. Experiments with radiographic instrumentation indicated that minor shape and thickness variations of the thin silver shell were responsible for significant nonsymmetric behavior in the post peak-compression expansion and explained why no jet had formed. I incorporated the shell thickness variations and slight aspherical dimensions into a fully 3-D simulation and saw the significant changes in expansion wrought by those asymmetries. Simulations of a Russian experiment—with thicker shells and easier-to-control tolerances—retained symmetry and were accurately represented by the simulation (including the jet).

**Studies of impact and fragmentation of ceramics and powders.** We moved on to powders and ceramics—but never had quite the same success as with metals—in part because adequate material models were lacking for ceramics, but also because a brittle fracture model was not available for these materials. However, the simulations very clearly demonstrated the compressive behavior observed in the experiments. We also explored various initiation schemes for the explosive charges. The metals were surrogates for more hazardous materials that might figure in explosive accident scenarios; on occasion, I provided estimates of source terms for first responder training exercises.

We also worked on ceramic surrogates for power supply materials such as PuO and $UO_2$. Understanding these materials became important when they were to be used for power supplies on satellites (e.g., the Mars Lander mission). I was involved at a low level—many simulations of these launch accident scenarios were done by Greg Bessette, with the goal to ensure as could be best assessed that no breach of the containment would occur. I was involved in other launch accident scenarios that we examined for NASA when post-shuttle options were evaluated to assess the shock and debris conditions that might exist at crew level from fuel explosions.

I used a P-$\alpha$ porous material model for the powders. The Soviets had reported extensive data for shock loading of porous fused silica for a wide range of under dense initial conditions. I had to make minor modifications to the quartz tabular model to include the reverse curvature at low pressures. With that table, I was able to show that the temperature rise in shocked porous quartz agreed with some limited published data and that, for many of the initial conditions, the code correctly showed final states with high temperatures and densities that were less than the initial values. I did not do nearly as well when tungsten carbide powders were shocked; the explosive experiments showed the powder being dispersed, whereas the P-$\alpha$ simulations led to a solid material consolidated behind the shock and very limited dispersal. I made extensive use of the P-$\alpha$ model in simulations with foam components. Where data were available, the model could represent the behavior quite well. In many situations, however, no data were available. In that case, an equation of state was chosen to represent the fully dense parent material, and an initial distension was used to be consistent with the reference density of the foam.

Other material studies included building models to represent the complex behavior of some ceramics for which Dennis Grady had obtained wave profile data from plate impact experiments. The models were basically elastic, with tailored moduli to capture the shock and release structure exhibited in the data. A brief diversion with Paul Taylor (which he later followed up in much more detail) was to examine the waves propagating through the skull and brain, looking for correlations to brain damage from head injuries.

**Simulation of impacts on concrete and foam microstructures.** Given improved computational capabilities, we were able to do some explicit modeling of concrete and foam microstructures and obtain very nice agreement with experimental wave profiles from Lalit Chhabildas and Bill Reinhart. To model the concrete structure, I used a random number generator to insert the aggregate as spheres or ellipsoids, which were then backfilled with mortar. For the foam, I inserted a solid and removed spherical shapes to create a discrete void structure with the correct total mass. These large 3-D problems were run with CTH, and the progression of the waves through the heterogeneous target structure could be readily followed; the spatial dependence of the shock emerging from the foam could also be observed with the model.

**Radiation deposition simulations.** Jeff Lawrence tutored me on the use of WONDY for radiation deposition problems to support reentry vehicle shock loading. The BUCKL code was used to determine the deposition profile based on the radiation characteristics and target materials; then that profile was inserted into WONDY to assess the nature of the induced shock waves and the consequent target motion. On similar problems, I worked with Bob Benham (and later Gary Rivera) to represent the impulse from radiation deposition with explosively or magnetically-driven flyers or directly with a light-initiated explosive in order to simulate underground test conditions without nuclear sources. Other deposition problems included nuclear reactor pulsed design scenarios, looking for the expansion characteristics of irradiated pellets and gases.

**Application of simulations to nuclear weapon systems.** One of the major weapon support efforts throughout my career was neutron generator standoff analysis. The first project was a critical need for the W80 system in the late 1970s that involved 1-D and 2-D simulations; we had considerable interaction with LANL, including the hydrodynamic test group to design requisite components. High-speed movies and shorting pins were used to characterize the component motion in the explosive environment and compare the data with the simulations. The simulations provided approximately one million dollars in savings for the program. By the time of the life extension program[104] for the W80, we were able to utilize 3-D models to redo some of the early analyses and demonstrate the accuracy of previous approximations (in collaboration with Dan Kletzli at Sandia's Livermore site). Experimental comparisons were done, with Karl Schuler and Mark Anderson providing the instrumentation on the LANL and LLNL hydrodynamic tests. By the end of my career, I had constructed computational models of major systems for a variety of applications that might arise. I also gave an annual lecture to the Weapon Intern classes on the application of computational techniques to nuclear weapon systems. Other activities included simulations of testing scenarios and looking for fragments that would form in non-nuclear accident scenarios. That topic area also included nearby explosion problems—that is, trying to assess the consequences of fragments from an exploding source impacting other explosive components in the vicinity. In some

[104] Editors' note: A life extension program is a program to repair or replace components of a nuclear weapon to extend the time that the weapon can safely and reliably remain in the Nation's stockpile.

cases, a debris footprint was of interest; the initial conditions of the fragments (i.e., the dimensions and velocity) extracted from a simulation were then fed into an aeroballistics code, and Monte Carlo techniques were used to track the fragments from the source to the ground to estimate the expected region in which fragments would land. Additional efforts were focused on modeling impact fuses—that is, calculating the transit time of shocks through nose cones and comparing the simulations to data that had been collected over the years.

**Application of simulations to conventional weapons.** A major effort in the conventional weapons area was chemical munition remediation. In addition to stockpiles of chemical munitions the Army needed to destroy, and a desirable mechanism was to render single munitions harmless that were unearthed on former battlefields or in test areas. Sandia had been commissioned to design such a unit, and, as part of the analysis, I was tasked with modeling a wide assortment of munitions, usually with a chemical inside a metal casing and a burster charge. If the latter were to detonate during decommissioning, casing fragments would be formed and accelerate outward from the charge. We (Reina Martinez and I) did many simulations to provide estimates of fragment dimensions and velocities; these characteristics were then used to evaluate the vulnerability of protective tank walls to penetration. Drawing sets from old Army manuals were used as a basis for building our models. CTH was used for the shock analyses and to determine the strain rates for the average fragment dimensions, to estimate the fragment sizes, and to then simulate the impacts of those fragments on tank walls. (See Kipp, Martinez, Hertel et al. 1999a; Kipp, Martinez, Benham et al. 1999b; Kipp and Martinez 2000.)

Another conventional weapon analysis project was that of the fluted liner-shaped charge. Gene Hertel was interfacing with Ernie Baker at Picatinny Arsenal in their attempts to model a fluted-shaped charge with CTH. When a shaped charge in a munition is fired from a rifle barrel, the charge spins and, if a smooth liner is used, the metal converging on the axis has such a large angular momentum that the jet is broken up, dispersing outward and ruining the performance of the charge. One way to compensate for this behavior, which has been known for many years, was to introduce fluted shapes in the liner—essentially overlapping "leaves" (the cross section has a serrated appearance). We figured out how to build a unit section of the liner with the insertion packages in the code (using Dave Crawford's newly developed Diatom library) and then repeat the insertion packages about a central axis to build a complete liner. This is clearly a 3-D problem, and we expended considerable effort to ensure that the liner geometry was accurately inserted into the model, along with the surrounding explosive charge and metal casing. Without any spinning motion, the flutes introduced their own spin, and the jet broke apart soon after forming. When we introduced spin, the flutes compensated correctly for the spin and a coherent jet was formed (although not quite as high in quality as a smooth, non-spinning liner produces). To our knowledge, this was the first time a simulation of a full device had been completed that showed this effect out to the extended jet. We could vary the spin rate to watch the change in jet quality and clearly see how the collapsing liner with the flutes was counter-rotating. We (Reina Martinez, Gene Hertel, myself, and the Picatinny staff) received the Neill Griffiths award at the 18th International Symposium on Ballistics in San

Antonio, TX, in November 1999 for our paper reporting this work (Kipp, Martinez, Hertel et al 1999a). For another application, I built a model of a rocket-propelled grenade and was able to simulate the full event, including the approach to a target, impact with the target, jet formation, and jet penetration into the target.

After September 11, 2001, I embarked on a wide-ranging series of analyses to address possible vulnerabilities of nuclear power plants and spent fuel storage areas. One of my assignments involved estimating the effects of large explosive charges on nuclear fuel transport containers. The limited amount of data provided at least some information on deformed containers. I built finely detailed models of several containers and simulated a variety of explosive configurations to evaluate the consequences to the container and fuel.

I pursued simulations of subscale penetrators into concrete targets, with impact conditions defined by a series of experiments at the Waterways Experimental Station. Deceleration and depth of penetration data were obtained, supplemented with triaxial static compression data from pre-shot samples of the concrete targets. Within the limits of the yield and fracture models available in CTH, a parameter set for the concrete was determined that fit the compression data; lacking was tensile fracture data, so a few iterations on the test configuration led to estimates (very small values), with the result that penetration data from most of the shots could be matched fairly well.

### A Few Miscellaneous Notes about Code Simulations

- In Eulerian code simulations, partially filled cells appear to be fragments, but the fracture model does not contain the physics to guarantee those apparent fragments actually represent a real size. That was part of the reason for our model: to provide an improved calculation of the fragment distribution. Some fragmentation analysis had to be done by hand, since CTH did not discriminate between volumetric expansion and thin-shell expansion in terms of the estimated fragment dimensions; so, for thin shells, I had to use the size as the in-plane measure and then estimate the thickness for the other important dimension of the fragment.
- The ALEGRA code had been developed with a magnetohydrodynamics capability (with Allen Robinson as the lead). I was involved in studies of shock compression of the magnetic field and the effect on subsequent material motion. The compressed field induces an additional magnetic pressure that must be released, and ALEGRA provided some keen insight into that phenomenon.
- I routinely ran a few test problems to make certain that codes accurately reproduced the expected shock behavior (e.g., for 1-D shocks I looked at amplitudes of shock jump conditions and explosive cylinder tests to compare with expansion data I had on hand for a variety of high-explosive types).
- I frequently used CKEOS to evaluate the ANEOS parameter sets, matching to shock Hugoniot data and enthalpy vs. temperature data to capture the solid-melt-vapor transitions. That was a utilitarian means to get operational equations of state for materials that were not yet fully modeled. I used Gerry Kerley's BCAT code to create the table accessed with CTH.

- I was able to show that, for a bare charge of TNT in air, the arrival times of the first shock were calculated fairly accurately, but a second less distinctive shock forms from reflections inside the fireball, and a third one as well. The latter two were not distinguishable in the CTH simulations, since the 1-D models did not have any turbulent mixing and secondary combustion physics. Nevertheless, the agreement with the primary shock was an indication that the explosive energy released and the tabular air equation of state were accurate.
- In situations where parameters were required for an explosive with little data available, I resorted to using the ideal gas model (as a degenerate JWL model); with known density and detonation velocity and an estimate of the Chapman–Jouguet pressure or gamma (typically about 3), I could make a reasonable simulation of an event. For a few situations, the explosive energy had to be released under less than detonation reaction conditions. This was accomplished by using the energy deposition option in the code; propellants (and gun powders) could be represented with this mechanism. I was also able to match extensive White Sands Missile Range data for liquid hydrogen and oxygen explosions with this method by capturing the rollover in peak pressure at points near the explosion. Of course, this was a means to an end without capturing the physics of the actual reaction processes.
- Condensation physics was not present, so although the code could calculate that the temperature increase was sufficient to cause a material to remain in the vapor state after full release from shocked states, it did not have a mechanism for recondensation—i.e., for agglomeration of gas into liquid droplets, leaving voids. Hence, large gaseous expansions could lead in some cases to cold, under dense materials, thereby limiting the usefulness of running simulations to late times.
- We had moved from an age of punched card decks to desktops interfacing with machines of all sizes; we were now doing simulations in at least 1-D and 2-D on our desktop machines that had required room-sized machines at the start of my career. As analyses expanded, I gradually accumulated a library of material model parameters and configurations in CTH. As an analogy, it was like the old punched card decks we used when I first arrived at the Labs; instead of copying the cards, though, now I could extract a configuration nearest to the one of current interest, modify the materials, adjust the geometry, and proceed with the simulation and analysis of the results. It was quite efficient for the walk-in work that I regularly received. Many of those requests relied on the existence of a reliable analysis tool for a problem, and the funds were not available to develop new models or to obtain new parameters for existing models; if I could provide the requesters with some significant insight into their puzzles, even with some approximations, it was of benefit. Those applications included small explosive experiments, shock loads into microelectromechanical (MEM) devices, missile destruct and stage separation mechanisms, linear-shaped charges, bullet penetration into targets, explosive and flyer plate configurations, overpressures in aircraft from small explosive charges, and mild-detonating fuse timing devices and shock interactions.
- By 2011, 3-D computing with CTH could accommodate an astounding billion cells, compared to what we started with: six million cells were all we could manage with CTH in the late 1980s on the Cray. We were addressing problems I never would have believed possible when I first signed on!

## Marcus D. Knudson

## (1998–Present[105])

Several events occurred in the late 1990s, as I was finishing up graduate studies in Shock Physics at the Institute for Shock Physics at Washington State University, which ultimately guided my choice to start my career at Sandia. First, Lawrence Livermore National Laboratory (LLNL) published a paper in *Science* on the landmark experiments that had recently been performed on deuterium at the Nova laser facility. Second, somewhat prompted by the success of the dynamic compression work at Nova, Don Cook, who at that time was the Director of the Pulsed Power Sciences Center at Sandia, convinced Jim Asay to evaluate the potential of performing meaningful dynamic compression experiments at the Sandia Z accelerator. Third, the Department of Energy established the Science-Based Stockpile Stewardship Program, necessitating national investment in such scientific fields as dynamic compression, which put me in the enviable position of being able to choose a career at any of the three national security laboratories. All three were attractive for their own reasons. Los Alamos National Laboratory (LANL) was probably considered the strongest "pure" shock physics effort, with its long history of a solid, gas gun-based experimental program on metals and energetic materials. LLNL was probably considered the "sexy" shock physics effort, with the emerging laser-driven shock program. Sandia was probably the most unorthodox of the three, having recently all but closed down their gas gun facility and being in the throes of trying to establish a high energy density shock physics program at the Z facility. I ultimately chose to join Sandia, to work with an incredible mentor in Jim Asay, and to have the unique opportunity to enter on the ground floor of what turned out to be an incredibly successful and rewarding endeavor. Needless to say, I have never regretted that decision.

When I first arrived at Sandia, in November 1998, the shock physics program at the Sandia Z accelerator was truly in its infancy. The team consisted of Jim Asay, Clint Hall, myself, and a few technicians. We had at our disposal a single continuous wave (CW) laser and three channels of VISAR for diagnostics. Experiments were being done in a "ride-along" mode; we had to beg and convince other experimentalists in the inertial confinement fusion (ICF) program to allow us to hang secondary hohlraums off the side of the return current cans in their wire-array experiments. In hindsight, this type of effort was doomed from the start. We know now the futility of trying to use the about 5-ns x-ray radiation burst from imploding z pinches to drive meaningful dynamic compression experiments. But back then we had to learn the hard way.

The Z accelerator was designed to perform pulse compression. Energy is stored in 36 Marx bank capacitors at the perimeter of the about 110-ft-diameter by 20-ft-tall tank of transmission oil and water. This energy is released and transmitted to the center of the accelerator via transmission lines with laser-triggered gas switches and self-breaking water switches, each designed to compress the discharge successively to a

[105] As of this writing, Marcus Knudson continues shock wave research on Z.

roughly 100-ns, 10–20 MA current pulse. The wire array is itself a final stage of pulse compression, which uses the current pulse to vaporize a cylindrical array of wires and implode them on axis, producing a roughly 5-ns burst of intense x-ray radiation. This x-ray radiation could then be used to drive large amplitude shock waves ablatively into materials of interest, in much the same way as intense laser radiation. Given the hohlraum temperatures that had been achieved at that time, we had expected to be able to attain approximately 5-Mbar pressures through x-ray ablation.

However, the z-pinch load on the Z accelerator proved to be at least one too many stages of pulse compression. We had felt that the resulting x-ray burst was insufficient as a source for dynamic compression. Rather, we desired a source capable of producing steady shock waves for durations of several tens of ns. We therefore spent nearly a year investigating the use of low-density, carbon-based foams (such as TPX®[106]) as a way of "conditioning" the radiation that reached a sample, such that the resulting radiation and foam tamper would support a significantly longer, steady shock. These efforts, while somewhat promising, ultimately proved to be inefficient and were unsuccessful. We were unable to produce any results that warranted publication.

Meanwhile, there was an interest from Rick Spielman, a pulsed power scientist who was very influential in the design of the Z accelerator, to use our VISAR capabilities to try to measure current density near the experimental load. The idea being that at small enough radii, the current density becomes large enough that appreciable conductor motion occurs during the discharge of the pulse that would be detectable using VISAR techniques. Jim Asay and Clint Hall were successful in demonstrating this type of load current measurement. As the temporal profile of the magnetic pressure mimicked that of the current pulse, the resulting electrode velocity exhibited a beautifully smooth acceleration indicative of a smooth ramp compression wave. Jim Asay was quick to identify the significance of this, and the use of pulsed power to perform a magnetically-driven isentropic (or quasi-isentropic) compression experiment—more commonly known as "ICE"—was born.

Early on it was recognized that Sandia would need to partner with one of the two weapon physics labs to accelerate the growth of this new dynamic compression capability. A relationship was struck with Art Toor, Bob Cauble, and David Reisman of LLNL. I'll assume that more detail of this partnership can be found in the recollections of Jim Asay and Clint Hall. I will only summarize that, through this partnership, significant progress was made in developing a modular, coaxial short-circuit load for performing ramp compression experiments on the Z accelerator.

The next significant breakthrough occurred during the 1999 American Physical Society meeting of the Division of Plasma Physics in Seattle, WA. Jim Asay, Clint Hall, and I were at the lobby bar of the Westin in Seattle, discussing the magnitude of the impulse these types of experiments delivered to the electrodes. Back of the envelope calculations suggested that the impulse was sufficient to launch relatively thin anode plates, roughly 1 mm in thickness, to significant velocities, well in excess

[106] Editors' note: TPX is the brand name for polymethylpentene (PMP), a thermoplastic polymer.

of 10 km/s. That evening I ran some simple hydrodynamic code simulations and the results were extremely encouraging. Upon returning from that meeting, we immediately began planning dedicated experiments on Z to test this new concept. Given that it takes roughly 3 to 4 months to design and fabricate hardware for Z experiments, the first experiments would not be performed until the spring of 2000.

In the meantime we presented the idea of magnetically accelerating flyer plates on Z at the 2000 Garwin review, an annual external review of Sandia's pulsed power program. I presented our plans on the first day of the 2-day review. I'll never forget the response of David Reisman, who was attending the meeting to present on the progress of the ICE platform the following day. He had been intrigued with the idea that we had presented, and that evening in his hotel room, he had run several MACH 2 calculations, including full magnetohydrodynamics. He came to the conclusion that it was a great idea, but unfortunately it would never work. The MACH 2 calculations indicated that no matter how thick the initial electrode, the magnetic field diffusion was so severe that the field would burn through the plate before it could reach a terminal velocity, and thus the plate would be vaporized prior to impact.

A few weeks later, we conducted the planned experiments. Given the fact that we only had a few VISAR channels at the time, we had to rely on shock breakout measurements. These were effectively reflectivity measurements in which a sudden loss of reflectivity occurred upon the arrival of the shock wave at the free surface, to obtain several transit time measurements in a wedged target to infer the shock velocity. I still remember sitting at home on the evening of that first flyer plate experiment analyzing the data, so anxious that my hands were literally shaking as I plotted the breakout time vs. distance. Lo and behold, despite the MACH 2 simulations, the data suggested that a significant portion of the plate was still at full density upon impact, implying that the diffusion rate in the simulation was too high. This marked the birth of the magnetically-accelerated flyer plate platform on Z.

Concurrently, Mike Desjarlais was beginning to explore the use of the relatively new quantum molecular dynamics (QMD) calculation capability to study equations of state and transport properties of warm dense matter and plasma states. At each time step, QMD uses quantum mechanics to solve for the electron density for a particular ion configuration and then uses classical mechanics to calculate forces on, and displacements of, the ions. This process is repeated for a long enough simulation time that steady states are achieved. Mike Desjarlais pioneered the use of the Kubo–Greenwood formulation with QMD to calculate electrical conductivity in condensed matter systems, beginning with aluminum. Using these techniques he discovered that the widely used Lee-More-Desjarlais conductivity model predicted conductivity that was about 20 % too low at temperatures just above melt. In simulations of the flyer plates, this resulted in a diffusion rate that was too fast compared to experiment. His revised conductivity model, which was fit to the QMD conductivities, in contrast was found to result in very good agreement between experiment and simulation.

Ray Lemke, who had begun approaching this problem from a simulation perspective using the magnetohydrodynamic (MHD) capability of ALEGRA, had independently identified the deficiency of the available conductivity models through his analysis of the experimental results. He worked with Mike Desjarlais to improve the conductivity models in ALEGRA and began an effort to develop a truly predictive, self-consistent, two-dimensional (2-D) simulation capability utilizing ALEGRA, which is the workhorse for experimental design and analysis to this day. Integral to this simulation capability is a circuit model for the entire vacuum section of the accelerator, with an element of the circuit being a 2-D ALEGRA simulation. This circuit can be driven by a Thevenin-equivalent voltage to represent the voltage at the insulator stack. Such a model enables one to predict the performance of an arbitrary short-circuit load accurately, while properly accounting for the effect of load deformation and the increases in inductance on the delivered current to the load. Also integral to this circuit model is the inclusion of a Z current flow loss and the eventual short that occurs at the convolute upstream of the short-circuit load.

This design capability has been used to develop a standard coaxial flyer plate load that has been used extensively over the past several years. The rectangular geometry features an offset cathode design that results in the two anode flyer plates reaching different terminal velocities, thereby enabling Hugoniot data to be obtained at two different pressure states on one Z experiment. Furthermore, the roughly 30-mm height of the flyer plate allows for multiple targets (typically up to five) to be fielded on each panel. By varying the cathode thickness and the machine charge voltage, flyer velocities ranging from about 7–32 km/s can be achieved. Later, using the stripline configuration (described below), we were able to launch an aluminum flyer plate to 46 km/s, which is in excess of 100,000 mph.

The flyer plate technique has proven to be extremely fruitful. Several problems of scientific interest have been addressed. Perhaps the most notable of these is the high-pressure response of liquid deuterium. This was the first set of experimental data to be published using the flyer plate technique, and, more significantly, these results resolved the controversy surrounding the high-pressure response of deuterium. Another notable result is the discovery that beryllium remains in the hcp (hexagonal close packed) phase at much higher pressure than previously thought. In fact, based upon the relative magnitude of the longitudinal elastic sound speed, beryllium either melts directly from the hcp phase or undergoes an hcp–bcc (body-centered cubic) transition at a pressure just below that of melt on the Hugoniot. Z flyer plate data also provide evidence for the existence of a diamond-bc8-liquid triple point along the diamond Hugoniot. Quartz has been extensively studied using Z, and it was determined that the Hugoniot response of quartz is stiffer than previously thought. This is quite significant given that quartz has become a widely used standard in multi-Mbar Hugoniot experiments. More recently, the flyer plate technique was used to demonstrate that two different equations of state for water that are commonly used in planetary modeling are far too compressible. Finally, the flyer plate technique has been used to obtain Hugoniot data on several cryogenic fluids including xenon, krypton, ethane, and krypton and ethane mixtures.

The circuit model for the vacuum section of Z developed by Ray Lemke has also been used extensively in the design of ICE loads. Through an effort led mainly by Jean-Paul Davis, a very robust method of ICE experimental design has been developed that relies on a strong synergy among experiment, theory, and simulation.

Briefly, one-dimensional (1-D) ALEGRA MHD simulations coupled to the DAKOTA optimization routine are used to determine the anode–cathode arrangement, electrode thickness, sample thickness, and time-dependent magnetic field required to produce a desired velocity profile at the rear of a sample of interest. The results of 2-D ALEGRA MHD simulations, which provide the changes in inductance and scale factor relating current density to magnetic field, are then used to convert the magnetic field profile to the desired load current pulse and, after accounting for current loss, to the desired machine current pulse. A transmission line model for the accelerator, called Bertha, is then used to identify a machine configuration (water switch settings, charge voltage, and laser trigger gas switch timings) necessary to produce the desired current. This method has been used time and again to design ramp compression experiments on Z.

Another significant advance for ramp compression experiments was the development of the stripline geometry. In the 2006 time frame, the year before the upgrade of Z, we were exploring the peak stresses that could be achieved through magnetic compression. To do this, we systematically reduced the size of the rectangular cathode stalk and the gaps between the anode and cathode at the load. The resulting load dimensions were a $9 \times 2$ mm cathode with a $17 \times 4$ mm anode box, leaving 1-mm gaps. It was quickly determined that the utility of this load was going to suffer due to the unrealizable alignment requirements of the cathode stalk within the anode box. Coaxial load geometries could be crudely thought of as two inductors in parallel. Misalignment results in differences in inductance, which in turn result in unequal distribution in power between the two panels. The unequal power distribution results in higher initial current density and therefore magnetic pressure, in the smaller anode–cathode gap side, which manifests itself as an apparent time shift between the velocity profiles.

We used the roughly year-long shutdown for the upgrade of the Z facility to evaluate the alignment requirements to enable high-precision ICE shots at multi-Mbar stresses. It became apparent early on that it would be next to impossible to achieve the required alignment fidelity at a facility such as Z. Careful analysis by Jean-Paul Davis revealed that the gap between the anode and the cathode would need to be uniform to within roughly 5 μm along the entire length of the load (about 30 mm) to ensure magnetic pressure uniformity. At that point we began to consider alternatives. Motivated by the load geometry of Veloce, the small pulser facility at Sandia, we embarked on the stripline design, which in principle can be thought of as a coaxial load with three anode panels removed. What is left is a two-panel arrangement where the cathode panel emerges through an opening in the anode plate and stands opposite the anode panel with a shorting spacer between the top of the anode and cathode panels. This arrangement allows samples to be mounted directly opposite each other, one on the anode and one on the cathode, with a single anode–cathode gap in between. This ensures that both samples experience the same magnetic field history.

The stripline geometry relieved the alignment requirements; however, it also resulted in a few technical complications that took some time to mitigate. Unlike the coaxial geometry, which fully contains the magnetic field within the enclosed anode structure, the stripline geometry allowed the magnetic field to encompass the area around the panels and samples. As a result, we had to come up with a tractable way of protecting the samples and the fiber optic cables used for the VISAR diagnostics. We protected the samples by completely covering the back of the panels with metal, effectively building a Faraday cage to exclude the magnetic field. The fiber optic problem proved to be more difficult and required several iterations over tens of shots to come up with a robust probe assembly that we refer to as the pitch-fork design. Furthermore, we learned the hard way that even though the fibers are encased in bicycle brake cable, the cathode probe cable had to be elevated off of the upper anode plate for several feet or the fibers would lose transmission over several hundred nanoseconds during the current pulse.

A second complication related to the unconstrained magnetic field is a current density gradient and thus pressure gradient, up the length of the stripline panels. Dave Seidel and others assisted us with EMPHASIS calculations of the load geometry to identify and mitigate this problem. The main source of the gradient was the transition from the radial feed of the magnetically insulated transmission lines to the vertical feed at the load. Since the magnetic field is unconstrained in this region, a significant current flow is induced on the sides and back of the cathode panel, with a current density that becomes lower along the length of the stripline. This aspect of the stripline geometry was mitigated by adopting tapered panels that widen near the short, with the exact taper identified through EMPHASIS calculations. The tapering has been shown through experiment to reduce the gradients significantly in the stripline design.

Both the coaxial and stripline geometries have been used to obtain valuable ramp compression and material strength data on a variety of materials. Perhaps the most notable ramp compression data have been performed on tantalum. During the same year in which Z was recommissioned, we were tasked with a level-1 milestone to perform a ramp compression experiment on tantalum to 4 Mbars. In a relatively short amount of time, we were able to reestablish short-circuit experiments on Z, commence experiments using the stripline geometry, resolve the diagnostic issues associated with the stripline, and establish the Bertha model to a high-enough fidelity to generate shockless compression of tantalum to 4 Mbars. Another notable study included a fairly complete investigation of the strength of LiF, a very common window material in dynamic material properties experiments, under ramp compression to 1 Mbar. A similar study of the strength of Be under ramp compression was performed to a similar stress level. High-stress ramp compression experiments have also been performed on aluminum, copper, gold, lead, and molybdenum.

In addition to the load development, other improvements have been made along the way to advance the magnetic compression capabilities on Z. The most notable of these improvements was the capability to perform high-fidelity pulse shaping. Initial experiments utilized what we now refer to as a synchronous long pulse; here, all 36 gas switches are fired synchronously (in fact, in the early days, a single laser split 36 ways triggered all of the lines simultaneously) and the water switches are directly shorted. In this configuration the current pulse essentially increases linearly in time over about 200–300 ns. For low-stress experiments, this is acceptable.

However, for higher-stress experiments, as well as high-velocity flyer plate launches, this linearly increasing current pulse results in shock formation at a relatively short propagation distance. Given that for most materials, the sound speed changes most rapidly at low stresses, a current pulse is required that will mitigate shock formation and maximize the propagation distance of the ramp compression wave. This was accomplished, in principle, by replacing the single laser trigger source with 36 individual lasers. In practice, significant improvement was also required to Bertha, the transmission line model used to predict machine performance. Jean-Paul Davis, Pat Corcoran, and others have done a tremendous job in continually improving the Bertha model and extending the pulse shaping extremes to very long and complex pulses. This work has been critical in the advancement of dynamic material property capabilities on Z.

Significant improvement in the diagnostics was also required to maximize the potential of Z for these types of studies. In particular, given that each Z experiment, whether ICE or flyer plate, could utilize up to 20 or more samples, it was critical that the number of VISAR channels be significantly increased. Relatively early on, within the first 2 years, we were able to increase the Sandia-owned VISAR channels from 3 to 14 by converting our two Valyn VISAR systems to multipoint VISARs. To increase the capability further, Jim Asay was able to leverage aboveground experimental capabilities by borrowing VISAR systems intended for Nevada operations. In this way, an additional 14 channels of VISAR and at least one more laser system would be transported to Sandia to support Z experiments during the downtime between subcritical experiments. Later, Jim was able to secure Nevada monies that were allotted to support experiments to fund members of the Special Technologies Laboratories (STL) in Santa Barbara, CA in developing and fabricating new-generation, multipoint VISAR systems and extremely fast photodiode detectors. Key contributors to this work were Bruce Marshall, Gerry Stevens, and Terry Davies, among others. This group was able to outfit Sandia with a state-of-the-art 38-channel, sub-200 ps-resolution VISAR system that is still in use today.

A second critical advance in the area of diagnostics was the repurposing of a short-pulse (5–10 ns) YAG laser that had been used for holography experiments at the STAR facility. As the number of VISAR channels grew, the CW laser became ineffective. It came to my attention that there was an unused Continuum-manufactured YAG laser at the STAR facility. We contacted Continuum to see if it was possible to convert this laser to a custom, long-pulse (5–10 μs) laser as a light source for velocimetry. Continuum was very gracious and offered to repurpose the system for about $80 k. The system, while it had its shortcomings (i.e., poor stability, low beam quality, etc.), increased our radiance at the sample by orders of magnitude and turned out to be necessary to provide sufficient light to enable us to use the fast photodiode detectors that had been designed by STL. We have since replaced the repurposed laser with a much more stable, higher beam quality Continuum system that has been our workhorse for the past several years.

Today we continue to push the envelope and develop new and exciting platforms to enable a wider variety of experiments in dynamic material properties. Particular areas currently in development that I am excited about are being enabled through more extreme pulse shaping. I'll briefly mention two here. First, an ability to produce a double-humped current pulse is paving the way for a shock-ramp platform that will open up the equation of state surface between the principal Hugoniot and the principal isentrope. In particular, one can now begin to design experiments in which the material is first shock melted and then ramp compressed back into the stability field of the solid, to investigate rapid resolidification. Second, we have been developing techniques using electroplating and diamond turning technologies to produce multi-layered flyer plates that can be used to perform double shock and shock and release experiments.

Looking back, it is staggering how much progress was made over the past decade. This progress was realized through the integration of various groups with expertise in shock wave physics, pulsed power, high energy density theory, computation and simulation, and optics and diagnostics. In addition to those I have specifically identified in these recollections, I would also like to acknowledge the following people who have played a key part in the advancement of material dynamics experiments on Z: hardware designers Chris Russell, Andy Maurer, Dustin Romero, and Devon Dalton; diagnosticians Randy Hickman, Jason Podsednik, Bruce McWaters, Kevin Youngman, Charlie Meyers, Devon Dalton, and Anthony Romero; scientists Dave Hanson, Jim Bailey, Greg Dunham, Chris Deeney, Dan Dolan, Seth Root, Chris Seagle, Justin Brown, Scott Alexander, Heath Hanshaw, Matt Martin, Becky Coats, and Mark Savage; and managers Chris Deeney, Tom Mehlhorn, Dawn Flicker, Thomas Mattsson, and Mark Herrmann. These are exciting times, and I look forward to the advances that will be realized in the years to come.

## Carl H. Konrad

## (1969–1998[107])

I was hired by Darrell Munson in 1969. Darrell, the Division Supervisor (i.e., level-1 manager), was starting a new gun facility to replace the explosive site at Building 9950 in Area Y. My first assignment was to report to Ray Reed at Building 9956 in Area Y. I was tasked to wire the instrumentation racks in the 9956 control room to the impact chamber, which would be located in the gun bay. I believe I ran 48 rigid 50 Ω (i.e., 50-Ω) cables to the racks and kept each cable in a given rack to the same length within less than one inch. At that time, the gun had not been delivered, and 9956 consisted of a 100-ft-long (30.5-m-long) gun bay and an attached bunker-type control room. Since 9950 was still certified as a 50-lb (22.7-kg) high-explosive site, a substantial bunker-style room was required for personnel protection. This distinction proved important later in the life of the facility.

Ray Reed was the staff member in charge of the startup of the gun, and Bob May was the staff member assigned as the facility supervisor for day-to-day operations. We were all assigned to the Shock Wave Phenomena Division, 5163. The powder gun was on order from General Motors (GM), located in Goleta, CA. The gun was delivered around the August to September time frame. The first shot (#50001) was conducted on September 19, 1969 and achieved a velocity of about 1 km/s. I believe GM was the only company making research guns of this type at that time. Both Los Alamos National Laboratory (LANL) and Lawrence Livermore National Laboratory (LLNL) had two-stage light gas guns that were fabricated at the GM Goleta facility. The Sandia powder gun filled a need for shock wave data in the velocity range from <1 to 2.2 km/s, which was difficult to achieve with two-stage light gas guns and was able to provide data on much larger targets.

The powder gun was also designed as a "convertible" gun. Incorporated into the initial rail design was the ability to convert the gun to a two-stage light gas gun as phase II of the facility plan. The thinking was that we would operate in the powder gun mode for 6 months and then convert it to the two-stage mode for the remaining 6 months. We worked in this mode for many years until we procured a second rail to support the powder gun in the same building. This made the gun bay very tight with the two guns side by side.

I was assigned to Area I before the new gun was delivered. I worked out of Building 805 at first and then moved to Building 855, which housed two gas guns

[107] Carl spent his career at Sandia in the shock wave field except for a short stint in the mid 1990s when he was in a department involved with environmental safety and health issues. Late in his Sandia career, Carl was asked to represent Sandia in the development of plans for the proposed Joint Actinide Shock Physics Experimental Research (JASPER) facility at the Nevada Test Site. The plan was to install a two-stage light gas gun at JASPER for Pu experiments. The JASPER plan served as an impetus to reactivate the STAR two-stage gun to obtain data on different velocity measurement techniques for JASPER. Bechtel ended up offering Carl a job to run the JASPER facility after he retired from Sandia in 1998; he accepted and moved to Nevada in March 2000.

owned by Barry Butcher's group. I supported Don Lundergan with his gas gun experiments. We worked on several projects, but the one I remember most was a target that consisted of thin stainless steel plates interleaved with EPON 828 Z® hardener. The main instrumentation was a velocity interferometer designed by Lynn Barker and Red Hollenbach. It consisted of a single frequency Spectra-Physics 119 laser. The laser had very low power that required the use of spectral mirrors on the target. Red Hollenbach designed and built large amplification pulsed photomultiplier (PM) tubes. The PM tubes drove 125 Ω cables connected directly to the plates of Tektronix 519 scopes. We had six or seven scopes and split them to monitor the two interference beams from the velocity interferometer. The scopes were connected in series, with one scope looking at the overall expected signal and the rest set to a faster sweep speed that would look at different segments of the data. Timing was critical and a precise fiducial was added to each scope. These experiments were conducted on the 125-Ω gun, named because of the impedance of the data lines. A second gas gun, the 50-Ω gun, was also located in Building 855. These guns had been used to perform shock wave studies on toxic beryllium at an earlier time. The catch tank area was a sealed bunker, and when the beryllium tests were completed, the bunker was cleaned and then painted with epoxy paint. The technicians at the 855 gas gun facility included Red Hollenbach (who was a TSA, aka a Technical Staff Associate), Kent Bowen, John Erni, Cliff Witten, Larry Kent, and myself.

I returned to the 9950 site around February 1971 as the electronics instrumentation technician. Bob May was then the engineer in charge of 9950. Bob Besecker was a mechanical technician in charge of the gun. He was a previous employee of GM and had worked on the GM guns at Goleta. Sandia hired Bob when they bought the gun because of his experience in this area. He was quite a character and at one time wanted to "ride the launch tube" during a maximum-charge powder gun shot. He held up the shot for about a half hour because I wouldn't fire the gun with him sitting on it. Dick Lingle was the electronics (instrumentation) technician. Dick also came from GM and was promoted to a TSA while at the STAR facility. I took Dick's place when I moved to STAR and he moved into Area I. Bob Hardy was hired in December 1969 and was also a mechanical technician.

## My First STAR Experiment

My first experiment on the powder gun was in support of Darrell Munson. Darrell was using a low-impedance Manganin® gauge as the primary instrumentation on the shot. I had never set up the instrumentation for this type of experiment, and Darrell handed me his shot folder and retired to a desk to do some work with the comment: "Give me a call when you are ready to shoot." I'm sure I spent longer than necessary to set up the experiment, but we got good data on all scopes and the correct velocity. From that time on, Darrell just sent out the shot folder for his experiments and did not personally come out for his shots.

During this experimental phase, we were doing equation of state shots using contact pins on the sample free surface to determine displacement after shock arrival and quartz gauge-type shots to directly measure the time-resolved stress history induced by the shock wave. Also we were using the abovementioned Manganin® gauges, both 50 Ω ones using Wheatstone-bridge-type instrumentation and low-impedance gauges of <1 Ω, which used a 500-V pulsed power supply and differential amplifiers in the scopes. The VISAR was introduced at STAR in the early to mid 1970s.

## Powder Gun Ballistic Impact Studies

**A brief history of powder gun studies.** During a period prior to the delivery of the two-stage gun, I believe staff members in our organization were not doing any experiments at STAR. Bob Hardy and I did not have any work to do or experiments to conduct. Darrell Munson, our Division Supervisor, called and asked me to contact a staff member named Scott (i.e., "Scottie"). Scottie was assigned to the Nevada Test Site (NTS) and had been tasked with designing a debris door for the underground tests. An underground test had been conducted that did not provide containment; hence, the debris door needed to be redesigned. What evolved was a series of experiments that kept Bob and me busy for the good part of a year. It was a good match; we needed something to do, and Scottie got complete use of our time, plus the full use of the STAR facility and machine shop. We had to develop a method to launch a 2.54-cm-diameter steel ball bearing to 2 km/s into the target Scottie provided. We mainly tested different geometric shapes. We had to modify the tank to accept x rays at the rear of the tank to record the dynamic impacts, measure the velocity, stop all projectile parts except the ball bearing, and provide the target for postmortem evaluation. That is the history, now the story.

We had completed the evaluation experiments and Scottie had developed the door design he believed would stop the ball bearing. Bob and I went out to help Scottie unload the targets from his vehicle in the parking lot at 9950A. If I remember correctly, the targets consisted of two thin 7075 aluminum plates separated by about 13–15 cm of steel-truss-grid honeycomb. I looked at Bob and he looked at me, and we told Scottie we would shoot through his design like a knife through hot butter. Obviously a discussion ensued, and we could not convince Scottie that his design would not work. Bob had access to a small-bore high velocity rifle. He left the discussion, got the rifle and loaded it with several bullets, and walked over to one of the targets we had set out in the parking lot. Scottie and I watched as he injected a bullet into the rifle and shot one of the targets. He blew a hole completely through it and Scottie's only comment was "That's not possible," at which point Bob loaded another shell and blew a second hole through the same target. That stopped the discussion as to whether the door design would work and changed it to "What was the velocity of the bullet?" and "Can you shoot steel bullets?"

Since Scottie was under a strict deadline, we did the following: (1) I developed and build a chronograph to measure the rifle's velocity, and (2) Bob took several bullets and mounted them in the mill and used a ball-end mill to machine a cavity into the lead bullet. Bob fitted a steel ball bearing slightly smaller than the rifle bore diameter into this cavity and reformed the lead around the ball bearing to hold it in place during acceleration. We then set up the chronograph and Scottie's target in the powder gun tank and started shooting. Scottie took the data and came back with a new door configuration with an added thin stainless steel plate in the middle of the honeycomb, and we were not able to penetrate it with the rifle. We tested this configuration on the powder gun and the door segment was able to stop the 2.54-cm ball bearing at 2 km/s. The back plate was distorted, but it did not fail. I assume this door design ended up in Nevada and was used on underground tests.

Also, in December 1971, we bought the two-stage conversion and had the ability to configure the facility as either a powder gun or a two-stage light gas gun. The first two-stage shot was conducted on January 21, 1972. Below is a table of the number and type of experiments conducted from 1973 to 1978.

Number of experiments at the STAR facility

| Year | 1973 | 1974 | 1975 | 1976 | 1977 | 1978[a] |
|---|---|---|---|---|---|---|
| Powder gun | 40 | 56 | 48 | 108 | 58 | 6 |
| Two-stage gun | 52 | 77 | 78 | 23 | 37 | 0 |
| Total | 92 | 133 | 126 | 131 | 95 | 6 |

[a]Accident occurred on November 1, 1977

We developed an evacuated ballistics range, which severely strained the gun bay of 9956. The final impact chamber was located outside the gun bay, and we had to cut a large half circle out of both south doors to close them around the range tubes. In addition, the impact chamber was shorter than what we needed and limited what we could accomplish. Developing the capability to conduct free-flight experiments required new instrumentation, sabot[108] development, and new shooting conditions. We built a ballistics range, which included two large range tanks for possible instrumentation locations and sabot stripping mechanisms, plus a range termination tank to house the target and the VISAR infrastructure as well as the interconnecting small range sections and vacuum interfaces. The addition of this range required the first expansion of the 9956 gun bay. An additional 100 ft (30.5 m) was added to the south end of the existing gun bay, thereby doubling its length. The width was also increased with the addition of a lean-to on the east side for use as a machine shop and to house the equipment used to hone the inner bore of the gun.

Our first series of experiments with this new ballistics range simulated a rain drop impacting a graphite target. Of the several problems we encountered in launching the small-diameter (2–6 mm) low-density polyethylene spheres, the most difficult was developing a sabot that would separate (or "strip") from the sphere without affecting the sphere's flight. Several new diagnostics also had to be developed. We overcame the following challenges in conducting these experiments:

1. developed a technique to measure the velocity of the sabot close to the muzzle,
2. used flash x rays to measure the projectile velocity in flight and provided a diagnostic for sabot development,
3. developed methods to detect the simulated raindrop and synchronize instrumentation prior to impact,

[108] Editors' note: A sabot is a device to ensure the correct positioning of a bullet in the barrel of a gun.

4. used alignment techniques to predict the impact location to within 1.5-sphere diameters,
5. adapted the VISAR to the ballistics target and developed submicrosecond triggering techniques,
6. developed tooling and techniques to manufacture precision sabots and machine precise cavities to allow the sphere to be captured and held during acceleration, but be released without affecting its free-flight trajectory, and
7. developed procedures to assemble the sabots and align the range component parts and target.

### A Nontypical Experiment or "Is the Plug Out?!"

We aligned all the range component parts using a technique of double focusing an alignment telescope by inserting a precision plug of various lengths into the muzzle end of the launch tube. The plug had a first surface mirror on the muzzle end. We would align a Keuffel & Esser Co. (aka K&E) telescope by first focusing on the center of the mirror, then projecting a grid on the surface of the telescope, and doubling the focal length of the telescope to focus on the projected grid. Using this method, we would align the telescope axis to the last few diameters of the launch tube, which we believed would determine the flight path of the sabot. On one shot after everything was aligned and we were ready to install the aligned target using the same telescope setting, I asked one of our contractors to get me the portable scope. I meant the oscilloscope located at the breech of the gun. He did not understand and brought me the K&E telescope instead. This, of course, ruined all the alignment of the range components. The range had to be opened up and the alignment process had to be done again. We always used a checklist to complete this process, and for this experiment, it had been filled out already. Upon restarting the alignment process, I did not start a new checklist, and after aligning the telescope again, I installed the target and closed up the range. I DID NOT REMOVE THE INVAR ALIGNMENT PLUG FROM THE LAUNCH TUBE MUZZLE, AND WE FIRED THE GUN. The first thing we noticed was that we did not get any velocity data, the x rays did not fire, and we did not get any triggers or VISAR data. We did get to cut about 15 cm off the muzzle of the launch tube to repair it.

During this time, a second gun rail was built to support the powder gun, which allowed us to have both guns operational at the same time. The two-stage gun was on the west side of the building and the powder gun was close to the center line of the building.

We had a major accident with the powder gun on November 1, 1977, that destroyed Building 9956. We were shut down for about 1 year. I believe the accident cost about $100,000 to repair 9956; the cost of the accident investigation was four or five times greater. The guns were not damaged, very little damage occurred to the instruments, and personnel were safe in the reinforced control room. (The control room was originally designed as an explosive bunker because the site was still an explosive site when 9956 was built.) This was a very difficult time and several things came out of the investigation that severely hampered our future operations. In no order of priority, the following two decisions were enacted:

1. Firing conditions for the two-stage light gas gun were frozen at the level prior to the accident, even though the gun was not involved in the accident. This decision meant we could not vary the piston mass, the maximum powder charge was determined by past experiments and limited the maximum gun velocity, and the hydrogen gas pressure was locked at one pressure, again limiting the gun operation. All this meant that the gun was not as versatile as it was designed to be.
2. The powder gun was similarly constrained, but we had considerably more data on that gun, so the constraints were not as much of a problem. We were required to develop a new primer for our powder charges. Of all the changes that were required, this was the most important. The original primer was a low-energy, black powder design. We were required to develop a high-energy, non-black powder primer, which improved safety.

## Going Back into Operation

The paperwork increased significantly, and all our procedures had to be formalized. To actually get the powder gun back into operation, it was decided that a pressure safety advisor would work with us to identify all possible problems. We spent months fixing trivial problems that he came up with, and when I finally challenged him, he told me, "I am too close to retirement to take the chance of letting you shoot this gun again." I sat down that afternoon and wrote a memo to Walt Herrmann, my level-2 manager, and copied John Galt (my Vice President) and personally guaranteed that the gun was operational, safe, and ready to start conducting experiments again. That memo was accepted, and we started experimental operations again. I think it is fair to say that the pressure safety engineer was very unhappy and would not sign my memo.

## Construction

The reconstruction following the accident actually started the expansion of the STAR facility into a world-class operation. In quick succession we added a new gun bay east of the original control room for the single-stage gas gun(s). A new powder gun bay, a breezeway, control room, optics clean room, target fabrication lab, dark room, and restroom were added. The TBF (Terminal Ballistics Facility) gun was added in the early 1980s. That gun was purchased from Hal Swift's company, Physics Applications. Ron Moody was in charge of getting the TBF gun operating. We made several modifications to the gun, but it never seemed to work up to its expected potential. I believe that the gun was not being operated with the required amount of gas in its pump tube. When we were able to verify that the pump tube was charged with the correct gas and pressure, the gun did, in fact, perform as expected.

### STARFIRE Project

The railgun project, STARFIRE, used a new Physics Applications two-stage light gas gun to inject a projectile at about 6.3 km/s into a railgun. A 1-MJ capacitor bank was used to drive the railgun. This project was part of a Star Wars project that began in the late 1980s. It was probably the most interesting project I worked on during my years at STAR. The STARFIRE team consisted of Ron Hawke as the lead scientist and Allan Susoeff as the technician in charge of boring the railgun to the correct diameter; both were from LLNL. At Sandia, Jim Asay was the overall project manager, I was the facility supervisor and worked on instrumentation and gun operations, and Clint Hall was the mechanical engineer for the projectile design. Gerry Sauve of EG&G worked on instrumentation and data collection and Randy Hickman of Ktech was the gun technician.

We were trying to increase the maximum velocity achievable to inject a projectile into the railgun at a velocity of at least 5 km/s. We injected the projectile at 6.3 km/s. We found that we had to develop a shooting condition that would achieve the correct injection velocity and not destroy the railgun insulators. Striking the correct arc to fire the railgun was like striking a match in a hurricane. Toward the end of the project, we achieved a measurable boost in the velocity of the projectile from the railgun.

### Misplaced Equipment

The technicians who worked at STAR were a very close knit group. We also enjoyed playing pranks on each other. One of the better pranks was played on Jeff Miller. Every year Jeff took a couple of weeks off and went back to Tennessee. One year he left his car parked at STAR to keep it safe while he was gone and so he would not have to pay for parking at the airport. We had an empty transportainer on the site, and Jeff's car would just fit into it. By just fit, I mean there was less than 12 inches of clearance on either side. We picked up Jeff's car with the forklift and carefully fitted it into the transportainer, took a photo of it, and closed the door. We placed the photo on Jeff's desk in plain sight and forgot about it. Jeff returned over a weekend and came out to the site to pick up his car. He could not find it, and he did not notice the photo of the car in the transportainer with all of us standing around it. He called Sandia security to see if they had it towed away and he may have filed a stolen car report. Monday morning came, we finally told Jeff to look on his desk where the photo clearly showed where his car was. We had a great laugh, although Jeff did not see the humor in it. We helped him get it out of the transportainer.

### STAR Facility Closure[109]

**Black Monday.** I think we all understood we were in trouble as our external customers were not extending their programs, but I was still caught off guard when Ed Barsis pulled the plug. Black Monday, December 12, 1994, is a chapter at STAR

[109] Editors' note: For additional details about the closure of STAR, see Jim Asay's recollection.

that I prefer not to dwell on. As the guys left the site, I remember I called Dave Cox to ask him if he had applied for the position to operate the 808 gun. He was not aware of the opening and applied on the last day they were accepting applications and got the job. In the end Clint Hall, Bill Reinhart, and I were the last three at the site and Lalit Chhabildas was the last staff member. We were trying to get in as many shots as we could before the final day. Considering there were only the three of us to do everything—target preparation, setup, data acquisition, cleaning and turning around the guns—I believe we accomplished a great deal that last month. We still had shots pending for Dennis Grady, but just ran out of time. Ed Barsis (the Director in charge of the facility) did not feel we were shutting down fast enough and sent Anita Vasey to the site to "assist" us. We were shooting at the time, and Anita started in our library and threw out many important papers and documents before I could stop her. We lost a lot of our history during this period. I may be wrong, but I always felt that Barsis wanted our organization to be in his group simply because he wanted our budget for his "Tera Flop" computer program. The STAR facility did not seem to me to be a good match with his organization. We were okay budget wise until our outside funds stopped, and we could not get our Sandia research budget back from Barsis.

**Safeguarding valuable STAR equipment.** I was not a part of this process, but as I remember it, Bill Reinhart moved all the important instrumentation and equipment from the site into a transportainer before the site was opened to the public for pillage and plunder. He had the transportainer moved to a parking lot outside Barsis' office and had Barsis billed for the move and the storage costs. When the site was reopened, STAR had the instrumentation necessary to begin operations again. *Thanks Bill!*

**Leaving STAR.** I left STAR before either Bill or Clint. I found a job with the Robotics Center (Center 9600). I was assigned to Org. 9601 under Linda Benavides as the ES&H (Environment, Safety, and Health) coordinator. My main job was as an interface between the Plant Engineering Department and the Center. Although the job was not technical in nature, it was very challenging. My job did not require a large ES&H effort, but the interface with Plant Engineering was very time consuming and intense. We were constructing Building 895, the Robotic Manufacturing Science and Engineering Laboratory (aka RMSEL) to house the majority of Center 9600. My main effort was to make sure every lab and office was configured to the user's requirements and versatile enough for future requirements. Since the building was to be a showcase for robotics research at Sandia, I required that Plant Engineering provide top-of-the-line equipment for the offices and labs. Sandia had a warehouse full of the old-style green telephones that they wanted to install in all offices and labs. Also, the office and lab chairs would be the gray generic nonadjustable style. I felt this equipment was inadequate and did the following.

To solve the chair problem, I contacted the Ergonomics Department and identified three chairs that were ergonomically adjustable and the same cost as the gray generic office chair. I set up a schedule and appointments for 9600 personnel to be fitted with the upgraded chairs. As for the telephones, I went out and purchased a rather expensive phone from Best Buy and delivered it to Plant Engineering and told them this is what I wanted installed in RMSEL. I knew Plant Engineering would not agree to the phone I had provided, but it did get them off the idea of the old green phones. They provided several phones from which to choose that were much less expensive than the one I had bought, and we ended up with a good, attractive functional system. It was my responsibility to ensure that every office and lab was completed and that the phone and computer hookups were active when the 9600 personnel were scheduled to move in. I believe I was 100 % successful in this effort.

After our personnel were moved in and the labs were set up, I set up a program for every lab to begin operation. ES&H was a "dirty" word to the staff members and technicians in the 1996 to 1997 time frame. Basically, it meant more and more paperwork for the staff members that took away from their primary work. I do not think I exaggerate when I say that I was the most unpopular person in 9600. I set the following program for every lab in RMSEL. I had the lab owners write their standard operating procedure for laboratory operations. This was their expertise, and they were the most qualified to determine those requirements, so they did not have a problem with that task. I personally wrote everything else to bring the labs into compliance with the ES&H regulations. This worked well and had two positive effects. I learned the operations of all the labs, important to me because I was also responsible for safety. It also made me much more popular to the 9600 personnel as they were not burdened with writing nontechnical documents.

Finally, and most importantly, I found out by accident that the Department of Energy (DOE) required that all new or modified labs had to be approved by DOE prior to beginning operations. Sandia had a department set up to help with this process, but they never contacted us. I identified the problem just prior to the start of the move. Several new labs in other Sandia buildings had been working for 3 years to get approval, and Sandia was being fined by DOE because those labs began operating without the DOE approval. I was given the task of getting this approval as soon as possible. Upon discussion with the Sandia department assisting with this process, I was informed that it was impossible to complete in my given time frame and probably would take years. They provided me with a ream of required boiler plate on top of the final paperwork that I would have to write for each laboratory. The process I had put together skipped this department and the DOE inspection requirement. My process stated that when I had a minimum of three labs ready to begin operating, I would schedule a walkthrough inspection by the following personnel: a subject matter expert, who was the department manager who owned the laboratory; a representative from Safety, Health Physics, ES&H, and Pressure Safety; a DOE observer if they wanted to attend; and someone from the Sandia department that had been set up to help with this process and any other organization that might be required to inspect laboratory operations. The department manager who owned the laboratory would approve the start of laboratory operations when all findings identified during the inspection had been corrected. This process bypassed the DOE signoff requirement, and all our laboratories were approved and placed in operation in a minimum period of time. I believe this process was the reason I received a promotion to Senior Member of Technical Staff.

Toward the end of my time at RMSEL, I was asked to assist with a LANL subcritical experiment at NTS as a subject matter expert on the VISAR. This was a very welcome relief from my ES&H duties since RMSEL was now in operation and my job was actually heading into the ES&H area. I was saved at this time by a request from Jim Asay to join his department to transfer technology developed at STAR to the Z machine in Area IV.

**Returning to STAR.** I joined the shock wave physics department that Jim Asay was forming in Area IV and was assigned to adapt the VISAR technology, which had been developed at STAR, to Z. If successful, this would give us a useful research tool on Z other than just providing assistance with the initial diagnosis of Z's performance. We conducted a proof-of-principle demonstration of the VISAR diagnostic on Z, and I then returned to STAR.

**JASPER.** I was asked to represent Sandia in the development of plans for the proposed Joint Actinide Shock Physics Experimental Research (JASPER) facility, to be located at NTS. The plan was to install a two-stage light gas gun at JASPER for Pu experiments. JASPER also needed data on different velocity measurement techniques. This gave us the justification to bring the STAR two-stage gun back on line at Sandia. We tested three in-bore velocity measurement techniques—MAgnetic Velocity Induction System (MAVIS), Optical Beam Reflectance (OBR), and flash x rays—in a slotted barrel extension and reported our findings to the JASPER team. The MAVIS and OBR techniques were rejected by the JASPER team, but are now standard techniques on the STAR two-stage guns. Bechtel offered me a job to run the JASPER facility and I accepted and moved to Nevada in March 2000.

# R. Jeffery Lawrence
# (1967–2005)
# Shock Physics: The Other Half

## Foreword

These notes are a fleshing out of an outline emphasizing the history of shock physics at Sandia, from its beginning in the late 1950s to the current time. The impetus for the outline was provided by Bob Graham's write-up of one side of that story (see below). As with all narratives of this type, the following is colored by my personal perspective, which is unique to my own experiences and first involved a 4-year tour with the Air Force and then nearly 40 years at Sandia. Only briefly mentioned are areas in which I was tangentially involved. Furthermore, my memory might be faulty or at least somewhat distorted. I was only peripherally involved in experimental shock physics, so the many important developments and advances in that arena will be left to others who are vastly more qualified than I.

## Very Early Background

Shock physics, in its modern form, evolved from early work at Los Alamos as part of the Manhattan Project. That work was an important part of the development of nuclear weapons. In follow-on efforts, but in some sense in parallel directions, scientists within the Department of Defense (DOD) and the national laboratories needed to study and understand these new weapons and their effects on military and civilian facilities and systems. These effects led to an increasing awareness of the need for research and development in shock physics, but in parameter spaces somewhat different from those directly connected with "bomb" physics.

The national laboratories and Sandia, in particular, already had efforts underway in several related fields. These were in areas needed for Sandia's core mission of "weaponization," or adding all the required "bells and whistles" to the new nuclear weapons in the development queue. At this time, the late 1950s and early to mid 1960s, Sandia had two different shock physics groups. The first is well described in a 2007 memo by Bob Graham on the occasion of the George Samara Memorial Symposium to commemorate George's life.[110] The research in George Samara's group concentrated on obtaining a fundamental understanding of the properties and behavior of materials under extreme dynamic loading conditions of pressure and temperature. The second group was directed more toward applied research—in particular, developing in-house experimental tools and facilities and parallel analysis capabilities to investigate questions about the dynamic response of weapon components and the materials within them. As time progressed, the data obtained from these efforts assisted in improving the design and subsequent implementation of many weapon components and related systems.

[110] Bob Graham, private communication, 2007.

## Early Contacts Between Sandia and the Air Force Weapons Laboratory

In the early 1960s, DOD made a great push to study and understand the detailed effects of nuclear weapons on military systems. The main group driving this effort was probably the Defense Atomic Support Agency (DASA), which was subsequently called the Defense Nuclear Agency (DNA) and is now the Defense Threat Reduction Agency (DTRA). DASA studied the key areas of vulnerability, survivability, and similar areas from the somewhat different perspective of lethality. After the cessation of atmospheric and high-altitude nuclear tests in the early 1960s, underground tests (UGTs) became the main source of these data. UGTs were generally of two types: weapon development tests, mostly run by Los Alamos and Livermore, and weapon effects tests, mostly under the auspices of DOD/DASA. There was some overlap, with some effects experiments conducted on development shots, and vice versa.

The Air Force Weapons Laboratory (AFWL) was the focus of much of the weapon effects test preparation, design, and execution, as well as post-test analysis. With the proximity of Sandia Laboratories on Kirtland Air Force Base (called Sandia Base at the time), along with Sandia's involvement in most phases of the UGT program, it was natural to establish a close working relationship. The main players at Sandia were Carter Broyles and his group, who supported many UGT activities at the Nevada Test Site. On the other side of the base at AFWL, Captain Donald Lamberson headed a group of young Air Force officers who did most of the pre-test and post-test analysis of one class of experiments—one-dimensional (1-D), pulsed x-ray-driven, and shock physics tests. The experiments provided active, time-dependent, shock wave profile measurements of 1-D coupons for basic and special materials and composites of great interest to the Air Force. The experiments used, for the first time, Sandia-developed quartz dynamic pressure gauges. The objectives of these and related experiments were to obtain actual dynamic and late-time data on both materials and systems to establish the vulnerability and lethality levels of missiles and weapon systems in the United States and elsewhere. Validation of theoretical calculations of dynamic material and system responses was also an important driver. Additional goals were to evaluate early concepts for hardening military systems against nuclear weapon effects. Another aim was to evaluate new active and passive experimental techniques, especially under the extreme conditions of the UGT environment. We learned early on, because of uncertainties associated with UGT environments, that if those tests were repeated at all, it should be three times—never two times!

During the mid 1960s, there was also broader interest in the vulnerability and survivability of missile systems, reentry vehicles, and even satellites. The principal threat was exposure to exoatmospheric nuclear detonations and specifically to the unrestricted output of about 3/4 of their total energy, in the form of low-energy pulsed x-rays. At the time, the Sandia group headed by Al Chabai (Dennis Hayes was one of the members!) was beginning to discuss many of these issues with the folks at AFWL. It was clear from these interactions that the tools to investigate these issues were mostly shock physics oriented, either directly related to shock generation and propagation or to the structural response from impulsive loads produced by pulsed radiation exposure. In fact, Chabai et al. pointed the way to the shock physics work underway at Sandia, which was then headed by Don Lundergan, Barry Butcher, and their colleagues.

### Hydrocodes

In those days, the main tools for experimental design and post-test analysis were the early shock physics hydrodynamic codes (aka "hydrocodes") that had evolved, however indirectly, from "bomb physics" codes at Los Alamos and Livermore. At the time, AFWL was using the PUFF code, which was being developed under the mentorship of a number of external folks. Two that come to mind were Olen Nance, a regular visitor from the RAND Corporation, and Mark Wilkins from Lawrence Livermore Laboratory.[111] Wilkins was instrumental in incorporating material strength into PUFF, and this was probably the first time strength effects were incorporated in any hydrocode used extensively to study weapon effects. That capability was needed because many UGT shock physics samples yielded stress pulses of relatively low amplitudes where these additional phenomena had a significant influence on the attenuation of the pulses propagating through the samples and gauges.

### Hydrocodes and Material Response

Because material response necessitated an extensive material property database, many labs were conducting experiments to obtain shock Hugoniot data on many different materials. Various compendia of these data were becoming available in journal review articles (e.g., describing early Los Alamos Scientific Laboratory[112] work) and in reports from all the national laboratories and their contractors. Many informal compilations were also put together by groups working in the field.

### Analytic Models

A number of significant developments occurred during this period. Analytic models for the dynamic impulse from pulsed radiation loads—for example, the original BBAY models—were developed under the AFWL-run RADS Program at AVCO Corporation, near Boston. The rationale was to find a simple approach for predicting impulse generation that did not require solving partial differential equations and was based on the conservation of energy and momentum. This particular description evolved from the work of Jack Whitener at the RAND Corporation in the 1950s. The BBAY acronym stands for the individuals involved in the effort: Hans Bethe

[111] Editors' note: Until 1971, our sister laboratory in Livermore was known as the University of California Radiation Laboratory.

[112] Editors' note: During this time period, our sister laboratory to the north was known as the Los Alamos Scientific Laboratory. Later, in 1979, its name would become Los Alamos National Laboratory.

(under a consulting contract to AVCO from Cornell), William Bade, John Averell, and Jerrold Yos. Dave McCloskey and Sam Thompson conducted follow-on work to modify the model further to refine the theory. The Modified BBAY (or MBBAY) model continues to be used today in the weapon effects and other communities for a variety of applications.

### Structural Response

Structural response was important because dynamic buckling of reentry vehicle, missile, and satellite structures was considered to be the primary mode of failure in response to exoatmospheric nuclear weapon attacks and thus was generally the limiting case for the vulnerability of these systems. Because of the different time scales of responses to dynamic loads, a material response of microseconds or less and a structural response of milliseconds or even longer are, to a large extent, independent. These two modes led to somewhat separate roles for material and shock wave response calculations with finite-difference hydrocodes and for structural response analyses, which typically used finite-element structural codes. For the latter, the simpler analytic models for impulse generation were used as initial conditions.

In the 1960s, these efforts were centered on theoretical and experimental structural response at AFWL, with most of the work actually done under contract at the Stanford Research Institute (SRI). SRI conducted many of the seminal experiments in buckling phenomenology and provided much of the theory behind pressure and impulse curves to describe the different modes of buckling phenomenology in a relatively simple fashion. Sandia also began to study these phenomena in the group headed first by Tom Lane and later by Lee Davison.

### Experimental Advances

I have neglected the many experimental developments associated with the various gas guns. These led to new techniques and capabilities and to unique high velocity and hypervelocity data. In turn, these capabilities generated new shock physics theories as well as a varied set of dynamic material properties, especially for materials of special interest to the Department of Energy (DOE) and DOD. In the applied shock physics groups, the people performing the theoretical, numerical, and experimental programs were in close proximity, both physically and organizationally, and thus the progress from new data to useful output for real applied problems sped up tremendously. During this process many new theories, models, and numerical techniques were developed and rapidly brought to bear on important applications.

### Evolution of the Group in the Mid 1960s

In the mid 1960s, Don Lundergan led the applied shock physics group. With the support of his management, he had the foresight to appreciate that the intimate connection between experiment, theory, and application was the best approach for attacking and solving the pressing research and development needs of the shock physics community. Don recognized that Walt Herrmann, then at the Massachusetts Institute of Technology (MIT) and at least partially under contract to the AFWL, had both a somewhat independent perspective (i.e., he was outside the Atomic Energy Commission weapons laboratories) and the technical depth to begin to address these issues. The arrival of Herrmann at Sandia was one of the steps in providing that close connection in the area of shock physics. The resulting organization was one, if not the first, to integrate these "A-to-Z" capabilities fully for shock physics in the United States.

The people involved in those years included Lynn Barker, Larry Bertholf, Olden Burchett, Barry Butcher, Charlie Karnes, Sam Key, Larry Lee, Darrell Munson, Ray Reed, and Karl Schuler. Jim Asay, Lalit Chhabildas, Doug Drumheller, Mike Furnish, Dennis Grady, Gene Hertel, Marlin Kipp, Jeff Lawrence, Jace Nunziato, and Floyd Tuler joined them somewhat later. Certainly others should be added to this list, both within the group and in many other organizations at Sandia.

## Early Code Development

With the arrival of Walt Herrmann, calculations of shock physics phenomena began in a more formal way. In fact, Herrmann's initial task was to establish an in-house computational capability to complement the experimental efforts. Rather than continue with the PUFF code, the group adopted a pair of codes, WONDY and TOODY. These were the 1-D and two-dimensional (2-D) Lagrangian codes that Herrmann originally developed at MIT under the name "RAVE" and which subsequently evolved into a series of codes named "WAVE I," "WAVE II," and so forth. These codes were designed using the more modern "structure" of modular units, which made them much more flexible as new features were implemented, such as improved material models and numerical techniques.

Because of the independent line of their development, the emphasis of the two codes was largely mechanical in nature. However, the well-accepted finite-difference and conservation algorithms were fully maintained. This allowed extreme loading conditions (e.g., hypervelocity impacts); however, the implementation stopped short of fully coupled radiation transport. As an example of these modifications, it was realized that a dynamic time-dependent radiation loading option for the targets was needed to address nuclear weapon effects. Because of the modular structure, this feature was implemented relatively easily in the codes.

**WONDY and TOODY**. The evolution of WONDY and TOODY led to extensive consulting between Sandia and AFWL (and its contractors) to incorporate additional dynamic phenomenology (e.g., the response of porous and rate-dependent materials) into PUFF, which was AFWL's production shock physics code. In fact, PUFF continued (and continues) to be used by the DOD contractor community, with many of its newer features following Sandia's lead. As with PUFF, the development of WONDY and TOODY was largely independent of the extensive shock physics codes at Los Alamos and Livermore, which were more driven by "bomb" physics rather than by "effects" phenomenology.

Because of its flexibility and modular design, WONDY was used during these years as a "test bed" for many new models, new numerical techniques, and other computer-oriented tasks (e.g., Input and Output). Examples of new equations of state and constitutive relations included material porosity, cumulative-damage fracture and spall, and various rate-dependent material responses. Rate dependence has its own characteristic time constant, different from that associated with the numerical stability of standard shock propagation analyses. The new technique of subcycling, necessary for stability of dual time-constant solutions, was therefore developed. The rate-dependent models were also an effective approach to studying complex, layered composites. Other innovations treated the electrical response of piezoelectric and ferroelectric materials as well as mesh rezoning for improved computational efficiency. With the huge increase in computer speed and storage over the decades, this approach has largely faded, at least for relatively simple 1-D problems. For 2-D and three-dimensional (3-D) codes, it is still important. Another aspect of WONDY's modular framework was that virtually any arbitrary equation of state or constitutive relation could easily be employed. The only restriction was that WONDY be formulated to return pressure as a function of density and energy.

**SWAP.** At the same time, Lynn Barker, with the assistance of Emily Young, was developing the SWAP code. SWAP, a method-of-characteristics code, ran much faster than any of the finite-difference codes and yielded clear mappings of propagation and attenuation of the various shock and pressure waves throughout the computational space. However, it was more difficult to interpret for complex problems where wave interactions often became very complicated. Multi-dimensional characteristics codes were flirted with, but received essentially no long-term support.

## Next Generation of Hydrocodes

**CHART, CSQ, and CTH.** In the 1970s and 1980s, Sam Thompson and his colleagues began migrating from the group that supported the Nuclear Regulatory Commission to the applied shock physics group. This was still a period of rapid shock physics code development at Sandia. Thompson brought his independently developed set of hydrocodes, which started with the 1-D versions, CHART and CHARTD. These evolved into the 2-D version, CSQ (CHARTD SQuared), and later into the 3-D code, CTH (CSQ to the Three Halves power). In contrast to WONDY and TOODY, these codes were Eulerian, as opposed to Lagrangian, and included radiation transport capabilities and their own equation-of-state formulations. In addition to Thompson, the developers included Mike McGlaun, Hank Lauson, James Peery, Steve Rottler, and others.

The 3-D code CTH is somewhat a special case. It has become a (maybe *the*) major workhorse hydrocode throughout the DOD and DOE communities and remains so today. Gene Hertel was the prime mover behind this evolution. He advertised, sold, and maintained the code for many years and continues to be one of the most knowledgeable individuals regarding its capabilities.

**ALEGRA.** As broader interest in shock physics developed, the need for a new generation of hydrocodes became evident, especially in the research-oriented arena. At Sandia the latest in a series of pulsed power accelerators, Z, has become a workhorse facility for studying nuclear weapon effects, issues of practical fusion energy via inertial confinement, and other facets of high energy density physics. Opening these expanded energy and power regimes has demonstrated the need to extend the modeling capabilities of our hydrocodes by incorporating features such as coupled magnetohydrodynamics, advanced radiation transport, and high-temperature and high-pressure properties of materials and plasmas. The code that has assumed the burden of these activities is ALEGRA. The historical hydrocode developers, in conjunction with the staff supporting Z, have worked together to develop this new and evolving tool; they continue to do so. In some sense this has been "closing the loop" with the earlier "bomb physics" efforts regarding high-energy phenomenology, but in a more modern context.

### Structural Response Studies

As mentioned earlier, the development and implementation of finite-element codes to study dynamic structural response were centered in Tom Lane's group, which was later headed by Lee Davison. Sam Key, who was still in the applied shock physics group, was instrumental in recognizing and addressing the many situations for which the time scales for material and structural response were not as different as had often been assumed, leading to the class of targets termed "stocky structures." Over the ensuing years, this effort led to the series of codes named HONDO, PRONTO, and their follow-on versions.

### Shock Physics Database

Because the field of shock physics was so fragmented at the time, with only one formal full academic curriculum in the country (at Washington State University in Pullman, WA), the source material for the field was also scattered. Access in some cases was as basic as word of mouth. Walt Herrmann realized as early as the late 1970s that correcting this situation would be of great value. He began the process of collecting, organizing, and cataloging the relevant material relatively slowly, but Marlin Kipp, and later Mike Furnish in the 1980s, eventually expanded that process.

These efforts resulted in a collection useful to the ever-growing cadre of folks working in shock physics, but who may not even be aware of this significant resource. The collection contains data and properties of many standard and specialized materials as well as information on the processes for their fabrication, diverse uses, and applications. Descriptions of old and new response models important for ongoing studies are also included. The Sandia Shock Compression Database consists of at least twelve space savers (typically five shelves each) of documents, books, reports, and other hard-copy versions of source material and other data. As a companion to this database, the Shock PHysics INdeX (SPHINX) consists of nearly twice the volume of references and additional related material. At present, the collection is in the basement of Building 962 in Area IV at Sandia and comprises 12,000 documents.

### Other Areas

Throughout this period, the multipronged shock physics group has continued to generate high-quality Hugoniot data sets for basic and exotic materials and composites. New wrinkles on dynamic material response have continued to be investigated, theoretically modeled, and incorporated into the shock physics codes. Examples have included variable strength, solid phase changes, advanced porosity, different forms of rate dependence, cumulative damage and fracture, and coupled piezoelectric and ferroelectric response. Sometimes these phenomena were important to the varied applications under investigation, and sometimes they were not. However, the close cooperation and rapid implementation of these technologies have continued to be an important feature of the applied shock physics group.

### UGT Support

Support for underground weapon effects tests was always an ongoing activity. Because of many newly developed materials for hardening and improved functioning of all sorts of military systems, UGTs were always an important driver for the group's experimental, theoretical, and modeling activities.

### Missing Links

These notes have touched on many research subjects important to shock physics and dynamic material and system response efforts at Sandia. However, some specific areas have been given only cursory attention, mainly because they were not major parts of my personal experience.

# Raymond W. Lemke

# (1989–Present[113])

I joined Tom Mehlhorn's group in 1999 because I wanted to work on z-pinch physics relevant to the inertial confinement fusion program in the Pulsed Power Sciences Center. I had spent my previous 10 years at Sandia in another group doing high power microwave (HPM) research and source development for the Air Force Phillips Lab (AFPL) under a work-for-others program. Prior to being hired at Sandia, I had been involved in HPM research for 5 years at AFPL as a Department of Defense employee. After 15 years of HPM research, I wanted new challenges. When I lobbied Tom to join his group, I told him I wanted to work on challenging problems that would require using areas of my educational background that had heretofore been underutilized—particularly, for radiation magnetohydrodynamics (RMHD).

I spent the first year in Tom's group trying to model nested wire-array dynamic hohlraum experiments performed on Z using the two-dimensional (2-D) RMHD code LASNEX. I had no prior experience with magnetohydrodynamic (MHD) codes. It quickly became clear to me that I had a challenging problem. With a PhD in physics and a BS in astrophysics, I had the prerequisites to quickly acquire a basic understanding of the physics of dynamic hohlraums. However, becoming adept at using a new physics code (i.e., LASNEX) is a challenge by itself, even if one is familiar with the physics. After a year working with LASNEX, I was starting to feel comfortable with the code. The experience I gained during that year would be applicable to another very challenging problem I would soon be asked to work on: computational modeling of dynamic material experiments using the relatively new (at that time) MHD code ALEGRA, which was then being developed at Sandia.

Near the end of 1999, Tom Mehlhorn requested that I try to model the magnetically-driven dynamic material experiments on Z using ALEGRA. Jim Asay, Clint Hall, and Marcus Knudson had recently proposed using the intense magnetic field produced by the current on Z to drive materials science experiments. In such an experiment, then referred to as isentropic compression experiment (ICE), the magnetic pressure is used to either accelerate a metallic flyer plate for shock physics studies or to ramp load a material to acquire off-Hugoniot data. At the time, a few experiments had been performed and more were planned. It was clear that a science-based design capability would be needed to take advantage of the energy available on Z. My task was to work with ALEGRA for this purpose. ALEGRA also needed a relevant application for validation purposes. Thus began a verification and validation effort focused on magnetically-driven dynamic material experiments. By the end of 2004, this effort produced a computational model with predictive capability that is now used extensively today in the Center to design and analyze dynamic material experiments.

[113] Ray Lemke was hired in 1989 but started research in shock compression science in 2000. As of this writing, he continues to model shock wave experiments on Z, particularly those with high-velocity flyer plates.

## The First Meeting

Very early in 2000, I met with Jim Asay, Marcus Knudson, Clint Hall, and Tom Haill (who was in my group and one of the ALEGRA developers) to discuss modeling of dynamic material experiments using ALEGRA. We discussed modeling some of the first ICE shots on Z that had been performed the previous December, which had square, short-circuit loads with a 20-mm×20-mm cathode and a 26-mm×26-mm anode. We also discussed modeling future shots that would use a load with a smaller cross section (an 11-mm×11-mm cathode and a 15-mm×15-mm anode) in order to increase the drive magnetic pressure. Using the large square load, a 400-μm-thick, 15-mm-diameter aluminum flyer plate had been accelerated to 7 km/s, the highest velocity at the time. Data that could be compared to the simulation results consisted of VISAR-measured flyer plate velocities. The currents measured on Z with B-dot diagnostics could be used to drive the simulations. Some near-term goals of the ALEGRA modeling were to simulate the small square load and determine (1) the current and initial flyer plate thickness required to produce an aluminum flyer plate with a peak velocity of 18 km/s and at least a 400-μm-thick aluminum solid at impact with the target and (2) the current and initial flyer plate thickness required to produce a 15-km/s, composite aluminum titanium flyer plate with at least a 300-μm-thick titanium solid at impact with a similar target. (3) After that, I would then simulate the flyer plate experiment that had been done using the large square load.

At the time of this meeting, very little was known about either the dynamics or the magnetic field topology of ICE loads fielded on the Z shots because the concept was new and a self-consistent modeling capability did not exist. Tim Pointon had done calculations using a three-dimensional (3-D) electromagnetic (EM), particle-in-cell (PIC) code in which the perfect conductors did not move. Those calculations showed that the magnetic field is uniform in height (i.e., the direction transverse to the flyer plate motion). Tom Haill had begun to build an ALEGRA input file for one-dimensional (1-D) Lagrangian simulations of the large square load, so I started my modeling effort with that input file. I will summarize here the key advances and events (not necessarily in chronological order) that led to the predictive model we use today to design and analyze material dynamics experiments on Z. I have extracted most of this information from my research notebooks.

## An Important Code Development

The 1-D ALEGRA MHD simulations of the ICE load require a geometric scale factor (S) that relates current to magnetic field; S is basically (but not exactly) the path length one would use when applying Ampere's law to calculate the magnetic field. Fortunately, the expertise I had gained using the 3-D, EM PIC code QUICKSILVER for HPM research could be applied to the dynamic material problem. Initially, I used QUICKSILVER to investigate the magnetic field topology in an ICE load and to calculate S for use in 1-D MHD simulations. The results of these initial 3-D EM simulations were very insightful. They confirmed Pointon's earlier result that the drive magnetic field (the pressure) is uniform in the direction of the current flow, transverse to the applied force (the direction of motion). From the applications perspective, this is important because an ICE load could contain multiple samples (flyer plates) along the height of a panel that would all experience the same magnetic pressure drive. From the modeling perspective, the uniformity of the magnetic field with height implied that the ICE load could be simulated self-consistently using 2-D MHD in a cross-sectional plane that bisects the flyer plate at a fixed height. Furthermore, in this 2-D plane, the magnetic pressure decreases monotonically with increasing distance from the center of the flyer plate (or material sample). This implied that flyer plates should become bowed in the direction transverse to both the direction of motion and the height of the load; that was an important consideration for fielding the wedge diagnostic being used at the time to measure the shock breakout time. The wedge slope must be oriented in the direction in which the flyer is not bowed; otherwise, the shock breakout timing would be biased by the bowing.

During the next few months, I ran 1-D, 2-D, and 3-D ALEGRA MHD simulations of the ICE load. By the end of 2000, I realized that a self-consistent, 2-D MHD simulation would not be possible unless a mechanism existed to energize the problem from the third dimension using a circuit representation of Z. I proposed this capability at ALEGRA meetings. By May 2001 Allen Robinson, then with the Computational Physics R&D Department at Sandia, had incorporated this capability into ALEGRA. Allen invented a very clever scheme for energizing the 2-D MHD simulation from the third dimension via a circuit that accounts for the effective height of the short-circuited ICE load. It's called the "e-normal return current boundary condition." That boundary condition made possible accurate, 2-D MHD simulations of isentropic compression experiments. When energized by a Thevenin-equivalent circuit model of Z, the 2-D ALEGRA simulations using the e-normal return current boundary condition self-consistently captured the effect of conductor motion and deformation on power flow into the ICE load. It is the key code development that produced the predictive model we use today.

Thus, between January 2000 and May 2001, until self-consistent 2-D MHD ALEGRA simulations of the ICE load became possible, I had to make due with 1-D simulation driven by a measured load current (provided by the B-dot diagnostics) to model the dynamic material experiments. As I would discover, these 1-D calculations could not be self-consistent, even if the load current was highly accurate, because the measured current includes time-dependent effects of conductor motion on the magnetic pressure drive that are not accounted for in the 1-D simulation. The 1-D simulations must be driven by the magnetic pressure drive experienced by the flyer plate, for example, in a self-consistent, multi-dimensional simulation. Hence, I tried some 3-D MHD ALEGRA simulations of the ICE load; although self-consistent, these calculations were too computationally intensive to be practical. The most useful data for code validation would not available until the end of November 2000.

## Critical Early Experiments for Model Development and Code Validation

The earliest dynamic material experiments were performed in December 1999. Although the data from these shots provided a useful starting point for MHD modeling, it was not sufficient to define the physics needed for model development. For this purpose, the data from Z Shots 575 (April 2000), 634 (September 2000), 658 (November 2000), and 668 (November 2000) would be critical for identifying the physical phenomena that govern the dynamics of the ICE load and affect the time-dependent velocities obtained from VISAR measurements.

The square load used in Z Shot 575 included aluminum samples with thicknesses of 400, 500, and 850 μm, all backed by LiF windows, for measuring the isentrope of aluminum. The 1-D simulations of the experiment driven by the measured load B-dot current produced reasonable agreement with all three Al/LiF interface velocities, but only after reducing the magnetic scale factor S calculated by QUICKSILVER by about 10 %. These results indicated that the ALEGRA MHD model, which used SESAME EOS (equation of state) 3700 for aluminum, was probably accurate; however, having to adjust the geometric scale factor S was evidence that the static value calculated by QUICKSILVER was inadequate for modeling these experiments. This deficiency would be resolved when self-consistent, 2-D simulations became possible.

Z Shots 634, 658, and 668 were flyer plate experiments. Z Shot 634 used a rectangular load to produce a 21-km/s, 850-μm-thick, Al flyer plate in an impact experiment (the highest velocity at the time); the VISAR was used to measure the flyer velocity. Z Shot 658 (with Greg Sharp as the principal investigator) was designed to deduce the rate of magnetic diffusion in aluminum by using the B-dot diagnostics to measure the appearance of the magnetic field on the surface opposite the magnetic drive side of flyer plates with thicknesses of 100, 200, 400, and 600 μm. Z Shot 668 was a symmetric shock impact experiment in which an 8.5-km/s, 925-μm-thick Al flyer impacted an aluminum target backed by a LiF window; the VISAR was used to measure the flyer and Al/LiF interface velocities.

Modeling these three flyer plate experiments using 1-D MHD simulations produced poor agreement with the measurements. In particular, the results for Shot 658 indicated that magnetic diffusion was 36 % faster in the calculations. The data from Shot 668 indicated unequivocally that the 8.5-km/s Al flyer plate was partly solid at impact, which was in stark contrast to the simulation result, which showed the flyer plate was completely melted. Indeed, at that time all ALEGRA simulations of flyer plate experiments produced flyer plates that were completely melted, in contrast to the experimental evidence. My efforts to find the cause of this discrepancy eventually led to the conjecture that the simulated electrical conductivity could be too low in melted aluminum, which was confirmed by Mike Desjarlais using the quantum molecular dynamics (QMD) code VASP.[114]

[114] Editors' note: G. Kresse and J. Furthmüller, "Efficient iterative schemes for *ab initio* total-energy calculations using a plane-wave basis set," *Physical Review B* **54**, pp. 11169–11186 (1996).

I came to this conclusion (after performing many test simulations) by modifying the value of a parameter in the LMD (Lee–More–Desjarlais) conductivity model for aluminum that controlled how much its electrical conductivity decreased when it melted. Consulting with Mike about this result, I learned that he had been working with the QMD VASP code and had modified it to calculate the electrical and thermal conductivities of materials; he had been doing so as part of an effort to model and to understand wire initiation experiments. Mike applied the modified VASP code to the aluminum flyer plate problem. His QMD calculations verified that, indeed, the electrical conductivity in melted aluminum is larger than the LMD model calculates for the same density and temperature. By May 2001 Mike had produced a new conductivity model for aluminum that incorporated the results from his QMD calculations. Using the improved model of aluminum electrical conductivity in my ALEGRA simulations of Z Shots 658 and 668 produced velocities in excellent agreement with the measured VISAR velocities. This discovery pertaining to the electrical conductivity of melted aluminum was a major advance on the path to developing a predictive MHD model.

Using the improved model of electrical conductivity in an ALEGRA simulation of Z Shot 634, driven by the raw load B-dot current, however, did not produce a flyer velocity that was in good agreement with the data at all times; the agreement was good at early times, but the discrepancy was significant at later times. This suggested some important phenomena were missing from the MHD model. The magnetically insulated transmission line (MITL) and load current diagnostics on Z Shot 634 indicated that significant current had been lost before the load, in contrast to the situation for Z Shot 668, which showed little current loss. I conjectured that the loss in Shot 634 occurred in the convolute and resulted in a short circuit that produced an exponential decay of the post-peak load current. However, since a simulation driven by the raw load B-dot current did not produce an accurate flyer velocity, I manually modified the post-peak load B-dot current with exponentials having different decay rates; using these to drive the MHD simulation, I was able to produce a flyer velocity in good agreement with the data. Aside from demonstrating the importance of current loss and the related short circuit in determining the flyer velocity, these calculations indicated that the measured load B-dot currents were not sufficiently accurate for MHD modeling of the flyer plate experiments. These discoveries were critical to the development of a realistic circuit model of Z and represented another significant advance on the path to developing a predictive MHD model for ICE.

### Predictive MHD Model, Circuit Representation of Z, and Dakota Optimization

Toward the end of 2001 the e-normal return current boundary condition (described above) had been incorporated into ALEGRA. At that time I began performing self-consistent, circuit-driven, 2-D MHD simulations of Z Shot 634 and using these calculations to develop a simple circuit model of Z. I modified a lumped-element,

Thevenin-equivalent circuit developed by Ken Struve (using the Bertha transmission line code) to include a time-dependent loss impedance that could account for current loss and a time-dependent crowbar (resistive) switch that accounted for the short circuit. I used the measured MITL current from Shot 634 and my modified load current obtained from the 1-D simulation (described above) to calculate the time-dependent loss impedance. This loss impedance showed an abrupt decrease toward zero during the time that current is lost, clear evidence that current loss results in a short circuit. I used this loss impedance and the relative amplitudes of the MITL and the modified load currents to surmise where the short circuit should be enforced, which I modeled with a separate, time-dependent crowbar switch. The crowbar switch is closed at the short-circuit time, which allows current to be lost to ground through the loss impedance. In this case, the time dependence of the current downstream of the short depends on the total downstream inductance and resistance.

In a discussion of this work with Dillon McDaniel (Sandia), he pointed out that I could use the stack voltage and current measurements on Z to derive a time-dependent voltage for my circuit-driven simulation of Shot 634, which I then did. Using this voltage in simulations of the shot with the new Z circuit model showed that the best agreement with the measured VISAR velocity was obtained by placing the short circuit on the upstream side of the convolute. At this location the inductance and resistance downstream of the short circuit are evidently sufficient to produce an accurate post-peak current decay rate. This was further support of my conjecture that time-dependent current loss occurs in the convolute and produces a short circuit, which must be accounted for in the design of plate impact experiments in which the flyer must be shocklessly accelerated.

The 2-D MHD model described above is the basis for the quasi-predictive ALEGRA model we use today to design and analyze dynamic material experiments. From 2002 to 2004 we designed experiments in which aluminum flyer plates were shocklessly accelerated to peak velocities up to 34 km/s (a corporate milestone), which was the maximum that could be produced on Z at that time. That model was also used to predict that flyer velocities greater than 40 km/s could be achieved after the Z accelerator was refurbished. This model is only quasi-predictive because it requires *a priori* knowledge of the current loss, something that is still lacking today. However, with a still growing database of accurate load currents, we are now able to estimate the time-dependent loss current with sufficient accuracy to design experiments. In order to obtain an accurate load current, that is, the current that actually reaches the ICE load and is not necessarily equivalent to the load B-dot current, we use the DAKOTA optimization code.

In 2004 my then manager, Tom Mehlhorn, sent me an article on DAKOTA optimization by Tony Giunta, one of the developers. It was clear to me that DAKOTA, in conjunction with ALEGRA, could be used to unfold the actual load current from a shot using the measured VISAR velocity. With Tony Giunta's help, I set up the first DAKOTA-ALEGRA optimization problem to unfold the load current from a shot. The calculation yielded a magnetic field drive that produced the measured flyer velocity to within the uncertainty of the measurement (i.e., less than 1 %). To get the actual current, I developed an iterative scheme using the 2-D MHD model.

Because the optimization uses the 1-D ALEGRA model, I developed a technique for calculating the initial scale factor S and the equivalent 1-D magnetic drive, both of which are required for the 1-D simulation, from the results of a 2-D simulation. The load current obtained in this way has proven to be a more accurate indicator of the actual current in the load than the current determined by the load B-dot measurement, which is frequently used as an initial guess at the actual current to start the optimization. This technique has evolved into a VISAR current diagnostic in which flyer plates are incorporated into an experiment (e.g., for a liner z-pinch) specifically to obtain an accurate load current. Furthermore, this technique is used in reverse to derive a current shape for either the shockless acceleration of a flyer plate or for dynamic loading of a material sample. The DAKOTA-ALEGRA optimization technique significantly advanced the dynamic material capability on Z.

### Refurbished Z Accelerator, 2007 to 2012

The refurbished Z accelerator (sometimes referred to as "ZR") was completed near the end of 2007; it has twice the energy of the old Z. Using the aforementioned predictive MHD model and associated techniques that were validated on Z, we made rapid progress in developing new platforms and experiments for ZR that took advantage of the larger energy and current. Some of the first shots on ZR were flyer plate experiments (Z Shots 1767, 1774, and 1775) designed to obtain an accurate load current (as described above). By April 2008 the first dynamic material experiments (Z Shots 1807 and 1810) were performed using a new stripline load that would produce the highest pressure for a given current. By September 2008 the compression isentrope of tantalum was measured to a peak pressure of 3.8 Mbars using a stripline load. By May 2009 a stripline load was used to accelerate a 900-μm-thick aluminum flyer plate shocklessly to a peak velocity of 46 km/s (the record to date on Z) in an impact experiment (Z Shot 1934) that generated a 20.7-Mbar shock in α-quartz. By October 2009 a stripline load was used to accelerate a composite aluminum-copper flyer plate shocklessly (with 800-μm-thick aluminum on the drive side followed by 200-μm-thick copper) to a peak velocity of 28 km/s (Z Shot 2006). Flyer velocity milestones achieved on both Z and ZR are compared with the velocity from Z Shot 634 (before pulse shaping techniques made shockless acceleration possible). Numerous successful flyer plate experiments have been fielded on ZR since 2009; the ones I mention here represent the state of the art.

### Acknowledgments

I am grateful to Jim Asay, Kyle Cochrane, Jean-Paul Davis, Mike Desjarlais, Tony Giunta, Tom Haill, Clint Hall, Marcus Knudson, Dillon McDaniel, Allen Robinson, Greg Sharp, Steve Slutz, Ken Struve, the ALEGRA developers group, and the Z-Team for their technical contributions to my work.

# C. Donald Lundergan

## (1956–1986[115])

## A Brief Description of the Early Days (1956–1966) of the Shock Wave Program at Sandia National Laboratories

The investigation of the propagation of shock waves in the elastoplastic pressure range had its beginnings in the late 1950s. The late 1950s and the early 1960s were a time of technological transition at Sandia—a transition that required the initiation of research into topics that would provide the tools for the laboratory to fulfill its mission of weaponizing advanced systems. A part of this transition was the initiation of studies into the propagation of stress waves in the region of interest to Sandia. The purpose of this memo is to recount that transition and relate some of the early days of shock wave studies.

The urgent task of Sandia Laboratories immediately after World War II was to build a stockpile of nuclear weapons, primarily bombs. The initial task confronting Sandia was to provide components and structures to meet operational requirements and to assure that the weapons would reliably detonate when required, but would be unquestionably safe under all other conditions. The early nuclear weapons were bombs carried by subsonic aircraft. The structural and environmental requirements for these early weapons did not substantially differ from conventional weapons. Accordingly, the application of extant engineering procedures and materials to the design of these early weapons was basically sufficient but required faultless application to meet stringent safety and reliability requirements. Innovative safety concepts and extensive reliability testing were initiated. Innovative designs such as parachutes to permit delayed delivery, weak-link and strong-link safety systems, and reliable barometric and contact fuzing capabilities were among the subsystems required.

The advent of supersonic and exoatmospheric missile delivery systems and their associated environments imposed new demands on the components and subsystems for which Sandia was responsible. To meet these demands required the understanding of the effects of these severe environments. One such demand was understanding the behavior of materials subjected to impacts at velocities associated with the new delivery systems and to radiation associated with space and countermeasures. One problem related to high velocity impacts was that the structure had to retain its integrity for a period after impact and fuze actuation to prevent deformation of critical structures or destruction of critical components. The problem associated with high energy radiation was material ablation and the resulting high-intensity shock waves that produced spallation that could preclude successful reentry.

To understand the effects of these environments and to resolve the related problems, shock wave research was initiated in the elastoplastic stress range. Leon Smith

[115] Editors' note: These are the dates Don Lundergan was employed at Sandia. After leaving the shock wave field in 1977, he participated in other Sandia programs and retired in 1986 after an illustrious career.

and Don Cotter asked why there were such inconsistencies as to the function of contact fuzes. The response, after some investigation, was that we could not predict the propagation and effect of the stress waves that resulted in failure of the structure. Depending upon subtle differences of impact, the fuze would provide timely initiation but, on other occasions, structural failure would negate detonation.

At that time, data existed on propagation of stress waves and material behavior at low strain rate boundaries such as those investigated using Hopkinson bars and comparable university laboratory equipment. The work at the national laboratories, in investigating the hydrodynamic behavior of materials, provided an upper boundary. Missing was the range of strain rates associated with the design needs of Sandia subsystems. Accordingly, the investigation of the elastoplastic behavior of materials was initiated at our laboratory.

To resolve this problem, experiments were conducted to determine the elastoplastic properties of the materials of interest and transform these properties into a format usable for analysis. To determine the material properties, a decision was made, circa 1958, to use a 4-inch-bore air gun that would accelerate a sabot[116] to velocities that, upon impact, would produce stress levels of interest. The gun accommodated targets of sufficient diameter to sustain one-dimensional strains for periods long enough to measure the elastic precursor and the following plastic wave and rarefaction wave. These events occurred prior to the unloading from the perimeter of the target. In addition, the gun permitted the use of various impedances and thicknesses of impact materials. However, a requirement was normal impact. This was achieved by accurate boring and polishing of the barrel and the design of the sabots that carried the impactor. The use of compressed air as an accelerant rather than explosives also avoided many operational restrictions, such as location and the safety regulations associated with the use of explosives. Here, it is appropriate to acknowledge the extraordinary qualifications of Roy "Red" Hollenbach.[117] His mechanical and electrical skills were essential to the success of this project. The air gun, a hand me down from Al Beck, who used it to test the effects of impacts on components, was refurbished and located in Area III. Because the gun was outside, the experimenter was exposed to the elements and to the basking rattlesnakes. But it was the start of a family of gas guns, all of which were housed in more comfortable locations.

Initially, the pin array developed by the Los Alamos shock wave group (GMX 6) was used to measure the velocity of impact and the deviation from normal impact. The Hugoniot equations of state for a variety of metals were determined. Bill Hartman, Jack Cannon, Jimmy Smith, Tom Looby, Wayne Brookshire, Bob Newman, and Bob Simpson all contributed to these early investigations. Lynn Barker invented

---

[116] A sabot is a device used to fire a projectile with accurate velocity and impact alignment control.

[117] Red recalls, "We started with nothing—a cast-off air gun mounted outside in Area III, no instrumentation, and two people that were not on the project full time. The dark room allocated to us was the women's unused wash room in Building 809. We progressed from firing the gun across the mesa to a sawdust tilled bunker that challenged us to find the test piece after each shot. Soon, our group seemed to really grow—more people, better facilities, lots of instrumentation, and more sophisticated measuring techniques."

measurement techniques to provide greater accuracy, with continuous rather than discrete data. Lynn, again with the able assistance of Red Hollenbach, went on to develop the slant wire technique in 1962 and then the laser interferometry instrumentation in 1964. He has continued to invent measurement techniques up to the present, including the velocity interferometer. These measurement techniques were essential to the success of the shock wave program.

The second requirement, processing the data into a usable form, has undergone equally extraordinary transformation. In place of the tedious, inaccurate hand calculations of the past, we now use three-dimensional analyses that are completed in seconds.

The initial calculations, around 1957, to determine the propagation of stress waves in a weapon structure were done by using known elastic wave velocities to hand plot the progression of stress waves and estimate the times of critical events. The calculations were both tedious and inaccurate because of the lack of computational techniques and material properties. The calculations, however, did indicate that a problem existed.

A year or so later, we had progressed to using limited data for a few materials and Marchant mechanical machines to do the calculations. Louise Cernich and Julie Bode deserve credit for their efforts in doing the calculations, which were obviously extremely tedious and subject to error. Now these calculations are done in seconds with the appropriate material properties and in three rather that one dimension. The transition from hand calculations to the current rapid analysis took place over a period of years.[118] Programs using punched tape were accomplished with the help of programmers, including Emily Young, who was in the shock wave program for many years.

As noted earlier, the time of transition required the addition of staff, schooled in the appropriate disciplines. Digital computing was coming of age, and problems requiring the results of shock wave research were mounting. During the early 1960s, Larry Bertholf, Charles Karnes, Walter Herrmann, Peter Chen, Barry Butcher, Karl Schuler, Darrell Munson, Floyd Tuler, Sam Key, and other well-qualified individuals were added to the organization. Their contributions added to the understanding and quantification of the time dependency for spallation and the propagation of stress waves in composites, laminates, and foams. These material characteristics and associated wave propagation codes provided the input and calculational techniques to solve problems in the weapons program, nuclear reactor safety, armor design, and many other areas.

Within a period of 15 years, the shock wave group had attained international recognition for its work. But, of most importance, the initial goal was realized of providing Sandia National Laboratories with the tools that are now being used to solve problems involving the dynamics of materials.

As a personal note, in addition to my esteem for all the people who made many noteworthy contributions to the field of shock wave physics, I would like to

---

[118] As an aside, it is a bit amusing that one of the earliest problems of interest was the cotact fuzing problem that is now being solved—just faster and better with all of our new capabilities.

acknowledge the foresight of Sandia management. The support of management provided the opportunity to enter into new fields of study—a study whose results vindicated their trust.

I again experienced the trust of management when I undertook a project to develop a new technique for detecting foreign underground nuclear tests. Once again, a "high risk, high payoff" endeavor was, fortunately, a success. My thanks this time are to Paul Stokes and Bob Clem for giving me the freedom to explore and my admiration is for Charles V. "Jack" Jakowatz and his group that developed the concept into an intelligence tool. I hope this freedom to explore still exists at the lab.

## An Addendum to Don Lundergan's Recollections

**Editors' comments**: Almost 30 years ago, in summer 1987, Lundergan put together a package of his old letters and memos dealing with his thoughts on the future needs for shock wave research at Sandia. The collection was sent to then Sandia Corporate Historian Necah Furman, along with a cover letter that gave a brief outline of how the various items contributed to planning and implementing the overall program. Some of the documents in the set date back to the late 1950s and early 1960s. Lundergan described several phases of the Sandia history that relate to the evolution of Sandia's shock physics research. The first, in the late 1950s, is characterized by the "discipline of dynamic material behavior in the elastoplastic range." A second phase, in the mid 1960s, is represented by the period during which the demands on Sandia scientists, engineers, and designers for improvements in weapon performance led to changes in the staff from those employing "textbook engineering" to those exploring application of advances in shock wave research and other fields for the design of weapon system components and for related programs.

In his September to December 1963 letters to Charlie Bild, at that time Director of Materials and Process Development, Lundergan outlined his vision for the future of the shock wave research and development program. He called out the need for both high- and low-pressure shock data to examine phenomena ranging from material and structural response to system-level vulnerability, survivability, and lethality analyses. The program would require extensive new laboratory-scale dynamic experimental capabilities, as well as sophisticated theoretical and numerical tools to analyze the data and predict how these phenomena extrapolate to real-world weapon scenarios. Lundergan's vision became especially important, as the various test ban treaties were then beginning to be implemented. What is remarkable is how many of his insights and recommendations have come to fruition throughout the succeeding decades.

We would like to include many of these items, but space limitations will not permit that. However, as an example of this material, we have appended a transcript of handwritten notes by Necah Stewart Furman. These notes are based on Lundergan's phone conversation with her on July 13, 1987, shortly after he had sent the memos and letters to her and a year or two after his retirement from Sandia. They recount some wide-ranging reminiscences of his early days and the evolution of the fledgling shock wave program. The overall topic of their conversation was the shock wave multi-lab research-and-engineering dichotomy.

## Edited Transcript of Lundergan–Furman Phone Call, July 13, 1987

The comments below are mostly in Lundergan's own words but, for continuity and clarity, we have edited and rearranged the paragraphs somewhat and divided the comments into separate subsections. In fact, some of his remarks are a little out of sequence because of the extemporaneous nature of the phone conversation. In a few instances, a phrase or sentence has been added in brackets for clarification:

In the late 1950s and early 1960s, there was a dichotomy between Sandia designers at fabrication and those at the research level—the purpose of the latter, I realized, was to aid those weapon designers. Charlie Bild also aided those design groups. The rest of the organization and Research Department was not in favor because they couldn't see the applicability to their own problems. Frank Hudson was chastised because they felt his work was inappropriate, but organizations in the Materials Directorate, 1100, run by Charlie Bild, felt some [of those activities] were appropriate, and motivations from design groups explained that demands being placed on the design organization [and] illustrated the need for transition. With that, Larry Bertholf performed a series of cause-of-failure development analyses using two-dimensional shock hydrodynamic codes applied to real situations and materials. [Those analyses actually occurred in the late 1960s and beyond.] However, at that time, the results were indeterminate.

Now, we needed to get to the stage where we knew the material properties that we could use to do those complex calculations. We were placing great demands on computers. The point is we were solving real problems arising from the results of research. Orval Jones, Director, and Dick Claassen [the first manager of the Physical Sciences Department] understood the problem. [This was eventually] solved by making Jones a Vice President and Herrmann a Director. They placed some key research people in the design organization to create more understanding, and they also had the knowledge to interpret [the needs and required approaches].

## Rationale for Sandia's First High-Precision Gun for Dynamic Material Property Studies

In 1956, I first came here from having taught physics and math at the university level. I was hired by Leon Smith and Don Cotter, 1260. [119] The problem to work on initially was the determination of whether or not contact fuzing would properly function. There were rather ambiguous results from the sled track. What came out of it? Bombs that were fuzed on impact came in at low velocity, which provided sufficient time to activate the fuze before actual deformation took place. [The generated impact often] caused a dud or deformation and it was not clear why. But as time went on, weapons and components were getting miniaturized, and the impact time interval became shorter; thus the situation would get even worse.

When you try to measure whether the design was appropriate by using sled tests—never repeat tests on the sled track; the results would be different each time—you get ambiguous results. This led to recognizing that we do not have the technology to make design calculations ahead of time, partly because we didn't have the materials and their material properties. Some work was done by using Hopkinson bars and basic strength of materials, but strain rates were too low and stresses were also too low. These tests could be done in Labs, but [Hopkinson bars were] not representative of the "real" world.

The other extreme was work done at Los Alamos where the pressures on materials were so intense they were treated as fluids. A Hugoniot equation (in the elastoplastic regime) was

[119] Editors' note: Department 1260 at that time was the Electrical Systems Department managed by L.D. Smith.

needed. One of our technicians, Red Hollenbach, thought of using an air gun under controlled conditions. I got a used gun. It was from Al Beck, who was employing it to determine effects of acceleration in components. It was used in Area II and in the Livermore Lab for acceleration designs for laydown weapons. But, it had not been used for this equation-of-state work.

Regarding that original gun, it was the first air gun utilized for EOS [equation of state] experiments, after being initially used by Al Beck. The gun cost $10,000.00, and, if amortized over the number of papers and reports written and the problems solved, it was one of the most economic investments Sandia ever made. For that first gas gun in Area III—I didn't appreciate the amount of energy we were working with—I didn't notify safety, and one of the plates landed near a safety person, so we rapidly got all sorts of catching devices in place!

We started out on aluminum.... Then, while we were working on contact fuzing, one of the Soviets mentioned a concern about hot x-ray problems. What is that? We came back and worked on understanding it [by looking at wave generation, propagation, and spallation, using the computer codes]. To study these effects, Barry Butcher developed one response model for foam.

At this time the group grew from section to division. Walt Herrmann [was promoted to] manager. Paralleling this [were other organizational changes], with Coyote Canyon under Orval Jones, and the materials groups under Charlie Bild.

## Sandia's Transition to a Multi-program Laboratory

[Some time in the 1970s], Bob Peurifoy asked one of the Departments he had and me to do a review [of Sandia's programs] because [e]nergy programs were coming into Sandia, and the bulk of the test program and most of the drafting and shop facilities were being paid for by the Defense programs. He asked Art [Jim Arthur] and me to determine how costs are distributed through the case system. Art and I talked to every case manager in the Lab and [we saw that] ... pure research ... and ... applied research ... went through the entire lab. [However, we were] surprised by the actually small amount of research going on throughout the Labs.

Research is not separate today as it was in the old days. We were a weapon laboratory—looked upon by Los Alamos and Livermore as the ones to do the work on their explosives. In 1972, I was asked by Charlie Bild to interface with the Laboratory Coordinator in one of the Divisions of ERDA [Energy Research and Development Administration[120]]; they had representatives from all laboratories to look at how they fit into a structure that we then had.

Peurifoy, Bill Nichol, and myself were asked to determine our contributions to the energy programs, particularly as it dealt with nuclear reactors. The AEC [Atomic Energy Commission] forced nuclear reactors on the public in an arrogant way. SNL had used reactors for the purpose of testing by generating environments to determine if components would survive [in nuclear weapon environments]—SNL was not a part of the transfer of using nuclear energy for generating power. But, much later, we did work to determine the safety of nuclear reactors. So, Peurifoy decided we could [begin to] support the Nuclear Regulatory Commission.

---

[120] Editors' note: ERDA was formed in 1975 as a result of the Energy Reorganization Act of 1974. ERDA assumed the functions of the Atomic Energy Commission (AEC) not assumed by the Nuclear Regulatory Commission. ERDA existed for only 2 years. In 1977 it was combined with the Federal Energy Administration to form the Department of Energy (DOE).

[121] Editors' note: This series of documents was a detailed "catalog" of Sandia's technical capabilities that, for many years, was updated regularly. These documents were a major tool in the lab's diversification (referred to in the next paragraph) involving various Department of Energy programs as well as reimbursable projects for other government agencies.

Peurifoy said, 'SNL does not have a way of informing other Labs of the many other disciplines of which we are capable'. So, I put together the 'Technical Capabilities of Sandia Laboratories' [121] with the help of Phil Mead. I called and got President Schwartz [actually, Morgan Sparks was the President at that time] to write a letter asking for cooperation.

The smallest number of technical areas that would pull together all of our disciplines—each of these capabilities—generic disciplines, [were] assigned to an individual to write up these capabilities. The end result was a set of 16 individual areas of capabilities: aerodynamics, terradynamics, material science, and so forth. [It was a] best seller—16,000 copies were distributed, with many outside requests, even from other countries, Israel, for example. So [we were in a] good position when we met with ERDA to present our representative disciplines. Result: I wrote a memo in 1973 suggesting that Sandia was in a position of the elite to be a multi-purpose laboratory, and we diversified. Survival through anonymity [the prevailing philosophy]—a low profile—that attitude had to change. The other labs didn't know anything about us. After this transition and after the change in the percentage of advanced-degree staff took place, our [shock wave] work became more in the area of applied physics.

## Some Interesting Anecdotes from that Older Period

A Don Cotter story—he likened [the] weapons program to a log in a turbulent river with ants on top of the log and, as the log would come to the top, the ants would yell, 'I can see; I'll drive now.'

In preparation for a Paris meeting in 1967, [I met] Max Williams, Dean of the College of Engineering at the University of Utah [1965 to 1973] who had worked on dynamic behaviors of polymers at Cal Tech. Max and I kept in contact, and we were given the responsibility to select papers for the meeting in Paris; the topic was High Pressure Physics—he asked me to chair the sessions and select the papers, so I called Orval Jones to help. Max came into my house, tossed the papers on a desk, and went into my bedroom and went to sleep. We were diligent and he submitted them to the Union of Applied Mechanics. The Committee of Atomic Physicists in Paris threw out [the] papers—[apparently there was a] problem related to [some of the subjects connected with information on the phenomenology of x-ray-generated impulse]. [We were] told by Security that we couldn't talk about the hot x-ray problem.

Stan Mitchell from London and Byork from Sweden were there and a fellow from Sweden asked when the decision was made for Sandia to take over shock wave physics.... Felipe Beatriz did not have permission to enter the United States. He wanted to talk to me. Security said he can't even come into the parking lot at Sandia. So I took him to dinner. He said your AEC is more stupid than ours. I got back to my office and Security said you can't even talk about open publications. In these early days of shock wave physics, we sought opportunities to get the job done, one way or another.

## Summary of Don Lundergan's Role in Sandia's Shock Wave Program

As a final point about Lundergan's role, Charlie Karnes recounts, in his recollections in this book, a conversation he had with retired Sandia Executive Vice President, Orval Jones:

> Two or three years ago I was talking to Orval Jones about the development of shock wave work at Sandia. He said that Don Lundergan deserves a lot of credit for Sandia's contributions in the field of shock waves because Don recognized the importance of shock wave computations [to Sandia's overall mission] and hired Walt Herrmann, and the rest is history.

# Peter C. Lysne

# (1966–1996[122])

I joined Sandia in the fall of 1966. My first assignment was in Mel Merritt's division, which was involved with planning field test events at the Nevada Test Site. This division also included Al Chabai and Bob Bass, both of whom stimulated my interest in geophysics. Rod Boade was another new hire, and we worked together for a few months before being transferred to Orval Jones' division to investigate low-amplitude shocks in porous materials. My first assignments from Orval were to publish my dissertation and to join with Walt Halpin in his study of shock propagation in a porous iron. Shortly thereafter, Lee Davison took Orval's job, and he encouraged the development of reverberation techniques to investigate states off the principal Hugoniot and the analysis of the nonlinear polarization effects within quartz gauges that compromised some reverberation experiments. Don Hardesty and I continued the off-Hugoniot work with the development of a complete equation of state of nitromethane. Later polarization work with Mark Percival led to studies of ferroelectric ceramics and inert materials, such as PMMA (polymethyl methacrylate) and epoxy. The polarization work continued even after I moved into geophysical programs because models were needed to explain the anomalous electrical effects observed in wet rocks.

The energy crisis of the mid 1970s caused many of us to branch out in our interests. Rod, Jim Johnson, and Al Stevens left Sandia, and I joined John Crawford, Tom Grissom, and Hugh Bivens in their work to build a neutron-activated logging tool for uranium. Later Hap Stollar, Carl Schuster, Bill Luth, and Dick Traeger enticed me into problems involving oil shale, tight gas sands, geothermal reservoirs, and scientific drilling programs. Interactions with industry, academia, the USGS (U.S. Geological Survey), and other national laboratories were essential to this work. Its culmination was the Geoscience Research Drilling Office with multiple national and international ties.

[122] Peter Lysne retired in 1996 from the Geothermal Research Department after leaving shock wave research in the late 1970s.

# J. Michael McGlaun

## (1976–2006[123])

## Shock Physics History: Outstanding People and Microprocessors

### Two-Dimensional Calculations

I hired into Sandia's Fireset Development group in 1976. I started working with Sam Thompson and John Freeman on a two-dimensional coupled shock and electromagnetics problem. We used Sam's Eulerian shock physics code CSQII. John Freeman was a manager in the inertial confinement physics group; he added electromagnetics to CSQII. John and Sam were my first contacts with senior Sandia scientists. Their work ethic and commitment to excellence made a lasting impression.

Sam Thompson started at Sandia in 1966. Sam developed an integrated suite of software that analyzed one-dimensional (1-D) and two-dimensional (2-D) strong shock problems. One of Sam's shock codes was the 1-D finite-difference Lagrangian radiation-shock physics code CHARTD. His other code was the 2-D CSQII finite-difference Eulerian shock physics code. Sam named the original version CSQ. Its applicability was limited because it allowed only two materials in a calculation. Sam named his second version CSQII; it allowed up to ten materials in a calculation. Sam developed an analytic equation-of-state package named ANEOS that both CHARTD and CSQII used. Sam also developed a powerful suite of pre- and post-processing graphical software to analyze the equations of state, time histories of Lagrangian and Eulerian points, and 1-D and 2-D calculations. Sam's software was powerful, was widely used, and had a reputation for being accurate, reliable, and well supported.

Shock wave analysts were using two other codes at Sandia when I arrived. WONDY was a 1-D Lagrangian finite-difference shock code supported by Marlin Kipp. Walt Herrmann had developed the original version. By 1976, he was in management and too busy to support codes. WONDY was widely used, was capable, and had an excellent reputation. TOODY was a 2-D Lagrangian finite-difference shock code supported by Jeff Swegle. Walt had also developed the original version of that code. Several outstanding scientists worked on TOODY, including Larry Bertholf, Jeff Lawrence, Darrell Hicks, and Billy Thorne. TOODY, like all 2-D Lagrangian shock codes, required a skilled user to analyze difficult problems. I analyzed a few problems with TOODY and was always impressed at the results obtained by some of its best users.

[123] Mike McGlaun moved to another organization in 1996 that was not directly involved with shock compression science. He retired from Sandia in 2006.

In 1978, I transferred to the Shock Physics group managed by Al Chabai. Much of my work was with Tom Bergstresser and Bill Davey, trying to understand explosively-operated craters. In 1979, I developed a post-processor for CSQII, called Crater, that would calculate the final shape of a crater generated by a buried explosive. CSQII performed the initial part of the calculation until the shock wave reflected off the soil surface and the ejecta were in free flight. Crater used restart files from a 2-D cylindrical CSQII calculation for initial conditions. Crater ballistically extrapolated the ejecta under the influence of gravity. It used slope stability criteria to calculate the final profile. The suite of codes worked well and the calculated crater shapes agreed with experimental data.

Running a calculation was a laborious process in the 1970s. We punched input on card decks. We walked the decks across the tech area ("sneaker-net"), submitted them across a counter, and then returned a few days later for the output. If we made an input error, we tried again. We ran calculations on Control Data Corporation (CDC) 6600s and 7600s. The process got easier when a remote batch facility was set up in a neighboring building. We could then read in card decks and receive printout in the next building, rather than walking across the tech area.

I started enhancing Sam's production codes in the 1980s. These enhancements included geological equations of state, boundary conditions, and second-order numerical convection. The improved numerical convection dramatically increased CSQII's accuracy.

Sandia promoted Sam to a manager position in the Nuclear Reactor Analysis group, so I began supporting the CHARTD and CSQII users. Most users were in the shock physics departments, e.g., Steve Rottler, Marlin Kipp, and Tim Trucano. Some were outside Sandia, e.g., Charles Anderson at the Southwest Research Institute.

Sandia got its first Cray around 1981. The 8 MB of memory were so much more than the CDC computers that I thought I had died and gone to heaven. The Cray exposed me to my first parallel constructs—loop-level vectorization. I spent many evenings analyzing loops and getting CSQII running efficiently on the Cray. Around this time, we started using Digital Equipment Corporation's Virtual Address eXtension (DEC VAX) computers to prepare the input decks and analyze the output files. We stopped using card decks!

After CSQII was running efficiently on the Cray, I left the shock physics group in 1982. I transferred to Sam's department and led a team building the MELCOR nuclear reactor analysis code. The accident at the Three Mile Island nuclear power plant spurred development of accident analysis models.

#### Three-Dimensional Calculations on the Crays

In the mid 1980s, Sam went back to a shock physics department as the manager. About this time, Sandia bought a Cray XMP 416 and management decided to build a three-dimensional (3-D) Eulerian shock physics code. I came back to the shock physics group in 1986 to design and build CTH. We decided to develop a tightly integrated suite that modeled 3-D, 2-D, and 1-D problems.

The 3-D database modeled a rectangular cuboid in x-y-z space. The 2-D database modeled a rectangular (x-y) or cylindrical (radius-y) space. The 1-D database modeled a linear (x), cylindrical (radius), or spherical (radius) space.

Sam Thompson, Millie Elrick (a contractor from Kaman Sciences Corporation), and I started building CTH. The size of the 3-D database was a challenge. The Cray XMP 416 had 128 MB of high-speed memory and 2 GB of slower-speed memory. The 3-D database would reside in the slower-speed memory. The memory management algorithm moved a 2-D x-y plane from the slower-speed memory to high-speed memory when the solution algorithm needed the data. Once the solution algorithm updated the data in the plane, the memory management algorithm wrote the 2-D plane back to the slower-speed memory. By carefully writing the algorithms, only a few 2-D planes needed to be in high-speed memory at one time. (This was an extension of how CSQII managed its 2-D databases on the small-memory CDC 6600s.) This memory management worked well. There was no appreciable penalty for moving the planes between the memories because we overlapped the input and output with calculations.

We wrote CTH in ANSI FORTRAN 77, which allowed the software to be more maintainable, readable, and portable than if we had written it in FORTRAN 66. New capabilities in FORTRAN 77 allowed us to do things that would have been very difficult in FORTRAN 66. For example, CHARACTER data allowed CTH to have a much more user-friendly input. The input was keyword driven rather than requiring the user to line up input in specific columns.

In 1987, CTH was routinely running 2-D calculations and important 3-D calculations. Fred Zeigler joined the development team and added stress deviators (material strength). Tragically, Fred died in a bicycling accident not long after he integrated his algorithms into CTH.

In the late 1980s, we used DEC VAX computers and International Business Machine Personal Computers (IBM PCs) and started using UNIX workstations. As UNIX workstations were phased in, the VAX computers were phased out. In the mid 1980s, Sandia's Crays used the Cray Time-Sharing System (CTSS) operating system. Los Alamos National Laboratory (LANL), in conjunction with Lawrence Livermore National Laboratory (LLNL), had developed CTSS. Later, Sandia transitioned from "home-grown," or proprietary, operating systems to the Cray-supported, UNIX-like UNICOS operating system.

CTH's outstanding user community was one of the key reasons CTH made rapid progress. I thought of Tim Trucano as a great "test pilot." He routinely stress-tested new features and provided invaluable feedback. Gerald Kerley, Stewart Silling, Marlin Kipp, and Dennis Grady were also outstanding users. They were extremely knowledgeable, routinely tested new features, and were always constructive. A few users started adding capabilities to CTH. Key external users like Charles Anderson's associates at the Southwest Research Institute and Kent Kimsey's associates at the Army Research Laboratory were knowledgeable and skilled users who also provided invaluable feedback.

This was a very productive and extremely exciting time. Many scientists inside and outside Sandia developed and documented improved material models, and we implemented many of them in CTH. Computational physicists developed and published improved numerical algorithms, and we integrated many in CTH. Code developers at LANL, LLNL, the United Kingdom's Atomic Weapons Establishment (AWE), and Sandia worked on 3-D codes and algorithms. We exchanged results at conferences and in-house presentations and with other external partners. In some cases, we exchanged software. For example, there was a particularly productive exchange with Tom Adam's code development activities at Los Alamos. The exchange of ideas and results accelerated everyone's activities. The users were ultimately responsible for the quality of algorithms in CTH. They would frequently bring in a calculation that did not meet their expectations. Interactive debuggers were now available on the Cray. These debuggers allowed us to probe individual computational cells to observe what was happening. (I started referring to interactive debuggers as interactive "understanders.") Some coding errors were uncovered. In other cases, we learned the algorithms were inadequate to model the phenomena. Using this process, we implemented the most accurate and robust algorithms.

In the late 1980s, Cray added another level of parallelism on their computers. A calculation could use more than one CPU.[124] The operating system could now spread the iterations of a vectorized loop over the available processors. CTH was able to achieve about a six-fold speedup on an eight-processor YMP.

In early 1990, Sam transferred back to the Nuclear Reactor Analysis area and I became the manager of a shock physics department. Steve Rottler became the leader of the CTH project. Sandia soon promoted Steve, so he moved on to other challenges. This was one of the busiest times of my life.

We started working with the Defense Advanced Research Projects Agency (DARPA). The Department of Defense (DOD) wanted an improved shock modeling capability for their laboratories and contractors. DARPA invested in shock codes at LANL, LLNL, and SNL and funded us to send CTH to key DOD users and add the features they needed. The users were very happy with CTH. In a year, DARPA was heavily investing in CTH. DARPA was a great partner that helped mature CTH and added features important to DOD. Our external user community exploded (no pun intended). By 1995, CTH had over 130 external users. DARPA's support was also politically important inside Sandia because their funding augmented the Department of Energy (DOE) funding, and they had a reputation for developing world-class technology.

## Three-Dimensional Calculations on Massively Parallel Computers

Calculations in 3-D were still slow and constrained to coarser meshes than desired. In 1987, Sandia demonstrated that a thousand CPU nCUBE massively parallel computer could speed up a physics calculation by a factor of 400 or more! By 1993, Sandia had an Intel Paragon massively parallel computer with 1872 CPUs and 37 GB of memory. We were convinced that the CTH algorithms would run effectively on these computers. This would allow us to perform much higher fidelity 3-D calculations in an acceptable run time.

[124] Editors' note: This acronym stands for Central Processing Unit.

In 1990, Allen Robinson and his team began developing a plan to move CTH to massively parallel computers. Our first effort was to write a special CTH version for massively parallel computers called PCTH. We were concerned with the size of memories on the computational nodes. The Paragon initially had 1344 nodes with 16 MB of memory and 528 nodes with 32 MB of memory. 16 MB was not a lot to hold the PCTH executable and a meaningful subset of the 3-D database. We limited the size of the PCTH executable by using only the core CTH 3-D algorithms rather than moving over all the 1-D, 2-D, and material algorithms. Debugging on a massively parallel computer was also a concern. We decided to write PCTH in C++ rather than in FORTRAN. We hoped to eliminate some coding errors using C++ features. We also wanted to learn more about using C++ for computer-intensive calculations.

Allen's team got a version of PCTH working that demonstrated excellent parallel speedup and good performance; his team also demonstrated that we could eliminate many software defects in a carefully written C++ program. The massively parallel computers were challenging to use at that time because of hardware and system software issues associated with all new computing environments. Mark Boslough, David Crawford, and Tim Trucano were some of the first users. In 1994, they analyzed the Shoemaker–Levy 9 comet impacts on Jupiter using 3-D calculations on the Intel Paragon. PCTH proved spectacularly successful. Astronomers could not observe the comet impacts because they were on the side of Jupiter facing away from Earth. Our calculations indicated an optical signature should be visible over the apparent visual edge (limb) of Jupiter. Astronomers could infer information about the impacts from that optical signature. Mark, Dave, and Tim received a Sandia Quality Award and published several papers on this work. It was a great demonstration of PCTH's capabilities and the power of massively parallel computers.

At the end of the PCTH project, we were comfortable with moving CTH to massively parallel computers. It was a lot of work. I believe Millie Elrick found the last modification required for the port. The CTH community owes a tremendous debt to Millie. She was one of the original developers, worked on the software longer than any of the other original developers, supported the external user community, and enhanced just about every part of the CTH suite. She was very, very good. By the mid 1990s, CTH ran on massively parallel computers and gave answers identical to the Cray version.

Sandia installed the ASCI Red supercomputer in 1996. It had 9072 CPUs and 594 GB of memory! CTH was ported over without undue difficulty. We were running much larger calculations than we ever thought possible and they were running fast. What a difference a decade makes. We went from 4 CPUs with 2 GB of memory to 9072 CPUs with 594 GB of memory!

The Army Research Laboratory purchased a massively parallel computer and quickly got CTH running. Now they could run high-fidelity 3-D calculations. DARPA's investment was still paying dividends to DOD. The DOD/DOE Memorandum of Understanding continued to support DOD users after DARPA's support ended.

### RHALE/ALEGRA Program

Tim Trucano and I began investigating a beyond-CTH shock physics code in the late 1980s. Sandia's inertial confinement fusion community needed capabilities that CTH did not provide. We believed the best underlying strategy would be to develop an arbitrary Lagrangian-Eulerian (ALE) code. The mesh in an ALE code could be Eulerian or Lagrangian or have an arbitrary motion.

James Peery, Kent Budge, and Mike Wong began developing a 3-D ALE code named RHALE in 1990. They studied the CTH algorithms as well as the algorithms in Sandia's PRONTO finite-element Lagrangian transient dynamics code. The plan was to use the best of both. They decided to use C++ rather than FORTRAN. Later they renamed the code ALEGRA. This was a very challenging project. ALEGRA demonstrated tremendous capability. It is a specialized code and not appropriate for all applications.

### The End of My Shock Wave Contributions

In 1996, I moved to another area at Sandia and ended my shock physics work. The shock codes remained in excellent hands. Gerald Kerley continued to improve the equations of state. David Crawford added adaptive mesh refinement. I stayed in touch with many of the researchers and am extremely proud of their accomplishments. I never ran another calculation or modified a code. My activities have been limited to explaining what we did, helping find bugs I authored, and writing histories.

## Annotated References to Sandia Code Development (1966–1992)

Below is a brief set of references about Sandia's shock codes. I have added brief annotations to some of the references. I started with the two codes Walt Herrmann brought to Sandia in the mid 1960s. I then included the codes Sam Thompson helped develop. I did not include codes developed after the 1990s such as ALEGRA. I did not include codes such as HONDO, PRONTO, or PRESTO that are not designed to be strong shock codes. I believe a WONDY and TOODY developer would collect a more complete set of references for those two codes. Hopefully the list below will be useful.

### TOODY Code

1. W. Herrmann, "A Lagrangian finite difference method for two-dimensional motion including material strength," Air Force Weapons Laboratory Technical Report WL-TR-64-107 (November 1964).

This is the earliest report I could find on the 2-D Lagrangian code that Walt developed when he was at Massachusetts Institute of Technology under contract to AFWL before coming to Sandia.

2. B.J. Thorne and W. Herrmann, "TOODY: A computer program for calculating problems of motion in two dimensions," SC-RR-66-602, Sandia Laboratories, Albuquerque, NM (January 1967).

Walt Herrmann moved to Sandia and brought his 2-D Lagrangian code with him. Billy Thorne worked with Walt on TOODY.

The next two computer programs are for pre- and post-processing the TOODY data. These are examples of the graphical analysis tools that were developed to examine the code results.

3. J.W. Swegle, "TOOHIST: A history post-processor for ARTOO," SAND88-1977, Sandia National Laboratories, Albuquerque, NM (September 1988).
4. J.W. Swegle, "TOOPLOT: A mesh plotting post-processor for ARTOO," SAND88-2074, Sandia National Laboratories, Albuquerque, NM (September 1988).

**WONDY Code**

1. W. Herrmann, P. Holzhauser, and R.J. Thompson, "WONDY: A computer program for calculating problems of motion in one dimension," SC-RR-66-601, Sandia Laboratories, Albuquerque, NM (February 1967).

Walt Herrmann also brought his 1-D Lagrangian code to Sandia. It was in continuous use through the 1990s. WONDY VII was documented in 1985 by Fred Zeigler. I am pretty sure Marlin Kipp continued to use WONDY for many more years—at least into the mid 1990s. Newer documents often described new capabilities and revised user input.

2. R.J. Lawrence, "WONDY IIIa: A computer program for one-dimensional wave propagation," SC-DR-70-315, Sandia Laboratories, Albuquerque, NM (August 1970).
3. M.E. Kipp and R.J. Lawrence, "WONDY V: A one-dimensional finite-difference wave-propagation code," SAND81-0930, Sandia National Laboratories, Albuquerque, NM (June 1982).
4. F.J. Zeigler, "WONDY VII: A vectorized version of the WONDY wave propagation code," SAND85-2338, Sandia National Laboratories, Albuquerque, NM (February 1986).

**CHARTD Code**

1. S.L. Thompson, "CHARTD: A computer program for calculating problems of coupled hydrodynamic motion and radiation flow in one dimension," SC-RR-69-613, Sandia Laboratories, Albuquerque, NM (November 1969).

This is the first document I could find of Sam Thompson's 1-D Lagrangian radiation hydrodynamics code. It used the diffusion approximation. Sam told me his first version didn't use the diffusion approximation and was called CHART. He replaced the radiation transport model with the diffusion approximation. Sam and others steadily improved CHARTD over the years. Newer documents described the new features and revised user input.

2. S.L. Thompson, "Improvements in the CHARTD radiation-hydrodynamic code I: Analytic equations of state," SC-RR-70-28, Sandia Laboratories, Albuquerque, NM (January 1970).

Sam developed his ANEOS equation of state. It is a thermodynamically consistent equation of state valid for multiphase materials. It did a good job for many materials for a long time.

3. S.L. Thompson and H.S. Lauson, "Improvements in the CHARTD radiation–hydrodynamics code II: A revised program," SC-RR-71-0713, Sandia Laboratories, Albuquerque, NM (February 1972).

Sam added a tabular equation of state option to CHARTD. I used them sometimes but they never really caught on. They were not as powerful as the tabular equations of state that Kerley developed.

4. S.L. Thompson and H.S. Lauson, "Improvements in the CHARTD radiation-hydrodynamic code III: Revised analytical equations of state," SC-RR-71-0714, Sandia Laboratories, Albuquerque, NM (March 1972).
5. S.L. Thompson and H.S. Lauson, "Improvements in the CHART D radiation-hydrodynamic code IV: User aid programs," SC-DR-71-0715, Sandia Laboratories, Albuquerque, NM (March 1972).
6. S.L. Thompson, "Improvements in the CHART D energy flow hydrodynamic code V, 1972/73 modifications," SLA-73-0477, Sandia Laboratories, Albuquerque, NM (August 1973).
7. S.L. Thompson, "CKEOS2 – An equation of state test program for the CHARTD/CSQ EOS package," SAND76–0175, Sandia Laboratories, Albuquerque, NM (May 1976).

This is a graphics system for checking equation-of-state data using the ANEOS equation of state. It would calculate data such as Hugoniots, phase diagrams, and isentropes. It was very powerful and extremely useful.

8. S.L. Thompson, "RSCORS—A revised SCORS plot system," SAND77-1957, Sandia Laboratories, Albuquerque, NM (May 1978).

Sam developed a graphics system to support his codes. It was very useful when we sent copies of the codes outside Sandia because it was easy to port to other graphics systems.

9. J.M. McGlaun, "New features and revised input instructions for CHARTD," SAND81-2581, Sandia National Laboratories, Albuquerque, NM (November 1981).

I added new features such as pressure-dependent yield surfaces for geological materials.

10. G.C. Padilla and S.L. Thompson, "RSCORS graphics system," SAND83-2639, Sandia National Laboratories, Albuquerque, NM (January 1984).
11. S.L. Thompson, "ANEOS analytic equations of state for shock physics codes input manual," SAND89-2951, Sandia National Laboratories, Albuquerque, NM (March 1990).

Sam continued to enhance ANEOS.

**CSQ Code**

1. S.L. Thompson, "CSQII—An Eulerian finite difference program for two-dimensional material response—Part 1. Material sections," SAND77-1339, Sandia National Laboratories, Albuquerque, NM (January 1979).

CSQII modeled problems with up to ten materials. CSQII also had diffusion approximation radiation transport. This report described the mechanics, not the radiation, algorithms.

2. J.M. McGlaun, S.L. Thompson, J.R. Freeman, "COMAG-III: A 2-D MHD code for helical CMF generators," in *Megagauss Physics and Technology*, edited by P.J. Turchi (Plenum Press, New York, NY), pp. 193–203 (1980).

This is an example of coupling CSQII to another code. John Freeman was a very smart manager in the inertial confinement fusion world. He developed and coupled an electromagnetics capability to CSQII. I worked with John and Sam on this code for a couple of years. It gave good agreement with experimental data.

3. J.R. Freeman, J.M. McGlaun, S.L. Thompson, and E.C. Cnare, "Numerical studies of helical CMF generators," in *Megagauss Physics and Technology*, edited by P.J. Turchi (Plenum Press, New York, NY), pp. 205–218 (1980).
4. J.M. McGlaun, "Crater: A computer code for calculating late stage crater formation," SAND80-1694, Sandia National Laboratories, Albuquerque, NM (February 1981).

This is another example of using CSQII to solve a more complicated problem. I wrote a post-processor for CSQII that calculated the final shape of explosively formed craters. It took a CSQII restart file and ballistically extrapolated the ejecta trajectories under a gravitational field. It gave good agreement with data. See the following report.

5. J.M. McGlaun, "Computer modeling of the pre-Gondola cratering experiments," SAND80-1695, Sandia National Laboratories, Albuquerque, NM (September 1980).
6. J.M. McGlaun, "Improvements in CSQII: An improved numerical convection algorithm," SAND82-0051, Sandia National Laboratories, Albuquerque, NM (January 1982).

This enhancement implemented the numerical method described in the following two papers by Bram van Leer. It dramatically reduced dispersion in the Eulerian remap step and dramatically improved accuracy.

7. B. van Leer, "Towards the ultimate conservative difference scheme. II. Monotonicity and conservation combined in a second-order scheme," *Journal of Computational Physics* **14**, pp. 361–370 (1974).
8. B. van Leer, "Towards the ultimate conservative difference scheme. IV. A new approach to numerical convection," *Journal of Computational Physics* **23**, pp. 176–199 (1977).
9. J.M. McGlaun, "Improvements in CSQII: A transmitting boundary condition," SAND82-1248, Sandia National Laboratories, Albuquerque, NM (June 1982).

This enhancement integrated recent numerical algorithms for non-reflective boundary conditions. It allowed a stress wave to exit the mesh with very little numerical reflection from the boundary.

10. S.L. Thompson, "CMP code maintenance package: Reference manual version 2.0," SAND85-0825, Sandia National Laboratories, Albuquerque, NM (April 1985).

The code systems were large and configuration management was a challenge. Sam developed his own configuration management system to manage the software inside Sandia and the versions we sent outside Sandia.

11. T.G. Trucano, D.E. Grady, J.M. McGlaun, "Fragmentation statistics from Eulerian hydrocode calculations," *International Journal of Impact Engineering* **10**, pp. 587–600 (1990).

This is an example of using a production code as a preprocessor for another problem, e.g., crater formation. Dennis Grady and Marlin Kipp did a lot of fragmentation work using CSQ and CTH output.

### CTH Code

1. J.M. McGlaun, S.L. Thompson, and M.G. Elrick, "CTH: A three-dimensional shock wave physics code," *International Journal of Impact Engineering* **10**, pp. 351–360 (1990).
2. D.L. Youngs, "An interface tracking method for a 3-D Eulerian hydrodynamics code," Atomic Weapons Research Establishment, AWRE/44/92/35 (April 1987).

David Youngs was a brilliant numerical scientist who worked at Great Britain's Atomic Weapons Research Establishment, the predecessor of the Atomic Weapons Establishment. He developed a 3-D Eulerian interface tracking method that was the best algorithm available and solved an extremely complex and difficult problem. LANL implemented it in their codes and it worked very well. LANL generously gave us their software, which helped us implement it in CTH and dramatically improved the accuracy.

# Stephen T. Montgomery
# (1975–2011)

I first visited the Albuquerque and Livermore sites of Sandia National Laboratories in December 1975, prior to finishing my thesis work at Purdue University. Because of my interest in interactions between mechanical and electrical phenomena, I accepted an offer to join a development group in Albuquerque to work on the analysis of explosively-driven ferroelectric power sources. While waiting on my security clearance, I studied the pioneering work conducted at Sandia on shock compression of dielectrics. I also became familiar with the basics of shock wave compression by reading review articles and a set of class notes prepared by Dennis Hayes.

Frank Neilson had proposed using the displacement current generated by rapid changes of electric polarization in materials under shock compression as a power source a few years prior to 1960. Research groups at Sandia conducted extensive experiments on a number of different materials, using explosive loading and gas guns, to study this idea in the following decade. For example, the nonlinear piezoelectric response in X-cut quartz crystals was used as a stress gauge for materials loaded in shock compression, and the depolarization of polycrystalline ferroelectric ceramics under shock compression was exploited to develop a variety of shock wave-activated power sources. Applied research, with the goal of an improved understanding of material polarization changes in ferroelectric ceramics, continued with Pete Lysne by conducting experiments and developing simple models for electrical response and with Pete Chen in developing more detailed coupled electromechanical constitutive models for use in numerical simulations. I was asked to work with Lysne and Chen to gain an understanding of what they were doing and to develop application models.

Chen and Lee Davison had developed a one-dimensional (1-D) response model for coupled mechanical and electrical response when the direction of material polarization and material motion was identical. This special configuration was called axial-mode operation. Jeff Lawrence and Davison built a numerical model for this special case in the 1-D wave propagation code WONDY. My group was interested in what was called the normal-mode configuration, for which the polarization direction was perpendicular to the direction of material motion. I converted the axial-mode response model in WONDY to a normal-mode response model and then applied the model to develop a complex device of interest to my group. The device relied on using a previously established explosive loading device to generate the desired shock compression for depolarization of ferroelectric ceramics. The culmination of this work was an "admiral's test" of the device with our vice president in attendance for that first test. To my great relief and that of the three levels of management above me, the electric output from the device was reasonably close to the result I predicted, and development engineers took over the project. After the demonstration of the ferroelectric device, I was asked to take primary responsibility for the firing system to be used in the Pershing II Earth penetrator. The Pershing II work was soon canceled and my career direction at Sandia changed.

These first few years at Sandia provided me with a working knowledge of shock wave physics, some experimental methods, material response modeling, and the numerical foundation of the WONDY wave propagation code. Pete Lysne and the development engineers in my group provided a nice introduction to gas gun impact, explosive shock generation, and the variety of power sources based on depolarization by shock compression. An important foundation for my future work at Sandia was built during the construction of material response models for ferroelectric ceramics with Pete Chen.

My experience in modeling and simulation in the firing systems group served me well when I transferred, in late 1979, to a computational mechanics group in an organization headed by Walt Herrmann. The other groups in his organization performed shock compression studies, computational physics simulations, and characterization of geological materials. At the time, there was considerable development work on the Waste Isolation Pilot Plant in southern New Mexico, and the oil embargo had motivated the creation of the Strategic Petroleum Reserve sited along the Gulf of Mexico in Louisiana and Texas. I was assigned to include transient creep models in the two-dimensional (2-D) finite-element code MARC and to validate these models by comparing code predictions to the measured creep closure of caverns excavated in potash deposits and brine flow from bore holes used to form oil storage caverns in salt domes.

After about 18 months, I moved into a computational mechanics and physics group, managed by Al Chabai, and was again modeling the response of materials and structures to shock waves. Initially I worked on characterizing the response of a porous material to sudden heating by radiation. I performed simulations of gas gun experiments that were being conducted by Lalit Chhabildas to study the spall strength of materials and their response under the rapid heating produced from exposure to an intense electron beam generated by the HERMES accelerator. I used the data from these experiments to characterize a porous material response model for WONDY.

I also performed numerous simulations on the 2-D Eulerian code CSQ that involved projectile impact into sands, hypervelocity impacts, and evaluation of an explosively-activated closure for the line-of-sight pipe in a proposed low-yield underground test bed. My work on that test bed allowed me to collaborate with staff in the underground test community at Sandia. I provided simulation support for scaled and full-size test activities conducted by Hal Walling, Les Hill, and Bob Bass. Many tests used conventional explosives to study the response of scale models. However, several tests at the Nevada Test Site used nuclear explosions to load models of the closure, and I needed to squeeze into protective clothing and a respirator to examine the excavated closure models. The test results indicated some validity of the calculations, and the feasibility of test bed modifications was demonstrated in an underground test.

I returned to the study of polarization changes by shock compression in ferroelectric ceramics in 1986. The groups at Sandia developing neutron generators were managed by Dennis Hayes, and he wanted to supplement the prototype testing approach with simulations to understand issues encountered during that development.

At the time, a systematic study of the mechanical and dielectric properties of ferroelectric ceramics used in components did not exist. Lalit Chhabildas initiated an experimental study of ceramic response under shock compression. I investigated methods to permit multi-dimensional simulations of the components. As a result, a simulation framework was developed in the SUBWAY code in collaboration with Pat Chavez. SUBWAY originally consisted of integration of a finite-element-based electric potential solution superimposed on the finite-difference grid of a TOODY simulation for wave motion in two dimensions. The code contained a simple model for transformation from the ferroelectric to anti-ferroelectric phase responsible for the polarization change in a ceramic under shock compression.

Computational capability was advancing, and three-dimensional (3-D) simulations were now feasible with the availability of the CDC, Cray, DEC, and SUN systems at Sandia. I decided to develop an integrated 3-D finite-element framework for SUBWAY and teamed with Lee Taylor to use the explicit finite-element code PRONTO to replace the functions of TOODY. Early 3-D simulations of the device provided mixed results. Some aspects, particularly the ability to simulate the electric field in the ceramic, were well received by the development group. However, while the electric output from the simulations was similar to the measured output, there was room for improvement.

In late 1990, significant organizational changes and directions of emphasis occurred at Sandia. I was assigned to an advanced visualization and mechanics group, where my time was divided among the development of Linux use conventions at the Laboratory, a transient arbitrary Lagrangian-Eulerian (ALE) finite-element response code, and improving SUBWAY's capability to support neutron generator design.

My interest in the ferroelectric ceramic work and application modeling motivated a transfer to the neutron generator group in summer 1992. Along with supporting several neutron generator development programs, I coordinated work to improve simulation capability by developing a fundamental understanding of ferroelectric ceramics and encapsulation materials under operational conditions.

Almost all studies of phase transformation in ceramics under compression used hydrostatic pressure, with the transformation occurring when the pressure reached a critical value. Ian Fritz noted significant differences from the hydrostatic response for samples he loaded in uniaxial stress and speculated that domain switching now accounted for almost all of the polarization change. This suggested that confinement would be important in driving the transformation in shock compression. Several limited experimental studies were initiated in collaboration with Jeff Keck between 1990 and 1996. A systematic series of experiments to characterize the role of stress on depolarization under shock compression was conducted at the Air Force Weapons Laboratory. These experiments showed that depolarization was rate dependent, and the axial stress intensity required to depolarize the ceramic completely was well in excess of the quasi-static hydrostatic transformation pressure. We also began a long collaboration with Dave Zeuch to characterize the quasi-static response of the ceramic systematically for various conditions of stress, electric field, and temperature. I also worked on a number of simple material response models for polarization changes of various dielectric materials such as quartz, ferroelectric copolymers, and non-transforming ferroelectric ceramics under shock compression during this period.

The Department of Energy began the stockpile stewardship initiative to reduce reliance on expensive tests and increase use of simulations to design and analyze nuclear weapons across the nuclear weapons complex. From 1996 to 2002, the stockpile stewardship and component development funds allowed a significant increase in experimental work and code development to improve the simulation capability.

My collaboration with Keck and Zeuch continued, and we were able to construct an advanced confinement vessel to allow the ceramic to be studied under complex states of stress and electrical bias across the temperature range of interest. The quasi-static loading experiments showed that the phase transformation in a ferroelectric domain was dependent upon the orientation of its spontaneous electric polarization in the stress field. This significant observation clarified many of the results from the comprehensive shock compression studies on mechanical and electric responses of the ceramic conducted by Bob Setchell, Lalit Chhabildas, Mike Furnish, and Mark Anderson. The shock compression experiments were challenging since, in addition to standard mechanical measurement diagnostics, voltages and currents from the ceramic were also measured. The experimental results provided basic data to improve material response models in the simulation code developed in collaboration with Rebecca Brannon and Joshua Robbins.

The stockpile stewardship funding also allowed implementation of algorithms and models developed for SUBWAY into a massively parallel ALE code called ALEGRA/EMMA. As an example of the simulation fidelity provided by parallel computation, a typical 3-D SUBWAY simulation I conducted in the early 1990s on a single processor used up to 400,000 finite elements. By the end of the 1990s, I was routinely conducting simulations with ALEGRA/EMMA on massively parallel computers with around five million elements, and several component simulations used element numbers ranging in tens of millions.

During this period, Keck and I developed a number of specialized shock compression test assemblies to help validate models in 3-D loading configurations. Some of these tests were conducted using proton radiography at Los Alamos to visualize component deformation under explosive loading. Additional testing, first by Bob Graham and then by Bob Setchell, used projectile impacts to study component and material response under different 3-D loading histories.

The effort to understand ceramic response was substantially completed in 2002. Additional studies at a much lower funding level were conducted from 2003 to 2008 to improve the understanding of dielectric properties, the effects of initial temperature, and composition variation on the shock compression response of the ceramic and encapsulation material in collaboration with Setchell, Anderson, and Dave Cox. Our primary interest was to understand the effect of temperature and composition variation on material response to support an initiative to quantify uncertainties in device performance.

In 2008 I moved to a solid mechanics group that was primarily concerned with structural system response. I helped code developers understand and verify equation-of-state models placed into a new family of finite-element codes (the SIERRA mechanics codes) and validated a coupling between an Eulerian hydrocode and the SIERRA codes to simulate blast loads on structures. In late 2009, the group responsible for impact fuzing wanted to develop a simulation capability. I joined this effort and again found myself working on stress-induced depolarization of ferroelectric ceramics.

I defined a project for simulation model development and experimental characterization for the ceramic material used as a sensing element by the fuze development group. A portion of the project involved integrating the ALEGRA/EMMA and SIERRA codes to allow simulation at a systems level. Since funding for the experimental characterization work was limited, I proposed a small testing program that relied on using ramp wave loading to study the material response. A test configuration using symmetric impact of fused silica plates to generate ramp loading of the ceramic was developed in collaboration with Jack Wise. We identified some unexpected electrical results as being due to a structural phase transformation between different ferroelectric structures. Jack showed this transformation was rate dependent with experiments on unpoled ferroelectric ceramic using ramp waves produced on Veloce. This work was still in progress when I transferred my responsibilities to others prior to my retirement in 2011.

## References (1977–2010)


1. P.J. Chen and S.T. Montgomery, "Normal mode responses of linear piezoelectric materials with hexagonal symmetry," *International Journal of Solids Structures* **13**, pp. 974–955 (1977).
2. P.J. Chen and S.T. Montgomery, "Boundary effects on the normal-mode responses of linear transversely isotropic piezoelectric materials," *Journal of Applied Physics* **49**, pp. 900–904 (1978).
3. P.J. Chen and S.T. Montgomery, "A macroscopic theory for the existence of the hysteresis and butterfly loops in ferroelectricity," *Ferroelectrics* **23**, pp. 199–207 (1980).
4. S.T. Montgomery, "Computer simulations of 75-mm projectiles impacting sand targets," SAND82-0831, Sandia National Laboratories, Albuquerque, NM (1982).
5. W.R. Wawersik, W. Herrmann, S.T. Montgomery, and H.S. Lauson, "Excavation design in rock salt – Laboratory experiments, material modeling and validations," in *Rock Mechanics: Caverns and Pressure Shafts*, edited by W. Wittke (Balkema, Rotterdam, 1982), pp. 1345–1356.
6. S.T. Montgomery, "Creep closure of an opening in a deep potash mine," in *Recent Advances in Engineering Mechanics and Their Impact on Civil Engineering Practice*, edited by Chen and Lewis (American Society of Civil Engineers, New York, NY, 1983), Volume II, pp. 1024–1027.

7. S.T. Montgomery, "Calculated loading on a fast acting closure," in *Proceedings of the Second Symposium on Containment of Underground Nuclear Explosions*, Kirtland Air Force Base, Albuquerque, NM, pp. 401–412 (1983).
8. S.T. Montgomery and P.J. Chen, "Influences of domain switching and dipole dynamics on the normal mode responses of a ferroelectric ceramic bar," *International Journal of Solids Structures* **22**, pp. 1293–1305 (1986).
9. S.T. Montgomery and P.F. Chavez, "Basic equations and solution method for the calculation of the transient electromechanical response of dielectric devices," SAND86-0755, Sandia National Laboratories, Albuquerque, NM (1986).
10. S.T. Montgomery, "Analysis of transitions between ferroelectric and antiferroelectric states under conditions of uniaxial strain," in *Shock Waves in Condensed Matter*, edited by Y.M. Gupta (Plenum, New York, NY, 1986), pp. 179–184.
11. D.H. Zeuch, S.T. Montgomery, and J.D. Keck, "Hydrostatic and triaxial compression experiments on unpoled PZT 95/5-2Nb ceramic: The effects of shear stress on the $F_{R1} \rightarrow A_O$ polymorphic phase transformation," *Journal of Materials Research* **7**, pp. 3314–3332 (1992).
12. D.H. Zeuch, S.T. Montgomery, and J.D. Keck, "Some observations on the effects of shear stress on a polymorphic transformation in perovskite-structured lead-zirconate-titanate ceramic," *Journal Geophysical Research* **98** (B2), pp. 1901–1911 (1993).
13. D.H. Zeuch, S.T. Montgomery, and J.D. Keck, "Further observations on the effects of nonhydrostatic compression on the $F_{R1} \rightarrow A_O$ polymorphic phase transformation in niobium-doped, lead-zirconate-titanate ceramic," *Journal of Materials Research* **9**, pp. 1322–1327 (1994).
14. S.T. Montgomery, R.A. Graham, F. Bauer, and H. Moulard, "Copolymer shock gauge response investigation with the fully coupled electromechanical code, SUBWAY," in *Proceedings from the Workshop on the Technology of Ferroelectric Polymers*, pp. 412–431 (1995).
15. S.T. Montgomery, R.A. Graham, and M.U. Anderson, "Return to the shorted and shunted quartz gauge problem: analysis with the SUBWAY code," in *Shock Compression of Condensed Matter—1995*, edited by S.C. Schmidt and W.C. Tao (American Institute of Physics, Melville, New York), AIP Conference Proceedings **370**, pp. 1025–1028 (1996).
16. D.H. Zeuch, S.T. Montgomery, and D.J. Holcomb, "The effects of nonhydrostatic compression and applied electric field on the electromechanical behavior of poled PZT 95/5-2Nb ceramic during the $F_{R1} \rightarrow A_O$ polymorphic transformation," *Journal of Materials Research* **14**, pp. 1814–1827 (1999).
17. D.H. Zeuch, S.T. Montgomery, and D.J. Holcomb, "Uniaxial compression experiments on lead zirconate titanate 95/5-2Nb ceramic: Evidence for an orientation-dependent, maximum compressive stress criterion for onset of the ferroelectric to antiferroelectric polymorphic transformation," *Journal of Materials Research* **15**, pp. 689–703 (2000).
18. L.M. Lee, S.T. Montgomery, and P.H. Jilbert, "Multi-element quartz shock gauge development," in *Shock Compression of Condensed Matter—1999*, edited by

M. D. Furnish, L. C. Chhabildas, and R. S. Hixson (American Institute of Physics, Melville, New York), AIP Conference Proceedings **505**, pp. 1015–1018 (2000).

19. R.E. Setchell, S.T. Montgomery, L.C. Chhabildas, and M.D. Furnish, "The effects of shock stress and field strength on shock-induced depoling of normally poled PZT 95/5," in *Shock Compression of Condensed Matter—1999*, edited by M.D. Furnish, L.C. Chhabildas, and R.S. Hixson (AIP, Melville, New York), AIP Conference Proceedings **505**, pp. 979–982 (2000).
20. S.T. Montgomery, R.M. Brannon, J. Robbins, R.E. Setchell, and D.H. Zeuch, "Simulation of the effects of shock stress and electrical field strength on shock-induced depoling of normally poled PZT 95/5," in *Shock Compression of Condensed Matter—2001*, edited by M.D. Furnish, N.N. Thadhani, and Y.M. Horie (AIP, Melville, New York), AIP Conference Proceedings **620**, pp. 201–204 (2002).
21. R.M. Brannon, S.T. Montgomery, J.B. Aidun, and A.C. Robinson, "Macro- and meso-scale modeling of PZT ferroelectric ceramics," in *Shock Compression of Condensed Matter—2001*, edited by M.D. Furnish, N.N. Thadhani, and Y.M. Horie (AIP, Melville, New York), AIP Conference Proceedings **620**, pp. 197–200 (2002).
22. M.D. Furnish, J. Robbins, W.M. Trott, L.C. Chhabildas, R.J. Lawrence, and S.T. Montgomery, "Multi-dimensional validation impact tests on PZT 95/5 and ALOX," in *Shock Compression of Condensed Matter—2001*, edited by M.D. Furnish, N.N. Thadhani, and Y.M. Horie (AIP, Melville, New York), AIP Conference Proceedings **620**, pp. 205–208 (2002).
23. S.T. Montgomery and D.H. Zeuch, "A model for the bulk mechanical response of porous ceramics exhibiting a ferroelectric-to-antiferroelectric phase transition during hydrostatic compression," in *Ceramic Engineering and Science Proceedings*, edited by E. Lara-Curzio and M.J. Readey (The American Ceramic Society, Westerville, Ohio) **25**, pp. 313–318 (2004).
24. R.E. Setchell, S.T. Montgomery, D.E. Cox, and M.U. Anderson, "Dielectric properties of PZT 95/5 during shock compression under high electric fields," in *Shock Compression of Condensed Matter—2005*, edited by M.D. Furnish, M. Elert, T.P. Russell, and C.T. White (AIP, Melville, New York), AIP Conference Proceedings **845**, pp. 278–281(2006).
25. M.U. Anderson, D.E. Cox, S.T. Montgomery, and R.E. Setchell, "Initial temperature effects on the shock compression and release properties of different alumina-filled epoxy compositions," in *Shock Compression of Condensed Matter—2007*, edited by M. Elert, M.D. Furnish, R. Chau, N. Holmes, and J. Nguyen (AIP, Melville, New York), AIP Conference Proceedings **955**, pp. 789–792 (2007).
26. R.E. Setchell, M.U. Anderson, and S.T. Montgomery, "Compositional effects on the shock-compression response of alumina-filled epoxy," *Journal of Applied Physics* **101**, 083527 (2007); Erratum: **102**, 119903 (2007).
27. J.C.F. Millet, D. Deas, N.K. Bourne, and S.T. Montgomery, "The deviatoric response of an alumina filled epoxy composite during shock loading," *Journal of Applied Physics* **102**, 063518 (2007).

28. S.T. Montgomery, "Effects of porosity and pore morphology on the elastic properties of unpoled PZT 95/5-2Nb," SAND2008-6185, Sandia National Laboratories, Albuquerque, NM (2008).
29. S.T. Montgomery, "Effects of composition on the mechanical response of alumina filled epoxy," SAND2009-6399, Sandia National Laboratories, Albuquerque, NM (2009).
30. B. Song, W. Chen, S.T. Montgomery and M.J. Forrestal, "Mechanical response of an alumina-filled epoxy at various strain rates," *J. Comp. Mat.* **43**, pp. 1519–1536 (2009).
31. T.J. Vogler, C.S. Alexander, J.L.Wise, and S.T. Montgomery, "Dynamic behavior of tungsten carbide and alumina filled epoxy composites," *Journal of Applied Physics* **107**, 043520 (2010).

# Bruno Morosin

# (1961–1997)

# Sandia's Shock-Induced Solid State Chemistry

Bruno came to Sandia in 1961 from the Microwave Department of Hughes Research, Hughes Aircraft as a staff member in the Solid State Physics Department. He was perhaps notable as the only supervisor being transferred five times between two departments, one dealing with the fundamental aspects of solid state physics while the other dealt with applied physics.

In the late 1970s, Robert (Bob) A. Graham and Lee Davison examined and reviewed the scientific literature dealing with the shock compression of solids. This comprehensive study eventually appeared in *Physics Reports* **55**, 255–379 (1979) as "Shock compression of solids" by Lee Davison and Bob Graham. Upon its completion, Bob Graham decided to look in more depth at Adadurov, Goldanski, and other Soviet scientists referenced in the *Mendeleev Chemical Journal* **18**, No. 1, 1973, and he was astonished to find a well-developed scientific effort reporting chemical changes and reactions. This was in contrast to the prevalent opinion of most U.S. experts who agreed that chemistry could not occur on the time scale of shock compression in normally inert solids. Many of the references were in Soviet journals that appeared without English translations. It was a major effort for Bob Graham to acquire the journal articles and have them translated. Pat Newman of the Sandia library staff provided much of the indispensable service needed for translations of the Soviet literature. In 1979, Bob Graham wrote an internal memorandum, "Chemical activity in shock-loaded solids," as well as subsequently going to Washington, D.C., to present a briefing to the National Materials Advisory Board on this well-developed Soviet scientific effort, existing essentially unknown to the remainder of the shock compression community. This triggered a briefing to the then current upper management, consisting of Sandia's President Morgan Sparks, Vice President Albert Narath, and the Director of Solid State Research, John Galt, on this Soviet shock compression-induced chemistry efforts. Such briefings were extended to other internal and external organizations.

In response, Sandia management decided to transfer Bruno Morosin from his Materials Division position under Jim Schirber's Solid State Department to become the Supervisor of the Shock Wave and Explosives Division in the Applied Physics Department under George Samara.

A core group was organized around various scientific disciplines. Within Samara's department, this consisted of Bob Graham for his physics involving shock loading as well as his explosive technology background, Gene Venturini for electron spin resonance and other magnetic properties, and Bruno Morosin for x-ray diffraction and general materials properties. In addition, other individuals at Sandia, such as Marty Carr for electron diffraction expertise related to defects, Ed Beauchamp for ceramics and their sintering properties, and Dave Webb for calculations of the spatial temperatures and pressures profiles due to the shock wave response within Bob Graham's containment fixtures, were interfaced as experts in their particular fields. This core group was tasked to aid Bob Graham in the evaluation of the Soviet literature on this subject, coined "shock chemistry," for what is essentially shock wave-induced solid state chemistry and materials modification and to propose an appropriate Sandia effort.

Initial research efforts were concentrated on several compound ceramics. Titania, alumina, and zirconia were studied for various applications by Ed Beauchamp, Bill Hammatter, and Ron Loehman in order to see if shock modification might offer an alternative or enhanced pathway to the desired effect for those current Sandia projects. Another example was an ongoing study of the mineral pyrite, namely, iron sulfide, on enhancing coal liquefaction, which appeared to offer some promise. Pyrite was shock loaded and employed in the liquefaction reactors to note whether the increase of defect levels might enhance the catalytic properties of the pyrite. Early on, it was established that the shock wave action and its inhomogeneous interaction of compression, relaxation, and materials mixing via local transport, as well as diffusion, could significantly alter ordinarily tough inert materials by introducing high numbers of defects.

Eventually, various professors and their students from universities were sought for particular capabilities lacking or oversubscribed at Sandia; these included Yuki Horie at (initially) North Carolina University on modeling shock processes, Jim Stewart at the University of Maryland on profile code development, Frank Williams at the University of New Mexico on catalysis, and Don Williamson for Mossbauer studies at the Colorado School of Mines. Pat Newman continued to provide the much needed translations of the Soviet literature as journal papers were uncovered. In addition, the assistance of many unmentioned individuals within Sandia and students at various universities provided the laboratory help required; these individuals became either coauthors or were acknowledged in the published journal papers.

Early in this effort on October 7, 1980, a one-day meeting with a Soviet scientist, Stephanov Batsanov, was organized by Jim Mote of the Denver Research Institute at the University of Denver. Jim Mote was one of the leaders in materials processing with high explosives and had previously been at Sandia in the Applied Physics Department. Besides Bob Graham, Bruno Morosin and Gene Venturini attended from Sandia with the aim of determining the current Soviet effort, if it could be assessed, and possibly determining if some unknown new technology was being developed. The outgoing Batsanov was accompanied by two others, who contributed nothing and did not appear to be scientists. Batsanov's earliest publications appeared in Soviet journals beginning in 1965 with somewhat meager content, and what he presented was already well delineated by Bob Graham's review of the Soviet literature.

It should be noted that Davison and Graham's previously mentioned review did include most of the early efforts by non-Soviet scientists of shock loading effects on materials. The early 1960s graphitic carbon-to-diamond work of Paul DeCarli and J. C. Jamieson, of F. P. Bundy, and eventually of DuPont's diamond grit multimillion dollar synthesis business enterprise using old underground Delaware mines, stemmed from the fundamental work at Los Alamos on controlled compression of condensed matter. Much of the work of the Soviets and other foreign investigators was reviewed in context with the early results of Sandia's program and was described in the *Annual Review of Materials Science* **16**, 315–341 (1986) by Graham et al.

By 1987, this core group had completed and presented eight short papers at the June 21–26 American Physical Society Shock Conference, dominating the shock chemistry area, and these papers were subsequently included in the conference proceedings. These included shock-induced reactions of specific inorganic compounds; defect formation and the resulting effects, such as enhanced catalytic action; the initial evaluation of line broadening of x-ray diffraction patterns; and enhanced sintering characteristics, all of which resulted from shock loading of selected single-phase inorganic compounds as well as a more complicated clay (nontronite), with its implication of evolution for planetary surfaces, namely, that of Mars. The major positive step demonstrated that, unlike the Soviet experiments, the Sandia results were very reproducible and consistent with the pressures and temperatures suggested by the preliminary calculations of Dave Webb. As time went on, these three-dimensional calculations were refined and samples were carefully taken from the recovery capsule as a function of location, subjected to various experiments, and shown to be consistent among various experimental shock configurations. In addition, Bob Graham, with the discussions and modeling results of Yuki Horie, was able to arrive at a conceptual chemical reaction model for shock chemistry, a process involving turbulent local mixing, possible massive defect generation, and a short, rapid transit pulsed temperature excursion—all of which were caused by the transit pressure pulse.

What proved surprising to several in the Solid State Physics Department were the consistent and reproducible results from these explosively-driven experiments. Shock-modification experiments were carried out with Bob Graham's containment fixtures, named "Baby Bear" and "Momma Bear," using standardized geometries and explosive loading arrangements. In these shock compression experiments, the peak pressure and temperature are controlled by the size and shape of the powder sample and the recovery fixture, in addition to the explosive loading system. In this standardized shock-recovery system, three explosive loading arrangements resulted in reproducible loading into five sample-fixture arrangements so that peak shock pressures from 4.5 to 27 GPa could be achieved. Then, by varying the density of the powder compacted over a typical range from 40 to 65 % solid density, the shock-induced temperature could be varied from 50 to 1000 C. Evidently, by carefully interfacing the components of the experiments, astonishingly reproducible shock-modified samples could be achieved. Bruno Morosin was queried numerous times by his colleagues in his former department on the reproducibility issue.

Initially, the sample capsules were opened, taking care to avoid contamination, but used post-shock essentially in their entirety. This probably caused some averaging of the sample characteristics. However, as Dave Webb's extensive two-dimensional (2-D) numerical calculations became more refined, detailed locations of the recovered samples were examined. The intact sample capsules were opened carefully under a binocular microscope in order to separate the compacted sample disk into smaller research samples for various analytical diagnostics; ruptured, but essentially recovered, capsules were treated in a similar fashion, but with some possible areas being discarded, especially if post-shock annealing of chemistry reactions was involved in that specific material. For example, ruptured capsules could possibly have introduced impurities that might easily confuse the electron paramagnetic resonance data. Not described was the use of the helium gun facility for some studies requiring peak pressures below the lower threshold of 4.7 GPa for the explosively-loaded system.

The three initial diagnostic tools were conventional x-ray patterns as well as the determination of line broadening of diffraction peaks, the intensity and shape of the electron paramagnetic resonance signatures, and the determination of dislocation densities of the transmission electron microscope images. Consistent and reproducible diagnostic results were obtained from the tiny, but carefully separated samples unloaded from various locations of the capsule as well as between different loading configurations when merged with values given by Webb's 2-D numerical calculations for the sample pressure and temperature.

From the very start, diffraction studies conducted on what were considered tough ceramic material, such as alumina, titania, various carbides and nitrides, yielded broadened diffraction lines more typical of those found for heavily cold-worked metals. Both crystallite or coherence size and residual lattice strain may be determined from such line broadening. Significant differences between the residual strains from ball milling and from shock modification were also apparent in these materials when examined. As expected, such sample modification affected the magnetic properties as well as any other measurement in which the local environment of component ions is important.

The importance of how x-ray diffraction profiles are collected and analyzed was pointed out early in the papers. Long-time scans of the diffraction peaks were required for good statistics for the points along the entire profile of the diffraction lines. The early Warren–Averbach analysis employed selective diffraction lines, but as most other peaks were also measured, it was realized that a more complete potential for analysis was present. Two computer codes were developed: XRAYL for producing idealized powder diffraction lines profiles from overlapping powder patterns and CRYSIZ for computing crystallite size and strain from broadening of powder diffraction lines. Eventually, with Cam Hubbard at Oak Ridge National Laboratory, documents were prepared with 3½-inch computer disks and published for external distribution. Furthermore, the study of lanthanum boride comparing shock-modified and various milling procedures, part of which was a graduate student's thesis at the University of Maryland using data collected at the National Institute of Standards and Technology, had the spin-off of establishing the pure material as one of the standards that the scientific community can obtain. Lanthanum boride's simple cubic pattern, large unit cell, and unusually sharp peaks allow excellent standardization of diffraction procedures and equipment as well as use as an internal standard. However, the eventual more gigantic step in this area came somewhat backward, probably from discussions Morosin had with computer code developers at other laboratories. This essentially allows, in crystal structure determinations and refinements, all the diffraction lines contained in the spectra to be treated as part of the least square refinement procedure using parameters for elliptical crystallite size and a strain tensor of the material being analyzed. Hence, the results of these parameter fits usually escape the notice of users of the computer codes since, for the most part, the values are considered as scaling parameters rather than as useful information on the crystallite nature of the substance under study.

One must recall that this was the era of President Reagan, the perceived Soviet threat, and the "Star Wars" initiative. Out of the shock chemistry effort came the concept of a radically different, long life, inert battery that could be activated upon impact for possible fuzing capabilities, that is, essentially a shock wave-activated battery.

Toward the end of 1986, superconductivity was discovered in copper-containing compounds with transition temperatures much higher than those in the metals and alloys employed at that time in various magnetic applications; this resulted in a large shift of the research effort at Sandia for many investigators, particularly those with a solid state physics background. The structure and magnetic response were two essential ingredients needing investigation for these new materials; hence, part of the core shock-induced solid state chemistry group suddenly became diluted. The intense era of high-temperature superconducting materials was followed by the discovery of the third form of carbon, coined "buckyballs," which again cut into the time the core group allocated to shock chemistry. Nevertheless, the core group carried out several intersecting studies, subjected superconducting ceramic materials to shock loading, and published the resulting effects. In addition, studies of MnO, SnO, and revisiting pyrite serve as examples of other studies that were conducted during this time frame.

In these later years, another unfortunate Sandia transition occurred. The implementation of more rigid budgeting resulted in the removal of the previous ease with which a researcher could "walk down the hall" and obtain a sample quickly that could be subjected to the experimental procedure.

In this reduced budget environment, a few experiments were conducted on materials thought to be important for other projects. The report that "red mercury" was a new explosive material led to much speculation as to its exact nature, from energy release of an isomer transition to the least exotic and to that of the compound itself because of its low decomposition temperature. Shock loading experiments were performed in 1990 at the Socorro explosive site to avoid the strict materials problems instituted at Sandia by the Department of Energy Secretary O'Leary and the "Tiger Teams."[125] Its decomposition under shock loading was demonstrated to be quite energetic and was initiated even for the lowest pressure experiments. The catastrophic rupturing of the containment fixtures even at the lowest pressures suggested that the helium gun facility should be used; however, that was impossible because of the potential for mercury contamination problems. Mitigating the shock impulse via foams, light flyers, etc. was discussed, but never undertaken.

[125] Editors' note: The term "Tiger Teams" at that time (the early 1990s) was used to describe the teams of experts that investigated each Department of Energy laboratory for environmental, health, and safety concerns.

As General McArthur stated in Congress upon his return at President Truman's request, "Old soldiers never die, they just fade away."

Fading away may appear to be the case for "shock chemistry," however, not for Bob Graham. He has continued to observe his surroundings and apply his knowledge of physics, chemistry, and materials science learned in his long career, wherever he was. On October 26–27, 2012, in Socorro at the annual meeting of the Four Corners Section of the American Physical Society, Bob presented a talk entitled "High Temperature and High Pressure Mix of Iron Oxides from the Impact Event at Bee Bluff Crypto-Meteorite Impact Crater of South Texas." This is one of the latest of his papers on the study of a geological feature that he observed while hiking near his home soon after retiring in Texas and his careful probing of rocks and formations discovered in the immediate area. Bob is currently writing a memoir, *Crystal Heart Beat: Sandia and Bell Labs and Lightning in a Bottle*, built around his long career at Sandia.

The intent of the above write-up is to give a perspective of Sandia's involvement in the area of the modification and solid state reaction of materials subjected to shock wave loading, coined as "shock chemistry." Those seeking a more complete understanding of the science can peruse the references where overall general reviews of the Sandia and Soviet efforts may be found. An exception was made concerning the description of Bob Graham's containment fixtures so that comments on line broadening of diffraction lines can be made more coherent and complete. Progress on line broadening was previously given in individual papers on specific materials as they were published; however, these have not been summarized to any extent. Moreover, the computer codes developed have not been indicated nor has the impact on the general science of crystallography been described.

# Darrell E. Munson[126]

# (1961–2003[127])

# History of Guns at Sandia National Laboratories

The history of the development and use of guns as engineering and research tools at Sandia covers a considerable period of time, beginning in the 1950s and extending to the 2000s with some fits and starts and periods of darkness. The need was in response to the Cold War struggle to maintain technical superiority over the Soviet missile threat. Much of this program hinged on the vulnerability of our missile systems. The need arose in the late 1950s and waned with the end of the Soviet threat and Star Wars in the 1980s. I will try to detail some of the history from a (minor) management viewpoint of one who was quite involved: Darrell Munson. However, in trying to reconstruct this history, it became apparent that one's memory (mine, especially) about past events is actually incredibly faulty. Considering that, I have tried to be as accurate as possible on dates, but some sequences may be difficult to define, and the problem with names remains serious.

## Early Non-Sandia History

This early history is probably not relevant to the subject of guns at Sandia but it does set the circumstances for a possible reason of my eventual involvement. After graduation from the South Dakota School of Mines and Technology (SDSM&T) with a BS degree in Process Metallurgy in 1954, I took a job with Allis Chalmers in West Allis, Wisconsin, a suburb of Milwaukee. While this seemed like a permanent situation, Professor Edwin Oshier, of the SDSM&T Mining Department and a graduate of Stanford University, asked me to apply for admission to that University.

After taking the SAT exams, to my surprise, I was admitted to Stanford and given a fellowship from U.S. Nickel Corporation (perhaps not the correct name) for tuition, room, and board. The first year was supplemented with a teaching assistant income. This was further augmented in the second year with work on a Navy research grant under the direction of Professor Robert Huggins to investigate the creep (long-term deformation of a metal as a function of stress and temperature) of internally oxidizable alloys.[N1] By this time I was living in a dormitory in a former military hospital facility in Menlo Park operated by the University. This situation was further augmented in the third year by part-time employment as a Night Clerk for Married Student Housing in the same Menlo Park facility. Several graduate

---

[126] Editors' note: The superscripts N1 to N12 in the text that follow refer to the Side Notes section at the end of Darrell's narrative. Because of the length of his explanatory notes, which are of substantial interest, these are provided as notes at the end rather than as footnotes at the bottom of each page. He retired in 2003 and died in 2014.

[127] Editors' note: Darrell was involved in shock wave research until 1978, when he transferred to a Geotechnology group at Sandia.

students in physics, mathematics, and I in physical metallurgy shared a set of rooms, formerly a projection booth and balcony above an unused auditorium. Then I became a part-time employee of Poulter Laboratory, a part of Stanford Research Institute, which also occupied part of the former military hospital complex. As a metallurgist, I polished metal witness plates for use in developing shaped-charge technology for penetrating oil well casings to enhance flow from the oil reservoir; these experimental tests were conducted at an extremely remote explosive facility in Calaveras Canyon.

After completing nearly all degree requirements, except for the final thesis presentation and taking the oral examination, I left Stanford in 1959 to take a position as assistant professor at Washington State College, WSC (which later became Washington State University, or WSU) in Pullman, Washington, in the Metallurgy Department under Professor Servet Duran, who was a fellow graduate student at Stanford while on leave from WSC.[N2]

## Early Sandia History

Western Electric sent a recruiting team to WSC, and while I enjoyed the teaching and social surroundings, it did not feel exactly right, so I went to interview with them. The Western Electric people sent their findings to Bell Laboratories, which at that time provided the managerial staff for a little known laboratory in New Mexico. That was of course Sandia Laboratories, the engineering branch of the nuclear weapons complex, which included the Los Alamos, New Mexico and Livermore, California laboratories. Sandia called me to arrange an interview at the New Mexico facility.[N3] As a result, I was offered employment and started in 1961 with the Metallurgy group under Charles Bild, the Director; Bill O'Neill, the Department Manager; Cecil Russell, the Division Supervisor[128]; and Keith Mead, the Section Leader. For 2 years this involved routine metallurgical problems, one of which was to obtain pure metal specimens for a group headed by Donald Lundergan, the Division Supervisor, by operating a compressed gas gun facility to study impact effects on bomb components, especially contact fuzes. This group was probably the first group to recognize the requirements for shock wave material studies to support the weapon programs.

## Initial Introduction to Gun Technology

Initially, the gun was part of an engineering support group for the weapon programs. After some use for contact fuzing studies, Lundergan acquired the gun, intending to use it for more intensive material studies of interest for weapons development. Both he and his group and the gun were then transferred to Bild's material directorate in

[128] Editors' note: Managers of divisions were called Division Supervisors at that time in Sandia's history. Currently, divisions are now called departments.

the late 1950s, just prior to my coming to Sandia.[N4] The intent was to study in detail the propagation of small amplitude stress waves in materials of interest caused by relatively high speed impacts. In this time frame, but perhaps somewhat later, another compressed gas gun of smaller diameter was built by Bob Graham in Area I; Graham's gun was in a research group while Lundergan's gun served only an engineering and phenomenological goal.

I was given a section supervisory position in Donald Lundergan's division, under Charles Bild, to oversee the 4-inch compressed gas gun operations. While the gun facility was in Area III, the offices and laboratory activities were in Building 806 in Area I. The office complex consisted of a large room separated into staff offices and offices for the supervisor and manager; another large room served as the laboratory. Technical staff included Bill Hartman, Lynn Barker, and Jack Cannon, while the laboratory staff included Red Hollenbach, Wayne Brookshire, Cliff Witten, and Bob Simpson, among others. Over the next few years, the group added technical staff, including Charles Karnes and then Barry Butcher in 1962.

Somewhat prior to this time, Bild's directorate had built an explosive facility[N5] in Area Y, a complex surrounding Area III. This consisted of a bomb-proof building (Building 9950) with a firing pad on the roof accessible by a dirt ramp at the rear of the building. The permissible charge was 25 pounds of high explosive. The facility consisted of the bomb-proof laboratory, an explosives storage igloo, and an explosives assembly building. The facility was not being used for its intended explosive tests on materials; instead, it had only a high-voltage exploding foil machine housed inside the building. At this time, the defense needs were to study the vulnerability of missile components for the weapon developers at shock wave stress levels above those produced by the 4-inch compressed gas gun. After some difficult discussions on the need for higher pressure impact material studies, I acquired this facility to add to the capability to produce material responses to very large impact shock waves.[N5] The Building 9950 operation was turned over to me in 1965. Within the laboratory building were three rooms, a long gallery with the high speed Tektronix oscilloscopes, Fluke pulse generators, and high voltage firing units. Two other larger rooms housed the Marx capacitor high voltage generator and the actual laboratory. The Marx machine used a sudden release of energy to propel a thin foil projectile onto a target. Although a skilled scientist, Ray Reed, tried to make it successful, it proved not to be. However, the high explosive capability was successful. The use of explosive lenses, using two explosives of different detonation velocities to produce a planar shock wave, allowed determination of the Hugoniots of many materials of interest to the weapon developers during the beginning of the era of ballistic missiles and their vulnerability to radiation from defensive warheads. These plane wave lenses were made at Pantex in Amarillo, Texas, therefore requiring some trips on my part to obtain them. Also, many trips were made to Los Alamos to learn these techniques from Bob McQueen's GMX6 group at their Ancho Canyon test site. Paul Matson was the explosives technician and Bob Biesecker was in charge of the specimen preparation.

However, in 1964 Walter Herrmann, a research associate at the Massachusetts Institute of Technology (MIT), came to Sandia to form a Deformation of Structures

Division to develop further his finite-difference numerical simulation "codes." Beginning with WONDY, these codes became one of the principal analysis tools for the weapon programs.[N6]

1966 was an interesting year; a massive reorganization took place with the elimination of all section supervisors and my return to staff member status, at my own request. In 1967 another major rearrangement took place to accommodate the increased demand for shock wave material studies. Bild had an old building (855) in Area I modified to accommodate both the 4-inch gun from Area III and a new 4-inch gun built for use in engineering studies. Butcher was in charge of the former and Larry Lee the latter.

That same year, 1967, further reorganization occurred, with Herrmann becoming the Department Manager and Lundergan, myself (promoted again), and Olden Burchett taking division management positions. I retained the Area Y Building 9950 Explosive Site. Somewhat unexpectedly, Lundergan stepped down to become a staff member. However, additional reorganizations took place in 1968. Lundegan's old group, now with Herrmann as the Department Manager, was moved in that year to the Solid State Research Directorate under Al Narath. Butcher, Karnes, Burchett, and I were Division Supervisors.

## The Two-Stage Light Gas Gun

In 1968 we became aware of some new innovative technology based on two-stage light gas guns[N7] that were being developed and were much superior to explosive methods for high speed impacts. Some interest developed at Sandia in the new technology, which I promoted, and as a result of a meeting with the Vice President of Research (name not recalled), Al Narath, and others, I made a pitch to acquire a two-stage light gas gun facility. Over the next year, I guided the procurement of a 3.5-inch gun from General Motors (GM) in Goleta, California, based on a gun constructed and operated by Sid Green and Arfon Jones. I was also responsible for both the conceptual and the final design of Building 9956 to house it. The project appeared as a line item in the Congressional Budget at something well over $100,000. The gun arrived in November 1969, was installed by GM staff, and fired its first test shot successfully before Christmas.

The gun building, 9956, was about 100 ft long and 20 ft wide and high enough to permit an overhead crane. Butting up to this building was a bomb-proof instrument room, which contained the oscilloscopes, pulse generators, and firing circuits. The 60-ft-long gun was aligned along the center line of the building and included an added 6-ft-diameter impact chamber. Sandia staff included Bob May, Ray Reed, Paul Matson, Bob Biesecker, Charles Kinsey, and the support of others who still manned the 9950 facility as well as some from other divisions. Somewhat later, Carl Konrad also joined the group. The gun could be used either as a two-stage light gas gun with a final stage diameter of 1.125 inch or as a simple powder gun with a 3.5-inch diameter. Joining these gun tubes was a tapered diameter reduction coupler section.

While the conversion between two-stage and powder guns was possible, it was a difficult and costly process. Consequently, another 3.5-inch tube was purchased in

1972 and placed against the west wall of the building as a complete, standalone powder gun. These two guns permitted shock wave material studies to around 100 kbars, field test impact studies, [N8] and measurement of the interior ballistics of the gun.

Many of these studies were made exceptional through the use of the interferometry techniques being developed by Lynn Barker.[N9]

## Other Changes to Expand the Capabilities of Sandia Testing

Sometime about 1972, another improvement in testing took place. Karl Schuler and I realized the need to devise a way to test stronger composite materials, typically ceramic or glass-filled epoxies or other plastics of interest used in missile heat shields, quasi-statically under high pressure. This required special testing equipment to determine the properties of these composites. Consequently, a rather new technology using a triaxial loading machine was required. In addition, shortly thereafter, the nation's energy needs brought the oil shale program to Sandia, where the shale was to be broken explosively in situ, in the bedded formation. The triaxial frame was also ideal for studies replicating the material response to explosive loading stresses of shale in situ.[N10]

The triaxial test frame was purchased from MTS Systems Corporation[N11] and a space was needed to house it. To accommodate this new test frame, Bild had space in Building 849 where the guns were housed. This building was a Quonset hut from the 1940s in Area I, which I said was too cold in the winter, too hot in the summer, and dirty always. Essentially, we built an interior frame building which was temperature controlled inside the hut, and the triaxial frame was installed here. Herb Sutherland joined my group at the beginning of a period of substantial change.

Other changes occurred in 1974 when the beginnings of Sandia's involvement in nuclear waste isolation began. (1974 was an unusual year, in which the fiscal year of the Federal government was changed by 3 months.) A planned repository in Kansas proposed by Oak Ridge National Laboratory run by Union Carbide proved unsatisfactory, and the Atomic Energy Commission asked Sandia to take up the problem. Sandia was provided with $400,000, which I was to use to establish a laboratory effort to begin a study of high temperature creep under triaxial conditions of rock salt, the geological formation that was suitable for a repository. We hired Wolfgang Wawersik, who had built such an apparatus at the University of Utah, to establish a similar one for Sandia. Originally, it was housed in Building 855 with our triaxial machine since they both required high pressure hydraulic fluid.

The need for fundamental studies of rocks, especially rock salt, for waste isolation programs precipitated further expansion, with the establishment of a Rock Mechanics Laboratory in a refurbished old metal building, 849, in Area I. I got priority internal monies to make the building useful.[N12] Building 849 required major modifications to be suitable for high pressure fluids, improved electrical wiring, and temperature controlled rooms. Two of the Wawersik test frames were installed in the building. I eventually added other major pieces of test equipment. The building was equipped further with a small machine shop and a specimen preparation laboratory. The two compressed gas guns remained in Building 855.

Further changes were to take place in 1978 as well, since economic problems resulted in a reduction of staff. After some difficulties, I stepped down to a staff member position. As a consequence, Barry Butcher got the two-stage gas gun facility.

## The Difficult Times

Within weeks of these changes in facilities management, an accident occurred at the two-stage gun facility in which the resulting pressure release damaged the interior of the gun building. After more than a year, the facility was restored. However, in further reorganizations, Butcher was replaced by Jim Asay. In a later rearrangement, Wawersik became the manager of the Rock Mechanics Laboratory. Much later, Butcher, in turn, replaced Wawersik as the leader of the Rock Mechanics Laboratory.

## Transition Period

In 1978, I went to work with Dave Northrop in a Geotechnical group working on subsidence over deep coal mines. We worked with the new ERDA (Energy Research and Development Administration) Energy Research Centers, which had formerly been part of the Bureau of Mines. I also began development of the Deformation Mechanism creep model for describing the constitutive behavior of salt. After 2 years in subsidence, I was given a 1-year appointment in 1980 to the Civilian Waste Repository group and the Department of Energy Headquarters in D.C. Wendell Weart was responsible for making this possible. Upon my return to Sandia, I joined the Waste Isolation Pilot Plant project under Weart. I was to be the Principal Investigator for the Thermal/Structural Interactions program. This involved the design and fielding of very large scale underground experimental rooms, which constituted a significant aboveground laboratory materials creep testing effort, the development of numerical finite-element calculational codes, and the verification of analysis methodologies relevant to the structural integrity of the site.

## The Future of the Two-Stage Gun

It was apparent that Jim Asay had big plans for the facility, renaming it the STAR facility. It did appear that the facility was to be involved in some advanced physics research. But, that is a story of which I have no real knowledge.

## Side Notes

**N1:** At this time, metallurgy was of very significant technical and scientific importance to the Nation. The advent of the jet engine led the way to global economic and military expansion. The principal effort was to develop metallic alloys and intermetallic compounds to improve the high temperature performance and operational life of turbine blades, since the engine efficiency improves as the operational temperature increases and cost decreases as the blade replacement time is extended. While many were searching for better materials, scientifically the question was to understand the creep (slow deformation) process at elevated temperatures. At Stanford, Professor Oleg Sherby, an outstanding scientist initially working with Professor John Dorn at the University of California at Berkeley, was my mentor. He was one of these rare people who required little if any sleep, and he wrote significant papers on his creep research while others were asleep. He certainly influenced my understanding and interest in the creep of metals and nonmetals, especially the role of micromechanical mechanisms. Eventually, the Metallurgy Department at Stanford morphed into the Materials Science and Engineering Department, reflecting the many wonderful applications in the field that we now enjoy. Professor Huggins became the originator at Stanford of one of the Centers for Materials Science funded by the Federal government (by the National Science Foundation) to provide facilities and scientists.

**N2:** As an introduction to Pullman in the heart of the Palouse wheat country, the newspaper headline was "Colton boy mauled by hogs." My duties were teaching four courses each quarter, starting a small research effort on the creep of metals through a small state grant to Professor Duran, and arranging and furnishing a new Nuclear Research Facility with benches, storage cabinets, desks, and other stuff built to my order by inmates of the state prison. I put in so much stuff that there was hardly any room left for students and professors to work.

**N3:** I arrived at the Albuquerque, New Mexico, airport on a beautiful, mild, soft air, autumn night. The terminal was in an old Mexican style wooden building with big beams and ceiling vigas. I was met by my hosts and taken to the Sundowner Inn. After 2 days of intensive interviews, I was picked up by my brother, Fred Oberg, who was working in Grants, New Mexico, at the Homestake Mining Company uranium processing plant as a process metallurgist. Coincidentally, he also was a graduate of SDSM&T.

**N4:** In 1963, I was promoted to Section Leader in charge of the 4-inch compressed gas gun under Lundergan. Other reorganizations occurred over the next 2 years. It was decided to eliminate two levels of management. As a result of the reorganization, I was offered the division for routine mechanical properties testing, which I turned down. After a short time thereafter (about 1965), Lundergan was promoted to a Department Manager, and I was promoted to the division that operated the 4-inch gun facility.

**N5:** Operating an explosive site required special staff, which included Bob Biesecker, Cliff Witten, and Paul Matson. Matson was an old Army sergeant from explosive ordinance who prepared all of the charges and fuzes. To aid him, I was the designated observer required by safety rules. As a result, I witnessed every explosive test. This continued as the practice for the two-stage gun with some later relief by Carl Konrad. Bob Hardy joined the group. Incidentally, later, perhaps in 1967,

Lundergan stepped down and Butcher took over his division. This department had Walter Herrmann as Department Manager with Charles Karnes, Barry Butcher, Olden Burchett, and me as division leaders. Eventually, Burchett was replaced by Al Chabai, I think.

**N6:** At this time, it was recognized that more powerful analysis techniques were essential to shock wave research. Consequently, Walter Herrmann was hired from MIT where he was a research associate involved in developing finite-difference shock wave analysis codes. Eventually, as stated above, he would be promoted to Department Manager and Larry Bertholf and Sam Key would join his group. The codes were initially WONDY and then TOODY.

**N7:** There had been several two-stage light gas gun in existence, of smaller diameter, in a couple of eastern U.S. groups (perhaps Picatinny Arsenal), at Cal Tech, and possibly one or two others elsewhere. The General Motors gun was, however, a major leap in capacity and was developed at Goleta to study compaction of porous materials. The principle of operation of the two-stage light gas gun is somewhat complicated. The gun consisted of a breech section, a 3.5-inch gun tube first-stage section, a tapered coupling section that reduced the internal diameter from 3.5 inch to the smaller diameter of the launch tube section, the second-stage launch tube section, and finally a catcher chamber. The breech section held the propellant. In operation, the first tube and the tapered section were filled with light gas (typically hydrogen), and the launch tube was under a vacuum. The gun used two different projectiles. The first, just ahead of the propellant charge, was a plastic-bodied projectile that compressed the light gas to an initial high pressure. In the tapered section, the plastic projectile extruded itself in the tapered section, thereby accelerating the projectile face and further compressing the gas. The tapered section was initially sealed off from the launch tube by a burst diaphragm, which released the high pressure light gas almost instantaneously into the second stage of the gun, to drive the second projectile. Although there are many possible configurations for the second projectile and the target, the second projectile usually was constructed with some variation of an impact plate component on the front face, a typical configuration for material studies. This projectile was accelerated down the tube to impact the target just off the end of the launch tube. Instrumentation consisted of projectile velocity, impact timing, and free surface velocity gauges. All of the debris was trapped in the catcher chamber by a laminate made of steel plates and aluminum honeycomb. In addition, high impact velocities required exceptionally high speed recording, so Tektronix 100-ns scopes with a 1-ns resolution were installed. These scopes were the top of the line in speed. A nice benefit to acquiring the gun was the trips to Goleta. On one trip, the field test people, who had a deep-sea oceanographic site off of Santa Cruz Island, let me visit the facility. I was flown there on their single engine plane by an ex-kamikaze pilot on to some questionable air strips after buzzing it to scare away the cows. I had lunch on their research ship, quite a unique honor.

**N8:** The field test people came to me because the fast gates intended to stop debris from a nuclear underground test from flying down the instrumentation tubes and destroying the gauges that were not working properly. We did a number of two-stage gun impact tests on the fast gates. Finally, I went to the Nevada test area to examine the results of an underground test. I was one of the first people underground for that test to look at the damage and suggest some possible remedies. The flight in a single engine plane to and from the site in strong cross winds over the ridges and canyons was exciting.

**N9:** Lynn Barker was one of the most talented and brilliant scientists I have ever known. He was in the Lundergan group when I joined it, then went to Columbia University under Sandia sponsorship, and then returned to Sandia to work with me again in my division. Shock wave studies using plate impacts require the measurement of the transit times through the specimen and the measurement of the movement of the free surface of the specimen. His early contributions were the use of a slanted wire resistor, which was progressively shorted out by the moving surface, a device that improved the resolution and, then, the interferometer systems. These elegant systems started with a simple interferometer and progressed to velocity interferometers for any surface. Resolutions were to the nanosecond level. Other clever ideas are too many to discuss here, except for his development of SWAP, a ray analysis computer code, for solving wave propagation problems.

**N10:** The testing of oil shale required specimens. To obtain these specimens, I traveled to Grand Junction, Colorado, a couple of times to select and have blocks of oil shale weighing tons shipped to Sandia.

**N11:** This was such a new and advanced concept that MTS Systems Corporation had never built a triaxial machine before. When we assembled it at Sandia, the manufacturer's name plate was upside down. Guess everyone had some learning to do.

**N12:** This involved conceptual and Title III design, which essentially raised the building shell and required pouring a new floor with trenches for piping and electronics, removing all the old wiring, installing new wiring and internal partitions, and building an attached high pressure pump shed. Eventually, the laboratory would house the original triaxial frame, a small four-post 100-kip[129] and a large four-post 400-kip test frame, plus a Hopkinson bar impact apparatus, and two triaxial creep frames. To accommodate the creep tests, the building was temperature controlled to within about 5° F. This was undoubtedly the most extensive remodeling of a shell of a building known at the time.

[129] Editors' note: A *kip* is a non-SI unit of force that equals to 1000 lb-force. It is used primarily by American architects and engineers to measure engineering loads. ("SI" refers to the International System of Units.)

# Ray P. Reed

# (1957–1961, 1967–1994[130])

# Stress Wave Related Activities at Sandia National Laboratories

I was the first of several University of Texas (UT) graduates introduced to Sandia National Laboratories in Albuquerque, by Professor E. A. Ripperger (Ripp) of the Engineering Mechanics Department, which later became and still is the Department of Engineering Mechanics and Aerospace Engineering. (Others included Robert Graham, Philip Stanton, Charles Karnes, Tom Priddy, and Herb Sutherland.) Ripp spent several summers at Sandia and other national labs. During one summer at Sandia, he designed the tall, inclined drop tower used in Area III for impact testing of weapon delivery systems.

## Pre-Sandia Years at University of Texas Austin

From the Engineering Mechanics Department of the fledgling Southwest Research Institute in San Antonio, TX, I had moved to the Military Physics Research Laboratory in UT Austin. There, at the Balcones Research Center, I designed and installed a punched-tape controlled-precision cam milling machine. While concurrently a student in the Engineering Mechanics Department under Ripp's guidance, I did research for a master's thesis on piezoelectric dynamic strain gauges. Later, Hans Baerwald would move from Clevite Corporation[131] to Sandia's Department 5130.[132]

## Sandia National Laboratories

After graduation, in 1957, I joined the newly formed Applied R&D Department (aka 5130) managed by Ken Erickson. (Temporarily bearded like a Viking, Ken and other employment interviewers were celebrating the 45th anniversary of New Mexico.) Shortly later, Ken left Sandia to form Kaman Nuclear, which, with three

[130] Editors' note: Ray left Sandia in the early 1960s to go to graduate school at UT Austin and did not intend to return to Sandia after that. Phone books from that era indicate he was not a Sandian from 1962 to 1966. But, as this recollection indicates, he did return.

[131] Editors' note: According to the Encyclopedia of Cleveland History, a site maintained by Case Reserve University (http://ech.cwru.edu/index.html), Clevite Corporation was founded as Cleveland Graphite Bronze in 1919 to make bearings and bushings for automobiles. After World War II, Clevite became heavily involved in defense contracts. The corporate name was changed to Clevite in 1952. In 1969, Clevite was acquired by Gould-National Batteries.

[132] Department 5130 was formed around 1956 with the name "Experimental Weapons Research" and managed by K. (Ken) W. Erickson. In 1957, the name was changed to "Physical Research" and managed by G. Hansche. Later, in 1971, the name became "Physics of Solids Research" under George Samara. The name and organization number continued throughout the rest of the department's history. In 1960 a sister "Physical Sciences Research" Department (5150) was formed with Richard Claassen as the first manager.

other divisions in 1968, became Kaman Sciences. He soon moved the start-up to Colorado Springs, CO. After Ken left, he recruited several of his former Sandia staff, including my then Division Supervisor, David C. Williams.

George Hansche then became the Department Manager. The three divisions under the department were led by George Anderson, Richard (Dick) Claassen, and Frank Neilson. A small group of the staff, led by Iben Browning, was tasked to discuss and propose possibilities for a separate advanced research and development (R&D) department. The Physical Sciences Department (aka 5150) was soon formed. Richard Claassen was chosen as its first Department Manager with a small cadre drawn from 5130. (About that time our president, Monk Schwartz, is said to have influenced Sandia to be an egalitarian organization by strongly discouraging the use of academic titles and white lab coats.) Our reduced department ranks were soon enhanced by hiring Ripp's student, Robert Graham, and a newly minted Cal Tech (California Institute of Technology) graduate, Orval Jones.

**The late 1950s at Sandia.** All small R&D, aeronautics, and field test organizations were housed in a few small offices and laboratories on the first floor of Building 802 and in cramped offices in the basement. The entire Sandia Credit Union was in a small Quonset hut near Building 800. The technical library was poorly organized and in a single small room in Building 800. Publishing papers or applying for patents was not encouraged or allowed. In that era, when the commercial and practical applications of nuclear explosions for large-scale excavation were being seriously considered, Warren Taylor and I studied explosive crater formation in scaled experiments in a specially designed building in Area III. Photographic instrumentation was by high-speed framing camera and overhead still camera. Small-scale explosive events were monitored by a lengthy, slender strain-gauge-sensed Kolsky pressure bar.

That era broadened the analysis by computer at Sandia. Large-scale computing was centralized in a small service organization. Locally, computer programs were submitted on IBM punched cards remote from the computer. Diane Martin, who had recently learned the new FORTRAN[133] language, programmed the analysis of my beginning experimental studies of wave propagation in tapered bodies. Computer programming then was not yet widely practiced by the staff members! There were no PCs, desktop calculators, or handheld calculators. (We had only the slide rule and abacus!)

**Shock wave research in the late 1950s.** At Sandia, exotic high pressure properties of weapon-relevant materials became of practical interest. A brief attempt to study very high pressure static properties of materials was initiated in 5130 by hiring a postdoc and acquiring a very expensive and massive static tetrahedral anvil device that was installed in Building 808 for the new hire. Unguided by a thesis advisor and frustrated, the new staffer shortly returned to academia. I don't know what happened to that monstrous apparatus.

[133] Editors' note: FORTRAN (for FORmula TRANslating System) was developed in the 1950s by IBM. In the 1990s the name was changed to Fortran.

Don Lundergan, in the Systems Development Directorate, had been performing dynamic material impact experiments with a compressed gas gun in a small building in Area III. Others, at an adjacent small-scale model of the terrain of the Nevada Test Site, used small explosive charges to explore the near-surface propagation of blast waves simulating air burst experiments. Following Lundergan's early work, attention was refocused on dynamic studies of material properties. Apparently, Lundergan's gas gun was moved from Area III to more convenient quarters nearer Buildings 802, 805, and 806 in Area I.

Separately, dynamic material studies were to be performed using an advanced compressed gas gun designed and implemented for Graham and others in 5130 by engineers Charles Muhlenweg and Sig Thundburg. Graham had worked on similar impact experiments as a student of Ripperger at UT. Orval Jones had related graduate student experience at the California Institute of Technology.

That gun was implemented to support dynamic phenomenological materials research. A precise scientific tool for that research was a newly conceived quartz crystal shock stress gauge. Quartz, used as a pressure sensor, was not novel. However, its development as a precise and elegant scientific tool was unique. The development of the "Sandia Quartz Gauge" required a special premium-grade, defect-free X-cut quartz as the piezoelectric sensor. Full development of this simple, yet sophisticated sensor would continue through years of study. Though patents were not yet in vogue at Sandia, the basic concept of the sensor was patented as the "Sandia Shock Stress Gauge" by Neilson, Benedick, and Graham. That precision gas gun facility and the sophisticated Sandia Quartz Gauge stress wave sensor became basic experimental tools for high-pressure studies in the department, using precise high velocity planar impact on subject materials. Independently, Lynn Barker, in the Systems Development Directorate, would begin to develop a different shock wave sensing system, the laser interferometry-based VISAR.

### Return to University of Texas Austin—Graduate School

This was during a time, as nuclear testing was paused, when the mission of Sandia, the other labs, and the Atomic Energy Commission seemed to be in decline. I decided to return to graduate school. I departed at the threshold of an age of profound shock wave experimentation and theory development.

With a grant to create and demonstrate concepts for microscale biological sensors, I returned to UT Austin to study engineering mechanics. Quite incidentally, the focus on tiny biosensors, just for convenient demonstration of the concept, involved thermocouple sensors of micron, cellular-scale, sensing dimension. Through that coincidental exposure to thermometry, I later discovered that the widely published traditional principles of thermoelectric thermometry were simplistic and fraught with pitfalls. Then, and even now, they are widely misrepresented at all technical levels, from casual tutorials to the deepest scientific material theory. Persistent misunderstanding still often invalidates commonplace costly and certified calibrations and degrades critical applications. A decade later, it led to my development of a general and authentic Functional Model of Thermoelectric Circuits and of special diagnostic circuits and instrumentation designed to avoid commonplace errors and flawed interpretation in thermometry.

The *incidental* thermometry activity by chance eventually enriched diagnostic activities in the areas of stress wave measurement, shock waves, and fossil fuel energy production. Working concurrently with Sandia and the American Society for Testing and Materials (ASTM) enriched both experiences and broadened my technical perspective.

The experience sensitized me to the many pitfalls of general measurement of stress waves as well as thermal waves and to an interest in *authenticity* of measurement as a general focus topic. In that connection, Sandia later allowed me to sponsor and conduct, in Albuquerque and in Las Vegas, a series of short courses focused on general measurement authenticity and diagnostics.

With my thesis completed and approved, I delivered it to the Dean's office. I casually walked from my university lab, visible via a fully exposed route around the foot of the tall UT "Texas Tower" on August 1, 1966, just as the infamous sniper, student Charles Whitman, already above on an observation deck of the tower, was setting up to begin his notorious shooting rampage. A half hour after my casual stroll, always visible from the observation deck, that mentally deranged shooter quickly killed 14 and injured 32 others around the tower base and across surrounding areas at distances to several hundred yards.

## Rehired at Sandia

I had not planned to rehire at Sandia. Fatefully, I was fortunate to return. At graduation, I interviewed several other potential employers, avoiding Sandia. A surprise phone call to interview at Sandia while I was interviewing elsewhere informed me that, in my absence, the lull in activities had reversed to an exciting new level of activity. The new urgency of the Lab mission had surged during the several years I was away. With acquired computer skills, a solid mechanics and dynamics background, and a continued interest in experimentation, I rejoined Sandia. I had withdrawn my accumulated contributions to the then-contributory Sandia retirement account to fund my schooling. As a rehire, I eventually finished funding the total amount again just as Sandia made the benefit noncontributory.

Don Lundergan was my interview host. Don had long recognized the emerging critical importance of shock wave technology to the Sandia mission. He had formally developed and proposed a comprehensive and very detailed plan for an entirely new department focused on all aspects—theoretical, computational, and experimental—of shock wave technology critically related to the Sandia mission. I joined the newly formed department Don had conceived and promoted rather than my prior R&D department. The new organization, as conceived by Lundergan, had emerged as a department of four divisions soon headed by Walter Herrmann, who was newly hired from the Massachusetts Institute of Technology. The department expanded quickly with its divisions led by supervisors Larry Bertholf, Olden Burchett, Barry Butcher, and Darrell Munson. That new department, the Solid Dynamics Department, as designed by Lundergan, was created especially for applied computational and experimental shock wave technology in a weapons directorate to complement the theoretical research efforts of the Physical Research Department.

**Research on the dynamic response of materials.** Munson's Shock Wave Phenomena Division focused on experimental stress wave applied research. The division first performed experiments by explosive loading at a newly constructed facility in Area Y.[134] There, I conducted tests using Capacitor Discharge Exploding Foil dynamic loading of materials. The dynamic response was monitored by advanced high-speed photography. Such a capacitor discharge-driven system had been used while at Northrop by Don Keller, who later founded Ktech.[135] (Very soon, Keller enlisted Larry Lee from our department to join his new company in Albuquerque.) The flyer facility was used to study dynamic fracture of complex advanced materials, such as boron-filament-reinforced composites.

Adjacent to the Area Y site, near Building 9950, Munson's division later built a large Butler building shelter[136] to house a new powder-driven gun to study higher velocity impacts. (Years later, a second-stage light gas gun, next to the powder gun, was added in this shelter.) A new adjoining control and instrument room housed an array of recording instruments and sheltered the operators and experimenters from possible harm. Bob May was active in the preliminary architectural setup of the facility. I developed the associated recording facility and instrumentation for the test chamber in preparation for the powder gun, which was then on order.

Sandia's first powder gun and its associated large experimental chamber were designed, fabricated, and first assembled by General Motors (GM) in Goleta, California (which is near Santa Barbara). From the enclosed muzzle of the gun, a projectile was fired into a walk-in, evacuated cylindrical containment chamber. Experimental targets were set up within the chamber for impact at the muzzle. The target chamber was barely large enough for experimenters to stand as they set up the experiments. Transverse ports allowed photographic and x-ray observation during impact, as well as electrical and VISAR monitoring of the tests.

As Sandia's representative, I monitored the single acceptance test firing of a projectile from the powder gun into its target chamber at the GM plant. Projectile impact on instrumented targets occurred at the muzzle. The volume and strength of the evacuated chamber were sufficient to capture and contain the expanded explosive products, remains of the impact-destroyed experiment, and debris from a test.

---

[134] Editors' note: Area Y was about 6 miles south of the main Sandia complex.

[135] Editors' note: Until recently, Ktech was a major technical support supplier to Sandia. In June 2011 Ktech Corporation was acquired by Raytheon and became part of the Raytheon Missile Systems division.

[136] Editors' note: A Butler building is a "temporary" corrugated metal structure that has a foundation. With the need for rapid expansion after World War II, Sandia relied on (and still relies on) such structures. Buildings 849 and 851, two Butler buildings installed in 1949, are still in use in Area I. See R.A. Ullrich, C. Martin, and D. Gerdes, *Historic Building Survey and Assessment Sandia National Laboratories New Mexico Site Albuquerque, New Mexico*, Vol. 1: Survey and Assessment, Sandia National Laboratories Report No. SAND2010-6117P (August 2010).

Access to the chamber and closure was at the rear via a heavy hinged circular door that was the full diameter of the chamber.

**The powder gun chamber.** A substantial catcher was required to stop high velocity debris from directly impacting the entry door. Peculiarly, the GM staff had not prepared a formal catcher for the projectile that was to be fired, without a target, in acceptance tests. Rather, they had improvised an ad hoc catcher from a massive surplus gun component. It had a tapered bore centered on the gun bore and was supported on a lab cart, secured by chains. The projectile was to be captured in the catcher. The sole acceptance firing was successful. The catcher was not. The test debris was contained in the chamber. However, the high velocity projectile impact propelled the improvised catcher excessively so the chamber and its rear access door were significantly dinged on first use—yet they remained functional and secure.

That dramatic acceptance test event foreshadowed a problem that occurred years later. The access door sealed vacuum and pressure. The door was secured by a ring of several dozen large bolts spaced around the circular perimeter. For years, following Sandia formal standard procedures, every bolt was secured. Eventually, the inconvenience and setup delay in securing the many bolts led to the progressive cautious reduction in the number of bolts actually used.

Earlier, I had been recruited by a field test organization to review and improve the design, fabrication, and analysis of their quartz shock stress gauges, which were employed on nuclear tests, and to develop other sensors and methods to use in severely noisy field environments rather than in the laboratory.

I eventually learned that, years later, in the most dramatic event of its service, too few bolts were secured on the chamber. The huge access door was blown open. By design and practice, there was no physical harm to the staff. However, much of the galvanized corrugated steel shell of the shelter building and portions of the sensor and control cable run to the instrumentation shelter were damaged. Decisively, the practical need to observe the appropriate, though inconvenient, standard operational procedures was convincingly illustrated. Lesson learned. Years later, the dramatic experience proved not to be unique to research guns at Sandia.

**Quartz gauge assemblies for nuclear field tests vs. the patented Sandia Quartz Gauge.** In my new organization, I learned that the patented Sandia Quartz Gauge design for laboratory use could not be used in nuclear field tests. Formerly, the field test gauges used underground had been of arbitrary design, consciously not following the critical proportions of the formal standardized laboratory "Sandia Quartz Gauge." Like that formalized sensor, the cylindrical sensor disks were of premium X-cut quartz but had different disk proportions and electrode geometries adapted to nuclear test constraints. Unlike the simple quartz disk laboratory sensors mounted bare to test specimens, a fieldable gauge assembly had to be encased for shielding, durability, and rapid installation in underground nuclear tests. Field data required measuring much longer pulse durations under very noisy conditions and could occupy only a severely limited overall diameter, crowded among many other experiments that were being installed. Enhanced sensitivity was needed for the noisy measurement environment.

The distinctive informal field test designs had been used in underground nuclear tests, yet it was always recognized that the signals were substantially distorted by the nonideal gauge geometry. Arbitrary deviation from the prescribed disk and electrode proportions had resulted in recorded waveforms that deviated significantly from the correct peak amplitude and wave shape. Building on those arbitrary designs, I standardized a family of "Sandia Field Test Quartz Gauges," different from the ideal laboratory "Sandia Quartz Gauge." New designs were limited to a few systematic, well-characterized design families. The revised designs unavoidably had nonideal step response, but those responses were predictable and characterized. Distortion in recording was reduced. Practical data correction was made possible. Complementary to the physical redesigns, I adapted a practical, though not ideal, new analysis to reduce the recorded waveform distortion. For each gauge series, we determined the reproducible characteristics by projectile impact using the workhorse gas gun facilities in Area I.

In the ideal "Sandia Quartz Gauge," only slightly tilted and sustained planar impact by a thick impactor results in a very narrow step to an ideal sustained flat plateau as the stress wave progresses through the quartz disk. That demonstrates the needed linear response. Rather, in a "Sandia Field Test Quartz Gauge," a sustained, nearly planar impact also results in an initial narrow step, but, consequentially, that step is followed by a nonlinear, progressively climbing amplitude to an abrupt end rather than a flat plateau. In some prior field test gauges, the relative terminal amplitude was more than double that of the initial step (an extremely nonideal response). Subject to a stress pulse loading of arbitrary form, as in field tests, the recorded pulse shape resembles neither the characteristic step response nor the imposed input pulse. Such a distorted response profoundly displays the dreaded nonlinear response. Critically, neither the recorded pulse shape nor the peak amplitude from the applied field test quartz gauges even approached faithful waveform reproduction.

The established mathematical sophistication and obscure pitfalls of deconvolution by conventional transform preclude its use by many. Convolution is a well-known mathematical operation. Traditionally, it is performed by rigorous classical mathematical transforms. Physical and analogous mathematical *convolution* is an inherently stable process that reproducibly represents the interaction of arbitrary forcing functions and system characteristics. Routine analysis software often includes the convolution transform as a familiar named function.

The complementary inverse operation, called *deconvolution*, attempts to deduce the input waveform from the observed distorted output and the associated known system characteristic. The deconvolution operation is notorious for its inherent tendency to dramatic instability. Casual use by the unwary of the canned transform function for the inverse deconvolution is prudently discouraged. Wild gyrations that result from application of deconvolution with inappropriate characterizing functions are viewed with horror by the casual user. Once bitten, most analysts avoid deconvolution as they would in leprosy.

The established characteristic step response of the Sandia Field Test Quartz Gauge convolved with an arbitrary stress input history results in a stable, accurate prediction of the waveform that ideally should be recorded. Fortunately, the standardized step response characteristics of the design set of Field Test Quartz Gauges used for analysis by deconvolution results in a stable and much improved correction of the distorted record to reveal an appropriate input waveform representation. The practical result is much improved, yet necessarily inferior to that from an ideal laboratory Sandia Quartz Gauge.

Signal distortion and the need for correction occur widely in measurements. Underground nuclear test conditions, in particular, involve additional significant cascaded signal distortion by recorders and lengthy multiple cables in series, each with a nonideal response characteristic. Signal distortion is affected by mixed, concurrent system characteristics. For such a commonplace practical application, I adapted and published, in a ten-part journal tutorial series, a unique *Algebra of Convolution and Deconvolution* that applies to multiple, cascaded fold and unfold operations that are commonplace in realistic field measurement. Coincidentally, this requirement to correct nonideal measurement system responses applies broadly to many other transient measurements in addition to stress measurements. The methods have been applied to a variety of data recorded by mechanical and thermal instruments with nonideal characteristics.

The simpler alternative for correcting Sandia Field Test Quartz Gauge records was algebraic. It was easily programmed and understood as a simple few-line algorithm. A comprehensive computer program incorporated the full set of standardized gauge and system characteristics. Applying the special algebra of cascaded convolution operations simplified the routine application of successive, realistic mixed convolution and deconvolution and the reduction of the test data from Sandia Field Test Quartz Gauges.

Later, to allow the field use of gauges of reduced diameter and greater sensitivity, Graham suggested that lithium niobate, which he had recently characterized, should be substituted for X-cut quartz in some developmental field test sensor designs. In certain crystallographic cuts, this grown crystalline material has a hydrostatic response, unlike X-cut quartz. Therefore, it could also be used in a variety of specialty stress wave sensors, particularly those designed for and used to document stress waves propagating from distributed explosives in many field experiments related to in-situ explosive rubblization of oil shale and coal for energy production.

**Fossil fuel research.** Triggered by the energy crisis of the 1980s, generous Federal emergency funds became available for fossil fuel energy research. Several major U.S. energy firms and a few opportunistic start-ups proposed speculative in-situ methods to produce liquid or gas energy products from underground resources. Historically, the evaluation of such tests in industrial energy exploration had been based on the surface-observed net product, rather than on the direct transient measurement of actual subterranean occurrences. Our field test organizations volunteered to the Department of Energy (DOE) to support diagnostic field instrumentation to evaluate the many novel, speculative geo-energy production schemes. The field test organizations partnered with other Sandia departments. Several field experiments in Colorado and Wyoming involved Rod Boade, George Davidson, new-hire Paul Hommert, and others. Field installation and recording were supported by the customized and specialized field test mobile instrumentation vans and staff used routinely in recording data at the Nevada Test Site.

Several experiments were performed on large field projects funded by the DOE that had been awarded to different bidders. Some bidders proposed to produce coal gas by underground in-situ combustion. Others proposed rendering liquid petroleum from oil shale by in-situ explosive rubblization (anticipating fracking technology), followed by in situ heating rather than excavation. In-situ combustion was proposed to produce a fraction of the resource economically. The characteristic of these proposals, the dynamic underground explosive and thermal process, could not be observed directly, so the actual fact of, or reason for, success or failure was indefinite. Sandia provided subsurface stress wave and thermometry instrumentation that revealed the actual remote subsurface transient stress and progressing thermal processes as they occurred. Techniques and instruments developed then were later applied to experiments on the underground storage of nuclear waste at Carlsbad and at the Nevada Test Site.

The processes involved the detailed monitoring of both explosively generated stress waves and diagnostic temperature. My contribution was to develop and apply special stress wave instrumentation of several underground explosion experiments and diagnostic thermal instrumentation of progressive combustion wave fronts to coal gasification experiments in Colorado and Wyoming. Sandia's diagnostic measurements used unconventional sensors and specially developed recording instrumentation. The data demonstrated that many optimistic predictions and superficial surface interpretations by the contracted experimenters were actually invalid. The energy development proposals were not as successful as could have been claimed without the Sandia diagnostics. With authentic measurement, the specific reason for any success or failure was concretely documented. Sandia's contributions were welcomed by our DOE sponsor, but not warmly appreciated by the unsuccessful experimenters.

**The science of measurement.** Measurement authenticity, not just accuracy, as a discipline became a broad topic of continuing interest to me. Sandia allowed me to host and present several short courses in Albuquerque and Las Vegas devoted to the Science of Measurement, as promoted by Peter Stein of Arizona State and Sandian Patrick Walter, whom Stein mentored en route to his Sandia-sponsored PhD in mid-career. The development of unique diagnostic stress waves and thermoelectric circuits, instruments, and test methods in fossil fuel experiments led to my several years of service, while at Sandia, as chairman of an ASTM international standards committee. My service resulted in internationally sharing—through papers, handbooks, and standards—concepts that were developed in the pursuit of the Sandia stress wave and diagnostic thermometry field experiments during the decades-earlier energy crisis and underground nuclear tests.

My long-time close association with Robert Graham, Bill Benedick, Larry Lee, and others in the Solid State Sciences Directorate, with their scientifically oriented material phenomenology studies, supported the varied applied measurement developments used by Sandia's field test organizations, from theory to application.

**A thin-film piezoelectric polymer and Francois Bauer.** At an American Physical Society symposium in Albuquerque, we first met the French scientist, Francois Bauer, who presented a paper on his thesis work with a new Japanese-discovered thin-film piezoelectric polymer material, polyvinylidene difluoride, or PVDF. Bauer was familiar with Graham and Benedick's work on X-cut crystalline quartz. Benedick spoke French. Bauer, then a graduate student, needed support for his thesis claim of reproducible dynamic properties from his special processing of PVDF. He convinced a skeptical Graham to test the claimed (but implausible) superior material properties using Sandia's precision gas gun facility. Graham's gas gun tests on Bauer's PVDF were convincing. Graham convinced (equally skeptical) me as well. From that encounter, formal symbiotic relations developed among Sandia, Ktech, and the Franco-Allemand (French-German) Institut de Recherches de Saint-Louis (ISL) in France.

Unlike solid-grown crystal piezoelectric materials, the very thin plastic film required special extrusion and polarizing to attain its reproducible piezoelectric properties. Bauer's critically important French-patented process yielded a precisely controlled and reproducible sensing material. ISL made the process accessible to Sandia and our contractor, Ktech. In collaboration with ISL, very thin sensitive shock stress sensors with a sensing area as small as 1-mm square were developed.

These cooperative relationships led to a pair of international workshops. With the workshop chairman, Bauer, I cohosted a multilingual PVDF Shock Sensor Workshop on November 6–8, 1990, in Saint-Louis, France. There were 72 attendees, with most from England and Europe. A corresponding PVDF Shock Sensor Workshop was held December 11–13, 1990, at Sandia in Albuquerque, cosponsored by Sandia and Ktech. That workshop was attended by 82 individuals representing 17 different, mostly U.S., organizations. Other PVDF-specialized workshops followed. Later, Larry Lee chaired a Workshop on the Technology of Ferroelectric Polymers from October 17–19, 1995. Additional workshops were held independently in France.

Our Sandia Field Test division developed standards, methods, designs, and specialized suppliers and maintained unique supplies of exceptional-quality defect-free quartz, lithium niobate, and PVDF film in uncut crystal boule form, as finished disks and sheets and as assembled gauges. Rigorous material and fabricating control were essential as commercial premium-grade materials are inadequate for the precision shock wave instrumentation. Sandia Field Test Quartz Gauges, as standardized, characterized, and used in underground nuclear test environments, were supplied as packaged test sensors for fielding by Benedick and Sandia experimenters in other directorates. As sometimes happens, as technology advances and matures, the Sandia test demand for these materials vanished. Precious, unique, and very costly materials remained unused (presumably, salvaged).

**Acknowledgments.** In retrospect, perhaps like other respondents who have offered their historic recollections, I have been fortunate to be associated with many exceptionally skilled associates and to have received the support of rarely credited, clever, dedicated, and capable engineers and technicians at Sandia who fabricated and performed experiments, analyzed data, implemented special designs, and offered perspective. Mark Anderson, John Birdsong, James Greenwoll, Red Hollenbach, George Ingram, Ken Kimball, Cliff Kinabrew, Robert Morris, Harold Roberts, and very many others made essential, if often inconspicuous, contributions to the progress of stress wave and thermometry testing at Sandia. More conspicuous was the generous support by Sandia management (e.g., Al Chabai, James Plimpton, and Joe Wistor) for the freedom to practice emerging technologies, theory, and science of stress wave pursuits and to allow the presentation of papers at national and international conferences. It has been interesting to observe the well-deserved advancement of several early associates who had been involved in Sandia's stress wave-related activities (e.g., Larry Bertholf, Richard Claassen, Ruth David,[137] George Davidson, Dennis Hayes, Paul Hommert, Orval Jones, Al Narath, Stephen Rottler, etc.).

## Post-retirement Activities

Following my retirement from the Applied Physics, Engineering, and Testing Center in June 1994, doing business as Proteun Services until 2004, I consulted nationally for several Sandia experimental departments and for the Sandia Primary Measurements Standards Department. I also contributed chapters to CRC/IEEE Press Publications (i.e., to *The Measurement, Instrumentation, and Sensors Handbook* and to *The Industrial Electronics Handbook*) and to Standards and Handbooks of ASTM. These later contributions reflected my measurement perspective developed earlier in stress wave-related activities at Sandia.

[137] Editors' note: Dr. Ruth David began her professional career at Sandia in 1975 with a bachelor's degree in electrical engineering. She served as the Director of Advanced Information Technologies at Sandia and, before that (from 1991 to 1994), as the Director of the Development Testing Center. In 1995, she left Sandia to become the Deputy Director for Science and Technology at the Central Intelligence Agency, and in October 1998, she was appointed President and Chief Executive Officer of Analytic Services, Inc., an independent, not-for-profit, public service institute that provides research and analytic support to a number of government agencies. In 2012 she was appointed a member of the National Science Board of the National Science Foundation.

# William D. Reinhart[138]

## (1999–Present)

I moved to Albuquerque in July 1981, with a job in hand at Ktech Corporation. I was assigned to the lateral impulse facility (Mag Flyer) at the Air Force Research Laboratory (AFRL). Mag Flyer also housed an 80-ft compressed light gas gun. My first day was quite unique and essentially repeated itself some years later. As I walked down the hall with Larry Lee (Vice President, Ktech) and Boyd Jennrete (Ktech manager for AFRL operations), it became clear that once again I would be fighting an uphill battle for respect. Technicians looked me up and down and, right from the start, assumed I was the hot-shot east coast recruit to teach them everything. Actually, I had never even seen instrumentation for experimental testing in my life. The control room was filled with oscilloscopes the size and weight of a small car, reel-to-reel recorders were spinning at incredible speeds, and panels of colored lights and racks of vacuum tubes were laden throughout the area. The lateral impulse facility was a large room containing capacitors and large aluminum blocks. But my first encounter with the lead engineer was something I could never forget. His first comment to me was, "So you are the one here to replace me? Well one day, all this will be yours!"

I believe it was in 1987 that I inquired about a position at STAR. I went to Carl Konrad, the manager of STAR, and asked if he thought there would be any problem with me working there. I had previously been at STAR to perform some tests and, again, encountered reluctance. STAR had just acquired state-of-the-art diagnostics in waveform recorders to obtain data digitally without cameras. I was familiar with them and researched how they worked and was able to use them for our STAR tests. This was the first and probably the only time they were used until I started work there in summer 1987.

STAR had a work environment to which I was not accustomed; it took me quite a few years to become an "accepted" worker among the technicians. I spent my time in the control room reading documents on instrumentation and reports and journals from the staff. I became an expert in instrumentation and interferometry—and I still do not remember anything about dropping a VISAR into a trench.

I was assigned to work with Lynn Barker to develop a three-stage gun. At that time, the intent was to launch a flyer plate to extreme velocities using very high gas pressures. Many issues had to be overcome, such as keeping the initial pressures low to avoid irreparable damage to the projectile yet significantly high to drive the flyer plate to high velocities without yielding the flyer plate. Though we never quite made it to an intact plate being launched, we did manage to ruin numerous barrels in these attempts.

I believe I started working with both Lynn and Lalit Chhabildas in 1988–1989 to follow up the gas launch to direct impacts on flyers with some type of graded-density impactor. But first, we studied what type of flyers would work best, as

[138] Bill Reinhart worked for a number of years as a contractor to the shock wave group at Sandia before finally becoming a Sandia employee in 1999.

appropriate candidates had to have adequate yield and spall strength to survive shock and quasi-isentropic loading at the high pressures. These impactors varied from simple PMMA (polymethyl methacrylate) and copper, to the Barker "pillow" (fabricated by spraying metallurgical powders and using sedimentation techniques so a smooth variation in shock impedance occurred), to the Chhabildas "plywood" (PMMA/Al/Ti/Cu), to the final configuration still in use, polymethylpentene (TPX)/Mg/Al/Ti/Cu/Ta. We had been using PMMA, but I found TPX gave the flyer a boost—something about the significant amount of hydrogen in the material—I am just guessing.

The new technique brought us into the 1990s and a new wave of customers. At this point, I was primarily working with Lalit at STAR on the two-stage gun. Looking at the records, over the next 4 years, I did over 50 tests a year on that gun alone, which had never before been accomplished. Additionally, I had taken over some of Carl's responsibilities and was essentially handling all diagnostics for nearly all STAR experiments. (This was the time of the ES&H[139] "Tiger Team" and major ES&H concerns laboratory wide, so Carl was basically swamped with facility paperwork.) About this time in the early 1990s, the statement by the Ktech engineer came up again, "One day all this will be yours." This time Carl made the comment, and it was definitely heard as the rest of the staff was reiterating this, even though I was a contractor.

However, as time went on, the STAR staff and I grew to respect each other since everyone (or most of them) had unique skills. Bob Hardy was probably one of the most resourceful, mechanically competent, experienced, and knowledgeable people I have ever encountered. He was greatly missed when he left, and STAR was significantly shorthanded in a facility that was heavily dependent on mechanical applications for operations. Clint Hall complemented Bob for not only his mechanical expertise but his ability to think through problems and document them (which, as years went by, became essential). Carl provided the stability and knowledge of the facility infrastructure for the diagnostics.

As everyone has mentioned, 1995 appeared to be the end of STAR. The rationale has been stated by others; the last days have been mentioned. What may not have been realized is that, at this time, I was not ready to throw in the towel. 1994 to 1995 had also brought the Internet and email to the STAR. I cannot remember exactly, but I did send off emails to our secretaries, managers, and director assistants—Lucille Verdugo, Anita Vasey, and others—about the status of STAR and offered up many options for operations continuing at a lower cost (to no avail), but I did manage to attract, probably not desired though, unwanted attention from our Director, Ed Barsis. I was in touch with Ktech about the possibility of taking over the facility, as well, and had made contact with the management at Phillips Laboratory. Ktech was not expressing too much interest since they had just gotten out of the AFRL gun and Mag Flyer test facility. But Phillips Labs was indeed interested. And they made contact with Ed Barsis, and I was not real sure of the real story—some hinted he

[139] Editors' note: ES&H is Environment, Safety, and Health.

said "Go ahead and take it;" other rumors were "no way" and that he would rather shut it down than give it away—I guess I may never know.

Regardless, the shutdown continued, but still I had not given up. The STAR staff was dissipated throughout the labs. Some went to the Sandia Ombudsman to complain that a contractor was still working, while employees were moved out. I believe Carl had a significant say on who stayed for the shutdown. After the last test (which was in May), I enlisted the help of Lucille and others, as they were dissatisfied with the way management handled all of us. I managed to obtain a couple transportainers and the use of about six large flatbed trucks. I had decided that I would not leave all the equipment that had value to the "vultures," so before the "STAR open house," I loaded all of what I thought was important onto trucks and into transportainers and "relocated" everything. The most valuable equipment (digitizers, computers, unique tooling) were loaded into a transportainer and parked behind Ed Barsis' building (not to his knowledge); two to three flatbeds went to Area II—the old Area II. I knew some people over there who said that one building was not set for demolition for another year. The flash x-ray instrumentation went to the new Building 905, in the attic or the mechanical area.

So, after all the important items were removed, the "garage sale" started and actually a lot of junk was removed—a good thing. Now my memory is a little vague, but Carl assumed responsibilities for ES&H in the robotics group, Clint went to the wind tunnel area, and I, well I, went to Building 808 with McGlaun—Dennis Grady was there as well. I had a HUGE office, yet I do not remember exactly what I did, but I do remember going over to the Explosive Components Facility and talking with Lloyd Bonzon at some point and asking him if he wanted STAR. I told him people from Phillips Lab were interested in taking the facility and wanted to know his thoughts. He surprised me by saying he would be damned if he would let an asset like STAR leave Sandia—so he took it! Again, the order of things that happened I cannot totally recollect, but Jim Asay, Don Cook, and many other people got involved on how to operate the facility in a campaign mode where it would come up to test, then go to a mothball state. We shut down STAR in about August 1995. So from August to December, I went from STAR, to Building 808, to Building 905 (as Lloyd brought me over to that building). When I arrived at 905, I had an office—well, it was an area where students (there were none) without clearances would be located—with my own bathroom. But here is where I started working on operating procedures, readiness assessments, and a plethora of other documents to bring guns back on line. I also got back into testing on the air gun in 905.

I believe that halfway through 1996, STAR reopened with testing on the Terminal Ballistics Facility gun. Don Cook gave us $25 k seed money, which wasn't much but helped a lot; Dennis Grady had some money for testing, and Lloyd Bonzon supplied me with ES&H support—i.e., a technician as needed. Lloyd was a unique manager since he had no background in our type of work, but he was able to get whatever he needed by getting the right people involved. He really did not care how it was perceived; he just did it—specifically, I was still a contractor and running a Sandia test facility without any Sandians present. During the beginning of the restart, I had at least one critic. He told me right after we had our first test that he did not know why I was doing this, as STAR will NEVER come back into full operation again as no one is really interested in the work we do out there—another catalyst for me.

Once operational, our group started reforming again; this was Jim Asay's doing and was followed by subsequent funding. I was able to bring John Martinez back (I have worked with John since 1982) and immediately did something I had always wanted at STAR—bringing consistency to the experiments. What I mean is that John would handle ALL setup of testing including, and especially, fabrication of projectiles. I had always thought that was a problem in the past: the projectiles were the product of generally nonqualified machinists and everyone was making them.

In 1999, Jim thought it would be best if I came over to his group as a Sandia staff member (MTS). When I brought this to Lloyd's attention (Jim had probably already talked to him), I asked him what he thought about moving over there. He surprised me by his response. First, he said, he wanted to maintain that any testing that could be done at 905 should be done on his gun rather than at STAR, and second, I should not go over as an MTS, but as a Senior Member of Technical Staff (SMTS). I believe that in September 1999, I was hired as a Sandian at the SMTS level in Jim's group.

From 1996 to 2007 (I think 2007 is when Lalit left), Lalit started incorporating me into his philosophy of performing experiments, obtaining the data and reducing, analyzing, and generating a report, followed by presenting the results. I have a different outlook (probably because of my ignorance in the subject field) in that I really must understand something before I can write or talk about the subject matter. This is the reason why it took me a painfully long time: not only must I understand, but I want to know *really* what of value the information is telling me and why it is important. I must not just have a simulation match the data, though I see the importance of having good models. Yet, I am not sure that, during my time, I have ever seen data generated at the facility go into a model, well, maybe once Kerley took some high-pressure data on the Ti6Al4V alloy and incorporated it, but that is all I can remember.

Now, this era (1999–2007) was incredible and, thanks to Lalit and Jim, is something I could tell my sons and daughters about the people I have met and talked with—well, they would not care as they would never have heard of most of them. But to me, it was very humbling to be able to talk, discuss problems, shake hands, and meet the likes of Jim Asay, Lalit Chhabildas, Lynn Barker, Dennis Grady, Alois Stilp, Datta Dandakar, Gennady Kanel, Stephan Bless, and many, many others! On a trip to the 50th International ARA (Aeroballistic Range Association) meeting, which was held at LLNL in 1999, I had the honor to see, shake hands with, and listen to him talk—**Edward Teller**. The interesting part was that I called my wife and told her this, and to my surprise, she was quite impressed and screamed at me over the phone, "Do you know who he is?! Why was I not invited to come along to this!"—which, of course, she was, but as usual she had declined.

Except for 1995 to 1996, I was never really concerned about my job or money issues for operations at STAR. During the leaner years (1996 to 2007), Lalit was able to get enough business for us to survive and, at times, thrive. When he left, I was worried. However, those years of traveling, reporting, and meeting people paid off. I was able to continue to obtain the business and keep STAR operational.

# George A. Samara

# (1962–2006)

**Editors' comment**: These two memorial tributes capture the seminal accomplishments George Samara made to the field of high-pressure science and, in particular, to understanding ferroelectric materials. The first was published in 2008 by the National Academies Press and the second in 2007 by Springer-Verlag.

**in Memoriam**

"For contributions to the understanding of dielectric, ferroelectric, and ferromagnetic materials applications."

**by Al Romig[140]**

Although George was born on December 5, 1936, in the small farming community of Jdeidet Marjayoun in southern Lebanon, his father was a United States citizen. He came to the United States as a 16-year-old, completed high school in Drumright, Oklahoma, and graduated from the University of Oklahoma with a B.S. in chemical engineering in 1958. He earned a doctorate in chemical engineering in 1962 from the University of Illinois at Urbana. This was under the direction of

[140] Al Romig prepared this memoriam at the request of the National Academy of Engineering for George A. Samara, who was also a member of the academy. See *Memorial Tributes: National Academy of Engineering* (The National Academies Press, Washington, D.C., 2008), Vol. 12, pp. 240–245, reprinted with permission from the National Academies Press. The first paragraph has been omitted from this version of Al Romig's memorium. Explanatory footnotes that are not part of the original have been added.

Professor Harry Drickamer, who was a hard task-master. His thesis work as well as continued collaborations with Drickamer covered a wide range of materials. These included pressure effects on the resistance of metalloids, simple fcc and bcc, and II–VI, III–V compounds to more exotic fused-ring aromatics. This collaboration with Drickamer continued into the 1970s. Later, other coauthors also enjoyed such long-enduring relationships, several extending over two and three decades. Many of these involved research topics resulting from informal questions and/or discussions at scientific meetings. Even though a list of coauthors exceeds 70, George was also the single author of nearly 100 journal papers and topical reviews.

Upon completing his degree at Illinois in 1962, he joined Sandia National Laboratories, but immediately fulfilled his military duty, spending the next 2 years as an ROTC[141] officer at the Institute of Exploratory Research for the U.S. Army Electronics Laboratory at Fort Monmouth, New Jersey. There, he was able to continue research involving pressure effects on solids, including the pressure homogeneity of the sample in the pressure apparatus, and developed various techniques to measure the physical properties under pressure, such as compressibilities, electrical conductivities, dielectric constants, and Curie points. This effort was with Armando Giardini and others at Fort Monmouth. Also included was a paper in the area of what became of major importance to not only the entire worldwide research field but also his personal research, as well as to Sandia—ferroelectrics. George's contributions to various aspects of ferroelectrics continued throughout his research career.

Returning to Sandia, George began and continued an active personal research effort which spanned 45 years on many areas besides ferroelectrics up to the final days of his life. Among such pending studies when he passed away were promising new ferroelectric materials based on lead lanthanum zirconate titanate and on terpolymers of vinylidene fluoride with other fluoroethylenes. George's numerous scientific papers included pioneering work on structural phase transformations, semiconductor physics, ferroelectrics, ferroelectric polymers, nanosized semiconductor clusters, defects, deep electronic levels, relaxation in crystalline solids and polymers, ionic transport, ceramics, photovoltaics, and MBE[142] and CVD[143] synthesis and processing. Among the invited reviews are three chapters in *Solid State Physics*: "The Study of Soft Mode Transitions at High Pressure" with Paul Peercy, "High Pressure Studies of Ionic Conductivity in Solids," and "Ferroelectrics Revisited—Advances in Materials and Physics."

Through this research effort, George became perhaps the world's most accomplished scientist in the use of high pressure for fundamental studies of electronic and structural properties of solid state materials. In addition, he was active in national and international organizations involving high pressure science. Among leadership roles he served are Vice President and [on] the Executive Committee of the International Association of High Pressure Science and Technology (AIRAPT);

[141] Editors' note: ROTC is the Reserve Officer Training Corps.

[142] Editors' note: MBE is molecular beam epitaxy.

[143] Editors' note: CVD is chemical vapor deposition.

chairman of the U.S. Department of Energy (DOE) Panel on High Pressure Science and Technology; chairman, 1974 Gordon Conference on Research at High Pressure; and co-chair and program chairman of the 1993 Joint AIRAPT/American Physical Society Conference on High Pressure Science and Technology.

In 1974, George was awarded the Ipatieff Prize (for scientists under the age of 40) of the American Chemical Society. In 1986, he was elected to the National Academy of Engineering. As an active member of the academy, he served on the Section 9 Peer Committee (2001–2003) and organized the regional meeting in Albuquerque (May 19, 2005) on *Solid State Lighting: The Next Revolution in Lighting*. He was a fellow member of the American Physical Society and the American Association for the Advancement of Science. His activities also included editorial or advisory boards for *Ferroelectrics*, *Reviews of Scientific Instruments*, *Journal of Physics and Chemistry of Solids*, and the Materials Research Society's *Journal of Material Research*.

Those of us who were familiar with George's unusually accomplished research career as well as his extraordinary organizational ability to develop a host of national and international materials research activities might also realize he possessed excellent management skills. At Sandia National Laboratories, he was initially promoted in 1967 and served in various management capacities involving oversight of research in condensed-matter physics, electronic materials and phenomena, chemistry research activities, advanced materials, and nanoscience.

During George's long and distinguished career at Sandia, he recruited the most promising young scientists in the country and then nominated them for invited talks, honors, and awards, promoting their career development. His interaction and support of these and other collaborating scientists led to worldwide recognition of excellence for various individuals and groups. He also served as a resource, sounding board, and mentor for generations of Sandians, many of whom came to occupy significant leadership positions both at Sandia or elsewhere. George had a unique combination of deep scientific understanding and insight, foresight, friendliness, and personal and organizational leadership abilities. He was widely recognized for his high professional, ethical, and scientific standards that have inspired all who had contact with him.

In addition to overseeing management of these research activities, George took on additional management responsibilities. During the last 13 years of his career, he served as manager of Sandia Natiional Laboratories Department of Energy (DOE) Basic Energy Sciences core research programs in materials science. He was largely responsible for developing the concept for the DOE Center of Excellence for the Synthesis and Processing of Advanced Materials, a coordinated, cooperative venture among 12 national laboratories and several industrial and university partners, and he served as its Director. In recent years, he was instrumental in developing the joint Sandia/Los Alamos Center for Integrated Nanotechnologies (CINT), one of five DOE Nanoscale Science Research Centers. In these efforts, he earned the admiration and respect from his colleagues at the other national laboratories and the materials research community for his unbiased even-handed coordination and leadership. Based on his technical management and leadership, George was awarded the American Chemical Society's Earle B. Barnes Award for Leadership in Chemical Management in 2000.

Finally, a few comments on his personal life, for amazing as his technical accomplishments were, George was much broader—a man for all seasons. In his younger days, he hiked the local New Mexico mountains as well as the midrange of the Sierra Nevada in California with various colleagues. He was a respectful opponent in tennis. He visited Lebanon through the years; there he met and married his wife, Helen. Those who lived in or visited Albuquerque enjoyed the Lebanese hospitality and food at their home. Fresh grapes, apricots, and particularly two varieties of figs were provided from his back yard. He enjoyed gardening and the arts, particularly music, and even wrote poetry for his church congregation. A loving and devoted husband and father, George leaves his wife Helen, daughter Victoria, son Michael, brother Emile, and sister Leila. Since George's passing, many have noted that he will not just be remembered for his accomplishments, of which there were many. More importantly, he will be remembered for the way he connected with other people, the generosity of his heart, his willingness to always be open to others, and his constant optimism. He was a brilliant scientist, a good citizen, mentor, and teacher. As Helen said, George was a gift to us all. He will be sorely missed.

**in Memoriam**

**by Bob Graham**[144]

George spent his almost 45-year scientific career at Sandia National Laboratories in Albuquerque and is especially known for the use of quantitative conditions of high pressure to better define the dependence of electronic properties on lattice compression. He received his PhD degree in Chemical Engineering in 1962 from the University of Illinois, Urbana, under the direction of Professor Harry Drickhamer.[145] Not only did George carry out an unusually accomplished research career, but in 40 years as a supervisor of physics and chemistry research activities, his interaction, insight, and support of his scientists led to worldwide recognition of excellence for his group in shock compression science. He is widely recognized for high professional, ethical, and scientific standards that have inspired all who had contact with him. As an ROTC officer, he spent 2 years at the Institute of Exploratory Research for the U.S. Army Electronics Laboratory at Fort Monmouth, New Jersey. He used his extraordinary organizational ability to help organize a host of national and international materials research activities. In high-pressure science, he served as the Vice President and [on the] Executive Committee of the International Association of High Pressure Science and Technology (AIRAPT); chairman of the U.S. Department of Energy (DOE) Panel on High Pressure Science and Technology; chairman, 1974

[144]This memoriam was published by Springer-Verlag in *Shock Waves* **7**, no. 3, pp. 213–214 (2007), reprinted with permission of Springer Science and Business Media. The first paragraph has been omitted and an explanatory footnote was added.

[145]Editors' note: Note that Professor Drickamer's name was misspelled in the original of this remembrance.

Gordon Conference on Research at High Pressure; and co-chair and program chairman 1993 Joint AIRAPT/American Physical Society Conference on High Pressure Science and Technology. He was active in the American Physical Society Topical Group on Shock Compression of Condensed Matter in official and consulting roles. Consistent with the focus of his group's orientation on materials science under deformation, his highly regarded shock compression science group uniquely focused on solid state physics and chemistry. The work drew from the physics capabilities from a broad range of scientific strengths.

George's 200 scientific papers included pioneering work on structural phase transformations, semiconductor physics, ferroelectrics, ferroelectric polymers, nanosize semiconductor clusters, defects, deep electronic levels, relaxation in crystalline solids and polymers, ionic transport, ceramics, photovoltaics, and MBE and CVD synthesis and processing. The work on ferroelectric polymers led to the development of nanosecond time resolution gauges. George was awarded the Ipatieff Prize of the American Chemical Society and was elected to the National Academy of Engineering. He was a fellow member of the American Physical Society and the American Association for Advancement of Science. His activities included editorial or advisory boards on *Ferroelectrics*, *Reviews of Scientific Instruments*, *Journal of Physics and Chemistry of Solids*, *The Journal of Materials Science*. Based on his technical management and leadership, he was awarded the American Chemical Society's Earle B. Barnes Award for Leadership in Chemical Management in 2000.

During the last 13 years of his career, George served as the manager of Sandia [National] Laboratories' DOE Basic Energy Sciences Materials effort. He served as the Director of the DOE Center of Excellence for the Synthesis and Processing of Advanced Materials, a coordinated, cooperative venture among 12 national laboratories and several industrial and university partners. He was largely responsible for developing the concept for the center. George played a leading role in formulating the DOE joint Sandia/Los Alamos, Center for Integrated Nanotechnologies. With his principal management interest in the DOE Basic Energy Sciences, George maintained a "secret" laboratory where he could have at least a little time with his high pressure science. True to his nature, he worked until the last.

## Karl W. Schuler

## (1967–1996[146])

I arrived at Sandia in April 1967. My background was theoretical continuum mechanics. I had never done any kind of experimentation but wanted to do so. One of my professors at Illinois Institute of Technology (IIT) suggested that, if I wanted to get into experimentation, I should definitely go to one of the national labs because they had lots of equipment ready to use. In contrast, if I went into a university, I would spend 2 or 3 years trying to get experimental apparatus in order to get a program started. Two Sandians at IIT working on their PhDs had also extolled the virtues of Sandia.

I was hired by Don Lundergan, who wanted me to work on wave propagation in polymeric materials. There were two reasons for this: the metal behavior at low shock pressures (<50 kbars) was being adequately covered, but many polymers involved in nuclear weapons had only been studied quasi-statically. Under shock loading their characterization was lacking, particularly for time-dependent effects. Secondly, Lynn Barker was developing the laser interferometer, and Plexiglas® looked like an ideal low-impedance window material.

Walter Herrmann was our Department Manager. The department, the Mechanical Behavior Department, was primarily located on the second floor in Building 805. I was given an office that was part of, what I guess you call it, a "bullpen." When you walked in from the second-floor hallway, you went down a central aisle which had cubicles on either side that were about eight-foot squared and had 5-ft-high walls. On the right side, the first cubicle was occupied by Floyd Tuler, the second by Larry Bertholf, the third by Ray Reed, and at the last one, by the window, was Barry Butcher. On the left side was a larger cubicle occupied by Sam Key and Zelma Besinger. Then there was a passageway that led into the Division Supervisor's office area. On the far side of that passageway was my cubicle, and next to my cubicle, again by the window, was Lynn Barker's cubicle.

I guess because of both the physical proximity and our joint interest in Plexiglas®, I worked closely with Lynn Barker. He was definitely a great mentor to me. I'll always remember that in his cubicle, there was a blackboard and along the top of the blackboard was written "I'm perfectly willing to be wrong." Very seldom was Barker wrong.

For the first couple months, I remember watching Lynn and Red Hollenbach set up shots in Building 855, which housed the gas gun. I also went out to the Area Y explosive site to watch explosive tests conducted by Darrell Munson. The gun facility in Area Y had not yet been built. I'm not exactly sure when I fired my first shot because the shot dates were typically placed on the front of the shot folders and I no longer have access to them. I do have a record of firing Shot 2110 in September 1967 on

[146] Karl Schuler left the shock wave program in 1978 and continued research in engineering applications until his retirement in 1996.

Plexiglas® that involved the use of the BRACIS (Beam Reflection At Center of Impact Surface) technique. This was a technique to determine the transit time of the shock wave by using a light beam that was totally internally reflected at the center of the impact surface of the transparent target. When the transparent projectile impacted, the light beam would continue on into the projectile, and one would observe a decrease in light level that could be interpreted as the shock wave initiation at the center of the impact surface. I'm not sure if this was the first shot to ever use the BRACIS technique, nor am I sure who came up with the idea, although it was probably Lynn. But I do remember that the first time it was tried, it was on one of my shots, and Lynn and Red were kind of hovering around in the control room. I don't think they thought it would work, but by golly it did. The usual technique, at the time, to determine when the central impact had occurred, was to have four pins around the periphery of the target carrying different voltage levels that would be shorted out as the projectile impacted. The average of these shorting times was the time of central impact. This technique worked well for materials that could be lapped very flat, but we found that trying to lap polymers like Plexiglas® flat left a lot to be desired. Without true flat surfaces, the uncertainty in the central time of impact was very large.

I'm sure that before Shot 2110, I had fired other shots. Some of these investigated the behavior of polyurethane and employed the other instrumentation techniques that the division used. These techniques included quartz gauges, slant-wire resistors, and displacement interferometers.

Using the interferometers, it was possible to resolve the fine structure of the stress waves in Plexiglas®. The loading shock typically consisted of a jump to a certain particle velocity followed by a gradual rise to the final particle velocity. By searching the continuum mechanics literature, I found papers by Alan Pipkin (a friend of my thesis adviser) that described the analysis of stress waves in nonlinear viscoelastic materials and predicted wave profiles similar to those that Lynn and I were observing. Thus, I was able to use Pipkin's work to analyze my experimental data. And by November 1969, I was able to submit a paper about the propagation of steady shock waves in Plexiglas®.

The extensions to the work on Plexiglas® came fast and furious. First, we did experiments where we backed up the Plexiglas® impactor with steel or aluminum. This produced a first shock followed by a second shock and allowed us to extend the range of the data. We also looked at the unloading waves in all these experiments. When the gun facility was completed in Area Y, I decided to extend the pressure range using the three-and-a-half-inch powder gun. In fact, I was the one who fired the first maximum shot on this gun (20 pounds of black powder). I seem to remember getting an impact velocity of 7000 ft/s. The next time somebody tried 20 pounds of powder in this gun, it was Bob Graham and it was his first shot on the powder gun. Unfortunately, the catcher had not been installed correctly in the impact tank. Furthermore, the door of the impact tank had not been properly secured. So, when Bob fired the shot, the rear door of the tank opened and a major disaster ensued. A lot of the sheet metal on the building was removed, along with all the lights and most of the instrumentation. In the wake of this disaster, I took the opportunity to drive our purchasing agent out to the site to show him what had happened. He was always complaining that I was trying to buy plant items using expense money. When he saw the destruction, he agreed that this was a hazardous environment and anything we bought in the way of instrumentation was at risk. In the future I never had any problems submitting purchasing requests through him, using whatever kind of money I had available.

But the real interest was in thin pulse attenuation in these polymeric materials. To look at thin pulse attenuation in a metal, one would take a thin impactor mounted typically on carbon foam. To accelerate thin pieces of polymer seemed very difficult because it was going to be very hard to find something to support a thin impactor. So I decided to turn the experiment around. What I did was put the specimen into the projectile off to one side and then, at the very center of the projectile, I put a prism and another prism behind the specimen; with this configuration, the interferometer beam went down the center line of the gun, made two reflections in the projectile, reflected off the back of the specimen in the projectile, and then went back out to the VISAR. The target consisted of a thin foil that was stretched using an O-ring over a hole in the target plate. Of course, the best foil to use was gold foil. So I found a supplier who would provide a 3-inch-wide, one-thousandths-of-an-inch-thick gold foil. What I didn't know was that every piece of precious material that arrived at Sandia receiving had to be inspected and weighed. The supplier assured me that he would ship my gold foil wrapped on a large-diameter cylinder so that it would stay nice and smooth since I needed it to be very flat.

When the shipment finally arrived at our lab, the gold foil looked like a used piece of tin foil. It had been wadded up in order to put it on the delicate balance to make sure that the fractions of an ounce of gold that I had bought were really there. I called the supplier and told him what had happened and he said no problem, I'll let you know the secret on how to make gold foil flat without a super expensive rolling mill. He told me to take a piece of the gold foil and put it on an optical flat, then put a drop of oil in the center of the foil, put another optical flat on top, and gently squeeze this sandwich in a press. The oil flows out and flattens and thins the foil. When I finished flattening the foil, I took it to the inspection department in the machine shop and had its thickness measured. The experiments worked well because the projectile did not rotate as it was launched. I was able to input pulses from a gold foil less than a few thousandths of an inch thick into various thicknesses of Plexiglas®.

My weak point was doing the paperwork at the end. **I was *not* a good or fast writer.** I always had a cartoon on my office wall depicting an outhouse with the caption "The job's not done till the paper work's finished." Fortunately, management recognized this and teamed me up with Jace Nunziato, who was a superb writer. Between the two of us, we pumped out quite a few papers. I'd do the experiments, reduce the data, make up some figures and tables, and discuss them with Jace. Together we would test already published theories with the data and scope out a paper.

Jace could start at 8 o'clock in the morning with a pad of yellow paper and a red felt tip pen. At 4 o'clock in the afternoon, he would throw the empty pen into the wastebasket and hand me the yellow pad fully covered with text for overnight proof reading so that it could be given to a secretary the next day. Jace's greatest coup was to show a collection of our papers to the editor of the *Handbook of Physics*, who asked us to write part of a volume. So now on my resume, I can claim to be the coauthor of a major monograph. After it was printed, we got a check from Springer, the publisher, which took forever to cash since it was for Deutschmarks. After it was cashed in dollars, we calculated that each of the four coauthors received a penny per word for our tome. So much for me making it as an author.

But my real problem in writing was the progress reports. Before I had been at Sandia for more than a few weeks, I had to write my first progress report. It went something like this: "But Don, I've only been here for eight weeks," I protested. "Yes, but in the research group we write progress reports. I've asked Lynn to help you write your first reports and I'm sure with his help you'll be done in no time," replied Don. "But Lynn, I haven't done anything," I insisted. Lynn replied, "Yes, you have. I know you've been in the library researching the literature. You also came to the lab and I showed you how to do a shot on the gas gun and how to use the laser interferometer, so you've started your experimental program. Why, we even discussed other experiments you plan to do." He continued, "So, 'Theories of Nonlinear Viscoelastic Wave Propagation' could be the title of your first report. For the second, I suggest 'Preliminary Experimentation' and, for the third, 'An Experimental Study of Shock Wave Propagation in Plexiglas®'."

I went back to my cubicle and began to write. I felt that never had so much been written on so little. This was before the days of word processors, so I had to scrawl everything out on paper. But I had been given three suggestions for progress reports which made me feel, after a while, that I might have actually done something in the 8 weeks I had been at the Labs.

After 2 days of scribbling, I had used up a whole pad of paper but my three reports were each less than one sheet of paper. I turned them into Don. Within an hour he came back with a number of suggestions that I quickly incorporated. I resubmitted them again to Don, who sent them up the line to Walter, my Department Manager. Walter immediately called me into his office and told me that I needed to have a first sentence or first paragraph in the reports to justify my existence.

My existence? He explained to me that in the research group, we always had to indicate to upper-level management how our research tied in with the Laboratory's objectives, and this was best done in the first sentence or paragraph of your report. That way, upper-level management could quickly glean that you were working on the right problems, and if they were interested, they could continue to read what you had done.

Also, he noted that I had one report that was about some kind of theoretical research and another report about my planned experimental program. He suggested that I should combine them and use the opening phrase, "A combined experimental and theoretical program has been started to…" In the future, I remembered this, and, in virtually every quarterly report I wrote in later years, I started with some variation of this phrase. Apparently, upper-level management loved studies that combined theory and experiment. A second point Walter made was that, once you came up with your justification sentence or paragraph, you shouldn't change it from report to report, because that would just confuse upper-level management. I got back to my cubicle to scrawl some more. When I was finally done, I had bragged profusely in a progress report that met the expectations of my management.

In the Department we only had to write progress reports every 6 months. But in the middle of 1968, a major reorganization occurred. We were moved into the research directorate and became the Solid Dynamics Research Department under Walt Herrmann, and I was required to write *quarterly reports*. At about the same time, Don Lundergan decided to step down and Barry Butcher became my Division Supervisor.

As good as Plexiglas® was for doing tests and being able to write journal articles, Plexiglas® was not really of interest to the weapons community. Sometime in 1968 I was put in a study group to look at the MC 2043 neutron generator. This experience brought me into direct contact with the kind of polymers of interest to the weapons community. In particular, loaded polymers were of great interest. These are polymers such as epoxy that are filled with various kinds of materials. Perhaps the most interesting of these was alumina-filled epoxy. I began to do shock wave experiments with this material. The most curious thing about this material was the great disparity between the loading wave velocities and the unloading wave velocities. The loading wave velocities were typically that of a polymeric material, but the unloading wave velocities were far higher than you would expect from a polymer. My explanation for this phenomenon was that initially the particles of aluminum oxide, which have a very high wave speed, are separated by the epoxy but, as the loading wave propagates through the material, the particles are brought into contact. Thus, the unloading wave is able to move through from one alumina particle to the next at a relatively high velocity. Specimens were prepared using different amounts of the alumina filler material, and I began a very extensive series of experiments. I also began reading papers on composite materials and mixtures and doing experiments on polymers loaded with other fillers. One of the most interesting fillers was epoxy loaded with glass microballoons. The material could sustain high hydrostatic pressures until the microballoons collapsed, at which point it would drastically decrease in volume.

As I began to accumulate shock wave data using the gas gun on these materials, I felt it would be desirable to have a triaxial test facility where specimens could be subjected to both a high hydrostatic pressure and a superimposed axial load. So, in the early 1970s, I began to order a pressure vessel and other equipment that could be built up into a hydrostatic triaxial test facility. In 1973, when the Arab oil embargo hit, energy research became very important. In fact, in 1974, Congress reorganized the Atomic Energy Commission into the Energy Research and Development Administration (ERDA). Sandia was instructed not only to continue support of nuclear weapons but to begin doing energy research. After much discussion, the research group at Sandia decided to investigate the production of liquid fuels from oil shale. The pressure vessel I had purchased to continue my research on polymeric material and mixtures suddenly became the centerpiece of a rock mechanics research lab devoted to oil shale research. We purchased a very large, stiff load frame to put the pressure vessel in so triaxial testing on oil shale could be done. I also began a series of experiments to look at the shock response of various grades of oil shale.

The rock mechanics lab was set up at the north end of Building 855. But, by 1977, I had essentially stopped working in shock waves and was heavily involved in rock mechanics. By 1978 we had completed many oil shale tests in the rock mechanics lab, and I had developed a very compact gauge to monitor the lateral expansion of oil shale samples in the triaxial testing machine. The work I had done on loaded polymers was taken over by Rod Boade, who ultimately coauthored a paper with me on alumina-filled epoxies. Although we planned to test alumina-filled epoxies and other filled epoxies in the triaxial test facility, the first priority was given to testing oil shale and other rock mechanics experiments such as those involved in the nuclear waste depositories in salt and at the Nevada Test Site at Yucca Mountain.

I did, however, continue my involvement with the neutron generator group; when I left the shock wave and rock mechanics groups in 1978 to become a member of the structural analysis group, I continued to support neutron generator standoff testing. In 1984, with the help of a lot of people who were still in the shock wave community, I fielded an instrumentation package for monitoring neutron generator standoff in weapon systems tests. I used the technique in various stockpile and proposed weapon systems from 1984 to 1996. After I retired in 1996, I was kept on as a consultant to continue advising the systems group on neutron generators.

My time in the shock wave group also significantly contributed to my health. In 1967, when Jeff Lawrence and I arrived, we were the only smokers in the department of 30+ people. There was evidence, such as stained fingers and old well-used ashtrays, that many others in the group had been smokers. However, because of the death of a member of the group from lung cancer, they had all quit! When Lynn bet me a dime that I could not stop smoking, I stopped. I still have the dime taped to a card inscribed with the words “Barker was wrong.”

## Herbert J. Sutherland
## (1970–2005[147])

I was hired into Sandia by Darrell Munson after graduating from the University of Texas (UT) at Austin (see Ray Reed's recollection). I received my master's degree under Professor Ripperger while working as a graduate assistant for Phil Stanton. Phil was at UT working on his PhD. He gave me my first introduction to Sandia. For my master's thesis, I used the explosively-driven flyer plate that was developed by Phil to examine a shock-induced phase transition in a polycrystal. For my doctorate, I examined acoustic wave propagation in a viscoelastic composite under Professor Calvit.

Darrell hired me primarily to examine shock waves in composite materials. I was paired with Doug Drumheller. (One departmental model for much of the research being conducted in this group paired an experimentalist with a theoretician. Working together was synergistic for both of us and proved to be a very fruitful model for producing advances in understanding and modeling.) My primary responsibility was on the experimental side and Doug's was on the theoretical side, but those lines were very loosely defined. Later we hired Marc Bedford from UT (Marc had been on my doctoral committee) as a consultant to aid in the analysis. We started with metal composites and moved on to viscoelastic composites. Doug and I shared an office in the "bullpen" for most of my tenure in the group.

When I first reported to Sandia, I did not have any experience working with gas guns. Don Lundergan took me "under his wing" and taught me the ropes. He let me observe one of his shots (start to finish) and then shepherded me through my first series of experiments on the 4-inch gas gun.

Over the years I collaborated with a large group of other researchers on various projects. In addition to my primary collaboration with Doug and Marc, I worked with Jace Nunziato, Karl Schuler, Joel Lipkin, Jim Kennedy, Gary Carlson, Pete Lysne, Peter Chen, and Rich Schmidt—just to name a few. All were very fruitful collaborations.

Doug, Marc, and I determined that the modeling of wave propagation in composites required a detailed knowledge of the geometric and material dispersion of the composite that could not be determined using shock waves alone. To examine dispersion in detail, I worked with Larry Kent and later with Dick Lingle, developing an ultrasonic (acoustic) wave propagation lab in our group. This capability was used on a number of projects in addition to the composite work.

In one case, Jace and I used ultrasonic waves to examine the mechanical degradation of a high explosive due to thermal aging. We were using elevated temperatures to conduct "accelerated life testing" of an explosive. To conduct these tests required monitoring, using ultrasonic techniques, of the explosive for approximately a week while maintaining the sample at a relatively high temperature. The first couple of

[147] Herb Sutherland left the shock wave group in 1978; he retired from Sandia in 2005.

days were critical and required measurements on a 24 hours basis. Jace "babysat" me (a minimum of two people was always required in the explosives lab when tests were being conducted) for a couple of nights. We brought in a couple of cots and a very loud alarm clock. I remember that Jace never got up, but he did stay awake while I did the measurements.

When funding became tight in the 1975 to 1976 time frame, Walt Herrmann transferred me from the work I was doing on the Shear Wave Velocity Interferometer to the department's reactor safety division. My work in ultrasonics guided Larry Kent and me in the development of an acoustic technique for monitoring the penetration of very hot liquid metals (including liquid sodium) into concrete (a composite with dispersion). The tests that Larry and I conducted at Sandia as part of a much larger test team led to an invitation from the KFK (Kernforschungszentrum Karlsruhe in Karlsruhe, West Germany, a German national laboratory for reactor research) to participate in similar experiments. I spent 2 months there as a visiting scientist adapting the techniques that Larry and I had developed to their experiments and serving as a member of the test team in conducting the experiments and analyzing their results.

For my last 4 years in this group and for an additional 6 years after that, I was an instructor in Sandia's Apprenticeship Program(s). I taught classes in algebra and trig, physics, and dynamics. All of the classes were held over the lunch hour or after work. Almost all of my students became full-time Sandia employees.

In 1978, I left the department. I moved to a recently created geomechanics department where I worked in the area of rock mechanics. In that research area, I had extensive collaborations with Darrell Munson and Karl Schuler. In 1985, I moved into the wind turbine development group, where I spent the next 20 years of my Sandia career.

# Samuel L. Thompson[148]

# (1966–1994)

# Tribute by Mike McGlaun: An Enduring Nationwide Impact

Samuel (Sam) L. Thompson developed state-of-the-art strong shock codes. He took the extra step of supporting a large user community both inside and outside Sandia National Laboratories. He strengthened the shock physics community throughout the Nation.

His codes stood the test of time because they were accurate and robust and many other people used them. Sam understood the importance of providing a complete shock physics analysis environment. This computational environment included material models and pre- and post-processing software. He worked exceptionally hard throughout his career to support the many users of his codes.

## Technical Contributions

Sam joined Sandia in 1966 after earning a PhD in astrophysics from the University of Kentucky. He began studying the response of materials to x-ray-driven thermo-mechanical impulses. Sam wrote the one-dimensional (1-D) **C**oupled **H**ydrodynamics **A**nd **R**adiation **T**ransport **D**iffusion (CHARTD) code soon after joining Sandia. It modeled strong shocks using the **AN**alytic **E**quation **O**f **S**tate (ANEOS) equation of state package. Sam developed the ANEOS package to model the complete multiphase equation of state. ANEOS provided equations of state that were valid from room conditions to very high pressures and temperatures. ANEOS did a remarkably good job of modeling many materials, including metals. Sam

[148] Sam had a varied career at Sandia. He worked on the development of hydrodynamic codes for shock wave simulations until he moved in 1980 into management involved with reactor safety analyses. He returned to the shock wave program in the mid 1980s and continued code development until about 1989. He died in 1994.

developed pre- and post-processing software to analyze the ANEOS equations of state and output from CHARTD. He even supported the underlying graphics package, RSCORS.[149] Sam somehow managed to find the time to write detailed reference manuals, some of which could serve as models for how to write good reference manuals today! He understood the importance of providing a robust and complete shock physics analysis environment from the beginning of his career.

Sam's most widely used code was **C**HARTD **SQ**uared (CSQ). CSQ is a two-dimensional (2-D) multi-material Eulerian code that modeled strong shocks, elasticity, plasticity, single-group diffusion radiation transport, and material fracture. Along with Daniel A. Matuska's HULL and Wally Johnson's OIL codes, CSQ helped define the direction that all general-purpose Eulerian shock wave physics codes would take in the 1980s and 1990s. Sam supported a large CSQ user community both inside and outside Sandia.

Sam manifested a fine craftsman's skill in his software. He developed his early codes in a challenging software and hardware environment. The challenges included FORTRAN 66 language limitations, card decks, batch processing, and multiday turnaround times. Documentation was challenging because he had to write the document by hand and then give it to another person to type with a typewriter! Despite this, he developed a complete environment for strong shock analysis.

Sam was also a highly respected analyst. Sandia management frequently pressed Sam into service to analyze very important problems. There were times in the 1970s and 1980s when management allowed only Sam to use Sandia's computers for these problems. Typically, Sam would use CSQ or CHARTD to analyze a time-critical problem for the weapons program. Management also recruited Sam for important non-weapons problems. One example was when he performed time-critical calculations following the Three Mile Island (TMI) reactor accident. Management pulled together a select team from across the lab to determine if a hydrogen detonation would breach the reactor's concrete containment structure. The team was able to tell the TMI Emergency Management that there was no danger of containment failure.

## Leader

Sam moved into management around 1980. His personal technical contributions decreased as he spent more time leading teams of developers and analysts. He handed off his code development and user support activities to other staff. He spent a few years working on nuclear reactor accident analysis following the Three Mile Island accident.

In the mid 1980s, Sam returned to the shock community when he assumed leadership of the department tasked with developing a three-dimensional (3-D) shock physics code. Sandia management believed that the new supercomputers (Cray XMPs) were capable of solving 3-D problems. The Computational Physics Research and Development Department developed the 3-D **C**SQ to the **T**hree-**H**alves (CTH) code. CTH is a 3-D, 2-D, and 1-D Eulerian code that models strong shocks, elasticity,

[149] RSCORS is an acronym for Revised Stromberg Carlson Optical Recoding System, which was the software and hardware system in the 1970s for generating graphs from computer runs.

plasticity, and fracture. Although Sam was the manager, he still found time to develop much of the generator program (CTHGEN), the post-processing program (CTHPLT), and parts of the CTH program. The team once again provided a complete shock physics environment.

In the early 1990s, the Defense Advanced Research Projects Agency (DARPA) became a strong supporter of CTH. They provided funding to add capabilities needed by the Department of Defense (DOD) and to support early DOD users. Sam's strong belief in a complete user environment paid benefits to the DOD users. The CTH team sends an external user a complete shock physics analysis computational environment. Many developers have enhanced CTH. They added improved material models and computational capabilities. CTH runs on massively parallel computers, thereby allowing multimillion cell meshes. David Crawford added adaptive mesh refinement, which provides superior shock resolution at an acceptable cost. CTH continues to be a very heavily used code inside and outside Sandia. Sandia distributes CTH to a large external user community. Sandia currently distributes CTH to approximately a hundred sites nationwide.

Many technical experts have used CTH to perform a variety of important, high-priority computational studies. High-fidelity, 3-D weapon safety calculations have helped to analyze complex weapon accident scenarios and minimize the number of expensive weapon safety experiments. Mel Baer enhanced CTH to model multiphase flow to study the combustion in the USS Iowa gun that exploded in 1989 with considerable loss of life. This was part of a Sandia team effort led by Dick Schwoebel. The team demonstrated that an over-rammed impact could ignite the propellant. The explosion was a tragic accident. Moreover, following the 1996 mid-air explosion of TWA flight 800 that resulted in the death of everyone onboard, Baer again used CTH to examine the possibility of a fuel–air explosion in the fuel tank. The National Transportation Safety Board concluded that the probable cause of the accident was a fuel–air explosion in a fuel tank. New requirements were developed for aircraft to prevent future fuel tank explosions. In another dramatic example, the massively parallel version of CTH analyzed the 1994 Shoemaker–Levy 9 comet impacts on Jupiter. Astronomers could not observe the impacts because they were on the side of Jupiter that faced away from Earth. The 3-D CTH calculations indicated an optical signature that should be visible over the limb of Jupiter. Astronomers could infer information about the impacts from the visible optical signature. Analysts still use CTH to analyze hypervelocity impact phenomena associated with asteroids. For example, Mark Boslough used it to analyze the 2013 near-Earth asteroid airburst explosion that entered the Earth's atmosphere near Chelyabinsk, Russia.

Sam has been a major influence in shock physics and code development since the 1960s. He was a strong individual contributor who transitioned to a respected leader. He developed a widely used strong shock modeling environment. Sending his codes outside Sandia strengthened the shock physics community nationwide. He set a standard for code capability, ease of use, and accuracy that endures today.

Sam Thompson died on November 3, 1994, following a long battle against a rare form of cancer.

# Wayne M. Trott

# (1978–2010)

# My Recollections of Shock Wave Research at Sandia

I came to work in shock wave studies in a more indirect way than many others. As it turns out, I was never assigned to an organization dedicated to shock physics research. Rather, I spent my time at Sandia in essentially two distinct organizations—the Laser Sciences Department from 1978 to 1989 and a group with a thermal and fluids emphasis in the Engineering Sciences Center from 1989 to 2010. My eventual participation in many shock physics projects over the years is thus a testament both to the highly multidisciplinary nature of shock wave studies in general and to the many opportunities for cross-disciplinary collaboration at Sandia. I have always felt that such opportunities are a hallmark and defining strength of research at Sandia. As will be evident in this recollection, my research assignments in shock physics involved a substantial number of distinct, diverse projects both large and small. Nearly always, the work was part time. I will not attempt to be exhaustive in describing the various projects. Instead, I intend to touch on the ones that I perceive as having the most impact overall.

My initial assignments at Sandia involved research into a candidate laser system for an inertial confinement fusion driver and the laser-induced decomposition of gas-phase mixtures of fuel and oxygen. Around 1980 I was encouraged by my supervisor, Jim Rice, to focus on energetic material decomposition. A joint effort eventually developed between our laser group and one of the component groups. George Perkins in the components area hired Anita Renlund as a principal investigator in his group, and a long-term collaboration resulted. Anita and I investigated various spectroscopic techniques (emission spectroscopy, Raman spectroscopy, time-resolved infrared spectral photography) for application to shock physics and shock chemistry problems. We initially focused on gas-phase energetic molecules, but soon concentrated on actual liquid and solid explosives in keeping with their higher programmatic interest and the potentially more direct ties to state-of-the-art theoretical modeling. Experimental techniques were generally first explored and developed with detonator drivers and then fielded on gas gun tests that allowed us to use well-characterized shock input. Probably the most quantitative results were achieved in Raman spectroscopic studies of shock-compressed TATB (triaminotrinitrobenzene) and nitromethane (Renlund and Trott 1988; Trott and Renlund 1989; Renlund and Trott 1990). We explored the response of these explosives as reflected by various pressure- and temperature-induced shifts in the molecular vibrational frequencies. Funding support for this fundamental work eventually died out and we were directed to more programmatic priorities. To my knowledge, follow-on work to our Raman studies on explosive materials has been minimal even though a quarter century has passed. The description of shock processes at the molecular level certainly involves enormous complexity, and this remains a challenging and largely underexplored area of research.

In parallel with the last few years of the spectroscopic work, my involvement began on issues associated with direct optical initiation of explosives. The concept of laser initiation of explosives had been explored off and on at Sandia since the early 1960s. The principal driver for the interest in this concept in the 1980s was the perceived safety advantage of an all-optical fireset for weapon applications. This idea was particularly attractive to Stan Spray in the Nuclear Weapons Safety area, and he provided early funding to explore this concept. Our early experiments involved yet another collaboration with the laser project (George Fisk) and component (Steve Sheffield, Bill Rogers) groups. Initial efforts involved interferometric studies of laser-driven flyers and direct irradiation of explosives (Sheffield et al. 1986). As it turns out, all of the researchers mentioned above took jobs elsewhere, and I was asked by Jim Rice to take over the work in this area.

In the first few years of my involvement, the work proceeded on several fronts, including the evaluation of the high-power handling capabilities of optical fibers, optimization of laser-driven flyers, and optical initiation of explosives by direct irradiation (with and without fiber optic delivery) as well as exploring the possibilities of laser-driven slapper initiation (Trott and Meeks 1990a, b). The earlier work (Sheffield, Rogers, and Castañeda 1986) had seemed to indicate that the laser energies to initiate secondary explosives reliably might be prohibitively high (>3 J) for practical fireset applications. But our experiments demonstrated that the perceived difficulties were primarily a matter of scale and, with suitably small spot sizes, the required energies for prompt initiation were much smaller—as low as 10 millijoules (Renlund et al. 1989). In these preliminary studies, we were also able to show that optical fibers of appropriate sizes (0.2–1 mm in diameter) could handle the corresponding irradiances to support this type of application.

At the same time this relatively low-level experimental effort was proceeding, study groups assembled by the systems groups evaluated the concept of optical initiation from nuclear safety and use-control perspectives and concluded that the concept was worth pursuing. Subsequent discussions involving both technical personnel (including at various times Bob Setchell, Kent Meeks, Phil Stanton, Anita Renlund, and myself) and management (including Don Schroeder, Dave Anderson, Jim Gerardo, and Dennis Hayes) resulted in collaborative funding of the Direct Optical Initiation project at Sandia. At this point, the work became distributed over several different organizations and eventually resulted in significant advances in understanding nearly all of the issues associated with the concept, including the miniaturization of robust, high-energy laser sources, optimization of optical fiber power handling capabilities, characterization of explosive response both to direct laser irradiation and optically-driven slapper initiation, etc. Parallel research at LANL also began to support this effort, perhaps most notably through the first demonstration of fiber-coupled, optical slapper initiation of high-density PETN (pentaerythritol tetranitrate) and HNS (hexanitrostilbene) explosives (Paisley 1990). My part in this effort became increasingly focused on the characterization and optimization of laser-driven flyers. In addition to its primary application to high explosive initiation, generating laser-driven flyers is of interest to shock physics generally since the optically-driven, high velocity foil may be used to impact a wide variety of materials, providing well controlled, short-pulse dynamic loading of the acceptor.

Over several years, we examined in detail various aspects of laser-driven flyer performance and integrity, collaborating both with experimentalists at other laboratories (Dennis Paisley at LANL and Alan Frank at LLNL) and analysts and theoreticians at Sandia (notably, Jeff Lawrence and Archie Farnsworth, Jr.). Experimentally, we evaluated numerous different flyer materials (e.g., Al, Ti, Mg) and composites (e.g., Al with an imbedded, thermally insulating layer of $Al_2O_3$) and explored many different facets of flyer generation phenomenology, including the laser absorption and driving plasma generation process, the extent of flyer ablation by the driving plasma, factors affecting flyer planarity, etc. (Trott 1994; Frank and Trott 1996). The collaboration with Lawrence and Farnsworth was particularly rewarding in making sense of the experimental results. Lawrence developed a simple but very robust model, based on the Gurney theory for explosively-driven plates, to predict the final velocity of the laser-driven flyers. With calibration from a limited set of experimental results for a given flyer material, the model was able to predict very accurately the experimentally observed trends arising from variations of laser fluence and pulse duration as well as flyer thickness and diameter (Lawrence and Trott 1993).

To address issues of flyer planarity as well as other more detailed aspects of the flyer launch process, Farnsworth incorporated a robust laser absorption model into the CTH hydrodynamic code; the simulations of the flyer acceleration history and flyer planarity were generated for comparison to our experimental results (Trott, Setchell, Farnsworth 2002). As in the case of Lawrence's work, Farnsworth's theoretical results led to a stimulating iteration of experimental development and simulation improvements that carried our understanding of the flyer generation process forward. Details of this synergistic work can be found in the references. This multiyear effort to provide the comprehensive, basic scientific underpinning for this approach to direct optical initiation was enthusiastically supported by the Technical Staff in the systems groups, most notably Bob Setchell and Kent Meeks. At a significantly later date, Setchell, Farnsworth, and I developed the aspects of this work further by demonstrating the utility of laser-driven flyer techniques for equation-of-state studies of microscale materials (Trott et al. 2002).

Our support of the Direct Optical Initiation project continued in parallel with significant changes in my line organization. The Laser Sciences organizations became increasingly involved in optoelectronics research. Accordingly, the flyer work (and other projects in which I was involved) proved to be better located in the Engineering Sciences Center. In the new organization, I was encouraged to continue this work (by John Cummings and later Art Ratzel), and in time I also developed new ties to experimentalists and analysts in that organization, as well as personnel in the Pulsed Power Sciences Center.

The workhorse optical diagnostic in the laser-driven flyer work was an Optically Recorded Velocity Interferometer System (ORVIS) that I "inherited" from Steve Sheffield after he left Sandia. We used this system as a point-measurement velocity interferometer (similar in many respects to the VISAR) throughout the course of the Direct Optical Initiation project work. Around 1996, I was encouraged by Jim Asay and Lalit Chhabildas to follow up on work reported by German and Russian researchers that used ORVIS in a line-imaging mode (Baumung et al. 1996). The opportunity to perform shock wave measurements in both a temporally- and spatially-resolved manner was of considerable interest to a wide variety of problems. Indeed, a functional design for a line-imaging VISAR had been demonstrated by Will Hemsing and coworkers at LANL several years earlier (Hemsing et al. 1990). Some of the disadvantages of this interferometer design included the need for a very powerful laser source (100 W or greater) and a complex light collection system involving multiple optical fiber bundles. The ORVIS adaptations used by Baumung et al. were somewhat simpler and required less investment in new hardware. Relatively early in our development of the line-imaging ORVIS, funding was obtained from the Laboratory Directed Research & Development (LDRD) program to pursue this line of research. With this support, we were able to develop the line-imaging ORVIS capability, and we systematically made improvements to the line-shaping optics, the collection optics, and the ORVIS image-processing routines and data analysis (Trott et al. 2001).

After coupling a line-imaging ORVIS setup to the gas gun impact facility at the Explosive Components Facility (ECF) in 1998, we began to apply the diagnostic to a wide variety of shock wave problems. This work is still ongoing at the ECF, and we have been continuously supported in this effort by the facility management, most notably Lloyd Bonzon and Leanna Minier. Also, over the years, additional interferometers based on our design have been developed for use at other Sandia facilities, including STAR (Shock Thermodynamics Applied Research), DICE (Dynamic Integrated Compression Experimental), and Z. My contributions to hardware development at these facilities as well as to the execution of various experiments have involved profitable and delightful collaborations with many Sandia researchers including Marcia Cooper, Alex Tappan, Bill Reinhart, Mike Furnish, Marcus Knudson, Tracy Vogler, Tommy Ao, Randy Hickman, and Justin Brown, in addition to Jim Asay and Lalit Chhabildas (as mentioned earlier). In recent years, we have also worked with Dana Dattelbaum, Steve Sheffield, Mark Short, and Scott Jackson of LANL.

While detailed discussion of the different problems addressed by line-imaging ORVIS is beyond the scope of this write-up, the materials and processes examined include but are not limited to:

- shock compression of heterogeneous explosives and inert surrogates such as HMX (a high melting explosive aka "octogen") and sugar to compare with mesoscale simulations (Baer and Trott 2002b; Trott, Baer, Castañeda et al. 2007),
- shock wave propagation through deterministic material geometries such as nitromethane-impregnated, geometrically regular samples of robocast aluminum oxide and highly ordered tin sphere lattices, also to compare with mesoscale simulations (Trott, Baer, Castañeda et al. 2006; Baer and Trott 2004),
- investigation of incipient spall in tantalum (Furnish, Chhabildas, Reinhart et al. 2009),
- investigation of edge-release effects in sapphire (Reinhart, Chhabildas, Trott, Dandekar 2002),

- investigation of multi-dimensional effects in shocked ALOX/PZT—that is, alumina-loaded epoxy/lead zirconate titanate (Furnish, Robbins, Trott et al. 2002),
- measurements of shock velocity resulting from hypervelocity flyer launch impact on the Z machine (Vogler, Trott, Reinhart et al. 2008),
- investigation of the response of interferometric windows under quasi-isentropic compression (Ao, Knudson et al. 2009a),
- high-pressure Hugoniot measurements using converging shocks (Brown, Ravichandran, Reinhart, Trott 2011),
- investigation of hot-spot formation in the initiation characteristics of nitromethane and particle mixtures (Dattelbaum, Sheffield, Stahl et al. 2010), and
- measurements of the response of the mining explosive ANFO, or AN/FO (ammonium nitrate/fuel oil), to low-stress planar impacts ( Cooper, Trott et al. 2012).

The daunting complexities of shock wave propagation in heterogeneous materials as well as multi-dimensional wave phenomena in nominally simpler materials seem to ensure the line-imaging ORVIS will find continued application in the future. Efforts to improve interferometer performance and the design of impact targets should prove profitable. There is undoubtedly significant room for improvement in the methods for image data analysis as well.

The foregoing has concentrated on my recollections in three principal areas: (1) development of spectroscopic diagnostics to track physical and chemical changes in materials under shock compression and energetic reaction, (2) evaluation and optimization of processes associated with the direct optical initiation of explosives with emphasis on the detailed processes associated with the launch and impact of laser-driven flyers, and (3) development and use of the line-imaging ORVIS diagnostic for a wide variety of shock compression studies. As indicated previously, budget constraints dictated that my investment in these efforts was most often part time, and my work was typically distributed over multiple projects (many of which were not shock physics-related). I have not touched on some additional contributions: e.g., projects in the energetic components groups to develop microenergetic processing and testing as well as efforts to develop a micromachined pressure transducer array detector. Discussion of these projects is best left to the appropriate principal investigator (in the examples cited, Alex Tappan and Dave Jones, respectively).

Lastly, it is important to acknowledge the excellent assistance of technologists who supported my efforts and those of my collaborators throughout. These include Hamp Richardson, Jaime Castañeda, Jill Miller, Dante Berry, Heidi Anderson, Dan Sanchez, John Liwski, and many others. I have tried to recall and acknowledge everyone who has made important contributions. I apologize in advance for any name that I may have omitted.

# Timothy G. Trucano
# (1971–1972, 1980–Present[150])

## Getting in the Door: 1971–1980

I have no formal training as a shock wave physicist and not much training as a physicist period. I'm a mathematician. How did I end up writing this recollection? The answer starts in the summer of 1971.

I had just completed my sophomore year at the University of New Mexico (UNM), majoring in mathematics and minoring in physics. I had completed a three-semester course in introductory college physics for physics majors. While I learned a lot of interesting stuff, in retrospect the most important thing I learned was from somebody who sat next to me for most of those semesters. He said Sandia had summer jobs for college students and I should think about applying. I had been on a high school tour as a senior but that was all I knew about Sandia. It had never occurred to me to try to get a summer job there.

I did apply and did get a summer job—and that's the beginning of the saga.

My summer of 1971 was a major harbinger of what was to come, but I didn't know it at the time. I was hired as a "Summer Student Development Associate" and the hiring manager was Lee Davison. The group was called something like the "Shock Wave Research Division" (first-level management units were called divisions in those days), although the formal hiring organization was the "Accounting Division." This group would have been a shock wave group in the "Physical Sciences Research Center" (physics) rather than in the "Engineering Sciences Center" (engineering). I found out after the fact that I got the job because a previously selected student ended up declining an offered job. I think this caught Lee by surprise, and somehow he ended up with me.

I needed a lot of growth as a person and a Sandia employee that summer. I don't remember accomplishing a single thing, but I certainly remember some of the people and the work to which I was exposed. I had to wait for a Q clearance, so I spent the first 2 weeks at an explosive firing site. My mentor that summer was Jim Kennedy; he was doing explosive shock initiation experiments out at the site. The site personnel modified the Standard Operating Procedures, while I was there with a line that said something like "Make sure Tim is in the blockhouse." I remember explosions from small sheets of Detasheet®, instrumented with fast oscilloscopes, and the resulting enormous quantities of Polaroid photos of oscilloscope traces. I also remember immediately becoming curious about waves, detonations, and so on in those first 2 weeks.

The culminating event at the site while I sat there was an enormous fuel–air explosive test (outside of Jim's shock initiation research interests) in which about 30 or 40 lb of TNT/PBX explosive lens launched an aluminum flyer plate into a big

[150]As of this writing, Tim left shock wave research around 2003 and, as of this writing, continues as a Senior Scientist in computational sciences.

gasbag. It was the largest explosion I have ever experienced. After the fact, we stared at the scorched ground and Jim immediately commented that the event had been a deflagration, not a detonation (imagine what a detonation would have been like!). More curiosity generated on my part—how did Jim know that? Deflagration? It turned out that the plastic bag would have been cut into triangular fragments (as I recall) if there had been a detonation because of the cellular instability associated with a propagating det (detonation) wave in the gas.

Another curious result was some primitive photometrics on that shot. I got to see a man named Bob Martis in action when I eventually got into the Tech Area—he provided digitization of the photometric records of the propagating deflagration in the event. He even showed me how to use his machine, and I tried to digitize some of the records myself, but that was a real art form people like Bob were good at, and I was awful at.

I also remember doing my first computing at Sandia while at the test site. I used an old Singer (I think) mechanical hand calculator while sitting in Bill Benedick's office. I eventually met Bill but not immediately. I was in his office because he was on vacation. I broke his calculator. Sigh. Was this a trend?

When I got into the Tech Area 2 weeks later, there was really no place to put me, so I crowded into Jim's office. Jim's office mate was T.J. Tucker. Suddenly I was hearing about exploding wires. This office was in Building 806, in the wonderful old physical sciences research complex of Buildings 805, 806, and 807. This place was heaven on Earth for me because of all the people and the opportunities to see how people actually did real work. How can I express my gratitude to Jim and T.J. for letting me sit in their small two-man office and learn from them? T.J. had a lovely lab right across the hall where he studied explosive wires, human capacitance (for reasons I was not told), and so on. It was called "Ye Olde Exploding Wire Laboratory" on a nameplate at its entrance (people weren't afraid to demonstrate public humor in their work in those days). I filed exploding wires on my list of things that would be interesting to learn more about. (Interestingly, exploding wires and their associated research questions are as important now around Sandia as they were in 1971!)

Jim and T.J.'s office had an ancient Wang electronic calculator system that could be used to do one-tenth of the things a modern $40 TI (Texas Instruments) or HP (Hewlett Packard) hand calculator can do. Jim asked me to program it to solve the Gurney equations. Failure—I was no computer programmer then and I've never been one. *However*, the Gurney equations were yet one more question in my mind, and they deepened my interest in understanding waves and explosives and such.

As I recall, Jim did some computing, probably using the always-important WONDY 1-D (one-dimensional) Lagrangian code or perhaps SWAP (Lynn Barker's 1-D characteristics code). I remember that T.J. ran a little code that solved a circuit equation-based empirical model for exploding wires, mainly as a data analysis tool, not as a research tool. In those days, the computer was a CDC 6600, and jobs were submitted as card decks. One took a card deck wrapped carefully in a black protective cover to a central point of the Buildings 805, 806, and 807 research complex, whereupon a courier took it to the central computing facility, and the deck was run. The output (mainly printout and hard-copy plots) was then delivered back to the central point. This sounds awful but one had a lot of time to think while waiting for computer results.

There was also a time-sharing system in 805, 806, and 807. It was a PDP 10 or something, with public terminals at specific junctions of the building complex. One of my favorite scares that I recall was centered on this rickety system. The same digitized records that Bob Marder created, written on *paper punch tape*, were loaded into the PDP system for data analysis. That meant Jim Kennedy gave me access to his personal account on the PDP so that I could upload some data one day (forget computer security implications). The system at that time had an endearing quality—when you logged out, it asked you if you wanted to KEEP your files. If you didn't answer the question, it simply deleted all of your files. Well, one day I was uploading data and the system died—that is, I couldn't log out or do anything. I immediately worried about losing all the data files Jim had stored on the system up to that point. I recall running around the hallways trying to find somebody who knew what to do. No luck. I returned to the terminal—and somebody else was already working on it. I thought I was going to die—but it turned out that Jim's files remained intact. I never wanted to look at a PDP again.

Another memory that summer was meeting Barry Butcher under trying circumstances. Jim went on vacation and someone had the idea of letting Barry (he was a manager then, I recall) try to find something for me to do. Poor Barry! The first thing he did was ask me to transfer a plot from a reference to a sheet of paper (using a Gerber scale), presumably so it could be used in a report or for some kind of further digitization. This is probably the kind of thing an engineering student learned to do within a few weeks of starting school. I was hopeless. In retrospect, it was really kind of funny because I was so hopeless at the task while in the middle of meeting one of the important shock wave people at Sandia. Remembering this incident, I was abashed for a while when I would run into Barry after I started at Sandia as a full-time staff member.

I returned to work with Lee, Jim, and T.J. the following summer (1972), having completed my junior year of college. By then I had started to learn some things at school about numerical methods for partial differential equations. Nothing that would allow me to write a code, but information that gave me better understanding of what a code like WONDY might be telling me. So, that summer I ended up spending a fair amount of time running WONDY and trying to help Jim with computational simulations. I was in my own office, which eased the pressure on Jim and T.J. but probably reduced the amount of information I was simply absorbing from their conversations and their general work environment.

Lee Davison was then managing, and Jim was a staff member in a group called "Explosive Research" or something. The lab had split off part of the original shock physics group to more thoroughly attack high explosive initiation problems and so on. During that summer I met Jace Nunziato for the first time, as he was a staff member in the group and would come and talk to Jim about the experimental data. Jace had just completed a major work with Karl Schuler on the shock response of PMMA (polymethyl methacrylate), which had general implications for understanding viscoelastic behavior in materials of interest at Sandia. (I believe Karl recalls this work in his own recollection.) They had published an article in a journal, an article that I tried to understand for years thereafter, and Jace (with his coauthors) was in the act of writing a major review article on viscoelastic response for the *Handbuch der Physik.*

On the computing front, remote access points were now available to the CDC computers. So, rather than have couriers haul card decks and deliver output, we were able to submit jobs through local card readers and receive printouts and hard-copy plots at local printers, while the couriers delivered more exotic outputs like microfiche to those local points. This reduced the amount of time spent thinking between job submissions. It was at the 806 remote access point that I first met John Van Dyke, who was then working on computational solid state physics. (John is still on roll in my current Center.) I continued to learn more about shock waves, continuum mechanics, and so on during the summer of 1972 (e.g., I recall reading a good deal of Bird, Lightfoot, and Stewart's book on transport theory). Also, the work I did with Jim Kennedy ultimately led to a manuscript continuing the study of the Gurney approximation for explosive product properties; Jim kindly made me a coauthor on that paper, titled "Limitations to the Efficiency of Chemical Explosives in Accelerating Metals." The paper was never published but is the first scientific manuscript with my name on it.

After 1972 I had no contact with Sandia until 1980. Among other things, Sandia had painful layoffs after that summer. I think it is unusual that the only student contact I had with Sandia was as an undergraduate.

I received a PhD in mathematics from UNM in the summer of 1980. I had originally planned to get my degree in 1979, and so I had gone through a rather halfhearted job search in fall 1978. By that time I knew I would be seeking employment in industry or at a lab and not an academic job. I recall interviewing with Arlin Cooper, a long-time Sandia recruiter at UNM. Our conversation went well until I started to look uncertain about getting my degree in the summer of 1979. I remember Arlin closing his notes with a very decisive gesture and suggesting that I interview again when I was certain of the degree. That happened to be the case in fall 1979. Fortunately, Arlin was in a forgetting or forgiving mood about my previous interview, and I did, in fact, complete my PhD and accepted a job offer at Sandia. The job offer came from the Computational Physics and Mechanics Division in the Engineering Sciences Center. My hiring manager was Al Chabai, my second-level manager was Walt Herrmann, and my Center Director was Orval Jones. Jim Asay had interviewed me, but that was more along the lines of giving me some information about the experimental effort. I also recall Larry Bertholf interviewing me, but Larry had moved to a different organization by the time I arrived.

## My First Year as a Full-Time Employee: 1980–1981

I consider myself to be remarkably lucky to have been hired by Sandia. First, I had no real training in shock wave physics. Second, although I had a PhD in mathematics, I had not worked directly on anything related to the mathematics of shock waves, compressible continuum mechanics, numerical methods, and so on. Third, what little anybody could remember about my two undergraduate summers at Sandia could not have been inspiring. Nonetheless, I was hired.

Of course, I was also lucky because it seems that I arrived in exactly the right place at exactly the right time to enable me to grow and be productive. Sandia's job offer demonstrated utter faith in the ability of someone to grow and prosper in an exciting and demanding work environment. The shock wave group I proceeded to work in (or with) for roughly the next 15 years was the most exciting and demanding environment I have experienced at Sandia, and most of the rest of my comments here will be devoted to explaining that. [As a side note, I'll mention that a couple of weeks after I hired in, Walt Herrmann's group moved to the east wing of the second floor of the Sandia administration building (802)—we were right under the executive office suite of the lab for the next 12 years.]

I believe that Walt Herrmann was mainly excited by my apparent devotion to wanting to learn new things (i.e., continuum mechanics, shock wave physics, and so on), and he may have felt that my mathematical background would ultimately be useful. Two specific initial tasks for me were to help with a two-dimensional (2-D) Lagrangian shock wave code called TOODY and to work on a shock wave project aimed at studying shear-banding phenomena in the shock response of metals. I never really got very far working *on* TOODY—as I've said I have never been a code *developer*. However, because my clearance was ready when I arrived in 1980 (something that disconcerted Al Chabai), I was immediately able to pick up some ongoing computational work involving penetration into geological materials using TOODY (with Darrell Hicks and Fred Norwood). So, the first real piece of computational work I did as a regular staff member was in the notoriously difficult area of penetration mechanics—little did I know! This effort went on for a while, allowing me to meet, among other people, Mike Forrestal, a long-time Sandian and a world expert in ballistic penetration phenomena.

The day I arrived, I was handed (1) a dictionary, (2) an HP calculator, (3) a Gerber scale, and (4) an SI[151] conversion guide. Thirty-one years later, I still use the conversion guide and the dictionary, and I have kept the Gerber scale as an archeological artifact. The HP calculator was gone within a few years. I indeed used the Gerber scale, and for years I couldn't help but think about Barry Butcher every time I touched the thing.

Darrell Hicks and Fred Norwood had pursued the computational work I was handed at that point. A funny story I can relate is that I needed to copy punch card input decks for TOODY that Darrell and Fred had developed for penetration simulations. An old-old-old machine in our office area *theoretically* duplicated punch card decks (I'd seen the thing back in 1971), but it seemed to me that all it really did was eat card decks and deposit green goo on the floor. I hand-duplicated Darrell's card deck (I think the *second* day I was on the job) at a punch card machine and then returned it to him. I happened to mention what I had done. His face turned ashen, and then he said "Well, if I'd known you were going to do that I never would have given you my deck!" There was a burst of laughter from his office mate, who was hidden behind

[151] Editors' note: An SI conversion guide refers to the guide to the International System of Units, or SI.

a desk partition. That was Marlin Kipp, with whom I would end up working for many years on a variety of weapon-related problems and who gave me invaluable help in learning how to work as a computational physicist and always served as an outstanding staff member to emulate. And I perfectly duplicated Darrell's deck.

My first weekend at the lab, I submitted enough penetration simulation computer runs to choke the computer system. Mike McGlaun was also helping me at that time with learning the mechanics of how to run jobs and so on and was one of the reasons I was able to start computing so quickly. Mike had not interviewed me—I always joked that if he had interviewed me, he would have wondered why they were hiring such a weirdo. On my first day, along with dictionaries and such, Mike met me and handed me a stack of CSQ (2-D Eulerian code) documentation and said he couldn't help me program TOODY, but he'd help in any other way he could. Mike ended up helping me until the day he retired and was ultimately an important mentor and is still a close friend. Mike would leave the group within 2 years to work with Sam Thompson on nuclear reactor safety codes.

After my initial weekend, when very little of my submitted work had run, Mike and I were at the cafeteria for lunch. He introduced me to Sam Thompson, the most important code developer at Sandia and Mike's mentor, someone who had been in Walt's group not too long before and who ultimately became a key figure in my own career. Sam was friendly enough, but he immediately demanded to know why I had submitted three times more jobs than could possibly be executed over the weekend, therefore clogging the computing queue and hurting others. I ended up doing a lot of computer-queue clogging during the course of my shock wave work.

The shear band project put me in direct contact with Jim Asay's experimental shock wave physics group. Doug Drumheller happened to be in that group at the time, and I had almost immediately made contact with him. I wanted to learn more about mixture theory and knew that Doug was actively doing research in that area and in related topics. I walked into Doug's office within a couple of days of my arrival. As Doug remembers it, I had one of his papers in my hands and was asking him about something that I thought was either a typo or an error. Doug wasn't used to people coming into his office having read the math in his papers that closely. Doug became the most important mentor of my career and remains my closest friend although he retired years ago. Doug introduced me to many people during my first couple of years, such as Karl Schuler, Darrell Munson, Pete Lysne, Herb Sutherland, Tony Chen, and many others. In this way I believe I was fairly rapidly and effectively absorbed into the shock wave culture at Sandia and very early developed an understanding and appreciation of the history and historical importance of the program.

I also met Dennis Grady in the context of the shear band work. Jim Asay had a very talented group of people looking at this problem and at related thermophysical localization problems involving the physics within shock fronts, its impact on and response to metal characteristics, and so on. Another person I remember meeting within a few months in the localization work was Al Romig, at that time a functioning metallurgy researcher, later to become an Executive Vice President at the Lab. (Orval Jones also achieved that level of management in his career. The shock wave program at Sandia has generated a lot of high-level Sandia managers.) As I learned more about the physics issues in shear localization, I also began to learn more of the deeper mathematical issues involved in attempting to explain it. This led me to do some joint work with Tim Burns, a mathematician and computational physicist in the shock group, on the stability issues involved in localization. I'd known Tim in graduate school at UNM—we were both in the mathematics department. Tim was a key reason I even had a chance to interview with the shock wave group, I believe. This work also led me to meet Steve Passman within my first year; I worked with him on the mathematical theory of localization for several years. This was probably the first project I was involved in that directly led me to learn a lot more mathematics as part of my job. Steve himself devoted a lot of effort to introducing me to people in the broader mechanics research community. I met many people and went to many meetings as a result of Steve's kindness and collegiality.

While I ended up doing work related to shear localization with Steve and Dennis and others, for several years the associated mathematical and physical problems remained very deep, and I really didn't contribute much when all was said and done. I did end up learning a lot, but more importantly I ended up working increasingly closely with Jim Asay and the people in his group.

We brought the penetration simulation work to a reasonable conclusion after about 1 year, and I went to my first American Physical Society shock wave meeting to present a brief paper on this work in the summer of 1981. I met a number of people at that meeting, including Steve Bless (at the University of Dayton at that time), who had ideas about better experimental diagnosis of some of the issues we were worrying about in our simulations. Steve got in contact with Jim and that led to a very long relationship between the two of them. I also met Professor Yogi Gupta for the first time—he had just replaced the recently retired Professor George Duvall at Washington State University, I think. Unfortunately, I never had the chance to meet George.

As it turns out, I also did some work in my first year at Sandia on computational simulations that provided information about potential brine migration problems at the Waste Isolation Pilot Plant. This had nothing to do with shock waves, but it was work that provided some funding coverage for me, and it was another learning opportunity. The project manager to whom I reported was Tom Hunter, who eventually became the President of the Lab.

### Learning and Doing: 1981–1991

Around early 1982 a conjunction of events led me to move internally into Jim Asay's group, the "Thermomechanical and Physical" division. I moved into his *experimental* group, with the very open-ended goal of providing as much *computational* support and insight as I could. Not only had I been directly working with Jim because of my involvement in shear banding, but I was also beginning to respond to other issues that he would bring up that I either might simulate or at least think about more mathematically. I was working more closely with Dennis Grady, and I had begun to work extremely closely with Doug Drumheller on constitutive modeling of alumina epoxy and its application to the computational simulation of components of interest to Sandia.

This move opened the floodgates for me. I suddenly wasn't able to turn around without bumping into interesting and challenging computational work that almost always promised to be put to severe experimental tests. Rather than go into all the details, I will explain the progress of this work from the high-level view of hypervelocity impact, which turned out to be the dominant theme of the work in Jim's group for the next decade. There were several reasons for that focus, including research problems that bordered on hypervelocity impact phenomena, such as a multiyear effort devoted to attempting to create high-pressure phase transitions in hydrogen (called the hydrogen metallization project). In general, Jim's group was probing the increasingly higher-pressure behavior of materials, of which hypervelocity impact phenomena provided both insight and application pull. Maybe the dominant factor in this period was the arrival of the SDI (Strategic Defense Initiative) Program and a large amount of funding for hypervelocity impact research pulled by a joint program with Los Alamos that we pursued for most of the 1980s.

The larger department that contained Jim's group underwent a significant management change as well. Walt Herrmann was promoted to Director of the Engineering Sciences Center. His replacement as the second-level manager of the shock wave group turned out to be Lee Davison.

Under Lee and under the driving conditions of the programs to which we were responding, including more Department of Defense (DOD) work as well, we made a further series of significant personnel additions. Lynn Barker returned to Sandia and Jim's group, and I ended up working on a large variety of research topics with Lynn. We also hired Gerry Kerley from Los Alamos, a world-class equation-of-state researcher. Lee convinced Sam Thompson to leave the nuclear reactor safety code development area and return to revive CSQ and begin to develop a serious three-dimensional (3-D) code capability, with a code that would be called CTH. When Sam returned, he brought Mike McGlaun back with him. I ended up working very closely with Sam and Mike during the rest of the decade as well, because it was their computational tools that enabled me to pursue most of the computational work that Jim's group needed. I believe Sam's first hire upon his return was Steve Rottler, who quickly developed an important role in the LANL-Sandia SDI project; Steve became the first formal PI (principal investigator) for CTH late in the decade, leading CTH to be the code that DARPA selected as the new-generation hydrodynamic code for DOD. Steve eventually became the Chief Engineer (Vice President) for nuclear weapons at Sandia. Sam also brought Luba Kmetyk into the group; Luba had a major role as a computational shock wave physicist using both the new generation of CSQ as well as CTH.

Because of my hypervelocity impact work, I also began working more closely with Lalit Chhabildas as the decade progressed. Finally, as a result of both the SDI program and certain DOD projects, I had the opportunity to work closely with a variety of LANL staff. Paul Yarrington had also returned to the shock wave group and was leading another group of computational shock wave physicists (including Kipp, Burns, Tom Bergstresser, Bill Davey, and others) in applying the modernized hydrodynamic code capabilities in CSQ and CTH to weapons-related problems. Paul's group applied these computational capabilities to address many shock wave physics problems, which provided a strong driver for further development of CSQ and CTH.

Because of the conjunction of funding, program drivers, research opportunities, and talent accumulated in the 1980s, the Sandia shock wave effort pursued every aspect of hypervelocity impact. These aspects included how to achieve hypervelocity impact, how to characterize it, how to simulate it both physically (i.e., experimentally) and computationally, and how to assess its consequences. This effort involved developing the experimental capabilities, theoretical understanding, computational capabilities, and specific applications—none of which had really existed prior to the 1980s. As part of the effort, Sandia became closely integrated with a revived national and international community in hypervelocity impact physics. For example, we developed a close relationship with Charlie Anderson and his group at the Southwest Research Institute.

A detailed recounting of my involvement with these varied activities would be inappropriately long. I will simply provide a brief description of three of the hypervelocity impact areas in which I worked, along with the names of some of my collaborators (but by no means *all* of those collaborators).

**Development of experimental capability to create high-pressure states in hydrogen (with Barker, Asay, Kipp, Grady).** This is a particular example of a larger-scope activity to create better experimental capabilities for generating high-pressure, low(er)-temperature isentropic or quasi-isentropic compressive states in materials. Along with the need for this with hydrogen, the effort was devoted to developing "soft-impact" techniques, such as Lynn Barker's "pillow" research and development. A specific research objective was to probe high-pressure phase transitions precisely. With considerable help from Marlin Kipp and Gerry Kerley, I explored aspects of both the experimental capability development and the simulation of material response subjected to quasi-isentropic compression. My work led Lalit Chhabildas to explore simplified soft-impact concepts and, ultimately, the reversal of these techniques to create higher-velocity launch capabilities, in particular the so-called HyperVelocity Launcher (the three-stage gun).

**A railgun research and development effort.** This effort was led by Jim Asay, in collaboration with Ron Hawke of LLNL and many others, including Lynn Barker. I conducted a computational study in collaboration with Lynn Barker on the initiation and growth of gouging in the fast sliding contact of materials. We also studied this issue in the context of Sandia's rocket sled track. The code developers (McGlaun, in particular) provided CTH modifications that allowed us to do this study and that ultimately enabled computational support for Chhabildas' work on the three-stage gun.

**Joint computational-experimental studies of hypervelocity impact phenomena.** These studies were conducted with McGlaun, Thompson, Kipp, Kerley, Asay, Barker, Grady, Chhabildas, and Jack Wise. We explored hypervelocity impact cratering, penetration, debris formation, and the effects on secondary structures. We looked at fundamental equation-of-state issues, such as impact vaporization and debris formation and propagation in both low- and high-density materials. Our work drove much of the CTH development as well as new experimental and diagnostic capabilities (such as a gun Grady developed that was particularly suitable for impact studies). We made a concerted effort to examine the mechanisms of shock vaporization, the formation of mixed liquid–vapor impact debris, and the details of debris propagation. Our work led to new insights in the formation of partially vaporized debris and its computational simulation, a problem being studied by many institutions at the time. Eventually, our efforts led to key spin-off studies by many people in the early 1990s, such as debris formation in radiation sources for extreme ultraviolet lithography (a microscale impact problem) and the successful prediction of phenomena associated with the impact of the Shoemaker–Levy 9 comet on Jupiter in 1994 (a giant macroscale impact problem). As a minor note, the first major 3-D calculation with CTH that I performed with Mike McGlaun involved the computation of the hypervelocity impact of two projectiles on a plate of material and the evolution of the subsequent penetration. Our ability to perform this calculation shocked Walt Herrmann when he saw it for the first time.

I ended the decade in close collaboration with Mike McGlaun to develop a proposal for a next-generation, multi-physics shock wave code. This code project, which is now called ALEGRA, started in 1990 and is still going strong at this time.

## Epilogue: 1990s and Beyond

After the 1980s, my work started to move in somewhat different directions, resulting in a relatively strong disconnect from shock wave physics research by the end of the 1990s. I want to place this in the context of my discussion above.

In the early 1990s, I was involved in work that represented, to some extent, a culmination of the shock group's work in the 1980s as well as a product of the organizational dislocation in Sandia's shock wave physics program. I mentioned above the start of the ALEGRA code development (circa 1990, with a lot of people), work on the Shoemaker–Levy 9 impact predictions (circa 1993–1994, with David Crawford, Mark Boslough, and Allen Robinson), and the analysis of debris formation in the initially conceived and implemented laser-radiation source for use in extreme ultraviolet (soft x-ray) lithography research and development (circa 1992–1994, with Dennis Grady, Glenn Kubiak, and Rick Stulen, which was overall a very large project).

The exclamation point of our 1980s work was the final perfection of the three-stage gun by Lalit Chhabildas and his experimental collaborators. I feel to this day that my best and most satisfying computational prediction involved my work with Chhabildas (circa 1993–1994) to predict the results of the first experiment on the completed gun. The simulation of the interior ballistics of the performance of the third stage, and the transition of the operation from the second to the third stage, exercised all of the computational capabilities that had been developed in CTH, while the experimental capability itself probably represented the essence of what we had really learned about hypervelocity impact during the 1980s. Our prediction of the final projectile velocity was within a few percent of the observations, and the shape of the formed projectile was very close to the experiment results. Perhaps of additional interest is that, by this time, the use of CTH was easily transferred to Lalit. Lalit was already using the code himself in conceptual design studies for the experiments. I interpret this as the ultimate confirmation of the vision Jim Asay had in bringing me into his group in early 1982—that the ever increasing sophistication and synergy of computation and experiment were the path to the future in shock wave physics. Now, experimenters themselves use sophisticated computational capabilities as a matter of course. Neither this mode of research nor the tools themselves existed when I first started working in shock wave physics.

I also want to say something about the organizational dislocation that occurred in the early 1990s. Experimental facilities at Sandia found themselves under great stress then because of funding reductions and programmatic uncertainties about future needs. When I was hired in 1980, the shock wave group included theory, computation, and experiment under a single second-level organization. This was a unique strength of our group. By the time we accomplished the tasks that I summarized above, the shock wave group was separating, with an experimental effort in a different organization than the computational effort. The overall computational group had been moved into a computational sciences center that was focused on advancing Sandia's capabilities in parallel computing. The computational shock wave work was an excellent driver for parallel computing. The experimental work was not, not the least because of the increasing funding challenges presented by the laboratory environment. Ultimately, the experimental group joined the Pulsed Power Sciences Center, where they are today.

Jim Asay was promoted to the position Lee Davison had held, and he had to deal with the above issue. The Lab then complicated matters further by eliminating the second level of management in the early 1990s as well. Ultimately, Jim assumed a programmatic role in the Lab's Nuclear Weapons (NW) program for several years.

The experimental story has a happy ending from my perspective. Jim left his NW programmatic role in the mid 1990s and returned to the experimental shock wave effort. He led the resurgence of the program in its new home in the Pulsed Power Sciences Center. One culmination of this phase of his leadership was implementing a program of shock wave experimentation on the Z machine. This is an entirely different story that I'll let Jim tell. I raise it here because my last shock wave related work was probably connected to this implementation. In the late 1990s, I helped simulate the first VISAR-instrumented experiments of radiatively-driven shock waves on Z, which were led by Jim Asay.

By the mid 1990s, I became increasingly immersed in research associated with verification and validation (V&V) of computational simulations. By the end of that decade, I was involved essentially full time with V&V and the Department of Energy ASCI (Accelerated Strategic Computing Initiative) program. That phase of my career, which is still ongoing, is really beyond the scope of this recollection. I will emphasize that the worldview I brought to V&V was strongly shaped by my approximately 15 years of concentrated computational simulation work with the Sandia shock wave physics program, especially my close engagement with experiments and experimenters.

I still feel, 31 years later, that I was given a fantastic and unique opportunity to participate in the Sandia shock wave physics program!

# Tracy J. Vogler
## (2001–Present[152])

My PhD dissertation project in engineering mechanics at the University of Texas (UT) at Austin involved microbuckling in composites loaded in compression. Prior to and following my defense, I interviewed for faculty positions at several universities around the country. When I didn't get any offers, I decided it was time to get out of the university environment for a while, so I started looking for jobs in industry and postdocs in various labs. Before I visited the Army Research Laboratory (ARL) in Aberdeen, Maryland, to interview for a postdoc with Bill Walters, the man who literally wrote the book on shaped charges, I had not even heard of a shaped-charge jet. I ended up going there, although I joined the impact physics branch doing crystal plasticity on titanium. Although I didn't do any impact or shock physics research while at ARL, I did become more familiar with high strain rate phenomena and got some exposure to shock physics through Datta Dandekar.

While the research at ARL was interesting, my girlfriend (later fiancé and wife) hated the Aberdeen area, so I started looking for a new job. A classmate from UT, Greg Bessette, was in one of the computational groups at Sandia. He helped connect me with Lalit Chhabildas, with whom he was working on missile defense problems. I interviewed and was impressed, even if I understood little of the technical details of the research being done at the STAR (Shock Thermodynamics Applied Research) facility or Z. I got married about a month after September 11 and started at Sandia in late 2001.

The first two projects I worked on were fragmenting tubes and the shock behavior of boron carbide. Both were done under the Joint Department of Defense (DOD)/Department of Energy (DOE) Munitions Technology Development Program, which brings DOD money to the National Nuclear Security Administration labs to work on problems of interest to both departments. To varying degrees, I have remained involved in this program throughout my career at Sandia thus far. Because shock physics was still new to me, I spent a lot of time learning the basics of the field. Besides reviewing some basic texts and papers in the field, I also watched tapes of Yogi Gupta's[153] shock physics class. It took me a year or two to come up to speed in the field, but Lalit and Jim Asay, who was my manager when I joined Sandia, allowed me that time to learn. Lalit especially was a tremendous mentor and resource for the work done in the field. Often, I would ask him about who had worked on a particular topic as a way to get started in learning about it; many times, the answer was that he had done it back in the 1980s or 1990s.

[152] As of this writing, Tracy Vogler works in an engineering mechanics group at Sandia in Livermore, CA, and continues his involvement with the shock wave work in the Pulsed Power Sciences Center.

[153] Yogi is a professor in the Department of Physics and the Director of the Institute of Shock Physics at Washington State University.

Since I was the only member of the experimental group with a background in solid mechanics, I became the lead person on the strength of materials. I used the shock-release and shock-reshock technique developed by Jim Asay and others to determine the strength of $B_4C$ and SiC as a function of pressure in a manner similar to what had been done for aluminum and other metals as well as for $Al_2O_3$ by Bill Reinhart and Lalit. Soon, I was using the same approach for Z experiments to measure strength under isentropic loading. Dennis Hayes and Jean-Paul Davis were working on ways to analyze ramp loading experiments to extract a loading isentrope, but it quickly became apparent that their approaches were not suitable to the unloading used to probe strength in the experiments because of the path dependence of elastic–plastic materials. Thus, a simple Lagrangian analysis approach was used. While this was not too unreasonable for aluminum, since it is well matched to the LiF windows used in the experiments, the approach was completely inadequate for high-impedance materials such as tantalum. Also, because our allocation of shots on Z for dynamic materials was relatively modest, only about two shots a year were devoted to obtaining strength data. This situation improved later when Jim Asay returned from Washington State and began doing experiments on Veloce, the small pulsed power system built in the mid 2000s. Several years later (around 2009), the hydrodynamic code LASLO was developed by Josh Robbins of Sandia for use on one-dimensional problems as a replacement for WONDY, which was the code used up to that point. With LASLO, it was possible to take some new approaches toward extracting strength from the Z experiments, such as iterative fits. When Justin Brown joined the lab in 2011 after finishing his degree at the California Institute of Technology (he had been an intern at Sandia's Dynamic Material Properties Department in high school and as an undergraduate at the University of New Mexico), he began the still-ongoing process of developing and improving these approaches.

In 2003, Lalit and Larry Larsen, an MD who happened to sit down the hall from us in Building 962, proposed an LDRD idea to study the shock kill of bioagents for missile interceptor applications. The idea was funded, and we began the effort to shock load and recover live bacteria. We modified the catcher tank of the air gun at STAR and, through several iterations of trial and error, designed a capsule that would hold the bacteria sample without leaking, at least up to a certain impact velocity. We studied *Yersinia pestis* and *Bacillus cereus*. The former was a genetically modified version of the bacterium responsible for plague, while the latter was closely related to *Bacillus anthracis*, commonly known as anthrax. While Larry assured us that both were harmless, I can't help but wonder if the technicians at STAR were still nervous in working with them, since they were the most likely to be exposed to them. From a technical standpoint, we found that at a certain pressure loading level, the survival of the bacteria abruptly fell by several orders of magnitude. Unfortunately, we were never able to obtain any more funding to continue the work or to do follow-on studies.

In working on the shock-release and shock-reshock technique, I extended a simple distributional model that Joel Lipkin and Jim Asay had developed in the 1970s. Unfortunately, the approach was almost entirely empirical and offered little in the way of insight into the problem. I came to believe that a modeling effort that addressed the behavior of the individual grains of a metal was needed to understand the material behavior and develop a realistic continuum model. Rich Becker at LLNL had published the first paper on wave propagation in a metal modeled with crystal plasticity in 2004 (Becker 2004). At Sandia around the same time, Wayne Trott had been working for a few years on a spatially-resolved velocity interferometer known as the Line VISAR (sometimes referred to as a Line ORVIS). In 2005, I proposed a Laboratory Directed Research and Development (LDRD) project that would incorporate both grain-scale crystal plasticity modeling and Line VISAR experiments in an effort to understand the dynamic behavior of metals. Unfortunately, the project fell short on both fronts, as our collaborator on the crystal plasticity simulations left Sandia, and the Line VISAR experiments proved more difficult than we had anticipated. While collaboration with John Clayton at ARL did produce a nice paper on tungsten alloys that included both experiments and two-dimensional (2-D) crystal plasticity calculations, the overall goal of using modeling and spatially-resolved simulations to improve our understanding of material behavior was not achieved. This idea has, to my knowledge, not been pursued to a successful conclusion yet.

The Line VISAR has remained tantalizingly promising yet extremely frustrating. For example, the work on tungsten alloys produced a distribution of spall strengths for the material, but attempts to obtain similar data on silicon carbide were not successful. Such data could have been extremely valuable to the stochastic failure model developed by Rebecca Brannon and her coworkers. The reality is that the instrument is, in general, complex and difficult to get to work well. While Wayne Trott was able to do this, most others have proved unsuccessful at it. Furthermore, the analysis of the streak images obtained from the experiments has proven problematic. Wayne utilized a simple line-out approach to the analysis of the data, whereas Tom Ao developed a Fourier transform approach that tended to smooth out the results. This raised the question of how much of the spatial variability observed in the Line VISAR measurements was due to real differences in behavior from point to point and how much was noise or an artifact of the measurement technique. Questions about repeatability of the measurements have also been raised since a thorough investigation of the Line VISAR for well-controlled phenomena has not been performed.

In the mid 2000s, I began working on the dynamic behavior of granular materials. Dennis Grady, who had retired from Sandia in the 1990s and now worked for Applied Research Associates, was also involved. Our initial effort involved shock compaction of granular tungsten carbide (WC) using a cell with multiple steps for shock arrival measurement. Since we used the VISAR on the different steps, we were able to resolve the propagating waveforms of the granular sample. The waves had long risetimes, from tens to a few 100 ns, and we learned two interesting things from them. First, the wave structure was nearly constant, unless attenuation was present, which indicated the waves were steady. To the best of my knowledge, this was the first verification of steady wave propagation in granular materials, and it meant that the use of jump conditions to analyze the results was valid. Second, when the strain rates were calculated based on the risetimes and plotted against the stress, the strain rate scaled with stress to the first power, i.e., proportional to stress. This behavior is quite different from the fourth power scaling, $\dot{\varepsilon} \propto \sigma^4$, which was observed by Swegle and Grady (1985) for fully dense materials, or from the second power scaling, $\dot{\varepsilon} \propto \sigma^2$, which was seen by Zhuang, Ravichandran, and Grady (2003) for layered composites. A few years later, we showed that nondimensional groups could be constructed that would collapse all of the experimental data on layered and granular/porous material onto a single curve. We also showed that the first power scaling originates from the need for mass to move across the gaps between particles, something that is not present in nonporous materials. Thus, granular WC displays a first power scaling, but granular WC in an epoxy matrix, which has a comparable bulk density and volume fraction of WC, displays a fourth power scaling.

Shortly after the first experiments on WC were completed, John Borg from Marquette University came to Albuquerque to spend a summer and begin a long-running collaboration on mesoscale modeling of granular materials. The approach was to use the CTH code to model individual grains of WC and simulate shock propagation through the material. The initial simulations were 2-D but were later extended to three-dimensional (3-D). Although mesoscale calculations of this type were not new, we simulated larger domains than most previous efforts and compared our results to the available experiments. We were able to match the experimental compaction curve by adjusting the strength parameter, and we were able to simulate the wave profiles reasonably well. The simulations also showed the first power scaling seen experimentally. Although the CTH mesoscale calculations perform reasonably well, they are nonphysical in not including realistic fracture or intergranular contact. To address some of these shortcomings, we began to explore other modeling approaches, notably the peridynamics formulation developed by Stewart Silling of the Computational Physics Department at Sandia. Work in this area has included collaboration with John Foster, formerly a Sandian and now at the University of Texas at San Antonio, and Chris Lammi, a student at the Georgia Institute of Technology.

As a graduate student at UT, I had heard Professor Rod Clifton talk about the pressure–shear loading technique as a means to measure the strength of materials at strain rates above those achievable with a Hopkinson bar. A slotted bore gun intended for oblique impact experiments had sat moribund at STAR for nearly two decades, until I was able to convince Clint Hall to fund the refurbishment of the gun around 2007. In fact, this gun dates to the infancy of shock physics at Sandia, as discussed elsewhere.[154] The process of bringing the gun online took much longer than I had expected, but eventually the team at STAR was able to get it working. We used VISAR probes oriented at an angle to measure both the normal and transverse velocity components of the target, as Lalit had done many years earlier. The first real project on which we utilized the oblique gun was measuring the shear stress transmitted by a granular sample, such as WC. It has also been used to study the strength of aluminum and PBX 9501.

[154] See the recollections of Jim Asay and Lalit Chhabildas.

Unfortunately, the results of the pressure shear experiments have sometimes been difficult to interpret. For example, for a couple of months, we were totally baffled by the transverse velocities measured at times when the shear wave could not have arrived at the free surface yet. Eventually, though, we realized that those transverse velocities were caused by the tilt of the impactor relative to the target, which led to an even more tilted longitudinal wave traveling through the target. When that wave reached the free surface, it also created the transverse velocity that we saw. Other aspects of the pressure–shear experiments continue to baffle us to this day. An interesting byproduct of the effort on the oblique impact gun is the magnetically applied pressure shear technique developed by Scott Alexander at Sandia. In what I view as an exceptional bit of thinking outside the box, Jim Asay conceived the idea in part, I think, as a result of discussions about the oblique impact gun.

In 2008, my wife took a faculty job in Stockton, CA, so I transferred to the Sandia California site. My manager in Albuquerque, Clint Hall, was very supportive during this time as was my new manager in California, Mike Chiesa. My group in California includes experimental solid mechanics, theoretical and computational solid mechanics, and molecular dynamics. While I fit in because of my educational background, most of my work is very different from what others in my department do. Shortly after I came to California, the Centers were reorganized, and my department became part of the Weapons Systems Engineering Center. Thus, from an organizational standpoint, I am quite close to systems groups and the CA gas transfer system organizations, so I have become much more familiar with nuclear weapons than was the case when I was in New Mexico. Since I came to CA, I have tried to justify my existence by becoming the main person at the site for hydrodynamic code (CTH) modeling, and I have supported projects on systems, such as the W80, W87, as well as other programs. I continue, though, to be involved in programs based in NM that involve gas gun and Z experiments, among other things.

As I write this, I am involved in new efforts, mainly related to the behavior of granular materials. With Seth Root, I am studying the high-pressure equation of state (EOS) of granular materials on Z as well as at STAR. This effort is tied to continuum EOS modeling by Dennis Grady and Gregg Fenton at Applied Research Associates as well as DFT (density functional theory) modeling by Kyle Cochrane, Luke Shulenburger, and Rudy Magyar at Sandia in NM. Also, Matt Lane in the Computational Materials and Data Science Department at Sandia in NM is performing molecular dynamics simulations to verify a hypothesis put forward by Dennis that porosity can enhance densification by triggering a phase transformation at a lower pressure. Finally, we are working to perform perturbation decay experiments similar to those performed by Andrei Sakharov and coworkers in the Soviet Union in the 1960s. Shock physics doesn't have a Nobel Prize winner, but the name Sakharov is one with strong meaning to me, having grown up during the Cold War. So, it's exciting to have the chance to extend work that he did.

I became involved in the American Physical Society Topical Group on Shock Compression of Condensed Matter after a few years at Sandia, and I served as secretary–treasurer from 2007 to 2009. In 2011, John Borg (Marquette), Jennifer Jordan (Air Force Research Laboratory-Eglin), and I co-organized the biennial shock conference in downtown Chicago. The location in the middle of the city worked out perfectly, and the meeting was the largest ever. We put special emphasis on attracting students and reaching out to our Chinese and Russian colleagues to facilitate their attendance. As part of the conference, we organized a side tour of Argonne's Advanced Photon Source, which was quite interesting. We also entered the mobile digital era, as the conference program was available on iPhone and iPod devices. Finally, the entire conference proceedings were made open access through the American Institute of Physics so that all the conference papers could be downloaded off their website for free.

As a result of coming to California, I have come to depend even more strongly on the team at the STAR gas gun facility, especially Bill Reinhart, Tom Thornhill, and Heidi Anderson, to get work done. I have always found them to be professional and careful in their work. Without them, it would have been impossible to remain involved in the experimental work on an ongoing basis. Of course, that does not mean that miscommunications or mistakes never occur, but we have been able to work quite well together despite me being a thousand miles from Albuquerque. I travel to Albuquerque fairly often to deal with things that are best handled in person, which helps keep all of us on the same page.

# Jack L. Wise

## (1979–Present[155])

## Personal Recollections of Shock Wave Research at Sandia

I first became aware of Sandia during the mid 1970s through discussions with recent graduates (John Cummings and Bob Setchell) and fellow students (Dale Berg and Glenn Laguna) who completed their doctoral studies in the Graduate Aeronautical Laboratories at the California Institute of Technology (Cal Tech). These contacts were all employed at some point by Sandia. I had first passed through New Mexico in September 1973 while en route to California for the beginning of my own graduate work at Cal Tech. Though my initial exposure to New Mexico was limited, I quickly perceived that it did, indeed, qualify as the Land of Enchantment on the basis of its unique scenic beauty, friendly people, history, culture, customs, cuisine, art, and architecture. I resolved then that I would return someday to spend additional time exploring more aspects of this captivating state; hence, as I neared the completion of my dissertation, the choice to pursue job opportunities with Sandia was clear and enthusiastic.

My interview trip to Sandia during the fall of 1978 was sponsored by Dave Northrop, who was seeking a new staff member to pursue research along the lines of that being done by Norm Warpinski regarding the fracturing of geological formations. During 2 full days at Sandia, I delivered a seminar regarding my thesis investigations of shock wave propagation in superfluid helium and met with management personnel from a number of technical groups. Although Jim Asay was not in attendance at my seminar, Dennis Hayes was on hand and thankfully suggested that Jim might want to be added to my interview agenda. This was accomplished, and I ultimately decided that Jim's group would be my first choice for employment at Sandia. A job offer was forthcoming for a staff position that was opened by virtue of Jim's earlier promotion to the Supervisor of the Thermomechanical and Physical Research Division, which was part of the Department managed by Walt Herrmann. After a successful thesis defense in June 1979, I began my Sandia career that summer. My Cal Tech thesis adviser, Hans Liepmann, strongly endorsed my choice of Sandia, telling me that "Now you're really going to see something!" He was absolutely right; the technical expertise and capabilities offered by Sandia far exceeded my grandest expectations.

When I started at Sandia, Jim's experimental staff included Lalit Chhabildas and myself; Dennis Grady was working at that time in a sister organization run by Barry Butcher. Our group also featured significant modeling capabilities, thanks to Douglas Drumheller and Tim Burns. The gun facility technician team, which Carl Konrad led, included Dave Cox and Bob Hardy. Dave was assigned to support my

---

[155] Jack Wise's shock wave research at Sandia was interrupted by a stint in geothermal research at Sandia from 1993–2005. Jack was promoted to Distinguished Member of Technical Staff in 2016.

testing activities. This was fortuitous for me: Dave and I had an excellent personality match, and we effectively contributed to a broad range of projects during our years together at the start of my technical career.

Now, midway through my 35th year at Sandia, I mentally divide this overall span into three distinct chronological phases. The first phase, which included the period from summer 1979 to December 1992, was dedicated to dynamic, high pressure experimental studies for the Thermomechanical and Physical Research Division and, later, the Experimental Impact Physics Division. During this time my first manager, Jim Asay, was promoted again and was then succeeded by Phil Stanton. Early on, I gained experience with laser velocity interferometry (i.e., the VISAR, which is an acronym for Velocity Interferometer System for Any Reflector) as well as with microwave interferometry. I also designed and performed recovery experiments with Jim Asay that allowed post-shock investigations of microstructural changes in brass and aluminum. I conducted impact testing on various pieces of weapon hardware, including contact fuze prototype designs, scaled reentry body nosetips, and embedded use-control devices. To support modeling and simulation efforts at Sandia, I obtained equation-of-state and wave profile data for a number of shock-loaded condensed materials, including lithium fluoride, nonhomogeneous carbon/carbon reentry body nosetip materials (with George Clark), glass (with George Raiser and Rod Clifton of Brown University and Dennis Grady, Stewart Silling, Paul Taylor, and Mike Furnish of Sandia), ceramics (with Dennis Grady), stainless steel (with Don Mikkola of Michigan Tech), beryllium (with Jim Asay and Lalit Chhabildas), copper, composites, and polymer-based foams. For the beryllium study, which constituted my inaugural involvement in a major body of technical work at Sandia, I benefited immensely from the guidance and collaboration of Jim Asay and Lalit Chhabildas. Later research on other materials earned a 1993 DOE (Department of Energy) Nuclear Weapons Program Recognition of Excellence Award for contributions as a member of the W88 Warhead Safety Analysis Team.

My initial work on optical properties featured measurements of the refractive index and Hugoniot equation of state for an aqueous solution of zinc chloride, which was being evaluated as a potential window material for velocity interferometer studies on high explosives of interest to Sandia and Los Alamos researchers. In a parallel study, I collaborated with Lalit Chhabildas to generate and publish the first optical calibration of lithium fluoride as a VISAR window for stresses ranging as high as 115 GPa; this was an exciting breakthrough for the shock wave community since it enabled wave profile measurements for embedded sample/window interfaces shocked to stresses far exceeding the limited ranges offered by the polymethyl methacrylate (PMMA), fused silica, and sapphire window materials that had been previously characterized by Lynn Barker and Red Hollenbach at Sandia. Bob Setchell provided significant insights for the analysis of the lithium fluoride data.

After his return to Sandia in the mid 1980s, Lynn Barker led an effort to generate high-pressure equation-of-state data for ramp-compressed samples of condensed (solid or liquid) hydrogen in order to detect the postulated metallic transition for this material. Based on my prior Cal Tech experience with experimental shock studies on superfluid helium, I supported the hydrogen work by developing a specialized cryogenic system for the preparation of condensed samples that were rapidly transferred from a cryostat to the impact chamber of the two-stage, light gas gun using Lynn's innovative Target Insertion Mechanism (TIM). I initially advocated deployment of a static, sacrificial sample system that would be mounted at the end of the gun's launch tube, similar to the approach used by Bill Nellis and his team at Lawrence Livermore. However, my recommendation was overruled when Lynn presented a convincing demonstration of his TIM concept using a homemade model. His concept, coupled with the suitably adapted cryogenic system, was successfully fielded on the two-stage gun; however, our results did not produce data pertinent to hydrogen metallization.

I also developed and implemented novel test methods for penetration and ceramic-armor studies done in conjunction with Dennis Grady and Marlin Kipp. This work included the first VISAR-based study of the dynamic, multiaxial impact response of confined and unconfined ceramic rods, as well as measurements of the rear-surface motion of ceramic and steel targets subjected to long-rod penetrator attack. This work benefitted from numerous technical discussions with Mike Forrestal. In a related effort, I completed tests and analyses to optimize the performance of the Terminal-Ballistics facility. Andy Piekutowski from the University of Dayton Research Institute provided invaluable guidance and consultation for this effort. In addition, I carried out shock vaporization experiments on aluminum, zinc, and porous carbon in collaboration with Gerry Kerley and Tim Trucano.

Near the end of my first career phase at Sandia, I was the ES&H (Environment, Safety, and Health) coordinator for the Solid Dynamics Research Department, overseeing the STAR (Shock Thermodynamics Applied Research) facility preparations and follow-up for the DOE Tiger Team[156] review. This assignment, which was all consuming at times, made me part of the Engineering Sciences Center's Compliance Initiative Appraisal Team, whose members also included Dan Aeschliman, Ken Erickson, and C.A. Davidson. With support from multiple third-party (non-Sandia) consultants, we inspected all offices and laboratory space utilized by personnel in the Center and issued a report detailing a total of 407 noncompliance findings that were ultimately resolved by the respective line organizations.

The end of the Cold War brought on the realignment of priorities at Sandia, and funding tightened for the traditional shock wave research program. In fact, the STAR facility was even shut down for a while. Just prior to this development, with encouragement from Doug Drumheller who had already changed his field of study,

[156] In 1990, Admiral James D. Watkins, the Secretary of Energy under George H.W. Bush, instituted a "Tiger Team" assessment process to identify, address, and rectify ES&H issues at the Department of Energy's research, production, and testing facilities. The independent Tiger Teams conducted environment, safety, and health and management compliance assessments at all DOE facilities. The Tigers arrived on April 15, 1991, and were at Sandia through May 24, 1991. The on-site visit was preceded by self appraisals by all Sandia organizations; following the visit, remediation plans were established to address the findings and concerns identified by the Tiger Teams.

I transferred to the Geothermal Research Department, where I began the second phase of my Sandia career in January 1993. Jim Dunn was the Department Manager at that time. In this new capacity, I assumed technical (design and analysis) and management responsibilities for projects involving laboratory and field tests related to sonic drilling, lost-circulation control, and Polycrystalline-Diamond-Compact (PDC) cutter and bit development. My work entailed the preparation of detailed program plans, reports, and updates for our DOE sponsors. I collaborated with, and supervised, other staff members and technicians who supported various elements of the drilling program. To meet our technology development goals, I established and maintained cooperative working relationships with multiple industry, university, and consultant partners; this effort involved drafting and overseeing contracts for university and industry support. I also generated descriptive web pages and brochures to advertise Sandia research activities and capabilities and conducted detailed technical reviews of DOE and NSF SBIR (National Science Foundation Small Business Innovation Research) proposals.

On the experimental side, I planned field tests and acquired extensive strain and acceleration data to characterize the dynamic behavior of sonic drilling systems. I also generated and documented extensive single-cutter data that served as a baseline for university efforts to model PDC cutter and bit dynamics. Dave Glowka and I conceived and implemented a novel transparent wellbore test configuration for lost-circulation studies. I also proposed, planned, and accomplished a fundamental parameter study for PDC cutters that was conducted with industry participation by the U.S. Synthetic Corporation. On another front, I led the development and implementation of a major Sandia and industry CRADA (Cooperative Research and Development Agreement) effort to assess the capabilities of current drag-bit technology via field demonstrations with commercial bits supplied by partnering manufacturers (ReedHycalog, Security DBS, Smith Bits, and Technology International) for operations in a hard-rock drilling environment. In all, my reported work on PDC cutter and bit technology garnered "best paper" awards at the 2002, 2004, and 2005 international meetings of the Geothermal Resources Council.

Following a nomination by Sandia management, I was appointed by the DOE HQ (Headquarters) Office of Geothermal Technologies to be the task leader for Annex VII ("Advanced Geothermal Drilling Techniques") of the International Energy Agency's (IEA) Geothermal Implementing Agreement (GIA). In this capacity, I coordinated multinational technical cooperation and exchanges by convening Annex VII meetings (e.g., in Pisa, Italy, during October 2004), participating in IEA-GIA Executive Committee meetings (e.g., in Antalya, Turkey, at the 2005 World Geothermal Congress), and managing the continued progress on Annex VII subtasks.

As geothermal research funding tapered off, members of the group were encouraged to seek opportunities for matrixing their time with other Sandia organizations. Clint Hall, who had periodically inquired about my potential interest in returning to the shock wave community, coordinated the establishment of a part-time research effort for me that involved wave profile measurements on various brazing alloys of interest to Scott Jones. With excellent support from Randy Hickman, Jeff Gluth, and the STAR facility team, we investigated the effects of annealing and preheating on the dynamic yield strength and spall failure of these alloys. This work triggered my full-time return to dynamic material studies in December 2004 when I started the third phase of my Sandia career by joining Clint as a member of the Shock and Z-Pinch Physics Department that was managed at that time by Chris Deeney. This group, which subsequently became the Dynamic Material Properties Department, has since been managed by Clint, followed by Dawn Flicker and, most recently, by John Benage.

Besides the braze alloy investigation, my more recent material studies have focused on samples of Kovar with Scott Jones and Clint Hall; tungsten carbide and alumina-filled epoxy composites with Tracy Vogler, Scott Alexander, and Steve Montgomery; a phenolic compound with Nathan Moore; lead zirconate titanate (PZT) with Steve Montgomery, George Clark, and Dan Jackson; and a metal oxide varistor composition, also with Montgomery, Clark, and Jackson. The PZT studies included experiments that simultaneously detected the time-resolved mechanical response and electrical output for this material under controlled ramp wave compression. I have also worked with Aaron Hall and Clint Hall to develop and characterize multi-constituent flame-spray depositions suitable for use as graded-impedance impactors or buffers for isentropic compression studies.

I have expanded on my earlier research on preheated braze samples by developing and utilizing a technique for preheating samples for ramp compression studies on the Veloce electromagnetic driver, which is operated by Randy Hickman and his DICE (Dynamic Integrated Compression Experimental) facility team. This technique has been employed in wave profile experiments that detected the temperature dependence of the α–β solid-solid transition for tin. Such high temperature work has since blossomed into a multiyear effort that I have led to develop a robust resistive preheat controller and associated load hardware to support dynamic material testing on Sandia's Z machine, as well as on the Veloce driver and gas guns located at the DICE and STAR facilities. This work has involved extensive collaboration among Sandia, NSTec (e.g., Morris Kaufman), and Ktech. Thus far, two preheat shots have been conducted on the Z machine for which initial sample temperatures of 500 and 656 C were achieved, respectively.

In collaboration with Sophie Chantrenne of Science Applications International, Jim Asay, Marlin Kipp, and Clint Hall, I have designed and fielded a sample soft-recovery assembly for ramp-loading experiments on the Veloce driver. This system has been applied to studies of nanostructure formation in joint Sandia investigations with Hongyou Fan and Randy Hickman. That research got a 2016 R&D 100 Award "as one of the 100 most technologically significant new products of the year."

Most recently, I have been involved in experimental studies of component function and damage under impact loading. This ongoing work has focused on an existing impact fuze design and on prototype assembly concepts for a radioisotope thermoelectric generator. In September 2013, a DOE/NNSA (National Nuclear Security Administration) Defense Programs Award of Excellence was presented to the W88 Contact Fuze Testing, Modeling, and Validation Team that I led together with George Clark and Dan Jackson.

In closing, I can only say that my time at Sandia has been both productive and gratifying. I have been extremely fortunate to be associated with the colleagues explicitly noted above, as well as many, many others who have collaborated in, or supported, my work. When I joined Sandia, I expected that my stay here would be limited to 5 years or so, before moving elsewhere. It's now been nearly 35 years… Time really does fly when you're having fun!

# List of Acronyms and Abbreviations

| | |
|---|---|
| 1-D | one-dimensional |
| 2-D | two-dimensional |
| 3-D | three-dimensional |
| AAS | American Astronomical Society |
| AEC | Atomic Energy Commission |
| AEDC | Arnold Engineering Development Center |
| AFPL | Air Force Phillips Laboratory |
| AFRL | Air Force Research Laboratory |
| AFWL | Air Force Weapons Laboratory |
| AGT | aboveground test |
| AGU | American Geophysical Union |
| AIRAPT | International Association of High Pressure Science and Technology |
| AIAA | American Institute of Aeronautics and Astronautics |
| ALE | arbitrary Lagrangian-Eulerian |
| ALEGRA | Arbitrary Lagrangian Eulerian General Research Applications |
| ANEOS | analytic equation of state |
| ANFO | ammonium nitrate and fuel oil |
| APS | American Physical Society |
| ARA | Aeroballistic Range Association |
| ARL | Army Research Laboratory |
| ASCI | Accelerated Strategic Computing Initiative |
| AT&T | American Telephone and Telegraph |
| AWE | Atomic Weapons Establishment |
| BBAY | Bethe, Bade, Averell, Yos |
| BKW | Becker–Kistiakowsky–Wilson |
| BN | Bechtel Nevada (corporation) |
| B-N | Baer–Nunziato |
| BRACIS | beam reflection at center of impact surface |
| BRL | Ballistic Research Laboratory |
| CDAR | Coupled Damage And Reaction |



J.R. Asay et al., *Impactful Times*, Shock Wave and High Pressure Phenomena,
DOI 10.1007/978-3-319-33347-2

| | |
|---|---|
| CDC | Control Data Corporation |
| CHARTD | Coupled Hydrodynamics and Radiation Transport Diffusion |
| CINT | Center for Integrated Nanotechnologies |
| CJ | Chapman-Jouguet |
| CLP | Corporate Lethality Program |
| CSQ | CHARTD SQuared |
| CTBT | Comprehensive Test Ban Treaty |
| CTH | CSQ to the Three Halves |
| DAC | diamond anvil cell |
| DASA | Defense Atomic Support Agency |
| DARPA | Defense Advanced Research Projects Agency |
| DDT | deflagration-to-detonation transition |
| DFT | density functional theory |
| DICE | Dynamic Integrated Compression Experimental |
| DMTS | Distinguished Member of Technical Staff |
| DNA | Defense Nuclear Agency |
| DOD | Department of Defense |
| DOE | Department of Energy |
| DTRA | Defense Threat Reduction Agency |
| EBW | exploding bridgewire |
| EEGS | Electrical Energy Gun System |
| ECF | Explosive Components Facility |
| EHVL | enhanced hypervelocity launcher |
| EOS | equation of state |
| ERDA | Energy Research and Development Administration |
| ES&H | Environment, Safety, and Health |
| FE | ferroelectric |
| FEM | ferroelectric model |
| GM | General Motors |
| GRL | Geophysical Research Letters |
| GRC | General Research Corporation |
| HARP | Hazards Assessment of Rocket Propellants |
| HE | high explosive |
| HEDP | high energy density physics |
| HEL | Hugoniot elastic limit |
| HNX | hexanitrostilbene explosive |
| HPC | high performance computing |
| HST | Hubble Space Telescope |
| HVIS | Hypervelocity Impact Symposium |
| HVL | HyperVelocity Launcher |
| IBM | International Business Machines |
| ICBM | intercontinental ballistic missile |
| ICE | isentropic compression experiment |
| ICF | inertial confinement fusion |
| IIT | Illinois Institute of Technology |

| | |
|---|---|
| ISL | Institut de Recherches de Saint-Louis |
| ISP | Institute of Shock Physics |
| ISS | International Space Station |
| IUTAM | International Union of Theoretical and Applied Mechanics |
| IVA | inductive voltage adder |
| JASPER | Joint Actinide Shock Physics Experimental Research |
| JCZ | Jacobs-Cowperthwaite-Zwisler |
| JANAF | Joint Army Navy Air Force (sometimes written as JANNAF to include NASA) |
| JMP | Joint Munitions Program |
| JSC | Johnson Space Center |
| JWL | Jones–Wilkins–Lee |
| KFK | Kernforschungszentrum Karlsruhe (Karlsruhe Institute of Technology) |
| LANL | Los Alamos National Laboratory |
| LDRD | Laboratory Directed Research and Development |
| LEO | low-Earth orbit |
| LIHE | light initiated high explosive |
| LMD | Lee-More-Desjarlais |
| LLNL | Lawrence Livermore National Laboratory |
| LTD | linear transformer driver |
| M&S | modeling and simulation |
| MAD | mutually assured destruction |
| MAPS | magnetically applied pressure shear |
| MAVEN | Model Accreditation via Experimental Sciences for Nuclear Weapons |
| MBBAY | Modified BBAY |
| MD | molecular dynamics |
| MDA | Missile Defense Agency |
| MF | magnetic flyer |
| MHD | magnetohydrodynamic |
| MIT | Massachusetts Institute of Technology |
| MITL | magnetically insulated transmission line |
| MOU | memorandum of understanding |
| MP | melting point |
| MPP | massively parallel processing |
| MSFC | Marshall Space Flight Center |
| NAE | National Academy of Engineering |
| NASA | National Aeronautics and Space Administration |
| NIF | National Ignition Facility |
| NIST | National Institute of Standards and Technology |
| NNSA | National Nuclear Security Administration |
| NNSS | Nevada National Security Site |
| NSWC | Naval Surface Weapons Center |
| NTS | Nevada Test Site |
| NRL | Naval Research Laboratory |
| ODE | ordinary differential equation |

| | |
|---|---|
| OMA | Optical Multichannel Analyzer |
| ORVIS | Optically Recorded Velocity Interferometer System |
| OSD | Office of the Secretary of Defense |
| P-α | Pressure-α (model) |
| P-λ | Pressure-λ (model) |
| PBFA II | Particle Beam Fusion Accelerator II |
| PBX | polymer-bonded explosive |
| PCTH | Parallel CTH |
| PDF | probability density function |
| PDV | photon Doppler velocimetry |
| PETN | pentaerythritol tetranitrate |
| PIC | particle in cell |
| PMMA | polymethyl methacrylate |
| PG | powder gun |
| PPG | Pacific Proving Ground |
| PVDF | polyvinylidene difluoride |
| PZT | Pb (lead) zirconate titanate |
| QMD | quantum molecular dynamics |
| R&D | research and development |
| RG | railgun |
| RHALE | Robust Hydrodynamics Arbitrary Lagrangian Eulerian |
| RPI | Rensselaer Polytechnic Institute |
| R-T | Rayleigh-Taylor |
| SCCM | Shock Compression of Condensed Matter |
| SCE | subcritical experiment |
| SDI | Strategic Defense Initiative |
| SDIO | Strategic Defense Initiative Organization |
| SITI | Sandia Instrumented Thermal Initiation |
| SL9 | Shoemaker–Levy 9 comet |
| SLIFER | Shorted Location Indicator by Frequency of Electrical Resonance |
| SNL | Sandia National Laboratories |
| SNM | special nuclear material |
| SPH | smooth particle hydrodynamics |
| SPR | Strategic Petroleum Reserve |
| SRI | Stanford Research Institute |
| SSP | Stockpile Stewardship Program |
| STAR | Shock Thermodynamics Applied Research |
| STScI | Space Telescope Science Institute |
| SWAP | Stress Wave Application Program |
| TARDEC | Tank Automotive Research, Development, and Engineering Center |
| TATB | triaminotrinitrobenzene |
| TBF | Terminal Ballistics Facility |
| THAAD | Theater High-Altitude Area Defense |
| TOE | third-order elastic |
| TMD | Theater Missile Defense |
| TMI | Three Mile Island |

| | |
|---|---|
| TIM | transient insertion mechanism |
| TP | triple point |
| TSG | Two-Stage light gas Gun |
| TTBT | Threshold Test Ban Treaty |
| UGT | underground nuclear test |
| UK | United Kingdom |
| U.S. | United States |
| VASP | Vienna Ab initio Simulation Package |
| VISAR | Velocity Interferometer System for Any Reflector |
| VSIP | Volunteer Staff Incentive Program |
| V&V | verification and validation |
| WDM | warm dense matter |
| WIPP | Waste Isolation Pilot Plant |
| WSU | Washington State University |
| XDT | explosive detonation transition |
| XRD | x-ray diode |
| YAG laser | neodymium-doped yttrium aluminum garnet (Nd:$Y_3Al_5O_{12}$) laser |
| Z | Z pulsed power facility |
| ZND | Zeldovich–von Neumann–Doring |

# Bibliography

**Editors' Note** Because of the extent of our bibliography, which contains nearly 1000 entries, it is somewhat unique in the following respects, as summarized below.

- The order of the entries in the bibliography is alphabetical by the first author chronologically by year. Entries in the same year for that same first author with one coauthor come next in the bibliography—that is, author 1 and author 2 (year). This is followed by entries in the same year for that same first author with multiple coauthors, *regardless of the alphabetical ordering of those coauthors*, using the customary a, b, and c notation for that specific year if there is more than one reference for that same first author with two or more coauthors—for example, author 1, author 2, and author 3 (year a); author 1, author 2, author 3 (year b); and author 1, author 2, author 3 et al. (year c).
- Our bibliography does contain some references by first authors that have the same last name but that are distinctly different individuals. An example of this occurs on the first page of our bibliography: entry in our bibliography and M. U. Anderson is the first author of the fifth entry in our bibliography. Another example is the four distinct first-author Browns (J. L. Brown, P. G. Brown, W. K. Brown, and W. T. Brown). Because of this, in the text of Part I and Part II of our book, you will find instances where three or four authors are listed, followed by the year—for example, (author 1, author 2, author 3, year) rather than our usual (author 1 et al., year).

 
J.R. Asay et al., *Impactful Times*, Shock Wave and High Pressure Phenomena,
DOI 10.1007/978-3-319-33347-2

T. Akashi, A.B. Sawaoka, R.A. Graham, The effect of shock compression on graphite-like boron nitride, in *Shock Waves in Condensed Matter*, ed. by Y.M. Gupta (Plenum, New York, NY, 1986), pp. 821–826

C.S. Alexander, L.C. Chhabildas, W.D. Reinhart, D.W. Templeton, Changes to the shock response of fused quartz due to glass modification. Int. J. Impact Eng. **35**, 1376–1384 (2008)

C.S. Alexander, J.R. Asay, T.A. Haill, Magnetically applied pressure-shear: A new method for direct measurement of strength at high pressure. J. Appl. Phys. **108**, 126101 (2010)

C.E. Anderson, T.G. Trucano, S.A. Mullin, Debris cloud dynamics. Int. J. Impact Eng. **9**(1), 89–113 (1990)

M.U. Anderson, R.A. Graham, D.E. Wackerbarth, Prediction and data analysis of current pulses from impact-loaded piezoelectric polymers (PVDF), in *Shock Compression of Condensed Matter*, ed. by S.C. Schmidt, J.N. Johnson, L.W. Davison (Elsevier, Amsterdam, 1990), pp. 805–808

M.U. Anderson, L.C. Chhabildas, W.D. Reinhart, Simultaneous PVDF/VISAR measurement technique for isentropic loading with graded density impactors, in *Shock Compression of Condensed Matter*, ed. by S.C. Schmidt, D.P. Dandekar, J.W. Forbes. AIP Conference Proceedings, vol. 429 (AIP, College Park, MD, 1998), pp. 841–844

M.U. Anderson, D.E. Cox, S.T. Montgomery, R.E. Setchell, Compositional effects on the shock compression and release properties of alumina-filled epoxy, in *Shock Compression of Condensed Matter*, ed. by M. Elert, M.D. Furnish, R. Chau, N. Holmes, J. Nguyen. AIP Conference Proceedings, vol. 825 (AIP, College Park, MD, 2006), pp. 789–792

J.A. Ang, Impact flash jet initiation phenomenology. Int. J. Impact Eng. **10**, 23–33 (1990)

J.A. Ang, G. Hauze, Impact of acceleration on barrel/launch package design. IEEE Trans. Magnetics **27**, 544–549 (1991)

J.A. Ang, L.C. Chhabildas, B.G. Cour-Palais, E.L. Christiansen, J.L. Crews, Evaluation of Whipple bumper shields at 7 and 10 km/s. AIAA Paper No. 92–1590 (1991)

J.A. Ang, A hypervelocity impact jet formation, in *Shock Compression of Condensed Matter*, ed. by S.C. Schmidt, R.D. Dick, J.W. Forbes, D.G. Tasker (Elsevier, Amsterdam, 1992), pp. 1019–1022

J.A. Ang, C.H. Konrad, C.A. Hall, A.R. Susoeff, R.S. Hawke, G.L. Sauve, A.R. Vasey, S.M. Gosling, R.J. Hickman, Hypervelocity projectile design and fabrication. IEEE Transactions on Magnetics **29**(1), 722–727 (1993a)

J.A. Ang, B.D. Hansche, C.H. Konrad, W.C. Sweatt, S.M. Gosling, R.J. Hickman, Pulsed holography for hypervelocity impact diagnostics. Int. J. Impact Eng. **14**, 13–24 (1993b)

J.A. Ang, B.D. Hansche, Pulsed holography diagnostics of impact fragmentation, in *High-Pressure Shock Compression of Solids II: Dynamic Fracture and Fragmentation*, ed. by L.W. Davison, D.E. Grady, M. Shahinpoor (Springer, New York, NY, 1996), pp. 176–193

T. Ao, J.R. Asay, J.-P. Davis, M.D. Knudson, C.A. Hall, High-pressure quasi-isentropic loading and unloading of interferometer windows on the Veloce pulsed power generator, in *Shock Compression of Condensed Matter*, ed. by M. Elert, M.D. Furnish, R. Chau, N.C. Holmes, J. Nguyen. AIP Conference Proceedings, vol. 955 (AIP, College Park, MD, 2007), pp. 1157–1160

T. Ao, J.R. Asay, S. Chantrenne, M.R. Baer, C.A. Hall, A compact strip-line pulsed power generator for isentropic compression experiments. Rev. Sci. Instrum. **79**, 013903 (2008)

T. Ao, M.D. Knudson, J.R. Asay, J.P. Davis, Strength of lithium fluoride under shockless compression to 114 GPa. J. Appl. Phys. **106**, 103507 (2009a)

T. Ao, R.J. Hickman, S.L. Payne, W.M. Trott, Line-imaging ORVIS measurements of interferometric windows under quasi-isentropic compression, in *Shock Compression of Condensed Matter*, ed. by M.L. Elert, W.T. Buttler, M.D. Furnish, W.W. Anderson, W.G. Proud. AIP Conference Proceedings, vol. 1195 (AIP, College Park, MD, 2009b), pp. 619–622

J.R. Asay, G.R. Fowles, G.E. Duvall, M.H. Miles, R.F. Tinder, Effects of point defects on elastic precursor decay in LiF. J. Appl. Phys. **45**(5), 2132–2145 (1972)

J.R. Asay, Shock-induced melting in bismuth. J. Appl. Phys. **45**, 4441–4452 (1974)

J.R. Asay, L.M. Barker, Interferometric measurement of shock-induced internal particle velocity and spatial variations in particle velocity. J. Appl. Phys. **45**(6), 2540–2546 (1974)

J.R. Asay, D.B. Hayes, Shock-compression and release behavior near melt states in aluminum. J. Appl. Phys. **46**, 4789–4800 (1975)

J.R. Asay, D. Hicks, D. Holdridge, Comparison of experimental and calculated elastic-plastic wave profiles in LiF. J. Appl. Phys. **46**, 4316–4322 (1975)

J.R. Asay, L.P. Mix, F.C. Perry, Ejection of material from shocked surfaces. J. Appl. Phys. **29**, 284–287 (1976)

J.R. Asay, Shock loading and unloading in bismuth. J. Appl. Phys. **48**, 2832–2844 (1977a)

J.R. Asay, Effects of shock wave risetime on material ejection from aluminum surfaces, SAND77-0731 (Sandia National Laboratories, Albuquerque, NM, 1977b)

J.R. Asay, Thick-plate technique for measuring ejecta from shocked surfaces. J. Appl. Phys. **49**, 6173–6175 (1978)

J.R. Asay, L.D. Bertholf, A model for estimating the effects of surface roughness on mass ejection from shocked surfaces, SAND78-1256 (Sandia National Laboratories, Albuquerque, NM, 1978)

J.R. Asay, J. Lipkin, A self-consistent technique for estimating the dynamic strength of a shock-loaded material. J. Appl. Phys. **49**, 4242–4247 (1978)

J. Asay, B. Butcher, C. Konrad, Internal pressure measurements on the Sandia powder gun, SAND79-2178 (Sandia National Laboratories, Albuquerque, NM, 1978)

J.R. Asay, L.C. Chhabildas, Some new developments in shock wave research, in *High Pressure Science and Technology – 1979*, (Proceedings of the VIIth International AIRAPT Conferences Part II), ed. by B. Vodar, P. Marteau (AIP, College Park, MD, 1980), pp. 958–964

J.R. Asay, L.C. Chhabildas, Determination of the shear strength of shock-compressed 6061-T6 aluminum, in *Shock waves and high-strain-rate phenomena in metals*, ed. by M.A. Myers, L.E. Murr (Plenum, New York, NY, 1981), pp. 417–424

J.R. Asay, L.C. Chhabildas, J.L. Wise, Strain rate effects in beryllium under shock compression, in *Shock Waves in Condensed Matter*, ed. by W.J. Nellis, L. Seaman, R.A. Graham. AIP Conference Proceedings, vol. 78 (AIP, College Park, MD, 1982a), pp. 427–431

J.R. Asay, L.C. Chhabildas, J.L. Wise, Viscoplastic response of beryllium under shock compression, in *High Pressure in Research and Industry – 8th AIRAPT and 19th EHPRG Conference Proceedings*, ed. by C.M. Backman, T. Johannisson, L. Tegnér (Arkitektkopia, Uppsala, 1982b), pp. 227–230

J.R. Asay, L.C. Chhabildas, G.I. Kerley, T.G. Trucano, High pressure strength of shocked aluminum, in *Shock Waves in Condensed Matter*, ed. by Y.M. Gupta (Plenum, New York, NY, 1986), pp. 145–150

J.R. Asay, G.I. Kerley, The response of materials to dynamic loading. Int. J. Impact Eng. **5**, 69–99 (1987)

J.R. Asay, T.G. Trucano, L.C. Chhabildas, Time-resolved measurements of shock-induced vapor-pressure profiles, in *Shock Waves in Condensed Matter*, ed. by S.C. Schmidt, N.C. Holmes (Elsevier, Amsterdam, 1988), pp. 159–162

J.R. Asay, T.G. Trucano, Experimental measurements of shock-induced vaporization in cadmium and lead, in *Shock Compression of Condensed Matter*, ed. by S.C. Schmidt, J.N. Johnson, L.W. Davison (Elsevier, Amsterdam, 1990), pp. 143–146

J.R. Asay, T.G. Trucano, R. Hawke, The use of hypervelocity launchers to explore previously inaccessible states of matter. Int. J. Impact Eng. **10**, 51–66 (1990)

J.R. Asay, C.A. Hall, C.H. Konrad, W.M. Trott, G.A. Chandler, K.J. Fleming, K.G. Holland, L.C. Chhabildas, T.A. Mehlhorn, R. Vesey, T.G. Trucano, A. Hauer, R. Cauble, M. Foord, Use of z-pinch sources for high-pressure equation-of-state studies. Int. J. Impact Eng. **23**, 27–38 (1999)

J.R. Asay, Isentropic compression experiments on the Z accelerator, in *Shock Compression of Condensed Matter*, ed. by M.D. Furnish, L.C. Chhabildas, R.S. Hixson. AIP Conference Proceedings, vol. 505 (AIP, College Park, MD, 2000), pp. 261–266

J.R. Asay, C.A. Hall, K.G. Holland, M.A. Bernard, W.A. Stygar, R.B. Spielman, S.E. Rosenthal, D.H. McDaniel, D.B. Hayes, Isentropic compression of iron with the Z accelerator, in *Shock Compression of Condensed Matter*, ed. by M.D. Furnish, L.C. Chhabildas, R.S. Hixson. AIP Conference Proceedings, vol. 505 (AIP, College Park, MD, 2000), pp. 1151–1154

J.R. Asay, M.D. Knudson, Use of pulsed magnetic fields for quasi-isentropic compression experiments, in *High Pressure Shock Compression of Solids VIII*, ed. by L.C. Chhabildas, L.W. Davison, Y. Horie (Springer, New York, NY, 2005), pp. 329–380

J.R. Asay, T. Ao, J.-P. Davis, C.A. Hall, T.J. Vogler, G.T. Gray, Effect of initial properties on the flow strength of aluminum during quasi-isentropic compression. J. Appl. Phys. **103**, 083514 (2008)

J.R. Asay, T. Ao, T.J. Vogler, J.-P. Davis, G.T. Gray, Yield strength of tantalum for shockless compression to 18 GPa. J. Appl. Phys. **106**, 073515 (2009)

J.R. Asay, T.J. Vogler, T. Ao, J. Ding, Dynamic yielding of single crystal Ta at strain rates of $\sim 5 \times 10^5$/s. J. Appl. Phys. **109**, 073507 (2011)

S. Attaway, S. Haniff, J. Stevenson, J. Wilke, Cielo CCC-1 summary: Lightweight, blast resistant structure development, SAND2011-6477P (Sandia National Laboratories, Albuquerque, NM, 2011)

M.R. Baer, J.W. Nunziato, A theory for deflagration-to-detonation transition (DDT) in granular explosives, SAND82-0293 (Sandia National Laboratories, Albuquerque, NM, 1983)

M.R. Baer, J.W. Nunziato, A Two-Phase Mixture Theory for the Deflagration-to-Detonation Transition (DDT) in Reactive Granular Materials. Int. J. Multiphase Flow **12**(6), 861–889 (1986)

M.R. Baer, R.J. Gross, J.W. Nunziato, E.A. Igel, An experimental and theoretical study of deflagration-to-detonation transition (DDT) in the granular explosive CP. Combust Flame **65**, 15–30 (1986)

M.R. Baer, Numerical studies of dynamic compaction of inert and energetic granular materials. J. Appl. Mech. **55**, 36–43 (1988)

M.R. Baer, J.W. Nunziato, Compressive combustion of granular materials induced by low velocity impact, in *Proceedings of the 9th International Detonation Symposium Office of Naval Research Report ONR 113291-7:293-305*, ed. by J.M. Short, E.L. Lee (Office of Naval Research, San Diego, CA, 1989)

M.R. Baer, A mixture model for shock compression of porous multi-component reactive mixtures, in *High-Pressure Science and Technology*, ed. by S.C. Schmidt, J.W. Shaner, G.A. Samara, M. Ross. AIP Conference Proceedings, vol. 309 (AIP, College Park, MD, 1994), pp. 1247–1250

M.R. Baer, P.W. Cooper, M.E. Kipp, Investigations of emergency destruction methods for recovered, explosively configured, chemical warfare munitions: Interim emergency destruction methods—Evaluation report, SAND95-8248 (Sandia National Laboratories, Albuquerque, NM, 1995)

M.R. Baer, Continuum mixture modeling of reactive porous media (Chapter 3), in *High-Pressure Shock Compression of Solids IV: Response of Highly Porous Solids to Shock Loading*, ed. by L. Davison, Y. Horie, M. Shahinpoor (Springer, New York, NY, 1996)

M.R. Baer, E.S. Hertel Jr., R.L. Bell, Multidimensional DDT modeling of energetic materials, in *Shock Compression of Condensed Matter*, ed. by S.C. Schmidt, W.C. Tao. AIP Conference Proceedings, vol. 370 (AIP, College Park, MD, 1996a), pp. 433–436

M.R. Baer, R.A. Graham, M.U. Anderson, S.A. Sheffield, R.L. Gustavsen, Experimental and theoretical investigations of shock-induced flow of reactive porous media, in *Proceedings of the 1996 JANAF Combustion Subcommittee and Propulsion System Hazards Subcommittee Joint Meeting* (Chemical Propulsion Information Analysis Center, Johns Hopkins University, Baltimore, MD, 1996b), pp. 123–132

M.R. Baer, Shock wave structure in heterogeneous reactive media, in *Proceedings of 21st International Symposium on Shock Waves* (University of Queensland, Great Keppel Island, 1997), pp. 923–927

M.R. Baer, M.E. Kipp, F. van Swol, Micromechanical modeling of heterogeneous energetic materials, in *Proceedings of the 11th International Detonation Symposium,* ONR 33300-5, ed. by J.M. Short, J.E. Kennedy (Office of Naval Research, Washington, D.C., 1998), pp. 788–797

M.R. Baer, Computational modeling of heterogeneous reactive materials at the mesoscale, in *Shock Compression of Condensed Matter*, ed. by M. Furnish, L. Chhabildas, R. Hixson. AIP Conference Proceedings, vol. 505 (AIP, College Park, MD, 2000), pp. 27–33

M.R. Baer, W.M. Trott, Mesoscale descriptions of shock-loaded heterogeneous porous materials, in *Shock Compression of Condensed Matter*, ed. by M. Furnish, N.N. Thadhani, Y. Horie. AIP Conference Proceedings, vol. 620 (AIP, College Park, MD, 2002a), pp. 713–716

M.R. Baer, W.M. Trott, Theoretical and experimental mesoscale studies of impact-loaded granular explosives and simulant materials, in *Proceedings of the 12th International Detonation Symposium, San Diego, CA*, ONR 333-05-2, ed. by J.M. Short, J.L. Maienschein (Office of Naval Research, Washington, D.C., 2002b), pp. 939–950

M.R. Baer, W.M. Trott, Mesoscale studies of shock loaded tin sphere lattices, in *Shock Compression of Condensed Matter*, ed. by M.D. Furnish, Y.M. Gupta, J.W. Forbes. AIP Conference Proceedings, vol. 706 (AIP, College Park, MD, 2004), pp. 517–520

J.E. Bailey, J. Asay, M. Bernard, A.L. Carlson, G.A. Chandler, C.A. Hall, D. Hanson, R. Johnston, P. Lake, J. Lawrence, Optical spectroscopy measurements of shock waves driven by intense z-pinch radiation. J. Quant. Spectrosc. Rad. Transfer **65**, 31–42 (2000)

J.E. Bailey, M.D. Knudson, A.L. Carlson, G.S. Dunham, M.P. Desjarlais, D.L. Hanson, J.R. Asay, Time-resolved optical spectroscopy measurements of shocked liquid deuterium. Phys. Rev. B **78**, 144107 (2008)

L.M. Barker, Measurement of free surface motion by the slanted resistor technology, SC-DR-610078 (Sandia National Laboratories, Albuquerque, NM, 1961)

L.M. Barker, Determination of shock wave and particle velocities from slanted resistor data, SC004611 (RR) (Sandia National Laboratories, Albuquerque, NM, 1962)

L.M. Barker, R.E. Hollenbach, System for measuring the dynamic properties of materials. Rev. Sci. Instrum. **35**, 742–746 (1964)

L.M. Barker, C.D. Lundergan, W. Herrmann, Dynamic response of aluminum. J. Appl. Phys. **35**(4), 1203–1212 (1964)

L.M. Barker, R.E. Hollenbach, Interferometer technique for measuring the dynamic mechanical properties of materials. Rev. Sci. Instrum. **36**(11), 1617–1620 (1965)

L.M. Barker, B.M. Butcher, C.H. Karnes, Yield point phenomenon in impact-loaded 1060 aluminum. J. Appl. Phys. **37**(5), 1989–1991 (1966)

L.M. Barker, Fine structure of compressive and release wave shapes in aluminum measured by the velocity interferometer technique, in *Behavior of Dense Media Under High Dynamic Pressures* (Proceedings of IUTAM Symposium), ed. by J. Berger (Gordon and Breach, New York, NY, 1968), pp. 483–504

L.M. Barker, R.E. Hollenbach, Shock wave studies of PMMA, fused silica, and sapphire. J. Appl. Phys. **41**(10), 4208–4226 (1970)

L.M. Barker, Velocity interferometer data reduction. Rev. Sci. Instrum. **42**(2), 276–278 (1971a)

L.M. Barker, A model for stress wave propagation in composite materials. J. Compos. Mater. **5**(2), 140–162 (1971b)

L.M. Barker, R.E. Hollenbach, A laser interferometer for measuring high velocities of any reflecting surface. J. Appl. Phys. **43**(11), 4669–4675 (1972)

L.M. Barker, VISAR data reduction, SLA-73-1038 (Sandia National Laboratories, Albuquerque, NM, 1974)

L.M. Barker, R.E. Hollenbach, Shock wave study of the $\alpha$-$\varepsilon$ phase transition in iron. J. Appl. Phys. **45**(11), 4872–4887 (1974)

L.M. Barker, K.W. Schuler, Correction to the velocity-per-fringe relationship for the VISAR interferometer. J. Appl. Phys. **45**(8), 3692–3693 (1974)

L.M. Barker, E.G. Young, SWAP-9: An improved stress wave analyzing program, SLA-74-0009 (Sandia National Laboratories, Albuquerque, NM, 1974) [This version supersedes an earlier report by Barker dated 1969]

L.M. Barker, P.J. Chen, W.A. Sebrell, Determination of the conditions of spallation in an impulsively loaded quartz phenolic-beryllium composite ring, SLA-74-0245 (Sandia National Laboratories, Albuquerque, NM, 1974a)

L.M. Barker, C.D. Lundergan, P.J. Chen, M.E. Gurtin, Nonlinear viscoelasticity and the evolution of stress waves in laminated composites: a comparison of theory and experiment. J. Appl. Mech. **41**, 1025–1030 (1974b)

L.M. Barker, α-phase Hugoniot of iron. J. Appl. Phys. **46**(6), 2544–2547 (1975)

L.M. Barker, High-pressure quasi-isentropic impact experiments, in *Shock Compression of Condensed Matter*, ed. by J.R. Asay, R.A. Graham, G.K. Straub (Elsevier, Amsterdam, 1984), pp. 217–223

L.M. Barker, T.G. Trucano, J.L. Wise, J.R. Asay, Experimental technique for measuring the isentrope of hydrogen to several megabars, in *Shock Waves in Condensed Matter*, ed. by Y.M. Gupta (Plenum, New York, NY, 1986), pp. 455–459

L.M. Barker, T.G. Trucano, J.W. Munford, Metal surface gouging by hypervelocity sliding contact, in *Shock Waves in Condensed Matter*, ed. by S.C. Schmidt, N.C. Holmes (Elsevier, Amsterdam, 1988), pp. 753–756

L.M. Barker, T.G. Trucano, A.R. Susoeff, Railgun rail gouging by hypervelocity sliding contact. IEEE Trans. Magnet **25**(1), 83–87 (1989)

L.M. Barker, L.C. Chhabildas, Gas-accelerated plate stability study, in *Shock Waves in Condensed Matter*, ed. by S.C. Schmidt, J.N. Johnson, L.W. Davison (Elsevier, Amsterdam, 1990), pp. 989–991

L.M. Barker, L.C. Chhabildas, T.G. Trucano, J.R. Asay, High gas pressure acceleration of flyer plates: Experimental techniques. Int. J. Impact Eng. **10**, 67–80 (1990)

L.M. Barker, The development of the VISAR, and its use in shock compression science, in *Shock Compression of Condensed Matter*, ed. by M.D. Furnish, L.C. Chhabildas, R.S. Hixson. AIP Conference Proceedings, vol. 505 (AIP, College Park, MD, 2000a), pp. 11–17

L.M. Barker, Multi-beam VISARs for simultaneous velocity vs. time measurements, in *Shock Compression of Condensed Matter*, ed. by M.D. Furnish, L.C. Chhabildas, R.S. Hixson. AIP Conference Proceedings, vol. 505 (AIP, College Park, MD, 2000b), pp. 999–1002

N.R. Barton, J.V. Bernier, R. Becker, A. Arsenlis, R. Cavallo, J. Marian, M. Rhee, H.-S. Park, B.A. Remington, R.T. Olson, A multiscale strength model for extreme loading conditions. J. Appl. Phys. **109**, 073501 (2011)

R.C. Bass, B.C. Benjamin, H.M. Miller, D.R. Breding, SLIFER measurement for explosive yield, SAND76-0007 (Sandia National Laboratories, Albuquerque, NM, 1976)

F. Bauer, R.A. Graham, M.U. Anderson, H. Lefebvre, L.M. Lee, R.P. Reed, Response of the piezoelectric polymer PVDF to shock compression greater than 10 GPa, in *Shock Compression of Condensed Matter*, ed. by S.C. Schmidt, R.D. Dick, J.W. Forbes, D.G. Tasker (Elsevier, Amsterdam, 1992), pp. 887–890

K. Baumung, J. Singer, S.V. Razorenov, A.V. Utkin, Hydrodynamic proton beam-target interaction experiments using an improved line-imaging velocimeter, in *Shock Compression of Condensed Matter*, ed. by S.C. Schmidt, W.C. Tao. AIP Conference Proceedings, vol. 370 (AIP, College Park, MD, 1996), pp. 1015–1018

R. Becker, Effects of crystal plasticity on materials loaded at high pressures and strain rates. Int. J. Plas. **20**, 1983–2006 (2004)

L.A. Behrmann, G. Dunbar, W. Wesloh, L.W. Davison, Reverse engineering of shaped charges, in *5th International Symposium on Ballistics* (Societe des Amis de l'ENSAE et de l'ENSTA Service Technique des Poudres et Explosifs, Toulouse, 1980), pp. 1–6

W.B. Benedick, Air guns and the use of air propelled projectiles. Sandia National Laboratories Technical Memorandum SCTM560079-51 (Sandia National Laboratories, Albuquerque, NM, 1956)

W.B. Benedick, Nitroguanidine explosive plane-wave generator for producing low amplitude shock waves. Rev. Sci. Instrum. **36**(9), 1309–1315 (1965)

W.B. Benedick, J.D. Kennedy, B. Morosin, Detonation limits of unconfined hydrocarbon-air mixtures. Combust. Flame **15**, 83–84 (1970)

W.B. Benedick, Detonation wave shaping, in *Behavior and Utilization of Explosives in Engineering Design* (Proceedings 12th Annual Symposium New Mexico Section of the American Society of Mechanical Engineers), ed. by L.W. Davison, J. Kennedy, F. Coffey (NM Section ASME, Albuquerque, NM, 1972), pp. 47–56

R.A. Benham, Light-initiated explosive for impulse measurements on structural members, SAND75-0516 (Sandia National Laboratories, Albuquerque, NM, 1976)

L.D. Bertholf, L.D. Buxton, B.J. Thorne, R.K. Byers, A.L. Stevens, S.L. Thompson, Damage in steel plates from hypervelocity impact. II. Numerical results and spall measurement. J. Appl. Phys. **46**, 3776–3783 (1975)

L.D. Bertholf, M.E. Kipp, Two-dimensional stress wave calculations of kinetic energy projectile impact on multi-layered targets, SAND76-9247 (Sandia National Laboratories, Albuquerque, NM, 1977)

G.C. Bessette, R.J. Lawrence, L.C. Chhabildas, W.D. Reinhart, T.F. Thornhill, W.V. Saul, Multi-dimensional hydrocode analysis of penetrating hypervelocity impacts, in *Shock Compression of Condensed Matter*, ed. by M.D. Furnish, Y.M. Gupta, J.W. Forbes. AIP Conference Proceedings, vol. 706 (AIP, College Park, MD, 2004), pp. 1323–1326

D.D. Bloomquist, S.A. Sheffield, Optically recording interferometer for velocity measurements with subnanosecond resolution. J. Appl. Phys. **54**, 1717–1722 (1983a)

D.D. Bloomquist, S.A. Sheffield, ORVIS, optically recording velocity interferometer system theory of operation and data reduction techniques, SAND82-2918 (Sandia National Laboratories, Albuquerque, NM, 1983b)

R.R. Boade, M.E. Kipp, D.E. Grady, A blasting concept for preparing vertical modified in situ oil shale retorts, SAND81-1255 (Sandia National Laboratories, Albuquerque, NM, 1981)

P.B. Bochev, C.J. Garasi, J.J. Hu, A.C. Robinson, R.S. Tuminaro, An improved algebraic multigrid method for solving Maxwell's equations. SIAM J. Sci. Comp. **25**(2), 623–642 (2003a)

P.B. Bochev, J.J. Hu, A.C. Robinson, R.S. Tuminaro, Towards robust 3D z-pinch simulations: Discretization and fast solvers for magnetic diffusion in heterogeneous conductors. Electron Trans. Num. Anal. **15**, 186–210 (2003b)

J.P. Borg, T.J. Vogler, Mesoscale calculations of the dynamic behavior of a granular ceramic. Int. J. Solids Struct. **45**, 1676–1696 (2008)

J.P. Borg, T.J. Vogler, Aspects of simulating the dynamic compaction of a granular ceramic. Model Simul. Mater. Sci. Eng. **17**, 045003 (2009a)

J.P. Borg, T.J. Vogler, The effect of water content on the shock compaction of sand, in *DYMAT 2009 – 9th International Conferences on the Mechanical and Physical Behavior of Materials under Dynamic Loading*, ed. by S. Hiermaier (EDP Sciences, Brussels, 2009b), pp. 1545–1552

J.P. Borg, T.J. Vogler, A. Fraser, A review of mesoscale simulations of granular materials, in *Shock Compression of Condensed Matter*, ed. by M.L. Elbert, W.T. Buttler, M.D. Furnish, W.W. Anderson, W.G. Proud. AIP Conference Proceedings, vol. 1195 (AIP, College Park, MD, 2009), pp. 1331–1334

J.P. Borg, T.J. Vogler, Rapid compaction of granular material: characterizing two- and three-dimensional mesoscale simulations. Shock Waves **23**, 153–176 (2013)

M.B. Boslough, R.A. Graham, Submicrosecond shock-induced chemical reactions in solids: First real-time observations. Chem. Phys. Lett. **121**, 446–452 (1985)

M.B. Boslough, E.L. Venturini, B. Morosin, R.A. Graham, D.L. Williamson, Physical properties of shocked and thermally altered nontronite: Implications for the Martian surface. J. Geophys. Res. **91**, E207–E214 (1986a)

M.B. Boslough, R.A. Graham, D.M. Webb, Optical measurements of shock-induced chemical reactions in mixed aluminum-nickel powder, in *Shock Waves in Condensed Matter*, ed. by Y.M. Gupta (Plenum, New York, NY, 1986b), pp. 767–772

M.B. Boslough, Shock-induced chemical reactions in nickel-aluminum powder mixtures: Radiation pyrometer measurements. Chem. Phys. Lett. **160**, 618–622 (1989)

M.B. Boslough, Shock modification and chemistry and planetary geologic processes. Annual Rev. Earth Planet Sci. **19**, 101–130 (1991)

M.B. Boslough, Thermochemistry of shock-induced exothermic reactions in selected porous mixtures, in *Proceedings of Explomet 1990 International Conference on Shock-Wave and High-Strain-Rate Phenomena in Materials*, ed. by M.A. Meyers, L.E. Murr, K.P. Staudhammer (Marcel Dekker, New York, NY, 1992), pp. 253–260

M.B. Boslough, J.R. Asay, Basic principles of shock compression (Chapter 2), in *High-Pressure Shock Compression of Solids*, ed. by J.R. Asay, M. Shahinpoor (Springer, New York, NY, 1993), pp. 7–42

M.B. Boslough, J.A. Ang, L.C. Chhabildas, W.D. Reinhart, C.A. Hall, B.G. Cour-Palais, E.L. Christiansen, J.L. Crews, Hypervelocity testing of advanced shielding concepts for spacecraft against impacts to 10 km/s. Int. J. Impact Eng. **14**, 95–106 (1993)

M.B. Boslough, L.C. Chhabildas, W.D. Reinhart, C.A. Hall, J.M. Miller, R. Hickman, S.A. Mullin, D.L. Littlefield, PVDF gauge characterization of hypervelocity—Impact-generated debris clouds, in *High-Pressure Science and Technology*, ed. by S.C. Schmidt, J.W. Shaner, G.A. Samara, M. Ross. AIP Conference Proceedings, vol. 309 (AIP, College Park, MD, 1994a), pp. 1833–1836

M.B. Boslough, D.A. Crawford, A.C. Robinson, T.G. Trucano, Watching for fireballs on Jupiter. Eos Trans. Am Geophys. Union **27**, 305–310 (1994b)

M. Boslough, E. Chael, T.G. Trucano, D.A. Crawford, Axial focusing of energy from a hypervelocity impact on Earth. Int. J. Impact. Eng. **17**, 99–108 (1995a)

M.B. Boslough, D.A. Crawford, T.G. Trucano, A.C. Robinson, Numerical modeling of Shoemaker-Levy 9 impacts as a framework for interpreting observations. Geophys. Res. Lett. **22**(13), 1821–1824 (1995b)

M.B. Boslough, D.A. Crawford, Impact-generated atmospheric plumes: observations on Jupiter and implications for Earth, in *Shock Compression of Condensed Matter*, ed. by S.C. Schmidt, W.C. Tao. AIP Conference Proceedings, vol. 370 (AIP, College Park, MD, 1996), pp. 1187–1190

M.B. Boslough, D.A. Crawford, Shoemaker-Levy 9 and plume forming collisions on Earth near-Earth objects, in *Annals of the New York Academy of Sciences*, ed. by J.L. Remo, vol. 822 (The New York Academy of Sciences, New York, NY, 1997), pp. 236–282

M.B. Boslough, D.A. Crawford, Low-altitude airbursts and the impact threat. Int. J. Impact. Eng. **35**, 1441–1448 (2008)

R.M. Bowen, P.J. Chen, Acceleration waves in chemically reacting ideal fluid mixtures. Arch. Ration. Mech. Anal. **47**, 171–187 (1972a)

R.M. Bowen, P.J. Chen, Acceleration waves in anisotropic thermoelastic materials with internal state variables. Acta Mech. **15**, 95–104 (1972b)

R.M. Bowen, P.J. Chen, J.W. Nunziato, Shock waves in a mixture of chemically reacting materials with memory. Acta Mech. **21**, 1–11 (1975)

R.M. Brannon, L.C. Chhabildas, Experimental and numerical investigation of shock-induced full vaporization of zinc. Int. J. Impact Eng. **17**, 109–120 (1995)

R.M. Brannon, S.T. Montgomery, J.B. Aidun, A.C. Robinson, Macro- and meso-scale modeling of PZT ferroelectric ceramics, in *Shock Compression of Condensed Matter*, ed. by M.D. Furnish, N.N. Thadhani, Y.M. Horie. AIP Conference Proceedings, vol. 620 (AIP, College Park, MD, 2002), pp. 197–200

N.S. Brar, Z. Rosenberg, S.J. Bless, Applying Steinberg's model to the Hugoniot elastic limit of porous boron carbide specimens, in *Shock Compression of Condensed Matter*, ed. by S.C. Schmidt, R.D. Dick, J.W. Forbes (Elsevier, Amsterdam, 1992), pp. 467–470

J.L. Brown, G. Ravichandran, W.D. Reinhart, W.M. Trott, High-pressure Hugoniot measurements using converging shocks. J. Appl. Phys. **109**, 093520 (2011)

J.L. Brown, C.S. Alexander, J.R. Asay, T.J. Vogler, J.L. Ding, Extracting strength from high pressure ramp-release experiments. J. Appl. Phys. **114**, 223518 (2013)

J.L. Brown, C.S. Alexander, J.R. Asay, T.J. Vogler, D.H. Dolan, J.L. Belof, Flow strength of tantalum under ramp compression to 250 GPa. J. Appl. Phys. **115**, 043530 (2014a)

J.L. Brown, M.D. Knudson, C.S. Alexander, J.R. Asay, Shockless compression and release behavior of beryllium to 110 GPa. J. Appl. Phys. **116**, 033502 (2014b)

P.G. Brown, J.D. Assink, L. Astiz et al., A 500-kiloton airburst over Chelyabinsk and an enhanced hazard from small impactors. Nature **503**, 238–241 (2013)

W.K. Brown, R.R. Karpp, D.E. Grady, Fragmentation of the universe. Astrophys. Space Sci. **94**, 401–412 (1983)

W.T. Brown, P.J. Chen, On the nature of the electric field and the resulting voltage in axially loaded ferroelectric ceramics. J. Appl. Phys. **49**(6), 3446–3450 (1978)

T.A. Brunner, Forms of approximate radiation transport, in *Nuclear Mathematical and Computational Sciences: A Century in Review, a Century Anew* (American Nuclear Society, La Grange Park, IL, 2003)

T.A. Brunner, C.J. Garasi, T.A. Haill, T.A. Mehlhorn, K. Cochrane, A.C. Robinson, R.M. Summers, ALEGRA-HEDP: Version 46, SAND2005-5996 (Sandia National Laboratories, Albuquerque, NM, 2005)

K.G. Budge, J.S. Peery, RHALE: A MMALE shock physics code written in C++. Int. J. Impact Eng. **14**, 107–120 (1993)

K.G. Budge, Verification of the radiation package in ALEGRA, SAND99-0786 (Sandia National Laboratories, Albuquerque, NM, 1999)

B.M. Butcher, L.M. Barker, D.E. Munson, C.D. Lundergan, Influence of stress history on time-dependent spall in metals. AIAA J. **2**, 977–990 (1964)

B.M. Butcher, J.R. Cannon, Influence of work-hardening on the dynamic stress strain curves of 4340 steel. AIAA J. **2**, 2174–2179 (1964)

B.M. Butcher, D.E. Munson, Influence of mechanical properties on wave propagation in elastic-plastic materials, in *Proceedings of the 4th International Detonation Symposium*, ONR ACR-126, ed. by S.J. Jacobs, D. Price (U.S. Naval Ordinance Laboratory, Silver Spring, MD, 1965)

B.M. Butcher, Computer program SRATE for the study of strain-rate sensitive stress wave propagation—Part I, SC-RR-650298 (Sandia National Laboratories, Albuquerque, NM, 1966)

B.M. Butcher, C.H. Karnes, Strain rate effects in metals. J. Appl. Phys. **37**, 402–411 (1966)

B.M. Butcher, Spallation in 4340 steel. J Appl. Mech. Ser. E **89**(1), 209–210 (1967)

B.M. Butcher, D.E. Munson, The application of dislocation dynamics to impact-induced deformation under uniaxial strain, in *Dislocation Dynamics*, ed. by A.R. Rosenfield, G.T. Hahn, A.L. Bement, R.J. Jaffe (McGraw Hill, New York, NY, 1967), pp. 591–607

B.M. Butcher, Spallation in 6061-T6 aluminum, in *Behavior of Dense Media under High Dynamic Pressures* (Proceedings of IUTAM Symposium), ed. by J. Berger (Gordon and Breach, New York, NY, 1968), pp. 245–250

B.M. Butcher, C.H. Karnes, Dynamic compaction of porous iron. J. Appl. Phys. **40**(7), 2967–2976 (1969)

B.M. Butcher, The description of strain-rate effects in shocked porous materials, in *Shock Waves and the Mechanical Properties of Solids*, ed. by J.J. Burke, V. Weiss (Syracuse University Press, Syracuse, NY, 1971), pp. 227–243

B.M. Butcher, Dynamic response of partially compacted porous aluminum during unloading. J. Appl. Phys. **44**, 4576–4582 (1973)

B.M. Butcher, L.A. Kent, L.M. Lee, A method for measuring unloading paths in partially compacted strain-rate insensitive porous materials, SLA-73-0152 (Sandia National Laboratories, Albuquerque, NM, 1973)

B.M. Butcher, M.M. Carroll, A.C. Holt, Shock wave compaction of porous aluminum. J. Appl. Phys. **45**, 3864–3875 (1974)

B.M. Butcher, A.L. Stevens, The shock wave response of Window Rock coal. Int. J. Rock Mech. Mining Sci. & Geomech. Abstr. **12**, 147–155 (1975)

R.K. Byers, A.J. Chabai, Penetration calculations and measurements for a layered soil target. Int. J. Num. Anal. Meth. Geomech. **1**, 107–138 (1977)

R.K. Byers, P. Yarrington, A.J. Chabai, Dynamic penetration of soil media by slender projectiles, in *International Journal of Engineering Science: Penetration Mechanics (Special Issue)*, ed. by A.C. Eringen, vol. 16 (Pergamon, Oxford, 1978), pp. 835–844

M.J. Carr, R.A. Graham, The effect of shock pressure and temperature on the deformation microstructure of rutile, in *Metallurgical Applications of Shock-Wave and High-Strain-Rate Phenomena*, ed. by L.E. Murr, K.P. Staudhammer, M. Meyers (Marcel Dekker, New York, NY, 1986), pp. 369–384

M.J. Carr, C.R. Hills, R.A. Graham, J.L. Wise, The effects of microstructure on the substructure evolution and mechanical properties of shock-loaded 6061-T6 aluminum and JBK-75 stainless steel, in *Shock Waves in Condensed Matter*, ed. by S.C. Schmidt, N.C. Holmes (Elsevier, Amsterdam, 1988), pp. 335–338

D. Carroll, E. Hertel, T.G. Trucano, Simulation of armor penetration by tungsten rods: ALEGRA validation report, SAND97-2765 (Sandia National Laboratories, Albuquerque, NM, 1997)

W.H. Casey, M.J. Carr, R.A. Graham, Crystal defects and the dissolution kinetics of shocked rutile, in *Shock Waves in Condensed Matter*, ed. by S.C. Schmidt, N.C. Holmes (Elsevier, Amsterdam, 1988), pp. 331–334

A.J. Chabai, Crater scaling laws for desert alluvium, SC-4391 (RR) (Sandia National Laboratories, Albuquerque, NM, 1959)

A.J. Chabai, D.M. Hankins, Gravity scaling laws for explosion craters, SC-4541 (RR) (Sandia National Laboratories, Albuquerque, NM, 1960)

A.J. Chabai, On scaling dimensions of craters produced by buried explosions. J. Geophys. Res. **70**, 5075–5098 (1965)

A.J. Chabai, R.J. Lawrence, E.G. Young, Elastic-plastic target deformation due to a high speed pulsed water jet impact, SLA-74-5227 (Sandia National Laboratories, Albuquerque, NM, 1974)

A.J. Chabai, C.W. Young, P. Yarrington, W.J. Patterson, R.K. Byers, Terradynamic technology – theory and experiment, in *Recent Advances in Engineering Science*, ed. by G.C. Sih (Lehigh University Publication, Bethlehem, PA, 1977), pp. 67–80

A.R. Champion, W.B. Benedick, Detection of strong shock waves with plastic tapes. Rev. Sci. Instrum. **39**(3), 377–378 (1968)

A.R. Champion, R.W. Rohde, Hugoniot equation of state and the effect of shock stress amplitude and duration on the hardness of Hadfield steel. J. Appl. Phys. **41**(5), 2213–2223 (1970)

S. Chantrenne, J.L. Wise, J.R. Asay, M.E. Kipp, C.A. Hall, Design of a sample recovery assembly for magnetic ramp-wave loading, in *Shock Compression of Condensed Matter*, ed. by M.L. Elbert, W.T. Buttler, M.D. Furnish, W.W. Anderson, W.G. Proud, AIP Conference Proceedings, vol. 1195 (AIP, College Park, MD, 2009), pp. 695–698

P.J. Chen, Growth of acceleration waves in isotropic elastic materials. J. Acoust. Soc. Am. **43**(5), 982–987 (1968a)

P.J. Chen, Thermodynamic influences on the propagation and the growth of acceleration waves in elastic materials. Arch. Ration. Mech. Anal. **31**(3), 228–254 (1968b)

P.J. Chen, M.E. Gurtin, E.K. Walsh, Shock amplitude variation in polymethyl methacrylate for fixed values of the strain gradient. J. Appl. Phys. **41**(8), 3557–3558 (1970)

P.J. Chen, The growth of one-dimensional shock waves in elastic nonconductors. Int. J. Solids Struct. **7**, 5–10 (1971)

P.J. Chen, M.E. Gurtin, Growth and decay of one-dimensional shock waves in fluids with internal state variables. Phys. Fluids **14**(6), 1091–1094 (1971)

P.J. Chen, M.E. Gurtin, On the use of experimental results concerning steady shock waves to predict the acceleration wave response of nonlinear viscoelastic materials. J. Appl. Mech. **39**(1), 295–296 (1972)

P.J. Chen, R.A. Graham, L.W. Davison, Analysis of unsteady waves in solids. J. Appl. Phys. **43**(12), 5021–5027 (1972)

P.J. Chen, Growth and decay of waves in solids, in *Handbuch der Physik*, Band **VIa** (3), ed. by S. Flugge (Springer, Berlin, 1973), pp. 303–402

P.J. Chen, J.W. Nunziato, On wave propagation in perfectly heat conducting inextensible elastic. J. Elast. **5**, 155–160 (1975)

P.J. Chen, J.E. Kennedy, Chemical kinetic and curvature effects on shock-wave evolution in shocked explosives, in *Proceedings of the 6th International Detonation Symposium*, ONR ACR-221, ed. by S.J. Jacobs, D.J. Edwards (Office of Naval Research, Washington, D.C., 1976), pp. 379–388

P.J. Chen, P.C. Lysne, H.J. Sutherland, Electrical responses of ferroelectric ceramics to dynamic loads of uniaxial strain, in *Propagation of Shock Waves in Solids* (The American Society of Mechanical Engineers, New York, NY, 1976a), pp. 73–78

P.J. Chen, L.W. Davison, M.F. McCarthy, Electrical responses of nonlinear piezoelectric materials to plane waves of uniaxial strain. J. Appl. Phys. **47**(11), 4759–4764 (1976b)

P.J. Chen, S.T. Montgomery, Normal mode responses of linear piezoelectric materials with hexagonal symmetry. Int. J. Solids Struct. **13**, 947–955 (1977)

P.J. Chen, S.T. Montgomery, Boundary effects on the normal-mode responses of linear transversely isotropic piezoelectric materials. J. Appl. Phys. **49**(2), 900–904 (1978)

P.J. Chen, M.F. McCarthy, T.R. O'Leary, One-dimensional shock and acceleration waves in deformable dielectric materials with memory. Arch. Ration. Mech. Anal. **62**(2), 189–207 (1978)

P.J. Chen, S.T. Montgomery, A macroscopic theory for the existence of the hysteresis and butterfly loops in ferroelectricity. Ferroelectrics **23**(1), 199–207 (1980)

P.J. Chen, T.J. Tucker, One dimensional polar mechanical and dielectric responses of the ferroelectric ceramic PZT 65/35 due to domain switching. Int. J. Eng. Sci. **19**, 147–158 (1981)

L.C. Chhabildas, H.M. Gilder, Thermal coefficient of expansion of an activated vacancy in zinc from high pressure self-diffusion experiments. Phys. Rev. B **5**, 2135–2144 (1972)

L.C. Chhabildas, A.L. Ruoff, The transition of sulfur to a conducting phase. J. Chem. Phys. **66**(3), 983–985 (1977)

L.C. Chhabildas, J.R. Asay, Rise-time measurements of shock transitions in aluminum, copper, steel. J. Appl. Phys. **50**(4), 2749–2756 (1979)

L.C. Chhabildas, H.J. Sutherland, J.R. Asay, A velocity interferometer technique to determine shear-wave particle velocity in shock-loaded solids. J. Appl. Phys. **50**(8), 5196–5201 (1979)

L.C. Chhabildas, J.W. Swegle, Dynamic pressure-shear loading of materials using anisotropic crystals. J. Appl. Phys. **51**(9), 4799–4807 (1980)

L.C. Chhabildas, J.R. Asay, Time-resolved wave profile measurements in copper to megabar pressures, in *High Pressure in Research and Industry* (Proceedings of 6th AIRAPT and 19th EHPRG International Conference), ed. by C.-M. Backman, T. Johannisson, L. Tegnér (Arkitektkopia, Uppsala, 1982), pp. 183–189

L.C. Chhabildas, R.D. Hardy, Pressure-shear loading techniques for material-property studies, SAND82-1546 (Sandia National Laboratories, Albuquerque, NM, 1982)

L.C. Chhabildas, J.W. Swegle, On the dynamical response of particulate-loaded materials I. Pressure-shear loading of alumina particles in an epoxy matrix. J. Appl. Phys. **53**(2), 954–956 (1982)

L.C. Chhabildas, J.L. Wise, J.R. Asay, Reshock and release behavior of beryllium, in *1981 Topical Conference on Shock Waves in Condensed Matter*, ed. by W.J. Nellis, L. Seaman, R.A. Graham. AIP Conference Proceedings, vol. 78 (AIP, College Park, MD, 1982), pp. 422–426

L.C. Chhabildas, Dynamic transverse particle velocity measurements using interferometric techniques, in *Proceedings of the SPIE*, vol. 427 (International Society for Optics and Photonics, Bellingham, WA, 1983), pp. 136–143

L.C. Chhabildas, D.E. Grady, Dynamic material response of quartz at high strain rates, in *High Pressure Science and Technology*, (Proceedings of the 9th AIRAPT International High Pressure Conference), vol 3, ed. by C. Homan, R.K. MacCrone, E. Whalley (AIRAPT, Albany, NY, 1984), pp. 147–150

L.C. Chhabildas, M.E. Kipp, Pressure-shear loading of PBX-9404, in *Proceedings of the 8th International Detonation Symposium*, NSWC MP 86-194, ed. by J.M. Short, W.E. Deal (Naval Surface Warfare Center, Dahlgren, VA, 1985), pp. 274–283

L.C. Chhabildas, J.M. Miller, Release-adiabat measurements in crystalline quartz, SAND85-1092 (Sandia National Laboratories, Albuquerque, NM, 1985)

L.C. Chhabildas, L.M. Barker, Dynamic quasi-isentropic compression of tungsten, in *Shock Waves in Condensed Matter*, ed. by S.C. Schmidt, N.C. Holmes (Elsevier, Amsterdam, 1988), pp. 111–114

L.C. Chhabildas, J.R. Asay, L.M. Barker, Shear strength of tungsten under shock- and quasi-isentropic loading to 250 GPa, SAND88-0306 (Sandia National Laboratories, Albuquerque, NM, 1988)

L.C. Chhabildas, L.M. Barker, J.R. Asay, T.G. Trucano, Relationship of fragment size to normalized spall strength for materials. Int. J. Impact Eng. **10**, 107–124 (1990)

L.C. Chhabildas, L.M. Barker, J.R. Asay, T.G. Trucano, G.I. Kerley, Sandia's new hypervelocity launcher, HVL, SAND91-0657 (Sandia National Laboratories, Albuquerque, NM, 1991)

L.C. Chhabildas, J.R. Asay, Dynamic yield strength and spall strength measurements under quasi-isentropic loading, in *Shock-Wave and High-Strain-Rate Phenomena in Materials*, ed. by M.A. Meyers et al. (Marcel Dekker, New York, NY, 1992), pp. 947–955

L.C. Chhabildas, L.M. Barker, J.R. Asay, T.G. Trucano, G.I. Kerley, J.E. Dunn, Launch capabilities to over 10 km/s, in *Shock Compression of Condensed Matter*, ed. by S.C. Schmidt, R.D. Dick, J.W. Forbes, D.G. Tasker (Elsevier, Amsterdam, 1992), pp. 1025–1031

L.C. Chhabildas, E.S. Hertel, S.A. Hill, Hypervelocity impact tests and simulations of single Whipple bumper shield concepts at 10 km/s. Int. J. Impact Eng. **14**, 133–144 (1993a)

L.C. Chhabildas, J.E. Dunn, W.D. Reinhart, J.M. Miller, An impact technique to accelerate flyer plates to velocities over 12 km/s. Int. J. Impact Eng. **14**, 121–132 (1993b)

L.C. Chhabildas, T.G. Trucano, W.D. Reinhart, C.A. Hall, Chunk projectile launch using the Sandia hypervelocity launcher facility, SAND94-1273 (Sandia National Laboratories, Albuquerque, NM, 1994)

L.C. Chhabildas, L.N. Kmetyk, W.D. Reinhart, C.A. Hall, Enhanced hypervelocity launcher: Capabilities to 16 km/s. Int. J. Impact Eng. **17**, 183–191 (1995)

L.C. Chhabildas, M.D. Furnish, D.E. Grady, Impact of alumina rods – A computational and experimental study. J. Phys. IV (Colloque) **7**(C3), 137–143 (1997)

L.C. Chhabildas, M.D. Furnish, W.D. Reinhart, D.E. Grady, Impact of AD995 alumina rods, in *Shock Compression of Condensed Matter*, ed. by S.C. Schmidt, D.P. Dandekar, J.W. Forbes. AIP Conference Proceedings, vol. 429 (AIP, College Park, MD, 1998), pp. 505–508

L.C. Chhabildas, W.D. Reinhart, Intermediate strain-rate loading experiments – technique and applications to ceramics, in *Proceedings of the 15th U.S. Army Symposium on Solid Mechanics*, ed. by S.C. Chou, K.S. Iyer (Battelle, Columbus, OH, 1999), pp. 233–240

L.C. Chhabildas, M.D. Furnish, W.D. Reinhart, Shock induced melting in aluminum: Wave profile measurements, in *Shock Compression of Condensed Matter*, ed. by M.D. Furnish, L.C. Chhabildas, R.S. Hixson. AIP Conference Proceedings, vol. 505 (AIP, College Park, MD, 2000a), pp. 97–100

L.C. Chhabildas, T.G. Trucano, R.M. Summers, W.D. Reinhart, J.S. Peery, D.A. Mosher, G.A. Mann, C.H. Konrad, M.E. Kipp, Experimental benchmark data for ALEGRA code validations, in *Shock Compression of Condensed Matter*, ed. by M.D. Furnish, L.C. Chhabildas, R.S. Hixson. AIP Conference Proceedings, vol. 505 (AIP, College Park, MD, 2000b), pp. 1011–1014

L.C. Chhabildas, W.M. Trott, W.D. Reinhart, J.R. Cogar, G.A. Mann, Incipient spall studies in tantalum—Microstructural effects, in *Shock Compression of Condensed Matter*, ed. by M.D. Furnish, N.N. Thadhani, Y. Horie. AIP Conference Proceedings, vol. 620 (AIP, College Park, MD, 2002), pp. 483–486

L.C. Chhabildas, W.D. Reinhart, T.F. Thornhill, G.C. Bessette, W.V. Saul, R.J. Lawrence, M.E. Kipp, Hypervelocity impacts on aluminum from 6 to 11 km/s for hydrocode benchmarking, SAND2003-1235 (Sandia National Laboratories, Albuquerque, NM, 2003)

L.C. Chhabildas, M.D. Knudson, Techniques to launch projectile plates to high velocities, in *High Pressure Shock Compression of Solids VIII*, ed. by L.C. Chhabildas, L.W. Davison, Y. Horie (Springer, New York, NY, 2005), pp. 143–200

L.C. Chhabildas, W.D. Reinhart, T.F. Thornhill, J.L. Brown, Shock-induced vaporization in metals. Int. J. Impact Eng. **33**(1-12), 158–168 (2006)

J. Clerouin, P. Renaudin, V. Recoules et al., Equation of state and electrical conductivity of strongly correlated aluminum and copper plasmas. Contrib. Plasma Phys. **43**(5-6), 269–272 (2003)

J. Clerouin, P. Renaudin, Y. Laudernet et al., Electrical conductivity and equation of state study of warm dense copper: measurements and quantum molecular dynamics calculations. Phys. Rev. B **71**, 064203 (2005)

K. Cochrane, M. Desjarlais, T. Haill, J. Lawrence, M. Knudson, G. Dunham, Aluminum equation of state validation and verification for the ALEGRA HEDP simulation code, SAND2006-1739 (Sandia National Laboratories, Albuquerque, NM, 2006)

K. Cochrane, T.J. Vogler, M.P. Desjarlais, T.R. Mattsson, Density Functional Theory (DFT) simulations of porous tantalum pentoxide, in *18th American Physical Society Shock Compression in Condensed Matter and 24th International Association for the Advancement of High Pressure Science and Technology Conference. Journal of Physics Conference Series*, ed. by W. Buttler, M. Furlanetto, W. Evans, vol. 500 (IOP Publishing, Bristol, 2014), 032005

G.W. Collins, L.B. Da Silva, P. Celliers, D.M. Gold et al., Measurements of the equation of state of deuterium at the fluid insulator-metal transition. Science **281**, 1178–1181 (1998)

J. Comley, B.R. Maddox, R.E. Rudd et al., Strength of shock-loaded single-crystal tantalum [100] determined using *in situ* broadband x-ray Laue diffraction. Phys. Rev. Lett. **110**, 115501 (2013)

M. Cooper, W. Trott, R. Schmitt, M. Short, S. Jackson, ANFO response to low-stress planar impacts, in *Shock Compression of Condensed Matter*, ed. by M.L. Elert, W.T. Buttler, J.P. Borg, J.L. Jordan, T.J. Vogler. AIP Conference Proceedings, vol. 1426 (AIP, College Park, MD, 2012), pp. 595–598

M.L. Corradini, D.S. Drumheller, Phenomenological Modelling of Steam Explosions, in *Proceedings of ANS/ENS Topical Meeting on Thermal Reactor Safety* (American Nuclear Society, La Grange Park, IL, 1980)

D.A. Crawford, M. Boslough, T.G. Trucano, A.C. Robinson, The impact of comet Shoemaker – Levy 9 on Jupiter. Shock Waves **4**(1), 47–50 (1994)

D.A. Crawford, M.B. Boslough, T.G. Trucano, A.C. Robinson, The impact of periodic comet Shoemaker-Levy 9 on Jupiter. Int. J. Impact Eng. **17**, 253–262 (1995)

J.C. Crowhurst, M.R. Armstrong, B.K. Knight, J.M. Zaug, E.M. Behymer, Invariance of the dissipative action at ultrahigh strain rates above the strong shock threshold. Phys. Rev. Lett. **107**, 104322 (2011)

R.T. Cygan, W.H. Casey, M.B. Boslough, H.R. Westrich, M.J. Carr, J.R. Holdren Jr., Dissolution kinetics of experimentally shocked silicate minerals. Chem. Geol. **78**, 229–244 (1989)

R.T. Cygan, M.B. Boslough, R.J. Kirkpatrick, Experimentally shocked quartz, NMR spectroscopy and shock wave barometry, in *Shock Compression of Condensed Matter*, ed. by S.C. Schmidt et al. (Elsevier, Amsterdam, 1990), pp. 653–656

R.T. Cygan, M.B. Boslough, Analysis of experimentally shocked minerals by NMR spectroscopy, SAND94-0294 (Sandia National Laboratories, Albuquerque, NM, 1994)

L.B. Da Silva, P. Celliers, G.W. Collins, K.S. Budil, N.C. Holmes et al., Absolute equation of state measurements on shocked liquid deuterium up to 200 GPa (2 Mbar). Phys. Rev. Lett. **78**, 483–486 (1997)

D.M. Dattelbaum, S.A. Sheffield, D.B. Stahl, A.M. Dattelbaum, W.M. Trott, Influence of hot spot features in the initiation characteristics of heterogeneous nitromethane, in *Proceedings of the 14th International Symposium on Detonation*, ONR 351-10-185, ed. by S. Peiris, C. Boswell, B. Asay (Office of Naval Research, Washington, D.C., 2010), pp. 611–621

J.-P. Davis, D.B. Hayes, J.R. Asay, P.W. Watts, P.A. Flores, D.B. Reisman, Investigation of liquid-solid phase transition using Isentropic Compression Experiments (ICE), in *Shock Compression of Condensed Matter*, ed. by M.D. Furnish, N.N. Thadhani, Y. Horie. AIP Conference Proceedings, vol. 620 (AIP, College Park, MD, 2002), pp. 221–224

J.-P. Davis, User manual for INVICE 0.1-beta: A computer code for inverse analysis of isentropic compression experiments, SAND2005–2068 (Sandia National Laboratories, Albuquerque, NM, 2005)

J.-P. Davis, S. Foiles, Experimental and computational study of the liquid–solid transition in tin, SAND2005-6522 (Sandia National Laboratories, Albuquerque, NM, 2005)

J.-P. Davis, C. Deeney et al., Magnetically driven isentropic compression to multi-megabar pressures using shaped current pulses on the Z accelerator. Phys. Plas. **12**, 056310 (2005)

J.-P. Davis, Experimental measurement of the principal isentrope for aluminum 6061-T6 to 240 GPa. J. Appl. Phys. **99**(10), 103512 (2006)

J.-P. Davis, D.B. Hayes, Measurement of the dynamic β-γ phase boundary in tin, in *Shock Compression of Condensed Matter*, ed. by M. Elert, M.D. Furnish, R. Chau, N.C. Holmes, and J. Nguyen, AIP Conference Proceedings, vol. 955 (AIP, College Park, MD, 2007), pp. 159–162

J.-P. Davis, CHARICE version 1.1 update, SAND2008-6035 (Sandia National Laboratories, Albuquerque, NM, 2008)

L.W. Davison, Propagation of plane waves of finite amplitude in elastic solids. J. Mech. Phys. Solids **14**, 249–270 (1966)

L.W. Davison, Perturbation theory of nonlinear elastic wave propagation. Int. J. Solids Struct. **4**, 301–322 (1968)

L.W. Davison, J.N. Johnson, Elastoplastic wave propagation and spallation in beryllium: A review, SC-TM-70-634 (Sandia National Laboratories, Albuquerque, NM, 1970)

L.W. Davison, Shock-wave structure in porous solids. J. Appl. Phys. **42**(13), 5503–5512 (1971)

L.W. Davison, A.L. Stevens, Continuum measures of spall damage. J. Appl. Phys. **43**(3), 988–994 (1972)

L.W. Davison, J. Kennedy, F. Coffey (eds.), *Behavior and Utilization of Explosives in Engineering Design* (Proceedings 12th Annual Symposium of New Mexico Section of the American Society of Mechanical Engineers, Albuquerque, NM, 1972)

L.W. Davison, A.L. Stevens, Thermomechanical constitution of spalling elastic bodies. J. Appl. Phys. **44**(2), 668–674 (1973)

L.W. Davison, Explosion containment devices: Design considerations, SAND74-0218 (Sandia National Laboratories, Albuquerque, NM, 1974)

L.W. Davison, A.L. Stevens, M.E. Kipp, Theory of spall damage accumulation in ductile metals. J. Mech. Phys. Solids **25**, 11–28 (1977)

L.W. Davison, M.E. Kipp, Calculation of spall accumulation in ductile materials, in *High Velocity Deformation of Solids*, ed. by K. Kawata, J. Shioiri (Springer-Verlag, Berlin, 1978), pp. 163–175

L.W. Davison, R.A. Graham, Shock compression of solids. Physics Reports **55**(4), 255–359 (1979)

L.W. Davison, Numerical modeling of dynamic material response, in *Shock Waves in Condensed Matter*, ed. by J.R. Asay, R.A. Graham, G.K. Straub (Elsevier, Amsterdam, 1984), pp. 181–186

L.W. Davison, *Fundamentals of Shock Wave Propagation in Solids* (Springer, Berlin, 2008)

M.P. Desjarlais, Practical improvements to the Lee-More conductivity near the metal-insulator transition. Contrib. Plasma Phys. **41**(2-3), 267–270 (2001)

M.P. Desjarlais, J.D. Kress, L.A. Collins, Electrical conductivity for warm, dense aluminum plasmas and liquids. Phys. Rev. E **66**, 025401 (2002)

M.P. Desjarlais, Density-functional calculations of the liquid deuterium Hugoniot, reshock, and reverberation timing. Phys. Rev. B **68**, 064204 (2003)

A. Dewaele, P. Loubeyre, Mechanical properties of tantalum under high pressure. Phys. Rev. B **72**, 134106 (2005)

J. Dietz, D.B. Hayes, Compilation of crater data, SC-RR-650220 (Sandia National Laboratories, Albuquerque, NM, 1965)

G. Dimonte, D.L. Youngs, A. Dimits et al., A comparative study of the turbulent Rayleigh-Taylor instability using high-resolution three-dimensional numerical simulations: the Alpha-Group collaboration. Phys. Fluids **16**(5), 1668–1693 (2004)

B.W. Dodson, M.B. Boslough, Techniques for recovery of shock-loaded samples, in *Shock Compression of Condensed Matter*, ed. by S.C. Schmidt, J.N. Johnson, L.W. Davison (Elsevier, Amsterdam, 1990), pp. 767–769

D.W. Doerfler, M.B. Vigil, The Cielo petascale capability computer: Providing large-scale computing for stockpile stewardship. *Stockpile Stewardship Quarterly*, vol. 3 (2), pp. 3–5 (2013)

D.H. Dolan, M.D. Knudson, C.A. Hall, C. Deeney, A metastable limit for compressed liquid water. Nat. Phys. **3**, 339–342 (2007)

D.H. Dolan, T. Ao, Cubic zirconia as a dynamic compression window. Appl. Phys. Lett. **93**, 021908 (2008)

D.H. Dolan, Accuracy and precision in photonic Doppler velocimetry (PDV). Rev. Sci. Instrum. **81**, 53905 (2010)

D.H. Dolan, C.T. Seagle, T. Ao, Dynamic temperature measurements with embedded optical sensors, SAND2013-8203 (Sandia National Laboratories, Albuquerque, NM, 2013a)

D.H. Dolan, R.W. Lemke, R.D. McBride, M.R. Martin, E. Harding et al., Tracking an imploding cylinder with photonic Doppler velocimetry. Rev. Sci. Instrum. **84**, 055102 (2013b)

D.S. Drumheller, A. Bedford, On a continuum theory for a laminated medium. J. Appl. Mech. **40**, 527–532 (1973)

D.S. Drumheller, C.D. Lundergan, On the behavior of stress waves in composite materials–Part II: Theoretical and experimental studies on the effects of constituent debonding. Int. J. Solids Struct. **11**, 75–87 (1975)

D.S. Drumheller, The theoretical treatment of a porous solid using a mixture theory. Int. J. Solids Struct. **14**, 441–456 (1978)

D.S. Drumheller, A. Bedford, On the mechanics and thermodynamics of fluid mixtures. Arch. Rat. Mech. Anal. **71**, 345–355 (1979)

D.S. Drumheller, A theory for the shock-loaded response of an alumina-filled epoxy mixture, in *Shock Waves in Condensed Matter*, ed. by W.J. Nellis, L. Seaman, R.A. Graham. AIP Conference Proceedings, vol. 78 (AIP, College Park, MD, 1982a), pp. 527–528

D.S. Drumheller, On the dynamical response of particulate-loaded materials: Part II – A theory with application to alumina particles in an epoxy matrix. J. Appl. Phys. **53**, 957–969 (1982b)

D.S. Drumheller, TOM MIX: A computer code for calculating steam explosion phenomena, SAND81-2520 (Sandia National Laboratories, Albuquerque, NM, 1982c)

D.S. Drumheller, M.E. Kipp, A. Bedford, Transient wave propagation in bubbly liquids. J. Fluid Mech. **119**, 347–365 (1982)

D.S. Drumheller, Wavecode constitutive models: Nonhomogeneous mixtures, SAND84-0713 (Sandia National Laboratories, Albuquerque, NM, 1984)

D.S. Drumheller, T.G. Trucano, L.C. Chhabildas, Wavecode constitutive models: Particulate-loaded composites, SAND84-0714 (Sandia National Laboratories, Albuquerque, NM, 1984)

D.S. Drumheller, Hypervelocity impact of mixtures. Int. J. Impact Eng. **5**, 261–268 (1987)

D.S. Drumheller, *Introduction to Wave Propagation in Nonlinear Fluids and Solids* (Cambridge University Press, New York, NY, 1998)

G. Dunham, J.E. Bailey, A. Carlson, P. Lake, M.D. Knudson, Diagnostic methods for time-resolved optical spectroscopy of shocked liquid deuterium. Rev. Sci. Instrum. **75**, 928–935 (2004)

J.E. Dunn, D.E. Grady, Strain rate dependence in steady plastic shock waves, in *Shock Waves in Condensed Matter*, ed. by Y.M. Gupta (Plenum, New York, NY, 1986), pp. 359–364

K.E. Duprey, R.J. Clifton, Pressure shear response of thin tantalum foils, in *Shock Compression of Condensed Matter*, ed. by M.D. Furnish, L.C. Chhabildas, R.S. Hixson. AIP Conference Proceedings, vol. 505 (AIP, College Park, MD, 2000), pp. 447–450

G.E. Duvall, R.A. Graham, Phase transitions under shock-wave loading. Rev. Mod. Phys. **49**(3), 523–579 (1977)

J. Eggert, M. Bastea et al., Ramp wave stress-density measurements of Ta and W, in *Shock Compression of Condensed Matter*, ed. by M. Elert, M.D. Furnish, R. Chau, N.C. Holmes, J. Nguyen. AIP Conference Proceedings, vol. 955 (AIP, College Park, MD, 2007), pp. 1177–1180

J.C. Eichelberger, D.B. Hayes, Magmatic model for the Mount St. Helens blast of May 18, 1980. J. Geophys. Res. **87**(B9), 7727–7738 (1982)

P. Embid, M. Baer, Mathematical analysis of a two-phase continuum mixture theory. Cont. Mech. Thermodynam. **4**, 279–312 (1992)

W.W. Erikson, E.S. Hertel Jr., M.J. Kaneshige, A.M. Renlund, A.C. Ratzel, Energetic materials research at Sandia National Laboratories, SAND2006-0806A (Sandia National Laboratories, Albuquerque, NM, 2006)

A.V. Farnsworth Jr., W.M. Trott, R.E. Setchell, A computational study of laser driven flyer plates, in *Shock Compression of Condensed Matter*, ed. by M.D. Furnish, N.N. Thadhani, Y. Horie. AIP Conference Proceedings, vol. 620 (AIP, College Park, MD, 2002), pp. 1355–1358

G. Fenton, D.E. Grady, T.J. Vogler, Intense shock compression of porous solids: application to WC and $Ta_2O_5$, in *Shock Compression of Condensed Matter*, ed. by M.L. Elert, W.T. Buttler, J.P. Borg, J.L. Jordan, T.J. Vogler. AIP Conference Proceedings, vol. 1426 (AIP, College Park, MD, 2012), pp. 1463–1466

D.J. Fogelson, L.M. Lee, D.W. Gilbert, W.R. Conley, R.A. Graham, R.P. Reed, F. Bauer, Fabrication of standardized piezoelectric polymer shock gauges by the Bauer method, in *Shock Waves in Condensed Matter*, ed. by S.C. Schmidt, N.C. Holmes (Elsevier, Amsterdam, 1988), pp. 615–618

J.W. Forbes, The history of the APS Shock compression of condensed matter Topical Group, in *Shock Compression of Condensed Matter*, ed. by M.D. Furnish, N.N. Thadhani, Y. Horie. AIP Conference Proceedings, vol. 620 (AIP, College Park, MD, 2002), pp. 11–19

J.W. Forbes, *Shock Wave Compression of Condensed Matter: A Primer* (Springer, Berlin, 2012)

M.J. Forrestal, D.E. Grady, K.W. Schuler, An experimental method to estimate the dynamic fracture strength of oil shale in the $10^3$ to $10^4$/s strain rate regime. Int. J. Rock Mech. Min. Sci. **15**, 263–265 (1978)

M.J. Forrestal, T.C. Togami, W.E. Baker, D.J. Frew, Performance evaluation of accelerometers used for penetration experiments. Exp. Mech. **43**(1), 90–96 (2003)

G.R. Fowles, Shock wave compression of hardened and annealed 2024 aluminum. J. Appl. Phys. **32**, 1475–1487 (1961)

A.M. Frank, W.M. Trott, Investigation of thin laser-driven flyer plates using streak imaging and stop motion microphotography, in *Shock Compression of Condensed Matter*, ed. by S.C. Schmidt, W.C. Tao. AIP Conference Proceedings, vol. 370 (AIP, College Park, MD, 1996), pp. 1209–1212

D.E. Fratanduono, T.R. Boehly, M.A. Barrios, D.D. Meyerhofer, J.H. Eggert et al., Refractive index of lithium fluoride ramp compressed to 800 GPa. J. Appl. Phys. **109**, 123521 (2011)

D.A. Fredenburg, T.J. Vogler, N.N. Thadhani, Meso-scale simulation of the shock compression response of equiaxed and needle morphology Al 6061-T6 powders, in *Shock Compression of Condensed Matter*, ed. by M.L. Elert, W.T. Buttler, M.D. Furnish, W.W. Anderson, W.G. Proud. AIP Conference Proceedings, vol. 1195 (AIP, College Park, MD, 2009), pp. 1341–1344

D.A. Fredenburg, N.N. Thadhani, T.J. Vogler, Shock consolidation of nano-crystalline 6061 aluminum. Mater. Sci. Eng. A **39**, 3349–3357 (2010)

J.R. Freeman, J.M. McGlaun, E.C. Cnare, Numerical studies of helical CMF generators, in *Megagauss Physics and Technology*, ed. by P.J. Turchi (Plenum, New York, NY, 1980), pp. 205–218

I.J. Fritz, R.A. Graham, Second-order elastic constants of high-purity vitreous silica. J. Appl. Phys. **45**(9), 4124–4125 (1974)

N.S. Furman, *Sandia National Laboratories: The Postwar Decade* (University of New Mexico Press, Albuquerque, NM, 1990)

M.D. Furnish, W.A. Bassett, Investigation of the mechanism of the olivine-spinel transition in fayalite by synchrotron radiation. J. Geophys. Res. – Solid Earth **88**(B12), 10333–10341 (1983)

M.D. Furnish, J.M. Brown, Shock loading of single crystal olivine in the 100-200 GPa range. J. Geophys. Res. – Solid Earth **91**(B5), 4723–4729 (1986)

M.D. Furnish, L.C. Chhabildas, Dynamic material properties of refractory materials – molybdenum, in *High strain rate behavior of refractory metals and alloys*, ed. by R. Asfahani, E. Chen, A. Crowson (The Minerals Metals & Materials Society, Warrendale, PA, 1992), pp. 229–240

M.D. Furnish, Recent advances in methods for measuring the dynamic response of geological materials to 100 GPa. Int. J. Impact Eng. **14**, 267–277 (1993)

M.D. Furnish, M.B. Boslough, G.T. Gray III, J.L. Remo, Dynamical properties measurements of asteroid, comet and meteorite material applicable to impact modeling and mitigation calculations. Int. J. Impact Eng. **17**, 341–352 (1995)

M.D. Furnish, J.L. Remo, Ice issues, porosity, and snow experiments for dynamic NEO and comet modeling, in *Near-Earth Objects: United Nations International Conference*, vol. 822 (New York Academy Sciences, New York, NY, 1997), pp. 566–582

M.D. Furnish, L.C. Chhabildas, W.D. Reinhart, Time-resolved particle velocity measurements at impact velocities of 10 km/s. Int. J. Impact Eng. **23**(1), 261–270 (1999)

M.D. Furnish, L.C. Chhabildas, R.E. Setchell, S.T. Montgomery, Dynamic electromechanical characterization of axially poled PZT 95/5, in *Shock Compression of Condensed Matter*, ed. by M.D. Furnish, L.C. Chhabildas, R.S. Hixson. AIP Conference Proceedings, vol. 505 (AIP, College Park, MD, 2000), pp. 975–978

M.D. Furnish, J.-P. Davis, M. Knudson, T. Bergstresser, C. Deeney, J.R. Asay, Using the Saturn accelerator for isentropic compression experiments (ICE), SAND2001-3773 (Sandia National Laboratories, Albuquerque, NM, 2001a)

M.D. Furnish, R.J. Lawrence, C.A. Hall, J.R. Asay, D.L. Barker, G.A. Mize, E.A. Marsh, M.A. Bernard, Radiation-driven shock and debris propagation down a partitioned pipe. Int. J. Impact Eng. **26**, 189–200 (2001b)

M.D. Furnish, J. Robbins, W.M. Trott, L.C. Chhabildas, R.J. Lawrence, S.T. Montgomery, Multidimensional validation impact tests on PZT 95/5 and ALOX, in *Shock Compression of Condensed Matter*, ed. by M.D. Furnish, N.N. Thadhani, Y. Horie. AIP Conference Proceedings, vol. 620 (AIP, College Park, MD, 2002), pp. 205–208

M.D. Furnish, M.E. Kipp, W.D. Reinhart, T.J. Vogler, W.W. Anderson, R.S. Hixson, Exploring pulse shaping for Z using graded-density impactors on gas guns (final report for LDRD Project 79879), SAND2005-6210 (Sandia National Laboratories, Albuquerque, NM, 2005)

M.D. Furnish, W.D. Reinhart, W.M. Trott, L.C. Chhabildas, T.J. Vogler, Variability in dynamic properties of tantalum: Spall, Hugoniot elastic limit and attenuation, in *Shock Compression of Condensed Matter*, ed. by M.D. Furnish, M. Elert, T.P. Russell, C.T. White. AIP Conference Proceedings, vol. 845 (AIP, College Park, MD, 2006), pp. 615–618

M. Furnish, T.J. Vogler, C.S. Alexander, W.D. Reinhart, W.M. Trott, L.C. Chhabildas, Statistics of the Hugoniot elastic limit from line VISAR, in *Shock Compression of Condensed Matter*, ed. by M. Elert, M.D. Furnish, R. Chau, N.C. Holmes, J. Nguyen. AIP Conference Proceedings, vol. 555 (AIP, College Park, MD, 2007), pp. 521–524

M.D. Furnish, L.C. Chhabildas, W.D. Reinhart, W.M. Trott, T.J. Vogler, Determination and interpretation of statistics of spatially resolved waveforms in spalled tantalum for 7 to 13 GPa. Int. J. Plast. **25**, 587–602 (2009)

J.M. Galbraith, L.E. Murr, A.L. Stevens, Electron microscopy of shock-loaded polycrystalline beryllium, in *32nd Annual Proceedings Electron Microscopy Society of America*, ed. by C.J. Arceneaux (Claitor's Publishing Co., Baton Rouge, LA, 1974), pp. 506–507

C.W. Gillard, G.S. Ishikawa, J.E. Peterson, J.L. Rapier, J.C. Stover, N.L. Thomas, Laser Velocimeter Development Program AD0834874 (Research and Development Division, Lockheed Missiles and Space Company, Inc., Palo Alto, CA, 1968)

S.F. Glover, L.X. Schneider, K.W. Reed et al., Genesis: A 5 MA programmable pulsed power driver for isentropic compression experiments. IEEE Trans. Plasma Sci. **38**(10), 2620–2626 (2010)

M.R. Gomez, S.A. Slutz, A.B. Sefkow, K.D. Hahn, S.B. Hansen, P.F. Knapp, P.F. Schmit, C.L. Ruiz, D.B. Sinars, E.C. Harding, C.A. Jennings, T.J. Awe, M. Geissel, D.C. Rovang, I.C. Smith, G.A. Chandler, G.W. Cooper, M.E. Cuneo, A.J. Harvey-Thompson, M.C. Herrmann, M.H. Hess, D.C. Lamppa, M.R. Nartin, R.D. McBride, K.J. Peterson, J.L. Porter, G.A. Rochau, M.E. Savage, D.G. Schroen, W.A. Stygar, R.A. Vesey, Demonstration of thermonuclear conditions in magnetized liner inertial fusion experiments. Phys. Plasmas **22**, 056306 (2015)

D.E. Grady, R.E. Hollenbach, High strain rate studies in rock. Geophys. Res. Lett. **4**, 263–266 (1977)

D.E. Grady, R.E. Hollenbach, K.W. Schuler, J.F. Callender, Strain rate dependence in dolomite inferred from impact and static compression studies. J. Geophys. Res. – Solid Earth and Planets **82**(8), 1325–1333 (1977)

D.E. Grady, R.E. Hollenbach, K.W. Schuler, Compression wave studies in calcite rock. J. Geophys. Res. **83**, 2839–2849 (1978)

D.E. Grady, Interrelation of flow or fracture and phase transition in the deformation of carbonate rock. J. Geophys. Res. **84**(B13), 7549–7555 (1979)

D.E. Grady, M.E. Kipp, The micromechanics of impact fracture of rock. Int. J. Rock Mech. Mining Sci. **16**, 293–302 (1979)

D.E. Grady, Shock deformation in brittle solids. J. Geophys. Res. **85**(B2), 913–924 (1980)

D.E. Grady, M.E. Kipp, Continuum modeling of explosive fracture in oil shale. Int. J. Rock. Mech. Mining Sci. **17**, 149–157 (1980)

D.E. Grady, Fragmentation of solids under impulsive stress loading. J. Geophys. Res. **86**, 1047–1054 (1981a)

D.E. Grady, Strain-rate dependence of effective viscosity under steady-wave shock compression. Appl. Phys. Lett. **38**, 825–826 (1981b)

D.E. Grady, Fragment size prediction in dynamic fragmentation, in *Shock Waves in Condensed Matter*, ed. by W.J. Nellis, L. Seaman, R.A. Graham. AIP Conference Proceedings, vol. 78 (AIP, College Park, MD, 1982a), pp. 456–459

D.E. Grady, Local inertial effects in dynamic fragmentation. J. Appl. Phys. **53**(1), 322–325 (1982b)

D.E. Grady, Analysis of prompt fragmentation in explosively-loaded uranium cylindrical shells, SAND82-0140 (Sandia National Laboratories, Albuquerque, NM, 1982c)

D.E. Grady, J.R. Asay, Calculation of thermal trapping in shock deformation of aluminum. J. Appl. Phys. **54**, 7350–7354 (1982)

D.E. Grady, J.R. Asay, R.W. Rohde, J.L. Wise, Microstructure and mechanical properties of precipitation hardened aluminum under high rate deformation, in *Material Behavior Under High Stress and Ultrahigh Loading Rates* (Sagamore Army Materials Research Conference Proceedings), ed. by J. Mescall, V. Weiss, vol. 29 (Plenum, New York, NY, 1983), pp. 81–100

D.E. Grady, Microstructural effects on wave propagation in solids. Int. J. Eng. Sci. **22**, 1181–1186 (1984)

D.E. Grady, M.E. Kipp, D.A. Benson, Energy and statistical effects in the dynamic fragmentation of metal rings. *Proceedings of the Conference of the Mechanical Properties of High Rates of Strain*, Inst. Phys. Conf. Series No. 70, 315–320 (1984)

D.E. Grady, M.E. Kipp, Geometric statistics and dynamic fragmentation. J. Appl. Phys. **58**(3), 1210–1222 (1985a)

D.E. Grady, M.E. Kipp, Mechanisms of dynamic fragmentation: factors governing fragment size. Mechanics of Materials **4**, 311–320 (1985b)

D.E. Grady, M.D. Furnish, Shock- and release-wave properties of MJ-2 grout, SAND88-1642 (Sandia National Laboratories, Albuquerque, NM, 1988)

D.E. Grady, Particle size statistics in dynamic fragmentation. J. Appl. Phys. **68**(12), 6099–6105 (1990)

D.E. Grady, M.D. Furnish, Hugoniot and release properties of a water-saturated high-silica-content grout, in *Shock Compression of Condensed Matter*, ed. by S.C. Schmidt, J.N. Johnson, L.W. Davison (Elsevier, Amsterdam, 1990), pp. 621–624

D.E. Grady, Dynamics of adiabatic shear. Journal de Physique IV, Colloque C3 Suppl., Vol. 1, 653–660 (1991)

D.E. Grady, Shock-compression properties of ceramics, in *Recent Trends in High-Pressure Research* (Proceedings of the International Conference on High Pressure Science and Technology, AIRAPT-XIII), ed. by A.K. Singh (Oxford and IBH Publishing, Oxford, 1992), pp. 641–650

D.E. Grady, Dynamic fracture and fragmentation, in *High-Pressure Shock Compression of Solids*, ed. by J.R. Asay, M. Shahinpoor (Springer, New York, NY, 1993), pp. 265–322

D.E. Grady, Dynamic failure in brittle solids, in *Fracture and Damage of Quasi-Brittle Structures*, ed. by Z. Bazant et al. (E&FN Spon Publications, London, 1994), pp. 259–273

D.E. Grady, Spall and fragmentation in high-temperature metals, in *High-Pressure Shock Compression of Solids II*, ed. by L. Davison, D.E. Grady, M. Shahinpoor (Springer, New York, NY, 1995a), pp. 219–236
D.E. Grady, Dynamic properties of ceramic materials, SAND94-3266 (Sandia National Laboratories, Albuquerque, NM, 1995b)
D.E. Grady, Shock wave compression of brittle solids. Mechanics of Materials **29**, 181–203 (1998)
D.E. Grady, N.A. Winfree, G.I. Kerley, L.T. Wilson, L.D. Kuhns, Computational modeling and wave propagation in media with inelastic deforming microstructure. J. Phys. IV **10**, 15–20 (2000)
D.E. Grady, N.A. Winfree, A computational model for polyurethane foam, in *Fundamental Issues and Applications of Shock-Wave and High-Strain-Rate Phenomena*, ed. by K.P. Staudhammer, L.E. Murr, M.A. Meyers (Elsevier, New York, NY, 2001), pp. 485–491
D.E. Grady, M.L. Olsen, A statistical and energy based theory of dynamic fragmentation. Int. J. Impact Eng. **29**, 293–306 (2003)
D.E. Grady, *Fragmentation of Rings and Shells – The Legacy of N. F. Mott* (Springer, New York, NY, 2006)
D.E. Grady, Fragment size distributions from the dynamic fragmentation of brittle solids. Int. J. Impact Eng. **35**, 1557–1562 (2008)
D.E. Grady, Dynamic fragmentation of solids, in *Shock Wave Science Technology Reference Library. Solids II*, ed. by Y. Horie, vol. 3 (Springer, New York, NY, 2009), pp. 169–276
D.E. Grady, Structured shock waves and the fourth-power law. J. Appl. Phys. **107**, 013506 (2010a)
D.E. Grady, Length scales and size distributions in dynamic fragmentation. Int. J. Fracture **163**, 85–99 (2010b)
D.E. Grady, Adiabatic shear failure in brittle solids. Int. J. Impact Eng. **38**, 661–667 (2011)
D.E. Grady, G. Fenton, T. Vogler, Equation of state and evidence of enhanced phase transformation for shock compression of distended compounds. Int. J. Impact Eng. **56**, 19–26 (2013)
D.E. Grady, Unifying role of dissipative action in the dynamic failure of solids. J. Appl. Phys. **117**, 165905 (2015)
D.E. Grady, Diffusion of dissipative correlation in the dynamic failure of solids (private communication, 2016)
R.A. Graham, Impact physics, SCR-59 (Sandia National Laboratories, Albuquerque, NM, 1958)
R.A. Graham, Piezoelectric behavior of impacted quartz. J. Appl. Phys. **32**(3), 555 (1961a)
R.A. Graham, Technique for studying piezoelectricity under transient high stress conditions. Rev. Sci. Instrum. **32**(12), 1308–1313 (1961b)
R.A. Graham, G.E. Ingram, W.D. Ingram, Performance of a high velocity powder gun. SC-4652 (RR) (Sandia National Laboratories, Albuquerque, NM, 1961)
R.A. Graham, Dielectric anomaly in quartz for high transient stress and field. J. Appl. Phys. **33**(5), 1755–1758 (1962)
R.A. Graham, O.E. Jones, J.R. Holland, Shock-wave compression of germanium from 20 to 140 kbar. J. Appl. Phys. **36**, 3955–3956 (1965a)
R.A. Graham, F.W. Neilson, W.B. Benedick, Piezoelectric current from shock-loaded quartz – A submicrosecond stress gauge. J. Appl. Phys. **36**(5), 1775–1783 (1965b)
R.A. Graham, O.E. Jones, J.R. Holland, Physical behavior of germanium under shock wave compression. J. Phys. Chem. Solids **27**, 1519–1529 (1966)
R.A. Graham, Impact techniques for the study of physical properties of solids under shock-wave loading. J. Basic Eng. Trans. ASME **89**, 911–918 (1967)
R.A. Graham, R.E. Hutchison, Thermoelastic stress pulses resulting from pulsed electron beams. Appl. Phys. Lett. **11**(2), 69–71 (1967)
R.A. Graham, D.H. Anderson, J.R. Holland, Shock wave compression of 30% Ni – 70% Fe alloys: the pressure-induced magnetic transition. J. Appl. Phys. **38**, 223–229 (1967a)
R.A. Graham, R.E. Hutchison, W.B. Benedick, Pulsed electron beam calorimetry utilizing stress wave measurements in solid absorbers, in *9th IEEE Annual Symposium on Electron, Ion, and Laser Beam Technology*, ed. by R.F.W. Pease (San Francisco Press, San Francisco, CA, 1967b), pp. 70–76

R.A. Graham, W.J. Halpin, Dielectric breakdown and recovery of X-cut quartz under shock-wave compression. J. Appl. Phys. **39**(11), 5077–5082 (1968)

R.A. Graham, G.E. Ingram, A shock-wave stress gauge utilizing the capacitance change of a solid dielectric disc, in *Behavior of Dense Media Under High Dynamic Pressure*, ed. by J. Berger (Gordon and Breach, New York, NY, 1968), pp. 469–482

R.A. Graham, O.E. Jones, A summary of Hugoniot elastic limit measurements, SC-R-68-1857 (Sandia National Laboratories, Albuquerque, NM, 1968)

R.A. Graham, Linear bulk modulus approximation for sapphire. J. Geophys. Res. **76**(20), 4908–4912 (1971)

R.A. Graham, W.P. Brooks, Shock-wave compression of sapphire from 15 to 420 kbar: The effects of large anisotropic compressions. J. Phys. Chem. Solids **32**, 2311–2330 (1971)

R.A. Graham, Determination of third- and fourth-order longitudinal elastic constants by shock compression techniques: Application to sapphire and fused quartz. J. Acoust. Soc. Am. **51**(5), 1576–1581 (1972a)

R.A. Graham, Strain dependence of longitudinal, piezoelectric elastic, and dielectric constants of X-cut quartz. Phys. Rev. B **6**(12), 4779–4792 (1972b)

R.A. Graham, Plasticity analysis in soil mechanics problems, in *Problems of Plasticity*, ed. by A. Sawczuk (Noordhoff International Publication, Leyden, 1973), pp. 392–396

R.A. Graham, R.D. Jacobson, Lithium niobate stress gauge for pulsed radiation deposition studies. Appl. Phys. Lett. **23**(11), 584–586 (1973)

R.A. Graham, Shock-wave compression of X-cut quartz as determined by electrical response measurements. J. Phys. Chem. Solids **35**, 355–372 (1974)

R.A. Graham, Piezoelectric current from shunted and shorted guard-ring quartz gauges. J. Appl. Phys. **46**(5), 1901–1909 (1975)

R.A. Graham, P.J. Chen, A new electrical to mechanical coupling effect for nonlinear piezoelectric solids. Solid State Commun. **17**, 469–471 (1975)

R.A. Graham, L.C. Yang, Inherent time delay for dielectric breakdown in shock loaded X-cut quartz. J. Appl. Phys. **46**(12), 5300–5301 (1975)

R.A. Graham, Pressure dependence of the piezoelectric polarization of $LiNbO_3$ and $LiTaO_3$. Ferroelectrics **10**, 65–69 (1976)

R.A. Graham, Second- and third-order piezoelectric stress constants of lithium niobate as determined by the impact-loading technique. J. Appl. Phys. **48**(6), 2153–2163 (1977)

R.A. Graham, J.R. Asay, Measurement of wave profiles in shock-loaded solids. High Temp. High Press. **10**(4), 355–390 (1978)

R.A. Graham, R.P. Reed (eds.), *Selected Papers on Piezoelectricity and Impulsive Pressure Measurements*, SAND78-1911 (Sandia National Laboratories, Albuquerque, NM, 1978)

R.A. Graham, Measurement of wave profiles in shock-loaded solids, in *High-Pressure Science and Technology*, ed. by K.D. Timmerhaus, M.S. Barber, vol. 2 (Plenum, New York, NY, 1979a), pp. 854–869

R.A. Graham, Shock-induced electrical activity in polymeric solids. A mechanically induced bond scission model. J. Phys. Chem. **83**(23), 3048–3056 (1979b)

R.A. Graham, Electrical activity in shock-loaded polymers, in *High Pressure in Science and Technology*, ed. by K.D. Timmerhaus, M.S. Barber (Pergamon, Oxford, 1980), pp. 1032–1039

R.A. Graham, Active measurements of defect processes in shock-compressed metals and other solids, in *Metallurgical Effects of High-Strain-Rate Deformation and Fabrication*, ed. by M.A. Meyers, L.E. Murr (Plenum, New York, NY, 1981), pp. 375–386

R.A. Graham, B. Morosin, B.W. Dodson, The chemistry of shock compression: A bibliography, SAND83-1887 (Sandia National Laboratories, Albuquerque, NM, 1983)

R.A. Graham, D.B. Webb, Fixtures for controlled explosive loading and preservation of powder samples, in *Shock Waves in Condensed Matter*, ed. by J.R. Asay, R.A. Graham, G.K. Straub (Elsevier, Amsterdam, 1984), pp. 211–216

R.A. Graham, M.J. Carr, Analytical electron microscopy study of shock synthesized zinc ferrite, in *Shock Waves in Condensed Matter*, ed. by Y.M. Gupta (Plenum, New York, NY, 1986), pp. 803–808

R.A. Graham, D.M. Webb, Shock-induced temperature distributions in powder compact recovery fixtures, in *Shock Waves in Condensed Matter*, ed. by Y.M. Gupta (Plenum, New York, NY, 1986), pp. 589–593

R.A. Graham, B. Morosin, Y. Horie, E.L. Venturini, M.B. Boslough, M.J. Carr, D.L. Williamson, Chemical synthesis under shock compression, in *Shock Waves in Condensed Matter*, ed. by Y.M. Gupta (Plenum, New York, NY, 1986a), pp. 693–711

R.A. Graham, B. Morosin, E.L. Venturini, M.J. Carr, E.K. Beauchamp, Shock-compression processes in inorganic powders, in *Metallurgical Applications of Shock-Wave and High-Strain-Rate Phenomena*, ed. by L.E. Murr, K.P. Staudhammer, M.A. Meyers (Marcel Dekker, New York, NY, 1986b), pp. 1005–1012

R.A. Graham, Shock compression of solids as a physical-chemical-mechanical process, in *Shock Waves in Condensed Matter*, ed. by S.C. Schmidt, N.C. Holmes (Elsevier, Amsterdam, 1988), pp. 11–19

R.A. Graham, L.M. Lee, F. Bauer, Response of Bauer piezoelectric polymer stress gauges (PVDF) to shock loading, in *Shock Waves in Condensed Matter*, ed. by S.C. Schmidt, N.C. Holmes (Elsevier, Amsterdam, 1988a), pp. 619–622

R.A. Graham, B. Morosin, D.M. Bush, Shock-induced melting of a KCl:LiCl eutectic powder as determined from electrochemical response measurements, in *Shock Waves in Condensed Matter*, ed. by S.C. Schmidt, N.C. Holmes (Elsevier, Amsterdam, 1988b), pp. 179–184

R.A. Graham, Issues in shock-induced solid state chemistry, in *Behavior of Dense Media under High Dynamic Pressures* (3rd International Symposium High Dynamic Pressures), ed. by R. Cheret (Association Francaise de Pyrotechnie, Paris, 1989), pp. 175–180

R.A. Graham, M.U. Anderson, F. Bauer, R.E. Setchell, Piezoelectric polarization of the ferroelectric polymer PVDF from 10 MPa to 10 GPa: Studies of loading-path dependence, in *Shock Compression of Condensed Matter*, ed. by S.C. Schmidt, R.D. Dick, J.W. Forbes, D.G. Tasker (Elsevier, Amsterdam, 1992), pp. 883–886

F.V. Grigoryev, S.B. Kormer, O.L. Mikhailova, A.P. Tolochko, V.D. Urlin, Experimental determination of the compressibility of hydrogen at densities 0.5–2 g/cm$^3$. JETP Lett. **16**, 201–204 (1972)

T.R. Guess, L.M. Lee, Spall strengths of five carbon materials. SC-DR-68-604 (Sandia National Laboratories, Albuquerque, NM, 1968)

T.A. Haill, C.J. Garasi, A.C. Robinson, ALEGRA-MHD: Version 4.0, SAND2003-4074 (Sandia National Laboratories, Albuquerque, NM, 2003) [Superseded by T.A. Haill, K.R. Cochrane, C.J. Garasi, T.A. Mehlhorn, A.C. Robinson, and R.M. Summers, ALEGRA-MHD: Version 4.6, SAND2004-5997 (Sandia National Laboratories, Albuquerque)]

T.A. Haill, K.R. Cochrane, C.J. Garasi, T.A. Mehlhorn, A.C. Robinson, and R.M. Summers, ALEGRA-MHD: Version 4.6, SAND2004-5997 (Sandia National Laboratories, Albuquerque, NM, 2005

T.A. Haill, T.A. Mehlhorn, J.R. Asay, Y.M. Gupta, R.J. Lawrence, C.J. Bakeman, J. LaFollett, A feasibility study for a fragment-producing chemical electrical launcher, *in 2007 16th IEEE International Pulsed Power Conference* (IEEE Piscataway, NJ, 2007), vol. 2, pp. 1753–1756

T.A. Haill, C.S. Alexander, J.R. Asay, Simulation and analysis of Magnetically-Applied Pressure-Shear (MAPS) experiments, *18th IEEE International Pulsed Power Conference* (IEEE Piscataway, NJ, 2011), pp. 1093–1098

T.A. Haill, T.R. Mattsson, S. Root et al., Mesoscale simulation of shocked poly-(4-methyl-1-Pentene) (PMP) foams, in *Shock Compression of Condensed Matter*, ed. by M.L. Elert, W.T. Buttler, J.P. Borg, J.L. Jordan, T.J. Vogler. AIP Conference Proceedings, vol. 1426 (AIP, College Park, MD, 2012), pp. 913–916

T.A. Haill, T.R. Mattsson, S. Root et al., Mesoscale simulation of mixed equations of state with application to shocked platinum-doped PMP foams, in *Proceedings of the 12th Hypervelocity Impact Symposium*, Procedia Engineering **58**, 309–319 (2013)

C.A. Hall, L.C. Chhabildas, W.D. Reinhart, Shock Hugoniot and release in concrete with different aggregate sizes from 3 to 23 GPa. Int. J. Impact Eng. **23**, 341–351 (1999)

C.A. Hall, J.R. Asay, W.M. Trott, M. Knudson, K.J. Fleming, M.A. Bernard, B.F. Clark, A. Hauer, G. Kyrala, Aluminum Hugoniot measurements on the Sandia Z accelerator, in *Shock*

*Compression of Condensed Matter*, ed. by M.D. Furnish, L.C. Chhabildas, R.S. Hixson. AIP Conference Proceedings, vol. 505 (AIP, College Park, MD, 2000), pp. 1171–1174

C.A. Hall, J.R. Asay, M.D. Knudson, W.A. Stygar, R.B. Spielman, T.D. Pointon, D.B. Reisman, A. Toor, R.C. Cauble, Isentropic compression of solids using pulsed magnetic fields. Rev. Sci. Instrum. **72**(9), 3587–3595 (2001a)

C.A. Hall, M.D. Knudson, J.R. Asay et al., High velocity flyer plate launch capability on the Sandia Z accelerator. Int. J. Impact Eng. **26**, 275–287 (2001b)

C.A. Hall, J.R. Asay, M.D. Knudson, D.B. Hayes, R.W. Lemke, J.-P. Davis, C. Deeney, Recent advances in quasi-isentropic compression experiments (ICE) on the Sandia Z accelerator, in *Shock Compression of Condensed Matter*, ed. by M.D. Furnish, N.N. Thadhani, Y. Horie. AIP Conference Proceedings, vol. 620 (AIP, College Park, MD, 2002), pp. 1163–1168

W.J. Halpin, O.E. Jones, R.A. Graham, A submicrosecond technique for simultaneous observation of input and propagated impact stresses, in *Symposium on Dynamic Behavior of Materials* (ASTM Special Technical Publications No. 336), (American Society for Testing and Materials, Philadelphia, PA, 1963), pp. 208–218

W.J. Halpin, R.A. Graham, Shock wave compression of Plexiglas from 3 to 20 kilobars, in *Proceedings of the 4th International Detonation Symposium*, ONR ACR-126, ed. by S.J. Jacobs, D. Price (Office of Naval Research, Washington, D.C., 1965), pp. 222–232

H.B. Hammel, R.F. Beebe, A.P. Ingersoll et al., HST imaging of atmospheric phenomena created by the impact of Comet Shoemaker-Levy 9. Science **267**, 1288–1296 (1995)

W.F. Hammetter, J.R. Hellmann, R.A. Graham, B. Morosin, Energy release and transformation of shock-modified zirconia upon annealing to 1550 degrees C, in *Shock Waves in Condensed Matter*, ed. by J.R. Asay, R.A. Graham, G.K. Straub (Elsevier, Amsterdam, 1984), pp. 391–394

W.F. Hammetter, R.A. Graham, B. Morosin, Y. Horie, Effects of shock modification on the self-propagating high temperature synthesis of nickel aluminides, in *Shock Waves in Condensed Matter*, ed. by S.C. Schmidt, N.C. Holmes (Elsevier, Amsterdam, 1988), pp. 431–434

D.L. Hanson, J.R. Asay, C.A. Hall, M.D. Knudson, J.E. Bailey, K.J. Fleming, R.R. Johnston, B.F. Clark, M.A. Bernard, W.W. Anderson, G. Hassall, S.D. Rothman, Progress on deuterium measurements on Z, in *Shock Compression of Condensed Matter*, ed. by M.D. Furnish, L.C. Chhabildas, R.S. Hixson. AIP Conference Proceedings, vol. 505 (AIP, College Park, MD, 2000), pp. 1175–1178

D.L. Hanson, R.R. Johnston, M.D. Knudson, J.R. Asay, C.A. Hall, J.E. Bailey, R.J. Hickman, Advanced cryogenic system capabilities for precision shock physics measurements on Z, in *Shock Compression of Condensed Matter*, ed. by M.D. Furnish, N.N. Thadhani, Y. Horie. AIP Conference Proceedings, vol. 620 (AIP, College Park, MD, 2002a), pp. 1141–1144

D.L. Hanson, M.D. Knudson, J.R. Asay, C.A. Hall, J.E. Bailey, R.W. Lemke, J.-P. Davis, R.B. Spielman, B.V. Oliver, D.B. Hayes, Precision shock physics capabilities for inertial fusion studies using the Z accelerator current drive, in *Inertial Fusion Sciences and Applications*, ed. by K.A. Tanaka, D.D. Meyerhofer, J. Meyer-ter-Vehn (Elsevier, Paris, 2002b), pp. 1091–1095

D.R. Hardesty, P.C. Lysne, Shock initiation and detonation properties of homogeneous explosives, SLA-74-0165 (Sandia National Laboratories, Albuquerque, NM, 1974)

D.R. Hardesty, An investigation of the shock initiation of liquid nitromethane. Combust. Flame **27**, 229–251 (1976a)

D.R. Hardesty, On the index of refraction of shock-compressed liquid nitromethane. J. Appl. Phys. **47**(5), 1994–1998 (1976b)

D.R. Hardesty, J.E. Kennedy, Thermochemical estimation of explosive energy output. Combust. Flame **43**, 45–59 (1977)

J.K. Hartman, J.L. Wise, R.A. Graham, R.O. Johnson, G.E. Clark, T.J. Burns, Microwave dielectric constant of shock-loaded lithium niobate, in *Shock Waves in Condensed Matter*, ed. by W.J. Nellis, L. Seaman, R.A. Graham. AIP Conference Proceedings, vol. 78 (AIP, College Park, MD, 1982), pp. 277–281

R.S. Hawke, Experiments on hydrogen at megabar pressures: Metallic hydrogen, in *Festkörperprobleme* **14**, ed. by H.J. Queisser (Springer, Berlin, 1974), pp. 111–118

R.S. Hawke, A.R. Susoeff, J.A. Ang, C.H. Konrad, C.A. Hall, G.L. Sauve, A.R. Vesey, Performance of hypervelocity armatures with replenished metal vapor plasmas, in *Third European Electromagnetic Launcher Symposium Proceedings*, UCRL-JC-106828 (Lawrence Livermore National Laboratory, Livermore, CA, 1991a)

R.S. Hawke, A.R. Susoeff, J.R. Asay, J.A. Ang, C.A. Hall et al., Railgun performance with a two-stage light-gas gun injector. IEEE Trans. Magnetics **27**, 28–32 (1991b)

D.B. Hayes, L. Kennedy, Unfolding of quartz gage records, SC-TM-690635 (Sandia National Laboratories, Albuquerque, NM, 1969)

D.B. Hayes, Wave propagation in a condensed medium with N transforming phases: Application to solid-I-solid-II-liquid bismuth. J. Appl. Phys. **46**, 3438–3443 (1975)

D.B. Hayes, D.E. Mitchell, A constitutive equation for the shock response of porous hexanitrostilbene (HNS) explosive, at Symposium on High Pressures, Commissariat a l'Energie Atomique, Paris, France, on August 22, 1978

D.B. Hayes, D.E. Grady, A thermal-viscous model for heterogeneous yielding in aluminum, in *Shock Waves in Condensed Matter*, ed. by W.J. Nellis, L. Seaman, R.A. Graham. AIP Conference Proceedings, vol. 78 (AIP, College Park, MD, 1982), pp. 412–415

D.B. Hayes, Unsteady compression waves in interferometer windows. J. Appl. Phys. **89**, 6484–6486 (2001)

D.B. Hayes, C.A. Hall, J.R. Asay, M.D. Knudson, Continuous index of refraction measurements to 20 GPa in Z-cut sapphire using pulsed magnetic loading. J. Appl. Phys. **94**, 2331–2336 (2003)

D.B. Hayes, C.A. Hall, J.R. Asay, M.D. Knudson, Measurement of the compression isentrope for 6061-T6 aluminum to 185 GPa and 46% volumetric strain using pulsed magnetic loading. J. Appl. Phys. **96**(10), 5520–5527 (2004)

J.R. Hellmann, K. Kuroda, A.H. Heuer, R.A. Graham, Microstructural characterization of shock-modified zirconia powders, in *Shock Waves in Condensed Matter*, ed. by J.R. Asay, R.A. Graham, G.K. Straub (Elsevier, Amsterdam, 1984), pp. 387–390

F. Herlach, J.E. Kennedy, The dynamics of imploding liners in magnetic flux compression experiments. J. Phys. D Appl. Phys. **6**, 661–676 (1973)

W. Herrmann, E.A. Witmer, J.H. Percy, A.H. Jones, Stress wave propagation and spallation in uniaxial strain. ASD-TDR-62-399 (Air Force Systems Command, 1962)

W. Herrmann, A Lagrangian finite difference method for two-dimensional motion including material strength. WL-TR-64-107 (Air Force Weapons Laboratory, 1964)

W. Herrmann, P. Holzhauser, R.J. Thompson, WONDY—A computer program for calculating problems of motion in one dimension, SC-RR-66-601 (Sandia National Laboratories, Albuquerque, NM, 1967)

W. Herrmann, Equation of state of crushable distended materials, SC-RR-66-2678 (Sandia National Laboratories, Albuquerque, NM, 1968)

W. Herrmann, Constitutive equation for the dynamic compaction of ductile porous materials. J. Appl. Phys. **40**(6), 2490–2499 (1969a)

W. Herrmann, On the dynamic compaction of initial heated porous materials, SC-DR-680865 (Sandia National Laboratories, Albuquerque, NM, 1969b)

W. Herrmann, Nonlinear stress waves in metals, in *Wave Propagation in Solids*, ed. by J. Miklowitz (American Society of Mechanical Engineers, New York, NY, 1969c), pp. 129–183

W. Herrmann, R.J. Lawrence, D.S. Mason, Strain hardening and strain rate in one-dimensional wave propagation calculations, SC-RR-70-471 (Sandia National Laboratories, Albuquerque, NM, 1970)

W. Herrmann, Constitutive equations for compaction of porous materials, in *Applied Mechanics Aspects of Nuclear Effects*, ed. by C.C. Wan (American Society of Mechanical Engineers, New York, NY, 1971), pp. 142–168

W. Herrmann, D.L. Hicks, E.G. Young, Attenuation of elastic-plastic stress waves, in *Shock Waves and the Mechanical Properties of Solids*, ed. by J.J. Burke, V. Weiss (Syracuse University Press, Syracuse, NY, 1971), pp. 23–64

W. Herrmann, Constitutive equations for the compaction of porous materials, SC-DC-71-4134 (Sandia National Laboratories, Albuquerque, NM, 1972)

W. Herrmann. On the evaluation of constitutive equations from experiment, in *Recent Advances in Engineering Science* **6**, (Proceedings of the Society of Engineering Science 10th Anniversary Meeting, 1973), pp. 297–307

W. Herrmann, J.W. Nunziato, Nonlinear constitutive equations (Chapter 5), in *Dynamic Response of Materials to Intense Impulsive Loading*, ed. by P.C. Chou, A.K. Hopkins (Air Force Materials Laboratory, Wright-Patterson AFB, OH, 1973)

W. Herrmann, Development of a high strain rate constitutive equation for 6061-T6 aluminum, SLA-73-0897 (Sandia National Laboratories, Albuquerque, NM, 1974)

W. Herrmann, R.J. Lawrence, The effect of material constitutive models on stress wave propagation calculations. J. Eng. Mater. Technol. Trans. ASME **100**, 84–95 (1978)

W. Herrmann, On constitutive modelling for the shock physicist, in *Shock Waves in Condensed Matter*, ed. by W.J. Nellis, L. Seaman, R.A. Graham. AIP Conference Proceedings, vol. 78 (AIP, College Park, MD, 1982), pp. 346–359

W. Herrmann, L.D. Bertholf, Explicit Lagrangian finite-difference methods, in *Computational Methods for Transient Analysis* (Mechanics and Mathematical Methods—Series of Handbooks), ed. by T. Belytschko, T. Hughes, vol. 1 (Elsevier, North-Holland, 1983), pp. 361–415

W. Herrmann, W.R. Wawersik, S.T. Montgomery, Review of creep modeling for rock salt, in *Mechanics of Engineering Materials*, ed. by C.C. Desai, R.H. Gallagher (Wiley, NY, 1984), pp. 297–317

D.G. Hicks, T.R. Boehly, P.M. Celliers, J.H. Eggert, S.J. Moon, D.D. Meyerhofer, G.W. Collins, Laser-driven single shock compression of fluid deuterium from 45 to 220 GPa. Phys. Rev. B **79**, 014112 (2009)

D.L. Hicks, Von Neumann stability of the WONDY wavecode for thermodynamic equations of state, SAND77-0934 (Sandia National Laboratories, Albuquerque, NM, 1977)

D.L. Hicks, F.R. Norwood, T.G. Trucano, TOODY-WONDY calculations of penetration events, in *Shock Waves in Condensed Matter*, ed. by W.J. Nellis, L. Seaman, R.A. Graham. AIP Conference Proceedings, vol. 78 (AIP, College Park, MD, 1982), pp. 544–547

M.L. Hobbs, M.R. Baer, B.C. McGee, JCZS: An intermolecular potential database for performing accurate detonation and expansion calculations. Propellants, Explosives and Pyrotechnics **24**(5), 269–279 (1999)

B.L. Holian, D.E. Grady, Fragmentation by molecular dynamics: The microscopic "big bang". Phys. Rev. Lett. **60**(14), 1355–1358 (1988)

K.G. Holland, L.C. Chhabildas, W.D. Reinhart, M.D. Furnish, Experiments of cercom SiC rods under impact, in *Shock Compression of Condensed Matter*, ed. by M.D. Furnish, L.C. Chhabildas, R.S. Hixson. AIP Conference Proceedings, vol. 505 (AIP, College Park, MD, 2000), pp. 585–588

A.C. Holt, M.M. Carroll, B.M. Butcher, Application of a new theory for the pressure-induced collapse of pores in ductile materials, in *Proceedings of the RILEM-IUPAC International Symposium on Pore Structure and Properties of Materials Part 5*, ed. by S. Modry (Academia, Prague, 1974), pp. 63–76

Y. Horie, R.A. Graham, I.K. Simonsen, Observations on the shock-synthesis of intermetallic compounds, in *Metallurgical Applications of Shock-Wave and High-Strain-Rate Phenomena*, ed. by L.E. Murr, K.P. Staudhammer, M.A. Meyers (Marcel Dekker, New York, NY, 1986a), pp. 1023–1035

Y. Horie, D.E.P. Hoy, I.K. Simonsen, R.A. Graham, B. Morosin, Shock synthesis of titanium aluminides, in *Shock Waves in Condensed Matter*, ed. by Y.M. Gupta (Plenum, New York, NY, 1986b), pp. 749–754

Y. Horie, M.E. Kipp, Modeling of shock-induced chemical reactions in powder mixtures. J. Appl. Phys. **63**(12), 5718–5727 (1988)

G.R. Hough, D.M. Gustafson, R.E. Thursby, Enhanced holographic recording capabilities for dynamic applications, in *Proceedings of the SPIE Ultrahigh and High Speed Photography,*

*Photons, and Velocimetry 1989 Conference*, ed. by P.A. Jaanimagi, vol. 1155 (SPIE, Bellingham, WA, 1990), pp. 181–188

H. Huang, J.R. Asay, Compressive strength measurements in aluminum for shock compression over the stress range of 4-22 GPa. J. Appl. Phys. **98**, 033524 (2005)

C.F. Huff, R.A. Graham, Pressure measurements very near an electrical arc discharge in a liquid using a lithium niobate piezoelectric transducer. Appl. Phys. Lett **27**(4), 163–164 (1975)

G.E. Ingram, R.A. Graham, Quartz gauge technique for impact experiments, in *Fifth Symposium (International) on Detonation*, ACR 184 (Office of Naval Research, Washington, D.C., 1970), pp. 369–386

B.J. Jensen, D.B. Holtkamp, P.A. Rigg, D.H. Dolan, Accuracy limits and window corrections for photon Doppler velocimetry. J. Appl. Phys. **101**, 13523 (2007)

J.N. Johnson, Shock waves in stress-relaxing solids, WSU SDL 66-01 (Washington State University, Pullman, WA, 1966)

J.N. Johnson, Basic Theory of irreversible thermodynamics with application to the anelastic solids, Internal Report No. 01-67 (Washington State University, Pullman, WA, 1967)

J.N. Johnson, W. Band, Investigation of precursor decay in iron by the artificial viscosity method. J. Appl. Phys. **38**(4), 1578–1585 (1967)

J.N. Johnson, A theory of rate-dependent behavior for porous solids: Steady-propagating compaction wave profiles, SC-RR-68-151 (Sandia National Laboratories, Albuquerque, NM, 1968a)

J.N. Johnson, Elastic precursor decay in quartzite for cylindrical and spherical flow. J. Appl. Phys. **39**(1), 290–296 (1968b)

J.N. Johnson, Single-particle model of a solid: The Mie-Grüneisen equation. Am. J. Phys. **36**(10), 917–919 (1968c)

J.N. Johnson, Constitutive relation for rate-dependent plastic flow in polycrystalline metals. J. Appl. Phys. **40**(5), 2287–2293 (1969)

J.N. Johnson, L.M. Barker, Dislocation dynamics and steady plastic wave profiles in 6061-T6 aluminum. J. Appl. Phys. **40**(11), 4321–4334 (1969)

J.N. Johnson, O.E. Jones, T.E. Michaels, Dislocation dynamics and single-crystal constitutive relations: Shock-wave propagation and precursor decay. J. Appl. Phys. **41**(6), 2330–2339 (1970)

J.N. Johnson, Shock propagation produced by planar impact in linearly elastic anisotropic media. J. Appl. Phys. **42**(13), 5522–5530 (1971)

J.N. Johnson, R.W. Rohde, Dynamic deformation twinning in shock-loaded iron. J. Appl. Phys. **42**(11), 4171–4182 (1971)

J.N. Johnson, An analysis of thermally-induced plane waves in elastic-plastic single crystals. J. Mech. Phys. Solids **20**, 367–380 (1972a)

J.N. Johnson, Calculation of plane-wave propagation in anisotropic elastic-plastic solids. J. Appl. Phys. **43**(5), 2074–2082 (1972b)

J.N. Johnson, Considerations for the calculation of shock-induced phase transformations in solids, SC-RR-72-0626 (Sandia National Laboratories, Albuquerque, NM, 1972c)

J.N. Johnson, Inelastic plane-wave propagation in anisotropic rocks. J. Geophys. Res. **79**(32), 4900–4907 (1974a)

J.N. Johnson, Wave velocities in shock-compressed cubic and hexagonal single crystals above the elastic limit. J. Phys. Chem. Solids **35**(5), 609–616 (1974b)

J.N. Johnson, Kinematic waves and group velocity: Application to natural and manmade environment. Am. J. Phys. **43**, 681–688 (1974c)

J.N. Johnson, J.R. Asay, D.B. Hayes, Equations of state and shock-induced transformations in solid-I, solid-II, liquid bismuth. J. Phys. Chem. Solids **35**, 501–515 (1974)

J.N. Johnson, L.E. Pope, Shock-wave compression of single-crystal beryllium. J. Appl. Phys. **46**, 720–729 (1975)

J.N. Johnson, Micromechanical considerations in shock compression of solids, in *High-Pressure Shock Compression of Solids*, ed. by J.R. Asay, M. Shahinpoor (Springer, New York, NY, 1993), pp. 222–240

L. Johnson, Sandia National Laboratories: A history of exceptional service in the national interest, SAND97-1029, ed. by C. Mora, J. Taylor, R. Ullrich (Sandia National Laboratories, Albuquerque, NM, 1997)

B. Jones, C.J. Garasi, D.J. Ampleford et al., Measurement and modeling of the implosion of wire arrays with seeded instabilities. Phys. Plasmas **13**, 056313 (2006)

G.E. Jones, L.D. Bertholf, J.E. Kennedy, Ballistic calculations of R. W. Gurney. Am. J. Phys. **48**, 264–269 (1980)

O.E. Jones, F.W. Neilson, W.B. Benedick, Dynamic yield behavior of explosively loaded metals determined by a quartz transducer technique. J. Appl. Phys. **33**(11), 3224–3232 (1962)

O.E. Jones, J.R. Holland, Bauschinger effect in explosively loaded mild steel. J. Appl. Phys. **35**, 1771–1773 (1964)

O.E. Jones, Piezoelectric and mechanical behavior of X-cut quartz shock loaded at 79 degrees K. Rev. Sci. Instrum. **38**(2), 253–256 (1967)

O.E. Jones, F.R. Norwood, Axially symmetric cross-sectional strain and stress distributions in suddenly loaded cylindrical elastic bars. J. Appl. Mech. Trans. ASME Ser. E **89**, 718–724 (1967)

O.E. Jones, J.R. Holland, Effects of grain size on dynamic yielding in explosively loaded mild steel. Acta Metall. **16**, 1037–1045 (1968)

O.E. Jones, J.D. Mote, Shock-induced dynamic yielding in copper single crystals. J. Appl. Phys. **40**(12), 4920–4928 (1969)

O.E. Jones, Shock waves and the mechanical properties of solids, in *Engineering Solids under Pressure*, ed. by H. Pugh, D. Li (Institution of Mechanical Engineers, London, 1971), pp. 75–86

O.E. Jones, R.A. Graham, Shear strength effects on phase transition pressures determined from shock-compression experiments, in *Accurate Characterization of the High Pressure Environment,* National Bureau of Standards Special Publication 326, ed. by E.C. Lloyd (U.S. Government Printing Office, Washington, DC, 1971), pp. 229–242

O.E. Jones, Metal response under explosive loading, in *Behavior and Utilization of Explosives in Engineering Design* (Proceedings 12th Annual Symposium New Mexico Section of the American Society of Mechanical Engineers), ed. by L.W. Davison, J. Kennedy, F. Coffey (NM Section ASME, Albuquerque, NM, 1972), pp. 125–148

O.E. Jones, Shock wave mechanics, in *Metallurgical Effects at High Strain Rates*, ed. by R.W. Rohde, B.M. Butcher, J.R. Holland, C.H. Karnes (Plenum, New York, NY, 1973), pp. 33–55

J.D. Kennedy, W.B. Benedick, Shock-induced polymorphic phase transformation in InSb. Bull. Am. Phys. Soc. **10**, 1112 (1965)

J.D. Kennedy, W.B. Benedick, Shock-induced phase transition in single crystal CdS. J. Phys. Chem. Solids **27**, 125–127 (1966)

J.E. Kennedy, Quartz gauge study of upstream reaction in a shocked explosive, in *Proceedings of the 5th International Detonation Symposium*, ONR ACR-184, ed. by S.J. Jacobs, R. Roberts (Office of Naval Research, Washington, D.C., 1970), pp. 435–445

J.E. Kennedy, Gurney energy of explosives: Estimation of the velocity and impulse imparted to driven metal, SC-RR-70-790 (Sandia National Laboratories, Albuquerque, NM, 1971)

J.E. Kennedy, Explosive output for driving metal, in *Behavior and Utilization of Explosives in Engineering Design,* Proceedings 12th Annual Symposium New Mexico Section of American Society of Mechanical Engineers, ed. by L.W. Davison et al. (New Mexico Section of American Society of Mechanical Engineers, New Mexico, 1972), pp. 109–124

J.E. Kennedy, Pressure field in a shock-compressed high explosive, in *Proceedings 14th Symposium (International) on Combustion* (The Combustion Institute, 1973), vol. 14, pp. 1251–1258

J.E. Kennedy, A.C. Schwarz, Detonation transfer by flyer plate impact, in *Proceedings 8th Symposium on Explosives and Pyrotechnics* (Franklin Institute, Philadelphia, PA, 1974)

J.E. Kennedy, J.W. Nunziato, Shock-wave evolution in a chemically reacting solid. J. Mech. Phys. Solids **40**, 107–124 (1976)

J.E. Kennedy, J.W. Nunziato, D.R. Hardesty, Initiation and detonation studies of condensed explosives using interferometric techniques. Acta Astronaut. **3**, 811–823 (1976)

L.W. Kennedy, O.E. Jones, Longitudinal wave propagation in a circular bar loaded suddenly by a radially distributed end stress. J. Appl. Mech. Trans. ASME Ser. E **91**, 470–478 (1969)

G.I. Kerley, Theory of ionization equilibrium: An approximation for the single element case. J. Chem. Phys. **85**, 5228–5231 (1986)

G.I. Kerley, Theoretical equation of state for aluminum. Int. J. Impact Eng. **5**, 441–449 (1987)

G.I. Kerley, Equations of state for calcite materials. I. Theoretical model for dry calcium carbonate. High Press. Res. **2**, 29–47 (1988)

G.I. Kerley, J.L. Wise, Shock-induced vaporization of porous aluminum, in *Shock Waves in Condensed Matter*, ed. by S.C. Schmidt, N.C. Holmes (Elsevier, Amsterdam, 1988), pp. 155–158

G.I. Kerley, Equations of state and gas-gas separation in soft-sphere mixtures. J. Chem. Phys. **91**, 1204–1210 (1989a)

G.I. Kerley, Theoretical model of explosive detonation products: tests and sensitivity studies, in *Proceedings of 9th International Detonation Symposium*, ONR 113291-7, ed. by J.M. Short, E.L. Lee (Office of Naval Research, Washington, D.C., 1989b), pp. 443–451

G.I. Kerley, Theory of calcite equation of state, in *Shock Waves in Condensed Matter*, ed. by S.C. Schmidt, J.N. Johnson, L.W. Davison (Elsevier, Amsterdam, 1990), pp. 613–616

G.I. Kerley, Equations of state for hydrogen and deuterium, SAND2003-3613 (Sandia National Laboratories, Albuquerque, NM, 2003)

K.Y. Kim, L.C. Chhabildas, A.L. Ruoff, Isothermal equations of state for lithium fluoride. J. Appl. Phys. **47**(7), 2862–2866 (1976)

M.E. Kipp, A.L. Stevens, Numerical integration of a spall-damage viscoplastic constitutive model in a one-dimensional wave propagation code, SAND76-0061 (Sandia National Laboratories, Albuquerque, NM, 1976)

M.E. Kipp, Calculation of borehole springing in oil shale (Rock Springs Site 6A), SAND77-1501 (Sandia National Laboratories, Albuquerque, NM, 1979)

M.E. Kipp, J.W. Nunziato, Numerical simulation of detonation failure in nitromethane, in *Proceedings of the 7th International Detonation Symposium*, NSWC MP 82-334, ed. by J.M. Short, S.J. Jacobs (Naval Surface Warfare Center, Dahlgren, VA, 1981), pp. 608–619

M.E. Kipp, J.W. Nunziato, R.E. Setchell, Hot spot initiation of heterogeneous explosives, in *Proceedings of the 7th International Detonation Symposium*, NSWC MP 82-334, ed. by J.M. Short, S.J. Jacobs (Naval Surface Warfare Center, Dahlgren, VA, 1981), pp. 394–407

M.E. Kipp, L.W. Davison, Analyses of ductile flow and fracture in two dimensions, in *Shock Waves in Condensed Matter*, ed. by W.J. Nellis, L. Seaman, R.A. Graham. AIP Conference Proceedings, vol. 78 (AIP, College Park, MD, 1982), pp. 442–445

M.E. Kipp, R.J. Lawrence, WONDY V—A one-dimensional finite-difference wave propagation code, SAND81-0930 (Sandia National Laboratories, Albuquerque, NM, 1982)

M.E. Kipp, D.E. Grady, Flaw nucleation and energetics of dynamic fragmentation, in *Shock Waves in Condensed Matter*, ed. by J.R. Asay, R.A. Graham, G.K. Straub (Elsevier, Amsterdam, 1983), pp. 159–162

M.E. Kipp, Modeling granular explosive detonations with shear band concepts, in *Proceedings of 8th International Detonation Symposium*, NSWC MP 86-194, ed. by J.M. Short, W.E. Deal (Naval Surface Warfare Center, Dahlgren, VA, 1985), pp. 35–41

M.E. Kipp, D.E. Grady, An application of geometric statistics to dynamic fragmentation, in *Shock Waves in Condensed Matter*, ed. by Y.M. Gupta (Plenum, New York, NY, 1986), pp. 435–439

M.E. Kipp, H.J. Melosh, A numerical study of the giant impact origin of the moon: The first half hour. Lunar Planet Sci. **18**, 491–492 (1987)

M.E. Kipp, R.E. Setchell, P.A. Taylor, Homogeneous reactive kinetics applied to granular HNS, in *Shock Waves in Condensed Matter*, ed. by S.C. Schmidt, N.C. Holmes (Elsevier, Amsterdam, 1988), pp. 539–542

M.E. Kipp, D.E. Grady, J.L. Wise, Planar-shock and penetration response of ceramics, in *Shock-Wave and High-Strain-Rate Phenomena in Materials*, ed. by M.A. Meyers, L.E. Murr, K.P. Staudhammer (Marcel Dekker, New York, NY, 1992), pp. 1083–1091

M.E. Kipp, D.E. Grady, J.W. Swegle, Numerical and experimental studies of high-velocity impact fragmentation. Int. J. Impact Eng. **14**, 427–1438 (1993)

M.E. Kipp, Target response to debris cloud incidence, in *High-Pressure Science and Technology*, ed. by S.C. Schmidt, J.W. Shaner, G.A. Samara, M. Ross, AIP Conference Proceedings, vol. 309 (AIP, College Park, MD, 1994), pp. 1849–1852

M.E. Kipp, W.D. Reinhart, L.C. Chhabildas, Elastic shock response and spall strength of concrete, in *Shock Compression of Condensed Matter*, ed. by S.C. Schmidt, D.P. Dandekar, J.W. Forbes, AIP Conference Proceedings, vol. 429 (AIP, College Park, MD, 1998), pp. 557–560

M.E. Kipp, R.R. Martinez, E.S. Hertel, E.L. Baker, B.E. Fuchs, C.L. Chin, Experiments and simulations of spinning shaped charges with fluted liners, in *18th International Symposium on Ballistics*, ed. by W.G. Reinecke, vol. 1 (Technomic Publishing Co, Lancaster, PA, 1999a), pp. 499–506

M.E. Kipp, R.R. Martinez, R.A. Benham, S.H. Fischer, Explosive containment chamber vulnerability to chemical munition fragment impact, SAND99-0189 (Sandia National Laboratories, Albuquerque, NM, 1999b)

M.E. Kipp, R.R. Martinez, Assessment of chemical munition fragment impact in an explosive containment chamber, SAND2000-0327 (Sandia National Laboratories, Albuquerque, NM, 2000)

M.E. Kipp, L.C. Chhabildas, W.D. Reinhart, M.K. Wong, Polyurethane foam impact experiments and simulations, in *Shock Compression of Condensed Matter*, ed. by M.D. Furnish, L.C. Chhabildas, R.S. Hixson, AIP Conference Proceedings, vol. 505 (AIP, College Park, MD, 2000), pp. 313–316

L.N. Kmetyk, L.C. Chhabildas, M.B. Boslough, R.J. Lawrence, Effect of phase change in a debris cloud on a backwall structure, in *High-Pressure Science and Technology*, ed. by S.C. Schmidt, J.W. Shaner, G.A. Samara, M. Ross, AIP Conference Proceedings, vol. 309 (AIP, College Park, MD, 1994), pp. 1829–1832

M.D. Knudson, D.L. Hanson, J.E. Bailey, C.A. Hall, J.R. Asay, W.W. Anderson, Equation of state measurements in liquid deuterium to 70 GPa. Phys. Rev. Lett. **87**, 225501 (2001)

M.D. Knudson, D.L. Hanson, J.E. Bailey, R.W. Lemke, C.A. Hall, C. Deeney, J.R. Asay, Equation of state measurements in liquid deuterium to 100 GPa. J. Phys. A: Math. Gen. **36**(22), 6149–6158 (2003a)

M.D. Knudson, D.L. Hanson, J.E. Bailey, C.A. Hall, J.R. Asay, Use of a wave reverberation technique to infer the density compression of shocked liquid deuterium to 75 GPa. Phys. Rev. Lett. **90**(3), 035505 (2003b)

M.D. Knudson, R.W. Lemke, D.B. Hayes, C.A. Hall, C. Deeney, J.R. Asay, Near-absolute Hugoniot measurements in aluminum to 500 GPa using a magnetically accelerated flyer plate technique. J. Appl. Phys. **94**, 4420–4431 (2003c)

M.D. Knudson, C.A. Hall, R. Lemke, C. Deeney, J.R. Asay, High velocity flyer plate launch capability on the Sandia Z accelerator. Int. J. Impact Eng. **29**, 377–384 (2003d)

M.D. Knudson, D.L. Hanson, J.E. Bailey, C.A. Hall, J.R. Asay, C. Deeney, Principal Hugoniot, reverberating wave, and mechanical reshock measurements of liquid deuterium to 400 GPa using plate impact techniques. Phys. Rev. B **69**(14), 144209 (2004)

M.D. Knudson, J.R. Asay, C. Deeney, Adiabatic release measurements in aluminum from 240- to 500-GPa states on the principal Hugoniot. J. Appl. Phys. **97**, 073514 (2005)

M.D. Knudson, M.P. Desjarlais, D.H. Dolan, Shock-wave exploration of the high-pressure phases of carbon. Science **322**, 1822–1825 (2008)

M.D. Knudson, M.P. Desjarlais, Shock compression of quartz to 1.6 TPa: Redefining a pressure standard. Phys. Rev. Lett. **103**, 225501 (2009)

M.D. Knudson, Megaamps, megagauss, megabars: using the Sandia Z machine to perform extreme material dynamic experiments, in *Shock Compression of Condensed Matter*, ed. by M.L. Elert, W.T. Buttler, J.P. Borg, J.L. Jordan, T.J. Vogler. AIP Conference Proceedings, vol. 1426 (AIP, College Park, MD, 2012), pp. 35–42

M.D. Knudson, M.P. Desjarlais, R.W. Lemke, T.R. Mattsson, Probing the interiors of the ice giants: Shock compression of water to 700 GPa and 38 g/cc. Phys Rev Lett **108**, 091102 (2012)

M.D. Knudson, M.P. Desjarlais, A. Becker, R.W. Lemke, K.R. Cochrane, M.E. Savage, D.E. Bliss, T.R. Mattsson, R. Redmer, Direct observation of an abrupt insulator-to-metal transition in dense liquid hydrogen. Science **348**, 1455–1460 (2015)

H. Kolsky, An investigation of the mechanical properties of materials at very high rates of loading. Proc. Phys. Soc. Lond. B **62**, 676–700 (1949)

C. Konrad, R. Hollenbach, Techniques for determining velocity and position of small hypervelocity spheres, SAND75-0624 (Sandia National Laboratories, Albuquerque, NM, 1975)

C. Konrad, R. Moody, Rear surface pin triggering technique, SAND86-0791 (Sandia National Laboratories, Albuquerque, NM, 1986)

C.H. Konrad, L.C. Chhabildas, M.B. Boslough, A.J. Piekutowski, K.L. Poormon, S.A. Mullin, D.L. Littlefield, Dependence of debris cloud formation on projectile shape, in *High-Pressure Science and Technology*, ed. by S.C. Schmidt, J.W. Shaner, G.A. Samara, M. Ross, AIP Conference Proceedings, vol. 309 (AIP, College Park, MD, 1994), pp. 1845–1848

C.H. Konrad, W.M. Trott, C.A. Hall, J.S. Lash, R.J. Dukart et al., Use of z-pinch sources for high pressure shock wave experiments, in *Shock Compression of Condensed Matter*, ed. by S.C. Schmidt, D.P. Dandekar, J.W. Forbes, AIP Conference Proceedings, vol. 429 (AIP, College Park, MD, 1998), pp. 997–1000

C.H. Konrad, W.D. Reinhart, L.C. Chhabildas, G.A. Mann, D.A. Mosher et al., Experimental benchmark data for ALEGRA code calculations, in *Shock Compression of Condensed Matter*, ed. by M.D. Furnish, L.C. Chhabildas, R.S. Hixson. AIP Conference Proceedings, vol. 505 (AIP, College Park, MD, 2000), pp. 1011–1014

R.G. Kraus, S. Root, R.W. Lemke, S.T. Stewart, S.B. Jacobsen, T.R. Mattsson, Shock thermodynamics of iron and impact vaporization of planetesimal cores. Nat. Geosci. **8**, 269–272 (2015)

G. Kresse, J. Hafner, Ab initio molecular dynamics for liquid metals. Phys. Rev. B **47**(1), 558–561 (1993)

G. Kresse, J. Hafner, Ab initio molecular-dynamics simulation of the liquid-metal – Amorphous-semiconductor transition in germanium. Phys. Rev. B **49**(20), 14251–14269 (1994)

G. Kresse, J. Furthmüller, Efficient iterative schemes for *ab initio* total-energy calculations using a plane-wave basis set. Phys. Rev. B **54**(16), 11169–11186 (1996)

R.D. Krieg, J.C. Swearengen, R.W. Rohde, A physically-based internal variable model for rate-dependent plasticity, in *Elastic Behavior of Pressure Vessel and Piping Compounds*, ed. by T.Y. Chang, E. Krempl (Elsevier, Amsterdam, 1978), pp. 15–28

R.W. Kulterman, F.W. Neilson, W.B. Benedick, Pulse generator based on high shock demagnetization of ferromagnetic material. J. Appl. Phys. **29**, 500–501 (1958)

J.M.D. Lane, G.S. Grest, A.P. Thompson et al., Shock compression of hydrocarbon polymer foam using molecular dynamics, in *Shock Compression of Condensed Matter*, ed. by M.L. Elert, W.T. Buttler, J.P. Borg, J.L. Jordan, T.J. Vogler, AIP Conference Proceedings, vol. 1426 (AIP, College Park, MD, 2012), pp. 1435–1438

J. Lankford, Mechanisms responsible for strain-rate-dependent compressive strength in ceramic materials. J. Am. Ceram. Soc. **64**(2), C33–C34 (1981)

R.J. Lawrence, WONDY IIIA: A computer program for one-dimensional wave propagation, SC-DR-70-315 (Sandia National Laboratories, Albuquerque, NM, 1970)

R.J. Lawrence, A nonlinear viscoelastic equation of state for use in stress propagation calculations, SLA-73-0635 (Sandia National Laboratories, Albuquerque, NM, 1973)

R.J. Lawrence, L.W. Davison, Analysis of nonlinear plane-wave propagation in piezoelectric solids, SAND77-0217 (Sandia National Laboratories, Albuquerque, NM, 1977)

R.J. Lawrence, J.R. Asay, The high pressure multiple shock response of aluminum, in *High-Pressure Science and Technology Sixth AIRAPT Conference*, ed. by K.D. Timmerhaus, M.S. Barber (Plenum, New York, NY, 1979), pp. 88–98

R.J. Lawrence, Enhanced momentum transfer from hypervelocity particle impacts. Int. J. Impact Eng **10**, 337–349 (1990a)

R.J. Lawrence, Stand-off shields for hypervelocity particles, in *Shock Compression of Condensed Matter*, ed. by S.C. Schmidt, J.N. Johnson, L.W. Davison (Elsevier, Amsterdam, 1990b), pp. 959–962

R.J. Lawrence, J.T. Kare, R.M. Zazworsky, D.K. Monroe, System requirements for low earth orbit launch using laser propulsion, in *Proceedings of the 6th International Conference on Emerging Nuclear Energy Systems* (ICENES 91). Fusion Technol **20**, 714–718 (1991)

R.J. Lawrence, A simple approach for the design and optimization of stand-off hypervelocity particle shields, in *AIAA Space Programs and Technologies Conference* (AIAA, Huntsville, AL, 1992a), pp. 24–27

R.J. Lawrence, The equivalence of simple models for radiation-induced impulse, in *Shock Compression of Condensed Matter*, ed. by S.C. Schmidt, R.D. Dick, J.W. Forbes, D.G. Tasker (Elsevier, Amsterdam, 1992b), pp. 785–788

R.J. Lawrence, W.M. Trott, Theoretical analysis of a pulsed-laser-driven hypervelocity flyer launcher. Int. J. Impact Eng. **14**, 439–449 (1993)

R.J. Lawrence, L.N. Kmetyk, L.C. Chhabildas, The influence of phase changes on debris-cloud interactions with protected structures. Int. J. Impact Eng. **17**, 487–496 (1995)

R.J. Lawrence, J.R. Asay, T.G. Trucano, C. Hall, Analysis of radiation-driven explosive flyers, in *Shock Compression of Condensed Matter*, ed. by M.D. Furnish, L.C. Chhabildas, R.S. Hixson. AIP Conference Proceedings, vol. 505 (AIP, College Park, MD, 2000), pp. 1079–1082

R.J. Lawrence, M.D. Furnish, T.A. Haill, T.G. Trucano, T.A. Mehlhorn, K.G. Budge, C.A. Hall, J.R. Asay, K.R. Cochrane, J.J. MacFarlane, Radiation-driven dynamic target response for dissimilar material jetting and for debris effects in partitioned pipes, SAND2001-1688C (Sandia National Laboratories, Albuquerque, NM, 2001) [Presented at III Khariton's Topical Scientific Readings International Conference, Sarov, Russia, February 2001]

R.J. Lawrence, T.A. Mehlhorn, T.A. Haill et al., Analysis of radiation-driven jetting experiments on Nova and Z, in *Shock Compression of Condensed Matter*, ed. by M.D. Furnish, N.N. Thadhani, Y. Horie, AIP Conference Proceedings, vol. 620 (AIP, College Park, MD, 2002), pp. 291–294

R.J. Lawrence, W.D. Reinhart, L.C. Chhabildas, T.F. Thornhill, Spectral measurements of hypervelocity impact flash. Int. J. Impact Eng. **33**(1–12), 353–363 (2006)

R.J. Lawrence, L.C. Chhabildas, Simple models for aspects of IFE shock mitigation. Fusion Sci. Technol. **52**(3), 494–498 (2007)

R.J. Lawrence, J.R. Asay, Y.M. Gupta, C.J. Bakeman, T.A. Haill, Fragment producing chemical-electrical launcher (FP-CEL): Feasibility study (Part I), SAND2008-7999 (Sandia National Laboratories, Albuquerque, NM, 2009a)

R.J. Lawrence, T.A. Haill, B.L. Freeman, Y.M. Gupta, Fragment producing chemical-electrical launcher (FP-CEL): Numerical analysis (Part II), SAND2008-8000 (Sandia National Laboratories, Albuquerque, NM, 2009b)

R.J. Lawrence, M.D. Furnish, J.L. Remo, Analytic models for pulsed x-ray impulse coupling, in *Shock Compression of Condensed Matter*, ed. by M.L. Elert, W.T. Buttler, J.P. Borg, J.L. Jordan, T.J. Vogler, AIP Conference Proceedings, vol. 1426 (AIP, College Park, MD, 2012), pp. 883–886

L.M. Lee, Some dynamic mechanical properties of pyrolytic boron nitride, SC-RR-67-2947 (Sandia National Laboratories, Albuquerque, NM, 1967)

L.M. Lee, Dynamic compaction of distended isotropic pyrolytic boron nitride, SC-RR-68-2 (Sandia National Laboratories, Albuquerque, NM, 1968)

L.M. Lee, R.P. May, T.R. Guess, Some dynamic mechanical properties of distended carbons. AIAA J. **8**(8), 1421–1428 (1970)

L.M. Lee, Shock response of distended CVD carbon felt, SC-RR-72-0814 (Sandia National Laboratories, Albuquerque, NM, 1972)

L.M. Lee, Low stress shock behavior of cellular concrete, SLA-73-0164 (Sandia National Laboratories, Albuquerque, NM, 1973a)

L.M. Lee, Nonlinearity in the piezoresistance coefficient of impact-loaded manganin. J. Appl. Phys. **44**(9), 4017–4022 (1973b)

L.M. Lee, Shock-induced index-of-refraction variations in PMMA, sapphire and lithium fluoride. Ktech Corporation Technical Report No. TR76–04 (Ktech Corporation, Albuquerque, NM, 1976)

L.M. Lee, A.C. Schwarz, Shock characterization of hexanitroazobenzene (HNAB), in *Proceedings of the 7th International Detonation Symposium*, NSWC MP 82-334, ed. by J.M. Short, S.J. Jacobs (Naval Surface Warfare Center, Dahlgren, VA, 1982), pp. 416–424

L.M. Lee, W.D. Williams, R.A. Graham, F. Bauer, Studies of the Bauer piezoelectric polymer gauge (PVDF) under impact loading, in *Shock Waves in Condensed Matter*, ed. by Y.M. Gupta (Plenum, New York, NY, 1986), pp. 497–502

L.M. Lee, R.A. Graham, F. Bauer, R.P. Reed, Standardized Bauer PVDF piezoelectric polymer shock gauge. J Phys Colloques **49**(C3), 651–657 (1988)

L.M. Lee, D.A. Hyndman, R.P. Reed, F. Bauer, PVDF applications in shock measurements, in *Shock Compression of Condensed Matter*, ed. by S.C. Schmidt, J.N. Johnson, L.W. Davison (Elsevier, Amsterdam, 1990), pp. 821–824

L.M. Lee, D.E. Johnson, F. Bauer, R.P. Reed, J.I. Greenwoll, Piezoelectric polymer PVDF application under soft x-ray induced shock loading, in *Shock Compression of Condensed Matter*, ed. by S.C. Schmidt, R.D. Dick, J.W. Forbes, D.G. Tasker (Elsevier, Amsterdam, 1992), pp. 879–882

L.M. Lee, S.T. Montgomery, P.H. Jilbert, Multi-element quartz shock gauge development, in *Shock Compression of Condensed Matter*, ed. by M.D. Furnish, L.C. Chhabildas, R.S. Hixson, AIP Conference Proceedings, vol. 505 (AIP, College Park, MD, 2000), pp. 1015–1018

Y.K. Lee, F.L. Williams, R.A. Graham, B. Morosin, Specific surface measurements of shock modified powders, in *Shock Waves in Condensed Matter*, ed. by J.R. Asay, R.A. Graham, G.K. Straub (Elsevier, Amsterdam, 1984), pp. 399–402

Y.T. Lee, R.M. More, An electron conductivity model for dense plasmas. Phys. Fluids **27**(5), 1273–1286 (1984)

R.W. Lemke, M.D. Knudson, A.C. Robinson, T.A. Haill, K.W. Struve et al., Considerations for generating up to 10 Mbar magnetic drive pressures with the refurbished Z machine (ZR), in *Shock Waves in Condensed Matter*, ed. by M.D. Furnish, N.N. Thadhani, Y. Horie, AIP Conference Proceedings, vol. 651 (AIP, College Park, MD, 2002), pp. 299–302

R.W. Lemke, M.D. Knudson, C.A. Hall, T.A. Haill, M.P. Desjarlais et al., Characterization of magnetically accelerated flyer plates. Phys. Plasmas **10**, 1092–1099 (2003a)

R.W. Lemke, M.D. Knudson, A.C. Robinson et al., Self-consistent, two-dimensional magnetohydrodynamic simulations of magnetically driven flyer plates. Phys. Plasmas **10**(5), 1867–1874 (2003b)

R.W. Lemke, M.D. Knudson, J.-P. Davis, D. Bliss, H.C. Harjes, Self-consistent, 2D magnetohydrodynamic simulations of magnetically driven flyer plate experiments on the Z-machine, in *Shock Compression of Condensed Matter*, ed. by M.D. Furnish, Y.M. Gupta, J.W. Forbes, AIP Conference Proceedings, vol. 706 (AIP, College Park, MD, 2004), pp. 1175–1180

R.W. Lemke, M.D. Knudson, D.E. Bliss, K. Cochrane, J.-P. Davis, A.A. Giunta, H.C. Harjes, S.A. Slutz, Magnetically accelerated, ultrahigh velocity flyer plates for shock wave experiments. J. Appl. Phys. **98**, 073530 (2005)

R.W. Lemke, D.B. Sinars, E.M. Waisman, M.E. Cuneo et al., Effects of mass ablation on the scaling of x-ray power with current in wire-array z pinches. Phys. Rev. Lett. **102**, 025005 (2009)

R.W. Lemke, M.D. Knudson, J.-P. Davis, Magnetically driven hyper-velocity launch capability at the Sandia Z accelerator. J. Impact Eng. **38**, 480–485 (2011)

R.W. Lemke, M.R. Martin, R.D. McBride et al., Determination of pressure and density of shocklessly compressed beryllium from x-ray radiography of a magnetically driven cylindrical liner implosion, in *Shock Compression of Condensed Matter*, ed. by M.L. Elert, W.T. Buttler, J.P. Borg, J.L. Jordan, T.J. Vogler, AIP Conference Proceedings, vol. 1426 (AIP, College Park, MD, 2012), pp. 473–476

R.W. Lemke, M.D. Knudson, K. Cochrane, M.P. Desjarlais, J.R. Asay, On the scaling of the magnetically accelerated flyer plate technique to currents greater than 20 MA, in *18th American Physical Society Shock Compression in Condensed Matter and 24th International Association for the Advancement of High Pressure Science and Technology Conference,* ed. by W. Buttler, M. Furlanetto, W. Evans, Journal of Physics Conference Series, vol. 500 (IOP Publishing, Bristol, 2014), 152009

J.F. Leon, R.B. Spielman, J.R. Asay, C.A. Hall, W.A. Stygar, P. L'eplattennier, Flux compression experiments on the Z Accelerator, in *Proceedings 12th IEEE International Pulsed Power Conference*, Vol. 1 (IEEE, Piscataway, NJ, 1999), pp. 275–278

H.E. Lindberg, A.L. Florence, Dynamic pulse buckling – Theory and experiment, DNA 6503H (Defense Nuclear Agency, Washington, DC, 1983) [Both Martinus Nijhoff Pubs., Dordrecht,

The Netherlands and Springer (in English) published versions of Lindberg and Florence's book in 1987]

J. Lipkin, M.E. Kipp, Wave structure measurement and analysis in hypervelocity impact experiments. J. Appl. Phys. **47**(5), 1979–1986 (1976)

J. Lipkin, J.R. Asay, Reshock and release of shock-compressed 6061-T6 aluminum. J. Appl. Phys. **48**, 182–189 (1977)

W. Lorenzen, B. Holst, R. Redmer, Metallization in hydrogen-helium mixtures. Phys. Rev. B **84**, 235109 (2011)

P. Loubeyre, S. Brygoo, J. Eggert, P.M. Celliers, D.K. Spaulding, J.R. Rygg, T.R. Boehly, G.W. Collins, R. Jeanloz, Extended data set for the equation of state of warm dense hydrogen isotopes. Phys. Rev. B **86**, 144115 (2012)

C.D. Lundergan, Contact fuzing study – First report, SC-TM-141-57 (12) (Sandia National Laboratories, Albuquerque, NM, 1957)

C.D. Lundergan, A method for measuring (1) the parameter of impact between two surfaces and (2) the properties of the plane shock waves produced, SC-4421 (Sandia National Laboratories, Albuquerque, NM, 1960)

C.D. Lundergan, The Hugoniot equation of state of 6061-T6 aluminum at low pressures, SC-4637 (RR) (Sandia National Laboratories, Albuquerque, NM, 1961)

C.D. Lundergan, J.H. Smith, Method of determining spall thresholds using one-dimensional shock waves, SC-DC-2629 (Sandia National Laboratories, Albuquerque, NM, 1962)

C.D. Lundergan, Spall Fracture, in *Proceedings of Symposium on Structural Dynamics under High Impulse Loading*, ASD-TDR-63-140 (Wright-Patterson Air Force Base, Dayton, OH, 1963), pp. 357–381

C.D. Lundergan, W. Herrmann, Equation of state of 6061-T6 aluminum at low pressures. J. Appl. Phys. **34**(7), 2046–2052 (1963)

C.D. Lundergan, Discussion of the transmitted waveforms in a periodic laminated composite. J. Appl. Phys. **42**(11), 4148–4155 (1970)

C.D. Lundergan, D.S. Drumheller, Dispersion of shock waves in composite materials, in *Shock Waves and the Mechanical Properties of Solids* (Proceedings of 17th Sagamore Army Materials Research Center Conference), ed. by J.J. Burke, V. Weiss (Syracuse University Press, Syracuse, NY, 1971a), Vol. 17, pp. 141–145

C.D. Lundergan, D.S. Drumheller, Propagation of stress waves in a laminated plate composite. J. Appl. Phys. **42**, 669–675 (1971b)

C.D. Lundergan, D.S. Drumheller, Propagation of transient stress pulses in an obliquely laminated composite, in *Dynamics of Composite Materials* (Proceedings of Applied Mechanics Division of the ASME), ed. by E.H. Lee (Springer, New York, NY, 1972), pp. 35–47

P.C. Lysne, W.J. Halpin, Shock compression of porous iron in the region of incomplete compaction. J. Appl. Phys. **39**, 5488–5495 (1968)

P.C. Lysne, R.R. Boade, C.M. Percival, O.E. Jones, Determination of release adiabats and recentered Hugoniot curves by shock reverberation techniques. J. Appl. Phys. **40**, 3786–3795 (1969)

P.C. Lysne, A comparison of calculated and measured low-stress Hugoniots and release adiabats of dry and water-saturated tuff. J. Geophys. Res. **75**, 4375–4386 (1970)

P.C. Lysne, Determination of high pressure equations of state by shock loading porous specimens. J. Appl. Phys. **42**, 2152–2153 (1971a)

P.C. Lysne, Equation of state of liquid $CCl_4$ to 16 kbar: A comparison of shock and static experiments. J. Chem. Phys. **55**, 5242–5246 (1971b)

P.C. Lysne, Nonlinear U(u) Hugoniots of liquids at low pressures. J. Chem. Phys. **57**, 492–494 (1972a)

P.C. Lysne, One-dimensional theory of polarization by shock waves: Application to quartz gauges. J. Appl. Phys. **43**, 425–431 (1972b)

P.C. Lysne, Dielectric breakdown of shock-loaded PZT 65/35. J. Appl. Phys. **44**, 577–582 (1973)

P.C. Lysne, D.R. Hardesty, Fundamental equation of state of liquid nitromethane to 100 kbar. J. Chem. Phys. **59**(12), 6512–6523 (1973)

P.C. Lysne, Prediction of dielectric breakdown in shock-loaded ferroelectric ceramics. J. Appl. Phys. **46**, 230–232 (1975)

P.C. Lysne, L.C. Bartel, Electromechanical response of PZT 65/35 subjected to axial shock loading. J. Appl. Phys. **46**, 222–229 (1975)

P.C. Lysne, C.M. Percival, Electric energy generation by shock compression of ferroelectric ceramics: Normal mode response of PZT 95/5. J. Appl. Phys. **46**, 1519–1525 (1975)

P.C. Lysne, Dielectric properties of shock wave compressed PZT 95/5. J. Appl. Phys. **48**, 1020–1023 (1976)

P.C. Lysne, C.M. Percival, Analysis of shock-wave-actuated ferroelectric power supplies. Ferroelectrics **10**, 129–133 (1976)

P.C. Lysne, Shock-induced polarization of a ferroelectric ceramic. J. Appl. Phys. **48**, 1024–1031 (1977)

P.C. Lysne, Dielectric properties of shock wave compressed PMMA and an alumina-loaded epoxy. J. Appl. Phys. **49**, 4186–4190 (1978a)

P.C. Lysne, Electrical response of relaxing dielectrics compressed by shock waves: The axial-mode problem. J. Appl. Phys. **49**, 4180–4185 (1978b)

P.C. Lysne, Dielectric relaxation in insulators slightly damaged by stress pulses. J. Appl. Phys. **54**, 3160–3165 (1983)

G.A. Lyzenga, T.J. Ahrens, Multi-wavelength optical pyrometer for shock compression experiments. Rev. Sci. Instrum. **50**, 1421–1424 (1979)

R.J. Magyar, S.S. Root, T.A. Haill et al., Equations of state of mixtures: Density functional theory (DFT): Simulations and experiments on Sandia's Z machine, in *Shock Compression of Condensed Matter*, ed. by M.L. Elert, W.T. Buttler, J.P. Borg, J.L. Jordan, T.J. Vogler, AIP Conference Proceedings, vol. 1426 (AIP, College Park, MD, 2012), pp. 1195–1198

M.R. Martin, R.W. Lemke, R.D. McBride et al., Analysis of cylindrical ramp compression experiment with radiography based surface fitting method, in *Shock Compression of Condensed Matter*, ed. by M.L. Elert, W.T. Buttler, J.P. Borg, J.L. Jordan, T.J. Vogler, AIP Conference Proceedings, vol. 1426 (AIP, College Park, MD, 2012), pp. 357–360

A.R. Mathews, R.H. Warnes, G.R. Whittemore, W.F. Hemsing, VISAR line-imaging interferometer. SPIE Proc. No. 1346, 133–140 (1990)

T.R. Mattsson, J.M.D. Lane, K.R. Cochrane, M.P. Desjarlais, A.P. Thompson, F. Pierce, G.S. Grest, First principles and classical molecular dynamics simulation of shocked polymers. Phys. Rev. B **81**, 054103 (2010)

T.R. Mattsson, S. Root, A.E. Mattsson, L. Shulenburger, R.J. Magyar, D.G. Flicker, Validating density functional theory (DFT) simulations at high energy-density conditions with liquid krypton shock experiments to 850 GPa on Sandia's Z machine. Phys. Rev. B **90**, 184105 (2014)

S. Mazevet, M.P. Desjarlais, L.A. Collins et al., Simulations of the optical properties of warm dense aluminum. Phys. Rev. E **71**, 016409 (2005)

J.M. McAfee, B.W. Asay et al., Deflagration to detonation in granular HMX, in *Proceedings of the Ninth International Detonation Symposium,* ONR 113291-7, ed. by J.M. Short, E.L. Lee (Office of Naval Research, Washington, D.C., 1991), pp. 265–279

J.M. McGlaun, S.L. Thompson, J.R. Freeman, COMAG-III: A 2-D MHD code for helical CMF generators, in *Megagauss Physics and Technology*, ed. by P.J. Turchi (Plenum, New York, NY, 1980), pp. 193–203

J.M. McGlaun, Crater: A computer code for calculating late stage crater formation, SAND80-1694 (Sandia National Laboratories, Albuquerque, NM, 1981a)

J.M. McGlaun, New features and revised input instructions for CHARTD, SAND81-2581 (Sandia National Laboratories, Albuquerque, NM, 1981b)

J.M. McGlaun, Improvements in CSQII: A transmitting boundary condition, SAND82-1248 (Sandia National Laboratories, Albuquerque, NM, 1982)

J.M. McGlaun, S.L. Thompson, M.G. Elrick, CTH: A three-dimensional shock wave physics code. Int. J. Impact Eng. **10**, 351–360 (1990)

J.M. McGlaun, J.S. Peery, E.S. Hertel, Shock physics code research at Sandia National Laboratories: Massively parallel computers and advanced algorithms, SAND96-0431C (Sandia National Laboratories, Albuquerque, NM, 1996)

J.M. McMahon, M.A. Morales, C. Pierleoni, D.M. Ceperley, The properties of hydrogen and helium under extreme conditions. Rev. Modern Phys. **84**, 1607–1653 (2012)

T.E. Michaels, Orientation dependence of elastic precursor delay in single crystal tungsten, Ph.D. thesis, Physics Department, Washington State University, Pullman, WA, 1972

J.M. Miller, L.C. Chhabildas, Low temperature experimental capability for use with gas guns, SAND85-0303 (Sandia National Laboratories, Albuquerque, NM, 1985)

J.C.F. Millet, D. Deas, N.K. Bourne, S.T. Montgomery, The deviatoric response of an alumina filled epoxy composite during shock loading. Appl. Phys. **102**, 063518 (2007)

S.T. Montgomery, Computer simulations of 75-mm projectiles impacting sand targets, SAND82-0831 (Sandia National Laboratories, Albuquerque, NM, 1982)

S.T. Montgomery, Creep closure of an opening in a deep potash mine, in *Recent Advances in Engineering Mechanics and Their Impact on Civil Engineering Practice*, ed. by W.F. Chen, A.D.M. Lewis, vol. 2 (American Society of Civil Engineers, New York, NY, 1983), pp. 1024–1027

S.T. Montgomery, Analysis of transitions between ferroelectric and antiferroelectric states under conditions of uniaxial strain, in *Shock Waves in Condensed Matter*, ed. by Y.M. Gupta (Plenum, New York, NY, 1986), pp. 179–184

S.T. Montgomery, P.F. Chavez, Basic equations and solution method for the calculation of the transient electromechanical response of dielectric devices, SAND86-0755 (Sandia National Laboratories, Albuquerque, NM, 1986)

S.T. Montgomery, P.J. Chen, Influences of domain switching and dipole dynamics on the normal mode responses of a ferroelectric ceramic bar. Int. J. Solids Struct. **22**(11), 1293–1305 (1986)

S.T. Montgomery, R. A. Graham, F. Bauer, H. Moulard, Copolymer shock gauge response investigation with the fully coupled electromechanical code, SUBWAY, in *Proceedings of the Workshop on the Technology of Ferroelectric Polymers* (1995), pp. 412–431.

S.T. Montgomery, R.A. Graham, M.U. Anderson, Return to the shorted and shunted quartz gauge problem: Analysis with the SUBWAY code, in *Shock Compression of Condensed Matter*, ed. by S.C. Schmidt, W.C. Tao, AIP Conference Proceedings, vol. 370 (AIP, College Park, MD, 1996), pp. 1025–1028

S.T. Montgomery, R.M. Brannon, J. Robbins, R.E. Setchell, D.H. Zeuch, Simulation of the effects of shock stress and electrical field strength on shock-induced depoling of normally poled PZT 95/5, in *Shock Compression of Condensed Matter*, ed. by M.D. Furnish, N.N. Thadhani, Y.M. Horie, AIP Conference Proceedings, vol. 620 (AIP, College Park, MD, 2002), pp. 201–204

S.T. Montgomery, D.H. Zeuch, Mechanical properties of engineering ceramics, composites and aerospace materials – A model for the bulk mechanical response of porous ceramics exhibiting a ferroelectric-to-antiferroelectric phase transition during hydrostatic compression, in *Ceramic Engineering and Science Proceedings*, Vol. 25, ed. by E. Lara-Curzio, M.J. Readey (The American Ceramic Society, Westerville, OH, 2004), pp. 313–318

S.T. Montgomery, Effects of porosity and pore morphology on the elastic properties of unpoled PZT 95/5-2Nb, SAND2008-6185 (Sandia National Laboratories, Albuquerque, NM, 2008)

S.T. Montgomery, Effects of composition on the mechanical response of alumina filled epoxy, SAND2009-6399 (Sandia National Laboratories, Albuquerque, NM, 2009)

R. Moody, C. Konrad, Magnetic induction system for two-stage gun projectile velocity measurements, SAND84-0638 (Sandia National Laboratories, Albuquerque, NM, 1984)

L.M. Moore, R.A. Graham, R.P. Reed, L.M. Lee, F. Bauer, T.W. Warren, Standardized piezoelectric polymer (PVDF) gauge for detonator response measurements, in *Behaviour of Dense Media Under High Dynamic Pressures* (Association Francaise de Pyrotechnie, Paris, 1989), pp. 35–43

L.M. Moore, R.A. Graham, Response of standardized PVDF piezoelectric polymer gauges to direct shock pressures between 8 and 32 GPa, in *Shock Compression of Condensed Matter*, ed. by S.C. Schmidt, J.N. Johnson, L.W. Davison (Elsevier, Amsterdam, 1990), pp. 813–816

B. Morosin, R.A. Graham, X-ray diffraction line broadening studies on shock-loaded $TiO_2$ and $Al_2O_3$, in *Shock Waves in Condensed Matter*, ed. by J.R. Asay, R.A. Graham, G.K. Straub (Elsevier, Amsterdam, 1984), pp. 355–362

B. Morosin, R.A. Graham, J.R. Hellmann, Monoclinic to tetragonal conversion of zirconia under shock compression, in *Shock Waves in Condensed Matter*, ed. by J.R. Asay, R.A. Graham, G.K. Straub (Elsevier, Amsterdam, 1984), pp. 383–386

B. Morosin, R.A. Graham, X-ray diffraction studies on shock-modified materials, in *Metallurgical Applications of Shock-Wave and High-Strain-Rate Phenomena*, ed. by L.E. Murr, K.P. Staudhammer, M.A. Meyers (Marcel Dekker, New York, NY, 1986), pp. 1037–1047

B. Morosin, E.L. Venturini, R.A. Graham, X-ray diffraction studies of shock-synthesized zinc ferrite, in *Shock Waves in Condensed Matter*, ed. by Y.M. Gupta (Plenum, New York, NY, 1986), pp. 797–802

B. Morosin, R.A. Graham, E.L. Venturini, D.S. Ginley, Shock-induced chemical synthesis of phases similar to the high temperature superconductor oxides, in *Shock Waves in Condensed Matter*, ed. by S.C. Schmidt, N.C. Holmes (Elsevier, Amsterdam, 1988a), pp. 439–442

B. Morosin, R.A. Graham, E.L. Venturini, M.J. Carr, D.L. Williamson, Shock-induced chemical synthesis of barium ferrites, in *Shock Waves in Condensed Matter*, ed. by S.C. Schmidt, N.C. Holmes (Elsevier, Amsterdam, 1988b), pp. 435–438

B. Morosin, R.A. Graham, S.S. Pollack, X-ray diffraction line broadening, in *Shock Compression of Condensed Matter*, ed. by S.C. Schmidt, R.D. Dick, J.W. Forbes, D.G. Tasker (Elsevier, Amsterdam, 1992), pp. 613–616

S.A. Mullin, D.L. Littlefield, L.C. Chhabildas, A.J. Piekutowski, Computational simulations of experimental impact data obtained at 7 to 11 km/s with aluminum and zinc, in *High-Pressure Science and Technology*, ed. by S.C. Schmidt, J.W. Shaner, G.A. Samara, M. Ross, AIP Conference Proceedings, vol. 309 (AIP, College Park, MD, 1994), pp. 1817–1820

D.E. Munson, L.M. Barker, Dynamically determined pressure-volume relationships for aluminum, copper, and lead. J. Appl. Phys. **37**(4), 1652–1660 (1966)

D.E. Munson, K.W. Schuler, Hugoniot predictions for mechanical mixtures using effective moduli, in *Shock Waves and the Mechanical Properties of Solids*, ed. by J. Burke, V. Weiss (Syracuse University Press, Syracuse, NY, 1971), pp. 185–201

D. Munson, R. May, Interior ballistics of a two-stage light gas gun, SAND75-0323 (Sandia National Laboratories, Albuquerque, NM, 1975)

D.E. Munson, R.R. Boade, K.W. Schuler, Stress wave propagation in $A1_2O_3$-epoxy mixtures. J. Appl. Phys. **49**, 4797–4807 (1977)

D.L. Munson, R.J. Lawrence, Dynamic deformation of polycrystalline alumina. J. Appl. Phys. **50**(10), 6272–6282 (1979)

S.A. Myers, C.C. Koch, Y. Horie, R.A. Graham, TEM of nickel aluminides produced by shock compaction, in *Shock Waves in Condensed Matter*, ed. by Y.M. Gupta (Plenum, New York, NY, 1986), pp. 755–759

F.W. Neilson, W.B. Benedick, The piezoelectric response of quartz beyond its Hugoniot elastic limit. Bull. Am. Phys. Soc. Ser. II **5**(7), 511 (1960)

F.W. Neilson, W.B. Benedick, W.P. Brooks, R.A. Graham, G.W. Anderson, Electrical and optical effects of shock waves in crystalline quartz, in *Les Ondes de Detonation*, ed. by G. Ribaud (Centre National de la Recherche Scientifique, Paris, 1962), pp. 391–419

W.J. Nellis, Metastable solid metallic hydrogen. Philos. Mag. B **79**(4), 655–661 (1999)

W.J. Nellis, S.T. Weir, A.C. Mitchell, Minimum metallic conductivity of fluid hydrogen at 140 GPa (1.4 Mbar). Phys. Rev. B **59**(5), 3434–3449 (1999)

W.J. Nellis, Metastable metallic hydrogen glass, UCRL-JC-142360 (Lawrence Livermore National Laboratory, Livermore, CA, 2001)

W.J. Nellis, Wigner and Huntington: The long quest for metallic hydrogen. High Press. Res. **33**(2), 369–376 (2013)

T.V. Nordstrom, R.W. Rohde, D.J. Mottern, Explosive strengthening of a Cu-Be alloy. Metall. Trans. **6**, 1561–1568 (1975)

F.R. Norwood, R.A. Graham, A. Sawaoka, Numerical simulation of a sample recovery fixture for high velocity impact, in *Shock Waves in Condensed Matter*, ed. by Y.M. Gupta (Plenum, New York, NY, 1986), pp. 837–842

J.W. Nunziato, W. Herrmann, The general theory of shock waves in elastic nonconductors. Arch. Ration. Mech. Anal. **47**, 272–287 (1972)

J.W. Nunziato, K.W. Schuler, E.K. Walsh, The bulk response of viscoelastic solids. Trans. Soc. Rheol. **16**, 15–32 (1972)

J.W. Nunziato, K.W. Schuler, Evolution of steady shock waves in polymethyl methacrylate. J. Appl. Phys. **44**, 4774–4775 (1973a)

J.W. Nunziato, K.W. Schuler, Shock pulse attenuation in a nonlinear viscoelastic solid. J. Mech. Phys. Solids **21**, 447–457 (1973b)

J.W. Nunziato, K.W. Schuler, D.B. Hayes, Wave propagation calculations for nonlinear viscoelastic solids, in *Computational Methods in Nonlinear Mechanics*, ed. by J.T. Oden et al. (The Texas Institute for Computational Mechanics, Austin, TX, 1974a), pp. 489–498

J.W. Nunziato, E.K. Walsh, K.W. Schuler, L.M. Barker, Wave propagation in non-linear viscoelastic solids, in *Mechanics of Solids*, ed. by C. Truesdell (Springer, Berlin, 1974b), pp. 1–108 [The second edition was published in 1984 by Springer, New York, NY]

J.W. Nunziato, Acceleration waves in elastic materials with two temperatures. Int. J. Nonlinear Mech. **10**, 137–142 (1975)

J.W. Nunziato, K.W. Schuler, E.K. Walsh, The influence of precompression on acceleration wave propagation in a nonlinear viscoelastic material. J. Appl. Mech. **42**, 731–732 (1975)

J.W. Nunziato, J.E. Kennedy, Shock-wave evolution in a chemically reacting solid. J. Mech. Phys. Solids **24**, 107–124 (1976)

J.W. Nunziato, J.E. Kennedy, D.E. Amos, The thermal ignition time for homogeneous explosives involving two parallel reactions. Combust. Flame **43**, 265–268 (1977)

J.W. Nunziato, D.S. Drumheller, The thermodynamics of Maxwellian materials. Int. J. Solids Struct. **14**, 545–558 (1978)

J.W. Nunziato, E.K. Walsh, J.E. Kennedy, A continuum model for hot-spot initiation of granular explosives, in *Behaviour of Dense Media under High Dynamic Pressures* (CEA, Paris, 1978a), pp. 139–148

J.W. Nunziato, E.K. Walsh, J.E. Kennedy, Behavior of one-dimensional acceleration waves in an inhomogeneous granular solid. Int. J. Eng. Sci. **16**, 637–648 (1978b)

J.W. Nunziato, M.E. Kipp, Numerical studies of initiation, detonation, and detonation failure in nitromethane, SAND81-0669 (Sandia National Laboratories, Albuquerque, NM, 1983)

J.W. Nunziato, Initiation and growth-to-detonation in reactive mixtures, in *Shock Compression of Condensed Matter*, ed. by J.R. Asay, R.A. Graham, G.K. Straub (Elsevier, Amsterdam, 1984), pp. 581–588

B.V. Oliver, Non-ideal MHD plasma regimes and their relevance to the study of dynamic z-pinches, MRC/ABQ-R-1947 (Mission Research Corporation, Albuquerque, NM, 1999)

D.L. Paisley, Laser-driven miniature flyer plates for shock initiation of secondary explosives, in *Shock Compression of Condensed Matter*, ed. by S.C. Schmidt, J.N. Johnson, L.W. Davison (Elsevier, Amsterdam, 1990), pp. 733–736

H.R. Pak, Y. Horie, R.A. Graham, Synthesis of nickel-aluminum alloys by shock compression of composite particles, in *Shock Waves in Condensed Matter*, ed. by Y.M. Gupta (Plenum, New York, NY, 1986), pp. 761–766

H.S. Park, N.R. Barton, J.L. Belof, K.J.M. Blobaum, R.M. Cavallo et al., Experimental results of tantalum material strength at high pressure and high strain rate, in *Shock Compression of Condensed Matter*, ed. by M.L. Elert, W.T. Buttler, J.P. Borg, J.L. Jordan, T.J. Vogler, AIP Conference Proceedings, vol. 1426 (AIP, College Park, MD, 2012), pp. 1371–1374

S.L. Passman, J.W. Nunziato, P.B. Bailey, E.K. Walsh, A mixture theory for suspensions, in *Rheology*, ed. by G. Astarita, G. Marrucci, L. Nicolais (Plenum, New York, NY, 1980), Vol. 2: Fluids, pp. 583–589

S.L. Passman, J.W. Nunziato, Shearing flow of a fluid-saturated granular material, in *Mechanics of Structured Media*, ed. by A.P.S. Selvadurai (Elsevier, Amsterdam, 1981), pp. 343–353

L.E. Pope, L.R. Edwards, The pressure dependence of the austenite start temperature in iron-nickel base alloys. Acta Metall. **21**, 281–288 (1973)

L.E. Pope, A.L. Stevens, Wave propagation in beryllium single crystals, in *Metallurgical Effects at High Strain Rates*, ed. by R.W. Rohde, B.M. Butcher, J.R. Holland, C.H. Karnes (Plenum, New York, NY, 1973), pp. 349–366

L.E. Pope, J.N. Johnson, Shock-wave compression of single-crystal beryllium. J. Appl. Phys. **46**(2), 720–729 (1975)

D.L. Preston, D.L. Tonks, D.C. Wallace, Model of plastic deformation for extreme loading conditions. J. Appl. Phys. **93**, 211–220 (2003)

G.F. Raiser, J.L. Wise, R.J. Clifton, D.E. Grady, D.E. Cox, Plate impact response of ceramics and glasses. J. Appl. Phys. **75**(8), 3862–3869 (1994)

R.P. Reed, Stress pulse-trains from multiple reflections at zone of many discontinuities. A notation for machine solution, SC-4462 (RR) (Sandia National Laboratories, Albuquerque, NM, 1962)

R.P. Reed, D.M. Schuster, Filament fracture and post-impact strength of boron-aluminum composites. J. Compos. Mater. **4**, 514–525 (1970)

R.P. Reed, D.E. Munson, Stress pulse attenuation in cloth-laminate quartz phenolic. J. Compos. Mater. **6**(2), 232–257 (1972)

R.P. Reed, J.I. Greenwoll, The PVDF piezoelectric polymer shock-stress sensor, SAND88-2907 (Sandia National Laboratories, Albuquerque, NM, 1989)

R.P. Reed, R.A. Graham, L.M. Moore, L.M. Lee, D.J. Fogelson, F. Bauer, The Sandia standard for PVDF shock sensors, in *Shock Compression of Condensed Matter*, ed. by S.C. Schmidt, J.N. Johnson, L.W. Davison (Elsevier, Amsterdam, 1990), pp. 825–828

W.D. Reinhart, L.C. Chhabildas, W.M. Trott, D.P. Dandekar, Investigating multi-dimensional effects in single crystal sapphire, in *Shock Compression of Condensed Matter*, ed. by M.D. Furnish, N.N. Thadhani, Y. Horie, AIP Conference Proceedings, vol. 620 (AIP, College Park, MD, 2002), pp. 791–794

W.D. Reinhart, L.C. Chhabildas, Strength properties of Coors AD995 alumina in the shocked state. Int. J. Impact Eng. **29**, 601–619 (2003)

W.D. Reinhart, L.C. Chhabildas, T.J. Vogler, Investigating phase transitions and strength in single crystal sapphire using shock-reshock loading techniques. Int. J. Impact Eng. **33**, 655–669 (2006)

W.D. Reinhart, T.J. Vogler, L.C. Chhabildas, Strength measurements on dry Indiana limestone using ramp loading techniques, in *Shock Compression of Condensed Matter*, ed. by M. Elert, M.D. Furnish, R. Chau, N. Holmes, J. Nguyen, AIP Conference Proceedings, vol. 955 (AIP, College Park, MD, 2007), pp. 1409–1412

W.D. Reinhart, T.F. Thornhill, L.C. Chhabildas, W.G. Breiland, J.L. Brown, Temperature measurements of expansion products from shock compressed materials using high-speed spectroscopy. Int. J. Impact Eng. **35**, 1745–1755 (2008)

D.B. Reisman, A. Toor, R.C. Cauble, C.A. Hall, J.R. Asay, M.D. Knudson, M.D. Furnish, Magnetically driven isentropic compression experiments on the Z accelerator. J. Appl. Phys. **89**(3), 1625–1633 (2001a)

D.B. Reisman, J.W. Forbes, C.M. Tarver, F. Garcia, D.B. Hayes, M.D. Furnish, J.J. Dick, Isentropic compression of high explosives with the Z accelerator, in *Proceedings of the 12th International Detonation Symposium*, ONR Report 333-05-2, ed. by J.M. Short, J.L. Maienschein (Office of Naval Research, Washington, DC, 2001b), pp. 343–348

J.L. Remo, M.D. Furnish, R.J. Lawrence, Plasma-driven z-pinch x-ray loading and momentum coupling in meteorite and planetary materials. J. Plasma Phys. **29**(2), 121–141 (2013)

A.M. Renlund, W.M. Trott, Spectrographic studies of shocked and detonating explosives, in *Shock Waves in Condensed Matter*, ed. by S.C. Schmidt, N.C. Holmes (Elsevier, New York, NY, 1988), pp. 547–552

A.M. Renlund, P.L. Stanton, W.M. Trott, Laser initiation of secondary explosives, in *Proceedings of the 9th International Detonation Symposium*, ONR 113291-7, ed. by J.M. Short, E.L. Lee (Office of Naval Research, Washington, DC, 1989), pp. 1118–1127

A.M. Renlund, W.M. Trott, Raman spectroscopic studies of shock-compressed nitromethane-d3, in *Shock Compression of Condensed Matter*, ed. by S.C. Schmidt, J.N. Johnson, L.W. Davison (Elsevier, New York, NY, 1990), pp. 875–878

P.A. Rigg, M.D. Knudson, R.J. Scharff, R.S. Hixson, Determining the refractive index of shocked [100] lithium fluoride to the limit of transmissibility. J. Appl. Phys. **116**, 033515 (2014)

A.C. Robinson, T.A. Brunner, S.K. Carroll, R.R. Drake et al., ALEGRA: An Arbitrary Lagrangian-Eulerian multimaterial, multiphysics code, in *Proceedings of the 46th AIAA Aerospace Science Meeting and Exhibit*, AIAA 2008–1235 (AIAA, Huntsville, AL, 2008)

A.C. Robinson, J.H.J. Niederhaus, V.G. Weirs, E. Love, Arbitrary Lagrangian–Eulerian 3D ideal MHD algorithms. Int. J. Num. Meth. Fluids **65**, 1438–1450 (2011)

R.W. Rohde, O.E. Jones, Mechanical and piezoelectric properties of shock-loaded X-cut quartz at 573 degrees K. Rev. Sci. Instrum. **39**(3), 313–316 (1968)

R.W. Rohde, J.R. Holland, R.A. Graham, Shock-wave-induced reverse martensitic transformation in Fe-30 pct Ni. Trans. Metall. Soc. AIME **242**, 2017–2019 (1968)

R.W. Rohde, Dynamic yield behavior of shock-loaded iron from 76 to 573 degrees K. Acta Metall. **17**, 353–363 (1969)

R.W. Rohde, R.A. Graham, The effect of hydrostatic pressure on the martensitic reversal of an iron-nickel-carbon alloy. Trans. Metall. Soc. AIME **245**, 2441–2445 (1969)

R.W. Rohde, Temperature dependence of the shock-induced reversal of martensite to austenite in an iron-nickel-carbon alloy. Acta Metall. **18**, 903–913 (1970)

R.W. Rohde, C.E. Albright, The influence of uniaxial tensile stress upon the austenite start temperature of an iron nickel carbon alloy. Scrip. Metall. **5**(2), 151–154 (1971)

R.W. Rohde, W.C. Leslie, R.C. Glenn, The dynamic yield behavior of annealed and cold-worked Fe-0.17 pct Ti alloy. Metallurgical Trans. **3**, 323–328 (1972)

R.W. Rohde, R.A. Graham, Stability of the magnetic phase transformation in shocked Fe-Ni alloys. Philos. Mag. **28**, 941–943 (1973)

R.W. Rohde, J.L. Wise, J.G. Byrne, S. Panchanadeeswaran, Microstructural-hardness correlations in shock-loaded and quasi-statically deformed 6061-T6 aluminum, in *Shock Waves in Condensed Matter*, ed. by J.R. Asay, R.A. Graham, G.K. Straub (Elsevier, Amsterdam, 1984), pp. 407–410

S. Root, J.R. Asay, Loading path and rate dependence of inelastic deformation: X-cut quartz. J. Appl. Phys. **106**, 056109 (2009a)

S. Root, J.R. Asay, Loading path dependence of inelastic behavior: X-cut quartz, in *Shock Compression of Condensed Matter*, ed. by M.L. Elert, W.T. Buttler, M.D. Furnish, W.W. Anderson, W.G. Proud, AIP Conference Proceedings, vol. 1195 (AIP, College Park, MD, 2009b), pp. 999–1002

S. Root, R.J. Magyar, J.H. Carpenter, D.L. Hanson, T.R. Mattsson, Shock compression of a fifth period element: Liquid xenon to 840 GPa. Phys. Rev. Lett. **105**, 085501 (2010)

S. Root, K.R. Cochrane, J.H. Carpenter, T.R. Mattsson, Carbon dioxide shock and reshock equation of state data to 8 Mbar: Experiments and simulations. Phys. Rev. B **87**, 224102 (2013a)

S. Root, T.A. Haill, J.M.D. Lane, A.P. Thompson, G.S. Grest, D.G. Schroen, T.R. Mattsson, Shock compression of hydrocarbon foam to 400 GPa: Experiments, mesoscale modeling, and atomistic simulations. J. Appl. Phys. **114**, 103502 (2013b)

S.E. Rosenthal, M.P. Desjarlais, K.R. Cochrane, Equation of state and electron transport effects in exploding wire evolution, in *Digest of Technical Papers PPPS-2001 IEEE Pulsed Power Plasma Science*, ed. by R. Reinovsky, M. Newton, vol. 1 (IEEE, Piscataway, NJ, 2001), pp. 781–784

S.D. Rothman, J.-P. Davis, J. Maw, C.M. Robinson, K. Parker, J. Palmer, Measurement of the principal isentropes of lead and lead-antimony alloy to ~400 kbar by quasi-isentropic compression. J. Phys. D: Appl. Phys. **38**, 733–740 (2005)

J.S. Rottler, S.L. Thompson, input instructions for the radiation–hydrodynamics code CHARTD, SAND87-0651 (Sandia National Laboratories, Albuquerque, NM, 1987)

A. Ruoff, L.C. Chhabildas, The sodium chloride primary pressure gauge. J. Appl. Phys. **47**(11), 4867–4872 (1976)

E.E. Salpeter, On convection and gravitational layering in Jupiter and in stars of low mass. Ap. J. Lett. **181**, L83–L86 (1973)

T.W.L. Sanford, G.O. Allshouse, B.M. Marder, T.J. Nash, R.C. Mock et al., Improved symmetry greatly increases x-ray power from wire-array z-pinches. Phys. Rev. Lett. **77**, 5063–5066 (1996)

G.S. Sarkisov, S.E. Rosenthal, K.R. Cochrane et al., Nanosecond electrical explosion of thin aluminum wires in a vacuum: Experimental and computational investigations. Phys. Rev. E **71**, 046404 (2005)

B. Schmitt, CTH strong scaling results for a one trillion zone problem to 156 million cores, SAND2013-8753 (Sandia National Laboratories, Albuquerque, NM, 2013)

K.W. Schuler, Propagation of steady shock waves in polymethyl methacrylate. J. Mech. Phys. Solids **18**, 277–293 (1970)

K.W. Schuler, The speed of propagation of release waves in polymethyl methacrylate, in *Proceedings of the 5th International Detonation Symposium*, ONR ACR-184, ed. by S.J. Jacobs, R. Roberts (Office of Naval Research, Washington, D.C., 1971), pp. 470–477

K.W. Schuler, J.W. Nunziato, The dynamic mechanical behavior of polymethyl methacrylate. Rheol. Acta **13**, 773–781 (1974)

K.W. Schuler, J.W. Nunziato, The unloading and reloading behavior of shock compressed polymethyl methacrylate. J. Appl. Phys. **47**, 2995–2998 (1976)

K.W. Schuler, P.C. Lysne, A.L. Stevens, Dynamic mechanical properties of two grades of oil shale. Intl. J. Rock Mech. Min. Sci. **13**, 91–95 (1976)

K.W. Schuler, R.A. Schmidt, Mechanical properties of oil shale of importance to in-situ rubblization, in *Proceedings of the American Nuclear Society Topical Meeting Energy and Mineral Resource Recovery* (ANS, La Grange Park, IL, 1977), pp. 381–391

D.M. Schuster, R.P. Reed, Fracture behavior of shock loaded boron aluminum composite materials. J. Compos. Mater. **3**, 562–576 (1969)

R.L. Schwoebel, *Explosion aboard the Iowa* (Naval Institute Press, Annapolis, MD, 1999)

A.B. Sefkow, S.A. Slutz, J.M. Koning, M.M. Marimak, K.J. Peterson, D.B. Sinars, R.A. Vesey, Design of magnetized liner inertial fusion experiments using the Z facility. Phys. Plasmas **21**, 72711 (2014)

R.E. Setchell, Index of refraction of shock-compressed fused silica and sapphire. J. Appl. Phys. **50**, 8186–8192 (1979)

R.E. Setchell, Ramp-wave initiation of granular explosives. Combust. Flame **43**, 255–264 (1981)

R.E. Setchell, Short-pulse shock initiation of granular explosives, in *Proceedings of the 7th International Detonation Symposium*, NSWC MP 82-334, ed. by J.M. Short, S.J. Jacobs (Naval Surface Warfare Center, Dahlgren, VA, 1982), pp. 857–864

R.E. Setchell, Effects of precursor waves in shock initiation of granular explosives. Combust. Flame **54**, 171–182 (1983)

R.E. Setchell, Grain-size effects on the shock sensitivity of HNS explosives. Combust. Flame **56**, 343–345 (1984)

R.E. Setchell, Experimental studies of chemical reactivity during shock initiation of hexanitrostilbene, in *Proceedings of the 8th International Detonation Symposium*, NSWC MP 86-194, ed. by J.M. Short, W.E. Deal (Naval Surface Warfare Center, Dahlgren, VA, 1986), pp. 15–25

R.E. Setchell, Microstructural effects in shock initiation of granular explosives, in *Proceedings of the International Symposium on Pyrotechnics and Explosives*, ed. by D. Jing (China Academic Publishers, Beijing, 1987), p. 635

R.E. Setchell, P.A. Taylor, A refined equation of state for unreacted hexanitrostilbene. J. Energ. Mater. **6**, 157–159 (1988)

R.E. Setchell, S.T. Montgomery, L.C. Chhabildas, M.D. Furnish, The effects of shock stress and field strength on shock-induced depoling of normally poled PZT 95/5, in *Shock Compression of Condensed Matter*, ed. by M.D. Furnish, L.C. Chhabildas, R.S. Hixson, AIP Conference Proceedings, vol. 505 (AIP, College Park, MD, 2000), pp. 979–982

R.E. Setchell, Refractive index of sapphire at 532 nm under shock compression and release. J. Appl. Phys. **91**, 2833–2841 (2002)

R.E. Setchell, Shock wave compression of the ferroelectric ceramic $Pb_{0.99}(Zr_{0.95}Ti_{0.05})_{0.98}Nb_{0.02}O_3$: Hugoniot states and constitutive mechanical properties. J. Appl. Phys. **94**, 573–588 (2003)

R.E. Setchell, Shock wave compression of the ferroelectric ceramic $Pb_{0.99}(Zr_{0.95}Ti_{0.05})_{0.98}Nb_{0.02}O_3$: depoling currents. J. Appl. Phys. **97**, 013507 (2005)

R.E. Setchell, S.T. Montgomery, D.E. Cox, M.U. Anderson, Dielectric properties of PZT 95/5 during shock compression under high electric fields, in *Shock Compression of Condensed Matter*, ed. by M.D. Furnish, M. Elert, T.P. Russell, C.T. White, AIP Conference Proceedings, vol. 845 (AIP, College Park, MD, 2006), pp. 278–281

R.E. Setchell, Shock wave compression of the ferroelectric ceramic $Pb_{0.99}(Zr_{0.95}Ti_{0.05})_{0.98}Nb_{0.02}O_3$: Microstructural effects. J. Appl. Phys. **101**, 053525 (2007)

R.E. Setchell, M.U. Anderson, S.T. Montgomery, Compositional effects of the shock-compression response of alumina-filled epoxy. J. Appl. Phys. **101**, 083527 (2007a)

R.E. Setchell, S.T. Montgomery, D.E. Cox, M.U. Anderson, Initial temperature effects on the dielectric properties of PZT 95/5 during shock compression, in *Shock Compression of Condensed Matter*, ed. by M. Elert, M.D. Furnish, R. Chau, N. Holmes, J. Nguyen, AIP Conference Proceedings, vol. 555 (AIP, College Park, MD, 2007b), pp. 193–196

S.A. Sheffield, J.W. Rogers Jr., J.N. Castañeda, Velocity measurements of laser-driven flyers backed by high-impedance windows, in *Shock Waves in Condensed Matter*, ed. by Y.M. Gupta (Plenum, New York, NY, 1986), pp. 541–545

S.A. Silling, P.A. Taylor, J.L. Wise, M.D. Furnish, Micromechanical modeling of advanced materials, SAND94-0129 (Sandia National Laboratories, Albuquerque, NM, 1994)

I.K. Simonsen, Y. Horie, R.A. Graham, Shock synthesis of nickel aluminides, in *Shock Waves in Condensed Matter*, ed. by Y.M. Gupta (Plenum, New York, NY, 1986), pp. 743–748

S.A. Slutz, M.C. Herrmann, R.A. Vesey, A.B. Sefkow, D.B. Sinars, D.C. Rovang, K.J. Peterson, M.E. Cuneo, Pulsed power-driven cylindrical liner implosions of laser preheated fuel magnetized with an axial field. Phys. Plasmas **17**, 56303 (2010)

J.H. Smith, L.M. Barker, Measurement of tilt, impact velocity, and impact time between two plane surfaces, SC-4728 (RR) (Sandia National Laboratories, Albuquerque, NM, 1962)

J.E. Smugeresky, T.T. McCabe, R.A. Graham, Effect of powder particle size and shape on the microstructure of explosively compacted stainless steel, in *Shock Waves in Condensed Matter*, ed. by S.C. Schmidt, N.C. Holmes (Elsevier, Amsterdam, 1988), pp. 411–414

B. Song, W. Chen, S.T. Montgomery, M.J. Forrestal, Mechanical response of an alumina-filled epoxy at various strain rates. J. Compos. Mat. **43**(14), 1519–1536 (2009)

R.B. Spielman, F. Long, T.H. Martin, J.W. Poukey, D.B. Seidel, W.A. Stygar, D.H. McDaniel, M.A. Mostrom, K.W. Struve, P. Corcoran, I. Smith, P. Spence, PBFA II-Z: A 20-MA driver for z-pinch experiments, in *Digest of Technical Papers 10th IEEE International Pulsed Power Conference*, ed. by W.L. Baker, G. Cooperstein (IEEE, Piscataway, NJ, 1995), pp. 396–404

R.B. Spielman, C. Deeney, G.A. Chandler, M.R. Douglas, D.L. Fehl, M.K. Matzen, D.H. McDaniel, T.J. Nash, J.L. Porter, T.W.L. Sanford, J.F. Seaman, W.A. Stygar, K.W. Struve, S.P. Breeze, J.S. McGurn, J.A. Torres, D.M. Zagar, T.L. Gilliland, D.O. Jobe, J.L. McKenney, R.C. Mock, M. Vargas, T. Wagoner, D.L. Peterson, PBFA Z: A 60-TW/5-MJ z-pinch driver, in *Dense Z-Pinches*, AIP Conference Proceedings, vol. 409 (AIP, College Park, MD, 1997), pp. 101–118

G. Stansbery, Space waste (Chapter 26), in *Waste: A Handbook for Management*, ed. by T.M. Letcher, D.A. Vallero (Elsevier, New York, NY, 2011), pp. 377–391

P.L. Stanton, R.A. Graham, The electrical and mechanical response of lithium niobate shock loaded above the Hugoniot elastic limit. Appl. Phys. Lett. **31**(11), 723–725 (1977)

P.L. Stanton, E.A. Igel, L.M. Lee, J.H. Mohler, G.T. West, Characterization of the DDT explosive, CP, in *Proceedings of the 7th International Detonation Symposium*, NSWC MP 82-334, ed. by J.M. Short, S.J. Jacobs (Naval Surface Warfare Center, Dahlgren, VA, 1982), pp. 865–876

D.J. Steinberg, S.G. Cochran, M.W. Guinan, A constitutive model for metals applicable at high-strain rate. J. Appl. Phys. **51**, 1498–1504 (1980)

D.J. Steinberg, C.M. Lund, A constitutive model for strain rates from $10^{-4}$ to $10^6$/s. J. Appl. Phys. **65**, 1528–1533 (1989)

D.J. Steinberg, Equation-of-state and strength properties of selected materials, UCRL-MA-106439 (Lawrence Livermore National Laboratory, Livermore, CA, 1996) [This is the revised version; an earlier version with the same report no. was published in 1991]

A.L. Stevens, L.E. Pope, Observations of secondary slip in impact-loaded aluminum single crystals (1). Scripta Metall. **5**, 981–986 (1971)

A.L. Stevens, F.R. Tuler, Effect of shock precompression on the dynamic fracture strength of 1020 steel and 6061-T6 aluminum. J. Appl. Phys. **42**(13), 5665–5670 (1971)

A.L. Stevens, O.E. Jones, Radial stress release phenomena in plate impact experiments: compression-release. J. Appl. Mech. Trans. ASME **39**, 359–366 (1972)

A.L. Stevens, L.W. Davison, W.E. Warren, Spall fracture in aluminum monocrystals: A dislocation-dynamics approach. J. Appl. Phys. **43**(12), 4922–4927 (1972)

A.L. Stevens, L.E. Pope, Wave propagation and spallation in textured beryllium, in *Metallurgical Effects at High Strain Rates*, ed. by R.W. Rohde, B.M. Butcher, J.R. Holland, C.H. Karnes (Plenum, New York, NY, 1973), pp. 459–472

A.L. Stevens, L.W. Davison, W.E. Warren, Void growth during spall fracture of aluminum monocrystals, in *Dynamic Crack Propagation*, ed. by G.C. Sih (Noordhoff, Leyden, 1973), pp. 37–48

A.L. Stevens, Residual mechanical properties of textured and spalled-damage beryllium, presented at AIAA/ASME/SAE 15th Structures, Structural Dynamics and Materials Conference, AIAA Paper No. 74-399 (AIAA, Reston, VA, 1974)

A.L. Stevens, P.C. Lysne, G.B. Griswold, Rock Springs oil shale fracturization experiment: Experimental results and concept evaluation, SAND74-0372 (Sandia National Laboratories, Albuquerque, NM, 1974)

P.H. Stolz, B.V. Oliver, Growth rates of the m=0 mode for Bennett equilibria with varying radial density and temperature profiles. Phys. Plasmas **8**(6), 3096–3098 (2001)

R.H. Stresau, J.E. Kennedy, Critical conditions for shock initiation of detonation in real systems, in *Proceedings of the 6th International Detonation Symposium*, ONR ACR-221, ed. by S.J. Jacobs, D.J. Edwards (Office of Naval Research, Washington, D.C., 1976), pp. 68–75

R.M. Summers, J.S. Peery, M.K. Wong, E.S. Hertel Jr., T.G. Trucano, L.C. Chhabildas, Recent progress in ALEGRA development and application to ballistic impacts. Int. J. Impact Eng. **20**, 779–788 (1997)

C.T. Sun, J.D. Achenbach, G. Herrmann, Continuum theory for a laminated medium. J. Appl. Mech. **35**, 467–475 (1968)

H.J. Sutherland, J.E. Kennedy, Acoustic characterization of two unreacted explosives. J. Appl. Phys. **46**(6), 2439–2444 (1975)

H.J. Sutherland, J.W. Nunziato, D.S. Drumheller, J.E. Kennedy, Wave propagation in unreacted, heterogeneous explosive materials, in *Proceedings of 12th Annual Meeting of the Society of Engineering Science*, ed. by M. Stern (Elsevier, Amsterdam, 1975), pp. 509–517

H.J. Sutherland, J.E. Kennedy, J.W. Nunziato, Behavior of the longitudinal acoustic velocity in PBX-9404 during thermal decomposition, SAND 77-0577 (Sandia National Laboratories, Albuquerque, NM, 1977)

J.C. Swearengen, R.W. Rohde, D.L. Hicks, Mechanical state relations for inelastic deformation of iron: The choice of variables. Acta Metall. **24**, 969–975 (1976)

M.A. Sweeney, History of z-pinch research in the U.S, in *Fifth International Conference on Dense Z Pinches*, ed. by J. Davis, C. Deeney, N.R. Pereira, AIP Conference Proceedings, vol. 651 (AIP, College Park, MD, 2002), pp. 9–14

J.W. Swegle, L.C. Chhabildas, Technique for the generation of pressure-shear loading using anisotropic crystals, in *Shock Waves and High Strain-Rate Phenomena in Metals*, ed. by M. Meyers, L. Murr (Springer, New York, NY, 1981), pp. 401–415

J.W. Swegle, D.E. Grady, Shock viscosity and the prediction of shock wave rise times. J. Appl. Phys. **58**(2), 692–701 (1985)

J.W. Swegle, D.E. Grady, Calculation of thermal trapping in shear bands, in *Metallurgical Applications of Shock-Wave and High-Strain-Rate Phenomena*, ed. by L.E. Murr, K.P. Staudhammer, M.A. Meyers (Marcel Dekker, New York, NY, 1986a), pp. 705–722

J.W. Swegle, D.E. Grady, Shock viscosity and the calculations of steady shock wave profiles, in *Shock Waves in Condensed Matter*, ed. by Y.M. Gupta (Plenum, New York, NY, 1986b), pp. 353–357

G.I. Taylor, The testing of materials at high rates of loading. J. Inst. Civil Eng. **26**, 486–519 (1946)

G.I. Taylor, The use of flat ended projectiles for determining yield stress I. Theoretical considerations. Proc. R. Soc. Lond. **A194**, 289–299 (1948)

L.M. Taylor, D.P. Flanagan, PRONTO-3D: A three-dimensional transient solid dynamics program, SAND87-1912 (Sandia National Laboratories, Albuquerque, NM, 1989)

P.A. Taylor, M.B. Boslough, Y. Horie, Modeling of shock-induced chemistry in nickel-aluminum systems, in *Shock Waves in Condensed Matter*, ed. by S.C. Schmidt et al. (Elsevier, Amsterdam, 1988), pp. 395–398

S.L. Thompson, CHARTD—A computer program for calculating problems of coupled hydrodynamic motion and radiation flow in one dimension, SC-RR-69-613 (Sandia National Laboratories, Albuquerque, NM, 1969)

S.L. Thompson, Improvements in the CHARTD radiation-hydrodynamic code I: Analytic equations of state, SC-RR-70-28 (Sandia National Laboratories, Albuquerque, NM, 1970)

S.L. Thompson, H.S. Lauson, Improvements in the CHARTD radiation-hydrodynamic code IV: User aid programs, SC-DR-71-0715 (Sandia National Laboratories, Albuquerque, NM, 1972)

S.L. Thompson, Improvements in the CHARTD energy flow hydrodynamic code V: 1972/73 modifications, SLA-73-0477 (Sandia National Laboratories, Albuquerque, NM, 1973)

S.L. Thompson, CKEOS2—An equation of state test program for the CHARTD/CSQ EOS package, SAND76–0175 (Sandia National Laboratories, Albuquerque, NM, 1976)

S.L. Thompson, CSQII—An Eulerian finite difference program for two-dimensional material response—Part 1 material sections, SAND77-1339 (Sandia National Laboratories, Albuquerque, NM, 1979)

S.L. Thompson, ANEOS analytic equations of state for shock physics codes input manual, SAND89-2951 (Sandia National Laboratories, Albuquerque, NM, 1990)

B.J. Thorne, W. Herrmann, TOODY: A computer program for calculating problems of motion in two dimension, SC-RR-66-602 (Sandia National Laboratories, Albuquerque, NM, 1967)

T.F. Thornhill, L.C. Chhabildas, W.D. Reinhart, D.L. Davidson, Particle launch to 19 km/s for micro-meteoroid simulation using enhanced three-stage light gas gun hypervelocity launcher techniques. Int. J. Impact Eng. **33**, 799–811 (2006)

T.F. Thornhill, W.D. Reinhart, L.C. Chhabildas, W.G. Breiland, C.S. Alexander, J.L. Brown, Characterization of prompt flash signatures using high speed broadband diode detectors. Int. J. Impact Eng. **35**, 1827–1835 (2008)

T.F. Thornhill, L.C. Chhabildas, W.D. Reinhart, Time-resolved optical signatures for Hugoniot state measurements in shock compressed composition B, in *Shock Waves in Condensed Matter*, ed. by M.L. Elert, W.T. Buttler, M.D. Furnish, W.W. Anderson, W.G. Proud, AIP Conference Proceedings, vol. 1195 (AIP, College Park, MD, 2009), pp. 404–407

S. Thunborg, G.E. Ingram, R.A. Graham, Compressed gas gun for controlled planar impacts over a wide velocity range. Rev. Sci. Instrum. **35**(1), 11–14 (1964)

W.M. Trott, A.M. Renlund, Pulsed-laser-excited Raman spectra of shock-compressed triaminotrinitrobenzene, in *Proceedings of the 9th International Detonation Symposium*, ONR 113291-7, ed. by J.M. Short, E.L. Lee (Office of Naval Research, Washington, D.C., 1989), pp. 153–161

W.M. Trott, K.D. Meeks, Acceleration of thin foil targets using fiber-coupled optic pulses, in *Shock Compression of Condensed Matter*, ed. by S.C. Schmidt, J.N. Johnson, L.W. Davison (Elsevier, New York, NY, 1990a), pp. 997–1000

W.M. Trott, K.D. Meeks, High-power Nd glass laser transmission through optical fibers and its use in acceleration of thin foil targets. J. Appl. Phys. **67**, 3297–3301 (1990b)

W.M. Trott, Investigation of the dynamic behavior of laser-driven flyers, in *High-Pressure Science and Technology*, ed. by S.C. Schmidt, J.W. Shaner, G.A. Samara, M. Ross, AIP Conference Proceedings, vol. 309 (AIP, College Park, MD, 1994), pp. 1655–1658

W.M. Trott, M.D. Knudson, L.C. Chhabildas, J.R. Asay, Measurements of spatially resolved velocity variations in shock compressed heterogeneous materials using a line-imaging velocity interferometer, in *Shock Compression of Condensed Matter*, ed. by M.D. Furnish, L.C. Chhabildas, R.S. Hixson, AIP Conference Proceedings, vol. 505 (AIP, College Park, MD, 2000), pp. 993–998

W.M. Trott, J.N. Castañeda, J.J. O'Hare, M.D. Knudson, L.C. Chhabildas, M.R. Baer, J.R. Asay, Examination of the mesoscopic scale response of shock compressed heterogeneous materials using a line-imaging velocity interferometer, in *Fundamental Issues and Applications of Shock-Wave and High-Strain-Rate Phenomena*, ed. by K.P. Staudhammer, L.E. Murr, M.A. Meyers (Elsevier, New York, NY, 2001), pp. 647–654

W.M. Trott, R.E. Setchell, A.V. Farnsworth Jr., Development of laser-driven flyer techniques for equation-of-state studies of microscale materials, in *Shock Compression of Condensed Matter, AIP Conference Proceedings*, ed. by M.D. Furnish, N.N. Thadhani, Y. Horie, vol. 505 (AIP, College Park, MD, 2002), pp. 1347–1350

W.M. Trott, M.R. Baer, J.N. Castañeda, A.S. Tappan, J.N. Stuecker, J. Cesarano, Shock-induced reaction in a nitromethane-impregnated geometrically regular sample configuration, in *Proceedings of the 13th International Detonation Symposium*, ONR-351-07-01, ed. by S. Peiris, R.M. Doherty (Office of Naval Research, Washington, D.C., 2006), pp. 308–318

W.M. Trott, M.R. Baer, J.N. Castañeda, L.C. Chhabildas, J.R. Asay, Investigation of the mesoscopic scale response of low-density pressings of granular sugar under impact. J. Appl. Phys. **101**(2), 024917 (2007)

T.G. Trucano, L.M. Barker, J.R. Asay, G.I. Kerley, Numerical studies of the dynamic isentropic loading of solid molecular hydrogen, in *Shock Waves in Condensed Matter*, ed. by Y.M. Gupta (Plenum, New York, NY, 1986), pp. 461–465

T.G. Trucano, J.R. Asay, Effects of vaporization on debris cloud dynamics. Int. J. Impact Eng. **5**, 645–653 (1987)

T.G. Trucano, J.R. Asay, L.C. Chhabildas, Hydrocode benchmarking of 1-D shock vaporization experiments, in *Shock Waves in Condensed Matter*, ed. by S.C. Schmidt, N.C. Holmes (Elsevier, Amsterdam, 1988), pp. 163–166

T.G. Trucano, D.E. Grady, J.M. McGlaun, Fragmentation statistics from Eulerian hydrocode calculations. Int. J. Impact Eng. **10**, 587–600 (1990)

T.G. Trucano, L.C. Chhabildas, Calculations supporting hypervelocity launcher development, in *High Pressure Science and Technology*, ed. by S.C. Schmidt, J.W. Shaner, G.A. Samara, M. Ross, AIP Conference Proceedings, vol. 309 (AIP, College Park, MD, 1993), pp. 1639–1642

T.G. Trucano, J.M. McGlaun, A. Farnsworth, Computational methods for describing the laser-induced mechanical response of tissue, in *Proceedings of Laser-Tissue Interactions V* (SPIE Proceedings Series 2134A), ed. by S.L. Jacques (SPIE, Bellingham, WA, 1994), pp. 179–203

T.G. Trucano, L.C. Chhabildas, Computational design of hypervelocity launchers. Int. J. Impact Eng. **17**, 849–860 (1995)

T.G. Trucano, D. Grady, G. Kubiak, M. Kipp, R. Olson, A. Farnsworth, Swords to plowshares: Shock wave applications to advanced lithography. Int. J. Impact. Eng. **17**, 873–890 (1995)

T.G. Trucano, K.G. Budge, R.J. Lawrence et al., Analysis of z pinch shock wave experiments, SAND99-1255 (Sandia National Laboratories, Albuquerque, NM, 1999)

R.F. Trunin, L.F. Gudarenko, M.V. Zhernokletov, G.V. Simakov, *Experimental Data on Shock Compression and Adiabatic Expansion of Condensed Matter* (Russian Federal Nuclear Center, Sarov, VNIIEF, 2001)

T.J. Tucker, J.E. Kennedy, D.L. Allensworth, Secondary explosive spark detonators, in *Proceedings of the 7th Symposium on Explosives and Pyrotechnics*, (Franklin Institute, Philadelphia, PA, 1971) [Also appeared as SC-R-713486. Sandia National Laboratories, Albuquerque, NM]

F.R. Tuler, B.M. Butcher, A criterion for the time dependence of dynamic fracture. Int. J. Fract. Mech. **4**(4), 431–437 (1968)

S.J. Turneaure, Y.M. Gupta, X-ray diffraction and continuum measurements in silicon crystals shocked below the elastic limit. Appl. Phys. Lett. **90**, 051905 (2007)

A.V. Utkin, G.I. Kanel, S.V. Razorenov, A.A. Bogach, D.E. Grady, Elastic moduli and dynamic yield strength of metals near the melting temperature, in *Shock Compression of Condensed Matter*, ed. by S. Schmidt, D. Dandekar, J. Forbes, AIP Conference Proceedings, vol. 429 (AIP, College Park, MD, 1998), pp. 443–446

E.L. Venturini, B. Morosin, R.A. Graham, Magnetic properties of shock-synthesized and furnace-reacted zinc ferrite, in *Shock Waves in Condensed Matter*, ed. by Y.M. Gupta (Plenum, New York, NY, 1986), pp. 815–820

E.L. Venturini, R.A. Graham, B. Morosin, Static magnetization and microwave loss in shock-modified ferrites, in *Shock Waves in Condensed Matter*, ed. by S.C. Schmidt, N.C. Holmes (Elsevier, Amsterdam, 1988), pp. 451–456

T.J. Vogler, T.F. Thornhill, W.D. Reinhart, L.C. Chhabildas, D.E. Grady, L.V. Wilson, O. Hurricane, A. Woo, Fragmentation of materials in expanding tube experiments. Int. J. Impact Eng. **29**, 735–746 (2003)

T.J. Vogler, J.R. Asay, A distributional model for elastic-plastic behavior of shock-loaded materials, in *Shock Compression of Condensed Matter*, ed. by M.D. Furnish, Y.M. Gupta, J.W. Forbes, AIP Conference Proceedings, vol. 706 (AIP, College Park, MD, 2004), pp. 617–620

T.J. Vogler, W.D. Reinhart, L.C. Chhabildas, Dynamic behavior of boron carbide. J. Appl. Phys. **95**, 173–4183 (2004)

T.J. Vogler, L.C. Chhabildas, Strength behavior of materials at high pressures. Int. J. Impact Eng. **33**, 812–825 (2006)

T.J. Vogler, M.Y. Lee, D.E. Grady, Static and dynamic compaction of ceramic powders. Int. J. Solids Struct. **44**, 636–658 (2007)

T.J. Vogler, J.D. Clayton, Heterogeneous deformation and spall of an extruded tungsten alloy: Plate impact experiments and crystal plasticity modeling. J. Mech. Phys. Solids **56**, 297–335 (2008)

T.J. Vogler, W.M. Trott, W.D. Reinhart, C.S. Alexander, M.D. Furnish, M.D. Knudson, L.C. Chhabildas, Using the line-VISAR to study multi-dimensional and meso-scale impact phenomena. Int. J. Impact Eng. **35**, 1844–1852 (2008)

T.J. Vogler, On measuring the strength of metals at ultra-high strain rates. J. Appl. Phys. **106**, 053530 (2009)

T.J. Vogler, C.S. Alexander, J.L. Wise, S.T. Montgomery, Dynamic behavior of tungsten carbide and alumina filled epoxy composites. J. Appl. Phys. **107**, 043520 (2010)

T.J. Vogler, C.S. Alexander, T.F. Thornhill, W.D. Reinhart, Pressure-shear experiments on granular materials, SAND2011-6700 (Sandia National Laboratories, Albuquerque, NM, 2011)

T.J. Vogler, J.P. Borg, D.E. Grady, On the nature of steady structured waves in heterogeneous materials. J. Appl. Phys. **112**, 123507 (2012)

E.K. Walsh, K.W. Schuler, Acceleration wave propagation in a nonlinear viscoelastic solid. J. Appl. Mech. **40**, 705–710 (1973)

S.L. Wang, M.A. Meyers, R.A. Graham, Shock consolidation of IN-100 nickel-base superalloy powder, in *Shock Waves in Condensed Matter*, ed. by Y.M. Gupta (Plenum, New York, NY, 1986), pp. 731–736

M.C. Wanke, A.D. Grine, M.A. Mangan, L.C. Chhabildas, W.D. Reinhart, T.F. Thornhill, C.S. Alexander, J.L. Brown, W.G. Breiland, E.A. Shaner, P.A. Miller, Advanced diagnostics for impact-flash spectroscopy on light-gas guns, SAND2007-0835 (Sandia National Laboratories, Albuquerque, NM, 2007)

W.R. Wawersik, W. Herrmann, S.T. Montgomery, H.S. Lauson, Excavation design in rock salt – laboratory experiments, material modeling and validations, in *Rock Mechanics: Caverns and Pressure Shafts*, ed. by W. Wittke (Balkema, Rotterdam, 1982), pp. 1345–1356

S.T. Weir, A.C. Mitchell, W.J. Nellis, Metallization of fluid molecular hydrogen at 140 GPa (1.4 Mbar). Phys. Rev. Lett. **76**, 1860–1863 (1996)

G.W. Wellman, K.W. Schuler, Structural consequences of railgun augmentation, IEEE Trans Magnet. **25**, 593–598 (1988)

E. Wigner, H.B. Huntington, On the possibility of a metallic modification of hydrogen. J. Chem. Phys. **3**(12), 764–770 (1935)

F.L. Williams, Y.K. Lee, B. Morosin, R.A. Graham, Catalytic activity of shock modified ZNO for CO oxidation and methanol synthesis, in *Shock Waves in Condensed Matter*, ed. by Y.M. Gupta (Plenum, New York, NY, 1986a), pp. 791–796

F.L. Williams, B. Morosin, R.A. Graham, Influence of shock compression on the specific surface area of inorganic powders, in *Metallurgical Applications of Shock-Wave and High-Strain-Rate Phenomena*, ed. by L.E. Murr, K.P. Staudhammer, M.A. Meyers (Marcel Dekker, New York, NY, 1986b), pp. 1013–1022

W.D. Williams, D.J. Fogelson, L.M. Lee, Carbon piezoresistive stress gauge, in *Shock Waves in Condensed Matter*, ed. by J.R. Asay, R.A. Graham, G.K. Straub (Elsevier, Amsterdam, 1984), pp. 121–124

D.L. Williamson, B. Morosin, E.L. Venturini, R.A. Graham, Mossbauer study of shock-synthesized zinc ferrite, in *Shock Waves in Condensed Matter*, ed. by Y.M. Gupta (Plenum, New York, NY, 1986), pp. 809–814

L.T. Wilson, D.R. Reedal, M.E. Kipp, R.R. Martinez, D.E. Grady, Comparison of calculated and experimental results of fragmenting cylinder experiments, in *Fundamental Issues and Applications of Shock-Wave and High-Strain-Rate Phenomena* (Explomet 2000), ed. by K.P. Staudhammer et al. (Elsevier, Amsterdam, 2001), pp. 561–570

J.M. Winey, J.N. Johnson, Y.M. Gupta, Unloading and reloading response of aluminum single crystals: Time-dependent anisotropic material description. J. Appl. Phys. **112**, 093509 (2012)

J.L. Wise, L.C. Chhabildas, J.R. Asay, Shock compression of beryllium, in *Shock Waves in Condensed Matter*, ed. by W.J. Nellis, L. Seaman, R.A. Graham, AIP Conference Proceedings, vol. 78 (AIP, College Park, MD, 1982), pp. 417–421

J.L. Wise, Refractive index and equation of state of a shock-compressed aqueous solution of zinc chloride, in *Shock Waves in Condensed Matter*, ed. by J.R. Asay, R.A. Graham, G.K. Straub (Elsevier, Amsterdam, 1984), pp. 317–320

J.L. Wise, L.C. Chhabildas, Laser interferometer measurements of refractive index in shock-compressed materials, in *Shock Waves in Condensed Matter*, ed. by Y.M. Gupta (Plenum, New York, NY, 1986), pp. 441–454

J.L. Wise, G.I. Kerley, T.G. Trucano, Shock-vaporization studies on zinc and porous carbon, in *Shock Waves in Condensed Matter*, ed. by S.C. Schmidt, R.D. Dick, J.W. Forbes, D.G. Tasker (Elsevier, Amsterdam, 1992), pp. 61–64

J.L. Wise, D.E. Grady, Dynamic, multiaxial impact response of confined and unconfined ceramic rods, in *High-Pressure Science and Technology*, ed. by S.C. Schmidt, J.W. Shaner, G.A. Samara, M. Ross, AIP Conference Proceedings, vol. 309 (AIP, College Park, MD, 1994), pp. 777–780

J.L. Wise, S.C. Jones, C.A. Hall, W.D. Reinhart, R.J. Hickman, J.W. Gluth, Effects of annealing and preheating on the impact response of selected braze materials, in *Shock Compression of Condensed Matter*, ed. by M.D. Furnish, M. Elert, T.P. Russell, C.T. White, AIP Conference Proceedings, vol. 845 (AIP, College Park, MD, 2006), pp. 686–689

J.L. Wise, D.G. Dalton, R.J. Hickman, M.I. Kaufman, S.A. Leffler, M.J. Jones, J.J. Lynch, A.C. Bowers, Sample preheating capability for dynamic material studies, *Bulletin of the American Physical Society*, Vol. 58, No. 7, June 2013, p. 86

E.P. Yu, B.V. Oliver, D.B. Sinars et al., Steady-state radiation ablation in the wire-array z pinch. Phys. Plasmas **14**, 022705 (2007)

E.P. Yu, M.E. Cuneo, M.P. Desjarlais et al., Three-dimensional effects in the wire array z pinch. Phys. Plasmas **15**, 056301 (2008)

F.J. Zeigler, WONDY VII: A vectorized version of the WONDY wave propagation code, SAND85-2338 (Sandia National Laboratories, Albuquerque, NM, 1986)

F. Zeigler, J.M. McGlaun, S.L. Thompson, T.G. Trucano, Computations of hypervelocity impact using the CTH shock wave physics code, in *Shock Waves in Condensed Matter*, ed. by S.C. Schmidt, N.C. Holmes (Elsevier, Amsterdam, 1988), pp. 191–194

D.H. Zeuch, S.T. Montgomery, J.D. Keck, Hydrostatic and triaxial compression experiments on unpoled PZT 95/5-2Nb ceramic: The effects of shear stress on the FR1→AO polymorphic phase transformation. J. Mater. Res. **7**(12), 3314–3332 (1992)

D.H. Zeuch, S.T. Montgomery, J.D. Keck, Some observations on the effects of shear stress on a polymorphic transformation in perovskite-structured lead-zirconate-titanate ceramic. J. Geophy. Res. **98**(B2), 1901–1911 (1993)

D.H. Zeuch, S.T. Montgomery, J.D. Keck, Further observations on the effects of nonhydrostatic compression on the FR1→AO polymorphic phase transformation in niobium-doped, lead-zirconate-titanate ceramic. J. Mater. Res. **9**(12), 1322–1327 (1994)

D.H. Zeuch, S.T. Montgomery, D.J. Holcomb, The effects of nonhydrostatic compression and applied electric field on the electromechanical behavior of poled lead zirconate titanate PZT 95/5-2Nb ceramic during the ferroelectric to antiferroelectric polymorphic transformation. J. Mater. Res. **14**(5), 1814–1827 (1999)

D.H. Zeuch, S.T. Montgomery, D.J. Holcomb, Uniaxial compression experiments on lead zirconate titanate 95/5-2Nb ceramic: Evidence for an orientation-dependent, maximum compressive stress criterion for onset of the ferroelectric to antiferroelectric polymorphic transformation. J. Mater. Res. **15**(3), 689–703 (2000)

S. Zhuang, G. Ravichandran, D.E. Grady, An experimental investigation of shock wave propagation in periodically layered composites. J. Mech. Phys. Solids **51**, 245–265 (2003)

# Index of Names of Individuals

## A


Abbott, Robert (Rob), 384
Aeschliman, Daniel (Dan), 591
Ahrens, Thomas (Tom), 332, 333, 336
Alexander, Scott, 217, 245, 301, 304, 375, 469, 587, 593
Allensworth, Dwight, 446
Allshouse, George, 401, 404, 407
Amos, Donald (Don), 117, 446
Anderson, Charles, 504, 505, 580
Anderson, David (Dave), 29, 568
Anderson, George, 51, 341, 344, 537
Anderson, Heidi, 571, 588
Anderson, Mark, 154, 174, 191, 192, 369, 370, 418, 458, 516, 545–546
Andrews, Joe, 421
**Ang, James (Jim)**, 132, 134, 177, 178, 254, **276–279**, 287, 289, 290, 366
Ao, Thomas (Tommy), 188, 228, 301, 304, 570, 585
Arthur, Jim (Art), 500
Asay, Blaine (B.W.), 308, 446
**Asay, James (Jim)**, 36, 50, 86, 91–93, 101, 102, 108, 112, 120, 128–135, 156, 171, 173, 188, 194, 195, 204, 207, 216, 217, 220–223, 228, 229, 239, 260, 269, 271, 276, 277, **280–305**, 320, 328, 335, 357, 361, 362, 369, 370, 372, 376, 377, 382, 383, 388, 400, 408–415, 419, 421, 425, 427, 437, 441, 462, 463, 468, 476, 479, 484, 488, 489, 494, 532, 549, 550, 569, 570, 575, 577, 578, 580, 582, 583, 584, 586, 587, 589, 590, 593
Ashcroft, Neil, 400


## B


Badruzzaman, Ahmed, 401
**Baer, Melvin (Mel)**, 117, 158, 160, 174, 184–188, 212, 216, 288, **306–319**, 452, 566
Bailey, James (Jim), 235, 255, 301, 469
Baker, Alan, 354
Baker, Ernest (Ernie), 158, 459
Band, William, 379, 380
Barker, Dennis, 387
**Barker, Lynn**, 8, 11, 25, 29, 34, 35, 40, 45, 57, 59, 60, 62, 64, 65, 66, 68, 75, 76, 85–91, 93, 95–97, 99, 105, 107, 112, 132–140, 146, 167, 174, 183, 215, 222, 227, 237, 264, 265, 276, 277, 283, 286, 287, 290, 291, 292, 305, **320–328,** 341, 343, 357, 358, 362–364, 380, 386, 393, 400, 409, 425, 431, 434, 435, 441, 471, 484, 485, 496, 529, 531, 535, 538, 547, 548, 550, 556, 561, 573, 579, 580, 590
Barr, Michael (Mike), 423
Barsis, Ed, 128, 169, 171–174, 196, 200, 278, 294, 330, 331, 367, 383, 410, 476, 477, 548, 549
Bass, Robert (Bob), 9, 23, 350, 420, 502, 514
Batsanov, Stephanov, 522
Bauer, Francois, 128, 151–153, 162, 544–545
Beatriz, Felipe, 501
Beauchamp, Ed, 149, 521–522
Beck, Albert (Al), 34, 35, 47, 496, 500
Becker, Richard (Rich), 585
Becker, Robert (Bob), 365, 371, 372
Becker, Steven (Steve), 385
Bedford, Marc, 562




J.R. Asay et al., *Impactful Times*, Shock Wave and High Pressure Phenomena,
DOI 10.1007/978-3-319-33347-2

Behrens, Richard (Rich), 312
Benage, John, 218, 262, 305, 593
Benedick, William (Bill), 10, 12, 28, 29, 34, 50, 54, 55, 64, 100, 123, 152, 153, 263, 286, 344, 357, 398, 436, 438, 448, 538, 544, 545, 573
Benham, Robert (Bob), 34, 458
Benner, Robert, 196, 278, 279
Benson, David (Dave), 185, 313
Benson, David A. (Dave), 144, 454
Benzley, Steve, 122
Berg, Dale, 589
Bergstresser, Thomas (Tom), 193, 268, 384–385, 504, 579
Berners-Lee, Tim, 199, 329
Bernier, Henri, 362
Berry, Dante, 571
Berry, Lew, 342
Bertholf, Larry, 29, 49, 72, 94, 122, 284, 425, 427, 431, 449, 451, 484, 497, 499, 503, 534, 539, 546, 556, 575
Besinger, Zelma, 556
Bessette, Gregg, 371, 457, 583
Biesecker, Robert (Bob), 40, 529, 530, 533
Bild, Charles (Charlie), 8, 35, 39, 108, 342, 425, 498–500, 528–531
Birdsong, John, 545–546
Bivens, Hugh, 502
Bless, Stephan, 550, 578
Bloomquist, Douglas (Doug), 188, 288
Boade, Rod, 502, 543, 561
Bochev, Pavel, 403
Bode, Julie, 497
Bonzon, Lloyd, 118, 174, 272, 324, 368, 370, 549, 570
Borg, John, 375, 586, 587
**Boslough, Mark**, 149, 173, 177, 182, 199–204, **329–340**, 366, 456, 507, 566, 581
Boughton, Bruce, 415
Bowen, Kent, 471
Boyer, William (Bill), 193, 301, 384
Bozman, Daniel (Dan), 391
Brannon, Rebecca, 155, 188, 191, 367, 516
Brookshire, Wayne, 342, 496, 529
Brown, Justin, 130, 148, 217, 245, 300, 301, 374, 375, 469, 570, 584
Brown, Michael (Mike), 382
Brown, William (Bill), 99, 393, 451
Browning, Iben, 537
Brownlee, Bob, 9, 22, 350
Broyles, Carter, 9, 76, 267, 481
Brunner, Thomas (Tom), 402–403
Budge, Kent, 207, 259, 403, 508
Budzinski, John, 389
Bundy, F.P., 522
Burchett, Olden, 50, 72, 484, 530, 534, 539
Burke, Michael (Mike), 193, 384
Burns, Tim, 578, 579, 589
**Butcher, Barry**, 6, 29, 37, 39, 49, 50, 65–67, 69, 72, 76, 85, 97, 102–103, 105, 111, 208, 216, 280, 286, 321, 322, **341–349**, 355, 356, 376, 425, 451, 471, 482, 484, 497, 500, 529, 530, 532, 534, 539, 556, 560, 574, 576, 589
Buttler, William (Billy), 392
Byers, Rupe, 449

C

Cannon, Jack, 341, 496, 529
Carlson, Gary, 562
Carlson, Ronald (Ron), 9, 420
Carr, Alan, 4
Carr, Marty, 149, 521
Carroll, Dan, 259, 371
Carroll, Sue, 259
Castaneda, Jaime, 571
Cauble, Robert (Bob), 239, 297, 463
Cernich, Louise, 497
**Chabai, Albert (Al)**, 6, 9, 21–23, 55, 75–76, 79, 100, 156, 174, 216, 267, **350–352,** 420, 421, 425, 451, 453, 481, 482, 502, 504, 514, 534, 546, 575, 576
Chael, Jerry, 193, 384
Chang, James (Jim), 291
Chantrenne, Sophie, 300, 304, 593
Chapman, Chap, 121
Chapman, Clark, 331, 337
Chavez, Al, 120
Chavez, Pat, 154, 515
Chen, Peter (Pete), 29, 49, 101, 107, 153, 210, 216, 425, 437, 497, 513, 514, 562
Chen, Tony, 577
**Chhabildas, Lalit**, 11, 61, 62, 93, 102–105, 129, 138–142, 147, 154, 173, 174, 177, 180, 181, 183, 186, 188, 190, 191, 199, 214, 216, 217, 221, 222, 226, 227, 247, 258, 267, 277, 285–287, 288, 296, 305, 329, 338, **353–376**, 383, 392, 409, 411–413, 415, 418, 419, 425, 458, 477, 484, 514–516, 547, 548, 550, 569, 570, 579–584, 586, 589, 590
Chiesa, Mike, 587
Chin, Chuck, 158
Chou, Anthony (Tony), 174, 368
Chyba, Christopher (Chris), 332
Cieslak, Michael (Mike), 95, 422

Claassen, Richard (Dick), 8, 25, 341, 499, 536, 537, 546
Clark, George, 590, 593
Clarke, Ed, 435
Clayton, John, 585
Clem, Robert (Bob), 498
Clifton, Rodney (Rod), 102, 103, 360, 586, 590
Coats, Rebecca (Becky), 469
Cochrane, Kyle, 217, 235, 239, 494, 587
Cogar, John, 372
Collins, Gilbert (Rip), 298
Connell, John, 142, 366, 367
Cook, Donald (Don), 207, 220, 239, 255, 295, 301, 304, 368, 369, 376, 410, 411, 462, 549
Cook, Wayne, 193, 384
Cooper, Arlin, 575
Cooper, Marcia, 216, 570
Cooper, Paul, 159
Corcoran, Patrick (Pat), 468
Corson, Dale, 341
Cotter, Donald (Don), 434, 496, 499, 501
Cowperthwaite, M., 309–310
Cox, David, 161, 173, 192, 283, 357, 368, 370, 477, 516, 589
Crawford, David (Dave), 173, 175, 200, 201, 331–332, 335, 338, 428, 432, 455, 459, 507, 508, 566, 581
Crawford, John, 502
Crews, Jeanne, 366
Crump, O. B., 443, 444
Current, Michael, 354

**D**
Dalton, Devon, 469
Dandekar, Datta, 374, 583
Daniels, Charlie, 123
Dattelbaum, Dana, 570
Davey, William (Bill), 166, 453, 504, 579
David, Ruth, 546
Davidson, C.A., 591
Davidson, George, 543, 546
Davies, Terry, 468
Davis, Jean-Paul, 217, 228, 300–302, 372, 465–468, 494, 584
Davis, William (Bill), 115, 443–444
Davison, Lee, 55, 67, 79, 98, 114, 117, 123, 128, 131, 149, 153, 156, 174, 183, 216, 289, 357, 382, 399, 400, 425, 436, 437, 443, 446–447, 451, 483, 486, 502, 513, 521, 522, 572, 574, 579, 582
DeCarli, Paul, 522
Decker, Daniel (Dan), 354
Deeney, Christopher (Chris), 217, 221, 228, 239–241, 255, 301, 303, 406, 412, 414, 419, 469, 593
**Desjarlais, Michael (Mike)**, 211, 217, 231, 233, 235–238, 301–303, **377–378**, 464–465, 491–492, 494
DeVault, Hunter, 9, 420
Dimos, Duane, 218, 305
Dolan, Daniel (Dan), 217, 271, 301, 417, 469
Dorn, John, 533
Dremin, Anatoly, 186, 314, 443
Drickamer, Harry, 552, 554
Drumheller, Douglas (Doug), 29, 96, 97, 165, 183, 216, 287, 307, 425, 449, 484, 562, 577, 578, 589, 591
Dunham, Greg, 469
Duran, Servet, 528, 533
**Duvall, George**, 36, 280, 305, 320, **379–381**, 399, 400, 421, 435, 578

**E**
Ek, David, 362
Elert, Mark, 392
Elrick, Millie, 431, 505, 507
Embid, P., 309
Erickson, Ken, 536, 591
Erni, John, 471

**F**
Fan, Hongyou, 593
Fang, Elliot, 277
Farnsworth, Archie, 166, 569
Ferry, Sylvia, 372
Fickett, Wildon, 115, 443–444
Fisk, George, 568
Fleming, Kevin, 222, 297, 385
Flicker, Dawn, 218, 262, 268, 301, 305, 382, 469, 593
Forrestal, Michael (Mike), 153, 576, 591
Fowles, G. Richard (Dick), 36, 284, 327, 380
Freedman, Jerry, 122
Freeman, John, 503, 511
Fuchs, Brian, 158
Furman, Necah Stewart, 3, 498
**Furnish, Michael (Mike)**, 62, 93, 99, 147, 154, 173, 186, 189, 191–193, 195, 196, 222, 228, 259, 267, 268, 287, 288, 294, 301, 368, 370, **382–392**, 484, 486, 516, 570, 590

**G**
Galt, John, 285, 356–357, 360–362, 475, 521
Garbin, Douglas (Doug), 384
Gerardo, James (Jim), 568
Giardini, Armando, 552
Gilder, Howard Michael (Mike), 353–354, 356, 357
Gilman, John, 280
Giunta, Tony, 493, 494
Glowka, David (Dave), 592
Gluth, Jeff, 416, 592
Gosling, Scott, 277, 278
Graboske, Harold (Hal), 129, 285, 362
**Grady, Dennis**, 61, 71, 91, 93, 99, 100, 105, 106, 131, 142–146, 166, 173, 174, 182, 183, 197, 216, 267, 286, 287, 291, 294, 295, 322, 351, 357, 358, 362, 369, 372, 374, 376, 382, **393–395**, 400, 418, 419, 453, 454, 457, 477, 484, 505, 512, 549, 550, 577, 578, 580, 585–587, 589–591
**Graham, Robert (Bob)**, 8, 10–12, 25, 28, 29, 34, 38–40, 53–56, 65, 83, 90, 91, 100–102, 124, 149–153, 168, 174, 184, 191, 214, 215, 216, 263, 271, 313, 329, 343, 344, 357, 359, 362, 370, 380, **396–400**, 425, 436, 438, 480, 516, 521–523, 526, 529, 536–538, 543, 544, 545, 554, 557
Gray, Rusty, 423
Green, Sid, 40, 530
Greenwoll, James, 545–546
Grissom, Tom, 502
Guenther, Arthur (Art), 280, 305
Guess, Thomas (Tommy), 29, 345
Gupta, Y. M. (Yogi), 222, 298, 305, 378, 381, 415, 578, 583
Gustafson, John, 196, 278, 279

**H**
**Haill, Thomas (Tom)**, 207, 228, 300, 304, **401–407**, 489, 494
Hall, Aaron, 593
**Hall, Clint**, 108, 132, 133, 139, 173, 174, 213, 217, 218, 221–226, 228–230, 239, 241, 243, 247, 286, 289, 290, 292, 294–297, 299, 300–303, 305, 368–370, 392, **408–419**, 423, 462, 463, 476, 477, 488, 489, 494, 548, 586, 587, 592, 593
Halpin, Walt (W.J.), 29, 502
Hammel, Heidi, 203, 335, 337, 340
Hankins, Dorris, 9, 420
Hansche, Bruce, 278
Hansche, George, 8, 25, 29, 536–537
Hansen, Gordon, 385
Hanshaw, Heath, 217, 235, 469
Hanson, David (Dave), 223, 235, 298, 301, 469
Hardesty, Donald (Don), 116–117, 446, 502
Hardy, Robert (Bob), 103–105, 109, 119, 216, 290, 347, 357–360, 376, 471, 472, 533, 548, 589
Harlow, Frank, 364
Harper, Fred, 456
Hartman, William (Bill), 341, 496, 529
Hawke, Ronald (Ron), 133–134, 235, 277, 289, 476, 580
**Hayes, Dennis**, 21–23, 44, 75, 77, 92, 93, 95, 106, 114, 115, 117, 120, 131, 147, 153, 154, 156, 173, 175, 190, 196, 204, 207, 216, 226, 228, 241, 244, 277, 282, 295, 300, 302, 399, 416, 419, **420–423**, 425, 427, 437, 450, 481, 513, 514, 546, 568, 584, 589
Hemsing, Willard (Will), 325, 385–387, 570
Henderson, Robert (Bob), 123
Herrera, Eugene (Gene), 385
Herrmann, Mark, 218, 305, 382, 469
**Herrmann, Walter (Walt)**, 8, 29, 40, 49–51, 63, 67, 68, 70–72, 77–79, 85, 86, 99, 108, 114, 154–156, 163, 171, 175, 212, 216, 217, 235, 264, 266, 275–277, 280, 285, 286, 293, 294, 305, 344, 356, 363, 365, 376, 382, 421, **424–426**, 427, 430, 432, 435, 437, 441, 443, 449, 475, 484, 486, 497, 499, 500, 501, 503, 508, 509, 514, 529, 530, 534, 539, 556, 560, 563, 575, 576, 579, 581, 589
**Hertel, Eugene (Gene)**, 155, 157, 158, 175, 266, 294, 366, **427–433**, 459, 484, 486
Hickman, Randy, 132, 228, 290, 301, 304, 416, 469, 476, 570, 592, 593
Hicks, Darrell, 29, 281, 425, 431, 450, 452, 503, 576
Hill, Les, 514
Hill, Scott, 366
Hitchcock, Tom, 182
Hixson, Robert (Rob), 382, 385, 423
Hockaday, Mary, 387
Holdridge, Diane, 281
Holland, J.R. (Reid), 29, 341, 436
Holland, Kathleen, 217, 221, 300–301, 371
**Hollenbach, Roy (Red)**, 8, 11, 29, 35, 44, 50, 57–59, 62, 66, 85–89, 91, 95, 99, 107, 136, 146, 227, 264, 286, 287, 296, 321, 323, 325, 328, 342, 393, **434**, 435, 441, 471, 496, 497, 500, 529, 545–546, 556, 590
Holman, G.T., 359

Holtkamp, David (Davy), 388, 390
Holtzhauser, Pat, 421
Hommert, Paul, 370, 372, 543, 546
Horie, Yuki, 149, 151, 522, 523
Hough, Gary, 277, 278
Hubbard, Cam, 524
Hudson, Frank, 499
Huggins, Robert, 527, 533
Huntington, Hillard (H.B.,) 234–235, 353–354, 356–357

I
Ingram, George, 29, 82, 545–546
Ito, Eichi, 383

J
Jackson, Dan, 593
Jackson, Scott, 570
Jakowatz, Charles V. (Jack), 498
Jamieson, J.C., 522
Jennrete, Boyd, 547
**Johnson, James (Jim)**, 29, 51, 59–60, 65, 81, 86, 91–93, 103, 104, 261, 269, 280, 282, 288, 322, 400, 421, 425, **435–437**, 502
Johnson, Leland, 3
Johnson, Wally, 565
Johnston, Robert (Bob), 298
Jones, Arfon, 40, 530
Jones, David (Dave), 571
**Jones, Orval**, 8, 25, 50, 51, 54, 64–65, 80, 98, 113, 114, 119, 156, 170, 174, 216, 271, 280, 360, 362, 399, 424, 425, 435, 436, **438–440**, 441, 442, 446, 449, 499–501, 502, 537, 538, 546, 575, 577
Jones, Scott, 592–593
Jordan, Jennifer, 587–588

K
Kanel, Gennady, 550
**Karnes, Charles**, 29, 49, 50, 57, 321, 342, 425, **441**, 484, 497, 501, 529, 530, 534, 536
Kaufman, Morris, 593
Keck, Jeff, 191, 515, 516
Keller, Don, 540
Kelly, Daniel (Dan), 371
**Kennedy, James (Jim)**, 29, 93, 113–117, 125, 187, 357, 422, **442–448**, 452, 562, 572, 574, 575
Kennedy, Jerry, 29
Kent, Larry, 471, 562, 563
Kerley, Gerald (Gerry), 216, 233, 292, 310, 331, 371, 455, 460, 505, 508, 510, 550, 579, 580, 591
Kerr, Richard (Dick), 336
Key, Samuel (Sam), 29, 49, 432, 436, 484, 486, 497, 534, 556
Kiefer, Mark, 391
Kim, Kuang Yul, 354
Kimball, Ken, 545–546
Kimsey, Kent, 505
Kinabrew, Cliff, 545–546
Kinchen, Richard, 357, 359
Kinsey, Charles, 40, 530
**Kipp, Marlin**, 98–100, 104, 105, 142–144, 158, 186, 197, 210, 216, 331, 357, 360, 372, 393, 425, 431, **449–461**, 484, 486, 503–505, 509, 512, 577, 579–591, 593
Kletzli, Daniel (Dan), 458
Kmetyk, Lubya, 177, 366, 579
**Knudson, Marcus**, 93, 136, 217, 221, 223, 225–234, 236–238, 241, 242, 245, 257, 298, 300–303, 305, 378, 412, 413, 416–418, 423, **462–469**, 488, 489, 494, 570
**Konrad, Carl**, 40, 102, 105, 108–110, 132, 137, 161, 173–175, 191, 221, 285, 290, 291, 294, 296, 347, 357–359, 365, 368, 370, 371, 376, 408, 410, 419, **470–479**, 530, 533, 547–549, 589
Kubiak, Glenn, 581

L
Laguna, Glenn, 589
Lamberson, Donald (Don), 76, 280, 305, 481
Lammi, Chris, 586
Lane, Matt, 587
Lane, Tom, 483, 486
Larsen, Larry, 373, 584
Lauson, Hank, 74, 450, 485
**Lawrence, R. Jeffery (Jeff)**, 29, 50, 73–75, 124, 153, 178, 179, 191, 216, 254, 371, 373, 391, 425, 431, 449, 458, **480–487**, 503, 513, 561, 569
Lee, Binky, 363
Lee, E. H., 437
Lee, Larry, 29, 39, 146, 170, 345, 361, 484, 530, 540, 544, 545, 547
Lee, Richard (Dick), 231, 302, 464, 492
Lee, Ronald (Ron), 445
Leifeste, Gordon, 218, 247, 262
**Lemke, Raymond (Ray)**, 228, 231, 271, 300, 302, 464, 465, **488–494**
Levy, David, 199, 329–330, 338

Lewis, John, 331
Lewis, Texas Jim, 435
Liepmann, Hans, 589
Lingle, Richard (Dick), 109, 471, 562
Lipkin, Joel, 105, 129, 269, 284, 357, 450, 562, 584
Liwski, John, 571
Looby, Thomas (Tom), 342, 496
**Lundergan, C. Donald (Don)**, 6, 8, 11, 25, 26, 29, 34–36, 38–41, 45, 49–52, 56, 61–63, 66, 68, 71, 72, 76, 85, 95–98, 102, 104, 111, 122, 140, 155, 175, 183, 217, 248, 263, 264, 271, 291, 296, 303, 321, 324, 327, 341, 342, 360, 380, 425, 434, 435, 441, 471, 482–483, **495–501**, 528–530, 533–535, 538–540, 556, 560, 562
Luth, Bill, 502
**Lysne, Peter (Pete)**, 29, 99–101, 116–117, 153, 162, 174, 216, 357, 436, 446, **502**, 513, 514, 562, 577

**M**

MacLow, Mordecai-Mark, 333
Mader, Charles (Chuck), 114, 442–443, 454
Magyar, Rudy, 217, 587
Mandalaywala, Chhabildas Chhaganlal, 375–376
Mandleco, Wally, 448
Marder, Robert (Bob), 574
Marshall, Bruce, 468
Marsh, Edward (Ed), 194, 385, 387, 389, 390
Martin, Diane, 537
Martinez, John, 132–133, 247, 295, 369, 550
Martinez, Reina, 158, 459–460
Martin, Matthew (Matt), 469
Matheson (Erik), 307
Matson, Paul, 40, 529, 530, 533
Mattsson, Thomas, 217, 218, 235, 239, 262, 301, 305, 469
Matuska, Daniel, 565
Matzen, M. Keith (Keith), 218, 260, 301, 305, 401
Maurer, Andy, 469
May, Robert (Bob), 11, 29, 40, 109, 110, 118, 216, 470, 471, 530, 540
Maybe, Ham, 341
McAfee, J. M., 308
McCloskey, David (Dave), 155, 483
McCoy, Nira, 384
McDaniel, Dillon, 217, 254, 373, 493, 494
McDowell, Jerry, 373
**McGlaun, J. Michael (Mike)**, 74, 75, 94, 156, 157, 200, 201, 207, 209, 216, 266, 277, 293, 330, 331, 367, 368, 401, 402, 427, 431, 433, 455, 485, **503–512**, 549, 564, 577, 579–581
McKee, G. Randall (Randy), 196, 240, 241, 303
McMillan, Charles, 277
McQueen, Robert (Bob), 344, 382, 400, 529
McWaters, Bruce, 469
Mead, Keith, 39, 528
Mead, Phil, 122, 501
Meeks, Kent, 568, 569
Mehlhorn, Thomas (Tom), 218, 296, 368, 382, 469, 488, 493
Meiners, Harry, 353
Melosh, Jay, 332, 455
Merritt, Mel, 502
Meyers, Charlie, 469
Michaels, Thomas (Tom), 51, 65, 436
Miller, Jeff, 177, 362, 366, 368, 476
Miller, Jill, 571
Minier, Leanna, 570
Mitchell, Dennis, 95, 115, 117, 216, 421, 422
Mitchell, Stan, 501
Mix, L. Paul (Paul), 284, 289
Mize, Greg, 194, 389, 390
**Montgomery, Steven (Steve)**, 93, 153–155, 190, 191, 209, 216, 266, 370, 422, **513–520**, 593
Montry, Gary, 196–197, 278, 279, 401
Moody, Ronald (Ron), 362, 368, 475
Moore, Gordon, 14, 15
Moore, Nathan, 593
More, Richard (Dick), 231, 302, 464, 492
Morel, Jim, 403
**Morosin, Bruno**, 34, 82, 149–151, 216, 329, 400, **521–526**
Morris, Robert, 545–546
Mosher, Dan, 191, 370
Mote, James (Jim), 65, 522
Moya, Jaime, 190, 370
Mueller, Daniel Carl, 346
Muhlenweg, Charles, 538
Mullin, Scott, 367
**Munson, Darrell**, 11, 29, 39, 40, 49, 50, 57, 64, 65, 72, 99, 108, 109, 111, 164, 216, 280, 342, 357, 362, 393, 425, 470–472, 484, 497, **527–535**, 539, 540, 556, 562, 563, 577
Murfin, Walter (Walt), 95, 422
Murphy, Byron, 420

**N**

Narath, Albert (Al), 25, 40, 113, 119, 294, 435–436, 442, 448, 521, 530, 546
Narayanamurti, Venkatesh (Venky), 356, 401
Neilson, Frank, 8, 10, 12, 13, 25, 28, 29, 31, 33, 50, 53, 54, 55, 64, 77, 83, 100, 153, 263, 264, 271, 341, 344, 351, 398, 513, 537, 538
Nellis, William (Bill), 129, 234, 285, 396, 400, 591
Nevers, John, 362
Newman, Pat, 521, 522
Newman, Robert (Bob), 342, 496
Nichol, Bill, 500
Northrop, David (Dave), 532, 589
Norwood, Fred, 576
Nunziato, Jace, 29, 68, 107, 114, 115, 117, 125, 167, 183, 184, 216, 306–307, 422, 425, 443–444, 446, 452, 484, 558, 562, 574

**O**

Oberg, Fred, 533
O'Brien, Miles, 338
O'Leary, Hazel, 385, 440, 525
O'Neill, William (Bill), 39, 528
Oschwald, Dave, 385–387

**P**

Passman, Steven (Steve), 216, 307, 578
Peery, James, 74, 156, 207, 266, 277, 293, 368, 485, 508
Percival, Mark, 29, 502
Perkins, George, 567
Perrett, William (Bill), 9, 45, 420
Perry, Frank, 284, 289
Peurifoy, Bob, 500, 501
Piekutowski, Andrew (Andy), 367, 591
Pipkin, Alan, 557
Plimpton, James, 546
Podsednik, Jason, 469
Pope, Larry, 65, 216, 436
Porter, John, 222, 297, 298, 303
Priddy, Tom, 536

**Q**

Quintenz, Jeffrey (Jeff), 208, 222, 298, 301

**R**

Raglin, Paul, 194–195, 222, 300, 305, 387
Ratzel, Arthur (Art), 311, 368, 569
**Reed, Raymond (Ray)**, 6, 11, 29, 40, 72, 77, 108, 109, 152, 162, 216, 470, 484, 529, 530, **536–546**, 556, 562
**Reinhart, William (Bill)**, 132, 139, 140, 142, 174, 177, 179, 191, 213, 221, 222, 247, 258, 277, 290, 294, 361, 365–371, 374, 376, 392, 409, 410, 458, 477, **547–550**, 570, 584, 588
Reisman, David, 224, 228, 229, 231, 236, 243, 300, 302, 413, 463–464
Remo, John, 339, 392
Renlund, Anita, 117, 174, 312, 567–568
Rice, James (Jim), 567, 568
Rice, Melvin, 400
Richardson, Hamp, 571
Ripperger, E.A. (Ripp), 397, 536–538, 562
Rivera, Gary, 458
Robbins, Josh, 155, 191, 516, 585
Roberts, Harold, 545–546
Robinson, Allen, 175, 200, 201, 207, 259, 293, 331, 335, 401, 403, 431, 452, 460, 490, 494, 507, 581
Robinson, Nicholas, 372
Romero, Anthony, 469
Romero, Dustin, 469
Romig, Albert (Al), 551, 577
Root, Seth, 93, 101–102, 217, 228, 301, 392, 417, 469, 587
Rothman, Steve, 300
Rottler, Stephen (Steve), 74, 157, 190, 257, 293, 365, 431, 485, 504, 506, 546, 579
Ruoff, Arthur, 354–356
Russell, Cecil, 39, 528
Russell, Chris, 469

**S**

Sagartz, Matt, 456
Sakharov, Andrei, 587
**Samara, George**, 8, 25, 51, 81, 90, 97, 168, 171, 173, 174, 216, 263, 271, 275, 357, 480, 521, 536, **551–555**
Sanchez, Dan, 571
Sandusky, Harold, 307
Sauve, Gerry, 132, 290, 476
Savage, Mark E., 469
Sawaoka, Akira, 362
Schmidt, Richard (Rich), 99, 562
Schmitt, Robert (Bob), 15, 198, 428
Schroeder, Donald (Don), 568
**Schuler, Karl**, 29, 49, 68, 85, 99, 100, 114, 132–134, 142, 159, 169, 216, 267, 290, 324, 425, 437, 443, 458, 484, 497, 531, **556–561**, 562, 561, 574, 577

Schuster, Carl, 502
Schwartz, Siegmund ('Monk'), 501, 537
Schwarz, Albert (Al), 47, 115, 117, 421
Schwoebel, Richard (Dick), 158–159, 356, 566
Scotti, James, 199, 329
Seagle, Chris, 217, 228, 301, 469
Seaman, Johann, 223, 297
Seaman, Lynn, 396
Seay, Glenn, 82, 113, 399, 435, 437
Setchell, Robert (Bob), 93, 116, 117, 120, 147, 149, 153, 154, 174, 187, 191, 192, 216, 368, 370, 444, 452, 516, 568, 569, 589, 590
Shahinpoor, Mohsen, 133
Shaner, John, 382, 393
Sharp, Greg, 491, 494
Shea, Neil, 354
Shear, Rodney (Rod), 193, 384
Sheffield, Steve, 115, 117, 188, 422, 568–570
Sherby, Oleg, 533
Shoemaker, Carolyn, 199, 204, 329–330, 338, 340
Shoemaker, Eugene (Gene), 204, 329, 330, 334, 338, 339, 451
Short, Mark, 570
Shulenburger, Luke, 217, 587
Sih, George, 449
Silling, Stewart, 505, 586, 590
Simpson, Robert (Bob), 342, 496, 529
Skews, B, 310
Slutz, Steve, 494
Smith, Albert (Al), 368
Smith, Carl, 9, 193, 216, 384
Smith, Jimmy, 496
Smith, Leon, 8, 34, 35, 495–496, 499
Sowell, Robett (Bob), 342
Spalding, Richard (Dick), 338
Sparks, Morgan, 439, 501, 521
Spielman, Richard (Rick), 213, 223, 228, 299, 412, 463
Spohn, Lawrence (Larry), 335
Spray, Stan, 568
Stanton, Philip (Phil), 81, 156, 172–174, 193, 216, 293, 367, 368, 410, 445, 536, 562, 568, 590
Steinberg, Daniel (Dan), 129–130, 285–286, 361–362, 383
Stein, Peter, 544
Stevens, Albert (Al), 65, 98, 99, 121, 216, 437, 451, 502
Stevens, Gerry, 468
Stilp, Alois, 362, 550
Stokes, Paul, 498
Stollar, Hap, 502
Struve, Ken, 493, 494
Stulen, Rick, 581
Stygar, William (Bill), 196, 303
Susoeff, Allan, 133, 277, 289, 476
**Sutherland, Herbert (Herb)**, 103, 165, 357, 360, 531, 536, **562–563**, 577
**Sweeney, Mary Ann**, 256
Swegle, Jeffrey (Jeff), 104, 131, 144, 183, 216, 288, 360, 431, 449, 454, 503, 586
Swift, Harold (Hal), 475

**T**

Takata, Toshiko, 332, 336
Tappan, Alex, 570, 571
Taylor, John (LANL), 280
Taylor, John M. (Sandia), 414, 415
Taylor, Lee, 154, 515
Taylor, Paul, 457, 590
Taylor, Warren, 537
Teller, Edward, 339, 550
Thompson, David, 384
**Thompson, Samuel (Sam)**, 74, 75, 95, 118, 155, 156, 167, 193, 216, 266, 276, 293, 366, 382, 422, 425, 427, 431, 450, 455, 483, 485, 503, 505, 508–512, **564–566**, 577, 579, 580
Thorne, Billy J., 29, 211, 425, 431, 503, 509
Thornhill, Tom, 247, 371, 588
Thundburg, Sig, 538
Toor, Arthur (Art), 224, 228, 239, 300, 413, 463
Traeger, Dick, 80, 502
**Trott, Wayne**, 93, 117, 174, 185, 186, 188, 216, 256, 288, 314, 315, 372, **567–571**, 585
**Trucano, Timothy (Tim)**, 114, 132, 134, 140, 141, 156, 157, 161, 172, 177, 200, 201, 207, 259, 284, 287, 290, 292–293, 297, 331, 332, 335, 338, 366–367, 394, 401–402, 425, 504, 505, 507, 508, **572–582**, 591
Trutnev, Yuri, 440
Tucker, Tillman (T.J.), 29, 445, 446, 573
Tucker, William (Bill), 182
Tuler, Floyd, 29, 49, 97, 451, 484, 497, 556

**V**

VanDevender, J. Pace (Pace), 401, 403–404
Van Dyke, John, 575
Vasey, Anita, 132, 174, 477, 548
Venturini, Eugene (Gene), 149, 521, 522
Verdugo, Lucille, 174, 548

**Vogler, Tracy**, 71, 93, 104, 183, 186, 188, 217, 228, 247–249, 261, 301, 361, 372, 374–376, 570, **583–588**, 593
Vortman, Luke, 9, 44, 80, 267, 420

**W**
Wackerle, Jerry, 355–356
Walker, Scott, 194, 389, 390
Walling, Hal, 514
Walsh, Ed, 68
Walsh, John, 400
Walsh, Robert (Bob), 29, 425
Walter, Patrick, 544
Walters, Bill, 583
Wanke, Michael, 373
Warpinski, Norm, 589
Wawersik, Wolfgang, 531, 532
Weart, Wendell, 532
Webb, David (Dave), 149–151, 521, 523–524
Weber, Laura. 435
Weingart, Richard (Dick), 445
Westmoreland, Sydney, 455
Wickham, Lewie, 435
Wigner, Eugene, 234–235
Wilke, Mark, 389
Williams, Andrew (Andy), 178, 277, 366
Williams, David C., 537
Williams, Frank, 522
Williams, Max, 501
Williamson, Don, 522
Wilson, Leonard, 183, 369, 375
Wilson, Leroy, 5
Winfree, Nancy, 371, 372
**Wise, Jack**, 62, 93, 135, 147, 155, 228, 244, 258, 285–287, 301, 361, 362, 417, 517, 580, **589–594**
Wistor, Joe, 546
Witten, Charles, 344, 471, 529, 533
Wong, Michael (Mike), 207, 370, 508

**Y**
Yarrington, Paul, 23, 99, 156, 166, 193, 216, 277, 293, 352, 367, 382, 384, 393, 425, 427, 453, 579
Yonas, Gerold (Gerry), 222, 298, 400
Young, Emily, 29, 75, 485, 497
Youngman, Kevin, 469
Youngs, David, 512

**Z**
Zahnle, Kevin, 332, 333
Zeigler, Fred, 452, 505, 509
Zeuch, David (Dave), 154, 515–516
Zwisler, W., 309–310

# Index of Terms

**A**

*ab initio* calculations, 217, 219, 231–233, 235, 236–239
  aluminum conductivity, 231, 236, 301, 377, 464, 492
  beryllium, 303
  carbon dioxide, 239
  diamond, 237, 238, 303
  deuterium, 232, 233, 235–237, 239
  hydrogen, 232, 233, 235
  polymers, 239
  water, 237, 238, 303
  xenon, 239, 303
aboveground testing (AGT), 9, 30, 77
Accelerated Strategic Computing Initiative (ASCI), 15, 196, 198, 217, 312, 367, 432, 433, 507, 582
acceleration wave, 107–108, 444
Advanced Photon Source (APS), 268, 270, 588
Advanced radiography, 193, 195, 196
Aeroballistics Range Association (ARA), 361, 362, 369–372, 374, 409, 550
Air Force Weapons Laboratory (AFWL), 75, 76, 78, 481–484, 508, 515
ALEGRA computer code, 16, 156, 190, 191, 204–207, 217, 218, 231, 259, 266, 271, 293, 368–370, 401–407, 433, 460, 464–466, 486, 488–494, 508, 581
ALEGRA/EMMA computer code, 516, 517
alumina, 131, 149, 183, 250, 251, 369, 374, 519, 520, 522, 524, 560
alumina-filled epoxy, 97, 104, 190–191, 288, 360, 372, 560, 561, 578, 593
aluminum, 42, 56, 57, 62–64, 67, 92–93, 98, 104–107, 114, 129, 131, 141, 147, 179, 187, 197, 221, 225–227, 231–232, 234–237, 244, 252, 281, 321, 324, 358, 364, 404–405, 421, 435, 441, 450–451, 455, 465, 467, 472, 491–494, 500, 573, 590–591
  strength properties, 129, 246, 363, 392, 586
Amdahl's Law, 278
American Telephone and Telegraph Co. (AT&T), 5, 7, 113, 341, 346, 440, 442
Ancho Canyon facility, 382, 529
ANEOS equation of state, 95, 422, 510, 511, 564
Area I, 10, 11, 38, 39–40, 98, 102, 109, 112, 173, 193, 224, 291, 343, 347, 410, 412, 470, 471, 529–532, 538, 540, 542, 549
Area II, 499, 549
Area III, 10, 31, 33–35, 38, 39, 108, 112, 328, 496, 500, 529, 530, 536–538
Area IV, 408, 416, 479, 487
Area Y, 31, 34, 39, 40, 108–109, 280, 327, 470, 529, 530, 540, 556, 557
Armando zero room, 390
Army Missile Defense Agency, 177, 179, 365, 371–374
Atomic Energy Commission (AEC), 5, 19, 484, 500, 501, 531, 538, 560
Atomic Weapons Establishment (AWE), 190, 192, 193, 384, 389–392, 445, 506, 512
Atomic Weapons Research Establishment, 445, 512



J.R. Asay et al., *Impactful Times*, Shock Wave and High Pressure Phenomena,
DOI 10.1007/978-3-319-33347-2

**B**
Baby Bear recovery fixture, 523
Bacchus/Barolo series of subcritical experiments, 391
backward analysis, 147, 244
Baer-Nunziato model, 184, 307–310, 312
ballistic holography work, 277
ballotechnics, 308
Bauer PVDF shock sensor, 151–153, 544–545
BCAT computer code, 460
Bechtel Nevada, 194, 384, 387, 389–391
Bell Telephone Laboratories, 5
beryllium Hugoniot, 231, 303, 465
biological shock experiments, 373, 538
Black Monday, 118, 171–207, 410, 411, 476–477
Boomerang subcritical test, 386
braze alloy studies, 601
BUCKL computer code, 458

**C**
carbon
  carbon-based foams, 223, 463, 558
  ramp loading, 219
  triple point, 219, 232, 237, 238, 378, 465
Cauchy-Green deformation tensor, 437
CDC computers, 14–15, 94, 154, 345, 427, 431, 504, 505, 515, 573, 575
ceramic, 149, 182, 183, 247, 250, 251, 361, 369, 370, 425, 522, 552, 555, 590, 591
  impact and fragmentation, 457
  strength of, 374–375, 395
Chapman-Jouget (CJ) detonation model, 186, 313, 461
CHARTD computer code, 74, 75, 77, 155, 266, 431, 485, 503, 504, 509–511, 564–565
CHEETAH code, 310
chemical reaction and chemical reactivity, 117, 149, 151, 399, 521, 523
Cielo computer, 15, 197, 403
Cimarron subcritical experiment, 194, 195, 387–388
CKEOS, 460
composites and mixtures, 7, 13, 26, 49, 73, 76, 95–97, 106, 149, 183–187, 215, 264, 266, 313–314, 316–317, 360, 425, 429, 481, 485, 487, 497, 531, 540, 562, 569, 583, 590, 593
compressive strength, 284–286, 292, 301, 304
concrete, 15, 95, 182, 247, 369–370, 372, 411, 422, 448, 458, 460, 563, 565
Congressional recognition of Z, 239, 304
constitutive relations, 26, 73, 311, 436, 485
Contact fuzes, 8, 11, 21, 26, 34, 39, 47, 52, 66, 263, 264, 496, 499, 517, 528, 590
containment for toxic materials, 97–98, 239–242, 285, 303, 413–414, 416–417
continuum mechanics
  damage mechanisms, 98
  energetic materials, 113–118
  modeling, 96, 557
cookoff modeling, 311–312, 319
copper
  Hugoniot, 57, 64
  isentropic loading, 299, 467, 494, 548
coupled damage and reaction (CDAR) model, 307
Crater computer code, 504, 511
cratering, 20, 21, 23, 105, 156, 267, 357, 420, 425, 511, 580
crater scaling
  strength, 21, 23, 156, 267, 420, 504, 537
Cray computers, 15, 154, 196, 197, 278, 345, 346, 455, 461, 504–507, 515, 565
cryogenic (capabilities and systems), 135, 222–223, 235, 291–292, 298, 300–301, 591
CRYSIZ computer code, 524
CSQ code, 14, 15, 74, 77, 94, 95, 118, 155–156, 158, 266, 422, 427, 431, 449–451, 485, 503–505, 511–512, 514, 565, 577, 579
CTH code, 15, 74, 118, 140, 144, 155–158, 160, 175, 179–180, 184–185, 191, 196–204, 207, 266, 293–294, 307–308, 310, 312–313, 331–332, 365, 366, 368–369, 373–375, 382, 392, 402, 422, 427–428, 431–433, 454–461, 485–486, 504–508, 512, 565–566, 569, 579–582, 586–587
Cygnus radiographic machine, 194–196, 390–391
cylindrical convergence
  EOS experiments, 220, 268, 271
  imploding wire array, 220–223, 297
  magnetic loading, 134, 206, 223–235

**D**
DAKOTA optimization software, 317, 466, 492–494
debris shields, 142, 180, 181, 200, 329, 366
Defense Nuclear Agency (DNA), 76, 99, 100, 140, 142, 177, 276, 351, 352, 363, 366, 367, 383, 393, 481

Defense Threat Reduction Agency (DTRA), 76, 140, 306, 318, 453, 481
deflagration, 113–115, 306, 573
deflagration-to-detonation transition (DDT), 184, 306–308, 311, 422
density functional theory (DFT), 235, 237, 406, 407, 587
depth of penetration, 334, 460
detonation
modeling of detonation, 117, 184, 186, 313–314, 317
modeling equation of state (EOS) products, 228
nitromethane, 452
PBX 9404, 114–116, 442, 444, 452
products, 318
shock dynamics of detonation, 350–352
shock induced, 184, 452
detonation wave, 7, 114, 116, 186, 313, 379, 443–444, 452
detonators, 7, 26, 34, 117, 184, 240, 241, 264, 307, 421, 445, 446, 567
deuterium Hugoniot, 219, 222, 232, 233–235, 236–237, 242, 298, 303, 377, 465
deuterium metallization, 93, 233, 234–235, 237
diamond, 251, 303, 355, 378, 382, 413, 465, 469, 522
diamond anvil cell (DAC), 130, 219, 234, 299
Diatom library, 459
diffusion
approximation, 77, 403–404, 509, 511, 565
current, 218, 228, 231
magnetic, 229, 230, 464, 491
thermal, 131, 132, 422
dirty-binder, 316
dislocation (theories and models), 51, 60–62, 65, 130, 131, 150, 184, 269, 270, 280, 281, 285, 322, 398, 435–438, 524
DOD/DOE MOU, 182–183, 186, 311, 368, 369–371, 374, 375
double shock, 469
Dynamic Integrated Compression Experimental (DICE), 189, 392, 416, 417, 570, 593

**E**

ejecta (of mass), 196, 268, 283–284, 289, 387–390, 455
Asay window, 388
thin foil technique, 112–113, 195
elastic constants
fourth order, 101–102
from ramp loading, 102
from shock loading, 102
third order, 101–102
elastic-plastic models and response, 13, 26–27, 42, 52–56, 58, 60, 63–65, 94–95, 131, 282, 284–285, 322, 436, 584
elastic-plastic shock structure, 6, 27, 129, 284
elastic precursor, 51, 60, 280–281, 327, 361, 362, 375, 435, 437, 496
electrical
conductivity, 231, 377, 407, 464, 491, 492
discharge in reactions, 407, 462–463
properties, 25
shorting pins, 10–12, 33, 35, 56, 63, 343, 385, 434, 458, 472, 557
electron paramagnetic resonance, 150, 524
electron spin resonance, 149, 521
encapsulants, 190–192
endothermic behavior at the shock front, 115–116, 443–444
energetic materials, 7, 13, 26, 101, 113–118, 152, 158, 174, 179, 182, 184, 188, 216, 248, 264, 266, 288, 306–309, 311, 316, 318, 398, 400, 428, 442, 462, 567
energy localization, 159, 185, 186, 313, 314, 319
energy release, 28, 114, 116, 150, 180, 193, 298, 443, 461, 525
Energy Research and Development Agency (ERDA), 19, 500, 501, 509–532, 560
enhanced catalytic action, 150, 523
equation of state (EOS), 6, 74
aluminum, 57, 63–64, 227, 235, 321, 404, 467
beryllium, 231, 303, 406
copper, 57–64, 321, 358, 467
deuterium, 219, 232–233, 236–237, 242, 298, 303, 377, 462, 465
diamond, 237
granite, 382
surface, 43, 235, 265, 268
equation of state/phase transitions
iron, 90–92, 94, 223, 224, 299, 324, 400, 412, 450
nitromethane, 116, 117, 446, 452, 502, 567, 570, 571
water, 219, 229, 237, 238, 303, 378, 411
xenon, 303, 407, 465
estane, 316
Eulerian hydrocode/hydrodynamic code, 73, 77, 94, 142, 155–158, 207, 266, 284, 402, 403, 429–431, 433, 449–451, 460, 485, 503–504, 512, 514, 517, 565

exoatmospheric missile delivery, 6, 27–29, 66, 495
exoatmospheric nuclear detonation, 75, , 481
exoplanets, 219, 231–232, 238
Explosive Components Facility (ECF), 114, 174, 183, 216, 218, 271, 311, 368–370, 410, 412, 418, 570
explosive plane wave generator, 33, 263

**F**
ferroelectric (FE)
  ceramics, 8, 10, 25, 101, 153, 154, 191, 216, 502, 513–518
  experiments, 151–152
  metals, 25
  models, 153–155
  polymers, 518, 544, 552, 560
ferromagnetic, 25, 398, 551
field test, 8, 23, 30, 45, 77, 441, 502, 531, 534, 537, 541–545, 592
finite difference, 14, 74–75, 77, 322, 341, 347, 349, 425, 430, 431, 483–485, 503, 515, 530, 534
finite-element, 154, 207, 312, 319, 346, 349, 402–403, 430, 432, 433, 483, 486, 508, 514–517, 532
fireset, 503, 568
flash x-rays, 292, 454–455, 473, 479, 549
flyers (laser-driven), 372, 568–569, 571
flyers (magnetically-driven, aka magnetic flyer), 43, 89, 218, 228–238, 243–244, 264–265, 268, 277, 292, 297, 301–303, 359, 369, 377, 381, 403, 405–406, 409, 413, 458, 464–469, 488–494
foams (porous materials), 7, 25, 26, 34, 42, 49, 69–71, 237, 281–282, 310–311, 345, 372, 375, 406–407, 435–436, 443, 457, 463, 510, 514, 534, 586
fourth power law, 61, 105–107, 112, 286, 358, 393, 422, 586
fracture, 66, 72, 73, 95, 97–99, 142–146, 157, 159, 177, 182–183, 251, 287, 289, 292, 342, 344, 349, 365, 393–394, 400, 429–430, 449–455, 457, 460, 485, 487, 540, 565–566, 586
fragment, 43, 69, 114, 135, 138, 142–146, 176, 180, 197, 247, 277, 278, 292, 345, 365, 406, 445, 453–455, 458–460, 573
fragmentation theories, 142–146
  analytic model, 143–145
  atomic fission (fragmentation model), 145–146
  expanding metallic cylinders, 183
  galaxy formation, 145
  general fragmentation law, 286
  molecular dynamics study, 146

**G**
geological materials, 8, 9, 13, 15, 23, 30, 99–100, 142, 157, 175, 182, 267, 287, 351, 363, 382, 393, 504, 510, 514, 531, 576
gold, 148, 269, 467, 558
graded density impactor, 134–140, 183, 227, 251, 265, 277, 286, 293, 364–365, 547
Grady-Kipp fragmentation model, 197
granular materials, 104, 111, 184, 187–188, 247–249, 306–309, 375, 585–587
Grüneisen ratio and MIE-Grüneisen equation of state, 157, 285, 328, 420
Guinan-Steinberg constitutive model, 129, 362
Gurney modeling, 445–446

**H**
Harry S. Truman letter to AT&T, 5
hazardous material, 193, 416, 450, 457
helium gas gun, 11, 36, 38, 40–42, 43, 112, 216, 265, 398, 524, 525
helium leak test for Z containment, 241, 414
  shock studies of superfluid helium, 589, 591
heterogeneous material, 184–190, 269, 288, 313–315, 372–373, 381, 400, 570–571
heterogeneous reactive media, 184, 312–314
heterogeneous spallation, 372
high explosive (HE), 10, 23, 30, 108, 117, 178, 264, 277, 387, 392, 421, 452, 460, 470, 522, 529, 562, 568, 574, 590
high strain rate, 61, 106, 107, 131, 143, 144, 157, 183, 226, 227, 250, 251, 286, 358–360, 370, 454, 499, 583
HMX explosive, 307, 314–317, 442, 570
Hopkinson bar, 6, 27, 63, 183, 226, 249–251, 369, 453, 496, 499, 535, 586
hot spots
  high explosives, 187–188
  hot spot formation, 183, 187, 319, 452, 571
Hubble telescope image of Jupiter impact, 15, 200–202, 334–338

Hugoniot, 10, 26, 31, 32, 37, 52, 53, 57, 102, 105, 116, 137, 182, 219, 222, 227, 229–233, 236–238, 251, 286, 297, 298, 303, 321, 322, 324, 351, 355, 377, 378, 381, 384, 387, 395, 406, 413, 416, 418, 444, 446, 460, 465, 469, 482, 487, 488, 496, 499, 502, 510, 529, 571, 590
Hugoniot elastic limit (HEL), 52, 53, 66, 105, 249–250
hydrodynamics codes, 14, 15, 72–74, 77, 93, 97, 104, 129, 142, 155–158, 198, 204–207, 229, 264, 266, 271, 284, 332, 351, 362, 371, 401, 404, 407, 427, 429, 430, 449, 464, 482, 483, 485
hydrogen equation of state (EOS), 232–234, 326, 580, 590–591
hydrogen metallization, 134–135, 138–139, 234–235, 291–292, 357, 579, 590–591
hypervelocity impact, 15, 94, 141, 178, 180, 199, 247, 287, 289, 329, 333, 428, 450–451, 484, 514, 566, 579–581
Hypervelocity Launcher (HVL), 11–12, 43, 134, 138–142, 176–181, 227–228, 230–231, 265, 268, 292, 364–367, 392, 409, 413, 580

**I**

ice (water), 202, 333, 336
ICE cube, 224–226, 240, 241, 244, 300, 413
ice giant planets, 378
ICF driver, 403
ignition and growth, 12, 113–117, 131, 144, 159, 160, 187, 206, 216, 308, 312, 315, 319, 443–444, 452
impact
  flash, 142, 176, 178, 179, 276, 277, 332, 373
  launchers, 27, 34–43, 264, 281
impedance matching
  transfer function approach, 147–148
in-bore projectile velocity measurements, 89
incipient spall, 97, 372, 570
index of refraction
  fused silica, 62, 323
  LiF, 62, 146–147, 285, 361, 392
  PMMA, 62, 146–147, 323, 370, 590
  sapphire, 62, 146–147, 216, 361
  zinc chloride solution, 590
inductance, 22, 446, 465, 466, 493
inductive voltage adder (IVA) radiography, 195
inertial confinement fusion (ICF), 205–206, 220, 221, 233, 296, 297, 304, 401–403, 406, 411, 462, 488, 508, 511, 567
infrared spectroscopy, 179, 247, 373, 567
initiation
  hot spot, 187
  nitromethane, 116, 446, 452, 461, 571
  thresholds, 315
Intel Paragon computer, 15, 172, 196, 200, 201, 331, 334, 431, 456, 506, 507
interferometry
  displacement, 57–58, 64, 320–323, 327, 421
  photon Doppler velocimetry (PDV), 390
  velocity, 11, 59, 62, 65, 68, 86–90, 105, 188, 251, 323, 590
  VISAR, 186, 194, 222, 299, 370
International Space Station, 175, 180, 181, 366
iron
  phase transitions, 90–91, 324, 450
  spallation, 66, 67
  yield, elastic, 89
isentropic compression experiment (ICE), 102, 130, 135, 137, 219, 224, 227, 229, 244, 292–293, 299, 300, 316–317, 326, 357, 363–364, 369–370, 401, 405, 406, 413, 416–417, 463, 466, 468, 488, 490, 493, 548, 571, 580, 584

**J**

JANAF library of explosives and energetic materials, 309
JASPER, 175, 371, 387, 470, 479
JCZ (Jacobs–Cowperthwaite–Zwisler) equation of state database, 310
JCZS database, 310
Johnson Space Center (JSC), 142, 180, 366
Jupiter evolutionary models, 235, 333
Jupiter impact by Shoemaker-Levy comet, 15, 175, 198–204, 266, 329–339, 507, 566, 581
JWL (Jones-Wilkins-Lee) model for explosives, 310, 454, 461

**K**

Kerinei/Krakatau subcritical experiment, 390
kinetic effects of shocks and phase transformations in models, 12, 92–93, 117–118, 142, 178, 185–186, 252–253, 266, 282, 324, 367, 373, 380, 398, 423, 452, 455, 456

kinetic energy kill for Theater Missile Defense, 175–180
kinetics of polymorphic phase transitions, 90, 92–95, 185–186, 231, 238, 299, 317, 324, 380, 398, 412, 450
Kirtland Field, 3, 4
Krakatau subcritical experiment, 195, 388–390
K/T dinosaur-killer, 200, 330, 331, 333
Ktech Corporation, 152, 174, 175, 295, 300, 361, 365, 368, 371, 392, 476, 540, 545, 547, 548, 601

**L**

LAMMPS molecular dynamics code, 407
laser-driven flyers, 372, 568–569, 571
laser-driven shock, 220, 232–234, 236, 270, 462
  deuterium, 232, 235–237
laser windows
  fused silica, 60, 62, 146, 323
  LiF, 62, 146–148, 234, 280, 285, 301, 361–364, 392, 467, 491, 584
  nitromethane, 117, 567
  PMMA, 62, 68, 146, 147, 323, 358, 361, 370, 590
  zinc chloride solutions, 590
LASNEX computer code, 488
lateral stress, 249
lead vaporization, 42, 141, 252
Ledoux nuclear test, 385
Lee-More-Desjarlais (LMD) conductivity model, 464, 492
light initiated high explosive (LIHE), 34, 445
Limited Test Ban Treaty, 29, 267
linear transformer driver (LTD), 196
Line VISAR, 13, 175, 186–190, 216, 231, 266, 268, 315–317, 374, 383, 389, 392, 585
  heterogeneous material studies, 175, 186–189, 268, 288, 315, 317, 372, 374, 385
  sugar shock experiments, 188–189, 314–316, 570
  tantalum spallation, 109, 190, 247, 372
lithium-niobate gauges, 101, 124, 266, 543, 545
Lyner facility, 193, 385

**M**

macroscale, 285, 314, 581
magnetically accelerated flyer plates, 43, 87, 228–233, 301–302, 405–407, 463–465, 491–492
magnetically applied pressure shear (MAPS), 12, 219, 245–246, 288, 304, 405–406
magnetohydrodynamic (MHD)
  computer codes, 16, 156, 231–232, 236, 244, 293, 300, 302, 401–407, 460, 464, 466, 485, 488–494
  experiments, 223–235, 239–240, 243–246, 299–300, 412, 415, 463–468
Manhattan Project, 3, 4, 6, 10, 19, 24, 27, 72, 399, 480
Manzano Base, 4
Mario subcritical experiment, 194–195, 388
MaRocco program, 195, 365, 388–389
Marshall Space Flight Center (MSFC), 142, 180, 366
massively parallel computers, 97, 155, 198, 202, 431–432, 506, 507, 516, 566
material measurement
  atomic scale, 145, 270
  continuum scale, 183, 270, 285, 287, 306–307, 309–310, 313–315, 319, 436, 437, 443, 451, 468, 556, 557, 575–576, 585, 587
  for cylindrical loading, 268
  for 3-D loading, 185, 187, 191, 516
  mesoscale, 175, 185–189, 216, 268, 270, 283, 285, 306, 312–319, 372, 375–376, 407, 570, 586
  for uniaxial strain loading, 28, 30–32, 35, 54, 56, 63–69, 71, 89–90, 94, 98, 103, 115, 129, 148, 153, 175, 223, 232–233, 238, 246, 250, 284
MAVEN experimental research, 190–192, 370–371
MAVIS velocity measurement, 479
maximum velocity for magnetically driven flyer plates, 43, 228, 231, 265, 302, 406, 465, 494
MELCOR nuclear reactor analysis code, 504
melting
  aluminium shock melting, 92–93, 392, 421
  beryllium, 465
  bismuth shock melting, 92, 281–282, 421
mesoscale
  modeling, simulation, 175, 183–186, 188, 270, 283, 312–318, 372, 375, 570, 586
metallic deuterium, 233–234, 237
metallic hydrogen, 134, 135, 234–235
metal-to-metal sliding contact, 134, 290, 430, 580
Meteor Crater, 330, 451
Michelson interferometer, 57–58, 64, 86, 215, 321–322, 327, 441

microenergetic processing and testing, 571
microscale, 184, 270, 285, 538, 569, 581
microwave loss measurements, 150
MINI JADE nuclear test, 351–352, 382
Missile Defense Agency (MDA), 177, 284, 365, 371, 428
mixture modeling of reactive porous media, 309
mixture theory, 184, 306, 308–310, 313, 344, 422, 577
molecular dynamics, 145, 146, 377, 407, 464, 491, 588–589
Momma Bear recovery fixture, 523
Monte Carlo techniques, 459
Moore's law, 14, 15
multiphase model, 184, 306, 310, 311
multipoint VISAR, 194, 468
multiscale modeling
  flow strength, 131
  foams, 187
  sugar, 188–190, 314–316, 570

**N**

National Aeronautics and Space Administration (NASA), 142, 177, 180, 333, 338, 365, 366, 457
National Defense Authorization Act of 1994, 190, 216, 267
National Ignition Facility (NIF), 218, 219, 298, 305
national missile defense, 128
National Nuclear Security Administration (NNSA), 9, 141, 172, 195, 216, 218, 221, 232, 239, 241, 242, 267, 270, 296, 301, 303–304, 412, 583, 593
National Security Technologies (NSTec), 391, 392, 593
nCUBE massively parallel computer, 196–197, 506
neutron generators, 34, 113, 154, 190, 383, 417, 422, 450, 458, 514, 515, 560, 561
Nevada National Security Site (NNSS), 6, 9, 22, 192, 194–196, 222, 239, 305, 350, 383, 391, 393
Nevada Test Site (NTS), 9, 22–24, 76, 77, 192–196, 222, 239, 255, 267, 293–294, 300, 301, 328, 350–351, 371, 383–385, 387, 389–391, 393, 414, 453, 470, 472, 479, 481, 502, 514, 535, 538, 543–544, 561
nitromethane explosive, 116–117, 446, 452, 502, 567, 570, 571
nuclear magnetic resonance, 149–150
nuclear testing, 9, 20, 21, 29–31, 267, 538
nuclear yield, 20–21, 23, 30, 267, 350, 453

**O**

oblique impact gun, 71, 248, 268, 291, 362, 587
oil shale, 99–100, 142–143, 146, 349, 393, 453, 502, 531, 535, 543–544, 560–561
one-shot power supplies, 25, 113
Optically Recording Velocity Interferometer System (ORVIS), 12, 13, 188, 266, 288, 372, 569–571, 585
Optical Multi-channel Analyzer (OMA), 178, 277
orbital debris, 142

**P**

P-$\alpha$ compaction model, 12, 70, 71, 375, 435, 457, 534
P-$\lambda$ compaction model, 38, 71, 375
Paragon massively parallel computer, 15, 172, 196, 200, 201, 331, 334, 431, 456, 506–507
Parallel CTH (PCTH) computer code, 200, 201, 203, 204, 367, 368, 431, 507, 544
parallel processing, 15, 128, 172, 197, 278, 294, 367, 382
paramagnetic resonance signatures, 150, 524, 566
particle velocity dispersion, 283, 372
PBFA, 30, 216, 220, 383
PBX 9404 explosive, 104, 114–116, 360, 442, 452
PBX 9501 explosive, 316–317, 586
penetration, 94, 100, 142, 176–177, 247, 311, 334–335, 338, 394, 433, 459–461, 563, 576–578, 580–581, 591
Pershing II Earth penetrator, 154, 513
PETN explosive, 310, 446, 568
phase transitions, 30, 59, 91, 247, 380, 398, 579, 580
  bcc crystals, 65, 66, 94, 231, 465
  ferroelectrics, 191, 264, 515, 518
  iron, 90, 94, 299, 400, 412, 450
  water, 238
piezoelectric
  experiments/modeling, 13, 51, 53–55, 100–101, 151, 397–398
pillow, 137–140, 292–293, 363–365, 548, 580

Piola-Kirchhoff stress tensor, 437
planetary physics, 149, 201, 204, 228, 238, 329, 332, 451, 455
  Jupiter, 175–176, 198–204, 235, 266, 329–340, 507, 566, 581
  Neptune, 219
plane-wave lenses, 10, 384, 529
plate breakup, 138, 292
plate impact, 28–29, 35–38, 42–43, 47, 52, 98, 135, 187, 220, 321, 362, 425, 445, 451, 457, 493, 535
plate impact launcher, 29, 42, 264
Plexiglas®, 68, 449, 556–560
Plowshare program, 23, 46, 420
plutonium
  containment experiments on Z, 239, 242–243, 303–304, 413
polymethyl methacrylate (PMMA), 62, 68, 95–97, 114, 140, 146, 147, 358, 361, 370, 443, 454, 455, 502, 548, 574, 590
polymorphic phase transformation, 88, 90, 94, 185, 231
polyurethane foam, 57, 187, 441, 557
polyvinyl-difluoride (PVDF) stress gauge, 56, 128–129, 151–153, 266, 352, 359, 369–370, 544–545
porous materials
  compaction, 7, 26, 42, 69–71, 88, 111, 237, 247–249, 281–283, 307–309, 311, 316, 324, 341, 370, 375, 457, 514, 534, 585–586
powder gun, 11, 12, 39–43, 102, 103, 105, 108–111, 118–119, 129, 139, 147, 161, 213, 265, 280–283, 285, 290, 347–348, 356–362, 364, 370, 374, 376, 382, 391, 447, 470–475, 530, 531, 540–541, 557
powder gun accident, 102, 110–111, 290, 348–349, 359–360, 447–448, 474–475, 532
preheating for shock wave experiments, 282, 304, 601
pressure-induced conductivity, 234
pressure-shear experiments, 90, 102–104, 129–130, 183, 245, 247–249, 266, 288, 304, 359–361, 363, 586–587
pressure-shear interferometer, 100, 248, 266, 288, 360
PRESTO computer code, 432, 433, 508
probabilistic models for reactive behavior, 319
projectile shape effect on debris, 142, 178, 366–367
PRONTO computer code, 154, 207, 402, 428, 432–433, 486, 508, 515
PUFF computer code, 482, 484
pulsed power
  for ramp and shock loading, 9, 43, 156, 217, 220–221, 241, 243, 268, 287, 295–296, 303, 406, 463, 582
Pyroceram®, 286
pyrometry, 376, 384–386, 388, 390
pyrotechnic mixtures, 308, 310
PZT 95/5, 104, 155, 190–191, 360, 370, 518–520, 571, 593

**Q**

quantum molecular dynamics (QMD), 12, 13, 217, 233, 236–239, 407, 464, 491, 492
quartz gauge
  configurations, 53, 54, 55
  displacement current, 31, 153, 513
  nonlinearity response, 77, 101, 542
  shock gauge, 25, 39, 62, 77, 151–153, 443
  stress gauges, 8, 25, 26, 32, 39, 62, 100, 153, 265, 436, 443, 513, 538, 541
  three-zone model, 13, 31, 53
  with underground testing, 267
quartz as a high-pressure standard, 215, 232
quasi-isentropic compression or loading, 130, 137, 138, 219, 227–228, 265, 291, 293, 326, 363, 412, 416, 418, 463, 548, 571, 580
QUICKSILVER computer code, 489, 491

**R**

radiation-driven shock, 297, 404, 405
radiography
  Cygnus machine, 194–196, 390–391
railgun, 43, 112, 128, 131–134, 138, 276, 277, 289–291, 302, 326, 365, 367, 407–409, 416, 476, 580
Raman spectroscopy, 567
ramp loading
  fused silica buffer, 108, 134, 136, 227, 286
ramp wave, 90, 102, 106–107, 112, 113, 117, 130, 134, 136–137, 147–148, 187, 217, 223, 224, 227, 228, 250, 269, 270, 286–287, 292–293, 299–304, 381, 423, 444, 517, 601
Rayleigh-Taylor instabilities, 131, 359, 405
reaction
  kinetics, 118, 422, 450, 450–453, 455
  spot initiation, 187

reactive
flow, 307, 319
Rebound I nuclear test, 192–195, 384–385
red mercury, 525
refurbished Z, 205, 231, 242, 246, 303, 493–494
release wave, 32, 42, 62, 91, 98, 231, 245, 252, 308, 322, 358, 359, 361, 363, 367, 371, 374, 375, 378, 383, 384, 386, 387, 437, 450, 452–454, 456, 457, 461, 469
reloading, 129, 187, 269, 284–285
resolidification, 469
reverberations, 66, 85, 96, 234, 363, 377, 444, 502
reverse ballistics, 383
RHALE computer code, 156, 207, 368, 402, 433, 508
ring expansion, 144, 454
RMHD code, 488
Rocco subcritical experiment, 194–195, 388–389
rock mechanics laboratory, 349, 531, 532, 560–561, 563
rock salt, 436, 531
RSCORS graphics package, 431, 510–511, 565

S

Sandia Base, 4, 5, 76, 481
Sandia field test quartz gauge, 77, 542, 543, 545
Sandia quartz gauge, 36, 77, 344, 397, 398, 538, 541–543
Sandia velocity interferometer, 59, 65, 68, 86–88, 90, 105
sapphire, 62, 82, 101, 131, 146, 216, 247, 361, 363, 372, 374, 570
index of refraction under ICE loading, 147, 590
scaling
explosive cratering, 20, 21, 23, 267, 420
nuclear yield estimates, 20, 21, 23, 267, 350
Science Campaign, 242, 415
Scooter explosive test, 23, 24, 46
Sedan crater, 23, 24
self-arcing railgun projectile, 133, 289–290, 409
self-consistent strength model, 129
semiconductor physics, 552, 555
Senate report on Z experiments, 239, 242, 344
SESAME equation of state, 233, 238, 491
shaped explosive charges, 157–158, 459–460, 528, 583
shear
deformation, 398, 452
shear band localization, 577–578
shock
compaction, 7, 26, 69, 88, 247, 248, 263, 283, 585
compression, 10, 12, 13, 16, 19, 24, 34, 51, 53, 54, 60–65, 87, 91, 102, 105–106, 113, 129, 149, 151–154, 175, 183, 184, 187, 189, 191, 192, 214, 215, 216, 223, 261, 263, 265, 267, 268, 280, 282, 285, 289, 295, 296, 299, 322, 324, 361, 379–381, 396, 397–400, 405–406, 429, 435, 437, 444, 460, 488, 503, 513–516, 521, 523, 554, 555, 570
initiation, 12, 113–117, 185, 187, 190, 216, 266, 313–315, 319, 422, 442–446, 568, 571–572
pyrometry, 376, 384–386, 388, 390
rise time, 60, 96, 105–107, 112, 131, 286, 322, 358, 422, 585
speed, 189, 386, 560
viscosity, 106, 357, 359–360, 393
artificial, 98, 184, 309, 312–313, 451, 452
measurements, 359–360
metals, 357
plastic, 357
4th power law, 393
shock attenuation
in porous materials, 71
in viscoelastic solid, 73–74
shock-change equation, 115, 443
shock chemistry, 149–151, 399, 522, 523, 525–527, 567
shock-induced detonation, 114–117, 186, 379, 442–444, 452, 461, 530, 572–573
Shock-induced melting, 92, 93, 146, 398
Shock-induced polarization, 101
Shock-induced reaction, 150, 452, 523
Shock-induced solid state chemistry, 149–151, 398–399, 521–526
Shock-induced vaporization, 30, 40, 42–43, 77, 88, 141–142, 176, 178, 180–181, 218, 220, 228, 247, 252–253, 284, 287, 289, 322, 367, 373, 580–581, 591
shock kill of biological agents, 373
shockless
acceleration, 135, 138, 139, 406, 494
loading, 102, 134, 136, 139, 140, 227–228
shock modification, 149, 522–524
shock physics database, 486–487

shock-ramp (loading or technique), 235, 268, 469
shock recovery, 97, 98, 149, 523
shock rise time, 60, 61, 64, 86, 105–107, 112, 131, 286, 322, 327, 357–358
shock temperature, 93, 137, 147, 237
Shock Thermodynamics Applied Research (STAR) facility, 9, 39, 40, 90, 105, 108–112, 129, 130, 133, 166, 172–175, 177, 179, 182, 189–191, 199–200, 216–218, 221, 222, 247–253, 278, 280, 287, 289, 290, 291, 294–295, 296, 301, 305, 328, 330, 357, 359, 367–371, 372, 374, 376, 382–384, 392, 408–409, 410, 412–413, 415, 418, 447–448, 468, 470–473, 475, 476–477, 479, 532, 547, 548–550, 570, 583, 584, 586, 587, 588, 591–593
  accident, 111, 290
  capabilities, 199, 201
  naming, 112, 328
shock-to-detonation transition in explosives, 452
Shoemaker-Levy comet SL9 impact on Jupiter, 199, 329
  S-L comet break up, 199–203, 329–338
SIERRA computer code, 312, 433, 517
slant-wire resistor, 11, 56–58, 63, 557
slapper (initiation, tests, or detonators), 319, 386, 391, 445, 568
SLIFER gauge
  determination of nuclear yield, 9, 21–23, 267, 350
sliding friction, 248
slotted-bore gun, 586
small-scale explosive events, 537
Smooth Particle Hydrodynamics (SPH) code, 332, 430
Socorro explosive site, 525
soft recovery, 244, 303
solidification, 135, 242, 291
sound speed
  aluminum, 42, 148, 464, 467
  beryllium, 285, 303, 344, 345, 357, 361, 406, 436, 465, 471, 590
  bismuth, 281
  LiF, 147
  tantalum, 148
space debris, 180–181, 200, 330, 366
spallation, 6, 27, 31, 49, 51, 65–67, 86, 88, 95, 97–98, 176, 190, 247, 322, 341, 372, 392, 495, 497, 500
  ceramics, 247
  iron, 66
  steel, 94
  tantalum, 383
spall strength, 27, 61, 66, 67, 89, 94, 138, 365, 384, 387, 514, 548
spark detonator, 446
SPARTAN radiation model, 403–404
Special Technologies Laboratories (STL), 417, 468
spectroscopy
  hypervelocity impacts, 247
  shocked deuterium, 235
SPH computer code, 332, 430
Splinter Gate, 447
spontaneous electrical polarization, 191, 515
Stagecoach subcritical experiment, 386, 387
STARFIRE railgun, 133, 276, 289–290, 408–409, 476
steady wave, 60, 61, 69, 96, 116, 309, 435, 452, 585
steam-air reaction, 448
stishovite, 383
Stockpile Stewardship Program (SSP), 9, 141, 190, 192, 193, 216–219, 221, 239, 242, 267, 296, 303, 304, 384, 462
Strategic Defense Initiative (SDI) aka 'Star Wars,' 127–128, 131–134, 142, 151, 157, 177, 276, 289, 363–365, 439, 476, 525, 527, 579
Stress Wave Applications Program (SWAP) code, 14, 75, 96, 322, 431, 485, 535, 573
stripline configuration or load, 147, 229, 231, 232, 243, 244, 245, 246, 304, 406, 465, 466, 467, 494
strip line ramp magnetic generator, 147, 406
subcritical experiment (SCE), 30, 192–196, 222, 267–268, 300, 384–392, 468
SUBWAY computer code, 16, 154, 155, 191, 266, 370, 515–516
sugar, 188, 189, 314–316, 570
SWAP computer code, 14, 75, 96, 281, 322, 378, 481, 485
T
Taj Mahal drift in U1a, 388
tantalum, 42, 105, 129–130, 136–137, 147, 178, 243, 245–246, 584
  compressive strength under ICE loading, 301
  mesoscale response, 372
  ramp compression, 467, 494
  spall, 384, 570
  Taylor wave, 36

temperature measurements, 237–238, 244, 269, 311, 544
Terminal Ballistics Facility (TBF), 111, 291, 295, 369, 475, 549, 591
Theater High Altitude Area Defense (THAAD), 177
Theater Missile Defense (TMD), 128, 142, 176–180
thermal trapping, 131
thermobaric explosives, 318, 319
thermoelastic coupling, 78
Thermos series of subcritical experiments, 390–391
Thoroughbred subcritical experiment, 195, 388
Three Mile Island (TMI) nuclear accident, 15, 95, 266, 422, 448, 504, 565
three-phase equation of state, 92, 282, 456
three-stage gun, 11, 43, 93, 134, 139, 179, 247, 252, 265, 366, 369, 371, 376, 547, 580–581
Threshold Test Ban, 23, 30, 420
TIGER code, 309, 310
Tiger Team, 525, 548, 591
time-resolved
  particle velocity measurements, 51–53, 56–62, 76, 105, 215, 284, 288, 361, 404, 435
  spectroscopy, 247
  stress wave measurements, 53
tin, 185–186, 315, 349, 570
TNT, 114, 298, 330, 331, 443, 461, 572
TOODY computer code, 14, 15, 74, 77, 94, 95, 98, 104, 154, 266, 284, 360, 425, 430–432, 449, 451–453, 484, 485, 503, 508, 509, 515, 534, 576, 577
TPX, 140, 223, 250, 463, 548
transmission electron microscopy, 150, 524
triaxial loading machine, 531, 535, 560–561
tungsten, (including strength), 65, 129, 134–139, 188, 291–292, 363, 364, 585
tungsten carbide, 248, 355, 457, 585, 593
Tunguska explosion in Siberia, 332, 339–340
turbulent local mixing, 151, 523
twinning, 65, 423, 436
two-phase mixture theory, 184, 306
two-stage light gas (that is, hydrogen) gun, 11, 12, 40–43, 93, 105, 108, 110, 112, 128, 129, 132–133, 134–135, 137, 139, 141–142, 178–180, 221, 228, 230, 234, 252, 265, 270, 277, 284, 285, 289–292, 302, 348, 356–359, 361–362, 364, 369, 374, 376, 382, 383, 408–409, 470, 473–476, 479, 530–532, 533–534, 548, 591

**U**

U1a underground test facility, 193, 195, 385, 388–391
underground tests (UGTs), 9, 20, 21, 23, 25, 29–31, 52, 55, 69, 75–77, 85, 99, 267, 276, 328, 345–349, 351, 383, 385–388, 420, 421, 449, 458, 472–473, 481–482, 487, 498, 514, 534
uniaxial strain loading, 227, 249, 251
uniaxial stress experiments on the gas gun, 183, 369
Unicorn subcritical experiment, 390
uniqueness of material models, 269
universal scaling relationship, 9
unpoled ferroelectric ceramic, 517
uranium
  ICE studies on Z, 3–4
U.S. Army Space and Strategic Defense Command, 178, 277
  USS Battle Ship Iowa explosion, 15, 128, 158–160, 175, 266, 306, 566

**V**

VANDAL code, 77, 421
vaporization
  shock-induced, 40, 42–43, 141–142, 218, 247, 252–253, 287, 367, 373, 580–581, 591
VASP (Vienna *ab initio* Simulation Package) code, 237, 407–408, 491, 492
Veloce pulsed power facility, 130, 218, 243–246, 268, 303–304, 317, 382, 406, 466, 517, 584, 601
verification and validation (V&V), 439, 488, 582
VISAR, 12–13, 62, 86–93, 98, 99, 103, 105, 112–113, 117, 134–135, 215, 265, 268, 291–292, 320, 323–325, 328, 385–392, 393–394, 590–591
  line imaging (aka the Line VISAR or Line ORVIS), 186–190, 216, 221, 231, 266, 268, 287–288, 315–317, 372, 569–570, 585
  particle velocity measurement, 88, 90, 223–225, 245–246, 248–249, 251, 268, 283–284
  pressure-shear measurements, 288, 304, 360–361, 372, 374–375, 586
  shock velocity measurement, 90, 195–196, 221–225, 230, 233, 282, 464, 582
viscoelastic materials, 13, 60, 67–69, 557

Vito subcritical experiment, 192–193, 388
vulnerability, 21, 25, 30, 40, 43, 66, 75–78, 97, 108, 252, 287, 347, 350–351, 421, 459, 481–483, 498, 527, 529

**W**

warm dense matter (WDM), 140, 233, 236, 464
Waste Isolation Pilot Plant (WIPP), 6, 346, 349, 514, 532, 578
water
  equation of state/Hugoniot, 219, 231, 237–238, 303, 378, 465
  freezing, 238
Wave structure, 25, 28, 32, 36, 38, 49, 52, 58, 62, 67–68, 99, 129, 184, 186, 284, 287, 299, 309, 312–313, 322, 381, 393, 398, 450, 585
weapon components, 4, 13, 14, 16, 20, 25, 26, 28, 29, 31, 34, 49, 52, 73, 93, 127, 190, 207, 263, 264, 267, 342, 480
WONDY computer code, 14, 68, 72, 73, 77, 97, 98, 144, 153, 226, 266, 285, 392, 421, 425, 430–432, 449–451, 453, 458, 484, 485, 503, 508, 509, 513, 514, 530, 534, 573, 584

**X**

X-cut quartz, 12, 13, 54, 62, 100, 102, 153, 363, 436, 513, 538, 541, 542–543
x-ray deposition, 42, 74–78, 97, 287, 350, 351, 391, 392, 421, 429, 462, 463, 481, 500, 501, 564
  shock-induced spall, 27, 65–67, 75–78", 97–98, 421, 481, 483
  stress generation, 55, 77, 101
x-ray diffraction studies, 149–150, 270, 521, 523, 524
x-ray-driven shock waves, 76, 481
XRAYL computer code, 524
x-ray vulnerability, 42–43, 77–78, 287, 351, 421

**Y**

Y-cut pressure-shear generator, 103, 104, 288, 360
Y-cut quartz, 103, 104, 248, 288, 360, 363, 437
yield measurements
  pressure-shear measurements, 288–289, 304, 359–360, 363, 406, 586, 587
  from wave profile, 280–282, 284, 285, 299, 301, 398, 438, 452, 457, 481, 590, 592
yield strength, 63–67, 74, 102, 106, 129, 131, 138, 143, 147, 157, 190, 226, 231, 245, 247, 250, 284, 321–322, 355, 358, 365, 372, 437–439, 548, 553, 593
  Hugoniot elastic limit (HEL), 52, 53, 63–66, 105, 250, 284
Yucca Flat weapons test basin, 193

**Z**

Z accelerator (Z, or Z machine), 30, 204–206, 217, 220–223, 230–237, 296–297, 303, 316, 369, 377, 411, 413, 462–463, 479, 493, 494, 571, 582, 593
Zapotec code, 428, 433
Z-Division at Los Alamos, 4–5
Zeldovich-von Neuman-Doring (ZND) model of detonation, 186, 313–314
Zirconia/zirconium dioxide, 149, 522
z-pinches, 205–206, 216, 223, 235, 304, 401, 403, 404, 407, 462
z-pinch source for shock drive, 205, 206, 207, 217, 221, 222, 223, 260, 297, 404, 463

9783319333458